Aerosol Sampling

Aerosol Sampling

Science, Standards, Instrumentation and Applications

JAMES H. VINCENT

Department of Environmental Health Sciences,
School of Public Health,
University of Michigan,
Ann Arbor, MI, USA

John Wiley & Sons, Ltd

Other Wiley Editorial Offices

John Wiley & Sons Inc., 111 River Street, Hoboken, NJ 07030, USA

Jossey-Bass, 989 Market Street, San Francisco, CA 94103-1741, USA

Wiley-VCH Verlag GmbH, Boschstr. 12, D-69469 Weinheim, Germany

John Wiley & Sons Australia Ltd, 42 McDougall Street, Milton, Queensland 4064, Australia

John Wiley & Sons (Asia) Pte Ltd, 2 Clementi Loop #02-01, Jin Xing Distripark, Singapore 129809

John Wiley & Sons Canada Ltd, 6045 Freemont Blvd, Mississauga, Ontario, L5R 4J3, Canada

Wiley also publishes its books in a variety of electronic formats. Some content that appears
in print may not be available in electronic books.

Anniversary Logo Design: Richard J. Pacifico

Library of Congress Cataloging-in-Publication Data:

Vincent, James H.
  Aerosol sampling : science, standards, instrumentation and applications / James H. Vincent.
     p. cm.
  Includes bibliographical references and index.
  ISBN-13: 978-0-470-02725-7 (cloth : alk. paper)
  ISBN-10: 0-470-02725-8 (cloth : alk. paper)
  1. Aerosols. I. Title.
  TP244.A3V563 2007
  628.5′30287 – dc22

                                                            2006036086

British Library Cataloguing in Publication Data

A catalogue record for this book is available from the British Library

ISBN: 978-0-470-02725-7 (HB)

Typeset in 10/12pt Times by Laserwords Private Limited, Chennai, India
Printed and bound in Great Britain by Antony Rowe Ltd, Chippenham, Wiltshire
This book is printed on acid-free paper responsibly manufactured from sustainable forestry
in which at least two trees are planted for each one used for paper production.

For Christine

Contents

Preface

My first book, *Aerosol Science: Sampling and Practice*, was published by John Wiley & Sons, Ltd in 1989. It had been conceived during a flight home to the United Kingdom from Australia in December of 1987, the long hours of pondering and doodling producing an outline that subsequently changed little as my notes evolved into a proposal and – ultimately – the book itself. Prior to that imposed period of reflection I had had no intention whatsoever of ever undertaking the arduous task of writing a major work on the scale of a book. But it dawned on me that aerosol sampling was a central aspect of the study, characterization and surveillance of atmospheric environments and that it needed to be addressed scientifically in an integrated way in order that the results of aerosol sampling in the real world would have any meaning. The book duly appeared at a time of burgeoning interest in aerosol sampling through the widening of the scope of health-related particle size-selective aerosol measurement, in particular the emergence of new criteria and standards, along with new instrumentation.

The field of aerosol sampling has moved forward dramatically during the past two decades, and the time has now come for a re-evaluation and update. A second edition was discussed. Eventually, however, it was decided – with the help of the publisher – that so much has changed, so much has happened, so much is better understood, that a new book is in order. Although written from scratch, *Aerosol Sampling: Science, Standards, Instrumentation and Applications* nonetheless clearly retains its origins in the earlier book, covering much of the same ground in the earlier parts, albeit in a way that has matured over the years of consideration and reconsideration. It still separates out the largely scientific from the largely practical elements since this most clearly defines, and best integrates, the scope of what has been achieved. In addition to a complete review of the latest science of aerosol sampling and the development of new instrumentation, major new ingredients in the latest book include extended chapters on aerosol sampling criteria and standards, new chapters on bioaerosols and direct-reading instrumentation, and a whole new section on field methods and applications. The first book had its origins in aerosol sampling science as it related to occupational exposure assessment. This is hardly surprising since, at the time, I was working at the Institute of Occupational Medicine in Edinburgh and my research team was consumed on a day-to-day basis almost entirely with such issues. But my move in 1990 to the University of Minnesota, and then on to the University of Michigan in 1998, launched me into the wider world of aerosols and aerosol measurement. So the new book presumes to provide a comprehensive review of the whole field of aerosol sampling, aimed at scholars, occupational and environmental hygiene practitioners, epidemiologists, engineers and instrument developers involved in all aspects of the study, development and application of aerosol exposure assessment methods.

A book like this demands a historical context, and I have attempted to provide one. It is an old cliché to say that 'how can we work out where we are going if we do not know where we have been?' Important work from the 1800s and early 1900s is referred to. Even Charles Darwin gets a mention. But the great majority of what is described comes from the explosion of interest in, and concerns about, occupational and living environments in the period post World War II. Indeed, it may be said that it was this period that gave birth to the disciplines that we now know as occupational and environmental hygiene.

Reading back what I have written, I see that the narrative in some respects has taken on the appearance of a 'memoir', not just by virtue of my own journey through the field over nearly three decades but also through my associations with a large number of people who, too, were players in what has taken place. This book is really about them. It follows, therefore, that I owe a great deal to the many people with whom I have interacted over the years. The list has to begin with C.N. Davies and W.H. Walton, the *doyens* of aerosol sampling science and its applications to aerosol exposure assessment. They had been together at the chemical and biological warfare research establishment at Porton Down in England during World War II, and had been among the pioneers of what we now know as aerosol science. After the war, Norman Davies, with the new British Occupational Hygiene Society, founded the *Annals of Occupational Hygiene*, bringing to it the strong scientific rigor that came from his academic grounding in physics. After Davies eventually moved on to found the *Journal of Aerosol Science*, Henry Walton took over the *Annals*, adding his own background in physics to further enhance the reputation of that journal as the leader in the field of what we can now refer to as 'occupational hygiene science'. In fact, it may safely be argued that Henry Walton was one of a small number of ground-breaking, true 'occupational hygiene scientists', his contributions over time reaching beyond the physics of aerosols and gases and into the realms of exposure assessment and occupational epidemiology. He was pivotal in the development of aerosol sampling methods and exposure assessment in the context of coalworkers' pneumoconiosis during his years with the National Coal Board and latterly the Institute of Occupational Medicine in Edinburgh. The obituary that appeared in *The Scotsman* after his passing in 2000 at the grand age of 87 contained the words 'It is not possible to estimate how many miners owe their lives and health to his innovations.' There could have been no more appropriate epitaph. Henry Walton was instrumental in persuading me to come to the Institute of Occupational Medicine when he retired in 1978, and that was the turning point that drew me into the field which is the subject of this book. Over the years I learned that, in his inimitable quiet way, he had been the source of many ideas now taken for granted in aerosol science, including the theories and development of aerosol dispersing devices such as the spinning disc generator, respiratory protection devices and testing methods, and particle size-selective sampling devices such as horizontal and vertical elutriators. He was the driving force behind the definition and measurement of *respirable aerosol* and latterly – and less well-known – *inhalable aerosol* and porous foam filtration. After his retirement, Henry was a regular and welcome presence in the Institute, and would often drop by my office for a 'chat'. It was during one of these, for example, that he was finally able to convince me how it was possible for the aspiration efficiency of an aerosol sampler to actually exceed unity! I last saw him when I visited him to discuss revisions to a paper we had written together for *Aerosol Science and Technology* on the history of occupational aerosol exposure assessment. It was a Saturday, the day of the funeral of Princess Diana, and his eyes were glued to the television as the event unfolded. We did not speak and, after a while, I left him to his thoughts. But in due course the paper duly appeared, sadly the only time our names appeared together on the printed page.

I am of course grateful to a very large number of others for their roles in my education about the science and practice of aerosol sampling, most of whom feature – many frequently – as the pages are turned. There are those in my own research teams over the years, including my colleagues David Mark, Alan Jones, Arthur Johnston, Rob Aitken, Harry Gibson and Gordon Lynch at the Institute of Occupational Medicine; then – later – my faculty colleague Gurumurthy Ramachandran and doctoral students Perng-Jy Tsai, Mark Werner, Terry Spear and Avula Sreenath at the University of Minnesota; and subsequently my doctoral students Sam Paik, Wei-Chung Su, Laurie Brixey, Yi-Hsuan Wu and Darrah Sleeth at the University of Michigan. Then there are the scientists with whom I frequently interacted, conversed, collaborated and shared research in Britain, including Richard C. Brown, Sarah Dunnett, Lee Kenny, Derek Ingham, Andrew Maynard, Trevor Ogden and Derek Stevens, to name just a few. There are the many others in Europe and Scandinavia, including Lorenz Armbruster, Paul Courbon, Peter Görner, Göran Lidén and Vittorio Prodi, and many others. I owe special words to Yngvar Thomassen for providing me with frequent opportunities to expound my thoughts at the many excellent symposia he organized, always in wonderful locations, as well as for his steady stream of ideas and suggestions about how the results of all our researches could be applied in the real world. I also specially want to mention Jean-Francois Fabries. Paul Courbon introduced us over 25 years ago at an aerosols conference in Minneapolis, and we maintained scientific contact over all the years that followed. He consistently made outstanding contributions to aerosol sampling science and its applications in occupational and environmental hygiene. It was a great loss to the whole aerosol science community when this charming and modest colleague was taken from us prematurely only a few weeks before this book was completed after a long and brave battle with illness.

Upon my arrival in the United States on a frigid New Year's Eve in 1990, my work and interests in aerosol sampling began a new phase, with expanded new opportunities for the funding of research and for the applications of what had been learned in the laboratory to field applications. I owe a special debt to Ben Liu, Virgil Marple, David Pui, Peter McMurry and others who encouraged – indeed facilitated – my move to the University of Minnesota, undoubtedly one of the 'meccas' of modern aerosol science. I am also especially grateful for my long associations with Paul Baron, David Bartley, Nurtan Esmen, Sergey Grinshpun, Martin Harper, Bill Hinds, Mort Lippmann, Walter John, Mike McCawley, Bob Phalen, Klaus Willeke, and many others. There are at least two of these – you know who you are – where our shared passion for music has exceeded even the joy of our scientific interactions! The work that some of us carried out in the ACGIH Air Sampling Procedures Committee resulted in the important monograph, *Particle Size-Selective Sampling for Particulate Air Contaminants* that is widely considered a working modern framework for aerosol-related exposure standards.

I am also grateful for my long associations with those in the various commercial enterprises who – along with names already mentioned – have been in the forefront of the realization of practical sampling instruments, including Debbie Dietrich, Bob Gussman, Pedro Lilienfeld, Eddie Salter, Gil Sem, and many others, including many that have kindly provided material for inclusion in this book.

The names I have listed above are all my friends. Overall, the list of distinguished aerosol scientists, occupational and environmental hygiene practitioners, and engineers who have contributed to what I have learned, and hence what is described in this book, is long. Sadly, many will have been inadvertently overlooked. To those who are not mentioned through my oversight, my thanks to you too and my sincere apologies.

Of course, the actual writing of a book is only the first step in the process of bringing a work to fruition. In this case, in addition to all the individuals and organizations that provided various materials in the form of photographs and other pictures, I am especially indebted to the whole editorial and production team at John Wiley & Sons, Ltd, notably Jenny Cossham, Lynette James and Richard Davies and their many unnamed colleagues. I thank them all for their professionalism, patience and understanding.

Finally, it will be wryly noted by any reader who themselves have undertaken a large work that the effort involves many hundreds of hours in solitude. Why do we undertake such projects? It is hardly the search for immortality. That achievement is reserved only for the likes of Einstein and Newton. For aerosol science, perhaps only Nikolai Fuchs has achieved that status. For the rest of us, the 99.999 %, we know that our contributions will fade soon after we have departed. What we hope for, therefore, is to be able to integrate our knowledge in order to create a small new step in a field which can be the basis of further steps in the future by others. It is with this in mind that our thoughts turn to our families who have been neglected so much for what might seem – ultimately – so little. In my case I am therefore grateful to my children, Jeremy, David, Heather and Claire, now all grown up, for their encouragement of my 'crazy' endeavour. Most especially, it could not have happened without the ongoing love and support of my wife Christine. In her 11 years as my editorial assistant when I was co-editor-in-chief (with Gerhard Kasper) of the *Journal of Aerosol Science*, she came to know the names – and frequently the faces – of very many of those same aerosol scientists that are featured in the book. Now, although her memory falters prematurely, she still remembers most of them.

Part A

Scientific Framework for Aerosol Sampling

1

Introduction

1.1 Aerosols

The scientific definition of an *aerosol* refers to it as a disperse system of liquid or solid small particles suspended in a gas, usually air. It applies to a very wide range of particulate clouds encountered terrestrially, including naturally occurring airborne dusts, mists, clouds, sandstorms and snowstorms, as well as the man-made smokes, fumes, dusts and mists that are found in our working and living atmospheric environments. It also applies to airborne particles of biological origin, including pollens, spores, bacteria and viruses. Although there are some examples of beneficial aerosols, including those specifically generated for medicinal and therapeutic purposes (e.g. inhalers like those widely used by people suffering from asthma) and those that are generated in closed systems for the purpose of specific engineering applications (e.g. carbon black manufacture, nanotechnology), the primary focus of this book is with aerosols that have the potential to be harmful to human health. This alone has been the subject of great – and continuing – interest. In this regard, many aerosols are regarded as pollutants and so are considered undesirable, especially if inhaled where they may be associated with adverse health effects in humans. It follows, therefore, that much occupational and environmental health research and practice is concerned with aerosols that are regarded as hazardous agents.

Figure 1.1 summarises some aerosol classifications. It also includes some examples of aerosol types encountered in occupational and environmental hygiene, reflecting wide ranges not only of chemistry and biology but also of particle size. As indicated, the range of scale is enormous, across all of which particles have the ability to remain airborne for significant durations in most environments and so be available for inhalation by humans. It will be a consistent theme throughout this book that particle size is one of the most important aerosol indices that influences the physical nature of aerosols – including their manner of transport in the air – and in turn the intensity of human exposure and dose. By a similar token, particle size is also extremely influential in all aspects of how aerosols are measured. But composition too is an important consideration since this is largely what drives toxicological effects in aerosol-exposed people. It may vary greatly throughout the particle size range, from location to location, and from one set of environmental conditions to another. This is especially the case for ambient atmospheric aerosols, which are subject to more widely varying conditions than workplaces, where aerosol composition, although often complex, is usually fairly constant. The continuing need to understand all these aspects of aerosols poses difficult and interesting challenges to the science and practice of aerosol sampling.

Aerosol Sampling: Science, Standards, Instrumentation and Applications   James Vincent
© 2007 John Wiley & Sons, Ltd

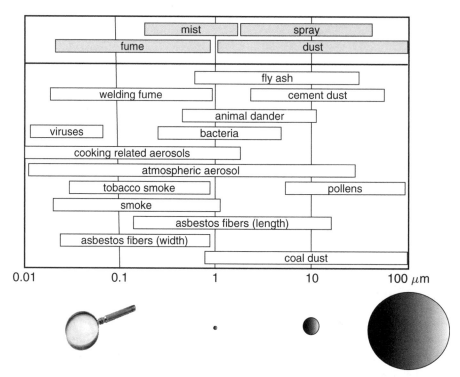

Figure 1.1 *Summary classification of aerosols, some types of interest in occupational and environmental health, and an indication of the relative range of scale of particle sizes*

1.2   Particle size

For a particle that is spherical, its definition is straightforward, requiring specification only of a single dimension, the particle *geometric diameter* (d). This is what would be obtained if the particle were to be observed and sized – by reference to an appropriate calibrated graticule – under a microscope. For a particle that is not spherical, however, the microscopist has a dilemma. There is no single defining dimension. Although spherical particles may be found in carefully generated laboratory aerosols, most aerosols in the real world – with the exception of mists and sprays – come into this second category. It therefore becomes necessary to find a metric of particle size that is not only consistent and accessible but is also relevant to the aspects of the behavior of the particles that are of specific interest. This leads to the concept of 'equivalent' particle size, for which there are several options. For example, *equivalent volume diameter* (d_V) is the diameter of a sphere that has the same volume as the particle in question, and is physically relevant to some aspects of particle motion, including diffusion. The *equivalent projected area diameter* (d_{PA}) and the *equivalent surface area diameter* (d_A), are both physically relevant to some aspects of how particles interact with electromagnetic radiation, including visible light. However, as will be seen later in Chapter 2 and beyond, none of these definitions is sufficient to fully describe the airborne behavior of the particle. From physical considerations, involving not only geometric size but also particle density and shape, particle *aerodynamic diameter* (d_{ae}) emerges as a property of great importance relevant to airborne behavior.

1.3 Elementary particle size statistics

Only rarely in practical cases do aerosols exist that consist of particles all of one size. When they do, they are referred to as 'monodisperse'. More generally, however, aerosols consist of particles of many different sizes, with particles usually belonging to populations with statistically continuous particle size distributions. These are referred to as 'polydisperse'. For the purposes of most of what follows in this book, a rudimentary outline of particle size statistics provides sufficient background. More detailed descriptions are available elsewhere (e.g. Fuchs, 1964; Hinds, 1999).

For particles whose size can be represented in terms of a single dimension (sometimes referred to as 'isometric'), the fraction of the total number of particles with diameter falling within the range d to $d + \mathrm{d}d$ may be expressed as:

$$\mathrm{d}n = n(d)\,\mathrm{d}d \tag{1.1}$$

where

$$\int_0^\infty n(d)\,\mathrm{d}(d) = 1 \tag{1.2}$$

and where $n(d)$ is the *number frequency* distribution function. Alternatively, the *mass frequency* distribution function $m(d)$ is given by:

$$\mathrm{d}m = m(d)\,\mathrm{d}d \tag{1.3}$$

and

$$\int_0^\infty m(d)\,\mathrm{d}(d) = 1 \tag{1.4}$$

Other forms of expression for the particle size distribution may similarly be described. They are all interrelated. Which form is actually used in practice depends on how the particle size is measured. For example, the counting and sizing of particles under a microscope yields distributions in the form of $n(d)$, while gravimetric methods, involving the weighing of collected samples on an analytical balance, yields them in the form of $m(d)$. As a general rule, when discussing particle size statistics, it is prudent to indicate the method by which the particle size analysis has been achieved.

It is often helpful in particle size statistics to plot distributions in the alternative cumulative form, for example:

$$\text{Fraction of mass with particle diameter less than } d,\ C_m(d) = \int_0^d m(d)\,\mathrm{d}(d) \tag{1.5}$$

A typical particle size distribution is given in Figure 1.2, shown both as mass frequency and cumulative functions, $m(d)$ and $C_m(d)$, respectively. This figure contains a number of important features. Firstly, the median particle diameter, at which 50 % of the mass is contained within smaller particles and 50 % is contained within larger ones, can be read off directly from the cumulative distribution [Figure 1.2(b) and (c)]. Secondly, the frequency distribution [Figure 1.2(a)] exhibits a strong degree of asymmetry such that the peak value of $m(d)$ lies at a value of d which is smaller than the median. The long tail in the distribution that extends out to relatively large particles is very common – indeed is ubiquitous – in aerosols in the real world. Skewed frequency distributions like the one shown in Figure 1.2 are often well represented by the log-normal function, given for example by:

$$m(d_{ae}) = \frac{M}{d_{ae}\sqrt{2\pi}\,\ln\sigma_g}\exp\left[-\frac{1}{2}\left(\frac{\ln d_{ae} - \ln\mathrm{MMAD}}{\ln\sigma_g}\right)\right] \tag{1.6}$$

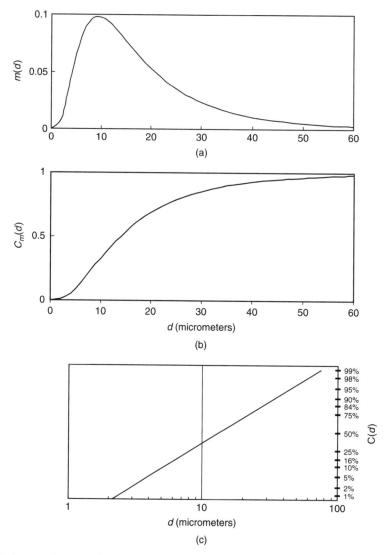

Figure 1.2 *Typical aerosol mass-related particle size distribution: (a) frequency distribution; (b) cumulative distribution; and (c) cumulative distribution on log-probability axes, indicative of a log-normal particle size distribution*

for the *mass* distribution in terms of particle aerodynamic diameter. Here MMAD is the commonly used acronym for the mass median aerodynamic diameter and σ_g is the geometric standard deviation reflecting the width of the distribution (or degree of polydispersity), while M is the total mass contained within the sample represented by distribution shown. Particle count and surface distributions may be also similarly represented. In Equation (1.6), σ_g is given by:

$$\sigma_g = \frac{d_{84\%}}{\text{MMAD}} = \frac{\text{MMAD}}{d_{16\%}} \tag{1.7}$$

But for calculation purposes, especially when the cumulative distribution is required, involving numerical integrations of $m(d)$, application of Equation (1.6) becomes unwieldy. The alternative expression:

$$C_m(d_{ae}) = \frac{\exp[a + b \cdot \ln(d_{ae})]}{1 + \exp[a + b \cdot \ln(d_{ae})]} \tag{1.8}$$

provides a very good approximation to the cumulative log-normal function and is much easier to use. Here:

$$b = 1.658/\ln(\text{MMAD}) \text{ and } a = -b \cdot \ln(\sigma_g) \tag{1.9}$$

For a perfectly monodisperse aerosol, $\sigma_g = 1$. More typically, for aerosols like those found in many workplace situations, σ_g ranges from 2 to 3. In such cases, when the cumulative distribution function is plotted on log-probability axes [as shown in Figure 1.2(c)], it appears as a straight line. Such behavior is easiest to interpret for scenarios where there are just single aerosol sources. But similar trends are also found for situations where there are multiple sources producing aerosols with essentially similar ranges of particle size. However, in situations where there are distinctly different types of aerosols generated into the same space, the individual particle size distributions characteristic of each source type can be clearly seen. The case illustrated in Figure 1.3 is typical of what might be found in a dusty workplace – for example, a coal mine – where, in addition to the dust sources associated with the primary mechanical mining operations, fine carbonaceous fume is also present in the form of diesel exhaust from mining vehicles. As will be seen later in this book, knowledge about particle size distribution can be extremely important in understanding the distribution of the deposition of inhaled particles to different parts of the respiratory tract. Knowledge of how chemical species are distributed throughout the overall particle size range is an important additional ingredient. Figure 1.3 therefore provides information that may be regarded as an aerosol 'fingerprint', containing much of the information needed for the estimation of health related dose and, in turn, risk.

The particle size distribution of an aerosol in the ambient atmosphere is highly relevant to aerosol sampling in the outdoors environment. Here, because of the very diverse range of sources together with the greater lifetime of some particles after they are released into the atmosphere, the picture is more complicated. Figure 1.4 contains a typical example, showing three clearly distinct regions. The first, the

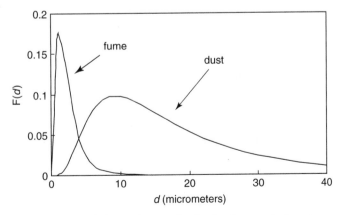

Figure 1.3 *Example of a typical bi-modal particle size distribution in a workplace (e.g. coal mining) where both dust and fume are being generated*

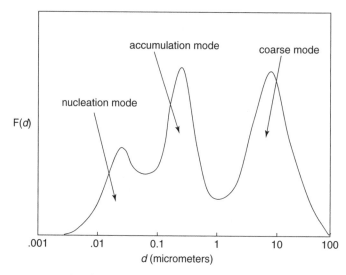

Figure 1.4   *Example of a typical particle size distribution for atmospheric aerosol*

coarse mode, is for the larger particles that are generally associated with mechanically generated airborne particles, including dusts from both natural and anthropological sources. The second, the *nucleation mode*, is for the finest particles that are generated as primary particles during combustion or other hot processes. Chemically, these are very different from those in the coarse fraction. In addition, following generation, they continue to evolve through both chemical and physical processes. Through some of these processes, involving condensation and coagulation, a third population of particles is produced, the *accumulation mode*. The transition between the nucleation and accumulation modes takes place for particle diameter around about 0.05 and 0.1 μm, and that between the accumulation and coarse modes around about 1.5 to 2.5 μm. As will be discussed later, these transition regions have important implications for aerosol sampling.

1.4   Aerosol measurement

Interest in aerosol sampling is stimulated by practical needs to understand, qualitatively and quantitatively, the properties of airborne particles in many occupational and ambient environmental situations. Such needs include the monitoring of emissions of particulate material to the atmosphere from industrial processes, assessment of the aerosol exposures of people for the purpose of epidemiology or risk assessment, and assessment of compliance with regulatory standards. In addition, there is the need in many situations to measure aerosols for the purpose of monitoring and controlling certain types of industrial process, including clean rooms. Interest in aerosol measurement across this range of situations expanded rapidly during the postwar period, fueled in particular by increasing public awareness of the problems associated with air pollution, especially in relation to health effects, leading in turn to the introduction of wide-ranging clean air acts, emission control limits, and air quality and occupational health standards in many countries. It has spawned a large – and still growing – body of scientific research, aimed at informing us about how to make the process of measurement as relevant as possible to the reason for making the measurement in the first place.

The property of most common interest is the aerosol mass concentration (c) defined as the mass of particulate material per unit volume of air. It is usually expressed in units of micrograms or milligrams per cubic metre ($\mu g\, m^{-3}$ or $mg\, m^{-3}$). Related properties of practical interest, used in some situations, include the number and surface area concentrations. Others include the distributions of particle size and chemical composition. The act of aerosol sampling involves the physical separation of particles from a given volume of the air so that they may be assessed in terms of these – and other – properties of interest.

Much of this book will be concerned with the process of aspiration whereby a known volume of the aerosol-laden air is extracted by sucking it (with the aid of pump) through one or more orifices in a solid casing which houses a *sensing region* which usually takes the form of a filter or some other substrate which may later be weighed, examined under a microscope or otherwise assayed. The aerodynamic processes by which airborne particles enter the sampler along with the air itself are – as we shall see – very complex, depending on many physical factors associated with the sampler itself as well as its surrounding environment and particle properties. The important practical question is: for given sampling conditions, how representative is the aspirated sample of the aerosol of interest in the atmospheric environment outside the sampler? In the first instance this requires definition of what is meant by 'representative', and this depends on the scientific rationale underlying the initial decision to carry out the sampling. Secondly, the sampled aerosol concentration and/or particle size distribution are strongly affected by the physical sampling process itself. Since this in turn is governed by physical processes dependent on particle size such that some particles will be sampled preferentially to others, the particle size distribution of the sampled aerosol will be biased and so will not properly reflect that of the aerosol at large. It follows that the measured aerosol concentration will also be biased.

The matters of *representative sampling* and the physics of the sampling process have both been the subjects of extensive discussion during recent decades. Industry and publicly funded research has been directed towards the generation of what has now become a large body of applied research involving applications of both fluid mechanics and aerosol mechanics. This in turn has enabled considerable progress towards the development of scientifically based practical methodologies.

1.5 Sampler performance characteristics

Scientific discussion of sampler performance at all levels can begin by referring to the diagram in Figure 1.5. It represents the air flow near a single-orifice sampler of arbitrary shape at arbitrary orientation with respect to a moving airstream. The sampler itself consists of an entry in a bluff body and an inner transition section leading to the sensing region, shown here as a filter on which the sampled particles are collected. The flow pattern indicated represents the motion of the air near the sampler. In particular, the limiting *streamline* – actually a *streamsurface* – is the invisible surface which separates the sampled air from that which is not sampled. To aid the discussion, the *incident* plane is defined as an arbitrary plane sufficiently far upstream of the sampler for the flow pattern there to be undisturbed by the presence or action of the sampler. The *sampling* plane is defined as the plane containing the sampler entry, and the *filter* plane defines the location of the sensing region inside the body of the sampler.

Under given particle size and air flow conditions, N_0 is the number of particles passing through that part of the incident plane contained within the limiting streamsurface. N_s is the number passing through the sampling plane, having arrived there directly whilst airborne at all times. N_r is the number passing through the sampling plane which have undergone impaction with and subsequently rebounded (or been blown off) from the external surfaces. It is important to note that both N_s and N_r may contain particles that were not included in the N_0 originally contained within the sampled air volume. Of the particles

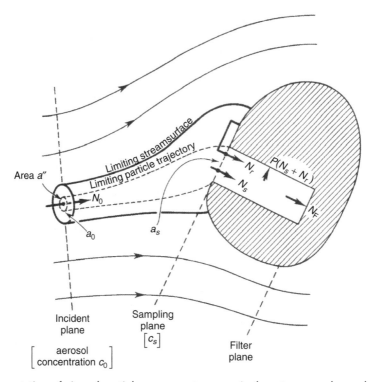

Figure 1.5 *Representation of air and particle movement near a single-entry aerosol sampler of arbitrary shape and orientation with respect to the external wind, identifying the important features needed on which to base a discussion about aerosol sampler performance. Reproduced with permission from Vincent, Aerosol Sampling: Science and Practice. Copyright (1989) John Wiley & Sons, Ltd*

crossing the sampling plane, some may be deposited on the walls of the transition section so that the number reaching the filter (N_F) is:

$$N_F = P(N_s + N_r) \tag{1.10}$$

where P is the fractional *penetration* of particles through the transition section. In what follows as this book proceeds, it will become apparent that sampler performance needs to be described by more than one parameter. Firstly, the *aspiration efficiency* (A) is the purely aerodynamic part of sampler performance, determined only by the air and particle motions outside the sampler. This is given by:

$$A = \frac{N_s}{N_0} \tag{1.11}$$

Next, the *entry* (or *apparent aspiration*) *efficiency* is:

$$A_{app} = \frac{N_s + N_r}{N_0} \tag{1.12}$$

Finally, the *overall sampling efficiency* (or *sampling effectiveness*) is

$$A^* = \frac{N_F}{N_0} = \frac{P(N_s + N_r)}{N_0} \longrightarrow \frac{PN_s}{N_0} = PA \text{ when } N_r = 0 \tag{1.13}$$

This last metric of sampler performance is of special importance when a particular fraction of the aspirated aerosol needs to be consistently selected. For example, as will be discussed in detail later in this book, this might be a sample that sets out to simulate a subfraction of particles inhaled by humans that deposit in a particular part of the respiratory tract. In such cases, the transition region between the entry and the filter is designed with regard to geometry, as well as fluid and aerosol mechanics, so that the penetration, P, is consistent and known.

Particle concentration in an aerosol is a scalar quantity that, for practical purposes (and except for extremely low concentrations), may be described in terms of a continuous spatial distribution. Consider again Figure 1.5. The limiting streamsurface encloses an area a_0 in the incident plane. The sampling orifice has area a_s, outside which all particles either strike the external wall of the sampler or pass by in the freestream. The limiting particle trajectory surface, inside which all particles enter the plane of the sampling orifice, encloses a corresponding area a''. Only for inertialess particles do the limiting streamsurface and limiting particle trajectory surface coincide. Then $a_0 = a''$. Otherwise $a_0 > a''$. The ideal situation is the one where there are no external or internal wall effects on sampler performance, in which case Equation (1.11) for aspiration efficiency may be rewritten as:

$$A = \frac{\int_{a_s} (cv)\, \mathrm{d}a}{\int_{a_0} (cv)\, \mathrm{d}a} \tag{1.14}$$

where c and v are local values (over local elementary areas $\mathrm{d}a$) for particle concentration and velocity, both distributed continuously across the sampling and incident plane respectively. Here, terms of the form (cv) are local particle fluxes, the numbers of particles flowing through unit area per unit time. Since, by continuity, the net flux of particles through a'' is the same as that through a_s, Equation (1.14) may also be expressed as

$$A = \frac{\int_{a''} (cv)\, \mathrm{d}a}{\int_{a_0} (cv)\, \mathrm{d}a} \tag{1.15}$$

It is obvious from Equation (1.15) that A is dependent on the spatial distribution of both particle concentration and velocity across the incident plane unless:

$$cv = \text{constant} \tag{1.16}$$

over a_s or a'', whichever is the greater. When Equation (1.16) is satisfied, Equation (1.15) reduces to:

$$A = \frac{a''}{a_0} \tag{1.17}$$

and this is the form that underlies most theoretical – and some experimental – assessments of aspiration efficiency. In terms of particle fluxes, Equation (1.17) leads to:

$$A = \frac{c_s U_s a_s}{c_0 U a_0} \tag{1.18}$$

where U is the freestream air velocity at the incident plane and U_s the mean air velocity across the sampling plane, and where – as already defined – c_s and c_0 are the mean particle concentrations at the sampling and incident planes, respectively. Continuity requires that:

$$U_s a_s = U a_0 \qquad (1.19)$$

So, for aspiration efficiency, Equation (1.18) becomes:

$$A = \frac{c_s}{c_0} \qquad (1.20)$$

It is important to reiterate that arrival at this familiar expression is entirely dependent on the assumption of uniform spatial distributions of particle concentration and freestream air velocity upstream of the sampler.

It follows from Equation (1.13) that the efficiency by which particles arrive at the filter is

$$A^* = \frac{c_F}{c_0} \qquad (1.21)$$

where c_F is the concentration of the particles reaching the filter.

In the last two expressions, c_s and c_F are the measured quantities of interest. The first depends on the aspiration efficiency of the sampler as reflected in A. This in turn is governed by the physics of the aspiration process so that for fixed geometrical and fluid mechanical variables, it is a function of particle size. As shown in Equation (1.13), the second depends both on A and on the similarly particle size-dependent efficiency P by which particles penetrate through the internal section of the sampler between the inlet and the filter.

For a polydisperse aerosol where the non-normalized mass-based particle size distribution is given by $M(d)$, c_s is given by:

$$c_s = \int_0^\infty A(d) M(d) \, \mathrm{d}d \qquad (1.22)$$

and c_F by:

$$c_F = \int_0^\infty P(d) A(d) M(d) \, \mathrm{d}d \qquad (1.23)$$

As will become apparent, Equations (1.20)–(1.23) are central to all discussions about the performance characteristics of aerosol samplers.

References

Fuchs, N.A. (1964) *The Mechanics of Aerosols*, Macmillan, New York.
Hinds, W.C. (1999) *Aerosol Technology: Properties, Behavior and Measurement of Airborne Particles*, 2nd Edn, John Wiley & Sons, Inc., New York.
Vincent, J.H. (1989) *Aerosol Sampling: Science and Practice*, John Wiley & Sons, Ltd, Chichester.

2

Fluid and Aerosol Mechanical Background

2.1 Fluid mechanical background

2.1.1 Introduction

The physics of air movement is fundamental to the behavior of suspended particles. At the microscopic level of an individual particle it defines the flow of air over and around the particle, and hence to the drag and lift forces acting on it. So the fluid mechanics of what happens at this level are relevant to the way in which a particle may move relative to the air itself. At the macroscopic level they are relevant to the behavior of larger-scale moving air systems in which aerosols move and in which they are translated and, in some cases, arrive at surfaces. This in turn is relevant to the airflow close to and around an aerosol sampler. It is not possible to have a discussion of the science of aerosol sampling without reference to both these aspects of fluid mechanics. In what follows, therefore, the aim is to provide a succinct framework of rudimentary fluid mechanical ideas and concepts to aid the discussion of aerosol transport and sampler characteristics which will form the main thrust of subsequent chapters. For detail in greater depth and breadth, the reader is directed elsewhere to the many specialized texts that are available (e.g. Schlichting and Gersten, 1999).

2.1.2 Equations of fluid motion

The starting point for describing the aerodynamic behavior of suspended particles is the set of equations representing the motion of the air itself. These are derived from applications of Newton's second law to a three-dimensional, incompressible elementary air volume (see Figure 2.1). That is:

$$\text{mass} \times \text{acceleration} = \text{sum of the forces acting} \tag{2.1}$$

in which the surface forces in question are *normal forces* associated with gradients in static pressure within the body of the fluid and *shearing forces* associated with gradients in fluid velocity and the attendant friction that occurs between adjacent layers of fluid moving at different velocities. By writing down expressions for these forces in three dimensions, a set of dynamic equations is obtained for describing the fluid movement, one for each of the three spatial dimensions. In addition, in order that

Aerosol Sampling: Science, Standards, Instrumentation and Applications   James Vincent
© 2007 John Wiley & Sons, Ltd

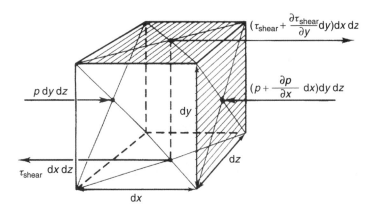

Figure 2.1 *Three-dimensional elementary air volume on which to base derivation of the fluid mechanical equations of motion. Reproduced with permission from Vincent, Aerosol Sampling: Science and Practice. Copyright (1989) John Wiley & Sons, Ltd*

mass is conserved, it is also necessary to ensure continuity. That is, the mass of air that flows into a given volume must be equal to that leaving it. The combination of dynamic and continuity equations provides the well-known Navier–Stokes equations, expressed generally as:

$$\rho \frac{D\mathbf{u}}{Dt} = -\mathrm{grad}\ p + \eta \nabla^2 \mathbf{u}$$

inertial   pressure   viscous
forces   gradient   shearing
forces   forces

$$\mathrm{div}\ \mathbf{u} = 0 \tag{2.2}$$

continuity

where the fluid is assumed to be incompressible, such that continuity of mass is equivalent to continuity of volume. Here, p is the local static pressure, $\mathbf{u}$ is the local instantaneous velocity vector, ρ the air density and η the air viscosity. The mathematical D-operator on the left embodies the *substantive acceleration* that comes from changes in local fluid velocity in both time and space.

The Navier–Stokes equations may be nondimensionalised by setting $\mathbf{U} = u/U$ and $X = x/D$, and similarly for the y and z directions. In addition

$$P_r = \frac{p}{p_r} \tag{2.3}$$

In the above U is a characteristic air velocity representing the flow system as a whole, D is a characteristic dimension and p_r is a reference static pressure. The set of equations of motion now becomes:

$$\rho \frac{D\mathbf{u}}{Dt} = -\frac{p_r}{D}\ \mathrm{grad}\ P_r + \frac{\eta}{\rho DU} \nabla^2 \mathbf{U} \tag{2.4}$$

from which it is seen that mathematical solutions for describing the flow are unique if the coefficient for the last term on the right-hand side is constant; that is, the nature of the flow is defined by the dimensionless quantity (*Re*) where:

$$Re = \frac{DU\rho}{\eta} \tag{2.5}$$

This is known universally as the *Reynolds number*, and its importance to aerosol science – and in turn to aerosol sampling – will soon become apparent.

2.1.3   Streamlines and streamsurfaces

Solutions of the Navier–Stokes equations can, in principle, allow determination of the pattern of the air flow for any set of boundary conditions and for a given value of *Re*. In practice, however, closed-form solutions are not generally accessible except for very simple flows, most of which are not ultimately of much use in practical applications. However, with the availability of powerful computers and appropriate software based firmly on the physical fundamentals outlined above, numerical solutions for most situations can now be obtained. There are many such routines developed specifically for research purposes by individual investigators. Some will be mentioned elsewhere in this book. Commercial routines are more accessible to the less-specialised user, but these are usually expensive and still require a significant level of expertise in engineering fluid mechanics in order to optimise the set-up and operation and, in turn, the quality of the results. Whether solutions are closed-form or numerical, flows with time-dependent behavior (e.g. oscillations such as vortex shedding) are especially difficult.

The endpoints of solutions of the Navier–Stokes equations may take many forms, depending on the desired goal of the enquiry. For present purposes, a solution that provides the shape of the flow pattern in a given situation is especially instructive. By way of illustration consider the simple case of a plane flow, two-dimensional in *x* and *y* only. Here, for steady flow, the Navier–Stokes equations from Equations (2.2) may be expanded into the form:

$$u_x \frac{\partial u_x}{\partial x} + u_y \frac{\partial u_x}{\partial y} = -\frac{1}{\rho}\frac{\partial p}{\partial x} + \frac{\eta}{\rho}\left(\frac{\partial^2 u_x}{\partial x^2} + \frac{\partial^2 u_x}{\partial y^2}\right)$$

$$u_x \frac{\partial u_y}{\partial x} + u_y \frac{\partial u_y}{\partial y} = -\frac{1}{\rho}\frac{\partial p}{\partial y} + \frac{\eta}{\rho}\left(\frac{\partial^2 u_y}{\partial x^2} + \frac{\partial^2 u_y}{\partial y^2}\right) \tag{2.6}$$

$$\frac{\partial u_x}{\partial x} + \frac{\partial u_y}{\partial y} = 0$$

The pressure term may be eliminated by differentiating the first equation with respect to *y* and the second with respect to *x* and then subtracting. In this process, terms of the form of the left-hand side of the third equation emerge which then vanish by virtue of continuity. If we now introduce the *stream function* (ψ), where:

$$u_x = -\frac{\partial \psi}{\partial y} \quad \text{and} \quad u_y = \frac{\partial \psi}{\partial x} \tag{2.7}$$

then the air flow pattern is fully described by the single equation:

$$\frac{\partial \psi}{\partial y}\frac{\partial \nabla^2 \psi}{\partial x} - \frac{\partial \psi}{\partial x}\frac{\partial \nabla^2 \psi}{\partial y} = \frac{\eta}{\rho}\nabla^4 \psi \tag{2.8}$$

which may be solved for $\{x, y\}$ for chosen constant values of ψ for given boundary conditions and *Re*. Here, each set of $\{x, y\}$ for a given value of ψ defines a fluid trajectory, or *streamline* (or, perhaps more

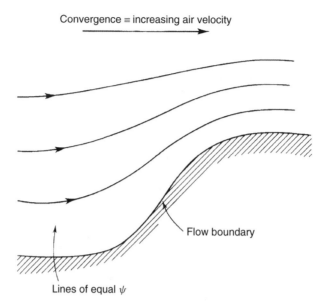

Figure 2.2   *Typical flow pattern over a surface to illustrate the nature of streamlines. Reproduced with permission from Vincent, Aerosol Sampling: Science and Practice. Copyright (1989) John Wiley & Sons, Ltd*

appropriately, *stream surface*). A typical streamline pattern is shown by way of illustration in Figure 2.2. It is required that there can be no net flow of fluid across such lines. Therefore the volumetric flow rate between adjacent streamlines having different values for ψ must remain constant. This in turn means that a divergence of the flow pattern is consistent with a decrease in fluid velocity; and vice versa for convergence. Qualitatively, it also means that, for a given flow configuration, the pattern of streamlines represents the trajectories of fluid-borne scalar entities moving in the absence of inertial or externally applied forces. The streamline pattern is therefore a vivid and instructive way of representing the nature of a fluid flow system, consistent with what would be observed in a flow visualisation, as for example might be achieved using a suitable visible tracer such as smoke.

Solutions for three-dimensional flow around a small spherical particle are particularly relevant in the context of this book. Some typical flow patterns for various ranges of *Re* are shown schematically in Figure 2.3. These show that, whereas the flow pattern at low $Re(< 1)$ is symmetrical upstream and downstream of the sphere, at higher *Re* (especially beyond 1) it tends to converge less rapidly on the downstream side. This is due to the increased inertial behavior of the flow at higher *Re*. The higher the value of *Re*, the greater will be the tendency of the flow in the wake of the sphere to separate (see below). Although this feature may not usually be applicable to the flow about aerosol particles, where in almost all practical cases *Re* is less than or does not exceed unity, it may be relevant to air movement around larger bluff obstacles (e.g. the body of an aerosol sampler) where *Re* is much higher.

2.1.4   Boundary layers

A *boundary layer* in a flow is defined as a region near a boundary (usually, but not always, a solid wall) where the influence of fluid viscosity is particularly important. For 'real' fluids, the physical concept of viscosity requires that the fluid velocity must fall to zero at the boundary itself. The boundary layer, as shown in the schematic example in Figure 2.4, is therefore seen to be the region within

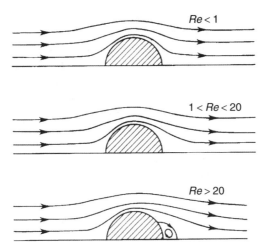

Figure 2.3 *General form of the air flow pattern near a spherical body for various ranges of Reynolds number (Re). Reproduced with permission from Vincent, Aerosol Sampling: Science and Practice. Copyright (1989) John Wiley & Sons, Ltd*

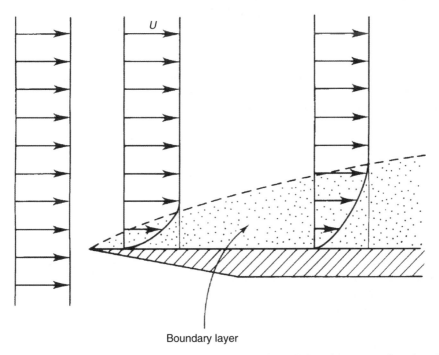

Figure 2.4 *Illustration of the flow in a boundary layer over a flat solid surface. Reproduced with permission from Vincent, Aerosol Sampling: Science and Practice. Copyright (1989) John Wiley & Sons, Ltd*

which the fluid velocity goes from zero–at the boundary–to the value corresponding to that which would be obtained by solving the Navier–Stokes equations for the zero-viscosity, so-called *inviscid*, case. However, as we penetrate deeper and deeper into the boundary layer, frictional forces become increasingly influential and, eventually at the wall itself, become dominant. In consideration of the general nature of the flow, the extent to which the effect of viscosity is significant depends on the thickness of the boundary layer in relation to the scale of the bulk flow. It may be shown that the boundary layer thickness for flow over a large body is relatively small compared with that for a small body. This, of course, is consistent with the earlier discussion about the Reynolds number–namely, that *Re* for a large body, for given fluid velocity, is greater than for a smaller body. It is a fair working assumption for the majority of aerosol sampling situations that the boundary layer at the sampler surface may be neglected and that potential flow models may be employed (see below). This in fact has been the assumption by many of the theoretical fluid dynamicists who have turned their attention to aerosol sampling.

For flow close to a solid surface, boundary layer properties are strong determinants of fluid behavior. For air flow over a smooth surface, energy is lost due to the friction in the boundary layer associated with the velocity gradient close to the surface. In the first instance, this is the source of the aerodynamic drag on bodies placed in the flow. But, other than for flows characterised by very small *Re* (*creeping flow with Re* $\ll$ 1), the loss of energy–and the associated change of local static pressure–may be such that the surface streamline can no longer remain attached to the surface, but rather may *separate*. For high enough *Re* this may happen even for relatively smooth, slowly curved surfaces, as in the flow over a spherical body. But even at relatively small *Re*, for flow over a surface with a sharp change in geometry, for example an abrupt step or a sharp-edged flat plate, the inertia of the flow in the boundary player may be such that the flow cannot remain attached to the surface. So it suddenly detaches. As shown in Figure 2.5, such flow separation is accompanied by flow reversal contained within a separation *bubble* which, depending on the flow geometry, may be more or less stable. Coherent vortex shedding is very prominent for two-dimensional flows, less so for axially symmetric flows. As will be seen later, considerations of such flow separation have some relevance to aerosol sampling.

2.1.5   Stagnation

Along a given streamline, ψ is constant and so the solution of the Navier–Stokes equations yields the well-known Bernoulli expression:

$$p + \frac{1}{2}\rho u^2 = \text{constant} \tag{2.9}$$

which states that the sum of the local static pressure and the local velocity pressure is constant. This is equivalent to the conservation of energy. In effect it states that the sum of kinetic and potential energy is constant provided that there are no friction losses. From Equation (2.9) we see that, as the velocity decreases, coinciding with the divergence of adjacent streamlines, the static pressure increases; and vice versa as velocity increases. *Stagnation* occurs when all of the velocity pressure is converted into static pressure and so the local velocity falls to zero. This is where the static pressure is highest. This usually – but not always – occurs at the surface of a bluff flow obstacle at points where a streamline intersects with the body and its continuation follows the surface of the body itself (see Figure 2.6). It is a particularly important concept in describing the flow near an aerosol sampler and will therefore feature repeatedly in future chapters.

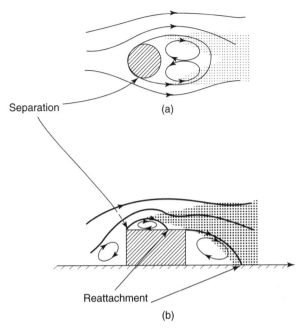

Separation

(a)

Reattachment

(b)

Figure 2.5   *Illustration of some typical separated flows: (a) sphere with a relatively stable wake cavity enclosed within the separation streamlines (or 'bubble'); and (b) surface mounted block, indicating separation and reattachment. The shaded areas represent turbulence that would be present if the Reynolds number (Re) is high enough. Reproduced with permission from Vincent, Aerosol Sampling: Science and Practice. Copyright (1989) John Wiley & Sons, Ltd*

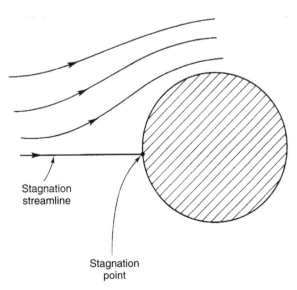

Stagnation
streamline

Stagnation
point

Figure 2.6   *Illustration of the phenomenon of stagnation in a typical streamline pattern over a body. Reproduced with permission from Vincent, Aerosol Sampling: Science and Practice. Copyright (1989) John Wiley & Sons, Ltd*

2.1.6   Potential flow

If friction forces are neglected, the flow is considered to be inviscid so that Equation (2.8) reduces to:

$$\nabla^2 \psi = 0 \qquad (2.10)$$

In addition it is useful to introduce the closely related concept of *velocity potential*, ξ, where:

$$u_x = -\frac{\partial \xi}{\partial x} \text{ and } u_y = -\frac{\partial \xi}{\partial y} \qquad (2.11)$$

which satisfies:

$$\nabla^2 \xi = 0 \qquad (2.12)$$

for inviscid flows. Described in this way, such flows are therefore widely referred to as *potential flows*, in which lines of constant ψ (streamlines) and constant ξ (equipotentials) form an orthogonal two-dimensional network.

   The particular value of the concept of velocity potential in relation to aerosol sampling is that it enables simple mathematical statements to be made about simple flows which may be combined in order to create more complex (and more relevant) ones. The example of a two-dimensional line sink is directly relevant to aerosol sampling. Consider the case where the flow rate is Q per unit length placed in a uniform freestream whose characteristic velocity is U. For the line sink, the potential field in $\{x, y\}$ coordinates, where the sink is located at the origin, is given by:

$$\xi_{sink} = -\frac{Q}{2\pi} \ln(x^2 + y^2)^{1/2} \qquad (2.13)$$

For the uniform (freestream) flow:

$$\xi_{freestream} = -Ux \qquad (2.14)$$

These individual potential fields may now be superposed to provide the overall potential field associated with the line sink placed in a uniform freestream. Thus:

$$\xi_{system} = \xi_{sink} + \xi_{freestream} \qquad (2.15)$$

which, with Equations (2.11), (2.13) and (2.14), yields the velocity components:

$$u_x = U - \frac{Qx}{2\pi(x^2 + y^2)}$$
$$u_y = -\frac{Qy}{2\pi(x^2 + y^2)} \qquad (2.16)$$

Much more complicated aspirating flows may similarly be constructed by superposing appropriate combinations of such simple solutions in much the same way.

2.1.7   Turbulence

Flows in which the layers of fluid – defined by the streamlines – slide smoothly over one another are referred to as *laminar*. However, when inertial forces become large enough in relation to viscous forces, the flow may tend to become unstable, or *turbulent*. This was first demonstrated experimentally for the flow of water in pipes by Sir Osborne Reynolds (1883), who found that the transition from laminar to turbulent flow is governed by a dimensionless quantity involving the diameter of the pipe, the fluid flow rate and the density and viscosity of the fluid itself. This is, of course, the Reynolds number (*Re*) already referred to, and it was found that the transition occurred at around $Re = 2300$. Since those earlier experiments, corresponding transitions from laminar to turbulent fluid flows have been examined by others for a wide variety of flow configurations, including internal flows in ducts and external flows of water, air and other fluids around bluff bodies, in boundary layers, etc.

The subject of turbulence presents considerable theoretical difficulties for the study of air movements. All that needs to be said here is that the physical manifestation of the instability associated with turbulence appears as randomly fluctuating fluid motions superimposed on the mean flow. To begin to construct a physical model for explaining the consequences of this, each velocity term in the Navier–Stokes equations may be replaced by its equivalent mean value plus the corresponding instantaneous fluctuation from that mean. The resultant set of equations for the instantaneous flow is complicated, but may be rendered manageable by time-averaging. Then it is seen that, as one of its consequences, turbulence is accompanied by an apparent sharp increase in the viscosity of the fluid. Another consequence is that there is a corresponding increase in the rate of mixing of fluid properties (e.g. momentum) and of fluid-borne material.

The structure of turbulence may be described in terms of the *spectral* characteristics of the fluctuating motions that are superimposed on the mean flow. Eddy-like flow formations are created at the large, low-frequency end of the spectrum, having a scale comparable with that of the flow system in question, such as the dimension of the pipe or of the bluff body, and their energy is derived from the mean flow. By the process of vortex stretching, these eddies are broken down into progressively smaller – and higher frequency – ones until the scale is reached where the energy may be dissipated viscously to generate heat. There are two views of the statistical properties of those fluctuating velocities, reflected in the Lagrangian and Eulerian *microscales* (λ_L and λ_E, respectively) which relate to the dimensions of the turbulence elements. The relationship between these quantities depends on how velocities in the fluid elements are correlated with one another. For instance, small λ_L/λ_E means that the velocities between successive elements are correlated; otherwise they are uncorrelated. This has important consequences in relation to the motion of a suspended particle in a turbulent flow as it encounters a succession of such elements.

Despite the physical complexity of turbulence, a useful, simplified picture may be obtained by describing it in terms of (a) a characteristic integral length scale (*L*) typical of the eddies containing the bulk of the turbulent energy (and which is closely related to the microscales referred to above), and (b) a characteristic 'energy' or mean square velocity fluctuation:

$$e = \overline{u_x'^2 + u_y'^2 + u_z'^2} \tag{2.17}$$

where u_x, u_y and u_z are the components of the instantaneous velocity fluctuation in the three directions indicated. If the corresponding velocities for the *mean flow* are U_x, U_y and U_z, respectively, then we may also define the *intensity* of turbulence (I_x) for each of the three directions. Thus:

$$I_x = \frac{\overline{(u_x'^2)}^{1/2}}{u_x} \tag{2.18}$$

describing the magnitude of the time-averaged velocity fluctuations of the turbulent motions as a proportion of the mean velocity for the *x* direction; and similarly for the *y* and *z* directions.

Mixing in turbulent flow may be described in terms of the transport of any scalar property, including airborne particles. The basic diffusivity of the turbulence itself may be expressed to a first-order approximation by:

$$D_t \approx Le^{1/2} \tag{2.19}$$

where it is seen to be a property of the turbulence associated with the fluid flow and is *not* a fundamental property of the fluid itself. Equation (2.19) is adequate in many cases for describing the turbulent diffusion of airborne scalar properties that do not exhibit inertial or buoyancy effects. It is therefore a useful starting point for consideration of the turbulent mixing of aerosols.

2.2   Aerosol mechanics

The science of aerosol mechanics is concerned with the mechanical properties and transport of airborne particles, and has been covered extensively in a number of well-known texts (Fuchs, 1964; Friedlander, 1977; Hinds, 1999). This chapter sets out to provide an overview of those aspects that are important for describing aspiration and other processes relevant to aerosol sampling.

2.2.1   Particle drag force and mobility

The drag force acting on a macroscopic body moving relative to the air is a net result of the distribution and magnitude of the local static pressure and shearing forces over its surface. These forces may be obtained from solutions of the Navier–Stokes equations of motion for the air flowing over the particle, with relevant boundary conditions. As early as 1851, Stokes solved the equations of fluid motion for the velocity and pressure distribution over the surface of the sphere of diameter d in a freestream where the velocity is v. He calculated that the net fluid mechanical drag force is a combination of the integrated pressure forces (one-third) and the integrated shearing – or friction – forces (two thirds). This led to the now-familiar expression of Stokes' law:

$$F_D = -3\pi d\eta v \tag{2.20}$$

for the conditions where the Reynolds number for the particle as given by

$$Re_p = \frac{dv\rho}{\eta} \tag{2.21}$$

is less than unity. This region ($Re_p < 1$) *is* now widely known as the *Stokes regime*.

Another quantity important for describing the airborne behavior of the particle, derived from Equation (2.20), is its *mechanical mobility* (μ) where:

$$\mu = \frac{1}{3\pi d\eta} \tag{2.22}$$

defines the ratio between the velocity of the particle and the external force associated with its motion.

2.2.2   Drag coefficient

The *drag coefficient* is given by

$$C_D = \frac{F_D/\text{area}}{\frac{1}{2}\rho v^2} \longrightarrow \frac{24}{Re_p} \text{ for } Re_p < 1 \tag{2.23}$$

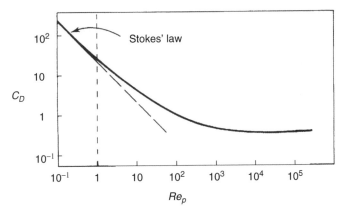

Figure 2.7 *Drag coefficient (C_D) for a spherical particle as a function of Reynolds number (Re_P) for the particle. Reproduced with permission from Vincent, Aerosol Sampling: Science and Practice. Copyright (1989) John Wiley & Sons, Ltd*

where the denominator on the right-hand side normalises the force per unit area on the sphere with respect to the velocity pressure in the fluid. For somewhat higher Re_p just outside the Stokes regime, the problem becomes more complicated because the inertial terms in the equations of fluid motion can no longer be neglected. Oseen (1910) modified Stokes' solution to take account of these conditions and obtained:

$$C_D = \frac{24}{Re_p}\left(1 + \frac{3}{16}Re_p\right) \text{ for } 1 < Re_p < 5 \tag{2.24}$$

The problem becomes yet more complicated for still larger $Re_p > 5$. The relationship between C_D and Re_p for a sphere over the full range of conditions of interest to most aerodynamicists is shown in Figure 2.7, including the case for much higher Re_p where the separated wake flow is turbulent and C_D tends to level off. However, Re_p values as large as this are largely irrelevant as far as the drag force on aerosol particles is concerned. Indeed, for airborne particles under many conditions of practical interest to aerosol scientists, Stokes' law as given by Equation (2.20) is an adequate working assumption.

From Equation (2.23) the general form for the drag force is:

$$F_D = \frac{C_D \pi d^2 v^2}{8} \tag{2.25}$$

which with Equation (2.21) becomes:

$$F_D = \frac{C_D Re_p}{24}(3\pi \eta d v) \tag{2.26}$$

where the term ($C_D Re_p/24$) on the right-hand side is the correction to allow for what is referred to as the 'non-Stokesian' regime.

2.2.3 Slip

The expression of Stokes law as embodied in Equation (2.20) operates under the assumption that the particle experiences the surrounding air as a continuum. In reality, this is not the case since the nature of the contact between the particle and the air is determined by interactions with individual gas molecules. In the extreme case where the particle is small in comparison with the mean free path (mfp) between gas molecules, the nature of particle motion relative to the air may be envisaged as taking the form of *slipping* in the vacuum that exists between successive collisions with gas molecules. This occurs in general for small particles at atmospheric pressure and, more specifically, for larger particles in some sampling situations where local static pressures are significantly lower than atmospheric. The drag force under such circumstances must therefore be modified to include the *Cunningham slip correction factor*, $C_c(d)$, so that:

$$F_D = -\frac{3\pi d\eta v}{C_c(d)} \tag{2.27}$$

where:

$$C_c(d) = 1 + \frac{\text{mfp}}{d}\left[2.34 + 1.05\exp\left(-\frac{0.39d}{\text{mfp}}\right)\right] \tag{2.28}$$

is one formulation that works well for d down to and even below 0.1 μm. Particle mobility (μ) must be modified accordingly. For air at standard temperature and pressure (STP), where mfp $\approx$ 0.1 μm, it is seen that the effect of slip only starts to become significant for $d \ll 1$ μm. For non spherical particles, it is a reasonable default in Equation (2.28) to express particle size in terms of the equivalent volume diameter, d_V.

2.2.4   General equation of motion under the influence of an external force

The starting point for all considerations of particle transport is the general equation of particle motion. It is given in vector notation by:

$$\text{Net force}, \, m\frac{\mathrm{d}v}{\mathrm{d}t} = \mathbf{F_E} + \mathbf{F_D} \tag{2.29}$$

where m is the mass of the particle in question, v its velocity relative to the air, $\mathbf{F_E}$ the external force (or combination of forces) acting and $\mathbf{F_D}$ the fluid mechanical drag force as before. In principle, exact solutions for particle motion in given flow and force fields can be obtained from Equation (2.29), and so the problem may be regarded as deterministic. For the spherical particle moving in the vertical (y) direction in air under the influence of gravity, Equation (2.29) becomes:

$$m\frac{\mathrm{d}v_y}{\mathrm{d}t} + \frac{C_D Re_p}{24}\frac{3\pi d\eta v_y}{C_c(d)} - mg = 0 \tag{2.30}$$

It is convenient in this important case that the external (driving) force and the particle drag (resisting) force are acting in one dimension only. Then Equation (2.30) reduces to:

$$\frac{\mathrm{d}v_y}{\mathrm{d}t} + \frac{v_y}{\tau} - g = 0 \tag{2.31}$$

where

$$\tau = \left[\frac{d^2\gamma C_c(d)}{18\eta}\right]\left(\frac{24}{C_D Re_p}\right) \tag{2.32}$$

in which γ is the particle density. In this expression, τ has dimensions of time. Equation (2.31) is a first-order linear differential equation which has the simple solution:

$$v_y = \tau g \left[1 - \exp\left(-\frac{t}{\tau} \right) \right] \tag{2.33}$$

for the case where $v_y = 0$ at time $t = 0$. This shows that the particle velocity under the force of gravity tends exponentially towards a terminal value, the *sedimentation velocity* or *falling speed*, given by:

$$v_y = \tau g \tag{2.34}$$

Typically for a spherical particle with $d = 20$ μm and density $\gamma = 2 \times 10^3$ kg m^{-3} settling in air at STP, we get $v_s \approx 0.02$ m s^{-1}. Inspection of Equation (2.21) reveals that $Re_p \approx 0.03$, confirming that Stokes' law is well satisfied. Particle velocity reaches $1/e$ of its final, terminal value at $t = \tau$. The quantity τ is therefore the time constant of the exponential process. From Equation (2.32), it is seen to be uniquely a property of the particle itself in the particular fluid, independent of the configuration or scale of the flow. It is referred to as the 'particle relaxation time'.

The same basic ideas may be applied to the cases of particles moving under the influence of other forces (e.g. electrostatic) or in flow fields where the drag force is not necessarily acting in the same direction as the external force. For very small particles where $d <$ mfp the preceding equations would of course need to include the slip correction.

2.2.5 Particle motion without external forces

The problem of a particle moving in air in the absence of an externally applied force is particularly relevant to aerosol sampling. Here the equation of motion:

$$\frac{d\mathbf{v}}{dt} = -\frac{1}{\tau}(\mathbf{v} - \mathbf{u}) \tag{2.35}$$

in which the velocity vector $(\mathbf{v} - \mathbf{u})$ represents the relative motion between the particle and the surrounding air. In the simplest case where the air is stationary, the solution for the particle velocity in the x direction is:

$$v_x = v_{0x} \exp\left(-\frac{t}{\tau} \right) \tag{2.36}$$

where v_{0x} is the initial velocity in that direction at time $t = 0$. Further integration yields the distance traveled by the particle by virtue of its inertia before it comes to rest, thus:

$$s = v_{0x}\tau \tag{2.37}$$

This quantity is known as the particle *stop distance*. Equation (2.35) may now be non dimensionalised by setting $\mathbf{U} = \mathbf{u}/U$, $\mathbf{V} = \mathbf{v}/U$ and $X = x/D$ for the x direction, and likewise for the y and z directions. Here, as before, U is a characteristic velocity of the system and D is a characteristic dimension. Equation (2.35) may now be manipulated to give:

$$\frac{U^2}{D}\frac{d\mathbf{V}}{dt} = -\frac{U}{\tau}(\mathbf{V} - \mathbf{U}) \tag{2.38}$$

leading to:

$$\frac{d\mathbf{V}}{dt} = -\frac{D}{\tau U}(\mathbf{V} - \mathbf{U}) = -\frac{C_D Re_p}{24} \frac{1}{St}(\mathbf{V} - \mathbf{U}) \tag{2.39}$$

where the dimensionless quantity:

$$St = \frac{d^2 \gamma U C_c(d)}{18 \eta D} \tag{2.40}$$

is known as the *Stokes number*. For values of Re_p within the Stokes regime ($Re_p < 1$) the pattern of particle motion obtained from solutions of Equation (2.39) depends only on St. But for particles where $Re_p > 1$, it will depend also on Re_p, for which corrections are available. The physical significance of St becomes apparent for situations where the flow is distorted (divergent and/or convergent) and the airborne particle does not reach dynamic equilibrium with its surroundings. This occurs, for example, in the air flow in the vicinity of a bluff obstacle, inside a bent duct or – notably, in the context of this book – close to an aerosol sampler. Here a suspended particle, as it moves through the system, experiences a continuously changing air velocity and in turn a drag force which is changing in magnitude and direction. For a distorted flow, it is apparent that the quotient D/U in Equation (2.39) is equivalent to a timescale which is characteristic of the flow distortion (τ_d). For the simplest case where $Re_p < 1$, then:

$$\frac{\tau U}{D} \longrightarrow St = \frac{\tau}{\tau_d} \tag{2.41}$$

For the case where τ/τ_d is small, the particle will tend to follow closely the motion of the surrounding air as it changes direction. Alternatively, when τ/τ_d is large, the particle is relatively slow to respond to the changes and so tends not to follow the air movement. Rather, it tends to continue in the direction of its original motion. Another view may be obtained by combining Equations (2.37) and (2.41) to show that St is the ratio of particle stop distance to the characteristic dimension of the flow distortion. Similarly to the preceding argument, the particle will tend to follow the air flow when s/D is small; and vice versa when s/D *is* large.

The result of all these considerations is the realization that St is an important measure of the ability of an airborne particle to follow the movement of the air, and that the airflow and particle trajectory patterns may differ to an extent dictated largely by the magnitude of St. This effect is illustrated qualitatively in Figure 2.8. It might be expected intuitively that the absence of external forces would simplify solution of the equation of particle motion. However, for flows where the local air velocity is changing everywhere, the problem is actually *more* complicated, and analytical solutions are not possible. In practice, therefore, the calculation of detailed particle trajectories must be carried out numerically on a stepwise, point-by-point basis.

The preceding discussion assumes the absence of gravity. In order to assess the relative magnitude of effects associated with gravitational forces in the presence of inertially controlled particle motion, a further dimensionless group is useful. The Froude number:

$$Fr = \frac{U^2}{gD} \tag{2.42}$$

is appropriate, being effectively the ratio between inertial and gravitational forces. The larger the value of Fr, the smaller the relative effect of gravity. However, if it is reduced to the stage where it becomes of the order of – or even smaller than – St, then gravity effects should not be ignored.

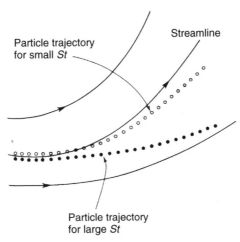

Figure 2.8 *Illustration of the effect of Stokes number (St) on the transport of particles in the flow about a bluff body. Reproduced with permission from Vincent, Aerosol Sampling: Science and Practice. Copyright (1989) John Wiley & Sons, Ltd*

2.2.6 Particle aerodynamic diameter

Consider again particles settling in air under the influence of gravity. From Equation (2.34), it is seen that all particles with the same relaxation time (τ) have the same falling speed. Thus for two spherical particles (subscripts 1 and 2) of different diameters (d) and different densities (γ), and in the absence of slip, their falling speeds in air are the same provided that:

$$\frac{d_1^2 \gamma_1}{C_D(d_1) Re_{p1}} = \frac{d_2^2 \gamma_2}{C_D(d_2) Re_{p2}} \tag{2.43}$$

from Equations (2.32) and (2.34). The *aerodynamic diameter* (d_{ae}) of a particle is defined as the diameter of a spherical particle of density $\gamma = \gamma^* = 10^3$ kg m^{-3} (equivalent to that of water) which has the same falling speed in air as the particle in question. Thus, from Equation (2.43), the aerodynamic diameter of a spherical particle of diameter d and density γ is given by:

$$d_{ae}^2 = d^2 \left(\frac{\gamma}{\gamma^*} \right) \left[\frac{C_D(d_{ae})}{C_D(d)} \right] \left[\frac{Re_p(d_{ae})}{Re_p(d)} \right] \tag{2.44}$$

For particles in the Stokes regime, drag and Reynolds number effects are negligible, in which case this last expression reduces to the more familiar:

$$d_{ae} = d \left(\frac{\gamma}{\gamma^*} \right)^{1/2} \tag{2.45}$$

So far, attention has been focused only on idealised, spherical particles. However, particles of extreme shape are often of considerable interest in the context of aerosol science, driven for example by the occupational and environmental health concerns about airborne asbestos fibers. Such fibrous particles are characterised by their long aspect ratio in terms of their length (l_f) to width (d_f). The equivalent particle aerodynamic diameter is highly dependent on these dimensions, as well as the orientation of the fiber as it falls under the influence of gravity. Cox (1970) theoretically derived the equations of motion for a fiber falling with its axis perpendicular and parallel to the direction of motion, respectively, yielding:

$$\left(\frac{d_{ae}}{d_f}\right)^2 \left(\frac{\gamma^*}{\gamma}\right) = \frac{9}{8}\left[\ln\left(\frac{2l_f}{d}\right) + 0.193\right] \tag{2.46}$$

$$\left(\frac{d_{ae}}{d_f}\right)^2 \left(\frac{\gamma^*}{\gamma}\right) = \frac{9}{4}\left[\ln\left(\frac{2l_f}{d}\right) + 0.807\right] \tag{2.47}$$

Stöber (1971) conducted experiments for long straight fibers of amosite asbestos, examining how they settled out by centrifugal forces in a centrifuge-based aerosol separator, and used the results to obtain the empirical relationship:

$$\left(\frac{d_{ae}}{d_f}\right)^2 \left(\frac{\gamma^*}{\gamma}\right) = 1.43\left(\frac{l_f}{d}\right)^{0.232} \tag{2.48}$$

which, by inspection, is seen to provide results that are quite close to Equation (2.46). This suggests that, under the conditions of the experiment in question, the fibers were settling with their axes perpendicular to their centrifugal motion.

For most aerosols of practical interest when it comes to aerosol sampling, particles are not spheres but range in shape from objects of low aspect ratio (*isometric*) to ones of high aspect ratio (*platelet* or *fibrous*). In order to generalise the preceding discussion to describe what happens for a particle of arbitrary shape, it is therefore necessary to introduce an additional quantity, the *dynamic shape factor* (κ). Thus, for particles in the Stokes regime:

$$d_{ae} = d_V \left(\frac{\gamma}{\gamma^*\kappa}\right)^{1/2} \tag{2.49}$$

where d_V is the *equivalent volume particle diameter*. In this expression, $\kappa = 1$ for spherical particles and $\kappa > 1$ for particles of all other shapes, becoming greater the more a particle departs from sphericity (or *isometricity*). For such particles, mechanical mobility, as originally expressed for spherical particles in Equation (2.22), is now re written as:

$$\mu = \frac{1}{3\pi\eta d\kappa} \tag{2.50}$$

Table 2.1 shows values of κ for some typical particles of the type that might be encountered in aerosol sampling situations. The values for the spherical and fibrous particles of the types indicated have been calculated or estimated from experimental measurements (Mark *et al.*, 1985; Hinds, 1999).

2.2.7   Impaction

The phenomenon of *impaction* is particularly important in the scientific discussion of aerosol sampling. Figure 2.9 shows a typical pattern of streamsurfaces about a simple bluff body in the form of a flat disk facing the freestream, where the air diverges to pass around the outside of the disk. Also shown is the corresponding pattern of trajectories for particles of given aerodynamic diameter (d_{ae}), where it is seen that some, by virtue of their inertia, intersect with the surface of the disk. The *limiting trajectory* is

Table 2.1 *Published values of dynamic shape factor (κ) for a range of nonspherical particles of simple shape (Mark et al., 1985; Hinds, 1999)*

Shape/type	Dynamic shape factor, κ
Spherical particle	1.00
Fiber (aspect ratio = 5)	1.34 (axis perpendicular to motion)
	1.06 (axis parallel to motion)
Quartz dust particles	1.36
Fused alumina particles	1.04–1.49
Talc particles (platelet)	1.88

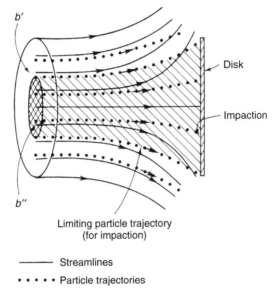

Figure 2.9 *Nature of particle impaction onto a flat plate facing directly into the wind, showing divergence of particle trajectories from air streamlines. Reproduced with permission from Vincent, Aerosol Sampling: Science and Practice. Copyright (1989) John Wiley & Sons, Ltd*

defined as the one inside of which all particles will impact onto the body and outside of which all will pass by. *Impaction efficiency* is defined as:

$$E = \frac{N_{imp}}{N_{inc}} \qquad (2.51)$$

where N_{imp} is the number of particles impacting onto the body in a given time interval and N_{inc} is the number of particles geometrically incident on the body in the same interval. By reference to Figure 2.9, for particles uniformly distributed in space in the uniform air flow approaching the body this amounts to:

$$E = \frac{b''}{b'} \qquad (2.52)$$

where b'' and b' are the areas projected upstream by the limiting trajectory surface and the body, respectively. If all the particles that strike the body stick, and are so are permanently removed from the flow, then Equation (2.52) also represents the efficiency of collection. For particles that obey Stokes' law ($Re_p < 1$) and for potential flow moving at velocity U about a body of characteristic dimension D, Equation (2.52) satisfies:

$$E = f(St) \tag{2.53}$$

where

$$St = \tau U/D \tag{2.54}$$

provided that, for reasons that will be given shortly, the particle geometric size (d) is small compared with the size (D) of the body. For most practical purposes in relation to aerosol sampling, this is a fair first-order approximation. But it has been suggested (e.g. Langmuir and Blodgett, 1946) that a relevant additional dimensionless group is the combination (Re/St). More generally, it seems sensible to write:

$$E = f(St, Re_p, Re) \tag{2.55}$$

in which Re is the Reynolds number defining the nature of the flow about the body itself. It is obvious from the preceding discussion that E will take values close to zero for St close to zero, will increase monotonically with St, then – at large values of $St (> 1)$ – tend to level off towards $E = 1$. Theoretical determination of the curve relating E and St for a given flow configuration is usually carried out by means of numerical solutions of the basic Equation (2.39), and many such calculations have been reported, starting as early as 1946 with those of Langmuir and Blodgett. There have also been many experimental studies confirming the predicted trend.

At this point, it is worth mentioning the interesting impaction regime corresponding to low pressure. Here, as discussed earlier, the Cunningham slip correction factor becomes progressively greater than unity as mfp for the air molecules increases beyond particle size. This means, for example, that, for $E = 0.5$, the value of d_{ae} required to maintain St *is* reduced, since this 'cut-off' value of d_{ae} falls as the operating pressure is reduced. This in turn raises possibilities for extending the range of the applications of the principles of impaction in practical sampling devices, as will be discussed in later chapters.

For a particle whose trajectory, as traced by the motion of its center, passes outside a bluff body but approaches closely enough, it may be collected by *interception*. For $d \ll D$, its effect on the behavior of E is negligible. But otherwise it is necessary to write:

$$E = f\left(St, Re_p, Re, \frac{d}{D}\right) \tag{2.56}$$

However, as far as aerosol sampling is concerned, interception may usually be disregarded.

2.2.8 Molecular diffusion

Particle motion has so far been assumed to be well ordered and, in theory at least, deterministic. In reality, however, even in apparently smooth (or laminar) air flow, aerosol particles exhibit random movement as a result of which there tends to be a net migration of particles from regions of high concentration to regions of lower concentration, irrespective of convection associated with any externally imposed air movement. Such motion is referred to as *molecular diffusion*, otherwise known as *Brownian motion*, and derives from the response of the particles to the thermal motion of the surrounding gas molecules. The flux of particles by this process may be described by the well-known Fick's law of classical diffusion. In the simplest, one-dimensional case, this is:

$$\text{Flux} = -D_B \frac{\partial c}{\partial x} \tag{2.57}$$

where D_B is the coefficient of Brownian diffusion. The resultant rate of change of local aerosol concentration is therefore:

$$\frac{\partial c}{\partial t} = D_B \frac{\partial^2 c}{\partial x^2} \tag{2.58}$$

whose solution for the simple one-dimensional case of N_0 particles released from $x = 0$ at time $t = 0$ gives the Gaussian form:

$$c(x, t) = \frac{N_0}{2(\pi D_B t)^{1/2}} \exp\left(-\frac{x^2}{4 D_B t}\right) \tag{2.59}$$

for the concentration distribution along the x direction at time t. The mean square displacement of particles from their origin is:

$$\overline{x^2} = 2 D_B t \tag{2.60}$$

The equivalent result for particle diffusion in three dimensions is:

$$\overline{x^2} = 6 D_B t \tag{2.61}$$

What is the significance of this to aerosol behavior in moving air? As discussed earlier, airflow streamlines represent the motion of inertialess particles in the absence of external forces and diffusion. However, when molecular diffusion is superimposed, the last two equations can be used to estimate the mean square excursion of particles from the streamlines on which they began – or, in the case of larger particles, from their ideal particle trajectories. Clearly the magnitude of the effect of molecular diffusion increases monotonically with D_B. Just how big is that effect? For small spherical particles in the Stokes regime, classical kinetic theory leads to:

$$D_B = \frac{k_B T}{3 \pi \eta d} C_c(d) \tag{2.62}$$

where T is the air temperature (in K), k_B is Boltzmann's constant ($= 1.23 \times 10^{10}$ J K^{-1}), and $C_c(d)$ is the Cunningham slip correction factor defined earlier in Equation (2.28). In this equation the numerator represents the thermal energy of the gas molecules that is transferred to the particles and the denominator represents the loss of particle energy due to viscous dissipation. Equation (2.62) therefore embodies the continual interchange of kinetic energy between the gas molecules and the particles, and vice versa. It shows that the effect of diffusion on overall particle transport is greater for smaller particles. Typically, for a spherical particle with $d = 1$ μm in air at STP, D_B is only of the order of 10^{-11} m^2 s^{-1}.

Strictly speaking, as already mentioned, Equation (2.62) only applies to spherical particles of geometric diameter d. For reasonably isometric nonspherical particles, for the particle diameter in this equation it is reasonable to represent d by the equivalent volume diameter, d_V, by virtue of its importance in the Stokes relationship for viscous forces acting on the particle (Williams and Loyalka, 1991). But now the particle dynamic shape factor should also be included. Particles of extreme shape are of special interest, especially in terms of their relevance to the lung deposition and sampling of very thin asbestos fibers. But then the orientation of the particle with respect to its motion at any point in time becomes important. Gentry *et al.* (1991) carried out an important series of experiments, using a diffusion battery, to measure D_B for thin chrysotile fibers characterised by a wide range of length (l_f) and width (d_f). The results showed that D_B was a complex function of l_f and d_f, and was much higher than predicted simply by using the equivalent volume diameter in Equation (2.62). It was also higher than predicted by other theories available at the time. Inspection of the results suggests that the closest agreement is achieved when the width of the fiber (d_f) is used as the particle size in Equation (2.62).

2.2.9   Turbulent diffusion

Earlier in this chapter the nature of the turbulent motions of the air itself was discussed. The problem of a particle of finite mass, that may exhibit finite inertia in response to the fluctuating turbulent motions, is even more complex. Yet it may be particularly important when it is considered that air movement in many aerosol sampling situations is turbulent. The concept of turbulence has already been characterised in terms of an integral *length scale* (L) and a *mean square velocity* (or energy) (e) associated with the fluctuating turbulent motions. For the fluid itself, the coefficient of turbulent diffusion may be expressed to a fair first approximation in terms of the product of the length scale and the root mean square turbulent velocity, as seen in Equation (2.19). For a suspended particle, its ability to respond to the eddying, turbulent motions of the surrounding fluid may be expected to depend on inertial considerations similar to those discussed earlier. Now, modified values of the length scale, mean square energy may be applied to the particle (L_p and e_p), and may be different to those for the fluid itself (L and e). For the diffusivity of a particle in a turbulent flow field, analogy with Equation (2.19) suggests:

$$D_{tp} \approx L_p e_p^{1/2} \tag{2.63}$$

Here, D_{tp} may differ significantly from D_t for the fluid alone. The early model of Peskin (1962), as described by Soo (1966), is useful. It was based on the experimental evidence of Soo *et al.* (1960) for two-phase turbulent flow in pipes, where it was noted that the correlation of velocities encountered by the particle was not the same as for the Lagrangian correlation of turbulence and that this effect was significant in reducing particle diffusivity. A relationship was proposed for the ratio between the local particle and fluid diffusivities (D_{tp} and D_t respectively) of the form:

$$\frac{D_{tp}}{D_t} \approx 1 - \left(\frac{\lambda_L^2}{\lambda_E^2} \right) \left(\frac{3k^2}{k+2} \right) \tag{2.64}$$

where λ_L and λ_E are the Lagrangian and Eulerian microscales, respectively, as introduced earlier. The dimensionless parameter:

$$k = \frac{\sqrt{\pi} Re_p \gamma d}{18 \eta \lambda_L \rho} \tag{2.65}$$

for a particle of diameter d and density γ in the Stokes regime, where Re_p here is the particle Reynolds number referred to the velocity fluctuations in the turbulent eddies. Equation (2.65) may be rewritten as

$$k = \sqrt{\pi} \left(\frac{d^2 \gamma}{18 \eta} \right) \left(\frac{e^{1/2}}{\lambda_L} \right) \approx \frac{\tau e^{1/2}}{\lambda_L} \tag{2.66}$$

This is similar to the related – and more accessible – parameter:

$$K_x = \frac{\tau (\overline{u_x'^2})^{1/2}}{L_x} \tag{2.67}$$

which, if it is plugged for k into Equation (2.64), provides a rough estimate of the coefficient of turbulent diffusion for particle transport in the x direction; and similarly for the other two directions. It is seen that K_x is equivalent to a Stokes number for a particle moving in the x direction in the distorted velocity field within an individual turbulent eddy.

From the above it is seen that, in general, the diffusivity of a particle relative to the fluid in turbulent flow is controlled by the two quantities K and λ_L/λ_E, whose relative importances can be interpreted as follows. For a given value of λ_L/λ_E and if $K \ll 1$, the particle will respond perfectly to the fluid motion, so that $D_{tp}/D_t = 1$. On the other hand, if $K \gg 1$, the particle will not respond well to the motion of the fluid, so that D_{tp}/D_t decreases. For a fixed but small value of K, and if λ_L/λ_E is large, a particle not following a given fluid element due to its inertia will encounter a succession of uncorrelated fluid velocities, and so its diffusivity will remain low. Conversely, if λ_L/λ_E is small, even though the particle might not follow perfectly the first fluid element it encounters, it will proceed to encounter successive correlated fluid elements such that it will appear to respond approximately 'perfectly'. As a result, it will exhibit high diffusivity regardless of the value of K. In the preceding discussion, understanding particle diffusivity as a function of K is relatively straightforward if the turbulence integral length scale and intensity are known. The ratio λ_L/λ_E is less easy, and no information is available for the sorts of distorted air flows relevant to aerosol sampling. For present purposes a fair working assumption is that:

$$\frac{D_{tp}}{D_t} = f(K) \tag{2.68}$$

What is the order of magnitude of D_{tp}? A rough estimate suffices to illustrate the importance of turbulent diffusion in relation to the corresponding molecular diffusion discussed earlier. Typical turbulence, relevant to air movement in a typical indoor environment, may have a length scale of at least a few centimeters and a root-mean-square velocity of a few centimeters per second. Assuming that D_{tp} and D_{tp} are approximately equal in magnitude, then, from Equation (2.68), D_{tp} is at least of the order of 10^{-3} m^2 s^{-1}. Clearly this is many orders of magnitude greater than the D_B for molecular diffusion indicated earlier.

References

Cox, R.G. (1970) The motion of long slender bodies in a viscous fluid: Part 1, General theory, *Journal of Fluid Mechanics*, 44, 791–810.

Friedlander, S.K. (1977) *Smoke, Dust and Haze: Fundamentals of Aerosol Behavior*, John Wiley & Sons, Ltd, New York.

Fuchs, N.A. (1964) *The Mechanics of Aerosols*, Macmillan, New York.

Gentry, J.W., Spurny, K.R. and Schörmann, J. (1991) The diffusion coefficients for ultrathin chrysotile fibers, *Journal of Aerosol Science*, 22, 869–880.

Hinds, W.C. (1999) *Aerosol Technology: Properties, Behavior and Measurement of Airborne Particles*, 2nd Edn, John Wiley & Sons, Ltd, New York.

Langmuir, I. and Blodgett, KB. (1946) US Army Forces Technical Report No. 5418.

Mark, D., Vincent, J.H., Gibson, H. and Witherspoon, W.A. (1985) Applications of closely-graded powders of fused alumina as test dusts for aerosol studies, *Journal of Aerosol Science*, 16, 125–131.

May, K.R. and Clifford, R. (1967) The impaction of aerosol particles on cylinders, spheres, ribbons and discs, *Annals of Occupational Hygiene*, 10, 83–95.

Oseen, C.W. (1910) Ober die Stokessche Formel und uber die verwandte Aufgabe in der Hydrodynamik, *Arkiv for Mathematik, Astronomi och Fysik*, 6, No. 29.

Peskin, R.L. (1962) The diffusivity of small suspended particles in turbulent fluids, Paper presented at the National Meeting of the AIChE, Baltimore, MD.

Ranz, W.E. and Wong, J.B. (1952) Impaction of dust and smoke particles, *Industrial and Engineering Chemistry*, 44, 1371–1381.

Reynolds, O. (1883) An experimental investigation of the circumstances which determine whether the motion of water shall be direct or sinuous, and of the law of resistance in parallel channels, *Philosophical Transactions of the Royal Society of London*, 174, 935–982.

Schlichting, H. and Gersten, K. (1999) *Boundary Layer Theory*, 8th Edn, Springer-Verlag, Berlin.

Soo, S.L. (1966) *Fluid Dynamics of Multiphase Systems*, Blaisdell, New York.

Soo, S.L., Ihrig, H.K. and El Kouh, A.F. (1960) Experimental determination of statistical properties of two-phase turbulent motion, *Transactions of the American Society of Mechanical Engineers, Journal of Basic Engineering*, 82D, 609–621.

Stöber, W. (1971) A note on the aerodynamic diameter and mobility of non-spherical aerosol particles, *Journal of Aerosol Science*, 2, 453–456.

Stokes, G.G. (1851) On the effect of the internal friction of fluids on the motion of a pendulum, *Transactions of the Cambridge Philosophical Society*, 9 (Pt 11), 8–106.

Vincent, J.H. (1989) *Aerosol Sampling: Science and Practice*, John Wiley & Sons, Ltd, Chichester.

Williams, M.M.R. and Loyalka, S.K. (1991) *Aerosol Science: Theory and Practice*, Pergamon Press, Oxford.

3

Experimental Methods in Aerosol Sampler Studies

3.1 Introduction

Studies of the characteristics of aerosol samplers require facilities and experimental methods that, as in other scientific fields, are chosen to provide the desired information to the level of quality desired and within the budgetary resources available. In recent decades, there has been a large expansion in the amount and degree of sophistication of equipment that is commercially available for aerosol research in general, including a bewildering array of aerosol generating and detection devices. Aerosol research may be an expensive activity. However, the history of aerosol sampling demonstrates that it is possible to conduct high-quality experiments with relatively simple equipment and instrumentation, provided that certain basic guidelines are adhered to. In this chapter, a review is given of a range of typical apparatus and methodology used by workers in this field.

3.2 Methodology for assessing sampler performance

The basic performance parameters of aerosol samplers were defined in Chapter 1. These provide the basis for the experiments that might be carried out to investigate the performance characteristics of aerosol samplers. Attention is now drawn to the distinction between the purely aerodynamic aspect as reflected in aspiration efficiency (A) and the more general performance as reflected in *apparent* aspiration efficiency (A_{app}) which allows for particles entering the sampler after having been in contact with its external surfaces. Choice of methodologies for assessing sampler performance – and in turn the design of experimental facilities and methodology – need to take account of this distinction.

3.2.1 The direct (trajectory) method

The *direct* (or *trajectory*) method is based on assessment of pure aspiration efficiency (A) as described by $A = a''/a_s$ (see Figure 1.5). It involves the experimental determination of particle trajectories near the sampler so that the limiting trajectory of particles which just enter through the plane of the sampling

Aerosol Sampling: Science, Standards, Instrumentation and Applications   James Vincent
© 2007 John Wiley & Sons, Ltd

orifice can be defined. Thus a'' is obtained directly. Meanwhile a_s is derived from knowledge of the sampling flow rate and the external wind speed.

Surprisingly, the method does not appear to have been widely applied, its use having being pioneered and used mainly by workers in the former Soviet Union in investigations of thin-walled sampling probes, notably by Belyaev and Levin (1972, 1974) and, subsequently, by Lipatov and his colleagues (1986). In their earlier version, Belyaev and Levin dispersed relatively monodisperse particles of *Lycopodium* powder into a wind tunnel and, for given sampling conditions, photographed the pattern of particle trajectories under strong flash illumination. Despite the fact that *Re* for the air flow in the wind tunnel was of the order of from 10 000 to 100 000, and hence that the air flow would almost certainly have been turbulent, Belyaev and Levin were able to trace limiting particle trajectories quite accurately. The later version of the method described by Lipatov *et al.* represented a substantial refinement. They used a small, laminar-flow wind tunnel, and a *vibrating-tip aerosol generator* to produce a stable, clearly visible stream of droplets that could be viewed under dark-field illumination using a pair of horizontal microscopes with cross-oriented optical axes. The principle of the experimental approach taken is illustrated in Figure 3.1 where aspiration efficiency is determined from:

$$A = \frac{\Delta^2 U}{\delta^2 U_s} \tag{3.1}$$

in which δ is the diameter of the sampling orifice and Δ the diameter of the 'tube' defining the limiting trajectory.

3.2.2   The indirect (comparison) method

The *indirect* (or *comparison*) method is based on assessment of aspiration efficiency as described by $A = c_s/c_0$ [see earlier Equation (1.20) and Figure 1.5]. This is the method by far the most widely used by experimenters for the determination of aerosol sampler performance. It involves exposure of the sampler to be tested to an aerosol of uniform spatial concentration in a flow field that is also spatially uniform. This method requires measurement of the concentration of aerosol passing through the plane of the sampling orifice (c_s) and that (c_0) in the undisturbed freestream. Determination of c_s is relatively straightforward, from knowledge of the mass of aerosol collected, sampling volumetric flow rate and sampling time. Determination of c_0 on the other hand requires *a priori* assumption about the performance of some other sampling device. It was established at an early stage in the history of aerosol sampling (see Chapter 5) that a thin-walled, cylindrical sampling tube facing the wind and operated isokinetically has $A = 1$ for all particle sizes, thus providing a valid and convenient means for the determination of c_0 in moving air. The situation is more difficult for calm air where, as will be seen later, models for accurately determining c_0 have been – and, indeed, remain – somewhat elusive.

Experimentally, it is necessary to expose both samplers – the test sampler and reference sampler, respectively – to the same aerosol-laden freestream. Researchers have achieved this by several means. One is to place the two samplers side by side, close enough together so that the aerosol concentration is the same upstream of each yet far enough apart to ensure that there is no fluid mechanical mutual interference. Another approach is to place the reference sampler upstream of, yet axial to, the test sampler, far enough away not to be influenced by any flow distortion associated with the test sampler. Here it is again assumed that the upstream presence of the reference sampler does not interfere with the performance of the test sampler. A third approach is a 'time-share' procedure, by which the test and reference samplers, respectively, are placed sequentially at the same test location. During intervals when either sampler is not in the test location, its sampling flow rate is interrupted. During a given experiment,

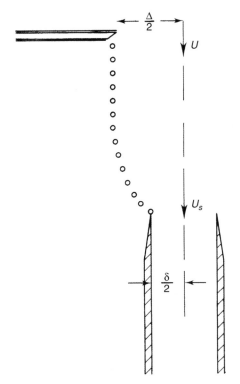

Figure 3.1 *Diagram to illustrate the principle of the direct method for assessing aerosol sampler aspiration efficiency. Reproduced with permission from Vincent, Aerosol Sampling: Science and Practice. Copyright (1989) John Wiley & Sons, Ltd*

several cycles of test and reference sampler interchange might take place. This method requires that, averaged over time, the freestream aerosol concentrations to which the two samplers are exposed are the same, imposing a stiff requirement on the test aerosol generation method. However, it has the advantage over the other variations of the indirect method that there is no possibility of aerodynamic interference between samplers.

3.2.3 Critique of the alternative methods

Of the two basic methods outlined above, the direct method has the advantage that it provides aspiration efficiency (A) unambiguously, and so provides the most ideal means for providing data against which to assess theory. However, it should ideally be carried out under smooth flow conditions so that the particle trajectories can be clearly viewed. Even then, it seems only to have been applied for studying the very simplest of sampler systems – small, thin-walled probes facing the wind. This, perhaps, reflects the degree of experimental difficulty involved in this approach.

The indirect method is more straightforward, and may be applied over a wide range of types of experiment. Its major disadvantage is that all particles that enter the sampling orifice are involved in the determination of aspiration efficiency. If these include some particles that have arrived there after having rebounded from the external surfaces of the sampler, then it is not the true aspiration efficiency

(*A*) that has been obtained but, rather, the *apparent* aspiration efficiency (A_{app}) (see Chapter 1). Under certain conditions, this could complicate the interpretation of experimental results.

3.3   Scaling relationships for aerosol samplers

In engineering science, dimensional analysis is widely used in order to describe the physical behaviors of systems using the smallest number of parameters and in a way that the description is independent of the system of units that might be applied. In fluid mechanics, Reynolds number (*Re*) is one such parameter. In aerosol mechanics, so too is Stokes number (*St*). In aerosol sampling science, the description of sampling efficiency – however it is defined – in terms of dimensionless groups allows systems of similar geometry to be scaled between different systems (e.g. large versus small, slow versus fast, etc.) whilst retaining similar behavior. If this can be achieved, then the results from experiments with a scaled system may be applied directly to other similar, perhaps larger, systems in the real world.

As will be seen as this book progresses, a set of primary variables for describing the aspiration efficiency of a single-orifice aerosol sampling system of given geometrical shape, along with their dimensions, contains:

- U, the air velocity in the freestream approaching the sampler (wind speed) (L T^{-1});
- U_s, the mean air velocity across the plane of the sampling orifice (L T^{-1});
- δ, the width of the sampling orifice (L);
- D, the width of the sampler body (L);
- d_{ae}, the particle aerodynamic diameter (L);
- θ, the orientation of the sampler with respect to the freestream (dimensionless).

Here L represents the dimension of length and T the dimension of time. It is assumed that the transporting medium is air, such that air density and viscosity are constant, and acknowledged that particle variables such as density are embodied within d_{ae}. It is easily recognised that the aspiration efficiency of the aerosol sampler can be expressed as a function of a set of dimensionless groups, thus:

$$A = \mathrm{f} \ \{Re, St, R, r, \theta\} \tag{3.2}$$

where gravity is neglected, and where

$$Re \propto DU \tag{3.3}$$

$$St \propto d_{ae}^2 U / \delta \tag{3.4}$$

$$R = U / U_s \tag{3.5}$$

$$r = \delta / D \tag{3.6}$$

More dimensionless groups would be required if the aim is to describe the effect of gravity and the more general sampling efficiency (*E*) that also includes particle transport inside the sampler after aspiration and on the way to the filter.

For aerosol sampling, regardless of the physical size of the sampler in question, the aspiration efficiency should remain the same as long as *Re*, *St*, *R*, *r* and θ are all also held constant (Kenny *et al.*, 2000). However, although this is strictly true, such an approach is unnecessarily restrictive. For example, if one wanted to simulate particle sizes up to 100 μm by using particles in the small-scale system of up to 20 μm, the sampler dimensions would have to be scaled down by a factor of 5 and the wind speed and sampling velocity scaled up by a factor of 5. This would give little flexibility in designing a small-scale system for the purpose of a laboratory experiment. Fortunately, however, there is some room for compromise. In general, the behavior of fluid mechanical systems tend to be relatively weak functions of *Re* if *Re* is large enough for inertial forces to dominate. Across the whole range of conditions of potential interest with respect to aerosol sampling, this might hold in some situations but not in others.[1] For practical purposes it is often a fair working assumption that behavior is only weakly dependent on *Re* for the purpose of designing laboratory-based aerosol sampler testing experiments. Then scaling may be based more simply on:

$$A = f\,\{St, R, r\} \tag{3.7}$$

and this opens up realistic opportunities for the use of smaller-scale laboratory systems.

With the above in mind, in order to keep R constant, U and U_s should be scaled by the same factor, k_U. The scaling factor for the sampler orifice, δ, may be scaled by the factor k_δ under the assumption that, to preserve geometric similarity, D will be scaled by the same factor. Since, as seen in Equation (3.3), St is proportional to the square of the particle size multiplied by U/δ, the basic scaling relationship for aerosol sampling is then given by

$$k_{dae}^2 = k_\delta / k_U \tag{3.8}$$

Such scaling laws allow the scaling not only of physical dimensions, wind speeds and flow rates but also, importantly, of the particle size in the test aerosol. The suggested scheme allows more flexibility in designing actual small-scale systems for laboratory research than would have been obtained from a more complete consideration of all the possible contributing factors. It provides a useful practical basis, provided that the investigator is aware of the possible limitations in specific cases.

3.4 Test facilities

Test facilities for aerosol sampler studies in the laboratory are designed and developed in order to simulate environments reflecting as far as possible those in the real world. But, since the real world is so complex and variable, it is inevitable that any such facilities must be approached with a significant element of compromise. It is therefore necessary to identify and prioritise the primary environmental variables that must be simulated. In the first instance, the ranges of the magnitude of the wind speed and direction must be defined for each given real-world scenario of interest. The structure of freestream turbulence (e.g. intensity and length scale) should also be included, at the very least as a secondary consideration. These considerations have featured in the best research in the field of aerosol sampler enquiry.

Air motions in the working section of test facilities should relate directly – or indirectly through the use of scaling relationships – to wind speeds known to exist in the environments of interest. Some occupational environments are characterised by quite high average wind speeds, up to as high as 4 m s^{-1}. Such values are frequently found, for example, in the highly ventilated workplaces of underground mining. It is notable that much of the earlier occupational hygiene research where such wind speeds

---

[1] For air at STP, it is a useful ad hoc tool to note that $Re = 7 \times 10^4 DU$. For a small sampler in a slowly moving air stream, where $D \approx 0.05$ m and $U \approx 0.5$ m s^{-1}, we get $Re \approx 200$. It rises to more than 2000 when $D \approx 0.30$ m and $U \approx 1$ m s^{-1}, conditions typical of some personal sampling situations where a small sampling device is mounted on the body of a person whose aerosol exposure is the subject of interest.

were simulated was driven by the need to address lung disease in dust-exposed mineworkers. The result was that much of the earlier aerosol sampler research was funded by agencies addressing those concerns, and so was carried out in facilities designed to reflect those types of environment. More recently, however, more information has emerged about the magnitudes of wind speeds in other types of workplace. Berry and Froude (1989) and Baldwin and Maynard (1998) of the United Kingdom Health and Safety Executive conducted surveys of wind speeds in a range of industrial workplaces, including the relative motions between workers and the surrounding air as they go about their tasks. They found that the measured wind speeds were surprisingly low, rarely exceeding $0.2-0.3$ m s^{-1} and were more typically less than 0.1 m s^{-1}. Similar results have been reported elsewhere. So it is now clear, at least for indoor occupational settings, that experimental facilities need to be able to simulate air movements all the way down to almost perfect calmness. The situation is somewhat different for outdoor air situations relevant to ambient air sampling, where the range of wind speeds is inevitably greater. Here, average wind speeds up to 10 m s^{-1}, or even higher, are relevant.

The realistic simulation of turbulence is more difficult. In the real world, intensities and length scales may often be greater than can be sustained inside the enclosed test sections of laboratory-based test facilities of the various types that will be described below. It then becomes a question of whether the effects of any such shortcomings are significant enough to change the interpretation of results. These are addressed in a later chapter of this book. As a general guideline for the design of laboratory test facilities, recognition of a possible role of turbulence should dictate a facility design that provides turbulence conditions that – at a minimum – are clearly defined and consistent.

3.4.1   Moving air

Moving air may be defined in the first instance for conditions where aerosol transport is governed largely by convection. That is, particles are assumed to be truly airborne, and so do not undergo any significant motion relative to the air associated with gravity. Purpose-built *aerosol wind tunnels* have been designed for this purpose.

The design of wind tunnels for aerodynamics research in general has a long history, and much of what has been learned may be applied to aerosol studies. However, in much experimental engineering fluid mechanics there is an expectation of very idealised conditions, some of which are very difficult to achieve in an aerosol wind tunnel, especially one built for aerosol sampler studies. One difficulty is that the injection of aerosol into the freestream of a wind tunnel working section will likely introduce undesirable and uncontrolled disturbance into the air flow. It is difficult to avoid such disturbances, but their effects can be minimised by careful design. Another difficulty is associated with the blockage ratio, reflecting the ratio of the frontal area of the model to the area of the working section. In a wind tunnel with an enclosed working section, the proximity of the solid wall causes an acceleration of the airflow as it diverges to pass around the body of the model. This will influence the boundary layer of the air movement close to the model, and in turn the transport of airborne particles. To minimise fluid mechanical effects arising from such blockage, aerodynamicists generally recommend that blockage should not exceed $10\,\%$ (Maskell, 1965), and perhaps in some situations should be even as low as $5\,\%$ (e.g. for aeronautical test studies). But for aerosol sampler studies the problem is compounded by the fact that large wind tunnel cross-sections make it difficult to achieve a spatially uniform test aerosol, providing an incentive to keep the cross-sectional area as small as possible. Compromises have to be made, taking all the considerations into account, and it has become customary in aerosol sampler research for blockage as high as $30\,\%$ to be thought of as satisfactory.

Wind tunnels used in aerosol sampler studies described in the literature vary greatly in size and configuration. When the design of such a facility for a new enquiry is being considered, there are certain general features and characteristics that need to be taken into account. Firstly, there is the question of size. For example, in some research dealing with small thin-walled probes facing the wind, very small wind tunnels – with cylindrical cross-sections as little as 10 cm in diameter – have been used (see Chapter 5). For such systems, the Reynolds number (*Re*) for the wind tunnel flow may be low enough for the air movement to be laminar. Smaller wind tunnels have the additional advantage over larger ones in that it is easier to obtain uniform spatial distribution of aerosol over the working region of interest. Also, because the overall volumetric flow rate is less, the amount of aerosol output which is required from the generator is less. More generally, however, practical aerosol sampler systems are too large to be tested in such miniaturised wind tunnels. A number of wind tunnels used in sampler studies over the years have embodied, in various ways, most of the desired features and so are worth describing here.

The miniature facility at Odessa State University in Ukraine (when it was the former Soviet Union) was built in the early 1980s by Professor Genady Lipatov and his colleagues for the investigation of the aspiration efficiency of thin-walled probes by the direct method (Lipatov *et al.*, 1986). It was designed to enable studies of small-scale, idealised sampling systems in experiments aimed at providing improved understanding of basic aerosol sampling processes, and was never intended for use in the study of practical sampling devices. As shown schematically in Figure 3.2, this tunnel was circular in cross-section and installed so that the air flow was vertically downwards. This last feature ensured that the effects of gravity on the trajectories of the test aerosols were minimised. The dimensions were small, with a working section diameter of just 3 cm, so that, for the range of air velocities studied, *Re* was low enough for the flow to be laminar. The contoured, contracting profile of the entry to the working section ensured uniform plug flow. Monodisperse liquid droplet test aerosols were generated for this system by means of a vibrating-tip generator and were injected into the airstream purely under the influence of gravity. By this combination of conditions, the air flow in the working section was not disturbed either by turbulence or by the injection of excess air. The result was a thin 'thread of droplets' that could be clearly visualised under appropriate lighting. By locating the position of this particle stream with respect to the sampler inlet, and incrementally moving one relative to the other, it was possible to

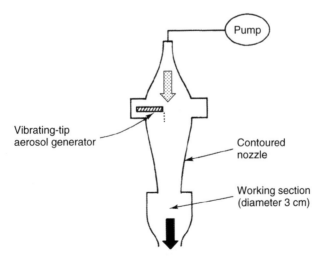

Figure 3.2 *University of Odessa miniature wind tunnel. Reprinted from Journal of Aerosol Science, 17, Lipatov et al., 763–769. Copyright (1986), with permission from Elsevier*

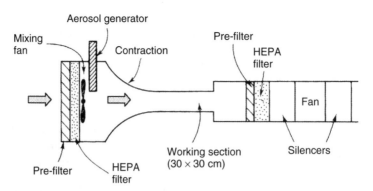

Figure 3.3   *University of Cincinnati wind tunnel. Reproduced with permission from Vincent, Aerosol Sampling: Science and Practice. Copyright (1989) John Wiley & Sons, Ltd*

locate the conditions under which the thread just entered the sampling orifice. This in turn provided the information needed to determine the aspiration efficiency of the sampler. Many years later, Witschger *et al.* (1997) described a somewhat similar cylindrical aerosol wind tunnel that was built at the French Institut National de Recherche de Sécurité (INRS) in Vandoevre, but with a larger working section of diameter 30 cm. This facility was intended for the study of small practical samplers, including the investigation of aspiration efficiency, wall losses, etc.

The wind tunnel at the University of Cincinnati in the United States featured in much of the work of Professor Klaus Willeke and his colleagues during the 1980s (Tufto and Willeke, 1982). As shown in Figure 3.3, this wind tunnel was an open-cycle machine of square cross-section. At its intake, air was drawn from the laboratory atmosphere through a high-efficiency (HEPA) filter to remove unwanted particles. Monodisperse test aerosols from a vibrating-orifice generator (described below) were introduced into the freestream immediately after this filter and were brought to a reasonably uniform spatial distribution by means of a slowly rotating mixing fan. The aerosol-laden freestream then passed through a 16:1 contraction down to the 0.3 m × 0.3 m square cross-section of the horizontal working region. This had the effect of reducing the overall intensity of turbulence in the air flow conveyed into the working section, and also served to further improve the aerosol spatial distribution. Air movement was achieved by means of a tubular centrifugal fan located downstream of the working section. The exhaust air was discharged back into the laboratory through a second HEPA filter. The performances of the aerosol sampler systems tested in this facility were characterised using the indirect method. This wind tunnel, like the others described above, was intended mainly for the study of simple, small, relatively idealised sampling systems.

The much larger wind tunnel at the Institute of Occupational Medicine (IOM) at Edinburgh in the United Kingdom was originally designed and built at the University of Nottingham for civil engineering studies (Brown, 1967). After it was relocated to Edinburgh in the late 1970s, it was modified to make it suitable for aerosol research, for which it was used for the next two decades in the study of full-scale practical sampling systems, including personal samplers mounted on the bodies of mannequins. As shown in Figure 3.4, it had an open-cycle, open-jet configuration, with a rectangular open working cross-section of about 1.5 m × 2.5 m, large enough to accommodate a life-size human mannequin. This configuration was such that, because of the absence of walls in the working section, potential blockage problems were eliminated. The horizontal air movement was generated by means of two 1.5 m diameter axial fans on the downstream side of the working region. Air was drawn in from the laboratory and discharged

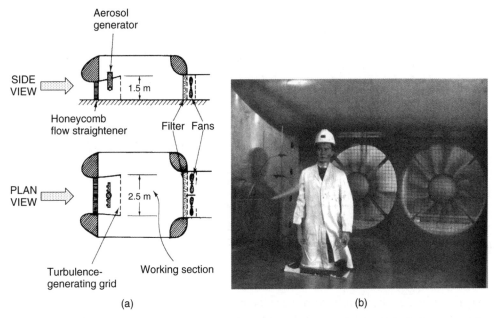

Figure 3.4 *Institute of Occupational Medicine (IOM) wind tunnel: (a) diagram of the lay-out of the tunnel; (b) photograph of the tunnel, also showing the mannequin that was used in some of the research. (a) Reproduced with permission from Vincent, Aerosol Sampling: Science and Practice. Copyright (1989) John Wiley & Sons, Ltd. (b) Reprinted from Aerosol Science for Industrial Hygienists, Vincent. Copyright (1995), with permission from Elsevier*

to outdoors through a filter. The airflow in the working region was turbulent with mean velocity in the range up to about 5 m s^{-1}. Nearly monodisperse test aerosols were mechanically generated from narrowly graded powders and dispersed through an array of nozzles into the wind tunnel upstream of the working section. For each powder grade and wind speed, the positions of the nozzles were adjusted in order to optimise the uniformity of the spatial distribution of aerosol in the working section, and this was achieved to within ±10 %. In order to facilitate mixing of the test aerosol, as well as control and define the level of turbulence, a square-lattice grid was placed across its entrance, in the manner prescribed by Baines and Peterson (1951) and others. Typically, a grid with a bar width of 2.5 cm and blockage less than 50 % provided turbulence with an intensity of about 7 % and a length scale of about 2 cm at about 1 m downwind of the plane of the grid. Importantly, it was found that, due to the action of the grid, the turbulence level in the working section was affected negligibly by the disturbance associated with the compressed air jet. In addition, there was no measurable influence on the spatial distribution of air velocity. This meant, therefore, that flow conditions in the working section were determined almost entirely by the grid and hardly at all by the aerosol injection process. Again the performances of the aerosol sampler systems tested in this wind tunnel were characterised using the indirect method.

During the 1980s, a similarly large wind tunnel was built by Professor William C. Hinds and his colleagues at the University of California at Los Angeles (UCLA) in the USA for use in studies of aerosol sampling and human inhalability, also using the indirect method. As later reported by Hinds and Kuo (1995), it was designed to provide uniform and stable concentrations of test aerosols up to 150 μm. It is shown in Figure 3.5 (Kennedy and Hinds, 2002). Like the IOM facility, this too was

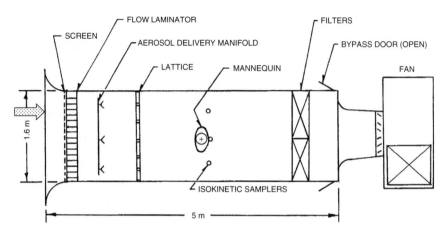

Figure 3.5 *University of California Los Angeles wind tunnel. Reprinted from Journal of Aerosol Science, 33, Kennedy and Hinds, 237–255. Copyright (2002), with permission from Elsevier*

large enough to accommodate a full-torso mannequin. But unlike the IOM machine it was a closed-jet facility, with an enclosed working section having a large square cross-section of 1.6 m x 1.6 m. Wind speeds were variable from less than 0.1 up to 2 m s^{-1}. Aerosol generation, again from narrowly graded powders, was achieved by an array of aspirating dust feeders which delivered aerosol to a reciprocating distribution manifold at the entrance to the working section. This arrangement enabled time-averaged aerosol concentrations in the working section to be uniform to within ±15 %. As in the IOM wind tunnel, a turbulence-generating grid was placed in the entrance to the working section, providing turbulence intensities in the center of the working section of from 3 to 14 %. The large wind tunnel built originally in the early 1990s by David Mark and his colleagues at Warren Spring Laboratory at Stevenage in the UK, later transferred to AEA Technology at Harwell, was also a closed-jet facility with working section large enough to accommodate a life-size mannequin (Figure 3.6). It featured in a large body of aerosol sampler research funded by the European Community that will be described later in this book. A similar machine was built by Mark Hoover and colleagues at the Lovelace Inhalation Toxicology Research Institute in Albuquerque, NM, USA in cooperation with scientists at Texas A&M University (Hoover *et al.*, 1996), with the specific goal of meeting US Environmental Protection Agency (EPA) guidelines for aerosol wind tunnel performance (Environmental Protection Agency, 1995). It had a 0.76 m × 0.76 m working section in which the air velocity could be varied continuously from 0.7 to 3.6 m s^{-1}.

The preceding wind tunnels were used in the work that contributed to the large body of aerosol sampling research that emerged during the 1980s and 1990s. Latterly, it had become an established philosophy in much such research that the testing and validating of aerosol samplers intended for practical use, in particular personal samplers of the type worn by people in order that their individual aerosol exposures might be assessed, should be carried out in wind tunnels with working sections large enough to accommodate life-size human mannequins. These are reflected, for example, in the recommendations of the Comité Européen de Normalisation (CEN) for standardised aerosol sampler testing (Comité Européen de Normalisation, 2001). However, this philosophy led to tension surrounding the requirement that much aerosol sampler testing would therefore always need to be carried out in large – and inevitably costly – facilities requiring laborious (and also costly) test procedures. So the search began for simpler and lower-cost approaches. Recent studies of the scaling relationships for

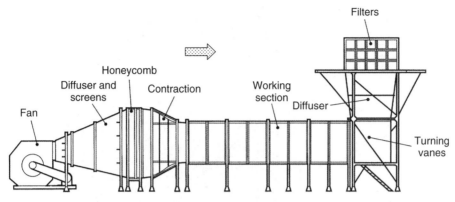

Figure 3.6   *The large wind tunnel at AEA Technology, Harwell, UK. Drawing courtesy of David Mark, Health and Safety Laboratory, Buxton, UK*

aerosol sampler performance, as described earlier in this chapter, pointed the way towards lower-cost, faster test methods, including a return to smaller and more manageable wind tunnel facilities. Such ideas provided the basis of the wind tunnel that was built at the University of Minnesota in the mid 1990s and the similar one that was built at the University of Michigan in the late 1990s. The more recent Michigan version is shown in Figure 3.7, featuring a test section with a square cross-sectional area of 0.3 m × 0.3 m, much smaller than the previously mentioned larger wind tunnels typically used to characterise aerosol samplers. In this machine, air was drawn through an inlet containing a bank of HEPA filters, passed through a honeycomb screen, and accelerated through a 6.25:1 contraction into the aerosol dispersion section. Test aerosol was introduced into the dispersion section and passed through a square-mesh turbulence grid, and then into the test section. Again the purpose of the turbulence grid was to facilitate mixing of the aerosol and to establish well-defined turbulence in the test section, yielding an intensity of from 3 to 4 % and length scale of from 2 to 3 cm in the middle of the working section. Downstream of the test section, the flow passed through a diffuser and was then discharged through another bank of HEPA filters. The fan was able to provide wind speeds in the test section ranging continuously from 0.1 to about 20 m s^{-1}. In this wind tunnel it was possible to study small, scaled-down versions of aerosol sampling systems and to relate the results to their full-scale equivalents. Again, for all the experiments that were reported for this system, the performances of the aerosol sampler systems tested were characterised using the indirect method.

The list of wind tunnel facilities for aerosol sampler studies given above is not exhaustive, and many others have been described in the literature. Most have features in common with one or more of those illustrated in this chapter.

3.4.2   Calm air

Unlike for moving air, in the ultimate limit of sampling under perfectly stationary ('calm') air conditions there is no externally provided convection that brings particles into the vicinity of the sampler system of interest. The only convection that is present is that derived from the aspirating action of the sampler itself, and this is only significant close to the sampler inlet. So, as Davies (1968) was the first to point out, the physical scenario is very different from that for moving air. Now, instead of convection, it is gravitational settling that brings the particles close to the sampling inlet.

Figure 3.7   *University of Michigan wind tunnel*

Aerosol sampling in perfectly calm air is highly idealised, and is probably never truly realised in the real world. That said, an understanding of the calm air situation is an important part of a complete picture of the physics of aerosol sampling over the full range of possible conditions. So it is important that research should be conducted for this important limiting situation. However, it is a fact that experiments under perfectly calm air conditions are very difficult to execute, one important aspect being the establishment and maintaining true calm air conditions, especially while test aerosol is being injected. This difficulty probably accounts for the relatively small amount of reported research in this area.

There have been a few notable contributions. The test chamber built at the IOM in Edinburgh during the late 1970s, and shown in Figure 3.8, is typical (Gibson and Ogden, 1977). It comprised a cube-shaped working section of dimension 1 m through which there was no net air flow. Test aerosols, generated by a variety of means, were injected into the space above the working section and mixed by means of four rotating, 18 cm diameter, dispersion disks. The particles entered the working section by settling under gravity through the 6 mm mesh and 50 mm thick honeycomb which served to minimise the penetration of air movement from the upper mixing volume to the lower working section. Local air velocities in this calm air chamber rarely exceeded 5 cm s^{-1}. A number of similar such calm air chambers have since been described. In most of the research using such test chambers, aerosol sampler performance was determined using the indirect method. However, there was a consistent weakness in those reported studies in that the methods that were employed to provide the reference aerosol concentration were likely biased uncertainly due to the general paucity of knowledge about aerosol sampler performance in calm air. There was, however, just one notable study by Yoshida *et al.* (1978) in which the direct method was used, in which the trajectories of particles near the sampler entry were visualised under suitable illumination, and then located quantitatively from photographs of particle tracks taken through a window in the side of the chamber. In this way, the efficiencies by which the particles entered the sampling test systems of interest were determined unambiguously.

This method was recently taken a stage further in the calm air test facility built in the author's laboratory at the University of Michigan (Su and Vincent, 2002). Development of this system extended what had been learned not only from the work of Yoshida *et al.* in terms of the general direct-method approach but also the previously mentioned thread of droplets approach of Lipatov *et al.* (1986) for

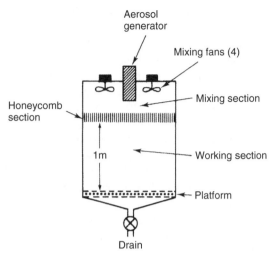

Figure 3.8 *Schematic of a typical calm air chamber of the type used in aerosol sampling studies. Reproduced with permission from Vincent, Aerosol Sampling: Science and Practice. Copyright (1989) John Wiley & Sons, Ltd*

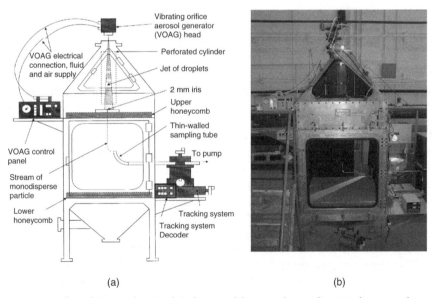

(a)   (b)

Figure 3.9 *University of Michigan calm air chamber used for sampler studies: (a) diagram of set-up; (b) photograph of the actual chamber. Reprinted from Journal of Aerosol Science, 33, Su and Vincent, 103–118. Copyright (2002), with permission from Elsevier*

generating the test aerosol. Here, therefore, as in the Lipatov *et al.* work for moving air, the experimental challenge was to locate the limiting condition when the droplet stream just entered the sampler inlet. As shown in Figure 3.9, it comprised three parts: an upper pyramid, a central main test chamber and a lower pyramid with a base. The upper and lower parts were separated from the main working section by honeycomb sections which provided for straight and slow air movement in the working section only as induced by the aspirating action of the sampler itself. An opening at the top of the upper pyramid provided the inlet for the aerosol injection. The main central chamber was a 1 m³ cube with one front glass door, two aluminum side-walls and one back glass window. The behavior of the falling particles was observed visually through the front glass door. On the right aluminum wall of the chamber was a narrow horizontal rectangular slot, providing access for the tube that carried the sampling system of interest. This slot was sealed by a rectangular rubber gland that permitted the tube, and in turn the sampler inlet, to move in both dimensions of the horizontal plane. Illumination of particles inside the test chamber was achieved through the rear glass window using an external lamp system. Monodisperse test aerosol was generated using a vibrating orifice aerosol generator (see below) and injected directly downwards into the upper pyramid of the facility. After evaporation of the volatile component, the resultant droplets of the desired stable liquid media approached the central part of the calm air chamber in the form of a conical spray. In order to achieve the desired ideal thread of droplets falling into undisturbed air inside the main test chamber, the primary droplet spray was collimated into a single droplet stream by the placement of a metal plate containing a central 2 mm iris through which a single, stable stream of droplets passed as it fell through the honeycomb.

3.4.3 Slowly moving air

It is now emerging that air in many indoor situations is neither calm, nor is it moving in the sense that has been discussed so far in the context of this book. In air which is moving slowly at velocities of the magnitude described earlier, particle transport has physical characteristics associated with both extremes. That is, particles move partially under the influence of an externally induced convection as well as under the influence of gravitational settling. The definition of 'slowly moving air' cannot be made simply in terms of air velocities, but must also reflect particle motion, in particular the effect of gravitational settling. Neither moving air nor calm air facilities of the type described above can adequately simulate such conditions. Recognising the importance of the slowly moving regime, a number of workers have therefore attempted to design experiments to characterise aerosol samplers accordingly.

In 2000, Fauvel and Witschger recognised the need to be able to carry out aerosol sampler studies in the laboratory but under realistic conditions (Fauvel and Witschger, 2000). For this, a test room was built at the French Institut de Protection et de Sûreté Nucléaire (IPSN) at Gif sur Yvette, having overall dimensions of 4 m × 3 m × 3 m. This facility was big enough to allow simulation of many actual workplace conditions, including specific ventilation arrangements. It was intended specifically for the study of both area (or static) and personal aerosol sampling systems. Soon afterwards, a facility was built at the University of Cincinnati that acknowledged the difficulty of studying the transport of large particles near life-sized bluff bodies in low wind speeds and attempted to address them directly (Aizenberg *et al.*, 2001). Figure 3.10 shows an open-throat wind tunnel designed such that, in the first instance, blockage effects were significantly reduced, in the same way that it was for the open-throat IOM wind tunnel described above. In the Cincinnati facility large particles with aerodynamic diameter in the range from 60 to 240 μm were delivered under gravity to the working section of the facility from above using a backwards-and-forwards oscillating spreader. Meanwhile the sampling system itself was mounted on a laterally oscillating frame. It was acknowledged that the instantaneous aerosol concentration distribution would not be uniform in such a system. However it was shown that, by careful adjustment of the rate and amplitudes of the oscillations of the moving components, reasonably uniform time-averaged aerosol concentration distributions could be achieved.

Very recently, a new facility was built at the University of Michigan that also attempted to address the problems with aerosol sampler experiments at such low wind speeds or large particle sizes. This

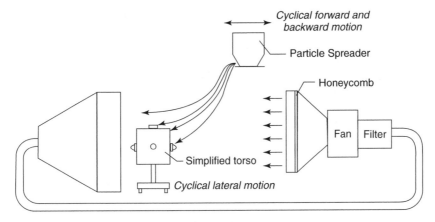

Figure 3.10 *University of Cincinnati low wind speed wind tunnel. Reprinted from Journal of Aerosol Science, 32, Aizenberg et al., 779–793. Copyright (2001), with permission from Elsevier*

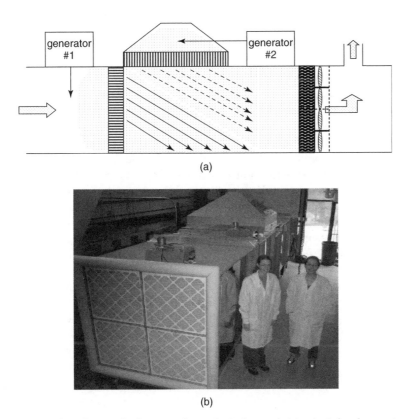

Figure 3.11 *University of Michigan ultralow-wind speed wind tunnel: (a) principle of operation (dashed arrows represent aerosol delivered from above from generator #1, solid arrows from the left from generator #2); (b) photograph of the wind tunnel itself*

facility was described as an 'ultra-low velocity wind tunnel', and combined features of both wind tunnels and calm air chambers of the types described above. As shown in Figure 3.11, mechanically generated aerosol was injected into a pair of dispersal chambers, one located directly upstream of the working section (as in the wind tunnels that have been described) and the other located above the working section (as in the calm air chambers that have been described). For a given test aerosol and given low wind speed, the injection of test aerosol into the respective dispersion chambers was 'balanced' in order to provide the optimum uniformity of test aerosol across the working section of the tunnel where the sampling systems of interest would be placed.

3.5   Test aerosol generation

For assessing the performance characteristics of aerosol samplers, the choice of type of test aerosol depends on the particular characteristic of sampler performance that is being studied. For example, if it is the purely aerodynamic part of performance (i.e. aspiration efficiency) that is of interest, then the only properties of the test aerosol that need to be defined are particle aerodynamic diameter (d_{ae}) and – possibly – state of electrical charge. The shape, density and substance of the individual particles are, in themselves, irrelevant. If, however, other aspects of sampler performance are the focus of attention (e.g. particle–wall interactions, both outside and inside the sampler), then specific particle properties such as shape, surface condition, chemical composition and, again, state of electrical charge may all be important additional factors.

For most experimental purposes it is required that test aerosols should be defined in terms of a single particle size. Monodisperse aerosols are therefore the ideal. The best – in the sense of being the closest to true monodispersity – invariably take the form of systems of idealized liquid or solid spherical particles since technical means for the generation of such aerosols with low geometric standard deviation ($\sigma_g \approx 1.00$) are available. Some of these will be described below. However, the extent to which 'true' monodispersity is actually required in a given experiment depends on the rate at which the particular physical process studied varies as a function of particle size. At one end of the scale, for example, impactors embody one of the most sharply particle size-dependent aerodynamic processes of interest to aerosol scientists. Indeed, they are designed with this very property in mind. Jaenicke and Blifford (1974) considered the monodispersity required of aerosols chosen for calibrating such devices and demonstrated that, in order to avoid significant broadening of the measured curve relating impaction efficiency (E) and d_{ae}, experiments should be carried out with narrowly dispersed aerosols with $\sigma_g < 1.05$. This is a stringent requirement, and is only met in practice with aerosols consisting of idealised liquid droplet or solid spherical particles generated under carefully controlled laboratory conditions. At the other end of the scale, however, there are aerosol processes, including many relevant to aerosol sampling, that are less sharply dependent on particle size. For these, such a high degree of monodispersity is not called for. Fuchs and Sutugin (1966) suggested that, for such cases, an aerosol with σ_g as large as or even greater than 1.20 may sufficiently monodisperse for practical purposes. This opens up the possibility of using test aerosols other than the idealised ones referred to, including narrowly graded, dry-dispersed powders.

3.5.1   Idealised test aerosols

Relatively monodisperse liquid droplet test aerosols with small σ_g may be generated by a variety of methods. The simplest, used in a number of aerosol sampling studies, involves the nebulisation of a suspension in a volatile liquid (e.g. alcohol) of previously prepared particles (e.g. polystyrene latex spheres), taking care to ensure that, as the resultant droplets are transported from the nebuliser to the test location, evaporation of the liquid phase has been completed and that agglomeration has not occurred. Here, therefore, the liquid plays a role only as the suspending medium. Other devices, however, have been developed for the generation of aerosols specifically of stable liquid droplet media.

The device widely used in earlier aerosol sampler studies was the *spinning disk aerosol generator* first described by Walton and Prewett (1949) and elaborated later by May (1966). It is still available commercially today (BGI Inc., Waltham, MA, USA). Its principle of operation is shown in Figure 3.12. It shows a flat, disk-shaped surface, typically with diameter about 50 mm, rotated about a vertical axis at high angular speed up to 70 000 revolutions per minute. In the version shown it was driven by compressed air, although there have been other versions where the disk was rotated by means of an electric motor. In all versions, a thin, continuous filament of the chosen liquid is delivered vertically – under gravity – to the center of the disk and is atomised as the liquid is thrown radially outwards under strong centrifugal forces. The physical process of atomisation involves the formation of a thin film of liquid over the surface of the disk and breakdown into individual droplets as the liquid escapes from the high velocity perimeter. Relatively monodisperse primary droplets are thus formed with geometric diameter:

$$d = \frac{2.5}{\omega} \left(\frac{\text{Surface tension}}{\gamma R_{disk}} \right)^{1/2} \tag{3.9}$$

where ω is the angular velocity of the disk, γ the density of the liquid in question and R_{disk} the radius of the disk. For particles generated in this way, it has been found that d ranged typically from about 20

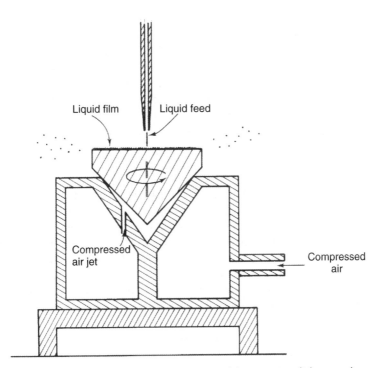

Figure 3.12 *Schematic to illustrate the principle of operation of the spinning disk aerosol generator. Reproduced with permission from Vincent, Aerosol Sampling: Science and Practice. Copyright (1989) John Wiley & Sons, Ltd*

to 100 μm with $\sigma_g \approx 1.10$. In the basic instrument, although the major proportion of the liquid mass is transformed into the primary particles described, a smaller proportion is released in the form of satellite particles of diameter about $d/4$. For the compressed air-driven version of the device shown in the figure, the disk formed part of a solid conical top which floated on an air bearing. By appropriate design of the air-flow configuration and choice of air pressure, such satellite particles may be separated inertially from the larger primary aerosols and so be removed from the test aerosol delivered to the test location.

Particles smaller than the range indicated may be obtained by using the spinning disk technique to generate the primary particles from solutions of the liquid in question in an appropriate solvent. The solvent component subsequently evaporates before the particles arrive at the test location. By the same technique, it is also possible to create solid monodisperse particles by generating the primary droplets from solutions of appropriate substances in appropriate solvents (e.g. polystyrene latex in toluene). In a related method, the spinning disk generator may be used to nebulise nondissolved particle suspensions, as already mentioned.

A later development for generating idealised test aerosols was the *vibrating-orifice aerosol generator* (VOAG) first described by Berglund and Liu (1973). This too continues to be available commercially (TSI Inc., St Paul, MN, USA). Its principle of operation is shown in Figure 3.13. In this apparatus, a filament of liquid is delivered under pressure through a small circular orifice of between 5 μm and 20 μm in diameter in a thin flat plate. The orifice plate is oscillated along the axis of the filament at frequencies in the range from 10 to 100 kHz by means of an electrically driven piezoelectric crystal. The filament is thus broken into individual droplets with geometric diameter:

$$d = \left(\frac{6Q_{liquid}}{\pi f}\right)^{1/3}$$

(3.10)

in which Q_{liquid} is the liquid flow rate in $cm^3\,s^{-1}$ and f the driving frequency in Hz. Aerosols are thus generated with d in the range from 0.5 to 50 μm and with σ_g as low as 1.01. The test aerosols produced by the last two methods are monodisperse enough to satisfy the experimental needs of most aerosol sampler studies. However, the delivery rate of test aerosol is low, particularly for the vibrating-orifice generator. Therefore such aerosol generation methods are generally suitable only for experiments at a relatively small scale where the total required volume of test aerosol is small or where extremely sensitive aerosol assessment is used (e.g. counting of individual particles).

3.5.2 Dry-dispersed dusts

Dry dispersion of bulk powder material has become a commonly used means of generating test solid-particle aerosols, driven in part by the commercial availability of sufficiently narrowly graded powders having very good consistency from batch to batch, even over many years of production. Most of the devices which have been developed for the aerosolisation of such powders have involved the metered steady delivery of powder to an aspirating air flow that lifted the particulate material and dispersed it as a cloud of unagglomerated particles that were then transported aerodynamically to the desired test location. In all of them, the objective of the metering has been to achieve aerosol delivery that was sufficiently stable for the application in question. Stability of the rate of aerosol delivery has been of greater or lesser importance depending on the application. For much aerosol sampling research, it is desirable to maintain a relatively constant aerosol flow not only from one sampling run to the next but also during an individual sampling run.

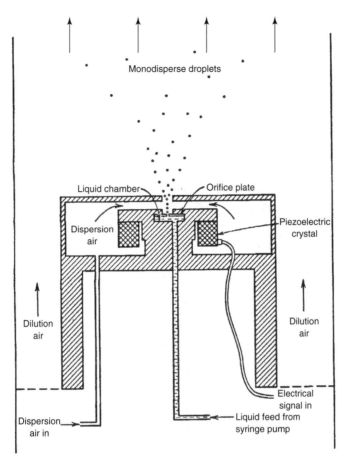

Figure 3.13 *Schematic to show the principle of operation of the vibrating-orifice aerosol generator (VOAG) (adapted from Hinds, 1999). Reproduced with permission from Vincent, Aerosol Sampling: Science and Practice. Copyright (1989) John Wiley & Sons, Ltd*

One device that emerged in the early days was the *Wright dust feed* (Wright, 1950). It is illustrated in Figure 3.14. Although versions of this device are still applied in current research, it is no longer available commercially. In this device a cake of the bulk powder of interest was formed by packing it into a cylindrical container. In practical operation, powder was scraped away from this cake by an electrically driven blade which rotated rapidly and advanced slowly by means of a fine worm thread. The scraping region was flushed with compressed air so that the particles released from the cake were entrained and carried to the outlet. Unwanted coarse agglomerates were intercepted en route by an impactor.

Rotating-table generators of the type first described by Hattersley *et al.* (1954) have also been widely used for many years. They come in various sizes and configurations, and have usually been custom built in-house for individual applications. Figure 3.15 shows the general configuration and principle of operation. Also shown is a photograph of a particular version that was constructed in the IOM laboratory

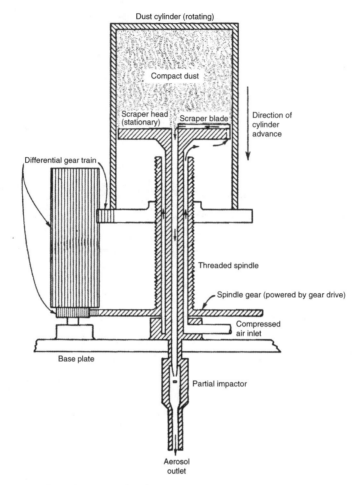

Figure 3.14 *Schematic to show the principle of operation of the Wright dust feed (adapted from Hinds, 1999). Reproduced with permission from Vincent, Aerosol Sampling: Science and Practice. Copyright (1989) John Wiley & Sons, Ltd*

to produce high dust output for aerosol sampler studies throughout the 1980s in the large wind tunnel described previously. In this particular device, the test powder of interest was loaded into each of three hoppers and was metered continuously to a shallow groove cut in a flat metal circular table which in turn was rotated by means of an electric motor. This delivery was assisted by an electrically driven stirrer which kept the powder well mixed and prevented blockage at the necks of the hoppers. Excess powder in the groove was scraped away so that reasonably constant metering of the powder to the groove was maintained. The powder carried by the groove passed under a succession of nozzles, one adjacent to each hopper, into which the powder was lifted under the strong suction provided by a compressed air-driven, Venturi-type aspirator. The dust thus made airborne was then subjected to strong shearing forces in the throat of the aspirator, serving to break up agglomerates before ejection into the test location.

Two metering dust feeders have been used in recent years successfully in aerosol sampler research applications that involved the rapid determination of sampler performance by means of direct-reading

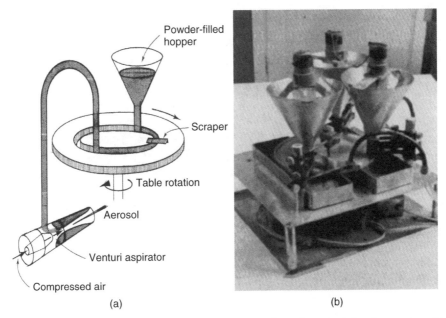

Powder-filled
hopper

Scraper

Table rotation

Aerosol

Venturi aspirator

Compressed air

(a)

(b)

Figure 3.15 *Rotating-table aerosol generator: (a) schematic to show the principle of operation; (b) photograph of version built at the Institute of Occupational Medicine (IOM) (reproduced from Mark et al., 1985). Reprinted from Journal of Aerosol Science, 16, Mark et al., 125–131. Copyright (1985), with permission from Elsevier*

particle size and counting instrumentation, and so required very constant aerosol delivery. One is a device originating from the United States National Bureau of Standards and known as the *NBS dust feeder* (Figure 3.16). In this device, the powder was metered from a vibrating hopper to a rotating serrated wheel, where the powder was delivered to the successive 'pockets' created by the individual serrations. The speed of the motor coupled to the serrated wheel was used to adjust the mass delivery rate in the range from about 0.01 to 1 g min^{-1}. The powder in each serration was aerosolised as it passed under a Venturi-type aspirator. This dust generator was at one time commercially available from BGI Inc. (Waltham, MA, USA) and was shown in many experiments to provide very good stability in aerosol delivery. But it has sadly been discontinued. A related device is still available commercially (at the time of writing) from Topas, Dresden, Germany (Model SAG410) and is quite widely used (Figure 3.17). In this device, the powder is delivered to a *dosing band* where the scraper ensures constant filling of each segment of a moving belt. The feed rate is controlled by adjusting the belt speed. The powder, thus metered, is aerosolised – as for the other devices mentioned – by means of a Venturi-type aspirator. For both the NPS and the Topas aerosol generators, experience has shown that, in order to optimise the stability of the aerosol delivery to the test section, the aspirator should be operated using clean, dry, filtered compressed air and the powder itself heated gently in an oven overnight prior to use and kept warm during operation by an infrared lamp directed at the hopper. These steps were necessary to reduce inter-particle adhesion effects due to moisture adsorption. When used carefully in this way, both generators have been shown to be capable of delivering nonagglomerated aerosol to a test system with mass injection-rate constancy to within ±10 % over a period of up to 2 h.

Fluidised-bed generators have become popular for some applications, partly because they are economical with the powdered material and partly because they provide highly effective means of breaking

Figure 3.16   *United States National Bureau of Standards (NBS) aerosol generator. Photograph courtesy of Robert Gussman, BGI Inc., Waltham, MA, USA*

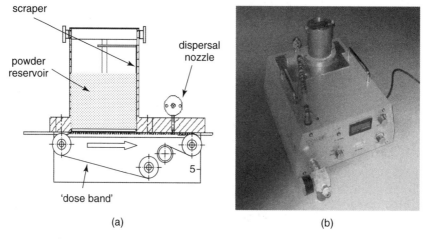

(a)	(b)

Figure 3.17   *Metering dust generator (Topas, Model SAG410, Dresden, Germany): (a) schematic to illustrate principle of operation; (b) photograph of commercially available instrument*

up agglomerates. Its principle of operation is shown in Figure 3.18. A version of the device originally described by Marple *et al.* (1978) remains commercially available today (TSI Inc., St Paul, MN, USA). Typically, the fluidised bed itself is composed of a cylindrical sump filled with 180 μm diameter bronze beads to a depth of about 2 cm. The beads are 'fluidised' by means of a jet of compressed air fed into the bottom of the bed through a sintered plate. Powder is delivered to the bed from a side reservoir through a close-fitting plastic sleeve by means of an electrically driven ball-chain conveyor (which returns to the reservoir through a second plastic sleeve). The particles aerosolised in this system are carried upwards in the vertical column above the fluidised bed itself, and are delivered from there to the test system.

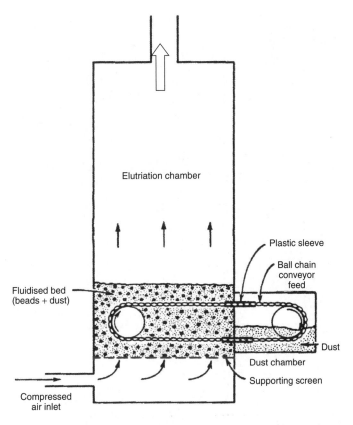

Elutriation chamber

Plastic sleeve

Ball chain
conveyor
feed

Fluidised bed
(beads + dust)

Dust

Dust chamber

Supporting screen

Compressed
air inlet

Figure 3.18 *Schematic to illustrate principle of operation of the fluidised bed aerosol generator. Reproduced with permission from Vincent, Aerosol Sampling: Science and Practice. Copyright (1989) John Wiley & Sons, Ltd*

3.5.3   Aerosol materials

Liquid droplet aerosols have been generated from a very wide range of fluid substances. The most commonly used have included water, uranine blue, di-2-ethyl-hexyl sebacate, oleic acid and di-iso-octyl phthalate. The most commonly used material for the idealised, monodisperse aerosols of solid spheres is polystyrene latex. Human serum albumen has also been used.

Over the years, a wide variety of dry-dispersed powders have been aerosolised for sampler evaluation, including silicon carbide, iron oxide, amorphous silica, zinc, ground limestone, glass beads and *Lycopodium* spores. One bulk material that has been found to be particularly suitable for aerosol sampler experiments, and has been used by many research groups, is fused alumina powder. Such fine ceramic powders are narrowly graded and are available commercially in a wide range of grades, originally for applications as abrasives for the smoothing and truing of optical lenses. Such powders are therefore readily available (e.g. as *Duralum*, Washington Mills Electrominerals, Manchester, UK). The photographs in Figure 3.19, taken from micrographs of a fine and a coarse grade, respectively, demonstrate the isometric nature and narrow size ranges of the particles in aerosols generated from such powders. Aerosols generated in this way may be virtually agglomerate-free provided that sufficient air pressure is supplied to the aspirator. For laboratory applications, airborne dusts of fused alumina have

Figure 3.19   *Scanning electron micrographs of particles collected from aerosols generated from two grades of narrowly graded powders of fused alumina (from Mark et al., 1985). Reprinted from Journal of Aerosol Science, 16, Mark et al., 125–131. Copyright (1985), with permission from Elsevier*

the advantage over many other mineral powders in that they present minimal inhalation-related health risks to personnel (Dinman, 1983). They have been widely used for aerosol sampling studies since the 1980s, especially because they generally meet the Fuchs–Sutugin criterion for *monodispersity*. Over the years, workers from a number of independent research groups have used the same powder grades from the same supplier in their own research and confirmed that the more recent calibrations have not differed significantly from the original ones. This is supported by information from the supplier that such 'optical grade' powders are manufactured to very high standards of industrial quality assurance

Table 3.1   *Mass median aerodynamic diameter (MMAD) (in μm) and geometric standard deviation (σ_g) for aerosols generated from a range of grades of narrowly graded fused alumina powders with grades as indicated (Mark et al., 1985)*

Grade designation	MMAD (σ_g)
F1200	6 (1.36)
F1000	9 (1.38)
F800	13 (1.30)
F600	18 (1.29)
F500	26 (1.21)
F400	34 (1.20)
F360	46 (1.20)
F320	58 (1.17)
F280	74 (1.19)
F240	90 (1.29)

and consistency, for example as specified by the Federation of European Producers of Abrasives (1993). The characteristics of such powders and their resultant aerosols are summarised in Table 3.1, where the quoted MMAD values and σ_g values are the ones originally published by Mark *et al.* (1985), still widely held to be applicable today.

3.5.4   Electric charge effects

It is well known that, during aerosol generation, particles become electrically charged. The mechanism of charging depends on the process of generation. For example, in the case of droplet formation, it arises from the charge that is made available as the surface of the bulk liquid material is reorganised. Reischl *et al.* (1977) measured the charge on 3 μm diameter droplets produced by means of the vibrating-orifice generator, and found it to be of net negative polarity with magnitude between 1 000 and 10 000 elementary electronic charges per particle. Whitby and Liu (1968) found that the median charge on 1.3 μm diameter spheres of polystyrene latex obtained from the nebulisation of liquid suspensions ranged from 10 to 100 elementary charges.

In the case of dry dust dispersal, charging arises from the combination of contact and triboelectric charging associated with particles making contact with one another during the process of being made airborne. The charge on a population of particles is dependent on the means of dispersal, and has been found to be essentially net neutral for mineral dust aerosols dispersed using the Wright dust feed and the rotating-table generator (Johnston *et al.*, 1987). The median magnitude of charge for 1 μm diameter particles ranges from about 10 to 40 elementary charges. For the fluidised-bed generator, however, it has been found that charge levels for some types of aerosol are not only higher but also asymmetrically distributed, and therefore are *not* net neutral.

The charge levels quoted above are for aerosols generated in the laboratory and delivered relatively quickly to the test location. They should be compared with the level of 'neutrality' corresponding to Boltzmann equilibrium which is about two elementary charges on a 1 μm diameter particle. This is what would be reached if the aerosol were allowed to 'age' for several minutes in the weakly ionised, bipolar air like that found in most terrestrial environments. Since such ageing is not possible in most aerosol experiments – including aerosol sampler studies – the possibility of complications introduced by the influence of particle charge should not be ruled out. Whilst for aspiration efficiency it transpires that charge effects may be neglected (see Chapter 7), this might not be the case for other aspects of sampler performance. Many researchers, therefore, consider it prudent to reduce the particle charge to a level close to Boltzmann equilibrium wherever possible. This is achieved by exposing the aerosol immediately after generation to a dense, bipolar cloud of ions. A radioactive source (e.g. krypton-85, thallium-204) is an effective means of providing such ions. So too is the sonic-jet corona discharge device described by Whitby (1961), although this appears to have fallen into disuse.

3.6   Reference methods

The indirect (or comparison) method described in Chapter 1 is the approach most commonly adopted by aerosol sampling researchers. In experimental measurements of the various indices of sampler performance by this means, a reference aerosol concentration measurement is required (for c_0) against with which to compare the concentration of the aerosol collected by the sampler of interest (c_s), as expressed earlier in Equation (1.20). This therefore requires a second sample, and in turn requires a reference sampler whose performance is known for the conditions pertaining to the test in question.

3.7   Assessment of collected aerosol

An important part of the process of testing an aerosol sampler is the assessment of collected aerosol. For dry-dispersed dusts, it is a relatively straightforward matter to determine the mass of collected particulate material by what are referred to as *gravimetric* procedures. Provided that the concentration, sampling flow rate and/or sampling time are high enough to provide a sufficiently large sample, assessment of on-filter mass may be carried out simply by means of an appropriate analytical balance. Aerosol deposited on the sampler wall can usually be assessed by brushing it off carefully (e.g. using a fine artist's brush) and either weighing it separately or, if desired, adding it to the on-filter mass. Wall deposits that cannot be recovered successfully in this way may be recovered by wash-off (e.g. using alcohol). In such procedures it is important to be aware of the weighing limitations, in particular the issues of limit of detection (LOD) and limit of quantitation (LOQ) as they might be influenced by instabilities associated with moisture uptake, on both the collected sampler and on the collection substrate (Smith *et al.*, 1997; American Society for Testing and Materials, 2000; Paik and Vincent, 2002).

In experiments where the amount of the collected aerosol is small, which is usually the case for idealised aerosols like those generated by, for example, the spinning disk or vibrating-orifice devices, direct weighing is often not possible. Other techniques are therefore employed. One approach is to tag the particles with an appropriate marker which can itself be assessed by sensitive means of detection. Tagging the particulate material with a fluorescent dye has been widely used, in which the collected material has been assessed by fluorimetry.

A few workers have used an alternative method which involves the counting of individual particles. This is particularly appropriate for aerosols consisting of solid spheres and where the amount collected is very small. One approach is to ensure that the aerosol to be assessed is distributed uniformly over the filter so that particle counting may be carried out by microscopy (optical or electron) over predetermined numbers of microscope fields of view, thus eliminating the need to assess all the particles on the filter. The emergence of computer-based, automatic scanning devices has eased the burden that such an approach had previously placed on human operatives. Some workers have extended this approach to polydisperse aerosols where particles are both counted and sized so that information about sampler performance may be obtained for a whole range of particle sizes from just one sample. For spherical particles of known composition, it is a simple matter to relate particle size obtained in this way to the particle aerodynamic diameter. Other workers (Armbruster and Breuer, 1982) have adopted a related approach in which the collected polydisperse particles have been simultaneously counted and sized by Coulter counter. In the use of this instrument, the collected particulate material is suspended in an appropriate electrolyte and the individual particles are counted and sized in terms of particle equivalent volume diameter as they flow through a small orifice and are detected by a change in the electrical resistance across the orifice due to the momentary presence there of the particle. For nonspherical particles, results may be obtained in terms of particle aerodynamic diameter by making appropriate corrections for particle density and dynamic shape factor.

In more recent years there has been a growth in experimental methods that have employed direct-reading optical photometers for determining aerosol concentration and more sophisticated discriminating instruments for counting and sizing individual sampled particles. For the latter, the *aerodynamic particle sizer* (APS) has been perhaps the most widely used. This instrument was first described in 1980 by Wilson and Liu, and soon afterwards was made commercially available and has remained so to this day (TSI Inc., St Paul, MN, USA). The APS senses and sizes individual particles based on their time-of-flight in a changing flow field and counts them based on light scattering. In this way, individual particle counts are placed into a large number of electronic bins representing narrow particle size bands within the range of d_{ae} from approximately 0.5 to 20 μm. In its use, data assembled in this way are managed by means of specialised software provided by the manufacturer. It is described in greater detail later in Chapter 20. Several versions have appeared, each one representing an improvement on the preceding one, with the elimination of various biases and artifacts along the way, including particle count coincidences associated with the presence of more than one particle in the sensing zone at any given time. Potential users should be aware of possible artifacts that were present in earlier versions (Armendariz and Leith, 2002; Stein *et al.*, 2002). The recommended range of particle concentration in order to minimise particle count coincidences is about 1000 particles cm^{-3}. For concentrations higher than this, dilution of the sampled aerosol is recommended and appropriate dilution equipment, designed specifically for use with the APS, is available from TSI Inc.

3.8 Aerosol sampler test protocols and procedures

Protocols for the laboratory testing, validation or investigation of the performances of aerosol sampling instruments strive to integrate the various components and methodologies – like those described above – in order that measurements of appropriate sampler performance indices may be made under conditions that are as relevant as possible to the real world in which the samplers are to be used. The search for methods that have the potential to be standardised has been of considerable interest in recent years, and research work towards this goal has been funded by the European Community and the National Institute for Occupational Safety and Health (NIOSH) and the EPA in the USA.

As already stated in Chapter 1, aerosol sampler performance characteristics relate to how sampling is carried out with reference to the dependence of particle aspiration and collection on particle size. As alluded to in Chapter 2, and will be discussed in greater detail in subsequent chapters, most particle size dependency of interest relates to particle aerodynamic diameter (d_{ae}). In that event, therefore, for the concentration of aspirated or collected polydisperse aerosols, we may rewrite Equations (1.22) and (1.23) as:

$$c_s = \int_0^\infty A(d_{ae}) M(d_{ae}) \, \mathrm{d}d_{ae} \qquad (3.11)$$

and

$$c_F = \int_0^\infty P(d_{ae})A(d_{ae})M(d_{ae})\,\mathrm{d}d_{ae} \longrightarrow c_F = \int_0^\infty F(d_{ae})M(d_{ae})\,\mathrm{d}d_{ae} \tag{3.12}$$

where $F(d_{ae})$ defines the particle size-selectivity of the sampler. The last equation is the most general form, reverting to Equation (3.11) when there are no losses of particles after they have passed through the inlet – that is, where $P(d_{ae}) \rightarrow 1$. Kenny and Bartley (1995) used this approach to develop the basis of a testing protocol for sampler performance evaluation. It is inevitable that, in the real world, where there is measurement error and system variability (i.e. between sampler specimens), there need to be statistical considerations in addition to the physical ones. To begin with, Kenny and Bartley defined three versions of Equation (3.12); namely:

$$\overline{c_{Fi}} = \int_0^\infty F_{actual}(d_{ae})M(d_{ae})\,\mathrm{d}d_{ae} \text{ for very many measurements for a single specimen, } i$$

$$\overline{c_F} = \int_0^\infty F_{actual}(d_{ae})M(d_{ae})\,\mathrm{d}d_{ae} \text{ for single measurements for very many specimens} \tag{3.13}$$

along with

$$c_{F,ideal} = \int_0^\infty F_{ideal}(d_{ae})M(d_{ae})\,\mathrm{d}d_{ae} \tag{3.14}$$

In these equations, the true mean $\overline{c_{Fi}}$ is estimated in practice as $\hat{c}_{Fi}$ and, similarly, $\overline{c_F}$ as $\hat{c}_F$. In addition, F_{actual} describes the actual performance curve, expressed as the selection efficiency as a function of d_{ae}. This is compared with F_{ideal}, the corresponding performance curve for an ideal sampler that exactly follows, without bias or error, a desired particle size-selection characteristic – for example one based on a prescribed health-based rationale. It is the relationship between these quantities that needs to be formalised in a standardised sampler testing protocol.

If the aerosol mass concentrations in the above are expressed in particulate mass per unit volume of air, then the mass bias associated with the sampler type relative to the ideal is given by:

$$\Delta = \frac{\overline{c_{Fi}} - c_{F,ideal}}{c_{F,ideal}} \tag{3.15}$$

Any physical variations between n different specimens (in the family i) of the sampler in question are described by the inter-specimen variance:

$$\sigma_i^2 \equiv \lim_{i \to \infty}\left[\frac{1}{n}\sum(\overline{c_{Fi}} - \overline{c_F})^2\right] \tag{3.16}$$

in which it is tacitly assumed that the inter-specimen variance is normally distributed with zero mean.

Since, in any given test sequence there is the inevitable experimental error, the estimate for $\overline{c_{Fi}}$ is given by:

$$\hat{c}_{Fi} = \overline{c_{Fi}} + \xi \tag{3.17}$$

where ξ is the evaluation error, reflecting the extent to which the estimated concentration is different from the true mean. This itself, over many measurements, is normally distributed, having zero mean and variance σ_{eval}^2. The variance of the estimated mean concentration from tests of n sampler specimens (in i) is then given by:

$$\text{Var}(\hat{c}) = \frac{1}{n}(\sigma_{eval}^2 + \sigma_s^2) \tag{3.18}$$

and that of the estimated bias by:

$$\sigma_\Delta^2 \equiv \text{Var}(\hat{\Delta}) = \frac{\text{Var}(\hat{c}_F)}{c_{F,ideal}^2} \tag{3.19}$$

Kenny and Bartley pointed out that, for this framework to be applicable in a practical testing situation, in particular to be able to assess how accurately a candidate sampler can collect the desired aerosol fraction, the challenge is to make measurements that can lead to $\hat{c}_F$, σ_i^2 and σ_{eval}^2. Such measurements would require experiments that can provide values of the terms inside the integral in Equations (3.13) for each of a number of discrete values of d_{ae} over the particle size range of interest. For example, the performance of a given candidate sampler, as reflected in $F_{actual}(d_{ae})$, may be determined for each particle size and the overall performance curve fitted by appropriate linear or nonlinear least-squares regression procedure.

For occupational settings, work like that described above has provided the foundation of, for example, the sampler testing protocol EN 13205 developed by the CEN entitled *Workplace atmospheres – assessment of performance of instruments for measurement of airborne particle concentrations* (Comité Européen Normalisation, 2002). In the CEN protocol, for any given sampler to be tested, the first step is a critical review of the sampling process for the instrument in question. This is intended to identify factors that may influence the performance of the sampler, including particle size, wind speed, aerosol composition, filter material, etc. This is essential in the process of sampler evaluation, determining under what conditions the sampler will need to be tested. Three options are then presented for the testing of samplers: (a) the laboratory testing of samplers with respect to sampling conventions; (b) the laboratory comparison of instruments; and (c) the field comparison of instruments. The second and third of these involve the identification and use of an existing aerosol sampler that is known and agreed to be capable of accurately sampling the appropriate aerosol fraction as a reference sampler. The test sampler of interest is then operated alongside the reference sampler, and the mass of aerosol collected in the test sampler is compared with that in the reference. In the laboratory, these experiments should be performed for the range of relevant environmental conditions, as identified in the critical review, and appropriately scaled. In the field, the comparisons should be carried out for as wide a range of conditions as possible pertaining to the field site(s) in question. For the latter, it is clear that the results can only be considered useful for future application at the same – or demonstrably similar – sites. The CEN protocol for laboratory testing contains a rather formal set of carefully prescribed ingredients:

- statement of principle
- description of the experimental test method
  - test conditions
  - test variables
  - particle size
  - wind speed
  - wind direction
  - aerosol composition
  - sampled mass
  - aerosol charge
  - specimen variability

- – flow rate variations
- – surface treatments
- experimental requirements
  - – environmental conditions (temperature and humidity)
  - – test aerosol
  - – choice of monodisperse or polydisperse
  - – measurement of particle size
  - – reference samples
  - – wind speed range and variability
  - – number of samplers to be tested simultaneously
  - – sampling pumps
- calculation methods
  - – nomenclature
  - – calculation of the actual sampled concentration
  - – calculation of ideal sampled concentration
  - – calculation of sampler bias
  - – application of a sampler correction factor
  - – calculation of uncertainty in the estimated sampler bias
  - – calculation of sampler accuracy
- test report.

In summary, the sampler performance is characterised fully for a representative range of particle sizes, wind speeds and sampler orientations, where sampling is usually orientation-averaged, as has been the case in the most of the research described in this report. The resultant data are used to construct a profile of overall performance – aspiration efficiency or sampling efficiency, depending on the recommended mode of operation of the sampler – as a function of particle aerodynamic particle diameter, d_{ae}. Then the performance profile is compared directly with the sampling convention (e.g. inhalability) by calculating the measured sampled mass fraction for a set of log-normal particle size distributions. This calculated mass is then compared with the equivalent results for a hypothetical ideal sampler that perfectly matches the convention. The two sets of calculated results are then used to construct a mathematical 'map' of the collected mass biases that would be found when the sampler is used over relevant ranges of conditions. Typically, for example, the bias may be described as a function of the particle size distribution, in which – if the distribution is log-normal – the variables affecting the bias are the mass median aerodynamic diameter (MMAD) and the geometric standard deviation (σ_g). This system is illustrated in the hypothetical bias map example shown in Figure 3.20, reflecting how well the sampler in question performs over the plane of the particle size distribution in terms of its ability to accurately measure the desired mass fraction. Other bias maps may be constructed for other combinations of variables affecting sampler performance. From such analyses, any given sampler may be classified according to its ability to sample more or less closely to the desired criterion. Further, EN 13205 suggests how, for a given sampler, a correction factor may be derived, provided that the bias is similar for all aerosol particle size distributions of interest and for all influencing variables. If those conditions are met, then application of the correction factor may facilitate use of the sampler in question.

As far as experimental methodologies are concerned, a number of aerosol sampler test protocols using direct-reading instruments have been proposed over the years, and the APS has featured prominently in many of these. The process involves successively exposing to a polydisperse test aerosol the sampler of interest and another that may be regarded as a reference sampler, and measuring the particle size distribution that passes through each using an appropriate direct-reading instrument like the APS. If it can be assumed that the aerosol concentration and particle size distribution in the test atmosphere

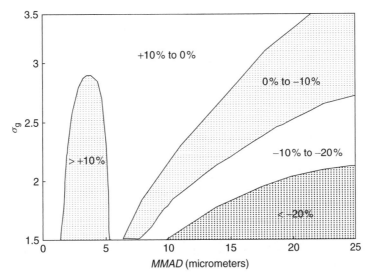

Figure 3.20 *Hypothetical bias map to illustrate the variation of the departure of sampled mass from the desired ideal as a function of particle size distribution as represented by mass median aerodynamic diameter (MMAD) and geometric standard deviation (σ_g)*

have not changed between the successive exposures of the two instruments, then the two successive particle size distributions provided by the APS – as reflected in counts recorded in each electronic bin of the instrument – provides the desired characteristic of the test sampler. This of course places strict requirements on the quality and consistency of the test atmosphere, especially on the aerosol generation equipment. So careful experimental design is required. The method of Kenny and Lidén (1991) employed the APS in this method for testing the pre-selection characteristics of samplers for fine aerosol fractions of the type often used by occupational hygienists. Since the influence of the air flow external to the sampler was not of interest, but rather the penetration properties of an internal section (e.g. a cyclone or horizontal elutriator), their studies were carried out in a calm air chamber. In the use of such an approach, it was clearly demonstrated that very reproducible data could be rapidly obtained for important aspects of sampler performance.

The approach to aerosol sampler testing is dependent on a number of factors. One is the size of the instrument to be tested. Many will be shown later in this book. Some that are developed for collecting aerosols in the ambient outdoor environment are very large indeed. Other devices may be small, but may require to be mounted on larger objects which, in themselves, become part of the overall sampling system. These may include the personal aerosol samplers that are widely used by occupational hygienists for determining the aerosol exposures of individual workers and are intended for use when worn somewhere on the body (e.g. on the lapel in what is often referred to as the 'breathing zone'). In all such cases, quite large wind tunnels or chamber facilities are required for full-scale testing. The recent emergence of reliable aerosol sampler scaling laws has begun to point the way towards reducing the size needed for such facilities. Brixey *et al.* (2002) described such an approach for testing aerosol samplers in wind tunnels, this time for testing the aspiration efficiencies of aerosol samplers in moving air, aiming to address the testing of samplers for the inhalable fraction. The method evolved out of the improved understanding that had been emerging about the scaling laws that could be applied to aerosol samplers, thus enabling studies to be carried out in quite small wind tunnels and for narrower ranges of particle sizes than in the real world (Ramachandran *et al.*, 1998). So, although the APS counts and sizes particles with d_{ae} only up to about 20 μm, application of the scaling laws referred to earlier permits identification of experimental conditions, including particle size range, that can in due course be related back to more realistic conditions at full-scale. Practical implementation of the approach described, however, was influenced by the awareness from previous work that the coupling between the external air flow and the internal air flow just inside the sampler entry is extremely complicated, relating to the fact that the external air flow undergoes a sharp transition during entry into the sampler, including boundary layer effects and – under certain conditions – flow separation and instability (Okazaki and Willeke, 1987; Hangal and Willeke, 1990; Ramachandran *et al.*, 2001; Sreenath *et al.*, 2002). By way of illustration, Figure 3.21 illustrates the *vena contracta* that results from the flow separation inside the inlet of a simple thin-walled sampling probe. More generally, the actual details of such flow interactions will be highly dependent on R, r and θ. So too will be particle losses by deposition in that region, and these have been shown to be considerable. There is no prospect of obtaining a general model for inlet losses that may be broadly applied to real-world aerosol samplers like those used in practical occupational and environmental hygiene. Brixey *et al.* expressed this inlet problem in terms of the transmission of particles through the various parts of the system, all the way from the test atmosphere in the wind tunnel down to the APS measuring instrument. Thus:

$$
\begin{aligned}
_T E &= \frac{_T c_{APS}}{c_0} = \frac{_T c_{APS}}{_T c_e} \cdot \frac{_T c_e}{_T c_s} \cdot \frac{_T c_s}{c_0} = {_T P_{tube}} \cdot {_T P_{entry}} \cdot {_T A} \\
_R E &= \frac{_R c_{APS}}{c_0} = \frac{_R c_{APS}}{_R c_e} \cdot \frac{_R c_e}{_R c_s} \cdot \frac{_R c_s}{c_0} = {_R P_{tube}} \cdot {_R P_{entry}} \cdot {_R A}
\end{aligned}
\tag{3.20}
$$

where the subscripts T and R refer to the test and reference samplers, respectively. In these expressions, P_{entry} is the penetration efficiency for particles passing through the flow transition region near the inlet and P_{tube} the penetration efficiency for particles passing through the length of the tube between the entry and the APS. The aerosol concentrations referred to at the various points within the system are as indicated in Figure 3.22. In Equations (3.20) it is seen that the APS counts can be used to determine

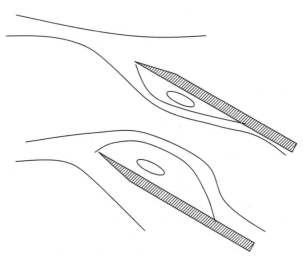

Figure 3.21   *Indication of the nature of the air flow near a simple sampling inlet, showing the presence of the vena contracta just inside the entry plane*

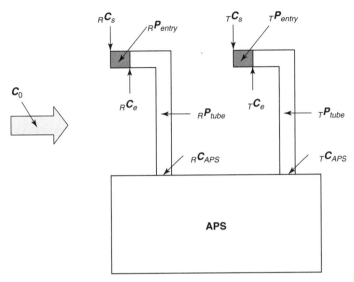

Figure 3.22 *Schematic diagram on which to base discussion of aerosol sampler performance testing procedure using a rapid data acquisition method*

aspiration efficiency of the test sampler ($_T A$) only if all the other terms are known or can be eliminated. The former can be achieved for $_R A$ by the choice of an appropriate reference sampler – for example, a thin-walled sampling tube facing the wind, whose aspiration efficiency is well-known from theory. The latter can be achieved for P_{tube} by ensuring that the physical dimensions of the tube between the entry and the APS are identical for each sampling line, so that $_R P_{tube}$ and $_T P_{tube}$ cancel. However, the same cannot be achieved for P_{entry} because, as already stated, this cannot be calculated. The solution that Brixey *et al.* adopted was to make artificial modifications to the entries of both the test and reference so that P_{entry} could be predicted after all. To do this they employed porous plastic foam media, based on their well-behaved filtration characteristics. Such media, described later in Chapter 8, have been considered in the past as possible pre-selectors for aerosol samplers, and their penetration characteristics are by now quite well understood and a reliable mathematical model is available (Vincent *et al.*, 1993). The approach taken by Brixey *et al.* was to place plugs of such media immediately inside the entries of both the test and reference samplers, with the intention that the flow inside the entry would now be determined entirely by the flow through the foam media, irrespective of the orientation of the sampler. In this way, therefore, the flow transition and disturbance described earlier are suppressed and $_R P_{entry}$ and $_T P_{entry}$ may both be calculated. Better still, if the foam plugs placed in the entries of the test and reference samplers, respectively, could be designed to be identical, $_R P_{entry}$ and $_T P_{entry}$ should then cancel. In that event, Equations (3.20) show that aspiration efficiency can be obtained from APS measurements, as intended.

Witschger *et al.* (1998) addressed the actual physical configuration of the sampling system, with special regard to the testing of personal samplers for the coarser inhalable fraction, for aerosols with particles with d_{ae} up to as high as 100 μm. Their approach involved experiments at full scale and with conventional aerosol sampling, collection and weighing procedures that have been the feature of many experiments like those that will be described as this book progresses. They suggested simpler experimental systems and procedures, including mounting the samplers on bluff bodies that were smaller and of simpler shape than the mannequins used by others.

More generally, the methods described above have been aimed primarily at the testing of aerosol samplers intended for applications in occupational environments. Lidén *et al.* (2000) noted that the idealized conditions in such laboratory scenarios are not fully reflective of true conditions in workplaces, and that protocols for field validation of laboratory tests should be included in any overall sampling testing program.

Similar guidelines to those described in the preceding have been proposed for the testing of samplers intended for use in monitoring the ambient environment. The United States led the way when the Environmental Protection Agency (Environmental Protection Agency, 1995 and later updates) developed protocols for the testing of candidate aerosol samplers for health-related fractions relevant to the US National Ambient Air Quality Standards (PM_{10} and $PM_{2.5}$). The aim was to ensure that, irrespective of which laboratory carries out the testing, consistent procedures should be carried out, meeting strict performance criteria. In that way, the experiments themselves may be carried out anywhere and by anyone in a manner that is consistent and the results are reproducible. Europe followed suit when CEN published its protocol EN12241 entitled *Air quality, determination of the PM$_{10}$ fraction of suspended particulate matter, reference method and field test procedure to demonstrate reference equivalence of measurement methods*. Both the EPA and CEN protocols place considerable emphasis on field testing procedures.

References

Aizenberg, V., Choe, K., Grinshpun, S.A., Willeke, K. and Baron, P.A. (2001) Evaluation of personal aerosol samplers challenged with large particles, *Journal of Aerosol Science*, 32, 779–793.

American Society for Testing and Materials (2000) Standard practice for controlling and characterizing errors in weighing collected aerosols, Standard D6552-00, American Society for Testing and Materials, West Conshohocken, PA, USA.

Armbruster, L. and Breuer, H. (1982) Investigations into defining inhalable dust. In: *Inhaled Particles V* (Ed. W.H. Walton), Pergamon Press, Oxford, pp. 21–32.

Armendariz, A.J. and Leith, D. (2002) Concentration measurement and counting efficiency for the aerodynamic particle sizer 3320, *Journal of Aerosol Science*, 33, 133–148.

Baines, W.D. and Peterson, E.G. (1951) An investigation of flow through screens, *Transactions of the American Society of Mechanical Engineers*, 73, 467–480.

Baldwin, P.E.J. and Maynard, A.D. (1998) A survey of wind speeds in indoor workplaces, *Annals of Occupational Hygiene*, 20, 303–313.

Belyaev, S.P. and Levin, L.M. (1972) Investigation of aerosol aspiration by photographing particle tracks under flash illumination, *Journal of Aerosol Science*, 3, 127–140.

Belyaev, S.P. and Levin, L.M. (1974) Techniques for collection of representative aerosol samples, *Journal of Aerosol Science*, 5, 325–338.

Berglund, R.N. and Liu, B.Y.H. (1973) Generation of monodisperse aerosol standards, *Environmental Science and Technology*, 7, 147–153.

Berry, R.D. and Froude, S. (1989) An investigation of wind conditions in the workplace to assess their effect on the quantity of dust inhaled, Internal Report IR/L/DS/89/3, Health and Safety Executive, London.

Brixey, L.A., Paik, S.Y., Evans, D.E. and Vincent, J.H. (2002) New experimental methods for the development and evaluation of aerosol samplers, *Journal of Environmental Monitoring*, 4, 633–641.

Brown, J.R. (1967) Low-speed, low-cost wind tunnel, *The Engineer*, 31 March, 481–483.

Comité Européen de Normalisation (2002) *Workplace atmospheres – assessment of performance of instruments for measurement of airborne particle concentrations*, CEN Standard EN 13205.

Davies, C.N. (1968) The entry of aerosols into sampling tubes and heads, *British Journal of Applied Physics*, 1, 921–932.

Dinman, B.D. (1983) Aluminium alloys and compounds. In: *Encyclopaedia of Occupational Health and Safety*, 3rd Edn, International Labour Organisation, Geneva, pp. 134–135.

Environmental Protection Agency (1995) *Ambient air monitoring reference and equivalence methods*, EPA 40 CFR 53, Sub-Part D, United States Environmental Protection Agency, Washington, DC, USA (see also *Proposed requirements for designation of reference and equivalence methods for PM$_{2.5}$ and ambient air quality surveillance for particulate matter*, US Federal Register, 13 December, 1996 (Volume 61, No. 241).

Fauvel, S. and Witschger, O. (2000) A test room for studying area and personal aerosol sampling, *Journal of Aerosol Science*, 31, S781–S782.

Federation of the European Producers of Abrasives (1993) *FEPA standard for bonded abrasive grains of fused aluminium oxide and silicon carbide*, FEPA 42-GB-1984 R1993.

Fuchs, N.A. and Sutugin, A.G. (1966) Generation and use of monodisperse aerosols. In: *Aerosol Science* (Ed. C.N. Davies), Academic Press, London, pp. 1–30.

Gibson, H. and Ogden, T.L. (1977) Some entry efficiencies for sharp-edged samplers in calm air, *Journal of Aerosol Science*, 8, 361–365.

Hangal, S. and Willeke, K. (1990) Overall efficiency of tubular inlets sampling at 0–90 degrees from horizontal air flows, *Atmospheric Environment*, 24, 2379–2386.

Hattersley, R., Maguire, B.A. and Tye, D.L. (1954) SMRE Research Report No. 103, Safety in Mines Research Establishment, Sheffield, UK.

Hinds, W.C. (1999) *Aerosol Technology*, 2nd Edn, John Wiley & Sons, Ltd, New York.

Hinds, W.C. and Kuo, T.-L. (1995) A low velocity wind tunnel to evaluate inhalability and sampler performance for large dust particles, *Applied Occupational and Environmental Hygiene*, 10, 549–556.

Hoover, M.D., Barr, E.B., Newton, G.J., Ortiz, C., Anand, N.K. and McFarland, A.R. (1996) Design and evaluation of an aerosol wind tunnel for applied industrial hygiene and health physics studies, Inhalation Toxicology Research Institute (ITRI) Annual Report, ITRI-148, Albuquerque, NM, USA.

Jaenicke, R. and Blifford, I.H. (1974) The influence of aerosol characteristics on the calibration of impactors, *Journal of Aerosol Science*, 5, 547–564.

Johnston, A.M., Vincent, J.H. and Jones, A.D. (1987) Electrical charge characteristics of dry aerosols produced by a number of laboratory mechanical dispensers, *Aerosol Science and Technology*, 6, 115–127.

Kennedy, N.J. and Hinds, W.C. (2002) Inhalability of large particles, *Journal of Aerosol Science*, 33, 237–255.

Kenny, L.C. and Bartley, D.L. (1995) The performance evaluation of aerosol samplers tested with monodisperse aerosols, *Journal of Aerosol Science*, 26, 109–126.

Kenny, L.C. and Lidén, G. (1991) A technique for assessing size-selective dust samplers using the APS and polydisperse test aerosols, *Journal of Aerosol Science*, 22, 91–100.

Kenny, L.C., Beaumont, G., Gudmudssen, A. and Koch, W. (2000) Small-scale aerosol sampler testing systems, *Journal of Aerosol Science*, 31, S406–S407.

Lidén, G., Juringe, L. and Gudmudsson, A. (2000) Workplace validation of a laboratory evaluation test for inhalable and 'total' dust, *Journal of Aerosol Science*, 31, 199–219.

Lipatov, G.N., Grinshpun, S.A., Shingaryov, G.L. and Sutugin, A.G. (1986) Aspiration of coarse aerosol by a thin-walled sampler, *Journal of Aerosol Science*, 17, 763–769.

Mark, D., Vincent, J.H., Gibson, H. and Witherspoon, W.A. (1985) Applications of closely-graded powders of fused alumina as test dusts for aerosol studies, *Journal of Aerosol Science*, 16, 125–131.

Marple, V.A., Liu, B.Y.H. and Rubow, K.L. (1978) A dust generator for laboratory use, *American Industrial Hygiene Association Journal*, 39, 26–32.

Maskell, E.C. (1965) A theory of the blockage effects of bluff bodies and stalled wings in a closed wind tunnel, *Aeronautical Research Council*, Reports and Memoranda No. 3400, Her Majesty's Stationery Office, London, UK.

May, K.R. (1966) Spinning-top homogeneous aerosol generator with shock-proof mounting, *Journal of Scientific Instruments*, 43, 841–842.

Okazaki, K. and Willeke, K. (1987) Transmission and deposition behavior of aerosols in sampling inlets, *Aerosol Science and Technology*, 7, 275–283.

Paik, S. and Vincent, J.H. (2002) Filter and cassette mass instability in ascertaining the limit of detection of inhalable airborne particulate, *American Industrial Hygiene Association Journal*, 63, 698–702.

Ramachandran, G., Sreenath, A. and Vincent, J.H. (1998) Towards a new method for experimental determination of aerosol sampler aspiration efficiency in small wind tunnels, *Journal of Aerosol Science*, 29, 875–891.

Ramachandran, G, Sreenath, A. and Vincent, J.H. (2001) Experimental study of sampling losses in thin-walled probes at varying angles to the wind, *Aerosol Science and Technology*, 35, 767–778.

Reischl, G., John, W. and Devor, W. (1977) Uniform electrical charging of monodisperse aerosols, *Journal of Aerosol Science*, 8, 55–65.

Smith, J.P., Bartley, D.L. and Kennedy, E.R. (1997) Laboratory investigation of the mass stability of sampling cassettes from inhalable aerosol samples, *American Industrial Hygiene Association Journal*, 59, 582–585.

Sreenath, A., Ramachandran, G. and Vincent, J.H. (2002) On the inlet characteristics of idealised blunt aerosol samplers at very large angles with respect to the wind, *Journal of Aerosol Science*, 33, 871–881.

Stein, S.W., Gabrio, B.J., Oberreit, D., Hairston, P., Myrdal, P.B. and Beck, T.J. (2002) An evaluation of mass-weighted size distribution measurements with the Model 3320 Aerodynamic Particle Sizer, *Aerosol Science and Technology*, 36, 845–854.

Su, W.C and Vincent, J.H. (2002) Experimental method to directly measure aspiration efficiencies of aerosol samplers in calm air, *Journal of Aerosol Science*, 33, 103–118.

Tufto, P.A. and Willeke, K. (1982) Dependence of particulate sampling efficiency on inlet orientation and flow velocities, *American Industrial Hygiene Association Journal*, 43, 436–443.

Vincent, J.H. (1989) *Aerosol Sampling: Science and Practice*, John Wiley & Sons, Ltd, Chichester.

Vincent, J.H., Atken, R.J. and Mark, D. (1993) Porous plastic form filtration media penetration characteristics and applications in particle size-selective sampling, *Journal of Aerosol Science*, 24, 929–944.

Walton, W.H. and Prewett, W.C. (1949) The production of sprays and mists of uniform drop size by means of spinning disc-type sprayers, *Proceedings of the Physical Society B*, 62 341–350.

Whitby, K.T. (1961) Generator for producing high concentrations of small ions, *Review of Scientific Instruments*, 32, 1351–1355.

Whitby, K.T. and Liu, B.Y.H. (1968) Polystyrene aerosols – electrical charge and residue size distribution, *Atmospheric Environment*, 2, 103–116.

Wilson, J.C. and Liu, B.Y.H. (1980) Aerodynamic particle size measurement by laser-doppler velocimetry, *Journal of Aerosol Science*, 11, 139–150.

Witschger, O., Wrobel, R., Fabries, J.F., Görner, P. and Renoux, A. (1997) A new experimental wind tunnel facility for aerosol sampling investigations, *Journal of Aerosol Science*, 28, 833–851.

Witschger, O., Willeke, K., Grinshpun, S.A., Aizenberg, V., Smith, J. and Baron, P.A. (1998) Simplified method for testing personal inhalable aerosol samplers, *Journal of Aerosol Science*, 29, 855–874.

Wright, B.M. (1950) A new dust feed mechanism, *Journal of Scientific Instruments*, 27, 12–15.

Yoshida, H., Uragami, M., Masuda, H. and Iinoya, K. (1978) Particle sampling efficiency in still air, *Kagaku Kogaku Ronbunshu*, 4, 123–128 (HSE translation No. 8586).

4

The Nature of Air Flow Near Aerosol Samplers

4.1 Introduction

In aerosol sampling, it is inevitable that the air that carries the particles of interest is in motion, as a result both of any externally driven wind (or *freestream*) and of the sucking (or *aspirating*) action of the sampler itself. In the preceding chapter it was shown how the ability of airborne particles to follow the motion of the air is determined not only by particle properties but also by the nature of the air movement. Before going on to discuss the aspiration characteristics of aerosol samplers, a picture of the nature of such air flows is therefore provided. In the terminology of experimental fluid mechanics, these may perhaps be regarded as *bluff-body flows with aspiration*. Without an appropriate appreciation of these it is not possible to make progress towards understanding aerosol sampling itself, neither its scientific nature nor the behaviors that are exhibited by practical sampling instruments on a day-to-day basis in real-world situations.

4.2 Line and point sink samplers

Line and point sink samplers are mathematically idealised systems in which the sampler has no body and no finite dimensions. The air that is sampled is simply drawn into an infinitesimally small orifice in the form of a line or point. Although such systems have no practical relevance as they stand, they do provide a useful starting point on which to build a picture of the flow near samplers more generally.

The simplest case of all is that for the three-dimensional potential flow near a point sink in calm air. The solution here is trivial, leading to the radial distribution of air velocity u_r with distance r away from the sink:

$$u_r = -\frac{Q_{point}}{4\pi r^2} \tag{4.1}$$

where Q_{point} is the sampling volumetric flow rate and the negative sign indicates that the flow is inwards, in the direction of r decreasing. The resultant streamline pattern is as shown in Figure 4.1(a), where

Aerosol Sampling: Science, Standards, Instrumentation and Applications   James Vincent
© 2007 John Wiley & Sons, Ltd

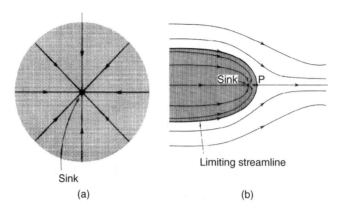

Figure 4.1 *Streamline patterns for two idealised aspirating flow situations: (a) from stationary air into a point sink; (b) from moving air into a line sink. Reproduced with permission from Vincent, Aerosol Sampling: Science and Practice. Copyright (1989) John Wiley & Sons, Ltd*

it is obvious that all the air in the system is eventually aspirated. The solution for the corresponding two-dimensional flow near an infinitely long line sink is equally trivial, giving:

$$u_r = -\frac{Q_{line}}{2\pi r} \tag{4.2}$$

where Q_{line} is now the sampling flow rate per unit length of the line. When a uniform freestream, representing a finite wind, is superimposed on the line sink, the flow pattern is accessible from potential flow theory. Now, as already shown in Chapter 2, the two-dimensional potential flow solution, in $\{x, y\}$ coordinates, yields a pair of equations – the earlier Equations (2.16) – that describes the air velocity field everywhere. The streamline pattern for this system is shown in Figure 4.1(b). One important feature is the pair of limiting streamlines that separates the sampled air (shaded volume) from that which passes by from left to right with the rest of the freestream. On the downstream side of the sink, these limiting streamlines come together at the stagnation point (say, P) where the velocity falls locally to zero. The position of P is easily shown to be:

$$x_{stag} = \frac{Q_{line}}{2\pi U} \quad \text{and} \quad y_{stag} = 0 \tag{4.3}$$

A similar set of solutions and flow pattern, this time in terms of three-dimensional streamsurfaces, is also easily obtained for the axisymmetric case of a point sink in a uniform freestream, giving:

$$x_{stag} = \left(\frac{Q_{point}}{4\pi U}\right)^{1/2} \quad \text{and} \quad y_{stag} = 0 \tag{4.4}$$

A number of workers (Levin, 1957; Stenhouse and Lloyd, 1974) have calculated flow patterns for the point sink in a uniform freestream, and gone on to use them as a basis for calculating particle trajectories – and, in turn, aspiration efficiency – for aerosol sampling in simple systems. Not surprisingly, comparisons with experimental data for real samplers with finite dimensions have exposed the obvious limitations of such idealised flow models. However, they confirm the value of this simple approach in providing first insight into the problem.

4.3 Thin-walled slot and tube entries

Real samplers have entries with finite dimensions for which the point or line sink models represent only rough approximations when those dimensions are very small. The next most complicated cases, albeit still highly idealised, are the two-dimensional potential flows near infinitely long, thin-walled slots and three-dimensional thin-walled tubes.

4.3.1 Facing the freestream

A thin-walled slot or tube-shaped sampler facing into the wind is the simplest system that can be studied. This is because, for these systems at this orientation, the wall of the sampler itself does not present any physical cross-section (or blockage) to the approaching flow. So any change in the flow pattern derives entirely from local changes in the air velocity associated with the difference in the undisturbed freestream air velocity and that as it enters the inlet of the sampler. These are therefore approachable systems from the theoretical standpoint.

Relatively straightforward solutions for the flow near a two-dimensional thin-walled slot have been obtained by a number of workers (Volushchuck and Levin, 1968; Bartak, 1974; Addlesee, 1980). These have been supplemented by electrolytic tank simulations by others (Vincent *et al.*, 1982). Although the two-dimensional thin-walled slot is, in itself, only of minor interest in relation to practical aerosol sampling, it provides a stepping stone to the more important three-dimensional case of the thin-walled cylindrical tube which features frequently in experiments and in other real-life situations. Extension of theory even to this simple next level substantially increases the degree of theoretical difficulty. Nonetheless, some early workers successfully achieved numerical solutions both for a moving freestream (Vitols, 1966) and for calm air (Agarwal and Liu, 1980). Others have determined the flow pattern experimentally (Rüping, 1968).

From this body of work, an overall qualitative picture of the flow near a thin-walled tube aspirating while facing into a uniform freestream may be constructed. Typical flow patterns are shown in Figure 4.2, and a number of interesting features are immediately apparent. Firstly, we again see the limiting streamsurface which divides the sampled from the unsampled air. As the freestream air velocity (U) falls, the limiting streamsurface encloses a larger and larger volume (shaded regions) until, eventually,

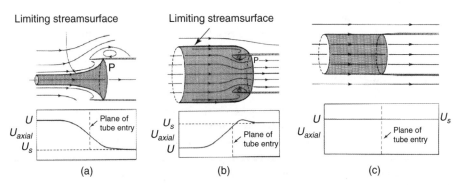

Figure 4.2 *Streamline patterns and axial air velocity distributions for flow in the vicinity of an aspirating thin-walled tube facing into the wind: (a) sub-isokinetic, with R > 1; (b) super-isokinetic, with R < 1; (c) isokinetic, with R = 1. Reproduced with permission from Vincent, Aerosol Sampling: Science and Practice. Copyright (1989) John Wiley & Sons, Ltd*

in the calm air limit (not shown), all the air in the system enters the sampling orifice. Secondly, the position of stagnation, which appears as a circular line around the circumference of the tube, is not fixed but moves over the surface of the tube as U varies relative to the average air velocity over the plane of the tube entry, U_s. An important parameter is therefore:

$$R = \frac{U}{U_s} \text{ where } U_s = \frac{4Q}{\pi \delta^2} \tag{4.5}$$

where Q is the sampling flow rate and δ is the diameter of the circular sampling orifice. Figure 4.2(a) shows the scenario for $R > 1$ corresponding to relatively low sampling flow rate. In the terminology of aerosol sampling using a thin-walled tube facing the wind, this is referred to as *sub-isokinetic*. Under these conditions, as demonstrated by Addlesee (1980) for the two-dimensional equivalent, stagnation occurs inside the tube, penetrating further into the tube as R increases. Associated with this changing pattern of the location of stagnation are regions of flow reversal on the inside of the tube. Addlesee pointed out that such behavior is well-known to fluid dynamicists for *intake flows*. In addition, as the reversing flow encounters the leading edge of the tube, separation (and subsequent reattachment) of the flow on the outside of the tube is likely. Finally, the air velocity in the approaching freestream along the axis of the tube decreases steadily, only leveling off at some distance into the tube. Figure 4.2(b) shows the corresponding scenario for $R > 1$ corresponding to relatively high flow rate. These conditions are referred to as *super-isokinetic*. Here, stagnation occurs on the outside of the tube, moving further away from the entry as R decreases. Flow reversal therefore occurs on the outside of the tube, and as the reversing flow encounters the leading edge of the tube, flow separation on the inside of the tube is likely. Now the effect is to cause a compression of the air flow streamlines close to the tube entry, with an associated increase in the axial air velocity there. This affect is referred to as a *vena contracta*. In this case, the axial air velocity along the approaching freestream increases steadily, continuing to increase even after the flow has entered the tube and experiences the *vena contracta*, finally falling back to a value that thereafter remains constant. Finally, Figure 4.2(c) describes the *isokinetic* limiting case where the air velocity of the freestream perfectly matches the air velocity over the plane of the sampling tube so that $R = 1$. Only here, does stagnation occur right at the leading edge of the tube. Now there are no flow distortions and the axial air velocity in the approaching freestream remains constant everywhere. This scenario has important practical implications to practical aerosol sampling, especially in that *isokinetic sampling* with thin-walled tubes is widely used for the accurate collection of aerosols from moving gas streams in industrial processes, stacks, etc. It also provides an important method for obtaining reference samples in laboratory-based aerosol sampler studies in wind tunnels.

4.3.2   Other orientations

The preceding describes the most idealised picture of the air flow close to an aerosol sampler and yet, as just mentioned, it has considerable practical relevance. Over the years, researchers and practitioners have been interested in the thin-walled tube at other orientations, partly in order to provide insights into how performance is degraded when the tube is not well aligned with the approaching freestream and partly to investigate the possibility of sampling at other angles. Unfortunately, the airflow pattern for orientations other than forwards-facing do not succumb as readily to theoretical treatment. One difficulty is that, as soon as the orientation moves away from the ideal forwards-facing, the body of the sampler begins to project finite cross-section – and hence blockage – to the incoming flow. Now the well-understood thin-walled tube flow picture that has just been described can only be taken as a

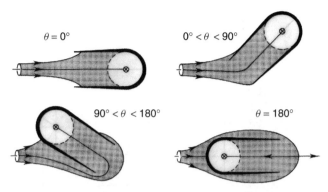

Figure 4.3 *Schematic to illustrate the changing nature of the air flow pattern near the entrance to an aspirating thin-walled tube at various orientations with respect to the wind, not to scale (Vincent et al., 1986). Reproduced with permission from Vincent, Aerosol Sampling: Science and Practice. Copyright (1989) John Wiley & Sons, Ltd*

starting point for considering the nature of the movement of air near a sampling system that, although it is very closely related, presents a significantly greater level of theoretical difficulty.

For the purpose of this discussion, consider first an idealised thin-walled sampling probe where the tube is assumed to be bent at right angles, with the short arm containing the sampler entry. Orientation of the probe with respect to the freestream is achieved by rotation of the tube about the axis of the long arm, which itself is held at right angles to the freestream. The expected general character of the flow pattern – by reference to the limiting streamsurface only – is suggested for a range of orientations of the tube inlet in Figure 4.3. Here, particular attention is drawn to the rapid changes that take place in the shape of the limiting streamsurface – in particular the positions of the lines of flow stagnation on the tube wall – for increasingly backwards-facing entry orientations. Here, following the ideas that began to emerge during discussion of the simpler forwards-facing case, the line of stagnation becomes increasing asymmetric as the angle of the axis of the tube with respect to forwards-facing, θ, increases. Then, for the most extreme case where $\theta = 180°$, the limiting streamsurface no longer terminates on the sampler body, but breaks away to fully enclose the sampler in a manner reminiscent of the case of the point sink shown in Figure 4.1(b).

4.4 Thick-walled tubes

In reality, ideal thin-walled probes of the type just described do not exist. Inevitably, the obstruction to the flow arising from the finite thickness of the tube wall modifies the nature of the flow. Walter (1957) was one of the first to consider this problem. By a combination of flow visualisation and air velocity measurements in a wind tunnel, he investigated experimentally the air movement near a thick-walled nozzle facing into the freestream. Typical results are shown in Figure 4.4. They show that the axial air velocity, U_{axial}, expressed here as a proportion of the freestream velocity, passes through a minimum at some distance in front of the entry. This differs distinctly from what happens for the ideal thin-walled tube illustrated in Figure 4.2. Walter found that the depth of this trough was greater for blunt-nosed tubes than for more streamlined ones. In addition, it became more pronounced as R increased in the range $R > 1$ yet, on the other hand, disappeared entirely for $R << 1$. Walter also noted that the depth of the trough increased with the relative thickness of the tube wall, and suggested that isokinetic sampling may be subject to error as a result. This view was later endorsed by Rouillard and Hicks (1978) who reported

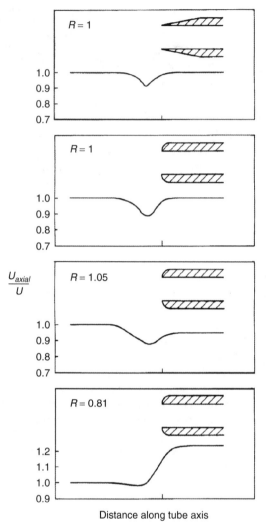

Figure 4.4   Axial distributions of air velocity (U_{axial}) along the axis of an aspirating thin-walled tube facing into the wind, showing the effect of velocity ratio (R) and the shape and thickness of the tube at the entry. Reproduced with permission from Vincent, Aerosol Sampling: Science and Practice. Copyright (1989) John Wiley & Sons, Ltd

experimental evidence that effects associated with tube blockage were transmitted to the upstream flow by creating stagnation-like regions in front of the tube.

4.5   Simple blunt samplers facing the wind

4.5.1   Two-dimensional blunt sampling systems

In his study of thick-walled tubes, Walter described a scenario where the blunt body of the sampler imposed significant blockage to the flow. More recently, attention has shifted towards the air flow

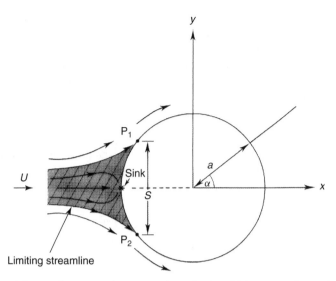

Figure 4.5 *Schematic of the air flow near the entry of a cylindrical blunt sampler with a thin line sink facing into the wind, showing the positions of flow stagnation P_1 and P_2 indicating where the dividing streamlines intercept the body of the sampler. Reproduced with permission from Vincent, Aerosol Sampling: Science and Practice. Copyright (1989) John Wiley & Sons, Ltd*

near aerosol samplers of wider practical relevance, where the sampler bodies are – geometrically and aerodynamically – even more 'blunt'. Once again, much can be learned by reference to more accessible two-dimensional flow systems. Ingham (1981) examined theoretically the potential flow of air near a cylindrical body placed normal to the freestream and with a line sink located at its leading edge. For the system shown in Figure 4.5, the solution for the velocity field is given by

$$
u_x = \frac{Q_{line}}{2\pi} \left[\frac{x}{(x^2 + y^2)} - \frac{2\left(x + \frac{D}{2}\right)}{\left(x + \frac{D}{2}\right)^2 + y^2} \right] + U\left[1 - \frac{D^2}{4}\frac{(x^2 - y^2)}{(x^2 + y^2)^2}\right]
$$

$$(4.6)$$

$$
u_y = \frac{Q_{line}}{2\pi} \left[\frac{y}{(x^2 + y^2)} - \frac{2y}{\left(x + \frac{D}{2}\right)^2 + y^2} \right] - U\left[\frac{D^2 xy}{2(x^2 + y^2)^2}\right]
$$

where D is the cylinder diameter. A most important feature is the pair of limiting streamlines which encloses the sampled air (shaded volume) and outside which all the air passes by the sampler with the rest of the freestream. Where these intersect with the surface of the cylinder defines the lines of stagnation that run parallel to the line sink. These are indicated in Figure 4.5 as P_1 and P_2 which, due to symmetry, are equidistantly spaced from the axis. From the above equations, the $\{x, y\}$ locations of P_1 and P_2 are given by:

$$
\cos\left(\frac{\alpha}{2}\right) = \left(\frac{Q_{line}}{4\pi\,DU}\right) \quad \text{and} \quad \alpha = 0
$$

$$(4.7)$$

where the angle α defines position on the surface of the cylinder. This equation shows that, when $Q_{line} = 0$, the system reduces to the simple case of flow around a cylinder without aspiration, for which stagnation occurs at $\alpha = 0$ and $180°$. When $Q_{line} = 2\pi DU$, the stagnation points are at $\alpha = 0, +90$ and $-90°$. When $Q_{line} > 4\pi DU$, there is one stagnation point only, at $\alpha = 0$.

If the line sink is now assumed to have finite width δ, then it is useful to define the dimensionless flow parameter (for two dimensions) as:

$$\phi_T = \frac{\delta U_s}{DU} \tag{4.8}$$

which, physically, is the ratio of the volumetric flow rate of the sampled air to that which is geometrically incident on the sampler body from far upstream. Here, both volumes are expressed per unit length of the cylinder. From Equations (4.7) and (4.8), the width (S) of the stagnation region enclosed by P_1 and P_2 is given by:

$$S = \frac{D\phi_T^{1/2}}{\pi^{1/2}} \longrightarrow 0.56D\phi_T^{1/2} \tag{4.9}$$

Similarly, a line sink in a wide flat plate placed normal to the freestream gives:

$$S = 0.80D\phi_T^{1/2} \tag{4.10}$$

Later, Dunnett and Ingham (1986) suggested that a more accurate representation of the flow field for this system could be obtained using the *linear boundary integral equation* (LBIE) method, requiring solution of Green's integral formula. It was noted that, unlike Ingham's earlier potential flow model, this approach would enable results to be obtained for a sampler of any shape, including those where no analytical solutions are available.

At about the same time as Ingham's original mathematical studies, electrolytic tank analog investigations were carried out to explore the flow fields for two similar problems – a flat-nosed two-dimensional sampler and another with the leading edges inclined backwards at $45°$, both facing the freestream (Vincent *et al.*, 1982). The aim was to explore the relationship:

$$S = B\phi_T^p b_T^q D \tag{4.11}$$

where

$$b_T = 1 - \left(\frac{\delta}{D}\right) \tag{4.12}$$

represents the geometric blockage ratio, B is the dimensionless *aerodynamic bluntness* parameter that relates to the shape of the sampler body, and p and q are coefficients to be determined. The results were found to satisfy the empirical expression:

$$S = B\phi_T^{1/2} b_T^{-1/7} D \tag{4.13}$$

where $B = 0.80$ for the two-dimensional flat-nosed sampler and $B = 0.48$ for the $45°$ sampler. Inspection of Equation (4.13) reveals that S is only weakly dependent on b_T and that the term in b_T is close to unity to within $\pm 10\%$ for $\delta/D < 0.5$.

Typical flow patterns from the electrolytic tank study for the flat-nosed sampler are shown in Figure 4.6. The first, for small $\phi_T \ll 1$ [Figure 4.6(a)], exhibits what has been described as the typical 'spring onion' shape in which the air flow pattern of the approaching freestream first diverges as it

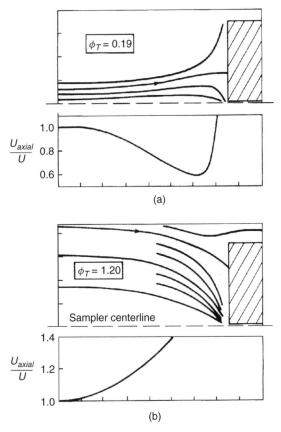

Figure 4.6 *Typical examples of the potential flow near the entries of two-dimensional, flat-nosed blunt samplers with high blockage ratio and facing into the wind, for contrasting values of the sampling ratio ϕ_T. Reproduced with permission from Vincent, Aerosol Sampling: Science and Practice. Copyright (1989) John Wiley & Sons, Ltd*

begins to experience the presence of the blunt sampler body and then converges as it comes increasingly under the influence of the suction associated with the flow into the sampling orifice. This observation of a two-part flow system will prove to be of considerable value later when the nature of the air flow near an aerosol sampler is used to evaluate aspiration efficiency. For this pattern, the axial air velocity distribution is consistent with the picture described by Walter and by Rouillard and Hicks, exhibiting a minimum just in front of the sampler. This may be taken to indicate the plane at which the shape of the sampled flow changes from divergent to convergent. By contrast, for higher sampling flow rate where $\phi_T > 1$ [Figure 4.6(b)], the distinction between the diverging (unaspirated) and converging (aspirated) flow components is less evident. However, in principle, it should still be regarded as a two-part flow system. The only difference between this and the one shown in Figure 4.6(a) is that the nature of the flow is now dominated by the flow into the sampling orifice so that the axial velocity distribution exhibits no minimum.

Typical similar results for a sampler with $45°$, backwards-facing external walls are shown in Figure 4.7. Qualitatively, the same trends are present as already noted for the flat-nosed sampler. However, comparison with the results in Figure 4.7 indicates that the flow distortion in front of the sampler is less

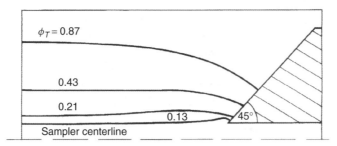

Figure 4.7   *Typical example of the dividing streamlines for potential flow near the entry of a two-dimensional blunt sampler with high blockage ratio and inclined leading edges, facing into the wind, for contrasting values of the sampling ratio ϕ_T. Reproduced with permission from Vincent, Aerosol Sampling: Science and Practice. Copyright (1989) John Wiley & Sons, Ltd*

marked. This is not surprising since the part of the flow associated with its deflection to pass around the sampler would be expected to be less divergent for the sleeker body configuration.

4.5.2   Axially symmetric blunt sampling systems

Potential flow theory for even relatively simple three-dimensional (e.g. axisymmetric) systems is substantially more complicated than for equivalent two-dimensional ones. However, Ingham and his colleagues, in an overall body of work that has evolved over the past two decades, have been successful in realising a mathematical approach to aerosol sampling for more complex systems. Ingham (1981) defined the sampling flow rate parameter for the sampler with axial symmetry as:

$$\phi_A = \frac{\delta^2 U_s}{D^2 U} \tag{4.14}$$

which is directly analogous to Equation (4.8) for the two-dimensional case. He went on to show that:

$$S = B\phi_A^{1/3} D \tag{4.15}$$

For a sampler of spherical shape and a small circular orifice facing the freestream, it was found $B = 0.70$. In Equation (4.15), S is now the diameter of the circular stagnation region on the sampler body surface enclosed by the limiting streamsurface where it intersects with the sphere. For the corresponding case of a flat disk-shaped sampler with a small circular central orifice facing the freestream, Ingham found $B = 0.92$.

   Vincent *et al.* (1982) extended the electrolytic tank study mentioned earlier to the example of the flat disk-shape sampler with a small central orifice. This was supported by wind tunnel studies in which the position of stagnation was located by measurements of the radial distribution of static pressure over the surface of the disk, and where – by virtue of the Bernoulli relationship – the maximum in static pressure was indicative of stagnation. The results for S/D as a function of ϕ – as suggested by Equation (4.15) – are shown in Figure 4.8 and they are seen to be in excellent agreement both with each other and with what Ingham had predicted. By way of illustration, Figure 4.9 shows a photograph of the same flow, visualised using fine balsa dust under strong slit illumination in the plane of the sampling orifice (Vincent and Mark, 1982).

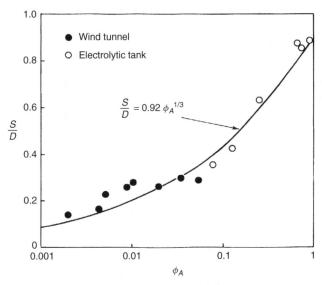

Figure 4.8 *S/D as a function of sampling ratio ϕ_A for an axisymmetric, flat-nosed, disk-shaped blunt sampler facing the wind, from both wind tunnel and electrolytic tank analog experiments, respectively. Reproduced with permission from Vincent, Aerosol Sampling: Science and Practice. Copyright (1989) John Wiley & Sons, Ltd*

Figure 4.9 *Photograph of the flow pattern near a disk-shaped blunt sampler with a central circular orifice facing the wind, obtained using balsa dust under slit illumination. Reproduced with permission from Vincent and Mark, in: Inhaled Particles V (Ed. W. H. Walton), Pergamon Press, Oxford, pp. 3–19. Copyright (1982) British Occupational Hygiene Society*

4.6   Blunt samplers with orientations other than facing the wind

4.6.1   A cylindrical blunt sampler

Figure 4.10 shows the general nature of the flow near a cylindrical blunt sampler with a narrow slot located at an angle θ with respect to the forwards-facing. Here, in addition to the *divergence* and *convergence* noted for the forwards-facing case, there is also *turning*. Ingham and Hildyard (1991) have studied this scenario mathematically and shown that the positions of stagnation can be found from solution of:

$$t^3\phi_T + t^2 2\pi \sin\theta + t(\phi_T - 4\pi\cos\theta) - 2\pi\sin\theta = 0 \tag{4.16}$$

where is ϕ_T is given by the earlier Equation (4.8), and:

$$t = \tan\left(\frac{\alpha}{2}\right) \tag{4.17}$$

which defines the angular positions of stagnation as given by α. Equation (4.16) is cubic in t and so has three roots, indicating the existence of the three lines of stagnation. In Figure 4.10, P_1 and P_2 are the primary stagnation lines towards the leading edge of the cylinder. In addition, there is a third stagnation line, P_3, towards the rear of the cylinder. By way of illustration, Figure 4.11(a) shows the angular positions of P_1 and P_2 calculated from the Ingham and Hildyard model for a range of slot orientations (θ) from $0°$ to close to $180°$ and for a single value of ϕ_T (see dotted lines). For $\theta = 0°$, stagnation splits symmetrically into two stagnation lines. But for increasing θ it is seen that P_1 moves with, but always just ahead of, the slot, so that α_{P1} is always only slightly greater than θ. Meanwhile for P_2, α_{P2} merges slowly towards $\alpha = 0°$ as θ increases. The positions of stagnation P_1 and P_2 are separated by the linear distance S, and this too is shown – also as a function of θ – in Figure 4.11(b). It is seen that S/D increases steadily as θ increases.

The positions of stagnation for this scenario were later studied experimentally by Sreenath *et al.* (1997) in a small wind tunnel similar to the one shown in Figure 3.7. Two test sampler models were built as the

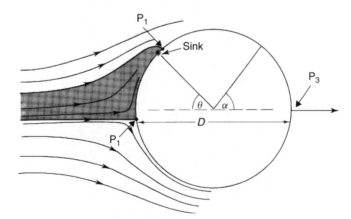

Figure 4.10   *General form of the streamline pattern near a two-dimensional cylindrical blunt sampler with its line sink entry placed at an angle θ with respect to the freestream. Here the angle α is used to denote position on the sampler body. Reproduced with permission from Vincent, Aerosol Sampling: Science and Practice. Copyright (1989) John Wiley & Sons, Ltd*

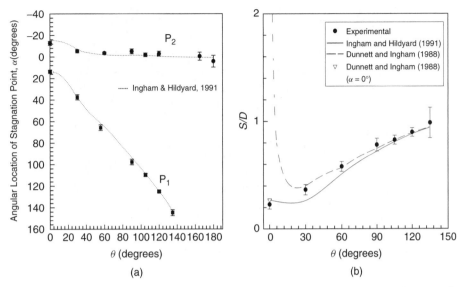

Figure 4.11 *Results from experimental flow visualisations of the flow near a cylindrical blunt sampler facing the wind: (a) angular positions of flow stagnation (α) as functions of slot orientation (θ); (b) width of the stagnation region as a function of θ (reproduced from Sreenath et al., 1997). Also shown are calculations from the numerical simulations of Ingham and Hildyard (1991) and Dunnett and Ingham (1988), as indicated. Reprinted from Atmospheric Environment, 31, Sreenath et al., 2349–2359. Copyright (1997), with permission from Elsevier*

subjects of the investigation. The first was a 3 cm diameter, 25 cm long cylinder with a 1.5 mm wide slot, which was mounted horizontally between the walls of the tunnel, using support rods such that the sampler could be easily rotated about its axis. End plates were used to reduce wall effects (Cowdrey, 1963). Smoke for flow visualisation was generated using a variation of the 'smoke-wire visualisation' technique that has been used extensively by aerodynamicists for many years, introducing minimal disturbance to the air flow and producing clear strands of smoke by which to track the streamlines of the flow. This method is based on the heating of a resistance wire that has been wetted with an appropriate oil. A typical flow visualisation obtained using this technique is shown in Figure 4.12. From such photographs, Sreenath *et al.* were able to determine the forwards-facing positions of stagnation P_1 and P_2 for a wide range of conditions. But location of P_3 was not possible. Experimental results for the angular positions of P_1 and P_2 obtained in this way are shown alongside those calculated by Ingham and Hildyard for $\phi_T = 0.25$ in Figure 4.11(a). Experiment and theory were found to be in good agreement. Similarly good agreement is shown in Figure 4.11(b) for S.

From all the preceding, it is interesting to consider what happens in the limiting, perfectly symmetrical case where $\theta = 180°$. Equation (4.16) does not provide a solution for the positions of stagnation for this case. But we are guided by what is known for angles approaching it. Here, to preserve symmetry, and to allow air to be aspirated into the slot inlet there can be only one position of stagnation on the cylinder itself – at the leading edge. On the downstream side, the limiting streamsurface must now enclose all the air that enters the slot, and so must be perfectly symmetrical and become detached from the cylinder. In that event, the downstream stagnation position is located where the flow divides between that which is aspirated and that which is not. The expected pattern of the time-averaged flow for this scenario is shown in Figure 4.13.

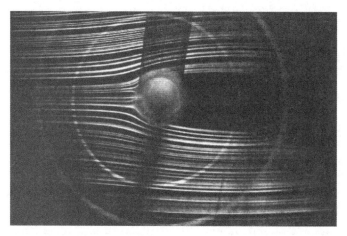

Figure 4.12   *Typical smoke tracer visualisation of the flow near a cylindrical blunt sampler with a slot entry facing the wind. Reprinted from Atmospheric Environment, 31, Sreenath et al., 2349–2359. Copyright (1997), with permission from Elsevier*

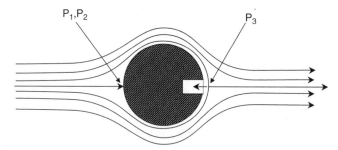

Figure 4.13   *Schematic to show the streamline pattern to be expected when a blunt sampler of simple shape is facing directly away from the freestream ($\theta = 180°$)*

4.6.2   Flow stability

For flow over a cylinder at Re values corresponding to those in the Sreenath *et al.* experiments ($Re \approx 1500$), it is expected that separation of the flow would occur at α close to $100°$. This would be accompanied by turbulence in the separation shear layers and by coherent vortex shedding motions. However, as seen from the experiments described above, the effect of the aspiration of air into the slot is to stabilise the boundary layer, and so 'postpone' the flow separation and any accompanying turbulence. For the cylindrical geometry discussed here, it is very well known that – at least in the absence of aspiration – such separation is accompanied by boundary layer instability so that the near wake flow is characterised by strong, coherent, periodic vortex shedding. The frequency of this vortex shedding (f) is usually expressed in terms of the dimensionless Strouhal number (Str) given by:

$$Str = \frac{fD}{U_0} \tag{4.18}$$

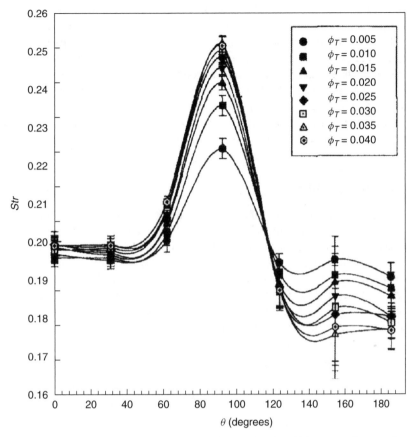

Figure 4.14   *Experiment results for Strouhal number (Str) for a cylindrical blunt sampler as a function of sampler orientation with respect to the wind (θ) for various values of sampling ratio (φ$_T$), reflecting the phenomenon of vortex shedding in the sampler wake. Reprinted from Atmospheric Environment, 31, Sreenath et al., 2349–2359. Copyright (1997), with permission from Elsevier*

Roshko (1954), and many others, have shown experimentally that $Str \approx 0.21$ for a cylinder without aspiration and for Re in the range from about 400 to greater than 2×10^5.

Such instabilities would inevitably influence aerosol transport. So Sreenath *et al.* also examined vortex shedding for the same aspirating cylindrical body already mentioned. For this, the fluctuating motions were detected using a fine hypodermic needle attached to a capacitive micromanometer, and fast Fourier transformation (FFT) methods were used to derive the frequency spectrum of the fluctuations. Figure 4.14 shows the results for Str as a function of θ for a range of ϕ_T for just one wind speed and one Re value. A complex pattern is seen. Firstly, for given ϕ_T, Str remained fairly constant at first as θ increases from zero. Then from about $\theta = 45°$, Str rose steeply, reaching a peak for θ at around 90°. Thereafter, it fell back even more steeply, levelling off at a value lower than that for $\theta = 0°$ as θ approached 180°. The magnitudes of both the rise and the fall in Str were greater as ϕ_T increased. Sreenath *et al.* explained these trends in terms of qualitative physical arguments along the following lines. For such flow scenarios, the location of one stagnation point remains roughly constant at $\theta = 0°$ as θ increases. The second stagnation point moves around with the slot and is located roughly a few degrees beyond θ. As θ increases, the angular location of the second stagnation point increases. The angular location of the separation point also increases, but proportionately less so. As a result, the angular distance between the stagnation point and the separation point decreases. This leads in turn to the formation of a smaller vortex and hence a corresponding increase in *Str* for θ in the range up to about 100°. However, when θ gets close to and beyond 104°, conditions are no longer favorable for flow separation from the body itself. At this point, it is suggested that the near-wake recirculation zone becomes detached and appears as a standing vortex just downstream from the cylinder. Now, since the boundary layer around the surface of the cylinder itself is essentially stabilised, the frequency of vortex shedding, and hence *Str*, is suddenly – and markedly – reduced.

4.6.3   A spherical blunt sampler

The three-dimensional flow at the surface of a sphere is more complicated, and needs to be characterised by three components of velocity, radial (V_r), tangential (V_α) and azimuthal (V_φ), respectively. This coordinate system is shown in Figure 4.15(a). When there is aspiration from a point sink or circular orifice located on the spherical body, the positions of stagnation are defined as *loci* on the surface of the sphere where all these three components are identical to zero. However, the radial component of velocity on the surface of the sphere is always trivially zero, so that stagnation zone *loci* can be determined by setting $V_\alpha = V_\varphi = 0$. Partial solutions to this were obtained mathematically by Dunnett and Ingham (1988) by setting only $V_\alpha = 0$, leading to:

$$
\begin{aligned}
&-\sin\alpha\cos\theta - \cos\alpha\sin\theta\cos\varphi + \\
&\frac{\phi_s}{3}\left[\frac{\sin\alpha}{(2+2\cos\alpha)^{3/2}} - \frac{1}{2\sin\alpha}\left(1 - \frac{1+\cos\alpha}{(2+2\cos\alpha)^{1/2}}\right)\right] = 0
\end{aligned}
\tag{4.19}
$$

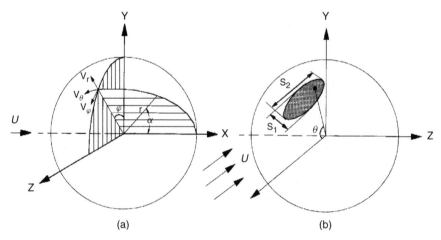

(a)   (b)

Figure 4.15   *Schematic to enable discussion of the flow about a spherical sampler with a small circular orifice at varying orientation with respect to the freestream: (a) coordinate system; (b) diagram to indicate the position and shape of the stagnation region on the surface of the sphere. Reprinted from Atmospheric Environment, 31, Sreenath et al., 2349–2359. Copyright (1997), with permission from Elsevier*

in which the dimensionless flux parameter for the sphere, ϕ_S, is given by:

$$\phi_S = \frac{\delta^2 U_s}{D^2 U} \tag{4.20}$$

where δ, as before, is the orifice diameter while D is the diameter of the sphere and U_S is the sampling velocity (averaged over δ). This is the same as the expression for ϕ_A given earlier in Equation (4.14). Dunnett and Ingham solved Equation (4.19) to obtain the two characteristic dimensions S_1 and S_2 of what is now an the oblate, egg-shaped stagnation region on the surface of the sphere enclosed by the condition of zero tangential velocity V_α, leading to:

$$S_1 = D \left(\frac{\phi_S}{3 \cos \theta} \right)^{1/2} \tag{4.21}$$

$$S_2 = D \sin \left[\frac{\phi_S}{24} \frac{\left(1 - \sin \frac{\theta}{2} \right)}{\sin^2 \frac{\theta}{2} \cos \frac{\theta}{2}} + \frac{1}{2} \sqrt{\frac{\phi_S}{3 \sin \theta}} + \frac{\theta}{2} \right] \tag{4.22}$$

The shape of the stagnation region in relation to the location of the sampling orifice is illustrated in Figure 4.15(b). Inspection of Equations (4.21) and (4.22) confirms that the shape of the stagnation region reverts to the expected circular shape when $\theta = 0°$.

Sreenath *et al.* (1997) extended their study of the cylindrical blunt sampling system described above to a similar experimental investigation of a spherical blunt sampler. This second body was a 6.35 cm diameter sphere with a 4 mm diameter circular orifice. It was mounted horizontally between the walls of the tunnel, again using a support rod arranged such that the sphere could be rotated about its horizontal axis. The sphere was oriented such that the inlet orifice was always in the $\varphi = 0$ plane. For the spherical body, the stagnation region was three-dimensional and could not be located in exactly the same way as for the cylindrical body. Although the maximum width of the stagnation region of the body of the sphere in the plane parallel to the freestream could be obtained from flow visualisation photographs in the same way as for the cylindrical sampler, a direct visual observation method was used to locate the extremities of the stagnation region in the transverse plane. In this way, quite accurate measurements were made of the widths S_1 and S_2. Some typical experimental results are shown in Figure 4.16 alongside the corresponding calculated values from Dunnett and Ingham. Here it is seen that agreement for S_1 [Figure 4.16(a)] was poor. While the measured values decreased as θ increased in the range up to about $\theta = 140°$ (beyond which measurements were not possible), the calculated results showed a steep rise followed by a sharp fall at $\theta = 0°$. The measured trend was more plausible. It was considered that this lack of agreement stemmed from the difference in the way that stagnation was defined. Dunnett and Ingham had defined it in terms of setting V_α to zero, whereas in the experimental study of Sreenath *et al.* the observational method used was equivalent to equating both V_α and V_φ to zero. By contrast, as seen in Figure 4.17(b), the original analytical expression for S_2 agreed quite well with the experimental results for θ greater than about 30°. This was as expected since the results did not depend on the additional assumption about V_φ. Better agreement for $\theta > 0°$ was later obtained by Dunnett (1999) in an improved mathematical model. At about the same time as the work of Sreenath *et al.*, Chung and Dunn-Rankin (1997) reported a similar study, also based largely on flow visualisation of smoke tracers. Their results for the location of the position of stagnation as defined above by S_2 were in similarly good agreement with the predictions of Dunnett and Ingham.

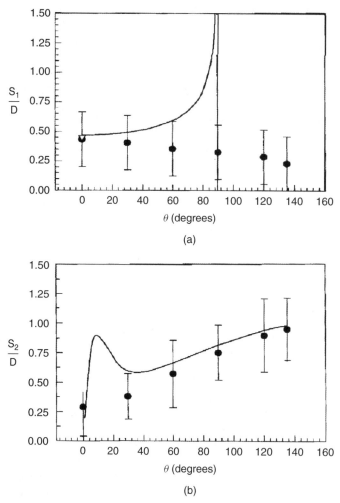

Figure 4.16   *Typical experimental results for the linear widths S_1 (a) and S_2 (b) of the enclosed stagnation region on the surface of the spherical blunt sampler for the entry located at varying angles (θ) with respect to the freestream, also showing (solid lines) the corresponding results of the numerical simulations of Dunnett and Ingham (1988). Reprinted from Atmospheric Environment, 31, Sreenath et al., 2349–2359. Copyright (1997) with permission from Elsevier*

From the theoretical and experimental studies of the air flow near a simple spherical blunt sampler with a single circular orifice, together with what is known about the flow about a sphere without aspiration, a general picture emerges for this system at various orientations that is equivalent to the one given earlier for the thin-walled tube (Figure 4.4) and for the cylindrical blunt sampler (Figures 4.10 and 4.13). Despite the likely relevance of such flows to many aspects of practical aerosol sampling, very little is known about them beyond what is described here. Here the nature of the flow at extreme orientations has more practical significance than for the thin-walled tube because most devices applied in real-world aerosol sampling are, in fact, blunt samplers and are expected to operate at a wide range of not-always-prescribed orientations. The near-wake vortex shedding mentioned earlier for the cylindrical

Figure 4.17   *Smoke tracer visualisation of flow near the ORB sampler (see Chapter 14), showing how – by virtue of the multiple-orifice arrangement of the sampling head – some of the air from within the sampler is re-entrained back into the freestream. Reproduced with permission from Vincent, Aerosol Sampling: Science and Practice. Copyright (1989) John Wiley & Sons, Ltd*

body is also likely to be present for the flow about a sphere (without aspiration), but is probably less marked than for two-dimensional flows (Fuchs *et al.*, 1979).

4.7   More complex sampling systems

From the picture of the flow which has been built up for the simple, single orifice sampling systems described above, it is possible to proceed – cautiously – to the construction of what is likely to happen when there is more than one orifice. One feature that emerges is that the number of positions of stagnation on the sampler body must increase, since the overall sampled air is now divided between multiple orifices so that the flow into each adjacent one must be separated at the surface of the sampler. It can easily be shown that the number of regions on the sampler surface enclosed by stagnation lines must be equal to the number of individual orifices.

Another interesting feature is that, even for orifices of identical size and shape, the net flow rate of air through each may not be the same, depending on their location with respect to the direction of the freestream and in turn the distribution of static pressure around the sampler body. Thus it is possible for higher local flow rates through some orifices to be offset by lower local sampling flow rates elsewhere. In fact, due to the presence of the bluff sampler body, there is the tendency for the local static pressure to be higher on the upstream side and lower – even negative – on the downstream side. Indeed, if the negative pressure in the near wake of the sampler is great enough, it is possible to envisage a scenario where there may even be a net *outflow* of air on that side. This phenomenon is illustrated in Figure 4.17 for an experimental prototype multi-orifice, omnidirectional sampler of a type described by Ogden and Birkett (1978). In the photograph, a smoke tracer has been released inside the body of the sampler, and the escape of smoke on the downstream side is clear confirmation of the 'negative aspiration' alluded to.

4.8   Effects of freestream turbulence

Most of the thinking about the nature of air movement near aerosol samplers contained in the preceding descriptions is based on the potential flow model. For laminar flow, this is reasonable if the characteristic Reynolds number is high enough. However, as already mentioned, in many practical aerosol sampling situations the freestream flow is not laminar but is turbulent. Since such turbulence is responsible for the diffusive transport of scalar properties of the flow, including momentum, it is necessary to ask the question: to what extent does freestream turbulence influence the mean flow pattern near an aerosol sampler? Solutions of the Navier–Stokes equations should then incorporate appropriate turbulence models, like those described by Launder and Spalding (1972). An experimental approach can provide useful guidance on how to think about the role of turbulence. Vincent *et al.* (1983) described some wind tunnel experiments for a simple disk-shaped blunt sampler of the type described earlier facing directly into the freestream. These were performed under conditions where freestream turbulence of well-defined characteristics, expressed in terms of length scale and intensity, respectively, was generated by the placement of square lattice grids across the entrance to the working section. These characteristics were varied by changing the grid bar width and distance downstream from the plane of the grid of the sampler location. In order to investigate possible changes in the flow pattern, measurements were made of S, the width of the stagnation region enclosed by the limiting streamsurface on the face of the sampler. This was achieved as before by plotting the radial distribution of the static pressure over the face of the sampler. The results showed no systematic variation in S for turbulence ranging widely in scale from substantially less than to substantially greater than the diameter of the sampler body with intensity up to as high as 15 %. Based on this limited evidence, it would appear that the freestream turbulence exerts little – if any – effect on the shape of the mean airflow pattern on the upstream side of an aerosol sampler.

References

Addlesee, A.J. (1980) Anisokinetic sampling of aerosols at a slot intake, *Journal of Aerosol Science*, 11, 483–493.

Agarwal, J.K. and Liu, B.Y.H. (1980) A criterion for accurate aerosol sampling in calm air, *American Industrial Hygiene Association Journal*, 41, 191–197.

Bartak, J. (1974) Dust sampling in flowing gases-two-dimensional case, *Staub Reinhaltung des Luft* (English translation), 34, 230–235.

Chung, I.-P. and Dunn-Rankin, D. (1997) Experimental investigation of air flow around blunt aerosol samplers, *Journal of Aerosol Science*, 28, 289–305.

Cowdrey, C.F. (1963) A note on the use of end plates to prevent three-dimensional flow at the ends of bluff cylinders, Aeronautical Report No. 1025, National Physical Laboratory, Her Majesty's Stationery Office, London, UK.

Dunnett, S.J. (1999) An analytical investigation into the nature of airflow near the sampling orifice of a spherical sampler, *Journal of Aerosol Science*, 30, 163–171.

Dunnett, S.J. and Ingham, D.B. (1986) A mathematical theory to two-dimensional blunt body sampling, *Journal of Aerosol Science*, 17, 839–853.

Dunnett, S.J. and Ingham, D.B. (1988) An empirical model for the aspiration efficiencies of blunt aerosol samplers oriented at an angle to the oncoming flow, *Aerosol Science and Technology*, 8, 245–264.

Fuchs, H.V., Mercker, E. and Michel, U. (1979) Large-scale coherent structures in the wake of axisymmetric bodies, *Journal of Fluid Mechanics*, 93, 185–207.

Ingham, D.B. (1981) The entrance of airborne particles into a blunt sampling head, *Journal of Aerosol Science*, 12, 541–549.

Ingham, D.B. and Hildyard, M.L. (1991) The fluid flow into a blunt aerosol sampler oriented at an angle to the oncoming flow, *Journal of Aerosol Science*, 22, 235–252.

Launder, B.S. and Spalding, D.B. (1972) *Mathematical Models of Turbulence*, Academic Press, London.

Levin, L.M. (1957) The intake of aerosol samples, *Izv. Akad. Naik SSSR SEr. Geofrz.*, 7, 914–925.

Ogden, T.L. and Birkett, J.L. (1978) An inhalable dust sampler for measuring the hazard from total airborne particulate, *Annals of Occupational Hygiene*, 21, 41–50.

Roshko, A. (1954) On the development of turbulent wakes from vortex streets, Technical Report No. 1191, National Advisory Committee for Aeronautics, Washington, DC.

Rouillard, E.E.A. and Hicks, R.E. (1978) Flow patterns upstream of isokinetic dust sampling probes, *Journal of the Air Pollution Control Association*, 28, 599–601.

Rüping, G. (1968) The importance of isokinetic suction in dust flow measurement by means of sampling probes, *Staub Reinhaltung des Luft* (English translation), 28, 1–11.

Sreenath, A., Ramachandran, G. and Vincent, J.H. (1997) Experimental investigations into the nature of airflows near bluff bodies with aspiration, with implications to aerosol sampling, *Atmospheric Environment*, 31, 2349–2359.

Stenhouse, J.I.T. and Lloyd, P.J. (1974). Sampling errors due to inertial classification. In: *Recent Advances in Air Pollution Control*, American Institution of Chemical Engineers (AIChE) Symposium Series No. 137.

Vincent, J.H. (1989) *Aerosol Sampling: Science and Practice*, John Wiley & Sons, Ltd, Chichester.

Vincent, J.H. and Mark, D. (1982) Applications of blunt sampler theory to the definition and measurement of inhalable dust, In: *Inhaled Particles V* (Ed. W.H. Walton), Pergamon Press, Oxford, pp. 3–19.

Vincent, J.H., Hutson, D. and Mark, D. (1982) The nature of air flow near the inlets of blunt dust sampling probes, *Atmospheric Environment*, 16, 1243–1249.

Vincent J.H., Mark, D., Gibson, H., Botham, R.A., Emmett, P.C., Witherspoon, W.A., Aitken, R.J., Jones, C.O. and Miller, B. (1983) Measurement of inhalable dust in wind conditions pertaining to mines, IOM Report No. TM/83/7, Institute of Occupational Medicine, Edinburgh.

Vincent, J.H., Stevens, D.C., Mark, D., Marshall, M. and Smith, T.A. (1986) On the aspiration characteristics of large-diameter, thin-walled aerosol sampling probes at yaw orientations with respect to the wind, *Journal of Aerosol Science*, 17, 211–224.

Vitols, V. (1966), Theoretical limits of errors due to anisokinetic sampling of particulate matter, *Journal of the Air Pollution Control Association*, 16, 79–84.

Volushchuck, V.M. and Levin, L.M. (1968) Some theoretical aspects of aerosol aspiration (large Stokes' numbers), *Izv. Atmospheric and Oceanic Physics* (English translation), 4, 241–249.

Walter, E. (1957) Zur problematik der Entnahmesonden und der Teilstromentnahme fur die Staubgehaltsbestimmung in Stromenden Gasen, *Staub Reinhaltung des Luft*, 53, 880–898.

5

Aerosol Aspiration in Moving Air

5.1 Introduction

As discussed in earlier chapters, aspiration efficiency is a primary index of aerosol sampler performance. It defines the efficiency by which particles from the ambient outside air enter through the inlet of the aspirating sampling device, and so is strongly governed by the nature of the airflow around and about the external body of the sampler. It is therefore highly influenced by the external environment which, in practical situations, is usually highly variable and something that cannot be controlled. However, an understanding of how particles move in the complex flow in the vicinity of the sampler can lead to models for estimating aspiration efficiency. As will be seen from what follows, such models have evolved to the point where they can explain the entry of particles into simple aerosol sampling systems. These can provide important insights into the performances of more complicated samplers like those that are found in real-world sampling.

Of course, aspiration efficiency is not the sole index of sampler performance. Other samplers, aiming to provide measures of particle size fractions of the ambient aerosol (e.g., as might be related to specific health effects), require knowledge of how particles are transported inside the sampler after aspiration. These will be discussed later. At this point, however, it is noted that aspiration efficiency has an important bearing on those other fractions too, since they are in fact subfractions of what has been aspirated.

In this chapter, attention is focused on scenarios where particles are transported near and around an aerosol sampler in a moving freestream. It is an underlying assumption that such transport is governed by convection and by inertial forces, and therefore that the effects of gravity and other forces may be neglected. This implies in particular that local air velocities in the systems of interest are always significantly greater than the velocities of particle motion falling under the influence of gravity. Such effects only become influential in calm or very slowly moving air, as will be discussed in a later chapter.

It is an important feature in all that follows in this chapter and the next that particles arriving at solid surfaces, for example the walls of sampling devices, are retained there. Generally speaking this is reasonable for the majority of practical aerosol sampling situations. Retention of deposited particles is based on the assumption that a fine particle experiences surface-derived adhesion forces that are greater than the fluid mechanical shearing – or other – forces that might be present to cause a particle not to stick in the first place, or to be dislodged after it has arrived. Nonetheless, there may be certain situations where this does not hold, and these will be discussed separately.

Aerosol Sampling: Science, Standards, Instrumentation and Applications   James Vincent
© 2007 John Wiley & Sons, Ltd

5.2 Thin-walled tube samplers

The thin-walled probe represents the simplest form of aerosol sampler because, when it is aligned with the flow facing the wind, it imposes little or no physical obstruction or blockage to the air movement. The case that has the greatest practical relevance is the cylindrical thin-walled tube. This is the one for which the most theoretical and experimental research has been conducted.

5.2.1 Qualitative picture of aerosol transport

In this chapter, the assumed nature of the airflow pattern is taken from the detailed description given in Chapter 4. In most of the theoretical developments which will be reviewed, however, it is the broad features – rather than the detailed structure – that form the main basis of discussion. For a qualitative description of the factors influencing the sampling process, Figure 5.1 shows the general form of the streamsurface pattern for a thin, cylindrical, sharp-edged tube aligned with and facing a moving airstream. The flow here is a simplified version of what was shown earlier in Figure 4.2, together with superimposed

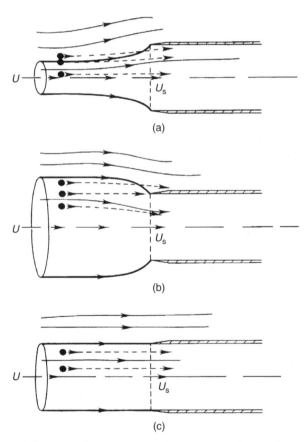

Figure 5.1 *General form of the streamline pattern and superimposed particle trajectories for an aspirating thin-walled cylindrical tube facing the freestream: (a) velocity ratio R < 1; (b) R > 1; (c) R = 1. Reproduced with permission from Vincent, Aerosol Sampling: Science and Practice. Copyright (1989) John Wiley & Sons, Ltd*

particle trajectories. When the aspiration rate is relatively low [Figure 5.1(a)], it is seen that the limiting streamsurface – dividing the aspirated from the nonaspirated air – is divergent. This is achieved when the mean velocity of the air passing through the plane of the tube entrance (U_s) is less than within the limiting streamsurface at a large distance upstream (U). Aspiration efficiency (A) may be expressed in terms of the net flux of aerosol particles passing through the plane of the tube entrance as a fraction of that contained within the limiting streamsurface upstream of the sampler.

For the case shown in Figure 5.1(a), the concentration of particles entering the tube is enriched by the arrival of particles within the sampled air volume by 'impaction' from the air outside. The result is that $A > 1$. This bias in the performance of the tube acting as an aerosol sampler becomes greater the lower the aspiration rate, and hence the more divergent the flow. It also becomes greater the larger the particle size, and hence the greater the tendency of the pattern of particle trajectories to differ from that of the streamsurfaces. For the case shown in Figure 5.1(b) where the aspiration rate is relatively large and so the limiting streamsurface is convergent, this is achieved when U_s is greater than U. Now the concentration of particles entering the tube is depleted by the fact that some particles originally inside the sampled air volume are lost by the same impaction process just referred to, except that the process now operates in reverse. The result is that $A < 1$, a bias that becomes greater the higher the aspiration rate (and hence the more convergent the flow) and the larger the particle size. Finally for the case shown in Figure 5.1(c), the aspiration rate is chosen so that the shape of the limiting streamsurface is neither divergent nor convergent and there is no distortion to the flow. This is achieved when U_s is equal to U. For such *isokinetic* conditions there can be no losses of particles from or gains to the sampled air volume by impaction. Now $A = 1$ exactly, independently of particle size. From this, it is clear why there has been so much practical interest in thin-walled aerosol samplers. If the freestream air flow is well defined in terms of velocity and direction, then a simple system is available which, by selection of the appropriate sampling flow rate, should in principle provide a truly representative sample of total aerosol.

These characteristics are summarised in Figure 5.2, where A is shown as a function of the velocity ratio $R = U/U_s$ for various particle sizes. Based on all that has been presented so far in this book, it seems reasonable to write down the initial functional relationship

$$A = \mathrm{f}\{R, d_{ae}\} \tag{5.1}$$

where here, and elsewhere unless stated otherwise, the effects of gravity and changes in Reynolds number for the air flow near the sampler are small enough to be neglected.

Now we proceed to examine the case where the tube is placed at a finite angle to the freestream (Figure 5.3). Here, the question of whether the limiting streamsurface is divergent or convergent is influenced not just by R but also by the angle (θ) which governs the extent to which the cross-sectional area of the tube is projected geometrically upstream. As θ increases, the air flow just upstream of the tube entrance will tend to be less divergent or convergent. This in turn suggests that, for a given value of d_{ae}, A may now be governed by a parameter of the form $R\cos\theta$, replacing R in the earlier discussion. In general:

$$A = \mathrm{f}\{R, d_{ae}, \theta\} \tag{5.2}$$

Since the earliest studies of thin-walled probes by Lapple and Shepherd in 1940, numerous scientific papers have been published which have served to underline the fact that aerosol aspiration for even the simple thin-walled tube is actually quite complex. In 1964, Fuchs remarked in reference to thin-walled probes that '... the problem of aerosol sampling has not yet been solved'. Even now, over 40 years further on, understanding is still not complete. Nevertheless, substantial progress has been made and the next few pages set out to describe in greater detail some of those achievements.

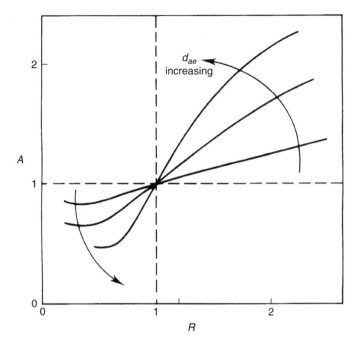

Figure 5.2   *General form of the aspiration efficiency (A) for aspirating thin-walled cylindrical sampling tubes facing the freestream as a function of velocity ratio (R) for various particle sizes. Reproduced with permission from Vincent, Aerosol Sampling: Science and Practice. Copyright (1989) John Wiley & Sons, Ltd*

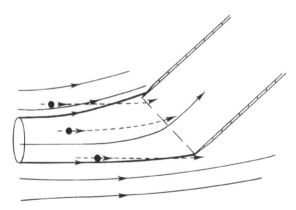

Figure 5.3   *General form of the streamline pattern and superimposed particle trajectories for an aspirating thin-walled cylindrical sampling tube placed at an angle to the freestream. Reproduced with permission from Vincent, Aerosol Sampling: Science and Practice. Copyright (1989) John Wiley & Sons, Ltd*

5.2.2   Impaction model for a thin-walled tube facing the freestream

Consider the thin-walled cylindrical tube aligned facing a flow in which the spatial distributions of aerosol concentration and freestream velocity upwind of the sampler are uniform. From the discussion in Chapter 1, we may reiterate that its aspiration efficiency (A) may be expressed as:

$$A = \frac{c_s}{c_0} \qquad (5.3)$$

where, as before, c_s and c_0 are the particle concentrations in the plane of the sampling orifice and in the undisturbed freestream, respectively.

Consider now a flow picture where the air aspiration rate is sub-isokinetic. Figure 5.4 identifies the important quantities necessary for constructing a physical model from which to determine aspiration efficiency. In addition to c_s, c_0, U and U_s, it also indicates the diameter of the cylindrical nozzle (δ), its cross-sectional area (a_s) and the area upstream contained within the sampled flow volume (a_0). The volume of air geometrically incident from upstream onto the tube entrance per unit time is $a_s U$. The volume actually aspirated by flowing through the tube entrance – and by definition through the upstream plane contained within the limiting streamsurface – is $a_s U_s$. Therefore the volumetric flow rate of air from the original $a_s U$ which now diverges to pass by outside the tube is $a_s (U - U_s)$.

In Figure 5.4 it is clear that all particles originally contained within the sampled air volume ($a_s U_s$) will enter the tube. Therefore c_s must be at least as large as c_0. However, in addition, there will be some particles from the air outside the limiting streamsurface that spill over into the sampled volume by impaction. The only difference from the impaction process described earlier in Chapter 2 is that the projection now takes place not onto a solid boundary but into a region in space bounded by the limiting streamsurface and the plane of the tube entrance. Since c_s must exceed c_0 by an amount equivalent to this spillover, the mass flow of particles into the tube may therefore be expressed as:

$$c_s a_s U_s = c_0 a_0 U + \beta c_0 a_s (U - U_s) \qquad (5.4)$$

where β here is a dimensionless impaction parameter equivalent to the impaction efficiency defined earlier (see Chapter 2) for particles impacting onto a simple bluff body. Since continuity requires $a_s U_s = a_0 U$, this reduces directly to:

$$A = \frac{c_s}{c_0} = 1 + \beta(R - 1) \qquad (5.5)$$

Since the spillover of particles described by the second term on the right-hand side of this expression cannot include particles which had not originally been geometrically incident on the tube entrance, β

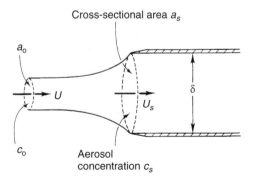

Figure 5.4 *Schematic of an aspirating thin-walled cylindrical sampling tube facing the wind, identifying the important variables needed for the development of a physical model for aspiration efficiency. Reproduced with permission from Vincent, Aerosol Sampling: Science and Practice. Copyright (1989) John Wiley & Sons, Ltd*

can take values only between 0 and 1. The same result as Equation (5.5) is derived for a flow pattern which is convergent (i.e. is super-isokinetic), the only physical difference being that particles spill *out of* – instead of into – the sampled air volume. Inspection of Equation (5.5) confirms trivially that $A \to 1$ under isokinetic conditions ($R = 1$), irrespective of particle size. Then the nature of β is not relevant. For nonisokinetic (or *anisokinetic*) conditions, however, the properties of β are very important, and so are central to much of what follows.

5.2.3   Physical definition of impaction efficiency for aerosol sampling

Badzioch (1959) pointed to the limiting case where the sampling flow rate falls to zero, where there is no aspiration, and suggested that the behavior of β should therefore resemble the equivalent quantity for simple, bluff body impaction. However, he also cautioned that the analogy with bluff body impaction is not exact even for a tube without aspiration. This is because the tube under such conditions represents a 'soft' obstacle in which the stationary air immediately in front of the tube entrance forms a 'cushion'. Badzioch reasoned that the impaction region of interest resembles not a disk-shaped obstacle, as in the limiting, zero-aspiration case, but rather a ring-shaped surface in the plane of the sampling orifice represented by the annular region between the superposed surfaces a_s and a_0 identified in Figure 5.4.

5.2.4   Experimental studies for thin-walled tubes facing the freestream

The impaction efficiency (β) may be determined experimentally by direct application of Equation (5.5), re-expressed as

$$\beta = \frac{A - 1}{R - 1} \tag{5.6}$$

where A and R are both quantities that may be measured experimentally. For A, this may be obtained by either indirect or direct methods (see Chapter 3), and R from the known freestream and sampling flow conditions. Ideally, in order to obtain the best results by either method, such experiments should be carried out under flow conditions as far away as possible from isokinetic. This is because experiments in which R is close to unity cannot produce meaningful results for β since the numerator and denominator on the right-hand side of Equation (5.6) both tend towards zero, so that the sensitivity of the estimation of β is less good.

Badzioch was one of the first to conduct extensive experiments for determining β along the lines indicated, using the indirect method to measure A. He employed a range of thin-walled tubes with internal diameter (δ) ranging from 6.5 to 19 mm, with wall thickness 0.6 mm, in a 20 cm diameter wind tunnel of circular cross-section capable of providing wind speeds up to about 3 m s^{-1}. Relatively monodisperse test aerosols of isometric particles, with MMAD values in the range of about 25–70 μm, were generated by mechanical dispersion of narrowly graded amorphous silica dusts and zinc powders. The masses of particulate material collected on filters in the test and reference samplers, respectively, were assessed gravimetrically to provide c_s and c_0. Sampling covered both super- and sub-isokinetic flow conditions, with R ranging from 0.22 to 4.62. Badzioch's original results are re-plotted in Figure 5.5 in the form of β as a function of St, a Stokes number for aspiration given by:

$$St = \frac{d_{ae}\gamma^* U}{18\eta\delta} \tag{5.7}$$

shown there as the open circles. Badzioch noted a possible additional dependency on tube diameter (δ), with β increasing monotonically with St for a given value of δ but also increasing with δ for a given

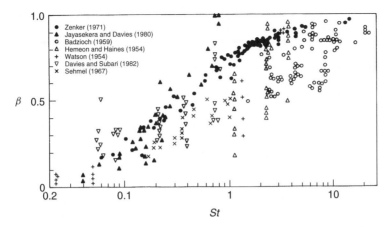

Figure 5.5   *Impaction efficiency (β) as a function of Stokes number (St) for an aspirating thin-walled cylindrical sampling tube, determined from the various earlier published data indicated, obtained using the indirect method. Reproduced with permission from Vincent, Aerosol Sampling: Science and Practice. Copyright (1989) John Wiley & Sons, Ltd*

St. Davies (1968) subsequently examined these trends and proposed that, for the range of conditions of Badzioch's experiments, β might be represented to a fair first approximation by the following empirical expressions:

$$\beta \approx 1 - \frac{1}{1 + 4St} \qquad \text{for } \delta > 2 \text{ cm}$$

$$\beta \approx 1 - \left[\frac{1}{1 - 0.6St \ln\left(1 - \frac{\delta}{2}\right)} \right] \qquad \text{for } 0.6 < \delta < 2 \text{ cm} \tag{5.8}$$

with δ expressed in centimeters. Badzioch's experiments have been widely discussed in the literature. A number of similar studies have also been reported by other workers, producing more results suitable for investigating the nature of β. The most important are listed in Table 5.1 along with summaries of the ranges of experimental conditions examined in each. Watson (1954) used a low turbulence wind tunnel in which he generated test aerosols of relatively monodisperse *Lycopodium* spores by atomisation of suspensions in water, with d_{ae} in the range from about 4 to 32 μm. Aerosol samples were collected using tubes with δ equal to 4.6, 7.0 and 10.5 mm, respectively, for R in the range from 0.43 to 2.25. At about the same time, Hemeon and Haines (1954) reported experiments in a pilot plant designed specifically for stack dust sampling studies. Test aerosols were generated by the mechanical dispersion of silicon carbide dusts, producing a range of d_{ae} from about 10 to 400 μm. Aerosol samples were collected using tubes with δ from about 0.3 to 1 cm for R in the range from 0.5 to 5. Later, Sehmel (1967) described experiments carried out using a 1 cm diameter tube, where monodisperse test aerosols of uranine blue with d_{ae} in the range from 1 to 28 μm were produced using a spinning disk generator, and sampling was carried out for R in the range from about 0.1 to 3. Soon afterwards, Zenker (1971) conducted experiments with test aerosols of relatively monodisperse glass spheres and ground limestone, using tubes with δ in the range from about 1 to 4 cm, and R varying from 0.4 to 1.7. Particle aerodynamic sizes per se were not specified directly in the paper, but were expressed in terms of stop distance so

Table 5.1   *Summary of ranges of experimental conditions for the most widely cited experimental studies of the aspiration characteristics of thin-walled cylindrical sampling tubes*

Reference	Method	δ(mm)	R	$d_{ae}(\mu m)$	$\theta(°)$
Watson (1954)	Indirect	5–11	0.4–2.3	4–32	0
Hemeon and Haines (1954)	Indirect	3–10	0.5–5	4–400	0
Badzioch (1959)	Indirect	6.5–19	0.2–4.6	25–70	0
Sehmel (1967)	Indirect	10	0.1–3	1–28	0
Zenker (1971)	Indirect	11–40	0.4–1.7	–	0
Belyaev and Levin (1972, 1974)	Direct	10	0.2–6	17–24	0
Yoshida *et al.* (1976)	Indirect	8–10	0.1–2.5	–	0
Durham and Lundgren (1980)	Indirect	0.5–0.7	0.5–2	1–20	0, 30, 60 and 90
Jayasekera and Davies (1980)	Indirect	5–14	1–11	16–22	0
Davies and Subari (1982)	Indirect	5–14	0.06–1	14–32	0, 90
Tufto and Willeke (1982)	Indirect	0.3–1	0.5–4	5–40	0, 30, 60 and 90
Lipatov *et al.* (1986)	Direct	1.8	0.03–2	34–80	0
Vincent *et al.* (1986)	Indirect	20–50	0.7–2	6–34	0, 45 and 90
Paik and Vincent (2002a)	Indirect	0.7–1	0.5–50	13–90	0

that the results may now readily be expressed in terms of *St*. Still later, Jayasekera and Davies (1980) and Davies and Subari (1982) performed experiments where monodisperse liquid droplet aerosols of di-2-ethyl-hexyl sebacate were generated by a spinning disk generator in the range of d_{ae} from 16 to 22 μm using a spinning top generator, and were collected using tubes with δ from about 0.5 to 1.4 cm with values of R in the range from 0.06 to 11. Jayasekera and Davies found that their experimental data provided a good fit with the inverse form of the aspiration efficiency equation, expressed as:

$$R = A + \left(\frac{A - 1}{-0.2A^2 + 6.9A - 1.9} \right) \frac{R}{St} \text{ for } R > 1 \tag{5.9}$$

The results of all the experimental studies summarised above are brought together alongside the original Badzioch ones in Figure 5.5. On closer inspection it is evident that the data of Zenker and of Davies and his colleagues were the more consistent and exhibited less scatter than those of Watson, Hemeon and Haines, and Badzioch. Those of Zenker were particularly good. Whilst no attempt is made in Figure 5.5 to bring out trends associated with variations in tube diameter (δ) or in velocity ratio (R), it is clear that the primary trend was the one involving *St*. Overall, these results provided no obvious support for Badzioch's original suggestion of a separate specific dependence on δ.

It is of interest at this stage to comment briefly on the scatter associated with the experimental results of the earlier studies and on the clear discrepancy between the results from the earlier studies and the later ones. Based on what became known later, it is now thought likely that the results in some of the earlier work may have been influenced by particle rebound or 'blow-off' from the external surfaces of the sampler and secondary aspiration into the tube entrance. This phenomenon, which will be discussed in greater detail in a later chapter, would have led to a tendency for measured values of *A* to be greater than their corresponding true values. In addition, it is thought that contributions to the scatter might also have arisen from failure on the part of some experimenters to correct properly for the losses of particles to the inside wall of the transition section between the tube entrance and the collecting filter.

From all the available information, the most appropriate general empirical expression for impaction efficiency (β) appears to take the form:

$$\beta = 1 - \left(\frac{1}{1 + GSt} \right) \tag{5.10}$$

where the coefficient G and its functional relationships with the various system parameters, particularly R, are now the subjects of interest. Unfortunately the indirect method for the experimental determination of sampler aspiration efficiency has been shown in the studies referred to so far to be insufficiently accurate for assessing the detailed nature of Equation (5.10). However application of the preferred direct method, originally developed and used primarily by workers in Russia and the Ukraine, *have* provided the desired insight. The most influential set of experiments of the period along these lines were conducted by Belyaev and Levin (1972, 1974), who examined the trajectories of relatively monodisperse *Lycopodium* spores in the vicinities of thin-walled sampling tubes facing the freestream in a wind tunnel, and so obtained measurements of aspiration efficiency by the direct method. This was achieved by means of a series of photographs of particle trajectory fields, from which it was possible to determine aspiration efficiency (A) directly without introducing any ambiguity arising from particle rebound, blow-off or internal wall losses. Experiments were conducted for ranges of R from 0.18 to 6.0, d_{ae} from 17 to 24 μm and St from 0.18 to 2.03. Values of β were estimated from the plotted aspiration efficiency results and are shown in Figure 5.6 as a function of St. As in the previous cases involving the use of

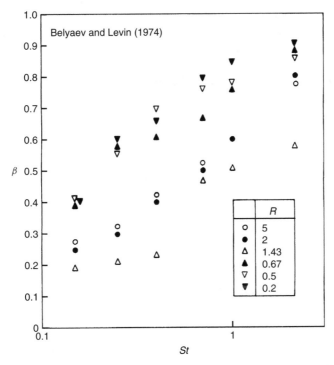

Figure 5.6 *Impaction efficiency (β) as a function of Stokes number (St) for various values of the velocity ratio (R) for an aspirating thin-walled cylindrical sampling tube, from experiments using the direct method by Belyaev and Levin (1972, 1974). Reproduced with permission from Vincent, Aerosol Sampling: Science and Practice. Copyright (1989) John Wiley & Sons, Ltd*

the indirect method, a strong upwards trend in β with increasing St *is* clearly apparent. In addition, however, despite the considerable scatter in the results, Belyaev and Levin found clear evidence of a consistent, systematic additional dependence on R, indicating the tendency for β to increase as R itself increases.

If Equation (5.10) is assumed for the behavior of β, then the trend observed in Figure 5.6 suggests that G decreases as R increases. Belyaev and Levin (1974) set out to examine quantitatively the form of the relationship between G and R, and proposed the empirical expression:

$$G = 2 + \frac{0.62}{R} \tag{5.11}$$

which they felt was sufficiently accurate '... for practical purposes'. They then proceeded to examine the experimental results of Zenker (1971) – considered to be the most accurate ones available at the time from the indirect method – in the light of this new expression, and found excellent agreement between experiment and theory for $0.4 < R < 1.7$.

Davies and Subari (1982) assessed their own indirect method measurements and concluded that Equation (5.11) led to an underestimate of A when $R < 0.2$. At first this view appeared to be supported by the results of Gibson and Ogden (1977). Later, however, Lipatov *et al.* (1986) re-examined this question in the light of careful new measurements using their more refined version of the direct method and for a wide range of R, including values below 0.2. Their results, obtained using a 0.2 cm diameter probe exposed to water droplet aerosols with d_{ae} in the range from 34 to 80 µm, are shown in Figure 5.7 in the form of A versus R for a single value of $St = 0.22$. These revealed a marked departure from the data of Davies and Subari and of Gibson and Ogden, especially when $R < 0.2$, providing strong, seemingly unambiguous, confirmation of the validity of Equation (5.11) over the full range of R investigated. They concluded that the model proposed by Belyaev and Levin '... may be regarded as universal'.

Lipatov and his colleagues went on to discuss the possible underlying physical reasons for the discrepancy between their results and those of Davies and Subari and of Gibson and Ogden. They suggested

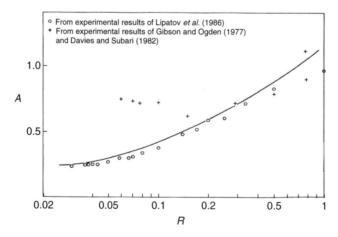

Figure 5.7 *Aspiration efficiency (A) as a function of velocity ratio (R) for the single value of Stokes number, St = 0.22, for an aspirating thin-walled tube, comparisons between different experimental data sets as indicated, also showing (solid line) the curve calculated from the model of Belyaev and Levin (1974). Reproduced with permission from Vincent, Aerosol Sampling: Science and Practice. Copyright (1989) John Wiley & Sons, Ltd*

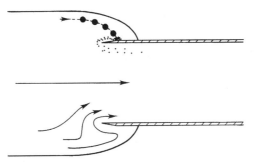

Figure 5.8 *Diagram to illustrate the 'splash' and break-up of droplets during impaction onto the surface of an aspirating sampling tube close to the entry, based on the suggestion of Lipatov et al. (1986) to explain the differences in experimental data like those shown in Figure 5.9. Reproduced with permission from Vincent, Aerosol Sampling: Science and Practice. Copyright (1989) John Wiley & Sons, Ltd*

that the excessive values of A found by the indirect method at low R might have been due to the disintegration of the liquid droplets in the test aerosols used as they collided at high velocity with the outer walls of the sampling tube adjacent to the entry. As illustrated in Figure 5.8, the splash fragments could have entered the tube and so been assessed as having been sampled.

From Table 5.1 it is seen that the experiments which have formed the basis of the discussion so far were carried out for a quite limited range of values for R, generally in the range below about 5. However, in more recent years, interest has stirred in studying aerosol aspiration for a much wider range of R. This is partly driven by the desire for new samplers designed to operate at lower sampling flow rates, in particular personal samplers for occupational hygiene applications that will be operated with small ultra-lightweight portable pumps. With this in mind, Paik and Vincent (2002a) conducted experiments to measure aspiration efficiency, by the indirect method, for thin-walled tubes of diameter $\delta = 0.7$ and 1 cm, respectively, and for R-values ranging from 0.5 to 50, far exceeding the range of any previous such studies. Nearly monodisperse test aerosols were generated in the range of d_{ae} from 13 to 90 μm from narrowly graded powders of fused alumina of the type described in Chapter 3. In this way, St ranged from 0.051 to 3.68. For each experimental data record (A_{meas}), it was compared with the corresponding value obtained using the Belyaev and Levin model (A_{B-L}) embodied in Equations (5.5), (5.10) and (5.11). The results of these comparisons are shown in Figure 5.9(a) in the form of a two-dimensional map that represents the bias, expressed as a percentage, calculated from:

$$\text{Bias}(\%) = \left(\frac{A_{B-L} - A_{meas}}{A_{meas}} \right) 100 \tag{5.12}$$

for each combination of R and St examined experimentally. The individual results were smoothed to generate the contours—or lines of constant bias – shown on the graph. This plot shows that the bias was always small when R was less than about 6, entirely consistent with the conclusions of the previous research. It shows an increasingly large positive bias in A_{B-L} compared with A_{meas} for R rising above 6. Paik and Vincent looked for a modified expression for G, and proposed:

$$G = 2 + \frac{0.62}{R} - 0.9R^{0.1} \tag{5.13}$$

It was found that this modified expression, when used with Equations (5.5) and (5.10), provided excellent agreement with all the available experimental results across the whole extended range of R and St, with

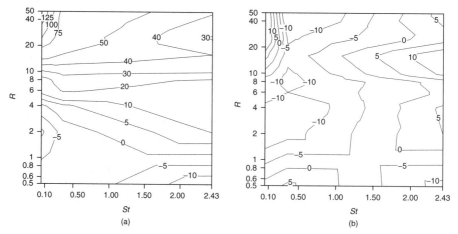

Figure 5.9 *Bias maps to show the percentage deviation of experimental data (Paik and Vincent, 2002a) for the aspiration efficiency (A) of an aspirating thin-walled tube from values calculated using two semi-empirical models, for extended ranges of Stokes number (St) and velocity ratio (R): (a) using the model of Belyaev and Levin (1974); (b) using the model of Paik and Vincent. Reprinted from Journal of Aerosol Science, 33, Paik and Vincent, 705–720. Copyright (2002), with permission Elsevier*

the exception of the few mentioned earlier where experimental artifacts are now thought to have existed (e.g. particle interactions with the sampler surfaces). This is clearly demonstrated in the new bias map shown in Figure 5.9(b).

5.2.5   Experimental studies for thin-walled tubes at other orientations

The majority of the body of research into the performance characteristics of thin-walled aerosol sampling probes has dealt with thin-walled tubes facing directly into the wind. Some early experimental studies of the directional dependence of aerosol samplers (Glauberman, 1962; Raynor, 1970) led to attempts to generate more general mathematical models, taking into account not only particle aerodynamic diameter, tube diameter, wind speed and sampling flow rate but also sampler orientation. These took the form of empirical curve-fitting exercises. However, a small number of studies have since been carried out which provide the basis for an improved, more general physical understanding of the aspiration process, including the effects of orientation. These too are listed in Table 5.1.

A very good experimental data set for thin-walled probes over ranges of conditions extended to include orientations other than facing the wind were obtained by Lundgren *et al.* (1978) and Durham and Lundgren (1980). They used liquid droplet test aerosols with d_{ae} in the range from about 1 to 11 μm produced by means of a spinning disk generator. Thin-walled sampling nozzles of diameter 0.5 and 0.7 cm, respectively, were placed at orientations (θ) of 0, 30, 60 and 90°, respectively. The velocity ratio (R) was varied in the range from 0.5 to 2. Soon afterwards, Tufto and Willeke (1982) reported experiments with monodisperse liquid droplet test aerosols of oleic acid in the range of d_{ae} from 5 to 40 μm generated by means of a vibrating-orifice generator. They used a single thin-walled sampler at orientations θ in the range from 0 to 90°, wind speed (U) in the range from 0.25 to 1 m s^{-1} and R in the range from 0.5 to 4. The experimental study of Davies and Subari (1982) already cited earlier also contained results for $\theta = 90°$. Later, Vincent *et al.* (1986) reported results of a study using relatively monodisperse test aerosols dispersed mechanically by means of a rotating table generator from narrowly

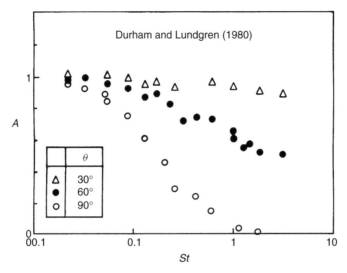

Figure 5.10 *Typical experimental data for aspiration efficiency (A) as a function of Stokes number (St) for an aspirating thin-walled tube placed at various angles (θ) to the freestream (data from Durham and Lundgren, 1980). Reprinted from Journal of Aerosol Science, 11, Durham and Lundgren, 179–188. Copyright (1980), with permission from Elsevier*

graded powders of fused alumina (see Chapter 3), producing aerosols with d_{ae} in the range from 6 to 34 μm. The test thin-walled samplers were larger than those used in the other work cited, with δ equal to 2 and 5 cm respectively, oriented at θ in the range from 0 to 180°. Wind Speeds (U) ranged from 1 to about 4 m s^{-1} and R from 0.67 to 2.00.

The experimental results from these studies all revealed, as expected, a strong dependence of aspiration efficiency on sampler orientation. Typical data, taken from Durham and Lundgren (1980), are given by way of illustration in Figure 5.10. They show A decreasing progressively as θ increased in the range up to 90°. In their earlier paper, Lundgren *et al.* (1978) attempted to relate this fall in A to the reduction in the effective area projected upstream by the sampling orifice.

5.2.6 Impaction model for other orientations

Using data from experiments like those just described, it became possible to develop an extended version of the model described earlier for the simpler forwards-facing case that can be applied to other orientations. As a starting point, Figure 5.11 indicates the limiting streamsurface pattern near a thin-walled cylindrical probe oriented at an angle (θ) with respect to the wind. Now the model must take into account the fact that the area projected upstream by the sampler body increases as θ increases in the range up to and including 90°. Following the earlier approach, the mass flow of particles entering the probe at orientation θ is given by:

$$c_s U_s a_s = c_0 U a_0 + \beta(\theta)(c_0 U a_s \cos\theta - c_o U a_s) \tag{5.14}$$

leading to:

$$A = \frac{c_s}{c_0} = 1 + \beta(\theta)(R \cos\theta - 1) \tag{5.15}$$

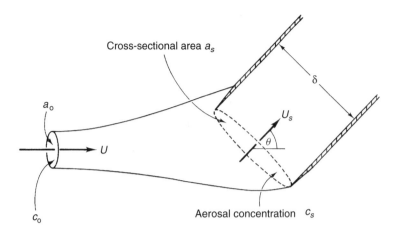

Figure 5.11 *Diagram of an aspirating thin-walled sampling tube placed at an angle (θ) to the freestream, identifying the variables needed for the development of a physical model for aspiration efficiency. Reproduced with permission from Vincent, Aerosol Sampling: Science and Practice. Copyright (1989) John Wiley & Sons, Ltd*

This is the form suggested by Lundgren and his colleagues. Now, however, it is to be expected that, whereas β was previously a function mainly of St and R, it now might also be a function of θ. To begin with:

$$\beta(\theta) = f(St_d) \tag{5.16}$$

where St_d is a new Stokes number that relates to particle moving in the nonsymmetric distorted airflow pattern shown in Figure 5.11. It may be expressed as:

$$St_d = \frac{\tau}{\tau_d} \tag{5.17}$$

where τ is the particle relaxation time and τ_d is the time scale associated with the flow distortion.

Modeling particle motion in this flow field may now proceed by reference to Figure 5.12, which indicates relevant velocity vectors. To estimate τ_d, the flow distortion may be represented as consisting of two parts. The first is associated with the divergence or convergence of the streamsurface pattern upstream of the probe as the air approaches the sampler. The timescale associated with this is τ_{d1} and may be represented by:

$$\tau_{d1} = \frac{\delta}{U \cos \theta} \tag{5.18}$$

The second part is associated with the turning of the flow as it approaches the plane of the sampling orifice, and this is superimposed on the divergence (or convergence) associated with the first part. The timescale associated with this part of the distortion is τ_{d2} and this may be represented in the first instance by:

$$\tau_{d2} = \frac{z}{U} \tag{5.19}$$

where z is a characteristic dimension associated with the turning effect. This may be estimated by considering the air velocity (U_z) that would be induced purely by the sampling action in the absence of

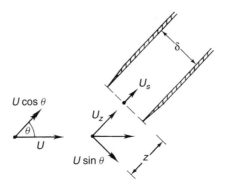

Figure 5.12 *Diagram to indicate the components of air velocity parallel and perpendicular to the axis of an aspirating thin-walled tube placed at an angle (θ) to the freestream. Reproduced with permission from Vincent, Aerosol Sampling: Science and Practice. Copyright (1989) John Wiley & Sons, Ltd*

any wind at this distance from the plane of the sampling orifice. Assuming a spherical flow field near the orifice, this provides:

$$4\pi z^2 U_z = \frac{\pi \delta^2 U_s}{4} \tag{5.20}$$

If z is taken to be the distance from the sampling orifice at which U_z is equal to the component of the freestream air velocity parallel to the plane of the sampling orifice, then $U_z = U \sin \theta$ so that, from Equations (5.19) and (5.20), we get:

$$\tau_{d2} = \frac{\delta}{4R^{1/2} U \sin^{1/2} \theta} \tag{5.21}$$

The processes of divergence/convergence and turning, respectively, take place in parallel, so that τ_{d1} and τ_{d2} may be combined using:

$$\frac{1}{\tau_d} = \frac{1}{\tau_{d1}} + \frac{1}{\tau_{d2}} \tag{5.22}$$

Finally, the preceding equations, along with Equation (5.10) and the earlier Equation (2.41) provide

$$St_d = St(\cos \theta + 4R^{1/2} \sin^{1/2\theta}) \tag{5.23}$$

This may now be plugged into Equation (5.15), leading to:

$$A = 1 + \left[1 - \left(\frac{1}{1 + G_\theta St(\cos \theta + 4R^{1/2} \sin^{1/2} \theta)}\right)\right](R\cos \theta - 1) \tag{5.24}$$

where G_θ corresponds to the earlier G, but takes form that is not yet known.

For $\theta = 0°$, Equation (5.24) reverts back to the simpler equation for a thin-walled tube facing the freestream, as given earlier by Equation (5.5). Further, $R = 1$ leads to the familiar result for isokinetic sampling, $A = 1$ always. In addition, it is interesting to note also that the condition for isokinetic sampling may be stated more generally, for tubes at all orientations $0 \leq \theta \leq 90°$, as:

$$R\cos \theta = 1 \tag{5.25}$$

For the interesting case where the tube is placed at right angles to the freestream, $\theta = 90°$, Equation (5.24) reduces to:

$$A_{90} = \frac{1}{1 + 4G_{90}St\,R^{1/2}} \tag{5.26}$$

Stevens (1986) arrived at the same expression by similar reasoning. On closer inspection, this reveals the interesting result that, for a fixed volumetric flow rate, A is independent of the probe dimensions. The functional form:

$$A_{90} = f\{St\,R^{1/2}\} \tag{5.27}$$

is consistent with that suggested by Tufto and Willeke (1982) and Davies and Subari (1982).

Vincent *et al.* (1986) compared the model described in Equation (5.24) with their own experimental results, as well as those of Durham and Lundgren (1980) and of Davies and Subari (1982). As shown in Figure 5.13, good agreement between experiment and theory was found for θ up to $90°$ for a constant value of $G_\theta \equiv G = 2.1$. Here it was noted that G did not after all contain a strong dependence on θ.

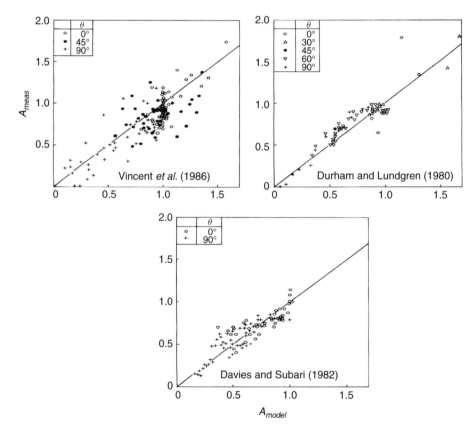

Figure 5.13 *Comparison between measured and calculated aspiration efficiency (A_{meas} and A_{model}, respectively) for an aspirating thin-walled tube placed at various angles to the freestream, for the three data sets indicated and for the blunt sampler model given by Equation (5.24) with G = 2.1 (Vincent et al., 1986). Reprinted from Journal of Aerosol Science, 17, Vincent et al., 211–224. Copyright (1986), with permission from Elsevier*

But it is likely that the data did not permit the level of sensitivity required to detect any such influence, if it did indeed exist. For the results that were obtained for the extreme orientation, $\theta = 180°$, there was – and still is – no prospect of creating a descriptive physical model, except to acknowledge that particle impaction onto the leading elbow of the downstream-facing sampling tube may play a significant role, so that the role of St remains highly significant, although for a different reason. A decade later, Tsai and Vincent (1993) inspected these data, along with corresponding data for blunt samplers (see below), and obtained a good fit using:

$$A_{180} = \frac{1}{(1 + 18R^{-1/3}St)} \tag{5.28}$$

indicating the dominant role of St and continuing influence of R.

Elsewhere, Hangal and Willeke (1990) described an alternative first approach towards developing a unified model for the aspiration efficiencies of thin-walled samplers placed at orientations that were generally forwards-facing. Soon afterwards, Grinshpun *et al.* (1993) created a more comprehensive descriptive physical model for the same scenario. As in all the preceding models, their rationale was to seek a system of empirical relationships that were based as far as possible on physical ideas, and embodied the appropriate combinations of variables, in particular as reflected in St and R. This system included the basic working expression in Equation (5.15) along with:[1]

$$\beta(\theta) = 1 - [1 + (2R + 0.62)St_i]^{-1} \text{ for } \theta = 0°$$

$$\beta(\theta) = \frac{\{1 - [1 + (2R + 0.62)St_{i,\theta}]^{-1}\}[1 - (1 + 0.55\lambda St_{i,\theta}R)^{-1}]}{1 - (1 + 2.62St_{i,\theta}R)^{-1}} \text{ for } 0 \leq \theta \leq 90°$$

$$\lambda = \exp(0.25St_{i,\theta}R) \tag{5.29}$$

$$St_{i,\theta} = St_i \exp(0.022\theta)$$

$$St_i = \frac{St}{R}$$

where St_i is the Stokes number for particle motion referred to the inlet velocity (U_s) and St is that for the freestream velocity (U). Grinshpun and his colleagues reported that this model was in good agreement with experimental data for airflow situations where U was significantly greater than the particle settling velocity, v_s. This work was part of a larger study to develop a general model that extended into the range where that is not the case, for which further discussion will follow in Chapter 6.

5.2.7 Mathematical models

The theories presented thus far for the aspiration efficiency of a thin-walled aerosol sampling probe facing the wind involve a combination of physical ideas and empirical reasoning which are based largely on the broad features of the air flow and particle trajectory fields. A number of other workers have attempted to develop more sophisticated mathematical models, the most notable features of which are the more rigorous, point-by-point description of the flow around the sampler entrance and of corresponding particle trajectories. One of the earliest such attempts was by Levin (1957), who calculated particle trajectories of particles moving in the idealised potential flow field near a point sink placed both in calm air, to which we will return in a later chapter, and in a uniform wind. For the latter, the important features are

---

[1] Also note the important Corrigendum by Grinshpun *et al.* (1994).

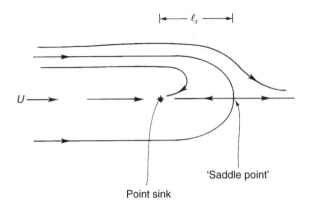

Figure 5.14 *Diagram to show particle trajectories for moving air near a point sink, upon which the model of Levin (1957) was based. Reproduced with permission from Vincent, Aerosol Sampling: Science and Practice. Copyright (1989) John Wiley & Sons, Ltd*

embodied in Figure 5.14, where the wind carries the particles into the region from whence they may be aspirated. The pattern of the air flow is characterised most notably by the 'saddle point' located at a distance l_s away on the downstream side of the point sink. Here the effect of the air velocity induced by the suction of the sink is just balanced by the wind velocity, so that:

$$U = \frac{Q}{4\pi \ell_s^2} \longrightarrow \ell_s = \left(\frac{Q}{4\pi U}\right)^{1/2} \tag{5.30}$$

where Q is the sampling volumetric flow rate. In order to describe the effect of inertia on particles moving in this flow field, Levin identified the dimensionless parameter:

$$k_L = \frac{\tau U}{\ell_s} \tag{5.31}$$

He started by setting $k_L \ll 1$ and then proceeded by assuming that the velocity of a particle moving in the flow field shown in Figure 5.14 is always approximately equal to the local vector sum of the wind speed and the local air velocity induced by the suction of the point sink. Thus he was able to calculate particle trajectories, and so – from the limiting trajectory – arrive at:

$$A = 1 - 0.8k_L + 0.08k_L^2 \tag{5.32}$$

By assuming that the point sink has finite dimensions, for example by setting the setting tube diameter at δ, this expression may now be written in terms of more familiar quantities, so that:

$$A = 1 - 3.2St R^{3/2} + 1.28(St R^{3/2})^2 \tag{5.33}$$

Even so, although this expression predicts the primary observed trends, it remains limited by the fact that the finite orifice and tube dimensions – and hence aerodynamic and particle transport processes in the region very close to the sampler – are not represented. Later, Voloshchuk (1971) modified Levin's equation specifically to contain the size of the sampling orifice, proposing:

$$A = 1 - 3.2 St R^{3/2} + 0.44 St^2 R \tag{5.34}$$

which was recommended so long as $64 St R > 1$ and $R < 0.25$.

Meanwhile, in 1966, Vitols published the results of calculations for the axially symmetric stream-surface pattern near a more realistic cylindrical probe, also based on solutions of the potential flow equations. Particle trajectories were determined numerically from solutions of equations of the form of the earlier Equation (2.38). From these, and from subsequent identification of the limiting trajectory, aspiration efficiency (A) was calculated corresponding to given experimental conditions. Vitols compared his theoretical results with selected data taken from the experimental sets available at the time from Hemeon and Haines (1954) and Badzioch (1959), and noted that agreement was fair. The theoretical results in Vitols's paper were originally presented in the form of A versus $1/R$ for given St. Now they may be replotted and examined to see to what extent they are consistent with the later empirical equations of Belyaev and Levin (1974) for impaction efficiency (β). The results are shown in Figure 5.15. Unfortunately the range of St examined was small and confined to relatively high values – between 2 and 6 – where β was close to unity. Therefore, Vitols's theory cannot be considered as having been properly tested. Furthermore, any effects due to changes in R were not detectable. Nevertheless, Figure 5.15 suggests that, for the range of conditions stated, the calculations of Vitols were actually in quite good agreement with the well-regarded results of Belyaev and Levin. Yoshida *et al.* (1976) later carried out numerical calculations based on similar ideas, and backed them up with a series of wind tunnel experiments for the more comprehensive range of St from 0.05 to 4. For the smaller

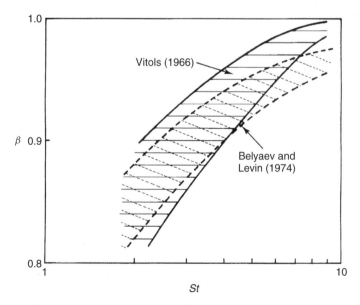

Figure 5.15 *Impaction efficiency (β) as a function of Stokes number (St) for an aspirating thin-walled tube facing the freestream, comparison between the theoretical calculations of Vitols (1966) and Belyaev and Levin (1974), respectively. Reproduced with permission from Vincent, Aerosol Sampling: Science and Practice. Copyright (1989) John Wiley & Sons, Ltd*

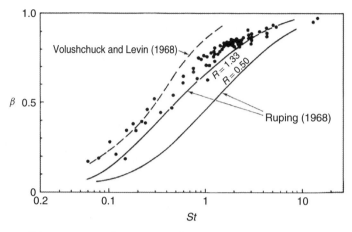

Figure 5.16   *Impaction efficiency (β) as a function of Stokes number (St) for an aspirating thin-walled tube facing the freestream, comparison between the experimental data of Zenker (1971) and the theoretical calculations of Volushchuck and Levin (1968) and Rüping (1968), respectively. Reproduced with permission from Vincent, Aerosol Sampling: Science and Practice. Copyright (1989) John Wiley & Sons, Ltd*

values of *St*, monodisperse test aerosols of methylene blue-uranine were produced using a spinning disk generator; for larger *St*, aerosols were generated by the mechanical dispersal of *Lycopodium* powder. Theory and experiment were reportedly in good agreement.

Rüping (1968) approached the problem from an intermediate standpoint in which he experimentally plotted the flow field in the vicinity of the thin-walled sampling tube using a device he referred to as a 'wedge probe'. Based on this measured flow field, he proceeded to calculate particle trajectories in a manner similar in principle to that of Vitols. Rüping then compared his calculated results for 'measurement error' (departure of *A* from unity) with the experimental data of Hemeon and Haines and of Badzioch, again the only ones available at the time. Agreement with these data was found to be fair. On the other hand, comparison with later, better data – particularly those of Zenker – was less impressive. The trend here, as reproduced in Figure 5.16, was for Rüping's theory to underestimate *β*. It is perhaps worth noting the observation of Belyaev and Levin (1974) in their own paper that Rüping's model *does* bring out a dependence of *β* on R – even if that trend is in the opposite direction to the one predicted by their own theory! In their model, Volushchuck and Levin (1968) began with the flow field near a thin-walled tube calculated for the case of zero wind and then superimposed the freestream air movement. They found that the semi-empirical expression for *A* given earlier by Equation (5.5) is a reasonable approximation and that *β* is a function of *St* only. Comparison between the resultant calculated curve for *β* as a function of *St* and the experimental results of Zenker is shown also in Figure 5.16, and agreement is seen to be generally fair.

A number of other attempts have been made to model the air flow mathematically and to calculate detailed particle trajectories. These include the work of Stenhouse and Lloyd (1979) who employed a flow approximation based on the potential field of a point source in a planar flow, an approach that they considered to be valid for small tube diameters. From particle trajectory calculations, they obtained results for aspiration efficiency which were reported to be in fair agreement with the experimental results of Belyaev and Levin (1974). A few years later, Dunnett (1990) mathematically modeled the behavior of aerosols near thin-walled probes for *θ* in the range from 0 to 90°, using the linear boundary integral equation (LBIE) method to solve numerically for the velocity potential in the airflow fields of interest and then using the resultant flow field to calculate particle trajectories from which aspiration efficiency could be obtained. Quite good agreement was obtained with the then-available experimental data. Later, she performed new numerical calculations for the forwards-facing case where R-values went as high as 50 (Dunnett, 2005). This was the scenario studied experimentally by Paik and Vincent (2002a), as discussed earlier, and Dunnett's calculations were in good agreement with those experimental results. In the process, she made an important contribution by identifying clearly the significant role in aspiration efficiency of the flow reversal that takes place inside the tube as the location of stagnation moves down inside the tube for large R-values. Earlier, Addlesee (1980) had commented that, for a particle to be successfully collected under such conditions, it would have to penetrate not just through the plane of the entry itself but also through the plane of stagnation inside the tube (and hence beyond the reverse flow region). Dunnett noted the same effect and argued that this might be the primary reason for the departure of the measured aspiration efficiency, as measured by Paik and Vincent, from the model of Belyaev and Levin (1974).

In 1989, Liu *et al.* published a set of results for aspiration efficiency from numerical simulations for aerosol collection by thin-walled tubes facing into the wind. In a follow-up paper, Zhang and Liu (1989) used these results as a set of error-free 'experimental' data to develop a modified set of the equations described above from Davies (1968), Voloshchuk (1971), Belyaev and Levin (1974) and Jayasekera and Davies (1980). This modified set [Equations (5.35)] is:

$$\beta \approx 1 - \frac{1}{1 + 3.8St} \text{ with Equation (5.5) for } 0.2 < R < 2,$$

for the Davies (1968) expression given earlier in Equation (5.8);

$$A = 1 - 3.2StR^{3/2} + 0.44St^2R \text{ for } 0.1 < R < 1 \text{ and } St < 1/64R, \tag{5.35}$$

for the Voloshchuk (1971) expression given earlier in Equation (5.34);

$$G = 2 + \frac{0.48}{R} \text{ with Equations (5.5) and (5.10) for } 0.1 < R < 10,$$

for the Belyeav and Levin (1974) expression given earlier in Equation (5.10); and

$$R = A + \left[\frac{A - 1}{-0.2A^2 + 6.9A - 1.9}\right]\frac{R}{St} \text{ for } R > 1,$$

for the Jayasekera and Davies (1980) expression given earlier in Equation (5.9); along with the expressions proposed by Liu *et al.* based on the same 'data', thus

$$G = \frac{1}{1 + 0.418St^{-1}} \text{ for } R > 1$$

$$G = \frac{1}{1 + 0.506R^{1/2}St^{-1}} \text{ for } R < 1 \tag{5.36}$$

both to be used with Equations (5.5) and (5.10). It is noted that, in the individual modified expressions contained in the set of Equations (5.35), some of the coefficients were changed from the originals. So too were the ranges of applicability, driven by the scope of the simulated 'data'. The modified expressions, as well as those of Liu *et al.* in the set of Equations (5.36), were all found to provide a fair fit with the available actual experimental data where available, and for the ranges of conditions indicated.

Earlier, when discussing the impaction model approach, the case of large R was noted. This regime is considered to be relevant to many possible sampling situations in occupational and environmental hygiene where it might be desirable to have a very low sampling flow rate, and hence low inlet velocity. Large R is also relevant to aerosol sampling from an aircraft, where now the relative velocity between the air and the sampler (effectively U) is very high. Krämer and Afchine (2004) studied this regime for a thin-walled tube facing the wind using a computer fluid dynamics (CFD) approach that employed a commercial package to solve the fluid and aerosol mechanical equations of motion and identify particle trajectories. In this way, they determined aspiration efficiency over a wide range of conditions. For the range of R corresponding to the conditions examined experimentally by Belyaev and Levin (1974), good agreement was found. But they, like Paik and Vincent (2002a) cited above, noted that the calculated A-values departed from the Belyaev and Levin model for larger values of R. Using the original Beylaev and Levin model as a starting point, Krämer and Afchine obtained a good fit with their calculated results using:

$$A = \left(\frac{1.26St + 0.27}{1.26St + 1.27}\right) R + 0.5 \left(\frac{1}{R}\right) \quad \text{for } St > 0.001 \text{ and } 5 \leq R \leq 140 \quad (5.37)$$

The overall approach taken by Krämer and Afchine was similar to that of Zhang and Liu (1989) in that it used numerical modeling to generate a set of error-free 'experimental' data which were then used to develop an explicit analytical equation. In turn, in both the Zhang and Liu and the Krämer and Afchine approaches, the philosophy adopted was the same as in the development of the impaction models described earlier. Thus a bridge was established between the impaction model approach and the mathematical approach.

The interesting case of a thin-walled cylindrical tube facing directly downstream, with $\theta = 180°$, is widely acknowledged to be a difficult one to analyze in the ways that have been proposed for the generally forwards-facing scenarios discussed above. But some such cases, including the closely related shallow-tapered tubes, were studied mathematically as part of the overall body of work from Ingham and his colleagues (Ingham *et al.*, 1995; Wen and Ingham, 1995). Here, unlike for most of their models for sampling in the forwards-facing scenarios where potential flow models were used, they based their approach on the initial description of a turbulent flow field. Reasonable agreement was obtained with the available experimental data. In these studies, as in many of the others cited above, the underlying approach was a simulation of the scenarios of interest based on numerical solutions of the physical equations of fluid and particle motion.

5.2.8 Conditions for 'acceptable' isokinetic sampling

From the preceding review of the development of thin-walled probe theory, a scientific framework exists by which the effects of departures from isokinetic sampling conditions on the amount sampled of a given aerosol may be assessed. A number of workers, notably Rüping (1968) and Belyaev and Levin (1974), have used the results of their analyses to examine the conditions for 'acceptable' isokinetic sampling for practical aerosol measurement in stacks and ducts. Figure 5.17(a) shows a pair of curves (the solid lines) relating R and St when $A = 0.95$ and 1.05, respectively. These might be considered to be reasonable bounds of acceptable performance. The hatched area contained between the curves therefore represents conditions for which $0.95 \leq A \leq 1.05$, within which sampling is achieved accurate to within $\pm 5\%$. This graph shows that, for small St, A will always be close to unity, regardless of the value of R. This is intuitively obvious because particles moving in this regime will tend to closely follow the air flow streamlines at all times.

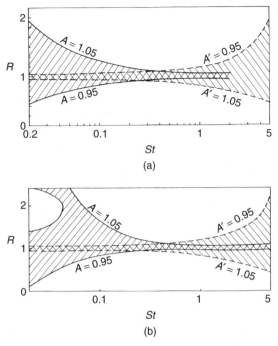

Figure 5.17 *Relationships for defining 'representative' sampling for an aspirating thin-walled tube, as defined by both 0.95 < A < 1.05 and 0.95 < A' < 1.05: (a) gravity neglected (based on Belyaev and Levin, 1974); (b) gravity included (based on Rüping, 1968). Reproduced with permission from Vincent, Aerosol Sampling: Science and Practice. Copyright (1989) John Wiley & Sons, Ltd*

For anisokinetic sampling and for large St, $A \rightarrow R$. This is because particles in this regime tend to continue to move in the direction of the freestream air movement and so will enter the tube if they are geometrically incident on it. Thus:

$$\frac{A}{R} \equiv A' \longrightarrow 1 \tag{5.38}$$

It is therefore apparent that we may plot a second pair of curves, this time for $A' = 0.95$ and 1.05, respectively. These are shown in Figure 5.17(b), where the hatched area between these two curves now represents conditions for which $0.96 \leq A' \leq 1.05$.

From the above it is seen that, in addition to the representative sampling which is achieved for all St provided that $R = 1$, we may also achieve representative sampling independently of R in the two extremes of St. From considerations like these, Rüping has thus referred to the 'fine dust method' (small St) and the 'coarse dust method' (large St), respectively, for practical aerosol sampling. For fine dust sampling, the representative airborne concentration (c_0) is obtained from:

$$c_0 = c_s = \frac{4M}{\pi \delta^2 U_s t} \tag{5.39}$$

where M is the mass sampled in time t. For coarse dust sampling, the same aerosol concentration is obtained from:

$$c_0 = \frac{c_s U_s}{U} = \frac{4M}{\pi \delta^2 U t} \tag{5.40}$$

For the intermediate range of St, or where the particle size distribution of the aerosol in question is wide, encompassing both fine and coarse particles, it is difficult to decide which of these techniques is the most suitable under given wind conditions. In practical sampling, the aerosols of interest are rarely monodisperse. Rather, they are invariably polydisperse, with the result that – even when the flow conditions are well known – the most important quantity St cannot be specified. Belyaev and Levin (1974) examined this problem and suggested that, in practice, an estimate of St on which a choice of which sampling method to adopt may be obtained in a simple initial experiment. This involves taking samples using two probes of the same diameter but operated at different sampling flow rates, one isokinetic and the other anisokinetic. By comparing the masses of particulate material collected by each probe, a rough estimate may be made of effective aspiration efficiency, and hence, from thin-walled sampler theory, a rough estimate of particle size.

5.3   Blunt samplers

In everything that has been discussed so far in this chapter, the tube wall has been assumed to be infinitesimally thin. However, as already described, the true thin-walled aerosol sampling probe does not exist in reality. The tube wall *must* have finite thickness, no matter how thin. The possibility of an effect on sampler performance due to wall thickness was first recognised by a number of workers (Walter, 1957; Belyaev and Levin, 1972; Rouillard and Hicks, 1978), and acknowledged to be associated with the additional flow distortion introduced by the finite aerodynamic blockage – or 'bluntness' – of the tube. We now proceed to examine the resultant effects on aspiration efficiency for aerosol sampling in moving air, extending some of the ideas presented in the preceding sections.

5.3.1   Impaction model for a blunt sampler facing the freestream

For simplicity, in order to maintain analogy with the thin-walled tube discussed above, the blunt sampler is assumed to take the form of an axisymmetric, disk-shaped sampler facing the freestream into which air is aspirated through a circular central orifice. Figure 5.18 identifies the important quantities necessary for constructing the same sort of model already described. As before, freestream and sampled mean air velocities are defined as U and U_s, respectively, with the velocity ratio $R = U/U_s$. The diameter of the sampling orifice is again δ, and now the diameter of the sampler body is D. The airflow in the system shown near the sampling orifice contains the diverging and converging features already introduced in Chapter 4. In the first instance, aspiration efficiency may be written down as:

$$A = f(St, R, r, \theta, B) \tag{5.41}$$

where the new term (r) is the dimension ratio given by:

$$r = \frac{\delta}{D} \tag{5.42}$$

in which D is the width of the sampler body. The other new term, B, is the aerodynamic bluntness of the sampler.

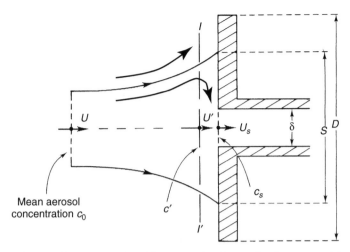

Figure 5.18 *Diagram of a simple axisymmetric blunt sampler facing the freestream, identifying the important variables needed to develop a physical model of aspiration efficiency. Reproduced with permission from Vincent, Aerosol Sampling: Science and Practice. Copyright (1989) John Wiley & Sons, Ltd*

The first efforts to develop an impaction model for the aspiration efficiency of a simple blunt sampler began in the early 1980s (Vincent and Mark, 1982; Chung and Ogden, 1986). For this, Figure 5.18 shows an important new quantity in the form of the diameter of the stagnation region enclosed on the sampler body by the limiting streamsurface. It also shows the two flow regions first described in Chapter 4 for the blunt sampler, the outer divergent part and the inner convergent part. Here, these are separated notionally by the intermediate plane I-I'. The intermediate mean air velocity across I-I' is U'. The corresponding mean aerosol concentration there is c'. It is assumed that the width of the region enclosed by the limiting streamsurface where the flow changes at I-I' is equivalent to S.

The approach taken is to examine particle motion in the two parts of the flow using ideas brought forward from the thin-walled sampler impaction model. This allows postulation of two quantities, A_1 and A_2 to describe the efficiency of the transmission of particles across each of the two regions. Thus:

$$A_1 = \frac{c'}{c_0} = 1 + \beta_1 \left(\frac{U}{U'} - 1 \right) \tag{5.43}$$

and

$$A_2 = \frac{c_s}{c'} = 1 + \beta_2 \left(\frac{U'}{U_s} - 1 \right) \tag{5.44}$$

where β_1 and β_2 are impaction efficiencies for particles moving in the two flow regions, each analogous to the single quantity β invoked earlier for the thin-walled probe. The overall transmission efficiency becomes the aspiration efficiency (A) where:

$$A = A_1 A_2 \tag{5.45}$$

It now remains to find a way to express the terms in Equations (5.43) and (5.44) which have not yet been evaluated. First, consider β_1. Again, by analogy with thin-walled sampler theory for particles with

$Re_p < 1$:

$$\beta_1 = f(St_1) \tag{5.46}$$

where

$$St_1 = \frac{d_{ae}^2 \gamma^* U}{18 \eta S} \longrightarrow St\left(\frac{\delta}{S}\right) \tag{5.47}$$

with

$$St = \frac{d_{ae}^2 \gamma^* U}{18 \eta \delta} \tag{5.48}$$

as before. Similarly, for β_2:

$$\beta_2 = f(St_2) \tag{5.49}$$

where

$$St_2 = \frac{d_{ae}^2 \gamma^* U'}{18 \eta \delta} \longrightarrow St\left(\frac{U'}{U}\right) \tag{5.50}$$

How might the equations for β_1 and β_2 be expressed more explicitly? As always, there are no analytical solutions for impaction efficiency – only empirical ones based on physical reasoning. Therefore, continuing the analogy with the thin-walled tube sampler, a pair of expressions is chosen of the form:

$$\beta_1 = 1 - \left(\frac{1}{1 + G_1 St_1}\right) \text{ and } \beta_2 = 1 - \left(\frac{1}{1 + G_2 St_2}\right) \tag{5.51}$$

Next, consider the intermediate velocity U'. By continuity:

$$U' = \frac{U_s \delta^2}{S^2} \tag{5.52}$$

The earlier Equation (4.15) for the axisymmetric aspirating body facing the freestream provides:

$$S = B \phi_A^{1/3} D \tag{5.53}$$

so long as the sampling orifice is substantially smaller than the size of the sampler body (i.e. $\delta << D$), with B the *bluntness* of the sampler already introduced. The second dimensionless quantity in Equation (5.53), ϕ_A, is the sampling ratio for the axisymmetric flow system in question, and is defined as (see Chapter 4):

$$\phi_A = \frac{\delta^2 U_s}{D^2 U} \longrightarrow \frac{r^2}{R} \tag{5.54}$$

where $r = \delta/D$. From the preceding, we may now summarise the resultant blunt sampler model for the disk-shaped sampler facing the wind in terms of the following set of working equations:

$$St = \frac{d_{ae}^2 \gamma^* U}{18 \eta \delta}$$

$$\phi_A = \frac{r^2}{R}$$

$$St_1 = St\left(\frac{r}{\phi_A^{1/3}}\right)$$

$$\beta_1 = 1 - \left(\frac{1}{1 + G_1 St_1}\right)$$

$$A_1 = 1 + \beta_1(\phi_A^{-1/3} - 1)$$ \qquad (5.55)

$$St_2 = St\phi_A^{1/3}$$

$$\beta_2 = 1 - \left(\frac{1}{1 + G_2 St_2}\right)$$

$$A_2 = 1 + \beta_2\left[\left(\frac{r}{\phi_A^{1/3}}\right)^2 - 1\right]$$

$$A = A_1 A_2$$

in which, based on the experimental data described in Chapter 4, B has been set at 1. Figure 5.19 illustrates the shape of some of the trends predicted by this model, with A plotted as a function of St for various R, D, S and B. It shows that A is close to unity when St is small, but tends towards R when St is large. It also shows that, for some conditions, the trend in A is not necessarily monotonic with St, reflecting the competing effects of the diverging and converging flow components in the region in front of the sampling inlet.

A similar set of equations may be constructed for the corresponding two-dimensional case of a long blunt sampler with a central slot inlet. For this system, the flow pattern is again as described in Chapter 4, and the equivalent of Equation (5.53) is:

$$S = B\phi_T^{1/2}D \qquad (5.56)$$

where $B = 0.80$ for a flat-nosed probe and $B = 0.56$ for a cylindrical probe of circular cross-section. Now, using the same nomenclature as above, but with δ and D becoming the slot and sampler *widths*, respectively, the set of working equations for determining aspiration efficiency corresponding to Equations (5.55) is:

$$St = \frac{d_{ae}^2 \gamma^* U}{18\eta\delta}$$

$$\phi_T = \frac{r}{R}$$

$$S = B\phi_T^{1/2}D$$

$$St_1 = St\left(\frac{\delta}{S}\right)$$

$$A_1 = 1 + \beta_1\left(\frac{S}{\phi_T D} - 1\right)$$ \qquad (5.57)

$$St_2 = St\phi_T\left(\frac{D}{S}\right)$$

$$A_2 = 1 + \beta_2 \left(\frac{\delta}{S} - 1 \right)$$

$$A = A_1 A_2$$

together with expressions for β_1 and β_2 of the form already given above and an appropriate choice for the coefficient B.

5.3.2   Experimental investigations of blunt samplers of simple shape facing the wind

In comparison with the case of the thin-walled probe, there are much less published experimental data for the aspiration characteristics of blunt samplers on which an assessment of the range of applicability of the above theoretical model may be carried out. From the theoretical point of view, the simplest blunt sampler scenario is the one where aerosols are sampled by a long two-dimensional cylinder with a slot-shaped sampling inlet facing directly into the wind ($\theta = 0°$). Despite its apparent simplicity,

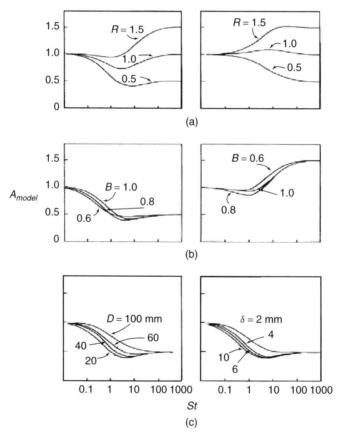

Figure 5.19   *Typical set of trends for aspiration efficiency (A) as a function of Stokes number (St) for various velocity ratios (R), bluntness values (B), body widths (D) and orifice widths (δ), calculated using the model given by Equations (5.55). Reproduced with permission from Vincent, Aerosol Sampling: Science and Practice. Copyright (1989) John Wiley & Sons, Ltd*

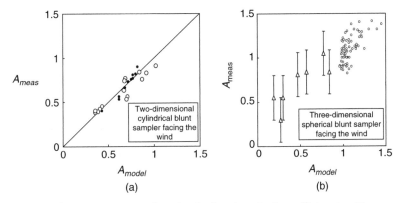

Figure 5.20 *Comparison between measured and calculated aspiration efficiencies (A_{meas} and A_{model}) for (a) two- and (b) three-dimensional blunt samplers of idealised simple shape facing the freestream, using the models described in Equations (5.57) and (5.55), respectively: (a) ○ experimental data, ● numerical data from Chung et al. (1994); (b) experimental data △ from Vincent et al. (1985) and ○ from Chung and Ogden (1986). Reproduced with permission from Vincent, Aerosol Sampling: Science and Practice. Copyright (1989) John Wiley & Sons, Ltd*

experiments are difficult for this system because the airflow into and around the sampler must be maintained as truly two-dimensional. This is especially difficult for the sampling airflow because it is important to achieve a uniform sampling air velocity across the whole length of the inlet slot. However, in their experimental study in a wind tunnel, Chung *et al.* (1994) achieved this to their satisfaction. They used the direct method (see Chapter 3) for determining aspiration efficiency (A), for monodisperse water droplet aerosols in the range of d_{ae} from 20 to 60 μm generated by means of a vibrating orifice generator. Photographs of particle tracks provided the basis for the determination of A. The results are shown in Figure 5.20(a), plotted to show the relationship between experiment and theory (A_{meas} and A_{model}, respectively) for this system as described by Equations (5.57), where – for the latter – the coefficients were chosen as $G_1 = 0.25$, $G_2 = 6.0$ and bluntness estimated as $B = 1.0$. Agreement was quite good.

In the 1980s, Vincent *et al.* (1985) and Chung and Ogden (1986) reported experimental results for axisymmetric disk-shaped blunt samplers facing the wind. Both were obtained using the indirect method. Chung and Ogden used monodisperse test aerosols of di-iso-octyl phthalate in the range of d_{ae} from about 3 to 13 μm produced by means of a spinning-disk generator, tagged with a fluorescent dye to enable the assessment of samples by fluorimetry. They used disk-shaped samplers of diameter (D) in the range from about 3 to 12 cm with orifice diameters (δ) in the range from 0.5 to 2 cm, sampling flow rates from 1 to 4 Lpm (liters per minute) and freestream velocities from 0.5 to 4 m s^{-1}. The experiments of Vincent *et al.* were carried out with relatively monodisperse test aerosols mechanically generated from narrowly graded dry powders of fused alumina with d_{ae} ranging from about 8 to 63 μm. Some experiments were also carried out using monodisperse aerosols of human serum albumen and wax with d_{ae} in the range from 10 to 40 μm, both dispersed by nebulisation of suspensions in ethanol. A single disk-shaped sampler was used, with $D = 4$ cm and $\delta = 0.4$ cm. For a single set of wind speed and sampling flow rate of 2 m s^{-1}, R was fixed at 0.16. Since the experiments of Chung and Ogden were carried out for $R > 1$ and our own for $R < 1$, the two sets of results are conveniently complementary. The results are shown in Figure 5.20(b), again plotted to show the relationship between experiment and theory where – this time – the model coefficients were chosen as $G_1 = 0.25$, $G_2 = 6.0$ and $B = 1.0$.

The error bars for the Vincent *et al.* data reflect the variability that was observed in multiple repeat runs. Again, agreement between experiment and theory was quite good.

More recently, interest in the simple, forwards-facing, axisymmetric blunt sampler was rekindled during the discussion of aerosol sampler scaling laws, driven by wider issues concerning the development of test methods for aerosol samplers of the types used for personal sampling by occupational hygienists. Paik and Vincent (2002b) set out to conduct a new experimental study of aspiration efficiency for disk-shaped blunt samplers with diameter (D) of 10 and 13 cm, with two dimension ratios ($r = \delta/D$) of 0.05 and 0.1, respectively, and velocity ratio (R) from 0.5 to 25. For the latter, the upper end was much higher than for the ranges of conditions studied earlier by Chung and Ogden and Vincent *et al.* As in their contemporary study for thin-walled tube samplers (Paik and Vincent, 2002a), they used nearly monodisperse test aerosols of fused alumina with d_{ae} ranging from about 13 to 90 μm. As before, they used the indirect method for determining aspiration efficiency. Figure 5.21(a) shows a set of contours in the plane of R and St, derived from the individual biases between A_{meas} and A_{model}, calculated point-by-point and smoothed to show lines of constant bias, where:

$$\text{Bias}(\%) = \left(\frac{A_{model} - A_{meas}}{A_{meas}} \right) 100 \tag{5.58}$$

for which A_{model} was obtained using Equations 5.55. Figure 5.21(a) reveals that the model increasingly overestimated A as R increases above about 2, while the magnitude of the bias was generally uniform across the range of St studied. Similar to the approach they adopted for the thin-walled sampler, Paik and Vincent sought in this later work to make empirical adjustments to the physical quantities in the model most sensitive to the variables most influential in the observed departures from the existing model. In

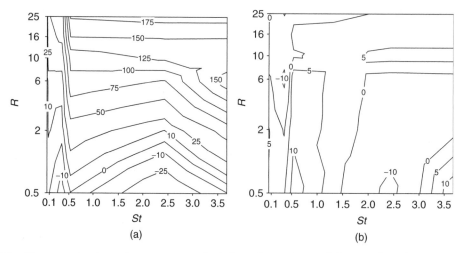

Figure 5.21 *Bias maps to show the percentage deviation of experimental data (Paik and Vincent, 2002b) for the aspiration efficiency (A) of a simple disk-shaped blunt sampler facing the freestream from values calculated using two semi-empirical models, for extended ranges of Stokes number (St) and velocity ratio (R): (a) using the original blunt sampler model given by Equations (5.55); (b) using the modified model given by Equations (5.55) with (5.59). Reprinted from Journal of Aerosol Science, 33, Paik and Vincent, 1509–1523. Copyright (2002), with permission from Elsevier*

so doing they chose:

$$\beta_1 = 1 - \left\{ \frac{1}{\left[1 + G_1 \left(\dfrac{R}{r} \right)^{G_2} St_1 \right]} \right\} \quad \text{and} \quad \beta_2 = 1 - \left\{ \frac{1}{\left[1 + G_3 \left(\dfrac{R}{r} \right)^{G_4} St_2 \right]} \right\} \qquad (5.59)$$

These were deliberately very similar to the original forms for β_1 and β_2 that Chung and Ogden tested, except that now the effects of R and r were explicitly included. The coefficients G_1, G_2, G_3 and G_4 were determined by nonlinear least squares regression, yielding $G_1 = 1.26$, $G_2 = -0.68$, $G_3 = 1.59$ and $G_4 = 0.40$ with $R^2 = 0.97$. The new bias map in Figure 5.21(b) shows that the revised model provided very good agreement with the experimental data over the whole range of conditions tested, including the results from the earlier studies cited. Finally, one particularly interesting – and potentially valuable – finding from the Paik and Vincent study was that aspiration efficiency appeared to be only weakly dependent on the dimension ratio, r. As will be discussed later, this could have considerable implications to the testing of small personal aerosol samplers of the type used by occupational hygienists.

5.3.3 Blunt samplers at other orientations

For thin-walled sampling tubes it has nearly always been the expectation that they would be applied in practical situations where the freestream would be quite well-defined in terms of velocity and direction, and that the tube would be placed to face directly into the freestream. This is the whole basis of what we have come to know as 'isokinetic sampling', and even this simple case has been widely applied in the sampling of aerosols from stacks and ducts. The expectations for blunt samplers more generally are quite different. Most aerosol samplers intended for practical applications in occupational and environmental hygiene are blunt and need to operate in wind conditions that are much less easily defined. Not only is the wind speed likely to be highly variable, but the orientation of the sampler inlet cannot be prescribed, and so may take any orientation between $\theta = 0$ and $180°$. Hence, there is considerable motivation towards finding predictive models over these ranges of conditions. But even to this day there is not a great deal of good experimental information on which to base a level of physical understanding matching that for simple blunt samplers facing the freestream. Early on, all that were available were the results of Ogden and Birkett (1977) for a full-sized, rather unidealised human head in wind tunnel experiments to determine the aspiration efficiency for the entry of particles through the mouth for orientations $\theta = 0$, 45 and $90°$, respectively. However, early attempts to use data like these towards modifying the impaction model for blunt samplers at angles other than directly forwards-facing were not very successful (Vincent, 1987).

Tsai and Vincent (1993) later returned to those earlier human head data of Ogden and Birkett, along with others from Armbruster and Breuer (1982), combining them with the thin-walled sampler data from Vincent *et al.* (1986), all of which contained data for orientations $\theta = 90°$ and $180°$. For $\theta = 90°$ they used as a starting point the form of expression already developed for the thin-walled sampler, as described by Equation (5.26). Now, however, an additional term involving r was added to account for the bluntness of the sampler body, leading to the suggestion:

$$A_{90} = \frac{1}{(1 + 4G_{90} r^{g_1} St R^{1/2})} \qquad (5.60)$$

where, as before, G_{90} was taken to be a constant coefficient and assigned the value 2.21. By nonlinear regression, a good fit to the entire available data set was found when $g_1 = -0.5$ so that:

$$A_{90} = \frac{1}{\left[1 + 8.84 \left(\dfrac{R}{r}\right)^{1/2} St\right]} \quad \text{with } R_{corr}^2 = 0.61 \tag{5.61}$$

which reverts back to the original form of Equation (5.26) when $r = 1$, corresponding to the thin-walled cylindrical sampler. The comparison between the calculated and experimental values for A_{90} is shown in Figure 5.22.

The situation for $\theta = 180°$ is more difficult because Equation (5.28) for the thin-walled sampler was based on a slighter physical idea – namely, only that A_{180} is governed by impaction onto the leading edge of the supported tube of the sampler. Tsai and Vincent proposed an extension of Equation (5.28), thus:

$$A_{180} = \frac{1}{(1 + 18 r^{g_2} R^{-1/3} St)} \tag{5.62}$$

which, as shown in Figure 5.23, provided quite a good fit to the whole data set when $g_2 = 1$ so that:

$$A_{180} = \frac{1}{(1 + 18 r R^{-1/3} St)} \quad \text{with } R_{corr}^2 = 0.71 \tag{5.63}$$

In a more recent experimental wind tunnel study to investigate a number of practical aerosol samplers that may generally be considered to be symmetric, so matching the criteria for application of the models

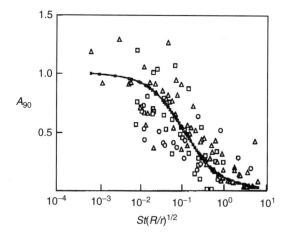

Figure 5.22 *Experimental data for aspiration efficiency (A_{90}) for thin-walled and blunt samplers placed with their orifices at an angle of 90° to the freestream, plotted as a function of $St(R/r)^{1/2}$ as suggested by Equation (5.60) (Tsai and Vincent, 1993); ◯ Ogden and Birkett (1977) for the human head, △ Armbruster and Breuer (1982) for the human head, and □ Vincent et al. (1986) for thin-walled cylindrical samplers. The solid line is calculated from the model in Equation (5.61). Reprinted from Journal of Aerosol Science, 24, Tsai and Vincent, 919–928. Copyright (1993), with permission from Elsevier*

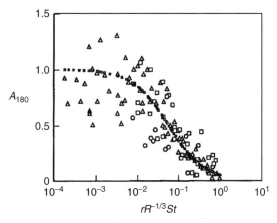

Figure 5.23 *Experimental data for aspiration efficiency (A_{180}) for thin-walled and blunt samplers placed with their orifices at an angle of 180° to the freestream, plotted as a function of $rR^{-1/3}St$ as suggested by Equation (5.62) (Tsai and Vincent, 1993); ○ Ogden and Birkett (1977) for the human head, △ Armbruster and Breuer (1982) for the human head, and □ Vincent et al. (1986) for thin-walled cylindrical samplers. The solid line is calculated from the model in Equation (5.63). Reprinted from Journal of Aerosol Science, 24, Tsai and Vincent, 919–928. Copyright (1993), with permission Elsevier*

described above for A_0, A_{90} and A_{180} respectively, Li and Lundgren (2002) obtained results that were generally in agreement with the models.

In conclusion from all the above, empirical models like those described models may provide the starting point for a physical discussion of the performances of aerosol samplers under less idealised, realistic conditions.

5.3.4 Mathematical and numerical approaches to blunt samplers

The approach described above towards describing and predicting aspiration efficiency for blunt samplers represents efforts to apply the successful impaction model approach that had been quite successful for the simple thin-walled sampling tube. It was more or less successful, the less so for extreme orientations of the sampling inlet with respect to the freestream. Other researchers have approached modeling from the alternative point of view of mathematical and numerical modeling. Ingham (1981) was the first to do so, for the simple case of a two-dimensional cylindrical sampler body with a line sink facing directly into the freestream ($\theta = 0°$). In addition to characterising the inviscid potential flow near the sampler and identifying positions of flow stagnation on the sampler body, he also carried out numerical calculations of particle trajectories leading to determinations of aspiration efficiency. The shapes of the trends predicted in Ingham's calculations were generally consistent with those exhibited by the few experimental data available at the time (Ogden and Birkett, 1977) and also with the ones calculated from the impaction model for the forwards-facing axially symmetric blunt sampler. Later, Ingham and Yan (1994) extended the cylindrical sampler model to a related geometry where a cylindrical sampler was attached to another, larger cylinder. The aim here was to simulate idealistically the situation where a small sampler is mounted on a large bluff body – like, for example, a small personal aerosol sampler mounted on the body of a person. The calculations were carried out only for $\theta = 0°$. It was shown that the size of the larger body had a significant effect on the overall aspiration efficiency of the sampling system. However, there were no data available against which Ingham and Yan could compare their calculations. It is notable that their main conclusion was inconsistent with the experimental observation of Paik and Vincent (2002b) that had suggested only weak dependencies on the size of the backing body.

Elsewhere, Chung and Dunn-Rankin (1992) performed numerical simulations for aerosol aspiration by a two-dimensional blunt sampler with its inlet facing directly forwards at $\theta = 0°$. This work differed from that of Ingham and his colleagues in that the flow around the sampler was taken to be viscous. Chung and Dunn-Rankin used a fully numerical routine to solve the Navier–Stokes equations directly. They demonstrated that the role of viscosity was reflected in a dependence of the flow field, and in turn, aspiration efficiency, on Reynolds number (Re) for the flow about the sampler body. From their results, however, it was shown that such dependencies were likely to be small for the ranges of Re expected in aerosol sampling situations.

In general, mathematical studies of three-dimensional blunt sampling systems have proved more difficult than for the two-dimensional ones that have been described. Dunnett and Ingham attempted to do so in their 1988 work, in particular using the analytical methods they had developed towards describing some macroscopic features of the airflow near a spherical sampler with its inlet at various orientations with respect to the freestream. Most notably, they determined the shape and dimensions of the stagnation region formed on the surface of the sampler body by the combined influence of the aspirated and nonaspirated airflow. They then used this information in a revised version of the impaction model as contained in Equations (5.55), modified to include the new stagnation region characteristics as now reflected in two dimensions (S_1 and S_2) instead of the original one (S). The new impaction model was used to calculate aspiration efficiency (A) and the results were compared with the human head data of Ogden and Birkett (1977). Although agreement between theory and experiment was quite good for forwards-facing ($\theta = 0°$), the results for other orientations were less encouraging. Later, Ingham and Wen (1993) took a different approach in their study of a disk-shaped body at $\theta = 0°$, in which the flow was assumed to be turbulent. Using conventional turbulence models like the ones described by Launder and Spalding (1972), they developed descriptions of both air and particle motions in turbulent flow near the blunt sampler system of interest. Their results showed a surprisingly weak dependence of A on St, not reflected in the experimental results of Vincent *et al.* (1985) and Chung and Ogden (1986).

5.3.5 Orientation-averaged conditions

We now refer back to the fact that, in real-world aerosol measurement situations, the sampler may instantaneously and arbitrarily take any orientation between $\theta = 0$ and $360°$ with respect to the external air movement. There will rarely be a situation where the distribution of the orientation over time can be identified. The simplest approach, and one which is the most valid for a scientific discussion, is where sampler orientation is averaged uniformly over time through the whole range of θ from 0 to $360°$. Ideally, aspiration efficiency averaged over all the orientations (A_{ave}) is given by:

$$A_{ave} = \frac{1}{2\pi} \int_{0°}^{360°} A(\theta)\mathrm{d}\theta \tag{5.64}$$

Unfortunately all the knowledge needed is not available at present. At best, as described above, reliable models exist only for A_0, A_{90} and A_{180} for sampling systems of quite simple, symmetrical shape. If it were possible to assume that A_0 is representative of aspiration efficiency over $-45° \le \theta \le +45°$, A_{90} over $+45° \le \theta \le +135°$, A_{180} over $+135° \le \theta \le +225°$ and A_{90} over $+225° \le \theta \le +315°$ (equivalent to $-135° \leq \theta \leq -45°$), then it would be possible to simply combine the models for A_0, A_{90} and A_{180} using:

$$A_{ave} = \frac{A_0}{4} + \frac{A_{90}}{2} + \frac{A_{180}}{4} \tag{5.65}$$

However, this is over-simplistic since the detailed distribution of aspiration efficiency with θ is likely to depend additionally on how St, R and r influence the shape of the air and aerosol flow patterns in the vicinity of the sampler that in turn must create a highly nonuniform dependence of A on θ. It is already known from the experimental studies that there is a tendency for aspiration efficiency to be nonuniformly biased by contributions from generally forwards-facing orientations – for example, as seen by inspection of the results of Durham and Lundgren (1980) shown in Figure 5.10. Therefore it is necessary to think in terms of a model of the general form:

$$A_{ave} = \phi_0 A_0 + \phi_{90} A_{90} + \phi_{180} A_{180} \tag{5.66}$$

where the coefficients ϕ_0, ϕ_{90} and ϕ_{180} are in fact 'weighting' parameters that are functions of St, R and r and do not necessarily add up to unity. Tsai *et al.* (1995) sought empirical relationships meeting these criteria and estimated the unknown constant coefficients contained within them by reference to experimental data for uniform orientation-averaged aspiration efficiency for the human head taken from the wind tunnel studies of Vincent and Mark (1982) and Vincent *et al.* (1990), and two rotating-head aerosol samplers – a small one sampling at 3 Lpm and a larger one sampling at 30 Lpm – taken from the wind tunnel studies of Mark *et al.* (1985) and (Mark *et al.* 1990). All the sampling systems in the studies quoted were geometrically uniform, and so were conveniently appropriate for the purpose of interest here, providing a total of 189 individual data records. More about these experimental studies will appear in later chapters of this book. Nonlinear regression of the data with respect to the postulated model yielded:

$$A_{ave} = \frac{A_0}{2} + \left(\frac{A_{90} - A_0}{181 R^{-2.31} r^{1.01} + 2} \right) + \frac{A_{180}}{2} \text{ with } R_{corr}^2 = 0.68 \tag{5.67}$$

applicable for θ averaged uniformly over 0 to 360°. In this expression, A_0 is calculated from Equations (5.55), A_{90} from Equation (5.59) and A_{180} from Equation (5.63). Equation (5.67) may be quite useful for evaluating the orientation-averaged performance of samplers that are relatively uniform in terms of the location of the sampling orifice on the blunt sampler body. Although this might be relevant to a number of practical systems, it still describes a scenario that is relatively simple in comparison with most of the aerosol sampling systems that are found in occupational and environmental hygiene practice. Even so, the model that has been described may at least be regarded as yet another stepping stone towards a fuller understanding of the physical nature of aerosol sampler performance.

References

Addlesee, A.J. (1980) Anisokinetic sampling of aerosols at a slot intake, *Journal of Aerosol Science*, 11, 483–493.

Armbruster, L. and Breuer, H. (1982) Investigations into defining inhalable dust, In: *Inhaled Particles, Vol. V* (Ed. W.H. Walton), Pergamon Press, Oxford, pp. 21–32.

Badzioch, S. (1959) Collection of gas-borne dust particles by means of an aspirated sampling nozzle, *British Journal of Applied Physics*, 10, 26–32.

Belyaev, S.P. and Levin, L.M. (1972) Investigation of aerosol aspiration by photographing particle tracks under flash illumination, *Journal of Aerosol Science*, 3, 127–140.

Belyaev, S.P. and Levin, L.M. (1974) Techniques for collection of representative aerosol samples, *Journal of Aerosol Science*, 5, 325–338.

Chung, I.P. and Dunn-Rankin, D. (1992) Numerical simulation of two-dimensional blunt body sampling in viscous flow, *Journal of Aerosol Science*, 23, 217–232.

Chung, I.P., Trinh, T. and Dunn-Rankin, D. (1994) Experimental investigation of a two-dimensional cylindrical sampler, *Journal of Aerosol Science*, 25, 935–955.

Chung, K.Y.K. and Ogden, T.L. (1986) Some entry efficiencies of disklike samplers facing the wind, *Aerosol Science and Technology*, 5, 81–91.

Davies, C.N. (1968) The entry of aerosols into sampling tubes and heads, *British Journal of Applied Physics*, 1, 921–932.

Davies, C.N. and Subari, M. (1982) Aspiration above wind velocity of aerosols with thin-walled nozzles facing and at right angles to the wind direction, *Journal of Aerosol Science*, 13, 59–71.

Dunnett, S.J. (1990) Mathematical modelling of aerosol sampling with thin-walled probes at yaw orientation with respect to the wind, *Journal of Aerosol Science*, 21, 947–956.

Dunnett, S.J. (2005) Numerical study of the aspiration efficiency of a thin-walled sampler facing the wind for high velocity ratios, *Journal of Aerosol Science*, 36, 111–121.

Dunnett, S.J. and Ingham, D.B. (1988) An empirical model for the aspiration efficiencies of blunt aerosol samplers oriented at an angle to the oncoming flow, *Aerosol Science and Technology*, 8, 245–264.

Durham, M.D. and Lundgren, D.A. (1980) Evaluation of aerosol aspiration efficiency as a function of Stokes' number, velocity ratio and nozzle angle, *Journal of Aerosol Science*, 11, 179–188.

Fuchs, N.A. (1964) *The Mechanics of Aerosols*, Macmillan, New York.

Gibson, H. and Ogden, T.L. (1977) Entry efficiencies for sharp-edged samplers in calm air, *Journal of Aerosol Science*, 8, 361–365.

Glauberman, H. (1962) The directional dependence of air samplers, *American Industrial Hygiene Association Journal*, 23, 235–239.

Grinshpun, S.A., Willeke, K. and Kalatoor, S. (1993) General equation for aerosol aspiration by thin-walled sampling probes from calm and moving air, *Atmospheric Environment*, 27, 1459–1470.

Grinshpun, S.A., Willeke, K. and Kalatoor, S. (1994) Corrigendum, *Atmospheric Environment*, 28, 375.

Hangal, S. and Willeke, K. (1990) Aspiration efficiency: unified model for all forward sampling angles, *Environmental Science and Technology*, 24, 688–691.

Hemeon, W.C.L. and Haines, G.F. (1954) The magnitude of errors in stack dust sampling, *Air Repair*, 4, 159–164.

Ingham, D.B. (1981) The entrance of airborne particles into a blunt sampling head, *Journal of Aerosol Science*, 12, 541–549.

Ingham, D.B. and Wen, X. (1993) Disklike body sampling in a turbulent wind, *Journal of Aerosol Science*, 24, 629–642.

Ingham, D.B. and Yan, B. (1994) The effect of a cylindrical backing body on the sampling efficiency of a cylindrical sampler, *Journal of Aerosol Science*, 25, 535–541.

Ingham, D.B., Wen, X., Dombrowski, N. and Foumeny, E.A. (1995) Aspiration efficiency of a thin-walled shallow-tapered sampler rear-facing the wind, *Journal of Aerosol Science*, 26, 933–944.

Jayasekera, P.N. and Davies, C.N. (1980) Aspiration below wind velocity of aerosols with sharp-edged nozzles facing the wind, *Journal of Aerosol Science*, 11, 535–547.

Krämer, M. and Afchine, A. (2004) Sampling characteristics of inlets operated at low U/U_0 ratios: new insights from computational fluid dynamics (CFX) modeling, *Journal of Aerosol Science*, 35, 683–694.

Lapple, C.E. and Shepherd, C.B. (1940) Calculation of particle trajectories, *Industrial and Engineering Chemistry*, 32, 605–617.

Launder, B.S. and Spalding, D.B. (1972) *Mathematical Models of Turbulence*, Academic Press, London.

Levin, L.M. (1957) The intake of aerosol samples, *Izv. Nauk. SSSR Ser. Geofiz.*, 7, 914–925.

Li, S.N. and Lundgren, D.A. (2002) Aerosol aspiration efficiency of blunt and thin-walled samplers at different wind orientations, *Aerosol Science and Technology*, 36, 342–350.

Lipatov, G.N., Grinshpun, S.A., Shingaryov, G.L. and Sutugin, A.G. (1986) Aspiration of coarse aerosol by a thin-walled sampler, *Journal of Aerosol Science*, 17, 763–769.

Liu, B.Y.H., Zhang, Z.Q. and Kuehn, T.H. (1989) A numerical study of inertial errors in anisokinetic sampling, *Journal of Aerosol Science*, 20, 367–380.

Lundgren, D.A., Durham, M.D. and Mason, K.W. (1978) Sampling of tangential flow streams, *American Industrial Hygiene Association Journal*, 39, 640–644.

Mark, D., Vincent, J.H. and Gibson, H. (1985) A new static sampler for airborne total dust in workplaces, *American Industrial Hygiene Association Journal*, 46, 127–133.

Mark, D., Vincent, J.H., Lynch, G., Aitken, R.J. and Botham, R.A. (1990) The development of a static sampler for the measurement of inhalable aerosol in the ambient atmosphere (with special reference to PAHs), Report No. TM/90/06, Institute of Occupational Medicine, Edinburgh.

Ogden, T.L. and Birkett, J.L. (1977) The human head as a dust sampler. In: *Inhaled Particles IV* (Ed. W.H. Walton), Pergamon Press, Oxford, pp. 93–105.

Paik, S. and Vincent, J.H. (2002a) Aspiration efficiency for thin-walled nozzles facing the wind and for very high velocity ratios, *Journal of Aerosol Science*, 33, 705–720.

Paik, S.Y. and Vincent, J.H. (2002b) Aspiration efficiencies of disk-shaped blunt nozzles facing the wind, for coarse particles and high velocity ratios, *Journal of Aerosol Science*, 33, 1509–1523.

Raynor, G.S. (1970) Variation in entrance efficiency of a filter sampler with air speed, flow rate, angle and particle size, *American Industrial Hygiene Association Journal*, 31, 294–304.

Rouillard, E.E.A. and Hicks, R.E. (1978) Flow patterns upstream of isokinetic dust sampling probes, *Journal of the Air Pollution Control Association*, 28, 599–601.

Rüping, G. (1968) The importance of isokinetic suction in dust flow measurement by means of sampling probes, *Staub Reinhaltung der Luft* (English translation), 28, 1–11.

Sehmel, G.A. (1967) Validity of air samples as affected by anisokinetic sampling and deposition within sampling line, In: *Proceedings of the Symposium on Assessment of Airborne Radioactivity*, Vienna, pp. 727–735.

Stenhouse, J.I.T. and Lloyd, P.J. (1979) Sampling errors due to inertial classification, In: *Recent Advances in Air Pollution Control*, American Institute of Chemical Engineers (AIChE) Symposium Series No. 137, AIChE, New York, NY.

Stevens, D.C. (1986) Review of aspiration coefficients of thin-walled sampling nozzles, *Journal of Aerosol Science*, 17, 729–743.

Tsai, P.J. and Vincent, J.H. (1993) Impaction model for the aspiration efficiencies of aerosol samplers at large angles with respect to the wind, *Journal of Aerosol Science*, 24, 919–928.

Tsai, P.J., Vincent, J.H., Mark, D. and Maldonado, G. (1995) Impaction model for the aspiration efficiencies of aerosol samplers in moving air under orientation-averaged conditions, *Aerosol Science and Technology*, 22, 271–286.

Tufto, P.A. and Willeke, K. (1982) Dependence of particulate sampling efficiency on inlet orientation and flow velocities, *American Industrial Hygiene Association Journal*, 43, 436–443.

Vincent, J.H. (1987) Recent advances in aspiration theory for thin-walled and blunt aerosol sampling probes, *Journal of Aerosol Science*, 18, 487–498.

Vincent, J.H. (1989) *Aerosol Sampling: Science and Practice*, John Wiley & Sons, Ltd, Chichester.

Vincent, J.H. and Mark, D. (1982) Applications of blunt sampler theory to the definition and measurement of inhalable dust, In: *Inhaled Particles V* (Ed. W.H. Walton), Pergamon Press, Oxford, pp. 3–19.

Vincent, J.H., Emmett, P.C. and Mark, D. (1985) The effects of turbulence on the entry of airborne particles into a blunt dust sampler, *Aerosol Science and Technology*, 4, 17–29.

Vincent, J.H., Stevens, D.C., Mark, D., Marshall, M. and Smith, T.A. (1986) On the aspiration characteristics of large-diameter, thin-walled aerosol sampling probes at yaw orientations with respect to the wind, *Journal of Aerosol Science*, 17, 211–224.

Vincent, J.H., Mark, D., Miller, B.G., Armbruster, L. and Ogden, T.L. (1990) Aerosol inhalability at higher wind speeds, *Journal of Aerosol Science*, 21, 577–586.

Vitols, V. (1966) Theoretical limits of errors due to anisokinetic sampling of particulate matter, *Journal of the Air Pollution Control Association*, 16, 79–84.

Voloshchuck, V.M. (1971) *Introduction to Coarse Aerosol Hydrodynamics*, Gidrometeisdat, St Petersburg (Leningrad) (in Russian).

Voloshchuck, V.M. and Levin, L.M. (1968) Some theoretical aspects of aerosol aspiration (large Stokes' numbers), *Izvestiya Atmospheric and Oceanic Physics*, (English translation), 4 241–249.

Walter, E. (1957) Zur problematik der Entnahmeesonden and der Teilstromentnahme fur die Staubgehaltsbestimmung in Stromenden Gasen, *Staub Reinhaltung der Luft*, 53, 880–898.

Watson, H.H. (1954) Errors due to anisokinetic sampling of aerosols, *American Industrial Hygiene Association Journal*, 15, 21–25.

Wen, X. and Ingham, D.B. (1995) Aspiration efficiency of a thin-walled cylindrical probe rear-facing the wind, *Journal of Aerosol Science*, 26, 95–107.

Yoshida, H., Osugi, T., Masuda, H. and Iinoyo, K. (1976) Particle sampling efficiency in still air, *Kagaku Kogaku Ronbunshu*, 2, 336–340 (HSE translation No. 8584).

Zenker, P. (1971) Investigations into the problem of sampling from a partial flow with different flow velocities for the determination of the dust content in flowing gases, *Staub Reinhaltung der Luft* (English translation), 31, 30–36.

Zhang, Z.Q. and Liu, B.Y.H. (1989) On the empirical fitting equations for aspiration coefficients for thin-walled sampling probes, *Journal of Aerosol Science*, 20, 713–720.

6

Aspiration in Calm and Slowly Moving Air

6.1 Introduction

In discussing the physics of the aspiration process in preceding chapters, it has been assumed that drag forces and inertia in convective air movement are the predominant influences on particle motion. The effects of gravity were neglected entirely. This is a fair assumption if the velocity of the moving air that brings the particles into the vicinity of a sampler is much greater than the falling speeds of the particles as they settle under the influence of gravity. There are, however, some situations where this might not hold. 'Calm air' refers to the special case where, ideally, the only air movement is that induced by the aspirating action of the sampler itself; that is, there is no external wind. Under such conditions, it is likely that gravitational settling will exhibit a significant effect on particle motion and hence on aspiration efficiency. Since this effect has not featured at all in the sampler theory which has already been described, 'calm air sampling' cannot be regarded simply as a limiting, zero-wind speed version of the general, moving air case. So it needs to be treated separately. In addition, because the local air velocities are very low in many indoor places where it is of interest to sample aerosols, it is also necessary to consider the intermediate range of conditions where the external air may be considered to be in motion, but where gravitational forces may also exert a significant influence on particle motion during sampling.

6.2 Sampling in perfectly calm air

6.2.1 Qualitative description

For a single-orifice sampler of arbitrary shape and orientation in calm air, the only air movement is that induced by the suction of air through the sampling orifice. The nature of the resultant air-flow pattern is shown (broken lines) in Figure 6.1, where it is seen that local velocities tend towards zero at large distances away from the sampler but increase progressively as the distance decreases, as reflected in the convergence of the streamlines. Typical particle trajectories in this flow system are also shown (continuous lines). In Figure 6.1(a), very small particles with negligible settling velocity and exhibiting negligible inertia follow the air-flow pattern closely. On the other hand, in Figure 6.1(b), larger particles at a large distance from the sampling orifice fall vertically, relatively undisturbed by the presence and

Aerosol Sampling: Science, Standards, Instrumentation and Applications   James Vincent
© 2007 John Wiley & Sons, Ltd

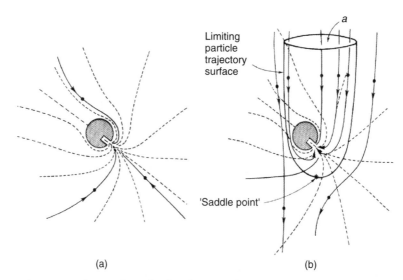

Figure 6.1   *General patterns of streamlines (broken lines) and particle trajectories (solid lines) near a sampler of simple shape with its entry located at arbitrary orientation with respect to the vertical: (a) negligible gravity and inertia; (b) larger particles with significant gravity and inertial effects. Reproduced with permission from Vincent, Aerosol Sampling: Science and Practice. Copyright (1989) John Wiley & Sons, Ltd*

action of the sampler. Those that approach close enough to the sampling orifice become increasingly influenced by the suction and so are drawn towards the sampler. For given particle size, depending on how rapidly the air velocity accelerates as the sampler is approached, inertial effects may or may not become significant.

In Figure 6.1(b), the net flux of particles entering the sampling orifice is the same as that passing through the area a above the sampler. For particles that exhibit negligible inertia, $a = a_0$. More generally, however, when inertial effects are not negligible, the net flux of particles entering the sampler is not the same as that passing through a_0 but rather that through some other area ($a = a''$, say), where $a'' \neq a_0$. The extent of the difference between a'' and a_0 is a measure of aspiration efficiency for calm air sampling, and depends on particle size, sampler shape and dimensions, and on sampling volumetric flow rate.

The surface described by the tube containing all the particles entering the orifice, and outside which all particle trajectories fail to enter, is effectively a limiting trajectory surface. It is equivalent to the ones already identified in earlier chapters for samplers in moving air. Unlike the moving air case, however, it is not now accompanied by a corresponding airflow limiting streamsurface. One important feature is the so-called 'saddle point' located immediately below the sampling orifice, defined as the position where the vertically downwards particle settling velocity is just balanced by the upwards air flow induced by the suction.

Aspiration efficiency (A) is defined initially in a manner similar to that for the moving air case, thus:

$$A = \frac{N_s}{N_0} \tag{6.1}$$

where N_s is the number of the particles sampled per second and N_0 is the equivalent number for inertialess particles. For the case shown in Figure 6.1(a), A is obviously unity. For the case shown in Figure 6.1(b), A is also unity provided that inertial effects are negligible. Otherwise it takes other values. From Figure 6.1(b), we have:

$$N_0 = c_0 a_0 v_s \tag{6.2}$$

where c_0 is the particle concentration at a large distance from the sampler and v_s is the particle settling velocity as defined earlier in Chapter 2. In addition:

$$N_s = c_0 a'' v_s \tag{6.3}$$

so that:

$$A = \frac{a''}{a_0} \tag{6.4}$$

Since we also have:

$$N_s = c_s Q \tag{6.5}$$

where c_s is the particle concentration entering the sampler and Q is the sampling volumetric flow rate, then it follows from Equations (6.3)–(6.5) that:

$$A = \frac{v_s a''}{Q} \longrightarrow \frac{c_s}{c_0} \tag{6.6}$$

and this is the basic working expression for the aspiration efficiency of a sampler in perfectly calm air.

One of the earliest attempts to understand the physics of the sampling of aerosols under calm air conditions was described by Davies (1968). He began by considering the air velocity due to aspiration by a single-orifice sampling head small enough to be regarded as a point sink. If the inner diameter of the tube is δ, then the mean velocity across the plane of the sampling orifice is:

$$U_s = \frac{4Q}{\pi \delta^2} \tag{6.7}$$

For gravitational forces to have negligible effect on aspiration efficiency, and so for orientation effects to be insignificant, it is necessary that $U_s \gg v_s$. On the other hand, for inertial effects also to be small:

$$s \ll \frac{\delta}{2} \tag{6.8}$$

where s is the stop distance of particles as defined earlier in Chapter 2. Here it relates to particle motion near the sampling orifice. The appearance of the factor of 1/2 on the right-hand side of Equation (6.8) arises because Davies chose to compare s with the orifice radius. He next identified the *dynamical sampling velocity*, the air velocity induced by the action of the sampler at a distance s away from the plane of the sampling orifice, as:

$$U_d = \frac{Q}{4\pi s^2} \tag{6.9}$$

based on the inverse square law variation of velocity with distance from a point sink. Substituting $s = U_d \tau$ leads to:

$$U_d = \left(\frac{Q}{4\pi \tau^2} \right)^{1/3} \tag{6.10}$$

where τ is the particle relaxation time as defined earlier in Chapter 2. For negligible inertial effects, the preceding equations provide:

$$\left(\frac{Q\tau}{4\pi}\right)^{1/3} << \frac{\delta}{2} \tag{6.11}$$

For negligible gravity effects, where $U_s >> v_s$, we have the corresponding expression:

$$\frac{\delta}{2} << \left(\frac{Q}{\pi g\tau}\right)^{1/2} \tag{6.12}$$

Equations (6.11) and (6.12) together provide a definition of the conditions under which both inertial and gravitational effects are negligible. In that event, there will be no changes in aerosol concentration as particles are drawn towards the sampling orifice, so that aspiration efficiency will be of the order of unity. For that to occur:

$$\left(\frac{Q\tau}{4\pi}\right)^{1/3} << \frac{\delta}{2} << \left(\frac{Q}{\pi g\tau}\right)^{1/2} \tag{6.13}$$

and this provides the basis of Davies' criterion for 'representative sampling' that will be discussed later. However, it does not tell us anything about aspiration efficiency when inertial and/or gravitational effects are *not* negligible. When neither inertial nor gravitational effects are negligible, then it is an important consideration that the concentration of particles in the aspirated air is modified due to particles striking the outer surface of the sampler.

In 1989, Grinshpun *et al.* published a review of a number of studies of sampling in calm air that had been carried out up to that point, including the ones by Kaslow and Emrich (1974), Davies (1977), Agarwal and Liu (1980), Belyaev and Kustov (1980), Lipatov *et al.* (1985) and Grinshpun and Lipatov (1986). It was noted that aspiration efficiency was consistently seen to be a function of Stokes number and the ratio between the particle settling velocity and the air velocity in the plane of the sampling orifice. From this and with the preceding physical picture in mind, a general function expression for aspiration efficiency for sampling in calm air is suggested along the lines:

$$A = \mathrm{f}(St_c, R_c, r, \theta, B) \tag{6.14}$$

This is analogous to the expression given earlier for moving air. Here, however, we have the modified Stokes number (St_c) and velocity ratio (R_c), the dimension ratio (r), orientation (θ, this time with respect to the upwards vertical) and sampler body shape (as represented by bluntness, B). The two new terms, St_c and R_c, are given by:

$$St_c = \frac{d_{ae}\gamma^* v_s}{18\eta\delta} \tag{6.15}$$

$$R_c = \frac{v_s}{U_s} \tag{6.16}$$

while the other terms are the same as before. Equation (6.14) embodies the fact that, for calm air sampling, it is the action of gravity, expressed in terms of v_s, that brings the particles into the vicinity of the sampler inlet, in contrast to the moving air case, where this is achieved by virtue of convection, and hence is dependent on U.

6.2.2   Experimental studies for sampling in perfectly calm air

Compared with the case for aerosol sampling in moving air, especially for thin-walled tubes, the amount of experimental work for calm air scenarios has been quite sparse. Part of the problem seems to be associated with the technical difficulties involved in conducting meaningful experiments under the required, quite stringent conditions. In particular, the introduction of a test aerosol into a calm air situation without disturbing the very environment that is of interest poses special challenges. Nonetheless, some useful experimental work has been reported.

In an early study conducted in a test chamber, Breslin and Stein (1975) used the indirect method to determine the aspiration efficiencies of a number of test samplers of simple geometry and a number of sampling flow rates. They used polydisperse test aerosols generated from dry powders using a Wright dust feed, and analyzed the collected samples on the basis of particle size by means of a Coulter counter. The results showed that aspiration efficiency was close to unity for conditions corresponding to small St_c, and decreased for large particle sizes corresponding to larger St_c. The most important earlier study was the one reported by Yoshida and colleagues in 1978. It was designed specifically to provide a definitive data set for the aspiration efficiencies of thin-walled tubes in perfectly calm air. Yoshida *et al.* employed a calm air version of the direct method used for moving air by Belyaev and Levin (1974) and Lipatov *et al.* (1986). Applied to the calm air system of interest this method eliminated the need for – and the difficulties of interpretation associated with – a reference sampler. The experiments were carried out in a $0.3 \times 0.3 \times 1.2$ m test chamber into which reasonably monodisperse aerosols of glass beads in the range of d_{ae} from 32 to 64 μm were dispersed by means of a mechanical feeder. Cylindrical test probes of diameter (δ) 7.5, 10, 16 and 20 mm were employed at sampling volumetric flow rates up to 80 liters min^{-1} (Lpm). Particle trajectories were photographed under strong illumination and, from the photographs, loci of particles corresponding to limiting trajectory surfaces were identified and, hence, aspiration efficiency (A) determined. It is noted, however, that the definition of aspiration efficiency employed by Yoshida and his colleagues differed slightly from that contained in Equation (6.6) in that a correction term was included to allow for the effect of sedimentation in the plane of the sampling orifice. However, this effect was negligible except in the few cases where particle settling velocity (v_s) was not small compared with mean sampling velocity (U_s). The experiments were designed to allow A to be represented as a function of St_c and R, and the results of Yoshida *et al.* are plotted in this way in Figure 6.2 for the case of upwards-facing sampling tubes. As shown in the results of Breslin and Stein, A fell steadily with increasing St_c, and was smaller for larger R_c.

In the years that followed, very little further work along these lines was reported, despite the practical importance of the calm air scenario. Quite recently, however, Su and Vincent (2002, 2003, 2004a) carried out experiments to determine the aspiration of samplers of simple shape – thin-walled and spherical – for orientations with respect to the vertical (θ) equal to 0, 90 and 180°, respectively The basic experimental approach was introduced earlier in Chapter 3. The investigators went to great lengths to ensure that the air was perfectly calm, so that the trajectory of a thin thread-of-droplets in the range of d_{ae} from about 40 to 70 μm – from a vibrating-orifice liquid-droplet aerosol generator – could be visualised clearly. By way of illustration, typical photographs obtained for the particle trajectories are shown in Figures 6.3–6.5, for both the thin-walled and spherical samplers for the various orientations listed. In the experimental set-up the location of the limiting streamsurfaces was facilitated by means of a micrometer-based tracking system that provided the means for moving the test sampler relative to the falling stream of droplets and hence for accurately measuring the dimensions required for the determination of a'' in Equation (6.6), leading in turn to A.

In addition to showing the particle trajectories, Figures 6.3 and 6.4 also indicate the shapes of the respective surfaces above the samplers through which all particles that were aspirated had to pass, defining a'' needed for the calculation of A. In Figure 6.3, for the upwards-facing tube, symmetry

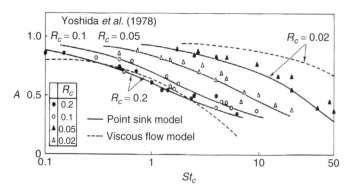

Figure 6.2   *Experimental data for aspiration efficiency (A) in calm air for upwards-facing, aspirating, thin-walled cylindrical tubes as a function of Stokes number (St_c) for a range of velocity ratios (R_c) as indicated: _____ numerical calculations based on the point sink model; --- numerical calculations based on the viscous flow model (from Yoshida et al., 1978). Reproduced with permission from Vincent, Aerosol Sampling: Science and Practice. Copyright (1989) John Wiley & Sons, Ltd*

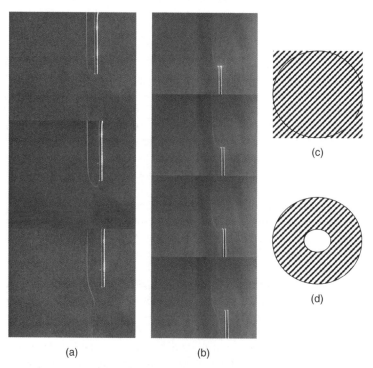

Figure 6.3   *Typical particle trajectories near a thin-walled sampling tube: (a) facing downwards; (b) facing upwards; (c) shape of the area (hatched) above an upwards-facing sampler enclosing particle trajectories for sampled particles; (d) likewise for downwards-facing sampler (Su and Vincent, 2004a). Reprinted from Aerosol Science and Technology, 38, 766–781. Copyright (2004), with permission from the American Association for Aerosol Research*

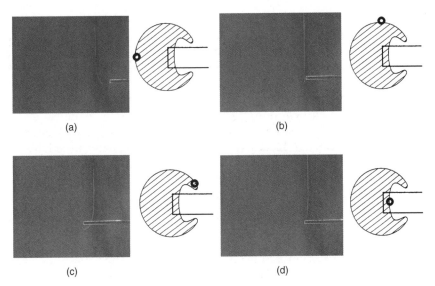

(a)   (b)

(c)   (d)

Figure 6.4   *Typical particle trajectories near a thin-walled sampling tube facing horizontally, showing trajectories of particles starting at different locations above the sampler. The drawings show the shape of the area (hatched) above the sampler enclosing particle trajectories for sampled particles, also indicating the starting points of the particle trajectories portrayed (Su and Vincent, 2004a). Reprinted from Aerosol Science and Technology, 38, 766–781. Copyright (2004), with permission from the American Association for Aerosol Research*

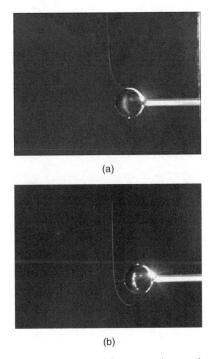

(a)

(b)

Figure 6.5   *Trajectories of particles near a spherical blunt sampler: (a) facing upwards; (b) facing downwards (Su and Vincent, 2003). Reprinted from Journal of Aerosol Science, 34, Su and Vincent, 1151–1165. Copyright (2003), with permission from Elsevier*

ensures that a'' was a circle. For the downwards-facing tube, a'' was similarly symmetrical but was modified by the presence of the shadow associated with the physical presence of the sampler body. Hence the shape of a'' was now annular. For the horizontal tube, the shape of a'' was no longer symmetrical. Now the shape was modified significantly by the asymmetrical presence of the tube so that a'' became closer to 'kidney-shaped'.

From inspection of a large number of trajectories like those shown in Figures 6.3–6.5, an extensive data set was obtained for A as functions of St_c. These are summarised in Figure 6.6 for the thin-walled samplers and Figure 6.7 for the blunt samplers. The results in Figure 6.6 show broadly similar trends to the results presented by Yoshida *et al.* Indeed, for the upwards-facing thin-walled tube where the same value of R_c was studied ($R_c = 0.1$), agreement with those earlier results (see Figure 6.2) was good. For the results in Figure 6.7 for the blunt samplers, the quantity B that is associated with the bluntness of the sampler here was represented simply by D/δ; that is, $1/r$.

6.2.3   Analytical models for aspiration efficiency in calm air

Unlike the case for moving air described in the previous chapter, the first efforts to develop a physical picture of aerosol sampling in calm air for the prediction of aspiration efficiency took the form of

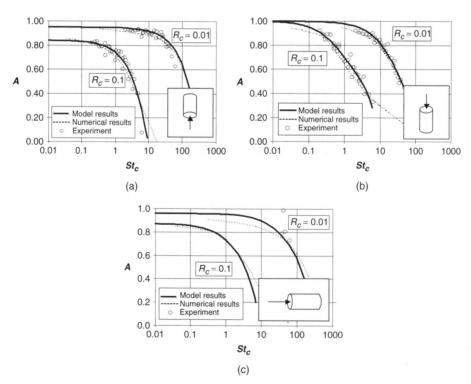

Figure 6.6   *Aspiration efficiency (A) for a thin-walled cylindrical sampling tube in calm air – orientations as shown – as a function of Stokes number (St$_c$) for various velocity ratios (R$_c$), experimental and numerical results as indicated (Su and Vincent, 2004a) and model calculations given by the solid lines (Su and Vincent, 2004b): (a) facing downwards; (b) facing upwards; (c) facing horizontally. Adapted from Journal of Aerosol Science, 35, Su and Vincent, 1119–1134. Copyright (2004), with permission from Elsevier*

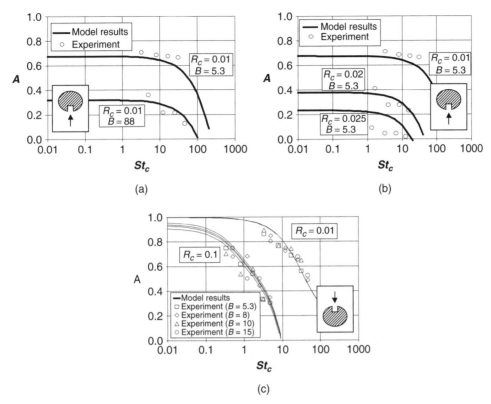

Figure 6.7   *Aspiration efficiency (A) for a spherical blunt sampler in calm air – orientations as shown – as a function of Stokes number (St_c) for various velocity ratios (R_c) and various bluntness values (B), experimental results as indicated and model calculations given by the solid lines (Su and Vincent, 2004b): (a) facing downwards, fixed R_c and different B; (b) facing downwards, fixed B and different R_c; (c) facing upwards. Adapted from Journal of Aerosol Science, 35, Su and Vincent, 1119–1134. Copyright (2004), with permission from Elsevier*

analytical models based in the first instance on mathematical descriptions of the airflow in the vicinity of the sampler. The earliest was Levin (1957) who, in addition to the moving air case referred to earlier in Chapter 5, also developed a mathematical model for the aspiration efficiency of a point sink in calm air, again based on determination of the airflow field from potential flow theory. The patterns of the resultant airflow streamlines and particle trajectories, respectively, are shown in Figure 6.8, providing a picture that is seen to be an idealised version of the more general case already shown in Figure 6.1. In order to normalise the equations of particle motion, Levin proposed the dimensionless *inertial parameter*:

$$k_L = \frac{\tau v_s}{\ell_s} \tag{6.17}$$

which is directly analogous to Equation (5.31) for his moving air scenario. In this expression, v_s replaces the wind speed U and ℓ_s is now the distance below the point sink of the *saddle point* that divides the aspirated from the nonaspirated airflow. For the latter, it identifies where the upwards suction velocity induced by the aspirating action of the sink just balances the particle's downwards falling speed, so

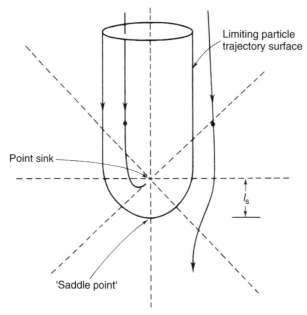

Figure 6.8   *General pattern of streamlines and particle trajectories near a point sink in calm air for particles with significant gravitational settling velocity, forming the basis of the theoretical model of Levin (1957). Reproduced with permission from Vincent, Aerosol Sampling: Science and Practice. Copyright (1989) John Wiley & Sons, Ltd*

that:

$$v_s = \frac{Q}{4\pi \ell_s^2} \qquad (6.18)$$

Equation (6.17) now becomes:

$$k_L = \left(\frac{4\pi \tau^2 v_s^2}{Q} \right)^{1/3} \qquad (6.19)$$

which, with the earlier equations for τ and v_s in Chapter 2, leads to:

$$k_L \propto d_{ae}^5 Q^{-1/2} \qquad (6.20)$$

For calm air sampling, Levin derived an expression for aspiration efficiency of the form:

$$A = 1 - 0.8k_L + 0.08k_L^2 \text{ for } k_L \ll 1 \qquad (6.21)$$

which is directly analogous to his expression in Equation (5.32) for moving air, simply replacing k_L with the version relevant to the calm air scenario. Inspection of Equation (6.21) shows that, for the applicable range of k_L, A decreases steadily as k_L increases, broadly consistent with the trend shown in the experimental results of Yoshida *et al.* It reveals that $A > 0.90$ for $k_L < 0.12$, a condition that has become known as Levin's 'dynamical limit'. While Levin's approach is useful in providing a preliminary quantitative physical picture of the factors influencing the performance of a sampler in calm air, it is limited. In particular, it was based on an infinitesimally small sampling orifice in an infinitesimally small sampler body. This may be accounted for by reorganising k_L in Equation (6.19) such that Equation (6.21) may be rewritten as:

$$A = 1 - 3.2 St_c R_c^{3/2} + 1.28 (St_c R_c^{3/2})^2 \tag{6.22}$$

and this now *does* incorporate the finite size of the sampling orifice. However, there is still no consideration of the possibility of particle impaction onto the solid external surface of the sampler.

In important yet unpublished work, Kaslow and Emrich (1973, 1974) carried out an extensive study to investigate a cylindrical tube sampling in calm air for orientations (θ) of 0, 90 and 180°. Their approach was built on that of Levin, but included recognition of the influence of finite size of sampling orifice and sampler body. The starting point was to divide the flow field for the air movement near the sampler into two regions: the first (outer) region at a distance sufficiently far away from the sampler that the point sink potential flow pattern was a good approximation, and the second (inner) region close to the sampler where the presence and finite dimensions of the sampler substantially modified the flow pattern. The flow pattern in the outer region was calculated theoretically using Levin's idealised, point sink, potential flow model. In the inner region, it was determined experimentally by photographing the illuminated trajectories of fine oil droplets in a way that enabled information to be obtained simultaneously about the shape of the streamsurface pattern and the magnitudes of the local air velocities. Particle trajectories in the outer flow region were calculated using Levin's method. In the inner region, numerical solutions of the equations of motion for particles in the measured flow field were performed. Thus limiting particle trajectory surfaces were determined for the orientations already referred to and for ranges of particle aerodynamic size and sampling flow rate. Extending these surfaces upwards far above the probe, the enclosed cross-sectional areas (a'') were identified for given sets of particle and sampling conditions, and were as shown above in Figures 6.3 and 6.4.

Kaslow and Emrich identified a useful pair of modified dimensionless groups, firstly Levin's inertial parameter k_L and secondly an *orifice size coefficient*:

$$m_Q = \delta \left(\frac{\pi g^{1/2}}{\sqrt{2} Q} \right)^{2/5} \tag{6.23}$$

which is seen to be independent of particle size. The parameters k_L and m_Q may be related to more familiar quantities, thus:

$$St_c = \frac{\tau U_s}{\delta} \longrightarrow 2 m_Q^{-3} k_L^{2/5} \tag{6.24}$$

and

$$R_c = \frac{v_s}{U_s} \longrightarrow \frac{1}{4} m_Q^2 k_L^{2/5} \tag{6.25}$$

Kaslow and Emrich carried out numerical calculations of aspiration efficiency using the flow fields they had developed for chosen values of m_Q and k_L and for the three sampler orientations already mentioned. The results were mapped in the $\{m_Q, k_L\}$ plane (see Figure 6.9), and these show that A was reasonably close to unity for all orientations provided that k_L did not exceed about 0.01 and m_Q was less than about 2. On closer inspection, these results are seen to be consistent with Davies' relationship in Equation (6.13).

Yoshida *et al.* (1978) carried out theoretical calculations of A in support of their own experimental studies described above. Two sets were performed, one based on solutions of the Navier–Stokes equations for viscous flow and the other based on the point sink potential flow model. The results

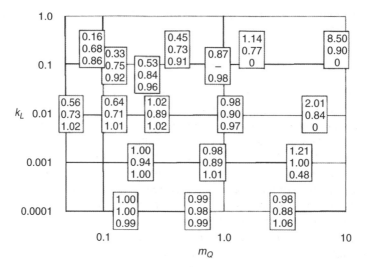

Figure 6.9 *Theoretical results of Kaslow and Emrich (1974) for the aspiration efficiency (A) of an aspirating, thick-walled cylindrical tube in calm air, for upwards, horizontal and downwards facing, respectively. Reproduced with permission from Vincent, Aerosol Sampling: Science and Practice. Copyright (1989) John Wiley & Sons, Ltd*

of these are plotted alongside their experimental data in Figure 6.2. Agreement between experiment and theory was very good for the viscous flow model, but less good for the point sink potential flow model, which progressively underestimated A as R_c decreased. This was explained in terms of the increasing role of impaction losses near the sampling orifice as sampling velocity effects became more prominent, in the same way that shortcomings in Levin's original theory were explained earlier. Later, Su and Vincent (2004a) also reported numerical simulations, based on solutions of the Navier–Stokes equations using a commercial package of the type which is now widely available (Fluent Inc., Lebanon, NH, USA). The results of these simulations are plotted alongside their experimental data in Figures 6.6 and 6.7, and agreement is seen to have been generally good. Indeed, the numerical results included ranges of conditions where it was not possible to obtain experimental data, and these may be regarded as complementary to – even extending – the experimental data.

Agarwal and Liu (1980) also investigated the problem for a thin-walled tube facing vertically upwards. In contrast to the method employed by Kaslow and Emrich, they based their calculations for particle trajectories on flow fields around the sampler obtained entirely from numerical solutions of the Navier–Stokes equations. They chose to present their results for aspiration efficiency (A) in terms of the $\{St_c, R_c\}$ combination of dimensionless parameters. Typical results are plotted in Figure 6.10(a) in the form of A versus St_c for various R_c. It is seen that, for the range of conditions indicated, A tended to decrease as both St_c and R_c increased. However, what is not revealed in this form of presentation is the fact that, at very low St_c consistent with near-zero sampling flow rates, particles may enter the sampling orifice directly by settling without help from the aspiration process. In the limit (Walton, 1954):

$$A \longrightarrow 1 + R_c \text{ as } St_c \longrightarrow 0 \qquad (6.26)$$

so that $A > 1$ is possible. This effect is illustrated in Figure 6.10(b) where the results of Agarwal and Liu are plotted in the alternative form of A versus R_c for various St_c.

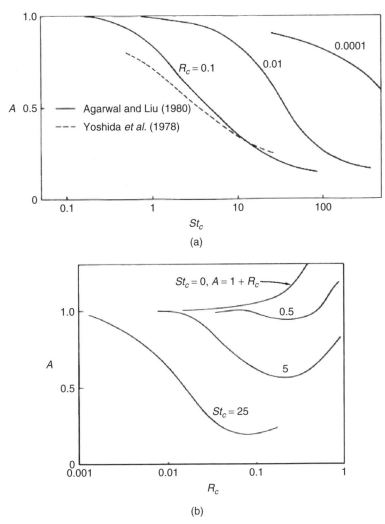

Figure 6.10 *Theoretical results of Agarwal and Liu (1980) for the aspiration efficiency (A) of an upwards-facing, aspirating, thin-walled tube in calm air, shown as (a) a function of Stokes number (St_c) for various velocity ratios (R_c) and (b) a function of R_c for various St_c. Also shown are some corresponding theoretical results from Yoshida et al. (1978). Reproduced with permission from Agarwal and Liu, American Industrial Hygiene Association Journal, 41, 191–197. Copyright (1980) American Industrial Hygiene Association*

More recently, Dunnett and Wen (2002) carried out a mathematical study of a simple thin-walled cylindrical sampler facing both vertically upwards and downwards. Two approaches were adopted. In one of them, the inviscid potential flow equations were solved numerically using the linear boundary integral equation (LBIE) method that Dunnett (1990) had described previously. In the other, a laminar flow model that took viscosity into account was applied in which the Navier–Stokes equations were solved using the control volume method together with a solution algorithm described by Patankar (1980). Dunnett and Wen showed that, when viscous effects were included, local air velocities close to the sampler inlet were slightly lower than for the inviscid flow case. In turn, the calculated A-values were somewhat lower. However, for most of the conditions examined they concluded that the potential flow model was quite accurate. Aspiration efficiency calculated in this way compared favorably with the experimental data of Yoshida *et al.* (1978) and Su and Vincent (2004a), respectively. Zaripov *et al.* (2000) have also made numerical simulations of particle transport near a tube facing vertically upwards, and reported similarly fair agreement. Later, Galeev and Zaripov (2003) studied the case of a spherical sampler with a single circular orifice for a range of orientations of the orifice with respect to the vertical. They reported good agreement with the results of Su and Vincent (2004a) for this sampling system with the orifice facing vertically upwards. This work was distinctive from the other mathematical studies cited by the way in which the finite size of the sampling orifice was represented. Here, it was approximated by means of an ensemble of individual point sinks so that the air velocity field for each could be expressed in analytical form and then they could all be superimposed to simulate an effective finite-sized sampling orifice.

6.2.4   Descriptive modeling of aspiration efficiency

The agency that brings the particles into the vicinity of the sampler inlet in calm air sampling is particle settling under the influence of gravity. Therefore, unlike the situation for moving air, where only St in Equation (5.41) is dependent on particle size, both St_c and R_c in the general functional expression in Equation (6.14) are dependent on d_{ae}. This means that St_c and R_c are inexorably coupled. It follows that there is no basis for developing a descriptive model for the aspiration efficiency of aerosol samplers in calm air along the lines of the impaction models developed for moving air situations. So other approaches are necessary. In the search for a descriptive model for aerosol sampling in calm air by means of thin-walled probes, Grinshpun *et al.* (1993) sought an equation for aspiration efficiency that was empirical but which embodied the physical parameters representing the effects of inertia and gravity as represented by St_c and R_c, as well as orientation θ. By reference to the experimental data then available, they arrived at:

$$A = \exp\left(-\frac{4St_c^{R_c^{1/2}+1}}{1 + 2St_c}\right) + R_c \cos\theta \text{ for } 0 \le \theta \le 90° \tag{6.27}$$

No model was offered for larger orientations since, as Grinshpun *et al.* noted, '. . . it is not commonly (used) in aerosol sampling from calm atmospheric environments'.

Using their own large new data set, Su and Vincent (2004b) attempted to establish a physically meaningful, more general semi-empirical description. The point sink model of Levin formed the starting point, and a new set of semi-empirical models in the form broadly expressed as:

$$A = A_{Levin} - IMP - BLOCK \tag{6.28}$$

where A_{Levin} is the ideal aspiration efficiency as expressed by Levin (1957) in the earlier Equation (6.21) or (6.22). This new expression acknowledged that departures from the ideal situation reflected in Levin's model were the result of combinations of inertial, 'impaction-like' effects (contained in IMP) and blockage, 'interception-like' effects that were mainly geometric in nature ($BLOCK$). Modeling began with an assessment of each of the scenarios described in the experimental data set (see Figures 6.6 and 6.7). For the thin-walled tube facing vertically downwards, particles do not impact onto the wall of the sampler for any value of St_c. So there is no inertial component, and:

$$A_{tube-down} = A_{Levin} - BLOCK_{tube-down} \tag{6.29}$$

where $BLOCK_{tube-down}$ can be a function only of R_c. The empirical expression:

$$BLOCK_{tube-down} = 0.5R_c^{1/2} \tag{6.30}$$

was found to provide a good fit. Taken together with Equation (6.22) this leads to:

$$A_{tube-down} = 1 - 3.2St_c R_c^{3/2} + 1.28(St_c R_c^{3/2})^2 - 0.5R_c^{1/2} \tag{6.31}$$

For the thin-walled tube facing vertically upwards there is no aspiration efficiency loss due to blockage, so that impaction onto the wall of the tube becomes the predominant loss mechanism, leading to:

$$A_{tube-up} = A_{point} - IMP_{tube-up} \tag{6.32}$$

Now it was expected that $IMP_{tube-up}$ would be a function of both of the coupled parameters St_c and R_c, as exhibited by the experimental data for this scenario. By trial and error, the expression:

$$IMP_{tube-up} = 0.12R_c^{-0.4}(e^{-p} - e^{-q}) \tag{6.33}$$

provided a good fit to the data when $p = 2.2R_c^{1.3}St_c$ and $q = 75R_c^{1.7}St_c$. With Equation (6.22), the above leads to:

$$A_{tube-up} = 1 - 3.2St_c R_c^{3/2} + 1.28(St_c R_c^{3/2})^2 - 0.12R_c^{-0.4}(e^{-p} - e^{-q}) \tag{6.34}$$

For the thin-walled tube sampling horizontally, aspiration efficiency was modified by the effects of both blockage and impaction. Aspiration efficiency was therefore given by

$$A_{tube-horiz} = A_{point} - BLOCK_{tube-horiz} - IMP_{tube-horiz} \tag{6.35}$$

in which the additional loss terms on the right-hand side were based in the first instance on the components for downwards and upwards sampling, respectively. It was assumed that

$$BLOCK_{tube-horiz} = h_1 BLOCK_{tube-down} \tag{6.36}$$

and

$$IMP_{tube-horiz} = h_2 IMP_{tube-up} \tag{6.37}$$

since the blockage effect applies to generally downwards-facing scenarios and the impaction effect applies to generally upwards-facing scenarios. In these last two equations, h_1 and h_2 are coefficients that reflect the balance between the two types of contribution, where $h_1 + h_2 = 1$. In general, this balance will depend on the orientation of the tube. In the limit of downwards sampling, $h_1 = 1$ and $h_2 = 0$, while for upwards sampling, $h_1 = 0$ and $h_2 = 1$. More generally, the balance between blockage and impaction derives from qualitative considerations of the envelopes of limiting particle trajectories defining particles 'just blocked' and 'just impacted', respectively. Combining Equations (6.35)–(6.37) with Equation (6.22) leads to:

$$A_{tube-horiz} = 1 - 3.2St_c R_c^{3/2} + 1.28(St_c R_c^{3/2})^2 - 0.4R_c^{1/2} - 0.024R_c^{-0.4}(e^{-p} - e^{-q}) \tag{6.38}$$

where Su and Vincent found that the coefficients $h_1 = 0.8$ and $h_2 = 0.2$ provided a good fit with the experimental results.

Moving on to the blunt sampler facing vertically downwards, they proposed an additional term to the expression that had been developed for the downwards-facing tube. Earlier they had shown (Su and Vincent, 2003) that for this scenario there was a decrease in aspiration efficiency due to the effect of blunt sampler geometry equivalent to $R_c(B^2 - 1)$. Here, therefore, it was proposed that:

$$A_{blunt-down} = A_{tube-down} - R_c(B^2 - 1) \tag{6.39}$$

so that, with Equation (6.31), aspiration efficiency is given by:

$$A_{blunt-down} = 1 - 3.2St_c R_c^{3/2} + 1.28(St_c R_c^{3/2})^2 - 0.5R_c^{1/2} - R_c(B^2 - 1) \tag{6.40}$$

which reverts back to the equation for the thin-walled probe under the same conditions when $B = 1$.

In the last case studied, for the blunt sampler facing vertically upwards, the observed experimental trends in aspiration efficiency [see Figure 6.7(c)] were noted as being very similar to those for the thin-walled tube, both in trend and magnitude. This was not surprising since the blunt body of the sampler body does not substantially block particles in the way that it does for the blunt sampler facing downwards. The model for the thin-walled tube shown in Equation (6.34) therefore provided a good starting point, and it was assumed that the small differences ascribable to the sampler bluntness may be accommodated by the inclusion of a term that shows a weak functional dependency on B and R_c. This suggested:

$$A_{blunt-up} = 1 - 3.2St_c R_c^{3/2} + 1.28(St_c R_c^{3/2})^2 - 0.12R_c^{-0.4}(e^{-p} - e^{-q}) - R_c^{3/2}(B^{1/2} - 1) \tag{6.41}$$

which provided a good description of what was observed experimentally. In this expression, the final term on the right-hand side meets the stated requirements, and also reverts back to the result for the idealised thin-walled tube case when $B = 1$.

The solid lines in the graphs in the earlier Figures 6.6 and 6.7 represent the values of A calculated from Equations (6.31), (6.34), (6.38), (6.40) and (6.41), respectively, and agreement is seen to be generally quite good across the whole range of experimental conditions studied. Such broad agreement suggested the possibility that a general single expression might be found that satisfies the five separate equations for the individual cases actually studied. To achieve that aim, Su and Vincent proposed:

$$A = 1 - 3.2St_c R_c^{3/2} + 1.28(St_c R_c^{3/2})^2 - {}_\theta h_1[0.5R_c^{1/2} - R_c(B^2 - 1)]$$
$$- {}_\theta h_2\{[0.12R_c^{-0.4}(e^{-p} - e^{-q})] - R_c^{3/2}(B^{1/2} - 1)\} \tag{6.42}$$

where, now, the coefficients h_1 and h_2 are expressed more generally for each angle θ with respect to the upwards-facing vertical, and they remain unknown except for the specific horizontal sampling case. From inspection of Equation (6.42), it is seen that:

- Thin-walled probe facing downwards: ${}_{180}h_1 = 1$, ${}_{180}h_2 = 0$, $B = 1$ so that $A_{calm-air} \rightarrow A_{tube-down}$.
- Thin-walled probe facing upwards: ${}_0h_1 = 0$, ${}_0h_2 = 1$, $B = 1$ so that $A_{calm-air} \rightarrow A_{tube-up}$.
- Horizontal thin-walled probe: ${}_{90}h_1 = 0.8$, ${}_{90}h_2 = 0.2$, $B = 1$ so that $A_{calm-air} \rightarrow A_{tube-horiz}$.
- Blunt probe facing downwards: ${}_{180}h_1 = 1$, ${}_{180}h_2 = 0$ so that $A_{calm-air} \rightarrow A_{blunt-down}$.
- Blunt probe facing upwards: ${}_0h_1 = 0$, ${}_0h_2 = 1$ so that $A_{calm-air} \rightarrow A_{blunt-up}$.

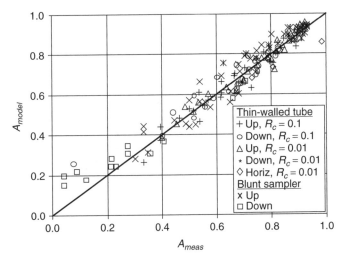

Figure 6.11 *Comparison between measured and modeled aspiration efficiency (A_{meas} and A_{model}) for various thin-walled and spherical blunt samplers in calm air with entry orifices placed at various orientations with respect to the vertical, experimental data from Su and Vincent (2003, 2004a) and calculated values based on the general model of Su and Vincent (2004b). Reprinted from Journal of Aerosol Science, 35, Su and Vincent, 1119–1134. Copyright (2004), with permission from Elsevier*

That is, the general Equation (6.42) conveniently reverts to the appropriate individual model for each of the scenarios studied. The full set of experimental data contained in Figures 6.6 and 6.7 are plotted alongside corresponding results obtained using this general model in Figure 6.11. The experimental and model values are seen to be in very good agreement ($R^2 = 0.95$) over the ranges of conditions studied. Although the model was developed on the basis of data for what is (still) a relatively narrow range of idealised scenarios, there is significant encouragement that a workable general model applicable to aerosol samplers like those used in practical situations might eventually be achievable.

6.2.5 Criteria for 'representative sampling' in calm air

The practical question of how to define the conditions for 'representative' sampling has been a major motivation for much of the work that has been done in relation to the collection of aerosols in calm air, especially in the earlier years. Davies (1968) extended what is contained in Equation (6.13) to suggest that $A \approx 1$ could be reasonably be achieved for practical purposes by placing factors of 5 in the inequalities. Thus:

$$5\left(\frac{Q\tau}{4\pi}\right)^{1/3} < \frac{\delta}{2} < 5\left(\frac{Q}{\pi g \tau}\right)^{1/2} \tag{6.43}$$

for the aspiration of particles of relaxation time τ at flow rate Q into a sampler with orifice diameter δ. Davies estimated that sampling for conditions falling within these bounds will give a departure in A from unity of no greater than about ± 0.05. From the earlier definitions of St_c and R_c, and noting the definition of τ in Chapter 2, these bounds may be expressed as:

$$St_c \leq 0.016 \text{ and } R_c \leq 0.04 \tag{6.44}$$

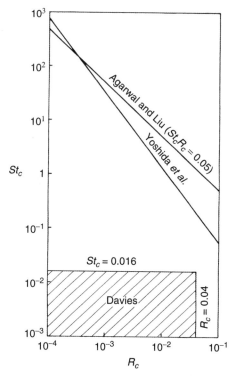

Figure 6.12   *Various criteria for 'representative' sampling in calm air (at the 95 % level of acceptance), plotted in the $\{St_c, R_c\}$ plane, from Davies (1968), Yoshida et al. (1978) and Agarwal and Liu (1980). Reproduced with permission from Vincent, Aerosol Sampling: Science and Practice. Copyright (1989) John Wiley & Sons, Ltd*

respectively. This defines the 'Davies criterion' for representative sampling. It is portrayed graphically in Figure 6.12. From such a diagram it is possible, in principle, to choose the sampling flow and sampling orifice diameter to yield $A \approx 1$ over whatever range of particle sizes is specified. Indeed, it has been quite widely applied in this way to relate to practical sampling situations.

Agarwal and Liu (1980) also sought to define 'acceptable sampling' in calm air. From their calculations of aspiration efficiency for the upwards-facing tube, as summarised earlier in Figure 6.10, they examined empirically the relationship $A = f(St_c, R_c)$. From their results, as summarised in this form in Figure 6.13, it is seen that:

$$A > 0.95 \text{ for } St_c R_c < 0.05 \tag{6.45}$$

This defines what has now become known as the 'Agarwal–Liu criterion' for calm air sampling, and this too is portrayed – alongside that of Davies – in Figure 6.12. Also shown in Figure 6.12 is the equivalent boundary determined by Yoshida *et al.* (1978) for upwards-facing sampling tubes, based on particle trajectories calculated in flow fields derived from solutions of the Navier–Stokes equations. It is seen from Figure 6.12 that the 'Yoshida criterion' is less restrictive than the Davies criterion, but not quite so relaxed as the Agarwal–Liu criterion.

Ter Kuile (1979) set out to determine the best combination of parameters for describing the aspiration efficiency of a sampler in calm air. He noted that the choice of the $\{St_c, R_c\}$ combination is unsatisfactory because both parameters depend on both sampler and particle properties. The practical problem is that it is

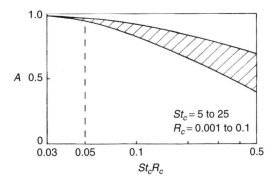

Figure 6.13 *Summary of the theoretical results of Agarwal and Liu for the aspiration efficiency (A) of an upwards-facing, aspirating, thin-walled tube, plotted as a function of $St_c R_c$, where the hatched area represents the range of data points over the ranges of St_c and R_c indicated. Reproduced with permission from Vincent, Aerosol Sampling: Science and Practice. Copyright (1989) John Wiley & Sons, Ltd*

not possible to characterise separately either the aerosol or the sampler using this form of presentation. Instead, therefore, Ter Kuile was drawn to the alternative combination of quantities very similar to that proposed by Kaslow and Emrich: namely, the tube size parameter m_Q exactly as described by Equation (6.23) and a particle size parameter k_Q given by:

$$k_Q = k_L^{1/5} \qquad (6.46)$$

where k_L is the Levin parameter as defined earlier. This preference was justified by the fact that, whereas m_Q depends solely on sampler properties, k_Q depends strongly on particle properties with only a very weak dependence on sampler properties, in the form of a one-tenth power of the sampling volumetric flow rate. Now, therefore, by plotting the conditions required for representative sampling in the $\{m_Q, k_Q\}$ plane, the particle size range of representatively sampled aerosols can be easily found for any sampling instrument. The form of the $\{m_Q, k_Q\}$ presentation is shown in Figure 6.14(a) in which it was suggested that, for representative sampling, $A > 0.95$. On this diagram are indicated the lines of constant St_c and constant R_c, for each of which – on the log–log form of presentation shown – there is a family of parallel straight lines of positive and negative slope, respectively. Also shown are the enclosed regions for the Davies, Agarwal–Liu, and Yoshida criteria. In addition, the 'dynamical limit' of Levin (1957), introduced earlier in association with Equation (6.21) is shown. Again it is shown that the Davies criterion is the most restrictive, while the Agarwal–Liu criterion is the least restrictive, particularly at large values of m_Q and k_Q. Figure 6.14(b) shows the corresponding plot for the more relaxed $A > 0.90$ level of acceptability.

6.3 Slowly moving air

Most of the preceding has been discussing the situation where the air is calm. Indeed, for some of the experiments described, the investigators went to great lengths to ensure that the air in the test chamber was *perfectly* calm – that is, with the exception of air motions induced by the aspiration action of the sampler itself. This is of course a highly idealistic scenario, one that is rarely encountered in the real world of occupational and environmental hygiene. Berry and Froude (1989) and Baldwin and

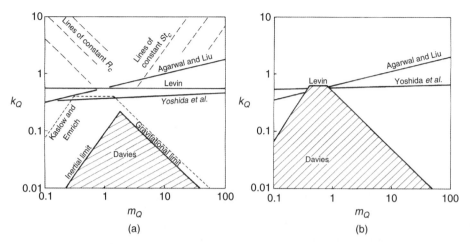

Figure 6.14   *Various criteria for 'representative' aerosol sampling in calm air, plotted in the {m_Q, k_Q} plane as suggested by Ter Kuile (1979): (a) at the 95 % level of acceptance; (b) at the more relaxed 90 % level of acceptance (from Levin, 1957; Davies, 1968; Kaslow and Emrich, 1974; Yoshida et al., 1978; Agarwal and Liu, 1980). Reproduced with permission from Vincent, Aerosol Sampling: Science and Practice. Copyright (1989) John Wiley & Sons, Ltd*

Maynard (1998) conducted surveys of wind speeds in a range of industrial workplaces, including the relative motions between workers and the surrounding air as they went about their tasks, and found as expected that the measured wind speeds were not zero. But neither were the air motions as rapid as those pertaining to most of the aerosol sampler research carried out in moving air (see Chapter 5). In fact it was found that air velocities were surprisingly low, rarely exceeding 0.2–0.3 m s^{-1} and more typically less than 0.1 m s^{-1}. With this in mind, it is an important question to ask: when is calm air calm? In other words, under what conditions may what has been learned from the research described in the preceding sections be relevant to practical situations?

6.3.1   Definition of calm air

Because the effects of gravity hardly feature in the theory of samplers in moving freestreams, yet gravity is so important in the calm air case, then one should not be regarded as a limiting case of the other. Nonetheless, Davies (1977) did attempt to extrapolate the model of Belyaev and Levin (1974), using their equations to arrive at:

$$A = \frac{1}{1 + 0.62 St_c} \text{ for } R = \frac{U}{U_s} \longrightarrow 0 \tag{6.47}$$

where it was implicitly assumed that aspiration efficiency was still governed by inertial effects. However, as pointed out by Gibson and Ogden (1977), this expression is unreliable because the experimental data upon which were Belyaev and Levin based their semi-empirical model extended only down to $R = 0.2$.

Ogden (1983) examined the extent to which calm-air ideas like those outlined above may be relevant to practical situations. The approach taken was to assume a model for aspiration efficiency in moving air and to determine the wind speed that, for a given sampling flow rate, causes A to fall to 0.90. For the purpose of this calculation, the simple moving-air model of Levin (1957) was considered to be sufficiently accurate. Reiterating Equation (5.31):

$$A = 1 - 3.2St R^{3/2} + 1.28(St R^{3/2})^2 \tag{6.48}$$

with St and R expressed as for the moving air case. By assuming low wind speed to the extent that the second term may be neglected, Ogden obtained:

$$\left(\frac{U}{U_s}\right)^{3/2} = \frac{1 - A}{3.2St} \tag{6.49}$$

leading to the condition:

$$U < \frac{0.1U_s}{St^{2/3}} \text{ in order to achieve } A > 0.90 \tag{6.50}$$

This now defines the wind speed below which the air may reasonably be considered to be 'calm' for the purpose of aerosol sampling considerations. Ogden pointed out that, in considering the question of representative sampling, this condition must first be met before applying any of the criteria referred to in the previous section. If it is not met at the outset, aspiration efficiency must be considered from the point of view of moving air as described in earlier chapters.

The boundary conditions for 'calm' air under the Ogden criterion may be plotted in terms of wind speed (U), particle aerodynamic diameter (d_{ae}), mean sampler entry velocity (U_s) and sampler orifice diameter (δ). Typical results are shown in Figure 6.15 for samplers with entry velocity $U_s = 1.2$ m s^{-1}, where it is seen that the limiting wind speed is highest for the smaller particles and decreases as particle

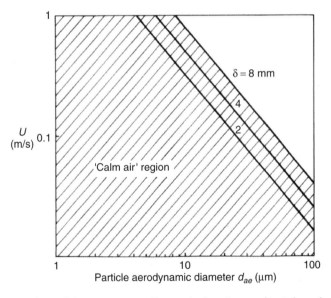

Figure 6.15 *Identification of conditions corresponding to 'calm air sampling', based on the theory of Ogden (1983), for samplers with entry diameter (δ) from 2 to 8 mm and having entry velocity (U_s) of 1.25 m s^{-1}. Reproduced with permission from Vincent, Aerosol Sampling: Science and Practice. Copyright (1989) John Wiley & Sons, Ltd*

size increases. It is interesting to note that doubling the sampling flow rate only increases the limiting wind speed by about 25 %.

6.3.2   Intermediate conditions

Physically, intermediate conditions may be defined as those where the external air is in motion but where the effect of gravity significantly modifies the trajectories of particles in the freestream. Grinshpun *et al.* (1993)[1] studied this important transition regime, beginning by comparing calculated aspiration efficiencies using their model – described in the set of Equations (5.29) in Chapter 5 – with a number of data sets for thin-walled sampling tubes facing the wind that spanned conditions all the way from moving air down to calm air. Figure 6.16 summarises the data comparisons for moving wind speeds down to values as low as about 5 cm s^{-1}. It shows clearly that the model accurately predicted the results when the wind speed (U) was much greater than the particle settling velocity (v_s), but that the agreement fell apart as U approached v_s and eventually went to zero. Grinshpun *et al.* therefore sought a version of their model that extended into this regime. They had already developed separate equations for the moving air and calm air cases, respectively, as contained in Equations (5.29) for the moving air case and Equation (6.27) earlier in this chapter for the calm air case. To accommodate the intermediate case, the transition between the extremes was accomplished by means of a pair of weighted functions that were empirical but – like their model equations – embodied physically sensible ideas and appropriate physical variables. In this approach they distinguished between two distinct orientations, one with respect to the upwards vertical for calm air sampling (referred to here as θ_1) and one the sampler makes with the moving freestream (referred to here as θ_2), both of which needed to be incorporated into the model as constructed. This therefore allowed for the possibility that the moving freestream might

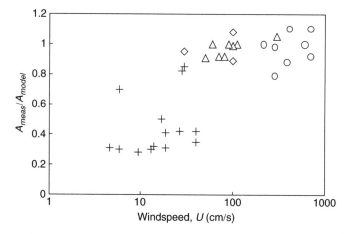

Figure 6.16   *Comparison between measured and modeled aspiration efficiency for a thin-walled sampling tube facing the wind as a function of wind speed (U) down to very low values, as summarised by Grinshpun et al. (1993). The individual sets of experimental data were taken from: ○ Belyaev and Levin (1974) and Belyaev et al. (1981); △ Lipatov et al. (1986, 1988) and Grinshpun et al. (1990); ◇ Hangal and Willeke (1990); + Grinshpun et al. (1989, 1990). Modeled results were based on the theory of Belyaev and Levin (1974). Adapted from Atmospheric Environment, 27A, Grinshpun et al., 1459–1470. Copyright (1993), with permission from Elsevier*

---

[1] Also note the important Corrigendum by Grinshpun *et al.* 1994.

not always be horizontal, as is the usual assumption. The resultant model took the form:

$$A = f_{moving} A_{moving, \theta = \theta_1} (1 + \Delta)^{1/2} + f_{calm} A_{calm, \theta = \theta_2} \qquad (6.51)$$

$$\uparrow \qquad\qquad\qquad \uparrow$$

Equations (5.28)   Equation (6.27)

for both θ_1 and θ_2 in the range from 0 to 90°. In this expression:

$$f_{moving} = \exp\left(-\frac{v_s}{U}\right) \text{ and } f_{calm} = 1 - \exp\left(-\frac{v_s}{U}\right) \qquad (6.52)$$

and Δ is a *particle velocity correction function* defined by Grinshpun *et al.* to link θ_1 and θ_2, given by:

$$\Delta = \frac{v_s}{U}\left(\frac{v_s}{U} + 2\cos(\theta_1 \pm \theta_2)\right) \qquad (6.53)$$

Calculations of A using this general model are shown in Figure 6.17 in comparison to some of the available experimental data – including calm air data from Belyaev and Kustov (1980) and Grinshpun *et al.* (1989, 1990), moving air data from Belyaev and Levin (1974), Lipatov *et al.* (1986) and Hangal and Willeke (1990), and slow-moving air data from Grinshpun *et al.* (1985, 1990). For the ranges of conditions indicated, agreement is seen to be very good.

The low wind speed regime between the most widely researched moving air scenario and the less-studied calm air one is very important for many practical indoor aerosol sampling situations. But it has received relatively little attention. The quest for appropriate experimental data is difficult because of the

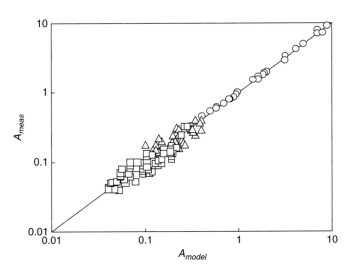

Figure 6.17 *Comparison between measured and modeled aspiration efficiency (A_{meas} and A_{model}) for thin-walled cylindrical samplers in calm air and moving air, with modeled results obtained using the general theory described by Grinshpun et al. (1993) and experimental data from: ○ Belyaev and Levin (1974), Lipatov et al. (1986) and Hangal and Willeke (1990) for moving air; △ Grinshpun et al. (1985, 1990) for slowly moving air; □ Belyaev and Kustov (1980) and Grinshpun et al. (1989, 1990) for calm air. Reprinted from Atmospheric Environment, 27A, Grinshpun et al., 1459–1470. Copyright (1993), with permission from Elsevier*

technical problems associated with achieving uniform, controlled experimental conditions. A number of studies have been reported that point the way to future progress. Roger *et al.* (1999) carried out studies of an annular slot sampler in a vertical wind tunnel where the wind speed in the freestream was only about 3 cm s^{-1}, for test aerosols with d_{ae} from about 10 to 50 μm. Experiments were carried out for different slot widths (h) equal to 1.1 and 2.5 mm and sampler overall diameter (D) of 30 and 50 mm. The results for aspiration efficiency (A) were plotted as a function of the dimensionless combination $(D/h)(v_s/U_s)$. Best fit with the experimental data was achieved using the empirical expression:

$$A = \exp\left[-0.38\left(\frac{D}{h}\right)^{0.61}\left(\frac{v_s}{U_s}\right)^{0.73}\right] \tag{6.54}$$

As shown in Figure 6.18, the association between A and $(D/h)(v_s/U_s)$ was clearly established. However, it is notable that the wind speed itself does not feature at all in Equation (6.54). Although the air was said to be moving, the actual wind speed was so low that the situation corresponded more closely to pure calm air according to the Ogden criterion. The same would likely have been true for other reports that have appeared that relate to aerosol sampling in the low wind speed regime (e.g. Maynard *et al.*, 1997; Aitken *et al.*, 1999; Kenny *et al.*, 1999; Witschger *et al.*, 2004).

At the time of writing, new research is under way to study both human inhalability and aerosol sampler performance for specific low wind speeds, some of it – for example – using new types of aerosol facilities like the ultralow-speed wind tunnel shown earlier in Figure 3.11. Such research is likely to produce a body of experimental data that will allow further development and extension of models like the one presented by Grinshpun *et al.* (1993).

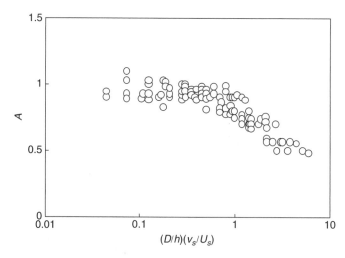

Figure 6.18   *Aspiration efficiency (A) for an annular slot sampler in air moving very slowly at about 3 cm s^{-1}, plotted as a function of (D/h)(v$_s$/U$_s$) as suggested by Equation (6.54), where D is the sampler body diameter and h the slot width (Roger et al., 1999). Reprinted from Journal of Aerosol Science, 30, Roger et al., S151–S152. Copyright (1999), with permission from Elsevier*

References

Agarwal, J.K. and Liu, B.Y.H. (1980) A criterion for accurate aerosol sampling in calm air, *American Industrial Hygiene Association Journal*, 41, 191–197.

Aitken, R.J., Baldwin, P.E.J., Beaumont, G.C., Kenny, L.C. and Maynard, A.D. (1999) Aerosol inhalability in low air movement environments, *Journal of Aerosol Science*, 30, 613–626.

Baldwin, P.E.J. and Maynard, A.D. (1998) A survey of wind speeds in indoor workplaces, *Annals of Occupational Hygiene*, 20, 303–313.

Belyaev, S.P. and Kustov, V.T. (1980) Sampling from calm air, *Trudy IEM*, 25, 102–108 (in Russian).

Belyaev, S.P. and Levin, L.M. (1974) Techniques for collection of representative aerosol samples, *Journal of Aerosol Science*, 5, 325–338.

Belyaev, S.P., Nikiforova, N.K., Smirnov, V.V. and Schelchkov, G.I. (1981) *Optics and Electronic Methods of Aerosol Investigation*, Energoizdat, Moscow (in Russian).

Berry, R.D. and Froude, S. (1989) An investigation of wind conditions in the workplace to assess their effect on the quantity of dust inhaled, Internal Report IR/L/DS/89/3, Health and Safety Executive, London.

Breslin, J.A. and Stein, R.L. (1975) Efficiency of dust sampling inlets in air, *American Industrial Hygiene Association Journal*, 36, 576–583.

Davies, C.N. (1968) The entry of aerosols into sampling tubes and heads, *British Journal of Applied Physics*, 25, 921–932.

Davies, C.N. (1977) Sampling aerosols with a thin-walled tube, In: *Atmospheric Pollution* (Ed. M. Benarie), Elsevier, Amsterdam, pp. 433–446.

Dunnett, S.J. (1990) Mathematical modelling of aerosol sampling with thin-walled probes at yaw orientation with respect to the wind, *Journal of Aerosol Science*, 21, 947–956.

Dunnett, S.J. and Wen, X. (2002) A numerical study of the sampling efficiency of a tube sampler operating in calm air facing both vertically upwards and downwards, *Journal of Aerosol Science*, 33, 1653–1665.

Galeev, R.S. and Zaripov, S.K. (2003) A theoretical study of aerosol sampling by an idealized spherical sampler in calm air, *Journal of Aerosol Science*, 34, 1135–1150.

Gibson, H., and Ogden, T.L. (1977) Entry efficiencies for sharp-edged samplers in calm air, *Journal of Aerosol Science*, 8, 361–365.

Grinshpun, S.A. and Lipatov, G.N. (1986) The initial composition distortions of aerosols being aspirated from the calm atmosphere, Abstract from the 14th All-Union Conference on Urgent Problems of the Physics of Aerosol Systems, Odessa (in Russian).

Grinshpun, S.A., Lipatov, G.N. and Dukatczenko, Z.M. (1985) The account of particle sedimentation for the correction of aerosol sample errors from downflows, *Meteorologia Gidrologia*, 9, 111–113 (in Russian).

Grinshpun, S.A., Lipatov, G.N. and Semenyuk, T.I. (1989) A study of sampling of aerosol particles from calm air into thin-walled cylindrical probes, *Journal of Aerosol Science*, 20, 1561–1564.

Grinshpun, S.A., Lipatov, G.N. and Sutugin, A.G. (1990) Sampling errors in cylindrical nozzles, *Aerosol Science and Technology*, 12, 716–740.

Grinshpun, S.A.,Willeke, K. and Kalatoor, S. (1993) A general equation for aerosol aspiration by thin-walled sampling probes in calm and moving air, *Atmospheric Environment*, 27, 1459–1470.

Grinshpun, S.A.,Willeke, K. and Kalatoor, S. (1994) Corrigendum, *Atmospheric Environment*, 28, 375.

Hangal, S. and Willeke, K. (1990) Overall efficiency of tubular inlets sampling at 0–90 degrees from horizontal aerosol flows, *Atmospheric Environment*, 24a, 2379–2386.

Kaslow, D.E. and Emrich, R.J. (1973) Aspirating flow pattern and particle inertia effects near a blunt, thick-walled tube entrance, Technical Report No. 23, Department of Physics, Lehigh University, PA, USA.

Kaslow, D.E. and Emrich, R.J. (1974) Particle sampling efficiencies for an aspirating blunt, thick-walled tube in calm air, Technical Report No. 25, Department of Physics, Lehigh University, PA, USA.

Kenny, L.C., Aitken, R.J., Baldwin, P.E.J., Beaumont, G.C. and Maynard, A.D. (1999) The sampling efficiency of personal inhalable aerosol samplers in low air movement environments, *Journal of Aerosol Science*, 30, 627–628.

Levin, L.M. (1957) The intake of aerosol samples, *Izv. Akad. Naik SSSR Ser. Geofizika.*, 7, 914–925 (in Russian).

Lipatov, G.N., Grinshpun, S.A., Semenyuk, T.I. and Sutugin, A.G. (1988) Secondary aspiration of aerosol particles into thin-walled nozzles facing the wind, *Atmospheric Environment*, 22, 1721–1727.

Lipatov, G.N., Grinshpun, S.A. and Shingaryov, G.L. (1985) Aerosol sampling from the calm air into upright cylindrical nozzle, *Meteorologia i Hydrologia*, 12, 99–102 (in Russian).

Lipatov, G.N., Grinshpun, S.A., Shingaryov, G.L. and Sutugin, A.G. (1986) Aspiration of coarse aerosol by a thin walled sampler, *Journal of Aerosol Science*, 17, 763–769.

Maynard, A.D., Aitken, R.J., Kenny, L.C., Baldwin, P.E.J. and Donaldson, R. (1997) Preliminary investigation of aerosol inhalability at very low wind speeds, *Annals of Occupational Hygiene*, 41, 695–699.

Ogden, T.L. (1983) Inhalable, inspirable and total dust, In: *Aerosols in the Mining and Industrial Work Environments* (Eds V.A. Marple and B.Y.H. Liu), Ann Arbor Science, Ann Arbor, MI., pp. 185–204.

Patankar, S.V. (1980) *Numerical Heat Transfer and Fluid Flow*, Series in Computational Methods in Mechanics and Thermal Sciences, Hemisphere Press, New York.

Roger, F., Fabries, J.F., Görner, P., Wrobel, R. and Renoux, A. (1999) Entry efficiency of an annular aerosol sampling slot in calm air, *Journal of Aerosol Science*, 30, S151–S152.

Su, W.C and Vincent, J.H. (2002) Experimental method to directly measure aspiration efficiencies of aerosol samplers in calm air, *Journal of Aerosol Science*, 33, 103–118.

Su, W.C. and Vincent, J.H. (2003) Experimental measurements of aspiration efficiency for idealized spherical aerosol samplers in calm air, *Journal of Aerosol Science*, 34, 1151–1165.

Su, W.C. and Vincent, J.H. (2004a) Experimental measurements and numerical determination of aspiration efficiencies for cylindrical thin-walled aerosol samplers in calm air, *Aerosol Science and Technology*, 38, 766–781.

Su, W.C. and Vincent, J.H. (2004b) Towards a general semi-empirical model for the aspiration efficiencies of aerosol samplers in perfectly calm air, *Journal of Aerosol Science*, 35, 1119–1134.

Ter Kuile, W.M. (1979) Comparable dust sampling at the workplace, Report F1699, Instituut voor Milieuhygiene en gezondheidstechniek, Delft, The Netherlands.

Vincent, J.H. (1989) *Aerosol Sampling: Science and Practice*, John Wiley & Sons, Ltd, Chichester.

Walton, W.H. (1954). Theory of size classification of airborne dust clouds by elutriation. *British Journal of Applied Physics*, Supplement 3, S29–S40.

Witschger, O., Grinshpun, S.A., Fauvel, S. and Basso, G. (2004) Performance of personal aerosol samplers in very slowly moving air when facing the aerosol source, *Annals of Occupational Hygiene*, 48, 351–368.

Yoshida, H., Uragami, M., Masuda, H. and Iinoya, K. (1978) Particle sampling efficiency in still air, *Kagaku Kogaku Ronbunshu*, 4, 123–128 (HSE translation No. 8586).

Zaripov, S.K., Zigangareeva, L.M. and Kiselev, O.M. (2000) Aerosol aspiration into a tube from calm air, *Fluid Dynamics*, 35, 242–246.

7

Interferences to Aerosol Sampling

7.1 Introduction

As discussed in detail in the preceding chapters, it is clear that the physical processes occurring during aerosol aspiration by aerosol samplers are controlled by a combination of inertial forces associated with particle motion in distorted air near the inlet to a sampler, and particle settling under the force of gravity. The former dominates under moving air conditions where local velocities of fluid mechanical convection are large compared with particle settling speeds. The latter dominates when the effects of convection are absent – as in calm air – or are small. The picture presented so far, however, is idealistic – an essential step towards understanding the behaviors of aerosol samplers under practical conditions. But it is incomplete. In the real world, flows are less ideal than those described, particles may experience forces other than inertial or gravitational, and particle interactions with solid surfaces may be complex such that they may or may not be retained upon impact. When these additional factors come into play, they represent interferences to the ideal behaviors of samplers. They are important in many practical aerosol sampling situations.

From time to time, in practical aerosol sampling, results are reported that do not fit the patterns predicted by theories like those already described. It is common for such observations to be explained away by broad reference to additional 'interfering factors' such as *air turbulence*, *electrostatic effects* or *wall effects*, often vaguely and without any physical or quantitative basis. Nonetheless, these are all plausible suggestions since, in most practical sampling situations, the external air flow *is* turbulent, particles and surfaces *are* electrically charged, and particles *do* blow or bounce off from external or internal walls. Interferences associated with such effects, leading to biases in sampling efficiency, may occur both outside the sampler during the process of aspiration and inside the sampler after aspiration and prior to arrival at the desired collection surface – filter or substrate. This chapter addresses both aspects.

7.2 Interferences during aspiration

Air turbulence, electrostatic effects and wall effects may all be influential during the processes involved in particle transport all the way up to the point where they pass through the plane of the inlet of an

Aerosol Sampling: Science, Standards, Instrumentation and Applications   James Vincent
© 2007 John Wiley & Sons, Ltd

aerosol sampler. These are external to the sampling device and so are largely uncontrollable. But it is important that an environmental or occupational hygiene practitioner is aware of their nature so that, at the very least, the results of sampling in the field may be interpreted intelligently when the need arises.

7.2.1   Effects of turbulence on aspiration

It is well known that turbulence can have a profound effect on the transport of airborne scalar entities – including aerosol particles – in unstable flows. One result of the superposition of the fluctuating turbulent motions of the air on the mean flow is the appearance of the phenomena of turbulent diffusion and associated mixing. As discussed earlier in Chapter 3, the physical description of these processes for aerosols is bound up with how well individual particles are able to follow the complex motions associated with the turbulent air movement. As far as aerosol sampling is concerned, the subject has been referred to scientifically by only a small number of workers (e.g. McFarland *et al.*, 1977; Ogden and Birkett, 1978; Addlesee, 1980; Jayasekera and Davies, 1980).

To a first approximation, turbulent diffusion may be thought of in terms of an extension of Fick's law of classical diffusion, in particular that there is a net flux of particles from regions of high to regions of low concentration. Consequently, turbulence tends to counteract the establishment of concentration gradients. How might this idea be related to aerosol samplers? For smooth air flow, we have described in previous chapters how, due to the inertial forces acting in the distorted streamline pattern near a sampler, particles may depart from the streamlines of the air flow. The departure of aspiration efficiency (*A*) from unity is a reflection of the resultant spatial nonuniformities in particle concentration. When turbulence is superimposed on the flow pattern, the resultant mixing through enhanced diffusion will act towards smoothing such concentration gradients and therefore to restore *A* towards unity. It is possible, in principle, to develop a theory for the effect of freestream turbulence on the performance of a blunt aerosol sampler that deals rigorously with particle motion in the mean and fluctuating components, respectively, of turbulent flow. A small amount of mathematical work along these lines has been reported for simple sampling systems (e.g. Wen and Ingham, 1995 for thin-walled samplers; Ingham and Wen, 1993 for a simple, disk-shaped blunt sampler), based on numerical solutions of the Navier–Stoke equations for the air motion.

A descriptive physical picture may be helpful. This may be constructed by referring back to the diagram in Figure 1.5, representing a blunt sampler with arbitrary shape and placed at arbitrary orientation with respect to the freestream. The streamsurface pattern represents the *actual* air motion if the flow is smooth and the *mean* air motion if the flow is turbulent. In the simpler case of the smooth flow that has formed the basis of most of the aerosol sampling ideas presented in this book so far, the limiting streamsurface defines exactly how the air is divided between what is sampled and what passes by outside the sampler body. For turbulent flow, however, the definition is less precise because there will be an exchange of individual packets of air across the limiting streamsurface. Now the limiting particle trajectory represents the statistically *most probable* surface describing the path of a particle that is just sampled. For the purpose of the present discussion, it is again useful to begin by writing down aspiration efficiency in terms of net particle fluxes. Thus:

$$A = \frac{N_s}{N_0} \tag{7.1}$$

where N_s and N_0 are as defined before. So too are the limiting streamsurface and limiting particle trajectory surface, respectively. In what follows, we examine the fluxes of particles in and out of the aspirated air volume in much the same way as during the earlier construction of the impaction models for thin-walled cylindrical and blunt samplers. It is recalled from Equations (1.14) and (1.15) that:

$$A = \frac{\int_{a_s} (cv)\,\mathrm{d}a}{\int_{a_0} (cv)\,\mathrm{d}a} = \frac{\int_{a''} (cv)\,\mathrm{d}a}{\int_{a_0} (cv)\,\mathrm{d}a} = A_0 \tag{7.2}$$

where c and v are the local particle concentration and velocity, respectively, and the (cv)-values represent particle fluxes. For smooth flow, these local values are steady so that A_0 may be assigned as the aspiration efficiency for zero turbulence. For turbulent flow, however, the local concentration and velocity also contain the fluctuating components c' and v'. Now the instantaneous particle flux in the direction of the mean air flow is given by:

$$Flux = (c + c')(v + v') \tag{7.3}$$

which by expanding and time-averaging reduces to:

$$\overline{Flux} = \overline{cv} + \overline{c'v'} \tag{7.4}$$

for the mean flux, where the two terms on the right represent transport by convection and diffusion, respectively. Terms linear in c' or v' have vanished by the process of time-averaging. For the direction (say, r) perpendicular to the local mean particle motion, we have a similar expression except that there is no transport by convection, so that:

$$\overline{Flux} = \overline{c'v_r'} \tag{7.5}$$

where v_r' is the instantaneous r-directed particle velocity. Equation (7.2) for turbulence flow now becomes:

$$A_t = \frac{\int_{a_s} (\overline{cv} + \overline{c'v'})\,\mathrm{d}a}{\int_{a_0} (\overline{cv} + \overline{c'v'})\,\mathrm{d}a} \longrightarrow \frac{\int_{a''} (\overline{cv} + \overline{c'v'})\,\mathrm{d}a - \int_{a^*} \overline{c'v_r'}\,\mathrm{d}a}{\int_{a_0} (\overline{cv} + \overline{c'v'})\,\mathrm{d}a} \longrightarrow \frac{\int_{a''} (\overline{cv})\,\mathrm{d}a - \int_{a^*} \overline{c'v_r'}\,\mathrm{d}a}{\int_{a_0} (\overline{cv})\,\mathrm{d}a} \tag{7.6}$$

in which it has been assumed that the flux by diffusion is much smaller than that by convection. The last term in the numerator represents the net lateral flow of particles by turbulent diffusion across the limiting trajectory surface, the surface area of which is a^*. Physically, Equation (7.6) embodies a simple model that involves two superimposed but competing particle transport processes. In the first, there is the reorganisation of the spatial distribution of particles near the sampler as the result of inertial effects associated with their movement in the distorted mean flow. Its primary effect is to try to establish particle concentration gradients in the region in question, the magnitude of which will be greater the more distorted the mean flow. In the second process, there is the competing turbulent diffusion which works to counteract the establishment of such concentration gradients.

Equation (7.6) combined with Equation (7.2) leads to:

$$A - A_0 \approx \frac{\int_{a^*} (\overline{c'v_r'})\,\mathrm{d}a}{\int_{a_0} (\overline{cv})\,\mathrm{d}a} \tag{7.7}$$

which expresses the effect of the turbulent motions on aspiration efficiency. The challenge now is to evaluate this expression and turn it into a manageable form that can be tested experimentally.

Fick's law may be applied to turbulent diffusion to a fair first approximation, so that the aerosol flux across the limiting streamsurface, representing aerosol transport into or out of the sampled air volume, may be written as

$$\overline{(c'v'_r)} \approx -D_{tpr}\frac{\partial c}{\partial r} \longrightarrow -D_{tr}\text{f}(K_r)\frac{\partial c}{\partial r} \tag{7.8}$$

where D_{tr} and D_{tpr} are the r-directed local turbulent diffusivities for the air and particles, respectively, as defined in Chapter 2. In Equation (7.8)

$$D_{tr} \approx L_r(\overline{u'^2_r})^{1/2} \text{ and } K_r \approx \frac{\tau(\overline{u'^2_r})^{1/2}}{L_r} \tag{7.9}$$

from Equations (2.19) and (2.67), in which u'_r represents the r-directed local air velocity fluctuation, as opposed to the equivalent particle velocity, v'_r. The first expression estimates the diffusivity of the air itself; the second is an inertial parameter akin to a Stokes number that represents the ability of a particle with relaxation time τ to respond to the turbulent fluctuating motions.

Using the preceding relations, and after some manipulation, Equation (7.7) can be resolved to:

$$\frac{A - A_0}{L_r(\overline{u'^2_r})^{1/2}} \approx \text{f}(K_r)\frac{\int_{a^*}\left(\frac{\partial c}{\partial r}\right)\text{d}a}{\int_{a_0}\overline{(cv)}\,\text{d}a} \tag{7.10}$$

Closer inspection reveals that the second bracketed term on the right is roughly equivalent to the impaction efficiency (β) introduced earlier for describing the physics of the aspiration process. Vincent *et al.* (1985) explored this relationship, estimating the right hand side by reference to terms in the blunt sampler impaction model described in Chapter 5. In this way they were able to reduce the equation further to:

$$\frac{A - A_0}{\Lambda\Omega} \approx \text{f}(K) \tag{7.11}$$

where

$$\Lambda = \frac{L(\overline{u'^2})^{1/2}}{DU} \text{ and } \Omega = \frac{LU}{(\overline{u'^2})^{1/2}D} \tag{7.12}$$

Here, Λ is equivalent to a dimensionless turbulent diffusivity and Ω expresses the relative magnitudes of the time scale associated with the fluctuating air motions and that associated with the mean flow distortion. In Equations (7.11) and (7.12), the turbulence characteristics are represented everywhere in terms of measurable quantities for the freestream. A more explicit form of the function f(K) may be roughly estimated by reference to the earlier Equation (2.64), leading to:

$$\frac{A - A_0}{\Lambda\Omega} \approx K\left(1 - \frac{3K^2}{K + 2}\right) \tag{7.13}$$

by which the general shape of the relationship may be ascertained.

The preceding provides a basis for experiments to test for the effects of turbulence on sampler aspiration efficiency, identifying the groups of variables that need to be controlled and others that need to be measured. Validation requires good experimental data. But such data are very sparse, perhaps because the desired experiments are difficult to execute. One study was carried out in the author's laboratory in which the aim was specifically to explore the functional relationship contained in Equation (7.11) (Vincent *et al.*, 1985). The experiments were performed in a large wind tunnel for a wide range of approximately isotropic turbulence conditions generated by means of an adjustable system of square-mesh, biplanar lattice-type grids, providing turbulence length scales (L) from about 2 to 10 cm and intensities from about 3 to 16 %. Monodisperse test aerosols of nebulised wax and human serum albumen with aerodynamic diameter $d_{ae} = 10$, 20 and 40 μm and narrowly graded aerosols of fused alumina with mass median values of d_{ae} ranging from 8 to 63 μm were employed. The experiments were carried out for one mean wind speed only, $U = 2$ m s^{-1}. Aspiration efficiency in the turbulent flow (A) was measured for a simple disk-shaped blunt sampler facing directly into a variable-turbulence freestream, where the disk diameter (D) was 40 mm and orifice diameter (δ) was 4 mm. The required reference concentration was obtained using a thin-walled cylindrical probe after it had been determined that its own aspiration efficiency was not significantly influenced by the turbulence when operated isokinetically. The results for the disk-shaped sampler were plotted in the form suggested by Equation (7.11) and are these are shown in Figure 7.1. There was considerable scatter in the data, especially for increasing K. However, the data plotted in this way exhibited a trend which was broadly consistent with Equation (7.13). The overall general conclusion from this work was that the effects of freestream turbulence were likely to be small except for large particles. However no such conclusion has been tested elsewhere – neither before nor since.

In other laboratory studies, Lutz and Bajura (1984) investigated the aspiration efficiencies of both thin- and thick-walled sampling tubes, conducting their experiments in a pipe where the turbulence was generated by an orifice plate axially at various distances downstream from the entrance to the pipe such that a range of turbulence conditions – defined in terms of intensity only – could be studied. For values of velocity ratio (R) from 0.3 to 3, and polydisperse test aerosols of glass beads with diameters ranging

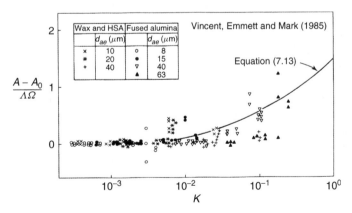

Figure 7.1 *Plot of $(A - A_0)/\Lambda\Omega$ as a function of the inertial parameter (K), as suggested by Equation (7.10), for a disk-shaped blunt sampler facing the freestream, for a single set of conditions with $U = 2$ m s^{-1}, $D = 40$ mm and $\delta = 4$ mm, and for a range of particles sizes as indicated (Vincent et al., 1985). The solid line is calculated from Equation (7.13) with an assumed relationship for K based on Equation (2.67). Reprinted from Aerosol Science and Technology, 4, Vincent et al., 17–29. Copyright (1985), with permission from Elsevier*

from 3 to 9 μm, Lutz and Bajura found a slight trend indicating the restoration of A towards unity, as predicted by theory. However they concluded that the effects of the turbulence on the aspiration efficiencies of the samplers tested under anisokinetic conditions were 'minimal'.

A small amount of work has been conducted to investigate the effects of turbulence on the performances of aerosol samplers of the type developed for practical applications. Most such samplers are substantially more complicated than the idealised ones described here, often having complex shapes and more than one entry orifice. It is therefore usually difficult to use them to draw conclusions of a fundamental nature. However they can, in certain circumstances, provide useful additional background. Vrins *et al.* (1984) carried out studies in the field, where they investigated the performances of a number of samplers designed for sampling suspended particulate matter in the ambient atmosphere. The tests were carried out at an outdoors test site where it was found that the turbulence generated by the upwind placement of wind shields (to simulate coal stock piles) was characterised by air velocity fluctuations with frequencies from 1 to 1000 Hz. Monodisperse test aerosols were produced upwind of the test site using spinning disk generators, and measurements of reference aerosol concentrations (c_0) upon which to base the determination of aspiration efficiency were made using 'Rotorod' samplers (see Chapter 17). The main conclusion was that, although aspiration efficiency itself did not appear to vary systematically with turbulence for any of the samplers tested, the variability in the results exceeded that arising purely from the analytical procedure. These findings are broadly in agreement with the other work reported above.

There are a few other reports that are helpful. In one study, Wiener *et al.* (1988) examined the effects of turbulence on the sampling efficiencies of thin-walled probes under anisokinetic conditions. The results, however, focused attention on the effects on particle deposition inside the probe, and did not contribute significantly to knowledge about aspiration efficiency per se. Kalatoor *et al.* (1995) from the same group reported a study on aerosol sampling by means of a thin-walled sampling tube from a harmonically varying freestream. Again, little light was shed on aspiration efficiency, although it was noted that there were significant increases in particle losses inside the sampler inlet associated with the fluctuating, turbulence-like motions in the freestream. Elsewhere, Gao *et al.* (1999) carried out a numerical study of a sampler with a circular slit inlet and their calculations predicted that overall sampling efficiency would be lowered as a result of freestream turbulence. They too showed increases in particle losses inside the sampler after aspiration, but did not show any marked effect on aspiration efficiency itself. Finally, Cain and Ram (1998) examined the highly specialised case of a cylindrical probe sampling in a very high-velocity turbulent freestream relevant to atmospheric aerosol sampling on an aircraft. Like the other work cited, they showed the effect of turbulence on particle losses in the shroud surrounding the inlet, but did not reveal much about aspiration efficiency.

The final conclusion from the small number of studies that have been reported is that, although it is physically plausible that there may be an effect of turbulence on the aspiration efficiency of a sampler, increasing with particle size and level of turbulence and for a greater departure of A_0 from unity (for the limiting smooth flow case), such an effect is likely to be small for most sampling situations. There does, however, appear to be an increase in variability of performance associated with the turbulence.

7.2.2   Effects of electrostatic forces on aspiration

Electrostatic forces may play a role in particle motion near a macroscopic body if (a) the particles are electrically charged but the body itself is uncharged, (b) the particles are uncharged but the body is charged, or (c) both the particles and the body are charged (Pich, 1966). As far as their relevance to the aspiration part of aerosol sampling is concerned, it may be assumed that the first two alternatives are of little interest since the electrostatic forces in those cases only come into play when the particles and the sampler are in close proximity with one another. Therefore, in the discussion that follows, it will be assumed that both the airborne particles and the sampler itself are charged.

The electrical forces acting on charged airborne particles are derived from the electric field (E_{el}) in the vicinity of the relatively large, charged sampler body, extending out from it but diminishing with distance away. Within this field, the local motion of particles carrying charge q *is* governed both by the combination of the inertial and gravitational forces (as before) and, now, the electrical force represented by:

$$F_{el} = q E_{el} \tag{7.14}$$

This may be incorporated into the basic equation of particle motion which, in turn, may be solved numerically to generate the pattern of particle trajectories. In this way it may be determined under what conditions the magnitude of the effect becomes significant in relation to the other forces acting and so makes a noticeable contribution to aspiration efficiency. But no such analysis appears to have been performed to date.

For a simple view, Equation (7.14) may be rewritten in terms of the electric charge carried by the sampler (q_s). Thus:

$$F_{el} \propto q q_s \tag{7.15}$$

so that the terminal velocity of the particle in the electric field is given by:

$$v_{el} \propto \frac{q q_s}{d} \tag{7.16}$$

The magnitude and sign of the particle charge depend on the mechanism by which the charge was acquired. Johnston and his colleagues carried out studies of the electric charge on particles in workplace aerosols in a range of industries (Johnston *et al.*, 1985) and during the mechanical generation of aerosols in the laboratory from dry powders (Johnston *et al.*, 1987a). It was shown that the charges on individual particles were usually and consistently distributed equally between positive and negative, so that the aerosols studied were net neutral. It was also shown that the median magnitude of charge was given by $|q| \propto d^n$, where n was found to take values between 1.2 and 2.5. This meant that:

$$v_{el} \propto q_s d^{n-1} \longrightarrow V_s d^{n-1} \tag{7.17}$$

where V_s is the electrical potential of the sampler. Since the magnitude of v_{el} in relation to other velocities in the aerosol aspiration system determines whether or not the effect of the electrical forces is significant, Equation (7.17) shows that any effect on aspiration efficiency would be greatest for larger particles.

Although there are many references in the literature to the potential problems in practical aerosol sampling arising from electrostatic forces, there are few experimental data upon which to base a good understanding of the physics of its effects on the aspiration process. There has been some interest in the problem from occupational hygiene researchers. Almich and Carson (1974) and Briant and Moss (1984) examined the performance of the 10 mm nylon cyclone widely used in the USA as a personal sampler for measuring the exposure of workers to fine 'respirable' particles. By comparing the results obtained using this instrument with those obtained using metal replicas, it was concluded that the charge that could accumulate on the nylon sampler during handling was responsible for reducing the amount of collected aerosol. Turner and Cohen (1984), in experiments in the laboratory with highly charged unipolar aerosols, reported the same trend for a 37 mm filter holder (of the type used in the USA for sampling 'total' aerosol) carrying charge of the same polarity as that of the aerosol. However, in field studies using the same samplers in a beryllium factory, no significant effects were found.

A number of studies were reported by Kuusisto (1979) and Lovstrand and Rosen (1980) where the main objective was to examine the effects of electrostatic forces on particle sampling under practical conditions. Most of the results were obtained for sampler potentials of the order of 1 kV, and effects on sampler performance were found to be generally small. However, in a few cases with sampler potentials as high as ± 5 kV, a significant increase was observed in the amount aspirated. Johnston *et al.* (1987b) carried out a laboratory study aimed at investigating electrostatic interference of aerosol sampling in direct relation to occupational hygiene. A series of experiments were performed in calm air to investigate the effects of electrostatic forces on the collection of particles by 25 mm diameter open-filter holders of the type used widely for sampling asbestos and other dusts. Test aerosols were mechanically generated from polydisperse silica gel powders by means of a Wright dust feed, where 90 % of the airborne material comprised particles with d_{ae} less than about 15 μm. The charge level on the dust was found to be net neutral, with the median magnitude of particle charge characterised by $|q| = 25d^{1.25}$, where d is expressed in micrometers. Particles were tagged with a fluorescent dye and the collected samples analysed fluorometrically. Potentials of the required magnitude and polarity were applied directly to the samplers by means of a high-voltage power supply, and were checked remotely and noninvasively using an electrostatic field mill (Chalmers, 1957). The ratio (R_{el}) of the amounts of aerosol collected by charged and uncharged samplers, respectively, was thus determined. Under the assumption that the effect of electrostatic forces on aspiration efficiency was determined by v_{el}, and hence by $V_s d^{1.25}$ as described by Equation (7.17), the results were plotted in the form of the bias, $(1-R_{el})$, as a function of $V_s d^{0.25}$. As shown in Figure 7.2, the effect of electrostatic forces was to decrease aspiration efficiency, the more so for larger particle sizes and higher electric potential on the sampler.

The question of course arises about the source and magnitude of the electric potential on the sampler. The magnitude of the potential required to produce a measurable effect is seen from Figure 7.2 to be of the order of kilovolts. Is this plausible? In fact it is. Gajewski (1980) showed that the electrical potential on a human body can, under the right circumstances of humidity, clothing worn, shoes worn, etc., reach the order of several kilovolts, derived from the friction and contact charging during ordinary day-to-day activities. Based on the research that has been described, and under appropriate conditions, such potentials are quite likely to influence the collection efficiencies of personal samplers worn on the bodies of human subjects.

The question of conducting versus insulating sampler materials deserves comment. The electric field near the sampler is determined solely by the charge on the sampler and on the sampler size and shape. However, it is the material of the sampler that governs its ability to hold such a charge. In the case of a highly insulating sampler, for example, high charge can easily be maintained. On the other hand, for a conducting sampler, the charge readily leaks away to earth through supporting structures. Therefore, in practical situations, the problem of electrostatic effects in relation to sampler performance is generally only significant for samplers made of insulating material. Smith and Bartley (2003) investigated the effect of sampler and mannequin conductivity during an experimental study in a wind tunnel of possible electrostatic effects on the performances of personal samplers. They showed that samplers made of nonconducting material exhibited lower sampling efficiency than those made of conducting material, placed on both conducting and nonconducting mannequins. In addition, the nonconducting mannequin provided lower sampling efficiency than the conducting one for both conducting and nonconducting samplers. From this and most other experiences in the laboratory and the real world, it is suggested that conducting materials should be used in sampler systems wherever possible.

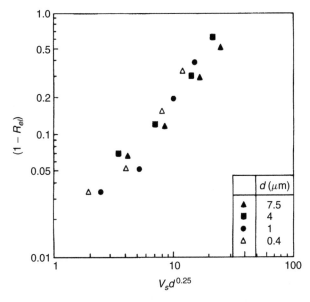

Figure 7.2 *Experimental results for the effect of electrostatic forces on sampler aspiration efficiency, where R_{el} is the ratio of aspiration efficiencies with and without electrostatic forces, respectively, and $V_s d^{0.25}$ is as suggested by Equation (7.17) with n = 1.25 based on measurements of electrostatic charge on workplace aerosols (Johnston et al., 1987). Here V_s is the electrostatic potential of the sampler in kV and d is particle diameter in micrometers. Reprinted from Johnston et al., American Industrial Hygiene Association Journal, 48, 613–621. Copyright (1987) American Industrial Hygiene Association*

7.2.3   External wall effects

As described in Chapter 1, *overall* sampler performance involves the aerodynamic transport of particles by aspiration directly into the sampling orifice together with other, secondary, processes involving the walls of the sampler, both external and internal. Pure aspiration is regarded as the ideal part of the process and, as discussed in the preceding chapters, this may be related mainly to the airflow pattern near the sampler and the aerodynamic diameter of the particles in the aerosol of interest, without the involvement of specific particle and sampler surface properties. Wall effects represent modifying influences to ideal sampler performance. They include combinations of deposition and rebound of particles at the solid surfaces of the sampler, involving aspects of air flow and particle properties over and above those that have appeared so far in the discussion of pure aspiration. Under certain conditions, such added complications may have an important bearing on the interpretation of observations made during experimental sampler research and practical investigations.

Here we consider effects arising from particle behavior outside the sampling orifice during processes leading up to aspiration. So far attention has been focused on the trajectories of particles that enter the sampling orifice and so are aspirated directly. However, the complete pattern of trajectories inevitably also includes those of particles that arrive at the solid walls of the sampler. Although these particles are not sampled by direct aspiration into the sampler, some of them *do* enter through the plane of the sampler inlet and so *are* removed from the freestream. From the discussion in Chapter 2 about impaction onto a solid body, the efficiency of this removal process depends on the shape of the pattern of airflow streamsurfaces and, if interception is negligible in comparison with impaction, on particle aerodynamic diameter, d_{ae}. It has been implicitly assumed in most theoretical and experimental investigations of the aspiration process that particles that arrive at the surface of the sampler in this way adhere permanently on arrival. Although this is easy to envisage for 'sticky' particles like liquid droplets, it is less obviously so for the coarse, gritty particles that are representative of many aerosols found in practical situations. Such particles are quite likely to rebound or be re-entrained from the surface upon or after arrival. If they do, those that are resuspended within the convergent, sampling flow contained within the stagnation region enclosed on the sampler surface by the limiting streamsurface will enter the sampling orifice. The result will be an increase in the overall concentration of particles entering the sampler and hence an apparent *increase* in aspiration efficiency. The magnitude of this effect depends on whether the process of rebound takes the form of particle 'bounce' or 'blow-off', on the degree of 'stickiness' of the particles, and on the magnitude and direction of the local aerodynamic forces acting on the particles.

The first experimental evidence for the effects of losses of particles from the external walls of sampling devices came from observations in the laboratory during trials with the 2.5 Lpm MRE (Mines Research Establishment) Type 113A gravimetric dust sampler. This is a device that was once used routinely for monitoring airborne dust in British coalmines, and consisted of a horizontal gravitational elutriator which admitted to a filter only the fine, so-called 'respirable', fraction (see Chapters 14 and 15). Entry of the dust-laden air to the elutriator took place through a number of horizontal narrow slits located flush with the blunt, flat-nosed front of the instrument. Ogden *et al.* (1978) and Mark (1978) independently carried out wind tunnel experiments to investigate the aspiration characteristics of this instrument facing into the wind. Ogden *et al.* used liquid droplet test aerosols of di-2-ethyl-hexyl sebacate and, for a given wind speed, found that measured aspiration efficiency, based on all the particles passing through the plane of the sampler entry, fell steadily with increasing particle size. This is a reasonable expectation for this system based on the blunt sampler theory described earlier in Chapter 5. In his experiments, Mark used aerosols generated from coal dust and observed the same trend for smaller particle sizes. However, for particles with particle aerodynamic diameter (d_{ae}) greater than about 10 μm, he found a sudden sharp increase in apparent aspiration efficiency. Comparing the results of these two experiments, it appears that the MRE sampler had oversampled the dry, gritty coal dust particles in relation to the 'sticky' liquid droplets. Based on the qualitative physical picture outlined above, this may be taken as clear evidence of an increase in the concentration of particles entering the sampler due to losses from the external walls of the device.

The next evidence came from wind tunnel experiments to investigate the aspiration characteristics of an idealised long cylindrical blunt body ($D = 125$ mm) placed transverse to the freestream with a single circular sampling orifice ($\delta = 4$ mm) facing into the wind (Vincent and Gibson, 1981). The experiments were carried out both using liquid droplet test aerosols of di-2-ethyl-hexyl sebacate and aerosols mechanically generated from narrowly graded powders of fused alumina. The results for apparent aspiration efficiency (A_{app}, as defined in Chapter 1) are shown as a function of d_{ae} in Figure 7.3. For the liquid droplet aerosols, the trend was for A_{app} to fall steadily with d_{ae}. This was as expected from blunt sampler theory for the prevailing conditions, indicating that A_{app} was probably close to the true aspiration efficiency (A). For the dry, gritty particles, the same trend was observed for d_{ae} below about 20 μm. But for larger particles and for the ungreased sampler bodies, there was again a sharp rise in A_{app}, indicating that $A_{app} > A$. But when the sampler body was greased, A_{app} fell back to close to the original trend. These and other experiments also produced direct visual evidence for the phenomenon in question. Figure 7.4 shows a typical picture of the dust deposit on the external surface of a 2 Lpm, single-hole personal sampler of the type that was once used by some British occupational hygienists. The sampler had been exposed to test aerosols generated from narrowly graded fused alumina powders. For the finer test aerosol the pictures show particles deposited uniformly over the leading face of the sampler, covering the surface right up to the edges of the sampling orifices, with no obvious indication of

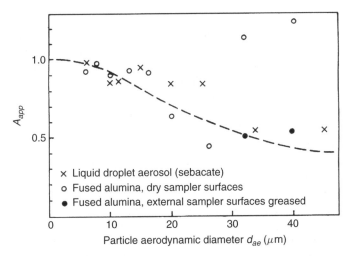

Figure 7.3 *Experimental data for apparent aspiration efficiency (A_{app}) as a function of particle aerodynamic diameter (d_{ae}) for a laboratory single-orifice, transverse cylindrical sampler for dry and liquid droplet aerosols, respectively, and for greased and ungreased external surfaces (Vincent and Gibson, 1981). Reprinted from Atmospheric Environment, 15, Vincent and Gibson, 703–712. Copyright (1981), with permission from Elsevier*

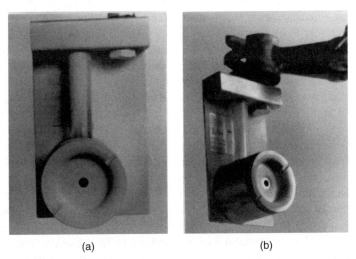

Figure 7.4 *Photographs of fused alumina dust deposits on the external surface of a small 2 Lpm sampler of the type that was once widely used as a personal sampler in the UK: (a) fine dust: (b) coarse dust. Particle blow-off is clearly apparent for the coarse dust (Vincent and Gibson, 1981). Reprinted from Atmospheric Environment, 15, Vincent and Gibson, 703–712. Copyright (1981), with permission from Elsevier*

particle losses from the sampler surfaces. For the coarser dust, however, the patterns of dust deposition show strikingly that the area immediately adjacent to the sampling orifice was now clear of deposited dust. It is reasonable to assume that particles reaching that part of the sampler surface by impaction had subsequently re-entered the air flow converging towards the sampling orifice, and so had been sampled. It had been thought that the phenomenon observed in the studies cited was associated solely with dry, gritty particles since, intuitively at least, it seemed that such particles should adhere to the sampler walls much less readily than the 'stickier' liquid droplets. However, Lipatov *et al.* (1986) noted a related phenomenon for the measured aspiration characteristics of thin-walled probes published by some other workers, even when liquid droplet test aerosols were used. The effect there was considered to be associated with the shattering of the liquid droplets as they impacted close to the leading edges of the probes under certain extreme conditions, in particular for very small velocity ratio R associated with high sampler inlet velocity. The final result was the same as for the dusts already discussed – that is, an apparent increase in aspiration efficiency.

The physical explanation for the rebound of particles from the external surfaces of samplers relates to the inability of such surfaces to retain impacting particles. One possibility is that particles may *bounce* on impact, in which case the phenomenon would be governed by whether or not the component of particle velocity normal to the surface on its arrival exceeds the critical rebound threshold value (Dahneke, 1971, 1973). This would be a function of particle size and surface properties, involving the *coefficient of restitution* which derives from the elastic flexuring of both the particle and the surface when they come together. The *elastic* bounce mechanism is well established for explaining particle loss in the high-velocity jets of impactors (e.g., Cheng and Yeh, 1979). As far as aerosol aspiration is concerned, however, the relative importance of this phenomenon may be assessed by referring back to the experiments described above. We might expect bounce to be most marked at locations on the sampler surface where the normal velocity components of impacting particles are greatest; that is, close to regions of aerodynamic stagnation. Inspection of the distributions of deposited dust over the leading surface of the sampler in the example shown in Figure 7.4 suggested that this was not the case. Rather the loss appeared to be greatest where the shear forces associated with local air motion over the sampler surface were greatest. It is therefore likely that particle bounce was not the most prominent mechanism there. A more likely mechanism in this scenario was *nonelastic* particle *blow-off,* where the fluid mechanical drag force on a particle momentarily deposited, and hence already in contact with the sampler wall in the sheared boundary layer flow, exceeded a critical value determined by the local friction force which was trying to retain the particle. Whether or not the observed reentrainment involved the coefficient of *dynamic friction* or of *static friction for sliding or rolling* depends on whether the particle in question was ever actually at rest in contact with the wall. Vincent and Humphries (1978) explored this effect for particles of fused alumina impacting onto a circular disk-shaped bluff body facing into the wind without aspiration. They conducted experiments in which the diameters of the regions where particles of different grades were and were not retained on the surface of the disk were measured by direct visual observation. The results were used to construct a model based on the re-entrainment model that Bagnold (1954) had proposed to explain the behaviors of particles of sand during the formation of dunes. Others have examined the same effect in studies of the formation of snow drifts. The main difference here was that, whereas Bagnold assumed gravity to be the main force holding the particle in contact with the surface, it was assumed for aerosol sampling to be associated with short-range, van der Waals interactions at the sampler surface. A notable result of the Vincent and Humphries study was that the process in question was a function not so much of the actual macroscopic sizes of the particles but, rather, of the dimensions of the small asperities that defined the microscopic surface geometry of the gritty particles of fused alumina. A related approach reported by Corn and Stein (1965), based on experiments with fly ash and glass beads, concluded that, in addition to factors like those already mentioned, relative humidity and electrostatic charge also played a significant part in the adhesion forces holding the particle and surface in contact. Furthermore, the action of turbulent eddies – penetrating into the boundary layer from the freestream – was suggested as part of the removal process.

It is certainly plausible that the blow-off phenomenon may account for what has been observed in some laboratory studies with aerosol samplers. Vincent and Gibson (1981) reported laboratory experiments that set out to examine specifically how well such ideas may be linked to the performance of a blunt aerosol sampler. Again, as in other research to clarify the physics of the sampling process, the simple example of a disk-shaped sampler with a single central orifice was chosen in versions with combinations of $D = 50$, 75 and 100 mm with $\delta = 2$, 3 and 4 mm, respectively. Each individual sampler was placed in a freestream where U was varied from 2 to 6 m s^{-1} and sampling flow rates ranged from 5 to 20 Lpm, and exposed to narrowly graded test aerosols of fused alumina with d_{ae} from 13 to 56 μm. This range of particle size corresponded to a range of equivalent volume particle diameter, d_V, from about 7 to 28 μm, thought to be a more appropriate, geometric metric of particle size for the discussion of particle blow-off effects. As in the case of the simple flat disk without aspiration, visualisation of the radial distribution of the deposit of particulate matter on the face of the sampler was the main experimental objective. Patterns like those evident in Figure 7.4(b) were even more vividly displayed. For the purpose of discussion, the diameter of the circular clear region around the sampling orifice was labeled as D''. Mark *et al.* (1982) noted the importance of the boundary of the stagnation region in this system, in particular that only particles re-entrained from the surface of the disk inside the region of diameter S – see Chapters 4 and 5 – would encounter an airflow pattern that drew them towards the sampling inlet. So, as shown in Figure 7.5, the results were plotted in the form of

$$\Phi \equiv \left(\frac{D'' - \delta}{D} \right) = f\{\phi_A\} \tag{7.18}$$

where the results contained a family of trends, one for each particle size, with Φ increasing with ϕ_A for given d_V and with d_V for given ϕ_A. These trends were entirely sensible, and it was obvious that the envelope for all these results should be defined by the stagnation boundary represented – that is, $D'' \to S$ as represented earlier in the Equation (4.15). It followed therefore that:

$$\left(\frac{D'' - \delta}{D} \right) \approx \phi_A^{1/3} - r \tag{7.19}$$

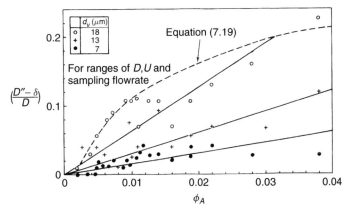

Figure 7.5 *Experimental data for (D''-δ)/D plotted as a function of the sampling ratio (φ$_A$) for an idealised disk-shaped blunt sampler facing the wind and exposed to aerosols of fused alumina for the mean particle sizes indicated, the dashed curve representing the envelope describing the situation where all particles striking the sampler inside the stagnation region are blown off and enter the sampling orifice (Mark et al., 1982). Reprinted from Aerosol Science and Technology, 1, Mark et al., 463–469. Copyright (1982), with permission from Elsevier*

where, as before, $r = \delta/D$. The envelope defined by this curve is also shown in Figure 7.5 and is seen to be consistent with the experimental results.

The scenarios described above are difficult to describe theoretically. However, Ingham and Yan (1994) studied the problem of the rebound of particles from the outer walls of an aerosol sampler from a mathematical standpoint, focusing attention on the microscopic level of the interaction between the particle and the sampler surface, in particular the respective roles of sliding or rolling motions in the rebound process. This was done by means of a detailed numerical description of the flow near the sampler, combined with mathematical descriptions of the physics of the nature of the forces on a particle in contact with the sampler surface. Ingham and Yan investigated both sliding ('static') and rotational models, and came to the conclusion that particle rotation was indeed the more likely explanation for the particle blow-off losses observed by Vincent and Humphries, Vincent and Gibson and Mark *et al.* (1982). As early as 1990, Wang had come to the same general conclusion for particle rebound scenarios close to those of interest here.

The empirical information contained in Figure 7.5 provides a useful basis for estimating the mass error associated with aspiration in the presence of blow-off. In the ideal case of no blow-off, the mass of aerosol sampled per unit time is given by:

$$m_s = Ac_0 \left(\frac{\pi \delta^2}{4} \right) U_s \tag{7.20}$$

where c_0 is the aerosol concentration in the undisturbed freestream and A is the sampler aspiration efficiency. In the presence of blow-off, it is necessary to add the mass of particles which first impact onto the surface of the sampler in the region between δ and D'' and are then re-entrained and blown into the sampling orifice. This mass may be estimated as:

$$m_r \approx Ec_0 U(D''^2 - \delta^2) \frac{\pi}{4} \tag{7.21}$$

where E is the efficiency of the impaction of particles onto the plane of the disc. The total mass of particles actually entering the sampling orifice is:

$$m = m_s + m_r \tag{7.22}$$

and the apparent aspiration efficiency (A_{app}) is given by:

$$m = A_{app} c_0 \left(\frac{\pi \delta^2}{4} \right) U_s \tag{7.23}$$

Combining the previous equations leads to

$$A_{app} \approx A + ER \left[\left(\frac{D''}{\delta} \right)^2 - 1 \right] \tag{7.24}$$

where, as before, $R = U/U_s$. The fractional oversampling error (Z) arising from particle blow-off is now:

$$Z = \frac{A_{app}}{A} - 1 = R \left(\frac{E}{A} \right) \left[\left(\frac{D''^2}{\delta} \right) - 1 \right] \tag{7.25}$$

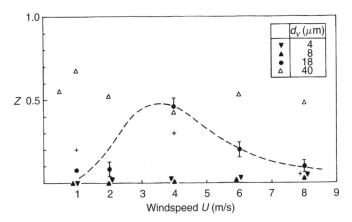

Figure 7.6 *Measured sampling error (Z) for the small single-hole sampler shown in Figure 7.4 as a function of wind speed (U) for dry fused alumina dust particles of various mean size as indicated, the dashed line indicating the broad trend expected from theory as represented by Equation (7.25) (Mark et al., 1982). Reprinted from Aerosol Science and Technology, 1, Mark et al., 463–469. Copyright (1982),with permission from Elsevier*

In this expression, A may be estimated from the blunt sampler model outlined in Chapter 5 and E from published data for impaction from a wide freestream onto a flat disk (e.g. May and Clifford, 1967). The last term may be estimated for given sampling conditions and particle size using data like those plotted in Figure 7.5. Mark *et al.* examined Z experimentally as a function of wind speed U for a range of particle sizes represented by d_V. The experiments were carried out in a small wind tunnel using the same single-hole personal sampler mentioned above, set up to face into the freestream and employing the same fused alumina test aerosols. Z was obtained by comparing the particulate matter aspirated by the sampler when the outer surfaces were clean and dry and when they were coated with grease to eliminate blow-off, respectively. The results are shown in Figure 7.6 and indicate biases as large as 50 %. The broken curve indicates the broad trend predicted by the model in Equation (7.25) and the graph shows that the results were in broad general agreement with the model. For the largest particle size, however, the results failed to exhibit the expected large values for Z, suggesting that some of those particles which were blown-off may not have entered the sampling orifice but overshot and re-entered the freestream.

In a separate experimental enquiry, Lipatov *et al.* (1988) reported the results of experiments to investigate sampling errors associated with particle rebound from the external sidewalls of cylindrical thin-walled sampling probes facing the wind. The experiments were carried out using probes with diameter (δ) from 5 to 10 mm and monodisperse *Lycopodium* particles of diameter 31 μm. Wind speeds (U) ranged from 0.9 to 4.5 m s^{-1} and the velocity ratio (R) from about 0.02 to 1. In this super-isokinetic range of R the flow pattern was such that, as described earlier in Chapter 4, the position of stagnation – separating the aspirated from the unaspirated airflow – was located on the outside wall of the tube downstream of the plane of the entry. As shown in Figure 7.7, particles rebounding after striking the wall between the plane of the entry and stagnation may, if their inertia is not too great in the downstream direction, be drawn back in the reverse flow and so be collected. Lipatov and his colleagues referred to this phenomenon as 'secondary aspiration'. Measurements were made of apparent aspiration efficiency (A_{app}) using the indirect method. By comparing the results with the true aspiration efficiency (A) obtained using their own direct method, as described in their earlier 1986 paper, they obtained a

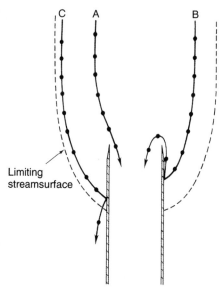

A Particle sampled by direct aspiration
B Particle sampled after rebound (by secondary aspiration)
C Rebounding particle with sufficient downstream inertia
  to avoid collection

Figure 7.7   *Diagram to illustrate the phenomenon of particle rebound from the external side wall of a thin-walled cylindrical sampling tube facing the wind. Reproduced with permission from Vincent, Aerosol Sampling: Science and Practice. Copyright (1989) John Wiley & Sons, Ltd*

rebound coefficient (A_r) where:

$$A_r = \frac{A_{app}}{A} \qquad (7.26)$$

The results for A_r are plotted in Figure 7.8 as a function of R for various values of U. They show a sharp rise in A_r as R fell below unity, the effect being the most marked for the lowest value of U. These trends were consistent with the fact that, in order for a rebounding particle to enter the sampling orifice, it had to overcome the downstream inertia associated with its motion in the freestream prior to impact with the side wall.

As mentioned earlier, interference effects during the process of aspiration are difficult to control. However, for external wall effects like those described above, there are some options. One method of dealing with secondary aspiration is to make the external surfaces of a sampler 'sticky' so that particles that are deposited are not easily re-entrained. This in fact was the approach taken in some of the experimental studies described. But there is more that can be done. Based on recognition that the process of rebound involves sliding or – more likely, according to Ingham and Yan – rolling motions associated with particle removal, the sampler inlet may be modified to 'intercept' such re-entrained particles. For example, Mark *et al.* (1985) and Mark and Vincent (1986) designed the inlets of two aerosol samplers intended for the collection of coarse inhalable particles by arranging for the inlet to incorporate a small lip that projected out slightly beyond the plane of the sampler surface. It was shown

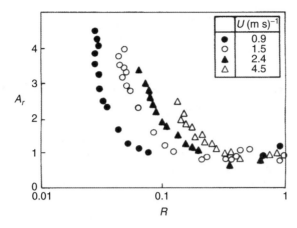

Figure 7.8 *Rebound coefficient (A_r) for a thin-walled cylindrical sampling tube facing the wind as a function of velocity ratio (R), for wind speed (U) in the range from 0.9 to 4.5 m s^{-1} for Lycopodium spores (Lipatov et al., 1988). Reprinted from Atmospheric Environment, 22, Lipatov et al., 1721–1727. Copyright (1988), with permission from Elsevier*

that the presence of such a lip enabled aerosol aspiration without any modification to basic aspiration efficiency, but acted so as to prevent re-entrained sliding or rolling particles from entering through the inlet.

7.3   Interferences after aspiration

The behavior of particles after they have passed through the plane of the sampler inlet after aspiration is important if the particles are not collected immediately but need to be transferred to a filter, some other collection substrate, or a sensing region. This behavior, especially as it relates to particle losses from the sampler airflow, is highly dependent on both the external and the internal shapes and dimensions of the sampler. It is also dependent on the complex fluid mechanical coupling that takes place between the external air movement and that inside the sampler.

The performance of a sampler in the context of practical aerosol measurement is usually thought of in terms of the overall efficiency with which particles in the freestream air are conveyed to a sensing region, typically a filter, inside the sampler body; that is, $A*$ as defined in Chapter 1. The conduit through which the aerosol must be transmitted between the plane of the entry and the filter may take the form, in its simplest case, of a straight cylindrical tube. In a somewhat more complicated case, the tube may be bent. More complicated still, it might take the form of a duct of nonuniform cross-section or a cavity of arbitrary shape.

7.3.1   Deposition losses inside a straight sampling tube

The simplest case of particle deposition in a straight cylindrical tube has been the subject of considerable research since it is relevant not only to aerosol sampling but to duct flows in general. The Reynolds number for flow inside a tube of internal diameter δ_{tube} at mean air velocity U_{tube} is:

$$Re_{tube} = \frac{U_{tube}\delta_{tube}\rho}{\eta} \tag{7.27}$$

The efficiency by which particles penetrate through the tube is given by:

$$P_{tube} = \frac{c_{exit}}{c_{entrance}} \tag{7.28}$$

where c_{exit} and $c_{entrance}$ are particle concentrations at the exit from and entrance to the tube, respectively. In the case where the Reynolds number for flow through the tube, Re_{tube}, is small enough (<2000) for the flow to be laminar, particle transport towards the wall primarily involves transport by gravitational settling and molecular (Brownian) diffusion. The latter has been studied theoretically by Gormley and Kennedy (1949) who showed that the effect on deposition increased with decreasing particle size, decreasing tube diameter and increasing residence time of particles during transport through the tube. It was independent of the orientation of the tube. Although any such effect is likely to be small for many systems of practical interest, it might become important in certain cases, for example in the sampling of ultrafine aerosols through very narrow tubes at low flow rates.

If diffusion is neglected, along with electrostatic effects, then gravitational forces become the dominating influence. Now orientation of the tube is very important. For a vertical tube where the flow is upwards, it is intuitive in the first instance that, if the particle settling velocity (v_s) is greater than the air velocity along the tube axis, no particles will be transmitted through the tube, so that $P_{tube} = 0$. But for all other orientations, P_{tube} takes values up to unity. For the case where the laminar flow is vertically downwards, $P_{tube} = 1$ always. The problem is perhaps most relevant to aerosol sampling when the tube is horizontal. This might apply, for example, to a tube that forms the sampling line between the entry of a thin-walled tube operated isokinetically and a filter holder in a stack sampling situation. In this scenario, particles move under the influence of gravity towards the lower surfaces such that the tube behaves as an elutriator (see also Chapter 8). Here, P_{tube} may be expressed initially as:

$$P_{tube} = f\{R_{tube}, r_{tube}\} \tag{7.29}$$

where

$$R_{tube} = \frac{v_s}{U_{tube}} \quad \text{and} \quad r_{tube} = \frac{\delta_{tube}}{L_{tube}} \tag{7.30}$$

and L_{tube} is the length of tube between the entrance and the exit. Here any dependence on Re_{tube} is neglected. The definitions in Equation (7.30) are consistent with similar nomenclatures in earlier chapters. Thomas (1967) showed that:

$$P_{tube} = \frac{2(ab + \sin^{-1}b - 2a^3b)}{\pi} \tag{7.31}$$

where

$$a = \left(\frac{3R_{tube}}{4r_{tube}}\right)^{1/3} \quad \text{and} \quad b = (1 - a^2)^{1/2} \tag{7.32}$$

An equivalent expression was found by Fuchs (1964), based on unpublished work by Natanson. It also appeared later in a paper by Pich (1972).

In order to test the theory, Thomas carried out experiments with a tube of diameter 10 mm and length 2.7 m, using a monodisperse, polystyrene latex test aerosol with $d_{ae} = 1.3$ μm. Flow rates were low enough for laminar flow conditions to prevail along the length of the tube. The results, presented in terms of penetration (P_{tube}) as a function of a^3, are shown in Figure 7.9 and agreement between experiment

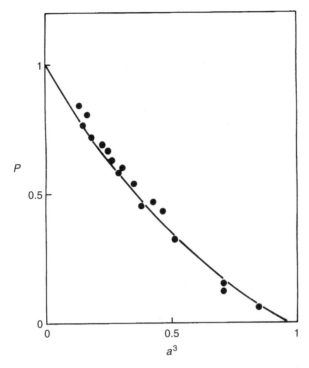

Figure 7.9 *Particle penetration through (P) through a straight cylindrical tube under laminar flow conditions, shown as a function of a³ as suggested by Equation (7.31) (Thomas, 1967). Reproduced with permission from Vincent, Aerosol Sampling: Science and Practice. Copyright (1989) John Wiley & Sons, Ltd*

and theory was very good. From this figure it is clear that, under certain conditions of particle size, sampling flow rate and tube dimensions, particle losses to the internal walls of the tube can be very large and so may well dominate overall sampler performance.

In the preceding, gravitational settling has been assumed to be a primary influence on particle transport towards the tube wall. This, however, is not the complete picture. Saffman (1965) provided a theoretical basis for an additional force on a particle arising from its motion in a flow field where there is a velocity gradient, as for example in a sheared viscous flow. The scenario is shown schematically in Figure 7.10. Here, put simply, the higher incident air velocity on one side of a particle gives rise to a lower pressure perpendicular to the flow, while the lower air velocity on the other side gives rise to a higher pressure. The result is a net force perpendicular to the direction of the flow. Particle rotation is also induced and this too contributes to the force. Saffman referred to this effect as 'lift', although – it should be noted – it is quite different from the more familiar aerodynamic lift phenomenon. Lipatov *et al.* (1989a) pointed out that this situation arises during particle motion in sheared laminar flow in a tube, and conducted experiments to investigate the resultant particle deposition on the wall of a tube for vertically upwards and vertically downwards flows. Test aerosols with d_{ae} of about 30 μm were generated from *Lycopodium* spores and penetration efficiency (P) was measured for ranges of L_{tube} from 0.1 to 0.25 m, δ_{tube} from 3 to 10 mm, and U_{tube} from 0.4 to 1.5 m s^{-1}. By way of illustration, Figure 7.11 shows some typical results for P as a function of U_{tube} for just one tube with $L_{tube} = 0.18$ m and $\delta_{tube} = 5.6$ mm. It is seen that deposition efficiency was significantly greater for the downwards than for the upwards

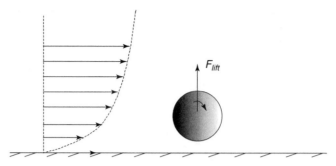

Figure 7.10   *Schematic to illustrate the principle of the Saffman lift force for a particle moving in the boundary layer close to a solid surface*

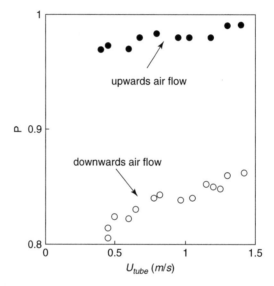

Figure 7.11   *Penetration (P) of particles in laminar flow through a vertical cylindrical tube associated with the Saffman lift force, shown as a function of the mean air velocity in the tube (U_{tube}) for both upwards and downwards air flow in the tube, for Lycopodium spores with d_{ae} about 30 μm and for L_{tube} from 0.1 to 0.25 m and δ_{tube} from 3 to 10 mm (data from Lipatov et al., 1989a)*

flows, indicating the role of particle gravitational settling in the relative motion between the particle and air flow. An important feature of the results of Lipatov *et al.* was the demonstration that, under certain conditions, the Saffman lift force could play a significant role in particle losses during transport through a tube. Fan *et al.* (1992) studied the problem analytically and, from the equations of motion for the particle transport in the tube, characterised the contribution of the lift force in terms of the dimensionless *lift number, Pl*, given by:

$$Pl = 0.65 \left(\frac{\gamma}{\rho} \right) \left(\frac{d}{\delta_{tube}} \right) \left(\frac{U_{tube}}{\sqrt{|\kappa| \nu}} \right) \tag{7.33}$$

where, following nomenclature introduced earlier, γ and ρ are the densities of the particle and the air, respectively, d is the particle geometric diameter, κ is here the rate of shear in the flow, and v the kinematic viscosity of the air. Fan *et al.* recommended that a functional dependence on *Pl* should be included in any discussion of deposition in tubes, along with R_{tube} and r_{tube} as already suggested – see Equation (7.29) – together with Re_{tube} as appropriate. They also noted that the lift force contribution applies not only in the body of laminar tube flow but also in the sheared region near the wall in turbulent flow. In practical aerosol sampling with this sort of configuration, there are indeed likely to be cases where Re_{tube} is large enough for the flow in the tube to be turbulent.

The effect of turbulence on the deposition of particles in tube flow has been dealt with extensively in the literature (e.g., Friedlander and Johnstone, 1957; Davies, 1965; Sehmel, 1967; Beal, 1970; Agarwal, 1972; Liu and Ilori, 1973; Liu and Agarwal, 1974; and others). In such cases, gravitational deposition has usually been neglected since it is negligible in comparison with that derived from the turbulent motions. The physical basis of turbulent deposition in the tube derives from the mixing of the airborne particles in the central 'core' of the tube and the 'free flight' projection of particles across the relatively quiescent layer immediately adjacent to the wall. Deposition may be described in terms of a *deposition velocity*, defined as the ratio of the number of particles deposition per unit area of the tube per unit time to the number initially present per unit volume in the bulk airflow. Thus, integrating particle deposition along the whole length of the tube leads to:

$$P_{tube} = \exp\left[-4\left(\frac{w}{U_{tube}}\right)\left(\frac{1}{r_{tube}}\right)\right] \tag{7.34}$$

where it is assumed that all particles are permanently removed from the flow once they have arrived at the tube wall. The challenge now is to obtain a suitable expression for w. Friedlander and Johnstone (1957) approached this by first assuming that the coefficient of turbulent diffusion for particle migration from the main flow in the tube was the same as that for the air itself. It was then assumed as the criterion for deposition that particles must diffuse to within one particle stop-distance (s) of the tube wall. For the purpose of obtaining s, the radially directed, root-mean-square velocity of turbulence in the core of the turbulent flow at the center of the tube was considered to be appropriate for describing the velocity of projection. For pipe flow, the empirical expression:

$$(u_r'^2)^{1/2} = 0.9U_{tube}\left(\frac{f}{2}\right)^{1/2} \tag{7.35}$$

was available based on the experimental work of others, where f is the well-known Fanning friction factor given by:

$$f = \frac{0.316}{4Re_{tube}^{1/4}} \tag{7.36}$$

The local radial flux of particles towards the tube wall j is given by Fick's law such that:

$$j = -D_{tp}(r)\frac{\partial \bar{c}}{\partial r} \tag{7.37}$$

where $D_{tp}(r)$ on the right hand side is the local coefficient of turbulent diffusion – at a radial distance r from the axis of the tube – for the particles, and the quotient is the radial gradient of the local time-averaged particle concentration. In order to arrive at a working analytical expression for w, Friedlander and Johnstone assumed the simplest case of turbulent deposition onto a flat wall from an infinite turbulent freestream. For this, they integrated Equation (7.37) from $r = \infty$ to $r = s$, at which point it was assumed that the concentration had dropped to zero. Thus:

$$w = \frac{|j|}{\overline{c}} \int_s^{\infty} \left[\frac{dr}{D_{tp}(r)} \right] \tag{7.38}$$

To evaluate this expression, s was calculated from Equation (2.37) along with Equation (7.35). For the viscous sublayer close to the wall, $D_{tp}(r)$ was estimated using an expression for the diffusion coefficient of air in the viscous sublayer:

$$D_{tp} \approx D_t = \left(\frac{\rho}{\eta} \right)^2 \left[\frac{r U_{tube} \left(\frac{f}{2} \right)^{1/2}}{14.5} \right]^3 \tag{7.39}$$

Elsewhere in the flow, other appropriate expressions were employed for D_{tp}, based on knowledge of the detailed characteristics of turbulent pipe flow.

Friedlander and Johnstone performed experiments to examine the turbulent deposition of particles of iron and aluminum in glass and brass tubes with δ_{tube} about 13 mm. Penetration (P_{tube}) was measured as a function of Re_{tube} in the range up to about 50 000 for particles with d_{ae} of the order of from 2 to 4 μm. Experimental values of w were then determined from the expression in Equation (7.34). The results are shown in Figure 7.12. Agreement between theory and experiment was fair, although there was a tendency for the measured values of w to fall somewhat below those predicted theoretically. The suggestion that this could have been caused by the loss of deposited particles from the wall by blow-off – by a mechanism similar to that described in the first part of this chapter – is supported by some additional results that were obtained for the inside surfaces of the tubes coated with glycerol jelly to increase particle-wall adhesion. For these, the experimental values of w were found to increase towards levels more consistent with theory.

Liu and Agarwal (1974) carried out similar experiments using a glass tube of diameter 12.7 mm and length 1.02 m. Liquid droplet test aerosols of olive oil were produced by means of a vibrating-orifice generator in the range of d_{ae} from about 1 to 20 μm. Like Friedlander and Johnstone, Liu and Agarwal obtained the turbulent deposition velocity (w) by plugging measurements of penetration into Equation (7.34). However, unlike Friedlander and Johnstone, they chose to plot their results in terms of:

$$w^* = \frac{w}{\left(\frac{f}{2} \right)^{1/2} U_{tube}} \tag{7.40}$$

where the denominator, the so-called 'friction velocity', served to nondimensionalise w. Values of w^* thus obtained were plotted as a function of the particle relaxation time, also nondimensionalised with respect to system parameters such that:

$$\tau^* = \frac{\tau \rho \left(\frac{f}{2} \right) U_{tube}^2}{\eta} \tag{7.41}$$

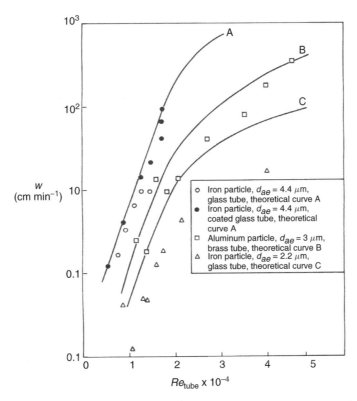

Figure 7.12 *Deposition velocity (w) as a function of Reynolds number (Re) for particles deposited inside a straight cylindrical tube by turbulent diffusion, comparison between experiment and theory (Friedlander and Johnstone, 1957). Reproduced with permission from Vincent, Aerosol Sampling: Science and Practice. Copyright (1989) John Wiley & Sons, Ltd*

This reflects particle size. The experimental results are shown in this form in Figure 7.13, revealing that deposition increased steadily with particle size and that w^* – unlike w in the plot of the data of Friedlander and Johnstone – did not depend significantly on Re_{tube}. Also shown in Figure 7.13 is the curve Liu and Agarwal calculated from their own version of theory as well as curves based on the theories of other workers, including that of Friedlander and Johnstone as already described, and those of Davies (1965), Beal (1970) and Liu and Ilori (1973). These theoretical models were constructed from the same basic diffusion model – except that of Davies which also includes the effects of Brownian diffusion for very small particles – and are seen to predict broadly the same trend. Where they differed primarily is in the way in which particles were assumed to be transported from the turbulent core of the tube to the point near the wall at which they undertook 'free flight' projection across the quiescent layer there. The only theory to depart substantially from the rest in this respect was that of Davies. Quite reasonably, he had assumed that the more appropriate turbulent velocity for the projection of particles towards the wall was that close to the wall, and not – as the other workers had assumed – close to the core of the turbulent flow in the tube. However, as Friedlander noted in his 1977 book (Friedlander, 1977), this did appear to lead to the prediction of excessively low deposition velocities. No explanation for the inconsistency has yet been offered.

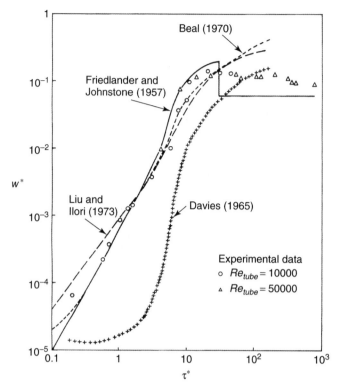

Figure 7.13 *Dimensionless deposition velocity (w*) for particle deposition in a straight cylindrical tube as a function of dimensionless particle relaxation time (τ*) – see Equations (7.40) and 7.41) – comparison between the experiment results of Liu and Agarwal (1974) and various theories as indicated. Reprinted from Journal of Aerosol Science, 5, Liu and Agarwal, 145–155. Copyright (1974), with permission from Elsevier*

7.3.2   Deposition losses inside a bent sampling tube

Agarwal (1972) considered the inertial losses of particles in a cylindrical sampling tube bent at right angles, and proposed a simple model based on the diagram in Figure 7.14. Here, impaction occurs as a result of the radial motion towards the outer wall of the tube of particles moving initially around the bend with the bulk air flow. Agarwal considered this system from the point of view of centrifugal forces, by analogy with cyclone theory. For further simplicity, uniform plug flow was assumed and deposition by turbulent diffusion and gravity were neglected. Despite the simplicity, quite good agreement with experiment was found (Vincent *et al.*, 1987). Cheng and Wang (1981) were the first to carry out a more rigorous theoretical analysis for the impaction of particles in a similarly bent tube, based on a more realistic description of the airflow in the tube. They used the fully developed laminar flow field model characterised by an inviscid core region and a boundary layer. They invoked the *Dean number*:

$$De = \frac{Re_{tube}}{(2R_{bend}/\delta_{tube})^{1/2}} \tag{7.42}$$

to define the nature of the velocity distribution over the cross-section of the tube in the bend for such flows. The higher the value of Re_{tube} and the tighter the bend, the greater the value of De and the greater

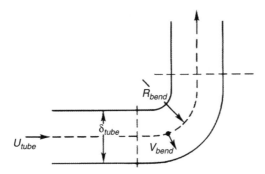

Figure 7.14   *Schematic of particle transport in flow in a 90° bent sampling tube upon which to base estimation of deposition under the combined influences of inertial and gravitational forces. Reproduced with permission from Vincent, Aerosol Sampling: Science and Practice. Copyright (1989) John Wiley & Sons, Ltd*

the degree to which the airflow is pushed towards the outside of the bend. From particle trajectories for transport in this sort of flow, Cheng and Wang calculated penetration (P_{bend}) as a function of St_{tube} for contrasting values of De, where:

$$St_{tube} = \frac{\tau U_{tube}}{\delta_{tube}} \tag{7.43}$$

is a Stokes number referred to aerosol transport in the tube. The results of these calculations are shown as the continuous curves in Figure 7.15. Later, Pui *et al.* (1987) reported some experimental measurements suitable for testing this theory. These were carried out using glass and steel tubes with δ_{tube} ranging from about 1 to 9 mm and monodisperse test aerosols of oleic acid produced by means of a vibrating-orifice generator. Results were obtained for ranges of St_{tube} and values of De closely corresponding to the ones examined by Cheng and Wang. These experimental results are plotted alongside the calculated ones in Figure 7.15. Agreement was good for the larger value of De but less good for the smaller one. Pui and his colleagues suggested, however, that the flow field model used by Cheng and Wang may not be applicable for De as small as 35.

7.3.3   Deposition inside a thin-walled tube facing into the wind

The description of particle deposition in the straight tube studies cited above does not quite take the form most relevant to aerosol sampling because it does not account for the transition – the coupling – that takes place between the external freestream and the airflow inside the tube. The more realistic case of particle deposition inside a cylindrical, thin-walled sampling tube facing into an aerosol-laden freestream has been investigated experimentally by Willeke and his colleagues (Okazaki *et al.*, 1987; Okazaki and Willeke, 1987). The experiments were carried out in a wind tunnel using horizontally placed thin-walled cylindrical sampling tubes with inlet diameter $\delta(\equiv \delta_{tube})$ equal to 3.2, 5.6 and 10.3 mm, respectively, and length (L_{tube}) equal to 200 and 300 mm. Wind speed in the freestream (U) was varied from 2.5 to 10 m s^{-1} and velocity ratio (R) from 0.5 to 4. Monodisperse test aerosols of oleic acid were produced by means of a vibrating-orifice generator, with d_{ae} in the range from 10 to 40 μm. The particles penetrating through the tube were counted by means of an optical particle detector. The ones deposited on the internal walls of the tube were subsequently washed off and assessed fluorometrically. Tube penetration was

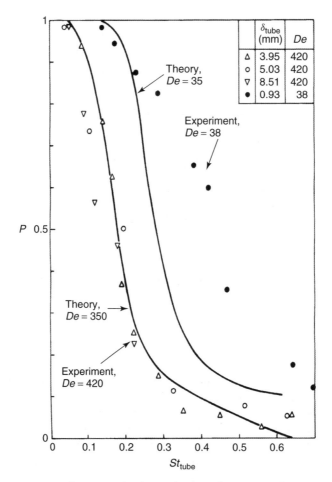

Figure 7.15 *Penetration (P) as a function of tube Stokes number (St$_{tube}$) for various Dean numbers (De), experimental data from Pui et al. (1987) for the range of 90° bent tubes with the various diameters indicated, solid curves based on the theory of Cheng and Wang (1981). Reproduced with permission from Vincent, Aerosol Sampling: Science and Practice. Copyright (1989) John Wiley & Sons, Ltd*

obtained experimentally from:

$$P_{tube} = \frac{N_{detector}}{N_{detector} + N_{wall}} \tag{7.44}$$

where $N_{detector}$ was the number of particles at the detector counter and N_{wall} the corresponding number obtained for the wall deposit. The results of these experiments were first examined as a function of the Stokes number for the sampler as defined in earlier chapters. Strong dependence was found with P_{tube} decreasing as St increased. This relationship was presumably associated with inertial effects associated with particle motion in the distorted air flow immediately adjacent to the tube entry. However, further dependencies on particle size, tube dimensions, wind speed and sampling flow rate were also clearly apparent. Under the assumption that gravitational deposition in the tube should play a significant role, the dimensionless tube gravitational deposition parameter (G_{tube}) was suggested as a suitable additional parameter, where:

$$G_{tube} = \left(\frac{L_{tube}}{\delta}\right)\left(\frac{v_s}{U_{tube}}\right) \qquad (7.45)$$

and where, now, $U_s \equiv U_{tube}$. Here G_{tube} reflects the ratio between the residence time of a particle passing along the whole length of the tube and the time associated with its motion under gravity towards – and deposition on – the wall. It is intuitive that the effect of gravitational deposition should increase as G_{tube} increases. In addition, close to the entry itself, the flow must inevitably be influenced by the nature of the boundary layer that grows inside the tube, and so will be a function of Re_{tube}, growing in thickness more rapidly the higher the value of Re_{tube}. It is therefore to be expected that effects associated with Re_{tube} would occur close to the tube entrance as particles encounter the slower-moving air in the boundary layer flow. In turn this would result in enhanced particle deposition in this region. This indeed was what was observed by Willeke and his colleagues in their experiments. Okazaki and Willeke (1987) proposed that:

$$P_{tube} = f\{K_{tube}\} = f\{St, G_{tube}, Re_{tube}\} \qquad (7.46)$$

and, from dimensional arguments, that:

$$K_{tube} = St^a G_{tube}^b Re_{tube}^c \qquad (7.47)$$

Best-fit with the experimental data was found for the combination $a = 0.5$, $b = 0.5$ and $c = -0.25$, from which it was shown that Equation (7.46) could be expressed more specifically as:

$$P_{tube} = f\left\{St, Re_{tube}, Fr, R, \frac{L_{tube}}{\delta}\right\} \qquad (7.48)$$

where now the effect of gravity is represented by the Froude number *(Fr)*, modified by the velocity ratio R to represent the flow condition at the entry, and including the aspect ratio of the tube (L_{tube}/δ). The experimental results of Okazaki and Willeke are plotted in Figure 7.16 in the form suggested by Equation (7.46) with the coefficients indicated. They displayed an excellent collapse which strongly supported the preceding physical arguments. Okazaki and Willeke found that the trend of P_{tube} with K_{tube} was described very well by the empirical expression:

$$P_{tube} = \exp(-4.7 K_{tube}^{0.75}) \qquad (7.49)$$

where P_{tube} may now be thought of in terms of two components, P_{entry} for the region of the tube where the effects of the coupling between the external and internal flows are felt and P_{inner} for the region of the flow inside the tube where the flow has stabilised and so is not dependent on the conditions at the entry. That is:

$$P_{tube} = P_{entry} + P_{inner} \qquad (7.50)$$

Soon afterwards, Wiener *et al.* (1988) from the same group extended the work to investigate the effect of a turbulent external freestream, again for a thin-walled tube facing directly into the wind. They noted in particular that the main effect of the turbulent motions in the external air approaching the sampling inlet was to increase the deposition inside the tube. A similar tendency was subsequently found by other researchers (e.g. Kalatoor *et al.*, 1995; Cain and Ram, 1998; Gao *et al.*, 1999). It was clearly related to the coupling between the external and internal flows, and the penetration of turbulence into – and eventual dissipation in – the tube.

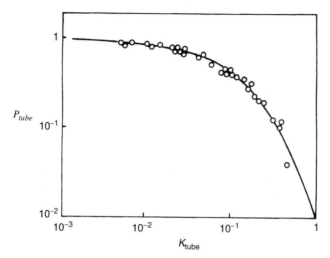

Figure 7.16   *Penetration (P_{tube}) as a function of deposition parameter (K_{tube}) as suggested by Equation (7.44), experimental data and theory (solid curve) from Okazaki and Willeke (1987). Reprinted from Aerosol Science and Technology, 7, Okazaki and Willeke, 275–283. Copyright (1987), with permission from Elsevier*

It was noted by Fan *et al.* (1992) that, during such consideration of the losses of particles during the entry of particles into a tube inlet in the work cited, the possible dependence on the Saffman lift force had not been included, despite its potential importance. It was included, however, in the theoretical study of Grinshpun *et al.* (1989) for laminar tube flows where $Re_{tube} < 1500$. They were able to demonstrate good agreement between their theory and experimental studies of particle inlet deposition in vertical cylindrical tubes, confirming the importance of taking this commonly neglected effect into account.

7.3.4   Deposition inside a thin-walled tube at other orientations

For orientations of the thin-walled tube other than forwards-facing, Hangal and Willeke (1990) noted the presence of the *vena contracta* just inside the tube entrance and its likely influence on particle deposition in that region. It is reasonable to expect that, as the angle of the sampler with respect to the freestream (θ) increases, the appearance of the *vena contracta* will become more marked and increasingly asymmetric. In addition, it is likely to become unstable and exhibit vortex shedding. This very complicated behavior must inevitably influence particle transport just inside the tube inlet.

Hangal and Willeke carried out experiments in the same facility described above by Okazaki and Willeke, and sought to use their experimental data towards the development of a unified performance model for the thin-walled tube aspirating at all forwards-facing angles ($0° \leq \theta \leq 90°$). The research included orientations with respect to not only the horizontal but also to the vertical, so that the effects of gravitational settling were also included. Special attention was given to what happens near the entry. For this, Hangal and Willeke identified possible contributions from both gravitational settling and from impaction onto the inner walls of the tube, such that:

$$P_{entry} = {}_{gravity}P_{entry} + {}_{impaction}P_{entry} \longrightarrow {}_{impaction}P_{entry} \qquad (7.51)$$

They then went on to express P_{entry} as:

$$P_{entry} = \exp[-75(I_W + I_V)] \qquad (7.52)$$

where I_W is the role of direct impaction onto the inner wall of the tube and I_V is the role of the *vena contracta*. These terms were given by:

$$I_W = St R^{1/2} \sin(\theta \pm \psi) \sin\left(\frac{\theta \pm \psi}{2}\right)$$

$$I_V = 0.09\left[(1 - R)\left(\frac{St}{R^3}\right)\cos\theta\right]^{0.3} \quad \text{for } R < 1 \qquad (7.53)$$

$$I_V = 0 \text{ for } R \geq 1$$

Here the angle ψ is the *gravity angle* first introduced by Hangal and Willeke and later simplified by Grinshpun *et al.* (1994) to:

$$\psi = \theta - \left[\sin^{-1}\left(\sin\theta - \frac{v_s}{U}\cos\theta\right)\right] \qquad (7.54)$$

This empirical model provided a good description of the experimental data. Later, Sreenath *et al.* (2001) conducted a new experimental study of particle losses associated with the inlet effect for a range of orientations $0 \leq \theta \leq 75°$. They used a version of the experimental method that involved the use of the direct-reading aerodynamic particle sizer (APS) to rapidly acquire large amounts of data. In those experiments, from measurements of the number concentrations of particles aspirated and then penetrating down to the APS, along with knowledge of the aspiration efficiency of the tube for various forwards-facing orientations from what were – by then – reliable models, it was possible to extract data for the efficiency of particles through the inlet. Sreenath *et al.* used them to construct an empirical model for the efficiency of penetration through the inlet region, taking the form:

$$P_{entry}(\theta) = \{1 - c_1 R^{c_2} St^{c_3}[1 - \exp(-c_4\theta)]\}[\exp(-c_5 K_\theta^{c_6})] \qquad (7.55)$$

where

$$K_\theta = 0.152 St \left(\frac{\cos\theta}{U_s^{2.5}R}\right)^{1/2} \qquad (7.56)$$

and where gravity was neglected. Nonlinear regression yielded a fair overall fit for the combination of coefficients $c_1 = 0.094$, $c_2 = 1.799$, $c_3 = 0.280$, $c_4 = 142.78$, $c_5 = 71.84$ and $c_6 = 1.695$. When $\theta = 0°$, the above model reverts to the model of Okazaki and Willeke (1987), as contained in Equation (7.49). Here, however, as seen in the models described above, the attempts to create semi-empirical mathematical descriptions of the losses of particles during the entry of particles into even very simple thin-walled tube sampling systems have resulted in models that are complicated and not particularly transparent.

7.3.5 Rebound of particles from internal walls

Earlier in this chapter there was detailed discussion of particle losses from external walls of the sampler body. Similar effects may occur inside a sampler after aspiration. This has not been widely studied, but it is to be expected that the physics will be similar to that for particle interactions with external surfaces. Lipatov *et al.* (1989b) acknowledged that such rebound may involve either *elastic* bounce or *nonelastic* blow-off as distinctly different phenomena. They described experiments to determine the rebound coefficient for particle losses inside the tube. As in their earlier studies of the rebound of particles from external surfaces, they again used *Lycopodium* spores with d_{ae} about 30 μm and cylindrical tubes of various diameters with greased and ungreased internal walls. They showed that particle losses for the ungreased tubes were negligible for U_{tube} up to about $2 \mathrm{~m \, s^{-1}}$, but increased sharply for velocities beyond that. They also showed that the rebound coefficient was also strongly dependent on Re_{tube}.

7.3.6 More complicated systems

The transition sections examined so far for linking the entry and filter, substrate or sensing regions of sampling systems have all been of relatively simple shape – cylindrical tubes of constant cross-section. However, sampling systems in the real world are often much more complicated, with internal geometries characterised by changes in cross-sectional dimensions and shape, as well as changes in flow direction. Experience for the thin-walled tube systems described above indicates that general models for inlet and other transmission efficiencies are not likely to be generalisable. Yet such particle losses, representing interferences to the performances of practical aerosol sampling systems, can lead to significant biases in measurement. In the real world, it is necessary that such biases should be accounted for. This requires that the various contributions to sampler performance like those referred to above should be characterised individually for each sampler by appropriate testing. Meanwhile, the results of research like that described will be useful in providing guidance for such testing – or, better still, in the design of sampling systems to minimise such effects.

7.3.7 Electrostatic effects

As mentioned earlier in this chapter, it is well-known that aerosols in many practical situations are electrically charged. This is especially the case for the 'relatively fresh' aerosols found in workplaces that have been charged during generation – for example, by contact or friction charging – and have not been airborne for long enough for the natural process of neutralisation by atmospheric charges (e.g. from cosmic rays, atmospheric processes). The time constant for such neutralization is thought to be of the order of about 20 min, long compared with the time of removal of aerosols and other air pollutants by industrial ventilation. By contrast, aerosols in the ambient atmosphere have usually been airborne for much longer and are 'relatively aged' and hence tend towards electric neutrality, as defined by Boltzman equilibrium (e.g. see Hinds, 1999). As far as the external aspiration process is concerned, as discussed already, particle charge is thought not to have a large effect on sampler performance. However, the same cannot necessarily be assumed for aerosols during transport in the relatively confined space *inside* a sampling system.

Liu *et al.* (1985) carried out a series of laboratory-based experiments to investigate the deposition of charged fine particles through a variety of coiled tubes of length 3 m, both of conducting copper and insulating plastic (various types), for various flow rates and tube diameters. Penetration through the sampling tube (P_{tube}) of monodisperse sodium chloride aerosol was measured as a function of particle geometric diameter (d) in the range from about 0.03 to 0.5 μm. Particle charge conditions studied included 'near-neutral' (corresponding to Boltzmann equilibrium, produced by exposing the particles to a bipolar ion cloud from a radioactive source of beta particles), 'singly charged' (produced using a mobility classifier) and 'multiply charged' (produced by diffusion charging in a cloud of unipolar ions). Typical results are shown in Figure 7.17 and are compared with calculations made on the basis of deposition by diffusion using the model of Gormley and Kennedy. The results for the conducting metal tube [Figure 7.17(a)] showed P_{tube} to be somewhat lower than expected on the basis of deposition by diffusion alone. However, the relatively weak dependence on particle charge seemed to rule out any contribution due to electrostatic forces. This is consistent with some observations reported by Thomas (1967). Liu and his colleagues concluded that the observed enhancement of deposition was associated

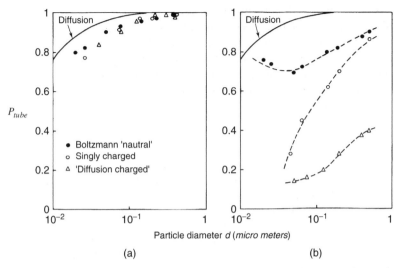

Particle diameter *d* (*micro meters*)

(a)   (b)

Figure 7.17 *Penetration (P_{tube}) of particles through sampling tubes as a function of particle diameter (d) for charged particles; (a) copper tube of length 3 m, diameter 4.3 mm, flow rate 1 Lpm; (b) polytetrafluoroethylene tube of length 3 mm, diameter 5 mm and flow rate 1 Lpm, wrapped in aluminum foil (Liu et al., 1985). Reproduced with permission from Liu et al., Annals of Occupational Hygiene, 29, 251–269. Copyright (1985) British Occupational Hygiene Society*

with the secondary flows induced in the tube due to its being coiled. For the plastic tubes, the results were strongly dependent on the type of plastic used. For polyvinylchloride (not shown), they differed very little from those for the copper tubing. However, for polyethylene and polytetrafluoroethylene (also not shown), P_{tube} was dramatically lower – in fact, so much so that close to 100 % deposition took place within the first few centimeters of tubing. It was suggested that the effect was associated with the charging of the tubing itself that had taken place during the flexing and bending which had been applied prior to sampling, possibly by a process related to, or analogous to, the piezoelectric effect. In such cases, however, Liu and his colleagues found that finite penetration could be restored somewhat by wrapping the tube in conducting foil. Even so, as shown in Figure 7.17(b), penetration was still substantially less than for the metal and poly(vinyl chloride) tubes.

A number of workers have considered theoretically the deposition of particles inside a conducting cylindrical tube due to electrical image forces and space charge forces. The main motivation for such work has been in relation to the electrostatic deposition of particles in the lung airways after inhalation. Liu and his colleagues examined their experimental results in the light of these models, and concluded that image and space charge effects had played no significant role in their experiments.

Over the years there has been concern about internal wall losses inside plastic samplers like those used by occupational hygienists in some countries for personal sampling applications. Puskar *et al.* (1991) investigated the problem for the widely used 37 mm plastic cassette through a series of experiments under field conditions in pharmaceutical industry workplaces. They reported very significant losses of particles in the complex cavity between the 4 mm inlet and the 37 mm filter and speculated that these were governed largely by electrostatic attractive forces. However, because of the lack of ability in this and other such studies in the field to control all potentially important sampling parameters, it was difficult to pin down the exact nature of the wall loss process.

Finally, Mark (1974) investigated problems associated with electric charge at the filter itself, observing that charged particles may be repelled electrostatically as they approached the filter and so may be deposited on the filter housing. Such charging, like that for the plastic tubes described above, probably arose from the handling of the filters prior to sampling. Mark discovered that the problem could be reduced by impregnating the filter prior to sampling with hyamine, thus raising the conductivity of the filter media sufficiently to prevent substantial charge accumulation.

References

Addlesee, A.J. (1980) Anisokinetic sampling of aerosols at slot intake, *Journal of Aerosol Science*, 11, 483–493.

Agarwal, J.K. (1972) The sampling of aerosols, Particle Technology Publication No. 208, Department of Mechanical Engineering, University of Minnesota, Minneapolis, MN, USA.

Almich, B.P. and Carson, G.A. (1974) Some effects of charging on 10 mm nylon cyclone performance, *American Industrial Hygiene Association Journal*, 35, 603–612.

Bagnold, R. (1954) *The Physics of Blown Sands and Desert Dunes*, Methuen, New York.

Beal, S.K. (1970) Deposition of particles in turbulent flow on pipe or channel walls, *Nuclear Science and Engineering*, 40, 1–11.

Briant, J.K. and Moss, D.R. (1984) The influence of electrostatic charge on the performance of 10-mm nylon cyclones, *American Industrial Hygiene Association Journal*, 45, 440–445.

Cain, S.A. and Ram, M. (1998) Numerical simulation studies of the turbulent airflow through a shrouded airborne aerosol sampling probe and estimation of the minimum sampler transmission efficiency, *Journal of Aerosol Science*, 29, 1145–1156.

Chalmers, J.A. (1957) *Atmospheric Electricity*, Pergamon Press, Oxford.

Cheng, Y.S. and Wang, C.S. (1981) Motion of particles in bends in circular pipes, *Atmospheric Environment*, 15, 301–306.

Cheng, Y.S. and Yeh, H.C. (1979) Particle bounce in cascade impactors, *Environmental Science and Technology*, 13, 1392–1395.

Corn, M. and Stein, F. (1965) Re-entrainment of particles from a plane surface, *American Industrial Hygiene Association Journal*, 26, 325–336.

Dahneke, B. (1971) The capture of aerosol particles by surfaces, *Journal of Colloidal and Interface Science*, 37, 342–353.

Dahneke, B. (1973) Measurements of bouncing of small latex spheres, *Journal of Colloidal and Interface Science*, 45, 584–590.

Davies, C.N. (1965) The rate of deposition of aerosol particles from turbulent flow through ducts, *Annals of Occupational Hygiene*, 8, 239–245.

Fan, B., McFarland, A.R. and Anand, N.K. (1992) Characterization of the aerosol particle lift force, *Journal of Aerosol Science*, 23, 379–388.

Friedlander, S.K. (1977) *Smoke, Dust and Haze*, John Wiley & Sons, Inc., New York.

Friedlander,. S.K. and Johnstone, H.F. (1957) Deposition of suspended particles from turbulent gas streams, *Industrial and Engineering Chemistry*, 49, 1151–1156.

Fuchs, N.A. (1964) *The Mechanics of Aerosols*, Macmillan, New York.

Gajewski, A. (1980) Electrostatic charges generated on the human body, *Ochr. Pr.*, 8, 17–19.

Gao, P.F., Dillon, H.K., Baker, J. and Oestenstad, K. (1999) Numerical prediction of the performance of a manifold sampler with a circular slit inlet in turbulent flow, *Journal of Aerosol Science*, 30, 299–312.

Gormley, P.G. and Kennedy, M. (1949) Diffusion from a stream flowing through a cylindrical tube, *Proceedings of the Royal Irish Academy*, 52A, 163–169.

Grinshpun, S.A., Lipatov, G.N. and Semenyuk, T.I. (1989) Aerosol separation within entrance regions of the upright channels (experiments and calculation), *Journal of Aerosol Science*, 20, 975–977.

Grinshpun, S.A., Chang, C.W., Nevalainen, A. and Willeke, K. (1994) Inlet characteristics of bioaerosol samplers, *Journal of Aerosol Science*, 25, 1503–1522.

Hangal, S. and Willeke, K. (1990) Aspiration efficiency: unified model for all forward sampling angles, *Environmental Science and Technology*, 24, 688–691.

Hinds, W.C. (1999) *Aerosol Technology: Properties, Behavior and Measurement of Airborne Particles*, 2nd Edn, John Wiley & Sons, Inc., New York.

Ingham, B.D. and Wen, X. (1993) Disklike body sampling in a turbulent wind, *Journal of Aerosol Science*, 24, 629–642.

Ingham, D.B. and Yan, B. (1994) Re-entrainment of particles on the outer wall of a cylindrical blunt sampler, *Journal of Aerosol Science*, 25, 327–340.

Ingham, D.B., Wen, X., Dombrowski, N. and Foumeny, E.A. (1995) Aspiration efficiency of a thin-walled shallow-tapered sampler rear-facing the wind, *Journal of Aerosol Science*, 26, 933–944.

Jayasekera, P.N. and Davies, C.N. (1980) Aspiration below wind velocity of aerosols with sharp-edged nozzles facing the wind, *Journal of Aerosol Science*, 11, 535–547.

Johnston, A.M., Vincent, J.H. and Jones, A.D. (1985) Measurements of electric charge for workplace aerosols, *Annals of Occupational Hygiene*, 29, 271–284.

Johnston, A.M., Vincent, J.H. and Jones, A.D. (1987a), Electrical charge characteristics of dry aerosols produced by a number of laboratory mechanical generators, *Aerosol Science and Technology*, 6, 115–127.

Johnston, A.M., Vincent, J.H. and Jones, A.D. (1987b), The effect of static charge on the aspiration efficiencies of airborne dust samplers: with special reference to asbestos, *American Industrial Hygiene Association Journal*, 48, 613–621.

Kalatoor, S., Grinshpun, S.A. and Willeke, K. (1995) Aerosol sampling from fluctuating flows into sharp-edged tubular inlets, *Journal of Aerosol Science*, 26, 387–398.

Kuusisto, P. (1979) Measurement with the filter method. *1. Investigation into electrostatic interference in sampling glass fibre dust*, The Swedish Board of Industrial Safety, Stockholm.

Levin, L.M. (1957) The intake of aerosol samples, *Izv. Akad. Naik. SSSR Ser. Geofiz.*, 7, 914–925.

Lipatov, G.N., Grinshpun, S.A., Semenyuk, T.I. and Sutugin, A.G. (1988) Secondary aspiration of aerosol particles into thin-walled nozzles facing the wind, *Atmospheric Environment*, 22, 1721–1727.

Lipatov, G.N., Grinshpun, S.A., Shingarov, G.L. and Sutugin, A.G. (1986) Aspiration of coarse aerosol by a thin-walled sampler, *Journal of Aerosol Science*, 17, 763–769.

Lipatov, G.N., Grinshpun, S.A. and Semeyuk, T.I. (1989a), Properties of crosswise migration of particles in ducts and inner aerosol deposition, *Journal of Aerosol Science*, 20, 935–938.

Lipatov, G.N., Semeyk, T.I., Grinshpun, S.A., Yakimchuk, V.I. and Skaptsov, A.S. (1989b), Interaction between coarse aerosol particles and inner walls of the upright tracts (adhesion, rebound and blow-off), *Journal of Aerosol Science*, 20, 939–941.

Liu, B.Y.H. and Agarwal, J.K. (1974) Experimental observation of aerosol deposition in turbulent flow, *Journal of Aerosol Science*, 5, 145–155.

Liu, B.Y.H. and Ilori, T.A. (1973) Inertial deposition of aerosol particles in turbulent pipe flow, Presented at the American Society of Mechanical Engineers (ASME) Symposium on *Flow Studies in Air and Water Pollution*, Atlanta, Georgia, June, 1973.

Liu, B.Y.H., Pui, D.Y.H., Rubow, K.L. and Szymanski, W.W. (1985) Electrostatic effects in sampling and filtration, *Annals of Occupational Hygiene*, 29, 251–269.

Lovstrand, K.-G. and Rosen, V. (1980) Interference effect of static electricity in particulate sampling by filtration methods, University of Uppsala Institute of High Voltage Research Report No. UURIE, Uppsala, pp. 136–180.

Lutz, S.A. and Bajura, R.A. (1984) Particulate sampling efficiency of thin- and thick-walled probes in highly turbulent flow streams. In: *Aerosols* (Eds. B.Y.H. Liu, D.Y.H. Pui and H. Fissan), Elsevier, Amsterdam, pp. 179–182.

Mark, D. (1974) Problems associated with the use of membrane filters for dust sampling when compositional analysis is required, *Annals of Occupational Hygiene*, 17, 35–40.

Mark, D. (1978) Instruments for sampling respirable airborne dust in coal mines: an investigation of their performance, M. Phil. Thesis, University of Nottingham.

Mark, D. and Vincent, J.H. (1986) A new personal sampler for airborne total dust in workplaces, *Annals of Occupational Hygiene*, 30, 89–102.

Mark, D., Vincent, J.H. and Witherspoon, W.A. (1982) Particle blow-off: a source of error in blunt dust samplers, *Aerosol Science and Technology*, 1, 463–469.

Mark, D., Vincent, J.H., Gibson, H. and Lynch, G. (1985) A new static sampler for airborne total dust in workplaces, *American Industrial Hygiene Association Journal*, 46, 127–133.

May, K.R. and Clifford, R. (1967) The impaction of aerosol particles on cylinders, spheres, ribbons and discs, *Annals of Occupational Hygiene*, 10, 83–95.

McFarland, A.R., Wedding, J.B. and Cermak, J.E. (1977) Wind tunnel evaluation of a modified Andersen impactor and an all-weather sampling inlet, *Atmospheric Environment*, 11, 535–539.

Ogden, T.L. and Birkett, J.L. (1978) An inhalable dust sampler for measuring the hazard from total airborne particulate, *Annals of Occupational Hygiene*, 21, 41–50.

Ogden, T.L., Birkett, J.L. and Gibson, H. (1978) Large particle entry efficiencies of the MRE -113A gravimetric dust sampler, *Annals of Occupational Hygiene*, 21, 251–263.

Okazaki, K. and Willeke, K. (1987) Transmission and deposition behaviour of aerosols in sampling inlets, *Aerosol Science Technology*, 7, 275–283.

Okazaki, K., Wiener, R.W. and Willeke, K. (1987) The combined effect of aspiration and transmission on aerosol sampling accuracy for horizontal isoaxial sampling, *Atmospheric Environment*, 21, 1181–1185.

Pich, J. (1966) Theory of aerosol filtration by fibrous and membrane filters. In: *Aerosol Science* (Ed. C.N. Davies), Academic Press, London, pp. 223–285.

Pich, J. (1972) Theory of gravitational deposition of particles from laminar flows in channels, *Journal of Aerosol Science*, 3, 351–361.

Pui, D.Y.H., Romay-Novas, F. and Liu, B.Y.H. (1987) Experimental study of particle deposition in bends of circular cross-section, *Aerosol Science and Technology*, 7, 301–315.

Puskar, M.A., Harkins, J.M., Moomey, J.D. and Hecker, L.H. (1991) Internal wall losses of pharmaceutical dusts during closed-face, 37-mm polystyrene cassette sampling, *American Industrial Hygiene Association Journal*, 52, 280–286.

Saffman, P.G. (1965) Inertial migration of a sphere in Poiseuille flow, *Journal of Fluid Mechanics*, 22, 385–400.

Sehmel, G.A. (1967) Validity of air samples as affected by anisokinetic sampling and deposition within sampling line. In: *Proceedings of the Symposium on Assessment of Airborne Radioactivity*, Vienna, pp. 727–735.

Smith, J. and Bartley, D. (2003) Effect of sampler and manikin conductivity on the sampling efficiency of manikin-mounted personal samplers, *Aerosol Science and Technology*, 37, 79–81.

Sreenath, A., Ramachandran, G. and Vincent, J.H. (2001) Experimental study of sampling losses in thin-walled probes at varying angles to the wind, *Aerosol Science and Technology*, 35, 767–778.

Thomas, J.W. (1967) Particle loss in sampling conduits. In: *Proceedings of the Symposium on Assessment of Airborne Radioactivity*, Vienna, pp. 701–712.

Turner, S. and Cohen, B.S. (1984) Effects of electrostatic charge on aerosol collection with polystyrene filter cassettes, *American Industrial Hygiene Association Journal*, 45, 745–748.

Vincent, J.H. (1989) *Aerosol Sampling: Science and Practice*, John Wiley & Sons, Ltd, Chichester.

Vincent, J.H. and Gibson, H. (1981) Sampling errors in blunt dust samplers arising from external wall loss effects, *Atmospheric Environment*, 15, 703–712.

Vincent, J.H. and Humphries, W. (1978) The collection of airborne dusts by bluff bodies, *Chemical Engineering Science*, 33, 1147–1155.

Vincent, J.H., Emmett, P.C. and Mark, D. (1985) The effects of turbulence on the entry of airborne particles into a blunt dust sampler, *Aerosol Science Technology*, 4, 17–29.

Vincent, J.H., Mark, D., Botham, R.A., Lynch, G., Aitken, R.J., Gibson, H. and Campbell, S. (1987) Realisation of a practical inspirable dust spectrometer, IOM Report No. TM/87/07, Institute of Occupational Medicine, Edinburgh.

Vincent, J.H., Mark, D., Gibson, H., Botham, R.A., Emmett, P.C., Witherspoon, W.A., Aitken, R.J., Jones, C.O. and Miller, B. (1983) Measurement of inhalable dust in wind conditions pertaining to mines, IOM Report No. TM/83/07, Institute of Occupational Medicine, Edinburgh.

Vrins, E., Hofschreuder, P., Ter Kuile, W.M., van Nieuwland, R. and Oeseburg, F. (1984) Sampling efficiency of aerosol samplers for large windborne particles. In: *Aerosols* (Eds B.Y.H. Liu, D.Y.H. Pui and H. Fissan), Elsevier, Amsterdam, pp. 154–157.

Wang, H.C. (1990) Effects of inceptive motion on particle detachment from surfaces, *Aerosol Science and Technology*, 13, 386–393.

Wen, X. and Ingham, D.B. (1995) Aspiration efficiency of a thin-walled cylindrical probe rear-facing the wind, *Journal of Aerosol Science*, 26, 95–107.

Wiener, R.W., Okazaki, K. and Willeke, K. (1988) Influence of turbulence on aerosol sampling efficiency, *Atmospheric Environment*, 22, 917–928.

8

Options for Aerosol Particle Size Selection After Aspiration

8.1 Introduction

So far in this book, attention has been focused primarily on the processes by which particles enter through the inlets of aerosol samplers, along with the processes post-aspiration by which particles are lost by deposition to internal walls before they can reach the collecting filter or substrate (or other sensing region). The latter were discussed from the point of view of interferences that might bias the actual performance of the sampler with respect to what is aspirated. These aspects are certainly important facets of aerosol sampler performance, and they need to be properly addressed in any design or application of a practical aerosol sampling system. But there are often further important practical aerosol sampling situations where the accurate representation of what is aspirated provides only the starting point for what is ultimately required in the real world of aerosol measurement. The other part relates to the particle size selection that may often be desired after aspiration in order that what is ultimately collected is representative of some finer fraction, as for example might be relevant to the aerosol that penetrates deep into the human respiratory tract. Such particle size selection in a sampling instrument should be capable of being designed specifically to match a given particle size-selection criterion. Knowledge of the range of possible physical mechanisms for selectively separating particles on the basis of their size can provide the basis of options for such instrumentation.

As will be seen later in this book, a given criterion for particle size selection may take the form of a continuous curve that defines the efficiency with which it is desired that a particle of each given size is collected across the whole range of interest. The performance of an instrument as a function of particle size, $E_i(d)$, that needs to match a given criterion i, $F_i(d)$, may be expressed in the first instance as:

$$E_i(d) = A(d)P_i(d) \equiv F_i(d) \tag{8.1}$$

where $A(d)$ is the aspiration efficiency of the sampler and $P_i(d)$ is the desired selection curve that is applied to aerosols after they have been aspirated in order that the accurate sampling of $F_i(d)$ may be achieved. Here, it is often the case that the criterion in question and the sampler performance parameters are expressed as functions of particle aerodynamic diameter, d_{ae}. But there are cases of

Aerosol Sampling: Science, Standards, Instrumentation and Applications   James Vincent
© 2007 John Wiley & Sons, Ltd

increasing interest, for fine and ultrafine particles, where the more appropriate metric of particle size is the geometric diameter, usually the equivalent volume diameter, d_V. In this chapter, physical options for achieving $P_i(d)$ are the new focus of attention. There is a wide range of such options for the selection of aerosol fractions based on their particle size, including gravitational elutriation (both horizontal and vertical), centrifugation, filtration by porous media, impaction and other inertial processes, diffusion, thermal and electrostatic precipitation, and optical and/or visual discrimination methods. These are reviewed in what follows.

8.2   Elutriation

Particle size selection by elutriation takes place by virtue of the fact that particles of different aerodynamic diameter settle vertically under the influence of gravity at different velocities, as described by the earlier Equation (2.34). In 1954 Walton provided a definitive scientific description of the elutriation process which has not been improved upon even to this day. Based on the settling behavior of particles moving in air flows in the absence of any inertial effects, Walton's theory applies both to vertical and to horizontal elutriators, both of which have been applied in the particle size-selection of particles for health-related aerosol sampling.

8.2.1   Vertical elutriation

Walton began by considering the scenario shown in Figure 8.1(a). This is very similar to the downwards-facing sampling system for calm air described earlier in Chapter 6 except that, here, inertial forces are neglected. This time, a filter is placed inside the sampling head shown, and provides the particle collecting surface. As before, the air flow is described by the dashed lines and the particle trajectories by the solid lines, but now it is gravitational settling that causes the one to depart from the other. If the filter is uniform and has typical pressure drop characteristics, the air flow will be uniformly distributed over its surface. Particles approaching the filter will reach the filter whenever the sampling velocity (U_s) is greater than or equal to the particle settling velocity, v_s. Walton described how the system of particle trajectories in this system is such that individual particle trajectories cannot cross one another. In this way, therefore, the moving particles may be considered to be contained within channels or 'tubes'. The external particle concentration is c_0 and that at the surface of the filter is $c(U_s - v_s)$, where the latter reflects the fact that the more rapidly falling particles fail to reach the surface. Integrating over all the 'tubes' entering the sampler, the total amount of aerosol reaching the filter surface is $Qc_0[1 - (v_s/U_s)]$, where Q is the overall sampling flow rate. The collection efficiency of vertical elutriation in this sampling system (E) is therefore given by:

$$E = \frac{Qc_0\left[1 - \left(\dfrac{v_s}{U_s}\right)\right]}{Qc_0} = 1 - \frac{v_s}{U_s} \quad \text{so long as } v_s \leq U_s \tag{8.2}$$

where the denominator describes what would reach the filter in the absence of gravitational forces. In turn, penetration efficiency, P, is given by:

$$P = 1 - E = \frac{v_s}{U_s} \quad \text{so long as } v_s \leq U_s \tag{8.3}$$

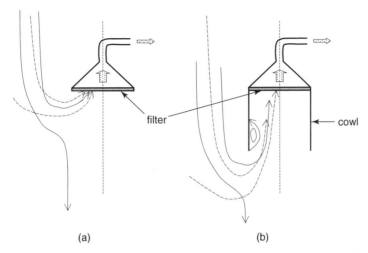

(a)   (b)

Figure 8.1 *Diagram to illustrate the principle of vertical elutriation: (a) downwards-facing open filter; (b) downwards-facing cowled filter. Airflow streamlines are the dashed lines, particle trajectories the solid lines*

Here it is seen that P falls to zero when $v_s = U_s$, indicating the existence of a sharp critical particle size above which nothing may be collected. Since the important variable in Equation (8.3) is the particle settling velocity, v_s, that critical particle size is expressed in terms of aerodynamic diameter (d_{ae}).

If the sampler shown in Figure 8.1(a) is now placed in moving air, the lines of air flow and particle trajectory are distorted from the picture shown. But, provided no inertial effects are introduced, the conditions at the surface of the filter remain the same, and so the collection efficiency is the same as that shown in Equation (8.2). One interesting outcome of Walton's 'tube-based' reasoning – for both calm and moving air – is that the deposition of particulate matter is uniformly distributed over the whole face of the filter. Walton went on to discuss a version of the vertical elutriator that is more commonly found in practice, where the presence of a vertical cylindrical shroud serves in the first instance to protect the filter from mechanical interference. As shown in Figure 8.1(b), the flow is inevitably more complicated in this system, for example by the possible introduction of flow separation inside the cylinder. But once again, in the absence of inertial effects, Walton's 'tube' approach requires that the arrival of particles at the surface of the filter remains as given by Equation (8.2), and so the result is the same as for the simpler system shown in Figure 8.1(a). Walton went on to examine other vertical elutriator configurations, including tapered cones, all with the same result.

Although vertical elutriators have been developed for practical applications in some occupational hygiene settings, notably in the cotton industry in the USA (e.g. Robert and Baril, 1984), there appears to be very little in the way of experimental evidence (e.g. in the form of data for P versus d_{ae}) to support Walton's elegant, yet simple, theory.

8.2.2 Horizontal elutriation

Walton extended his theory to the passage of an aerosol in laminar flow through a straight horizontal channel where, here too, particle motion is controlled by gravitational settling. The principle is shown in Figure 8.2(a), where the airflow streamlines are again represented by dashed lines and particle trajectories

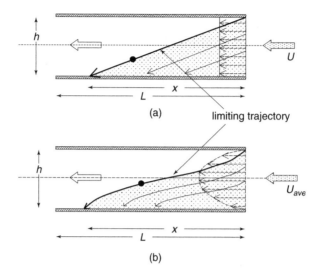

Figure 8.2 *Diagram to illustrate the principle of horizontal elutriation: (a) idealised uniform flow; (b) more realistic parabolic laminar flow. Airflow streamlines are the dashed lines, particle trajectories the solid lines*

by solid lines. Walton assumed in the first instance that the air flow was uniformly distributed over the cross-section of the channel. The dimension x is the distance along the channel as measured from its entrance. In this system, the particles follow straight trajectories towards the lower surface of the channel at angles given by $\tan^{-1}(v_s/U)$, where U is the uniform air velocity. At distance x, there are no particles above the depth $v_s x/U$, but a uniform concentration of particles below. This is shown in Figure 8.2(a), defined by the limiting particle trajectory. From this scenario, the penetration (P) of particles leaving the channel at $x = L$ is seen trivially to be

$$P = 1 - \frac{v_s L}{Uh} \text{ for all } \frac{v_s L}{Uh} \leq 1 \tag{8.4}$$

where h is the height of the channel. This means that, for $v_s L/Uh > 1$, all particles reach the floor of the channel so that $P = 0$. Again, therefore, as for the vertical elutriator, there is a sharp cut-off, this time corresponding to $v_s L/Uh = 1$.

Walton took the argument a stage further to discuss the situation where the velocity profile in the channel is not uniform. This is more realistic. As shown in Figure 8.2(b), provided that the Reynolds number for the channel flow is low enough for laminar flow to prevail, the velocity profile of the two-dimensional horizontal air flow indicated is parabolic. For this scenario, Walton used his now familiar reasoning to show that:

$$P = 1 - \frac{v_s L}{U_{ave}h} \text{ for all } \frac{v_s L}{U_{ave}h} \leq 1 \tag{8.5}$$

where U_{ave} is the air velocity averaged over the cross-section of the channel. Further, Walton showed that Equation (8.5) applies regardless of the actual shape of the air velocity profile.

Experimental data from which to validate the theory for horizontal elutriation are available from the work that was carried out during the development of specific particle size-selective sampling instruments. Those specific instruments will be described later in this book. For the present, Figure 8.3 summarises the

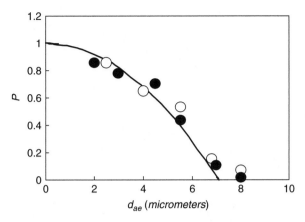

Figure 8.3   *Penetration (P) of a rectangular horizontal elutriator as a function of particle aerodynamic diameter (d_{ae}): comparison between theory from Walton (1954), shown as solid line, and experimental data from Wright (1954) and Dunmore et al. (1964), shown as open and solid circles, respectively*

available experimental data for elutriator penetration (P) as a function of particle aerodynamic diameter (d_{ae}) reported by Wright (1954) and Dunmore *et al.* (1964). Wright used test aerosols generated into a calm air chamber from polydisperse glass spheres, counting and sizing (under an optical microscope) particles that settled out onto glass cover slips, both for aerosol which had passed through the elutriator and for aerosol which had not. Dunmore *et al.* adopted a similar approach, except that they used liquid droplet aerosols produced from methylene blue. Also shown in the figure is the curve calculated directly from Walton's theory. In general, experiment and theory were in good agreement, with perhaps the only significant departure being that some particles larger than the predicted 'cut-size' were seen to penetrate through the elutriator.

The theoretical and practical simplicity of the horizontal elutriator shown in the preceding provided encouragement for the historical development of both sampling criteria and particle size-selective sampling instrumentation, as will be seen in later chapters.

8.3   Filtration by porous foam media

Walton was also influential in the early suggestions that the relatively poor filtration performance characteristics of porous plastic foam media might be useful for particle size-selective sampling. The porous plastic foam media in question are formed from reticulated polyurethane with a structure consisting of a matrix of bubbles which, pierced at their points of contact, create an open, three-dimensional lattice of connected short elements of approximately triangular cross-section. These media are produced commercially in large quantities for wide ranges of industrial and military applications, including sound and vibration isolation, rough filtration, padding, etc. Their applications in occupational and environmental health and hygiene were – and remain – secondary to the manufacturers. That said, however, the consistency during production of their geometrical and other physical properties, expressed in terms of packing fraction and the number of pores per linear dimension, provide excellent opportunities for useful applications in particle size-selective aerosol sampling. They are available from several sources, in different grades and sheet thicknesses.

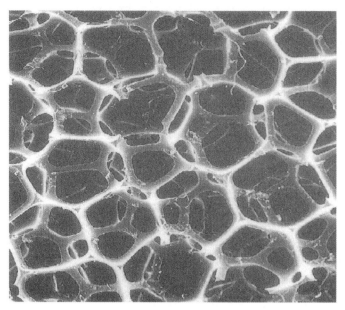

Figure 8.4   *Electron micrograph of reticulated polyurethane foam media of the type used for particle size-selective aerosol sampling. Reprinted from Journal of Aerosol Science, 24, Vincent et al., 929–944. Copyright (1993), with permission from Elsevier*

An electron micrograph of a typical sample of porous plastic foam media is shown in Figure 8.4. Here the short solid elements of the structure may be regarded as somewhat analogous to the fibrous elements that make up more traditional filtration media. It is this analogy that provides the basis for developing a model for porous plastic foam filtration which may then be applied to particle size-selective aerosol sampling. Foam media like the sample shown in Figure 8.4 may be characterised in terms of the following geometric parameters:

- nominal porosity (Po), expressed conventionally (by the manufacturers) in terms of 'pores per inch' (or ppi);
- volume (packing) fraction (σ_{foam}), expressed as the portion of the overall volume occupied by the solid material;
- the effective 'fiber' width, d_f.

For a fibrous filter of overall thickness t in the flow direction, penetration (P) is first written down in the general form:

$$P \equiv \exp\{-\alpha t\} \tag{8.6}$$

where α is the efficiency of collection per unit of filter depth, referred to as the *layer efficiency* (Brown, 1993), given by:

$$\alpha = \left(\frac{4}{\pi}\right)\left(\frac{\sigma_{foam}}{d_f}\right) E_{fiber} \tag{8.7}$$

in which E_{fiber} is the single fiber collection efficiency embodying the physics of the filtration process (Pich, 1966).

From the experimental studies of Brown (1980), Gibson and Vincent (1981) and Wake and Brown (1991) a relatively simple picture of the filtration process emerged where the primary forces acting on particles are gravitational and inertial, with gravity predominant at low air velocities through the media and inertia predominant at higher velocities. Other mechanisms such as interception, diffusion and electrostatic forces were found to be negligible in comparison with these for the particle size ranges studied in the works cited. By way of illustration, Figure 8.5 shows some of the data reported by Gibson and Vincent from measurements in a calm air chamber, using spinning disk-generated monodisperse aerosols of di-(2-ethylhexyl)sebacate, of the filtration efficiency of cylindrical plugs of foam media of thickness $t = 25$ mm and pore sizes of 30 and 60 ppi, respectively. These data, plotted here in the form of penetration (P) as a function of air velocity at the face of the plug (U_{face}), show very clearly the gravity and inertia-controlled regions – to the left and right, respectively, of each graph – and the

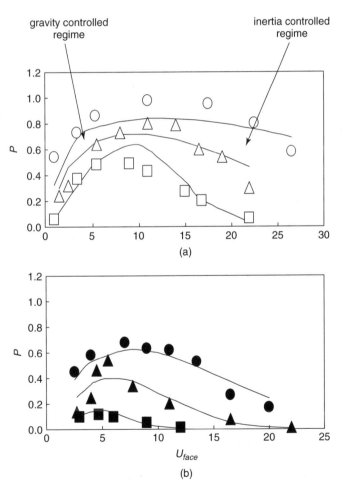

Figure 8.5 *Penetration (P) of a cylindrical plug of porous plastic foam media as a function of air velocity at the face of the plug (U_{face}) (results from Gibson and Vincent, 1981): ○● $d_{ae} = 7$ μm; ▲△ $d_{ae} = 9$ μm; □ ■ $d_{ae} = 12$ μm. (a) 30 ppi foam; (b) 60 ppi foam. Adapted with permission from Gibson and Vincent, Annals of Occupational Hygiene, 24, 205–215. Copyright (1981) British Occupational Hygiene Society*

transition that took place between them. With such data in mind, Equations (8.6) and (8.7) may be combined to give:

$$-\left(\frac{d_f}{t}\right)\ln P = kE = f\{St_{foam}, N_{gfoam}\} \equiv y \tag{8.8}$$

where the coefficient k embodies the packing fraction which, for present purposes, is considered to be constant, and y becomes the subject of interest. In Equation (8.8), we have the inertial parameter:

$$St_{foam} = \frac{d_{ae}^2 \rho^* U_{face}}{18\mu d_f} \tag{8.9}$$

in the form of a Stokes number similar to the ones already introduced elsewhere in this book, and the gravitational parameter:

$$N_{gfoam} = \frac{d_{ae}^2 \rho^* g}{18\mu U_{face}} \tag{8.10}$$

Vincent *et al.* (1993) defined y in terms of the empirical relationship:

$$y = a\, St_{foam}^{\ b} + cN_{gfoam}^d \tag{8.11}$$

where a, b, c and d are coefficients that were statistically fitted to the experimental data of Gibson and Vincent (1981), yielding:

$$y = 54.86St_{foam}^{2.382} + 38.91N_{gfoam}^{0.880} \tag{8.12}$$

and hence:

$$\ln P = -\frac{t}{d_f}\left(54.86St_{foam}^{2.382} + 38.91N_{gfoam}^{0.880}\right) \times 10^{-3} \tag{8.13}$$

In this expression, all the variables shown are expressed in SI units. In addition to its consistency with the Gibson and Vincent data, this model was found also to be in good agreement with the experimental results of Wake and Brown (1991). It was later compared with a significant body of new data from two different laboratories and was further confirmed as providing a good prediction of foam penetration (Kenny *et al.*, 2001).

As already mentioned, foams like those described are conventionally classified in terms of nominal porosity (*Po*) in ppi (the number of pores intersected per linear inch). From microscopy, this quantity was found to be well described by:

$$d_f = 9.633 \times 10^{-3} \cdot Po^{-1.216} \tag{8.14}$$

where d_f is expressed in SI units. A linear correlation between fiber diameter and cell (or pore) diameter also exists, in the form:

$$d_f = 8.86 \times 10^{-6} + 0.076 \times \text{cell diameter} \tag{8.15}$$

where both d_f and cell diameter are expressed SI units (Kenny *et al.*, 2001).

Inspection of the model reveals, not surprisingly, that the penetration of larger particles is lower than for smaller ones. In Figure 8.6, calculated P is plotted as a function of d_{ae} for two typical combinations of variables, and it is seen to fall from unity for d_{ae} close to zero, tailing off towards zero at larger values of d_{ae}. For the same foam plug dimensions and air flow rate, the penetration of the 30 ppi foam

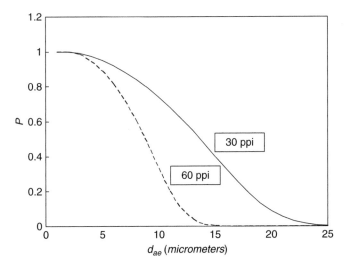

Figure 8.6 *Calculated penetration of typical foam media (P) as a function of particle aerodynamic diameter (d_{ae}), for foam plug width 25 mm, thickness 25 mm and flow rate 2 Lpm, and for foam grades 30 and 60 ppi, respectively*

is seen to be greater than for the 60 ppi foam. The trends shown in Figure 8.6 confirm the potential usefulness of porous plastic foam media as pre-selectors for particle size-selective aerosol sampling.

At the higher end of the inertial regime (e.g. to the right-hand side of the graphs in Figure 8.5), the velocity of particles travelling inside the foam media may be great enough that particles fail to be retained on contact with the foam media surface during collection, involving blow-off mechanisms like those touched on earlier in Chapter 7. This effect has been noted by Chen *et al.* (1998) and others, along with the suggestion that the problem can be addressed in practical situations by the judicious use of thin grease coatings. Preparation of foam media prior to sampling in this way has been shown to enhance the adhesion of particles to the extent that the problem can virtually be eliminated.

8.4   Centrifugation

Particle size selection by centrifugation is directly analogous to gravitational settling, where the force of gravity is replaced by the centrifugal force on particles derived from the rapid rotation of the body of air containing them. In this way, depending on the angular velocity of the externally applied rotation and the dimensions of the flow system in question, the force may be significantly greater than for gravitational settling. The basic principle is shown in Figure 8.7, where the sampled air flow passes through a rotating chamber. In this system the volume of air itself contained within the chamber is also rotated so that the particles migrate under the influence of the centrifugal force towards the outer wall of the chamber. By analogy with Equation (2.34), the velocity of migration of airborne particles towards the wall is given by:

$$v = \tau(\omega^2 r) \equiv \tau \left(\frac{U_r^2}{r} \right) \tag{8.16}$$

where, as before, τ is the particle relaxation time, ω the angular velocity of the rotation, U_r the linear (or circumferential) air velocity that is characteristic of the rotating flow and r the radius of the rotation.

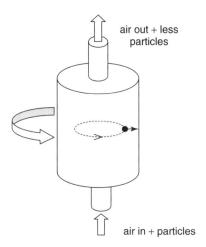

Figure 8.7   *Diagram to illustrate the principle of centrifugation, showing the rotation of the body of sampled air contained inside the device and the resultant force on airborne particles towards the inside wall*

The term in brackets on the right is analogous to the acceleration due to gravity in gravitational settling. It is seen from Equation (8.16) that larger particles are again collected more efficiently than the smaller ones. In this system, if the sampled flow passing through the rotated body of air is laminar and otherwise well-defined, the spatial distribution of collected particles on the interior collecting surface may be used to determine the particle size distribution. Indeed, for such idealised conditions, aerosol centrifuges have been developed as aerosol spectrometers, as will be discussed in a later chapter. Alternatively, by appropriate design and setting of operating parameters, the particle size-related aerosol penetration may be matched to a specific particle size fraction.

The cyclone is a class of centrifugation-based devices that is commonly used in aerosol sampling systems. It operates on the basis that rotation of the body of air instantaneously contained within the cylindrical body of the device is achieved by the asymmetric introduction of the aerosol at the inlet. Such rotation provides the source of the centrifugal force that drives particles towards the wall where they may be separated from the flow. The version that is most widely used for aerosol sampling purposes is the one shown in Figure 8.8, commonly known as the *reverse-flow cyclone*. Figure 8.8(a) shows the general nature of the air flow in this system, indicating how the air enters into the plane of the page, tangentially to the body of air inside the body of the cyclone. This leads to strong rotation of the contained air volume. Particles in this rotating flow migrate towards the inside wall of the cyclone under centrifugal forces. Once they arrive there, they are either collected or fall to the bottom of the cyclone in the quiescent boundary layer close to the wall. The bottom of the cyclone is sealed, so the airflow must reverse – now depleted of particles – and return upwards through the central core of the rotating air, exiting through the outlet at the top. Inevitably the air motion inside this system is extremely complex, exhibiting not only the general patterns indicated but also turbulent and various secondary motions that are highly dependent on the specific inlet and internal geometries. Earliest models to predict cyclone performance were based on considerations of the residence time of particles inside the body of the cyclone, as dictated by particle migration towards the wall and how this related to the overall residence time of the volume of air rotated within the body of the cyclone. In this way it was possible to estimate that all particles equal to or greater than a certain size arrived at the wall (and so were assumed to be collected) and hence that all particles smaller than this penetrated through the cyclone and emerged at

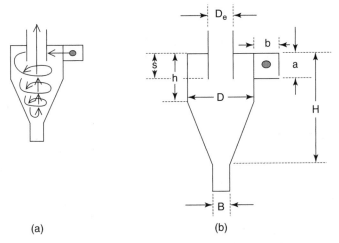

(a)   (b)

Figure 8.8 *Schematic of the reverse-flow cyclone: (a) to illustrate the general air movement inside the cyclone; (b) to show the important dimensions required to develop a model for penetration or collection efficiency*

the exit. This pointed to a sharp cut in penetration at a given particle size, specifically aerodynamic diameter. Rosin *et al.* (1932) were the first to analyze this scenario, and developed a model based on the set of dimensions shown in Figure 8.8(b). This may be summarised by the expression:

$$_{50}d_{ae}, = k_1 \left[\frac{R}{U_s N} \left(1 - \frac{R}{D} \right) \right]^{1/2} \tag{8.17}$$

where $_{50}d_{ae}$ defines the particle aerodynamic 'cut-size' for the cyclone at which penetration falls to 50 %. Here, k_1 is a coefficient that embodies physical quantities such as the density of water and the viscosity of air and other constants, R is the average distance of the particle in the inlet from the axis of the cyclone, $1/2(D - b)$, and N is the number of revolutions of the airflow within the cyclone given by:

$$N = \frac{TU_s}{\pi D} \tag{8.18}$$

Here T is the residence time of the gas stream as it passes through the body of the cyclone, estimated from V/Q where Q is the volumetric flow rate and V the volume of the cyclone. Rosin *et al.* calculated V from:

$$V = \frac{\pi}{4} \left[\left(\frac{H - h}{D - B} \right) \left(\frac{D^3 - B^3}{3} \right) + D^2 h - D_e^2 s \right] \tag{8.19}$$

Later, Dalla Valla (1952) suggested that the preceding formulation was overly complicated and proposed more simply that an empirical value of N varying between 0.5 and 3 would be appropriate. Similar formulations to that of Rosin *et al.* were later developed by Davies (1952), Barth (1956), Rietema (1961), Dietz (1981) and others. The form of Equation (8.17), as well as the other suggested equations for $_{50}d_{ae}$, suggest that the performance of the reverse-flow cyclone may be defined in terms of a Stokes number-like quantity, and some workers have expressed it as such (Lidén and Gudmundsson, 1997). However, this is somewhat misleading because the physics of collection of particles inside the cyclone is closer to centrifugation – and hence to elutriation – than to the inertial behavior that is usually linked with the Stokes number.

Later models took into account the more complex features of the air motion inside the cyclone, including diffusion and boundary layer effects (Beeckmans, 1972; Kim and Lee, 2001). Ma *et al.* (2000) carried out a numerical study of a number of small cyclones of the type used as pre-selectors in aerosol sampling applications, modeling the air motion in terms of incompressible, fully developed, turbulent flow. Some of the results for one of the cyclones studied are shown in Figure 8.9 in the form of penetration (P) as a function of particle aerodynamic diameter (d_{ae}). Ma *et al.* compared their results with some experimental data reported by Kim and Lee (1990) that had been obtained for polydisperse aerosols of polystyrene latex dispersed by means of an atomiser and where an aerodynamic particle sizer (APS) had been used to determine penetration for individual particle sizes. They also included in the comparison data calculated from the empirical models of Barth (1956) and Dietz (1981). What is seen in Figure 8.9 is that, far from there being a sharp cut, P varied quite strongly with d_{ae}, in much the same way as for elutriation and foam filtration. More specifically it is seen that the numerical results agreed quite well with the experimental ones in predicting d_{ae} at $P = 0.5$, equivalent to $_{50}d_{ae}$ in the preceding discussion. Otherwise, however, agreement between any of the models and the experimental data was poor. From this and other such comparisons, it is inescapable that the performance of the cyclone is a very complicated function of all the geometrical variables shown in Figure 8.8(b), so much so that a general predictive model is very difficult to achieve (Ranz, 1984).

Perhaps the most useful modeling exercises were the ones carried out by Chan and Lippmann (1977) and, later, Lidén and Gudmundsson (1997). In the earlier study, Chan and Lippmann examined the available experimental data for the performance characteristics of a number of American cyclones for selecting the fine respirable fraction – both large, high flow rate devices intended as static samplers in the ambient atmosphere as well as workplaces, and small, low flow rate ones intended as personal samplers. They showed that penetration (P) as a function of d_{ae} could be described by the universal empirical expression:

$$P = 0.5 - 0.5 \tanh \left[B \frac{d_{ae}^2}{KQ^{2n}} + (A - 2B) \frac{d_{ae}}{KQ^n} + (B - A) \right] \qquad (8.20)$$

where A, B, K and n are coefficients that were subsequently obtained by fitting the expression to the measured performances of several contrasting types of cyclone (Lippmann and Chan, 1979). Lidén and

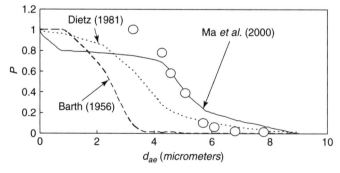

Figure 8.9 *Results for penetration (P) versus particle aerodynamic diameter (d_{ae}) for a typical small cyclone of the type used for aerosol sampling: ○ experimental data (Kim and Lee, 1990); —— numerical results (Ma et al., 2000); – – – model results (Barth, 1956); · · · model results (Dietz, 1981)*

Gudmundsson used statistical multilinear regression methods to examine the relative roles of all the cyclone variables. Their goal was to develop a framework for scaling the various parameters influencing cyclone performance and to identify which ones were the most important. For the parameters indicated in Figure 8.8(b), they showed that the ratio $_{50}d_{ae}/D$ was a function of a Reynolds number based on the inlet velocity (U_s) and the annular distance between the vortex finder and the cyclone wall, $1/2(D - D_e)$. In addition they showed that the steepness of the penetration curve was mainly a function of the ratio of the cyclone body diameter to the vortex finder diameter (D/D_e), independently of Reynolds number.

8.5 Impaction

8.5.1 Conventional impaction

The mechanism of impaction was introduced earlier in Chapter 2. Some notable work on the physics of impaction was reported in the early 1950s by Davies and Aylward (1951), Davies *et al.* (1951) and Ranz and Wong (1952), where the nature of the impaction of particles onto a surface from an impinging particle-containing air jet was explored. Thus emerged the device that has since become widely known as the 'impactor'. The principle is shown schematically in Figure 8.10.

The level of interest of the application of impactors in modern aerosol sampling owes much to the large body of work of Professor Virgil A. Marple and his colleagues at the University of Minnesota, beginning with his seminal work of 1970 (Marple, 1970) and continuing to this day. Marple, however, in his historical review of impactors noted the existence of practical impactors even as early as 1860 with the *aeroscope* of Pouchet (1860)[1], followed soon after by related devices by Maddox (1870) and Miquel (1879). But it is now generally recognised that it was Marple himself that first provided the first, full, definitive theoretical basis. Figure 8.10 indicates not only the principle of particle collection in a single-jet impactor but also the primary dimensions needed in the first instance to define collection

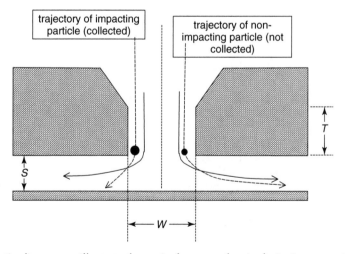

Figure 8.10 *Schematic diagram to illustrate the main features of a single-jet impactor: W is the impactor jet width, S is the jet-plate spacing, and T is the length of the impactor jet throat (or nozzle)*

---

[1] It is noted, however, that there are reports of a impactor-based device introduced by Maddox and Cunningham even as early as the 1840s (e.g. see Gregory, 1973).

efficiency. Here, W is the impactor jet width, S the jet-plate spacing, and T the length of the impactor jet throat (or nozzle), defining the fluid mechanical conditions for the impactor system shown. The jet may be circular or rectangular. For the latter, the length of the jet into the plane of the page (not shown) is L where $L \geq 10\, W$. Although the airflow in this system is apparently much simpler than for the cyclone described in the preceding section, there are again no formal analytical solutions, either for the air motion or for the particle trajectories. However, Hinds (1999) described a useful simple empirical model that embodied some aspects of the overall physical nature of particle transport in the system shown. Figure 8.11 shows a modified, simple version of the air and particle flow upon which to base such a model. Assuming symmetry, only one half of the system needs to be discussed. It is similar to the system shown in Figure 8.10, but now the air flow streamlines are idealised such that they take the form of concentric quarter-circular arcs. The forces acting on particles moving in this simplified flow field are assumed to be centrifugal in nature so that, by analogy with Equation (8.16) for centrifugation, the particle shown departs from the streamline at radial velocity:

$$v_r = \tau \left(\frac{U^2}{r} \right) \tag{8.21}$$

where r is the radial distance of the particle from the center of the circular arc of the air flow of velocity U. The radial displacement of the particle from its original streamline is Δ given by the product of the radial outwards velocity and the time that the particle spends in the quarter-circle arc, so that:

$$\Delta = \tau \left(\frac{U^2}{r} \right) \left(\frac{2\pi r}{4U} \right) \longrightarrow \frac{\pi \tau U}{2} \tag{8.22}$$

Since the particle path shown is the limiting trajectory, then the penetration efficiency (P) of the two-dimensional impactor system shown is given by:

$$P = 1 - \left(\frac{2\Delta}{W} \right) \longrightarrow 1 - \frac{\pi \tau U}{W} \longrightarrow 1 - \pi \, St_{impactor} \tag{8.23}$$

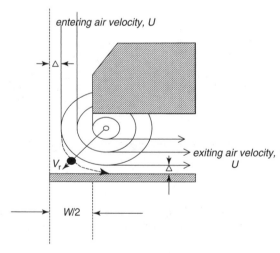

Figure 8.11 *Simplified version of the scenario shown in Figure 8.10 (right hand side only for symmetrical system) on which to base an empirical model for impactor penetration*

where

$$St_{impactor} = \frac{\tau U}{W} \qquad (8.24)$$

in which $St_{impactor}$ is a Stokes number reflecting inertia-dominated particle motion in the changing flow field in the impactor jet.[2] The inconsistency in arriving at a Stokes number dependency after an analysis that depends on centrifugal motion is analogous to the similar inconsistency already noted in some discussions of cyclone behavior. Here, however, the dependency on $St_{impactor}$ is entirely appropriate because it is clear that, in reality, the physical nature of particle transport in an impactor is indeed controlled by inertia. But Hinds was careful to advise that an analysis like that described would not be suitable for predicting actual values of penetration through the impactor.

In seeking a more rigorous and complete model, Marple (1970) – and later Rader and Marple (1985) – performed numerical calculations for single-jet impactor systems for both circular and two-dimensional rectangular-shaped jets, and obtained curves of collection efficiency (E) as a function not only of $St_{impactor}$ but also other dimensionless quantities embodying the physics of the air and particle movement and the dimensions of the impactor system. In the scientific literature for impactors, performance is usually portrayed in the form of collection efficiency (E) as a function of particle aerodynamic diameter (d_{ae}). But for present purposes and the sake of consistency with the rest of this chapter, it is illustrated in Figure 8.12 in the form of penetration (P) versus d_{ae}. While for most applications it is desirable that such curves should be as sharp as possible, in reality the shape of the curve shown indicates that some particles with particle aerodynamic diameter greater than $_{50}d_{ae}$ pass through the impactor system while others less than $_{50}d_{ae}$ are collected. A corresponding large body of experimental data has provided both qualitative and quantitative confirmation of the numerical simulations. This led Marple and Willeke (1976) to apply those numerical results to the development of an engineering framework by which impactors may be designed for specific applications having predictable $_{50}d_{ae}$-values and the

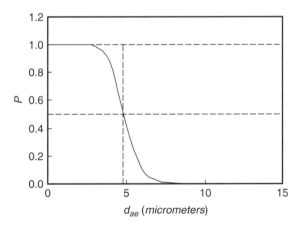

Figure 8.12 *Typical performance curve for a single-jet impactor of the type studied numerically by Marple and his colleagues, shown in the form of penetration (P) as a function of particle aerodynamic diameter (d_{ae})*

---

[2] For consistency in this book, $St_{impactor}$ is referred to the diameter or width of the jet. But it is noted that in some literature it is referred to the radius or half-width. Therefore in applications of this model, or in reviewing its descriptions in other literature, readers should take care to check that the right version is used.

desired sharp cuts. They proposed:

$$\sqrt{C_c(d)}_{50}d = \left(\frac{18_{50}St_{impactor}\pi\eta W^3}{4Q\gamma} \right)^{1/2} \quad \text{for round jet impactors} \quad (8.25)$$

and

$$\sqrt{C_c(d)}_{50}d = \left[\frac{18_{50}St_{impactor}\eta \left(\frac{L}{W}\right)W^3}{Q\gamma} \right]^{1/2} \quad \text{for rectangular jet impactors} \quad (8.26)$$

where $_{50}St_{impactor}$ is the Stokes number corresponding to $_{50}d_{ae}$ and Q is the air flow rate through the nozzle. Equations (8.25) and (8.26) allow the convenient calculation of $_{50}d_{ae}$ so long as the slip factor can be assumed to be unity. Otherwise the slip factor, $C_c(d)$ takes account of aspects of particle motion where particle size is small compared with the distance between collisions with gas molecules. In this, the geometric particle diameter, d, is required. It follows that the specific value $_{50}d$ corresponds to $_{50}d_{ae}$. In addition, of course, as indicated earlier in Equation (2.40), $C_c(d)$ should also appear in $_{50}St_{impactor}$. In the context of impactors the phenomenon of slip applies not only to small particles at atmospheric pressure but also to larger particles at the lower pressures that are found in some impactors. Therefore, unlike for most other situations and devices described in this book, the slip correction cannot be routinely neglected. Unfortunately, when it is needed, the calculation of $_{50}d_{ae}$ becomes much less convenient since, as shown earlier in Equation (2.28), $C_c(d)$ itself contains terms involving particle size. Marple and Rubow (1986) described an iterative procedure by which the desired $_{50}d_{ae}$ could be extracted from Equations (8.25) and (8.26) if the local static pressure in the impactor jet was known.

Referring back to Figure 8.10, in particular to the aspect ratios of the important dimensions indicated, several studies have shown that the shape of the collection curve is relatively insensitive to the ratio T/W provided that it is small enough to avoid the development of a parabolic velocity profile in the jet as it exits the nozzle (e.g. Mercer, 1969; Marple and Liu, 1974). This requires T/W less than about 5, and it is generally recommended that $1 < T/W < 2$. By contrast, the same authors noted that $_{50}d_{ae}$ was quite sensitive to the ratio S/W. Marple and Rubow (1986) recommended that, in order that small changes in geometry should not significantly influence $_{50}d_{ae}$, $S/W \geq 1$ for circular jets and $S/W \geq 1.5$ for rectangular jets (where $L \geq 10W$). They also noted that the sharpness of the cut was influenced by the jet Reynolds number (based on W for circular jets and $2W$ for two-dimensional rectangular jets), and that the best results were obtained for values between about 500 and 3000.

During the process of impaction, it is seen that particles arrive at the collecting surface at relatively high velocity. This raises the problem that some particles, especially solid, gritty ones, may fail to be retained on impact. Whether or not this happens depends on many factors, including not only the velocity of impacting particles and the particle size, type and surface properties, but also the properties of the collecting surface. All of these have a strong bearing on the ability of the particle to adhere to the collecting surface. As discussed in Chapter 7, particle losses during aspiration by aerosol samplers appear to be due to 'blow-off', associated with particle surface properties and the nature of local air flow in the boundary layer of the surfaces involved. During impaction, however, where the velocities of particles on impact with collecting surfaces are relatively high, the process is more in the nature of 'bounce' or 'rebound', involving the elastic properties of both the particle and collecting surface (Cheng and Yeh, 1979). When such particle losses occur, collection efficiency is degraded. In practice such effects are reduced by the treatment of collecting surfaces to enhance adhesion (e.g. the application of grease).

8.5.2 Low pressure and micro-orifice impaction

Conventional impactors, operating at close to atmospheric pressure, cannot provide useful particle size selection below about 0.4 μm (Hering and Marple, 1986). But low pressure and micro-orifice impactors have been found to allow that range to be significantly extended. Stern *et al.* (1962) were the first to investigate the performance of impactors at low pressure, and showed that the Stokes number may continue to account for collection efficiency even when particles are small compared with the mean free path (*mfp*) between gas molecules. It has since been found that low pressure impaction allows for the possibility of $_{50}d_{ae}$-values even as low as 0.05 μm, using nozzles of the same width as for conventional impactors of the type described above. In practice the desired low $_{50}d_{ae}$ may be obtained by the use of a vacuum pump to draw air through the nozzle at high flow rate, generating large pressure drop such that impaction takes place at much lower pressures (e.g. as low as 0.03 atm). It is the greatly increased slip factor that accounts for the low $_{50}d_{ae}$. In a separate, related development it was found that particles down to as small as 2 nm and 5 nm could be successfully sampled by expanding the aerosol into an evacuated region leading to 'hypersonic impaction' (e.g. Fernandez de la Mora and Schmidt-Ott, 1993).

Micro-orifice impactors are a more recent innovation, differing significantly from the simple impactor systems described so far in that the nozzle plate contains a large number of very small orifices – as many as 2000 with widths as low as 60 μm, achieved by chemical etching – providing the desired flow rate at relatively low pressure drop. Values of $_{50}d_{ae}$ below 0.1 μm have been achieved in this way (Marple *et al.*, 1981). The collection of ultrafine particles in this range has been of increasing interest in recent years, driven by concerns about health effects specifically related to such small particle size (Brown *et al.*, 2000). In this regard, in another impactor application, de Juan and Fernandez de la Mora (1998) showed how, by focusing the aerosol in an impactor jet, it was possible to achieve sharp size selection particles all the way down to 0.01 μm by means of an impaction beam.

Strictly, the characterisation of particle size in terms of aerodynamic diameter at such small values is inconsistent with the widely accepted original definition based on the falling speed of particles under the influence of gravity (see Chapter 2). Under normal conditions, the motion of particles in this size range is dominated by diffusion. However, extension of the definition of particle aerodynamic diameter in the manner indicated is useful in many practical fine-particle applications (e.g. the study of diesel fume, welding fume, smog, etc.).

8.5.3 Virtual impaction

Virtual impaction takes place in a system somewhat similar to the impactor that has been described, but where the passive collecting surface is replace by an active collection probe. The principle of this process is shown in Figure 8.13. The principle is very similar to that for the conventional impactor except that, now, the particles are impacted onto the plane of a sampling probe, rather than directly onto a solid surface. In the axisymmetric system shown, the air that emerges from the impaction nozzle is divided into the *major flow* that is extracted radially and the *minor flow* that is drawn into the probe, typically between about 5 and 10 % of the total flow. To a greater or lesser extent depending on inertial forces, the larger particles have the greater tendency to enter the probe part of the system. By the choice of nozzle and probe dimensions, together with the magnitudes of the major and minor flows, the efficiency with which particles enter the probe – and are hence selected from the original aerosol – may be adjusted. One advantage of virtual impactors over conventional ones is that the problem of particle bounce is eliminated.

Marple and Chien (1980) reported a theoretical study of virtual impactor efficiency, and shown that, although the shape of the performance curve was basically similar to that for conventional impactors, *P* did not reach unity for the smaller particles. Rather, as shown in Figure 8.14, it leveled off at a

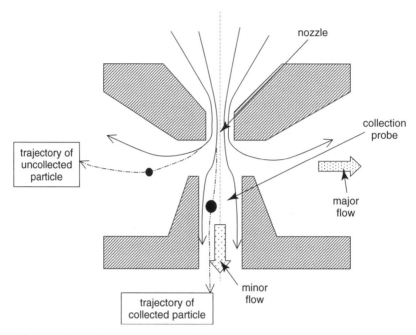

Figure 8.13   *Schematic diagram to illustrate the principle of virtual impaction*

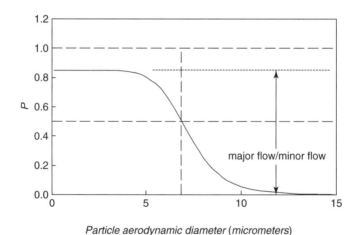

Particle aerodynamic diameter (micrometers)

Figure 8.14   *Typical curve showing penetration (P) as a function of particle aerodynamic diameter (d_{ae}) for a virtual impactor*

value equal to the ratio of the major flow to the total flow. This shape has been widely confirmed experimentally. The offset effect has been eliminated in some virtual impactor systems by providing a central core of aerosol-free air in the nozzle (e.g. Chen *et al.*, 1986). Since this is the part of the air flow that contains the minor flow, then the smaller particles are now not contained in the probe flow, so that P does approach unity zero as d_{ae} continues to decrease. In general, the sharpness of the cut for virtual impactors is less than for conventional ones. For some applications in particle size-selective sampling this property in itself might have some advantages over conventional impactors.

8.6 Diffusion

As discussed in Chapter 2, particles in the size range below about 0.1 μm are not strongly influenced by gravitational and inertial forces so that their behavior in relation to most practical contexts are not best represented by aerodynamic diameter. Rather, particle motion is usually dominated by diffusion, so that equivalent volume diameter (d_V) becomes a more appropriate metric of particle size.

8.6.1 Deposition by diffusion in laminar flow through tubes

In Chapter 7 particle deposition from laminar flow in tubes was discussed in the context of its contribution to unwanted biases in aerosol sampling. Gormley and Kennedy (1949) had shown that the efficiency of deposition in a cylindrical tube (E) increased with decreasing particle size, decreasing tube diameter and increasing residence time of particles during transport through the tube. Fuchs (1964) and Ingham (1975) also reported the results of theoretical studies for the same idealised system. Soon afterwards, Sinclair and Hoopes (1975) published experimental results for the penetration of particles through tubes of the form of 'collimated holes' for the range of particle size from 10 nm up to about 0.1 μm. From comparison of the various theoretical expressions that had previously been proposed, Soderholm (1979) suggested that the most accurate representation of penetration (P) as a function of the appropriate system variables was:

$$P = 1 - 5.5\mu^{2/3} + 3.8\mu + 0.81\mu^{4/3} \tag{8.27}$$

for $\mu < 7.22 \times 10^{-3}$ and

$$P = 0.82 \ \exp(-11.5\mu) + 0.098 \ \exp(-70.1\mu) + 0.033 \ \exp(-179\mu) + 0.015 \ \exp(-338\mu) \tag{8.28}$$

for $\mu \geq 7.22 \times 10^{-3}$. In these expressions:

$$\mu \frac{4D_B L}{\pi D^2 U_{ave}} \longrightarrow \frac{D_B L}{Q} \tag{8.29}$$

in which L is the length of the tube, D its diameter, U_{ave} the average air velocity through it and Q the volumetric flow rate. As before, D_B is the coefficient of Brownian diffusion, expressed as a function of particle size, specifically d_V. It was noted from Equation (8.29) that, for given flow rate, penetration was independent of the tube diameter.

Scheibel and Porstendörfer (1984) carried out experimental studies of deposition in batteries of narrow capillary tubes with very fine test aerosols generated using a condensation-type aerosol generator together with a differential mobility analyzer to obtain the desired particle size and monodispersity. For the tube systems studied, the tube length was varied. Some of their experimental results are summarised in Figure 8.15, combining data for both charged and uncharged particles, for which no difference was found. The data set shown here is for a bundle of 484 tubes, each of length 9.3 cm, and for a total flow rate (through the whole set of tubes) of 2 Lpm, providing 0.004 Lpm per tube. The general observed trend was that P increased steadily as d_V increased in the range from a few nanometers up to about 0.1 μm. It is seen from the graph that the experiments were in quite good agreement with the Soderholm model.

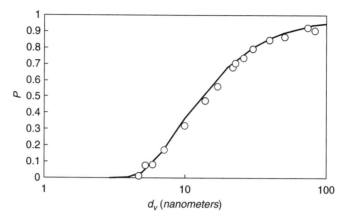

Figure 8.15   *Penetration (P) as a function of particle equivalent volume diameter (d_V) for deposition by diffusion in a tube; solid line from the theory of Soderholm (1979), experimental data from experiments reported by Scheibel and Porstendörfer (1984) for a battery of tubes of length 9.3 cm each with flow rate 0.004 Lpm*

Two other tube bundles were studied. In one, with tubes longer than the one represented in Figure 8.15 (results not shown), even better agreement with theory was found. But in the other, for shorter tubes, the experimental data (also not shown) fell significantly below the corresponding theoretical values. Scheibel and Porstendörfer interpreted the discrepancy between experiment and theory for the shorter tubes in terms of entrance effects at the tube inlet and the distance taken for the laminar flow parabolic profile inside the tubes to become fully established.

8.6.2   Deposition by diffusion in flow through screens

Fine particles may be deposited by diffusion from an aerosol that is passed through a wire-mesh screen. One view is that this is an extension of the tube model discussed above, where now the flow through each of the openings in the mesh is equivalent to that through a short tube. Cheng and Yeh (1980), however, treated the screen as equivalent to a slice of fibrous media, and on this basis – for a single screen – proposed the expression

$$P = \exp\left[-\frac{10.8\sigma_{screen}t}{\pi(1 - \sigma_{screen})d_W^{5/3}} \left(\frac{D_B}{U_{face}} \right)^{2/3} \right]$$

(8.30)

based on the so-called 'fan-model' that assumed the filter elements to comprise rows of parallel cylinders. It is directly analogous to Equation (8.6) shown earlier for particle penetration through porous plastic foam media. Here, however, σ_{screen} is the volume (or packing) fraction of the screen, t the screen thickness (analogous to the thickness of the foam filter media), U_{face} the air face velocity just ahead of the screen, d_W the width of the screen wire elements (analogous to the foam 'fiber' thickness for the foam media), and D_B the coefficient of Brownian diffusion. As before, it is the latter that contains the particle size, expressed in terms of equivalent volume diameter, d_V.

Scheibel and Porstendörfer extended their experimental study to include also this scenario and their results for a single screen with $d_W = 50$ μm, $t = 85$ μm, $\sigma_{screen} = 0.28$ and $U_{face} = 0.024$ m s^{-1} are shown in Figure 8.16 alongside the corresponding curve calculated from the model of Cheng and Yeh. Agreement was quite good, especially for larger particles with d_V above about 10 nm.

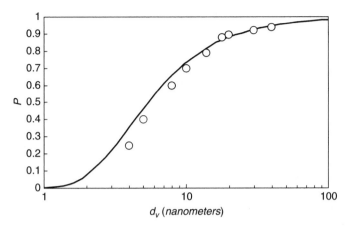

Figure 8.16 *Penetration (P) as a function of particle equivalent volume diameter (d_V) for deposition by diffusion in a single screen; solid line from the theory of Cheng and Yeh (1980), experimental data from experiments reported by Scheibel and Porstendörfer (1984) for a screen with wire width $d_W = 50\ \mu m$, thickness $t = 85\ \mu m$ and face velocity $U_{face} = 0.024\ ms^{-1}$*

8.7   Other particle size-selective mechanisms

Most aspects of particle motion in air are highly dependent on particle size, and so provide opportunities for applications in particle size-selective aerosol sampling. The ones that have found the most widespread use in this way have been described above. Other inertial devices, beyond the conventional and virtual impactors described above, have been developed, differing in the way the particles are injected into the distorted air flow and are collected, some of which will be described later in relation to aerosol spectrometers. Other options are also available.

8.7.1   Electrostatic precipitation

Electric charging of airborne particles, for example in a corona discharge, allows for particle collection that is strongly dependent on particle size. This process is complicated, however, because both the charging and the collection processes are particle size-dependent. Electrostatic precipitation has been used for large-scale industrial gas cleaning since the early 1900s, having the advantages over other particle collecting systems that the resistance to the gas flow is very low yet the collection efficiency can be very high (e.g. White, 1963). Yeh (1993) reviewed electrostatic precipitation as it may be applied to aerosol sampling and detection, and showed that it can produce an especially useful option for particles in the size range below about 1 μm. The processes of particle charging, involving both diffusion charging (by which ions arrive at the particle surface by diffusion, especially effective for small particles) and field charging (by which ions arrive under the influence of an external electric field, especially effective for larger particles), are sufficiently well understood that the distribution of electric charges given to particles of given size can be quite well predicted for given corona discharge conditions. In turn, the subsequent motion of particles in electric fields of well-defined intensity and direction may be predicted. This may be very useful in aerosol measurement. It has been used in some sampling systems simply as an effective low-pressure drop means to collect particles onto a substrate. Most importantly, however, it has found special applications in sampling instruments – or spectrometers – intended for the determination of particle size distribution, especially for small particles below 1 μm in diameter extending into the ultrafine region well below 0.1 μm.

8.7.2   Thermal precipitation

Thermophoresis derives from the interaction of particles with gas molecules whose random (or thermal) motions are a strong function of the gas temperature. As already noted earlier in the discussion of Brownian diffusion (see Chapter 2), particles exchange momentum in collections with surrounding gas molecules. A particle in a strong temperature gradient may experience greater momentum exchange with gas molecules impacting from the high temperature side than from the low temperature side. The result is a net force, and in turn particle motion, in the direction from high to low temperature. For small particles with $d < mfp$, the derivation of the force is straightforward, based only on the random motions of both the particles and gas molecules, respectively, depending neither on the particle size itself nor its composition. For large particles with $d \gg mfp$, the particle sees the surrounding gas as a continuum and the force depends both on particle size and particle composition. The latter occurs by virtue of the role of the thermal conductivity of the particle in distributing heat through the body of the particle. The deposition of particles in a temperature gradient – for example between a pair of plates in a narrow rectangular channel – takes place by virtue of their resultant thermophoretic drift towards the cooler surface. The graph in Figure 8.17 shows some calculated values for the thermophoretic velocity (v_T) as a function of particle size for particles whose thermal conductivity relative to that of air is 10, typical of many aerosols found in workplaces and the ambient environment. The curves are plotted for a temperature gradient of 1000 K cm^{-1}, typical of the range found in many of the sampling instruments – *thermal precipitators* – that have applied the principles of thermophoresis in particle collection. In Figure 8.17, particle size is expressed in terms of equivalent volume diameter (d_V), consistent with the earlier

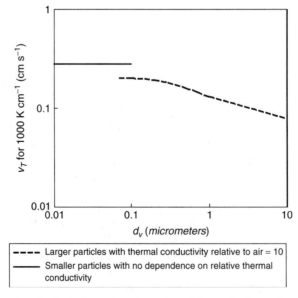

Figure 8.17   *Thermophoretic velocity (v_T) as a function of particle equivalent volume diameter (d_V) of particles in a temperature gradient of 1000 K cm^{-1} (calculated from equations summarised by Hinds, 1999)*

discussion about the choice of particle size metric for diffusion-related phenomena. Two relationships are shown. The first for very small particles indicates that v_T is totally independent of d_V. The second, for larger particles where the thermal conductivity of the particles comes into play, shows a relatively weak dependence on particle size compared with the other processes mentioned above. Hinds (1999) pointed out that, under typical conditions in a thermal precipitator, thermophoretic particle velocities may exceed gravitational settling velocities for d_V below about 0.1 μm, and thermophoretic fluxes may exceed diffusive fluxes for d_V greater than about 0.1 μm. Devices operating on such principles were first proposed as collectors for aerosol samplers as early as the 1930s (e.g. Green and Watson, 1935). In the past few years, interest has been shown in this mechanism for the collection of ultrafine particles (Maynard and Brown, 1991).

8.7.3 Optical processes

Beyond aerosol mechanics, particles may be selected according to their size by both optical and visual means. The scattering of light by suspended particles in air is defined by the way in which electro-magnetic radiation interacts with dielectric media, as first described in the late 1800s by James Clerk Maxwell (Maxwell, 1888). Applied to aerosols, it is a complex function of the wavelength of the light, the angle of both illumination and detection, and particle properties such as geometric size, shape and composition. For aerosol measurement, the interaction between a beam of light and the particles may be viewed in terms of either the *extinction* (or, conversely, the *transmittance*) of the beam or the light that is scattered into a given solid angle (e.g. as reviewed by Hodkinson, 1966). A full understanding of the underlying physics and its applications to these scenarios for specific conditions provides a wide range of options for particle size-selective aerosol measurement. A major advantage of optical detection methods is that the quantities in question (e.g. concentration of sampled aerosol, particle size distribution) may be obtained in real time, and this is where they have received greatest interest.

References

Barth, W. (1956) Berechnung und Auslegung von Zyklonabscheidern auf Grund neuerer Untersuchungen, *Bund der Ingenieure für Wasserwirtschaft, Abfallwirtschaft und Kulkurbau*, 8, 1–9.

Beeckmans, J.M. (1972) A steady-state model of the reverse-flow cyclone, *Journal of Aerosol Science*, 3, 491–500.

Brown, L.M., Collings, N., Harrison, R.M., Maynard, A.D. and Maynard, R.L. (2000) Ultrafine particles in the atmosphere, *Philosophical Transactions of the Royal Society of London A*, 358, 2563–2565.

Brown, R.C. (1980) Porous foam size selectors for respirable dust samplers, *Journal of Aerosol Science*, 11, 151–159.

Brown, R.C. (1993) *Air Filtration: An Integrated Approach to the Theory and Applications of Fibrous Filters*, Elsevier, Oxford.

Chan, T.L. and Lippmann, M. (1977) Particle collection efficiencies of air sampling cyclones: an empirical theory, *Environmental Science and Technology*, 11, 377–382.

Chen, B.T., Yeh, H.C. and Cheng, Y.S. (1986) Performance of a modified virtual impactor, *Aerosol Science and Technology*, 5, 369–376.

Chen, C.C., Lai, C.Y., Shih, T.S. and Yeh, W.Y. (1998) Development of respirable aerosol samplers using porous foams, *American Industrial Hygiene Association Journal*, 59, 766–773.

Cheng, Y.S. and Yeh, H.C. (1979) Particle bounce in cascade impactors, *Environmental Science and Technology*, 13, 1392–1395.

Cheng, Y.S. and Yeh, H.C. (1980) Theory of a screen-type diffusion battery, *Journal of Aerosol Science*, 11, 313–320.

Dalla Valla, M.M. (1952) *United States Technical Conference on Air Pollution* (Ed. L.C. McCabe), McGraw-Hill, New York, p. 341.

Davies, C.N. (1952) The separation of dust and particles, *Proceedings of the Institute of Mechanical Engineering*, 1B, 185–198.

Davies, C.N. and Aylward, M. (1951) The trajectories of heavy, solid particles in a two-dimensional jet of ideal fluid impinging on a plate, *Proceedings of the Royal Society of London*, B64, 889–911.

Davies, C.N., Aylward, M. and Leacey, D. (1951) Impingement of dust from air jets, *AMA Archives of Industrial Hygiene and Occupational Medicine*, 4, 354–397.

de Juan, L. and Fernandez de la Mora, J. (1998) Sizing particles with a focusing impactor: effect of the collector size, *Journal of Aerosol Science*, 29, 589–599.

Dietz, P. (1981) Collection efficiency of cyclone separators, *American Institute of Chemical Engineers Journal*, 27, 888–892.

Dunmore, J.H., Hamilton, R.J. and Smith, D.S.G. (1964) An instrument for the sampling of respirable dust for subsequent gravimetric assessment, *Journal of Scientific Instruments*, 41, 669–672.

Fernandez de la Mora, J. and Schmidt-Ott, A. (1993) Performance of a hypersonic impactor with silver particles in the 2 nm range, *Journal of Aerosol Science*, 24, 409–415.

Fuchs, N.A. (1964) *The Mechanics of Aerosols,* Macmillan, New York.

Gibson, H. and Vincent, J.H. (1981) The penetration of dust through porous foam filter media, *Annals of Occupational Hygiene*, 24, 205–215.

Gormley, P.G. and Kennedy, M. (1949) Diffusion from a stream flowing through a cylindrical tube, *Proceedings of the Royal Irish Academy*, 52A, 163–169.

Gregory, P.H. (1973) *Microbiology of the Atmosphere*, 2nd Edn, John Wiley & Sons, Inc. New York.

Green, H.L. and Watson, H.H. (1935) Physical methods for the estimation of the dust hazard in industry, Medical Research Council Special Report No. 199, HMSO, London.

Hering, S.V. and Marple, V.A. (1986) Low-pressure and micro-orifice impactors, In: *Cascade Impactor* (Eds J.P. Lodge and T.L. Chan), American Industrial Hygiene Association, Akron, OH, pp. 103–127.

Hinds, W.C. (1999) *Aerosol Technology*, 2nd Edn, John Wiley & Sons, Ltd, New York.

Hodkinson, J.R. (1966) The optical measurement of aerosols, In: *Aerosol Science* (Ed. C.N. Davies), Academic Press, London, pp. 287–357.

Ingham, D.B. (1975) Diffusion of aerosols from a stream flowing through a cylindrical tube, *Journal of Aerosol Science*, 6, 125–132.

Kenny, L.C., Aitken, R.J., Beaumont, G. and Görner, P. (2001) Investigation and application of a model for porous foam aerosol penetration, *Journal of Aerosol Science*, 32, 271–285.

Kim, J.C. and Lee, K.W. (1990) Experimental study of particle collection by small cyclones, *Aerosol Science and Technology*, 12, 1003–1015.

Kim, J.C. and Lee, K.W. (2001) A new collection efficiency model for small cyclones considering the boundary-layer effect, *Journal of Aerosol Science*, 32, 251–269.

Lidén, G. and Gudmundsson, A. (1997) Semi-empirical modelling to generalise the dependence of cyclone collection efficiency on operating conditions and cyclone design, *Journal of Aerosol Science*, 28, 853–874.

Lippmann, M. and Chan, T.L. (1979) Cyclone sampler performance, *Staub Reinhaltung der Luft*, 39, 7–12.

Ma, L., Ingham, D.B. and Wen, X. (2000) Numerical modelling of the fluid and particle penetration through small sampling cyclones, *Journal of Aerosol Science*, 31, 1097–1119.

Maddox, R.L. (1870) On the apparatus for collecting atmospheric particles, *Monthly Microscopical Journal*, 1, 286–290.

Marple, V.A. (1970) *A Fundamental Study of Inertial Impactors*, PhD Thesis, University of Minnesota, Minneapolis, MN, USA.

Marple, V.A. and Chien, C.M. (1980) Virtual impactors: a theoretical study, *Environmental Science and Technology*, 8, 648–654.

Marple, V.A. and Liu, B.Y.H. (1974) Characteristics of laminar jet impactors, *Environmental Science and Technology*, 8, 648–654.

Marple, V.A. and Rubow, K.L. (1986) Theory and design guidelines. In: *Cascade Impactor* (Eds. J.P. Lodge and T.L. Chan), American Industrial Hygiene Association, Akron, OH, pp. 79–101.

Marple, V.A. and Willeke, K. (1976) Impactor design, *Atmospheric Environment*, 10, 891–896.

Marple, V.A., Liu, B.Y.H. and Kuhlmey, G.A. (1981) A uniform deposit impactor, *Journal of Aerosol Science*, 11, 333–337.

Maynard, A.M. and Brown, L.M. (1991) The collection of ultrafine aerosol particles for analysis by transmission electron microscopy, using a new thermophoretic precipitator, *Journal of Aerosol Science*, 22, S379–S382.

Maxwell, J.C. (1888) *An Elementary Treatise on Electricity*, 2nd Edn, Clarendon Press, Oxford.

Mercer, T.T. (1969) Impaction from round jet, *Annals of Occupational Hygiene*, 12, 41–48.

Miquel, M.P. (1879) Etude sur les poussierres organisees de l'atmosphere, Annales d'Hygiene Publique et de Medecine Legale, January, 226.

Pich, J. (1966) Theory of aerosol filtration by fibrous and membrane filters, In: *Aerosol Science* (Ed. C.N. Davies), Academic Press, London, pp. 223–285.

Pouchet, M.F. (1860) Micrographie atmospherique, *Compte Rendu des Séances de l'Academie des Sciences*, 16 April, 748–750.

Rader, D.J. and Marple, V.A. (1985) Effect of ultra-Stokesian drag and particle interception on impaction characteristics, *Aerosol Science and Technology*, 4, 141–156.

Ranz, W.E. (1984) Wall flows in a cyclone separator: a description of internal phenomena. In: *Aerosols* (Eds B.Y.H. Liu and D.Y.H. Pui), Elsevier, New York, pp. 631–634.

Ranz, W.E. and Wong, J.B. (1952) Impaction of dust and smoke particles, *Industrial and Engineering Chemistry*, 44, 1371–1381.

Robert, K.R. and Baril, A. (1984) Sampling for respirable cotton dust, *Journal of the American Oil Chemists Society*, 10, 1553–1558.

Rietema, K. (1961) *Cyclones in Industry* (Eds K. Rietema and C.G. Verver), Elsevier, Amsterdam, pp. 46–85.

Rosin, P., Rammler, E. and Intelmann, W. (1932) Grundlagen und Grenzen der Zyklonentstaubung, *Zeitschrift Ver. Deutsch. Ing.*, 76, 433–437.

Scheibel, H.G. and Porstendörfer, J. (1984) Penetration measurements for tube and screen-type diffusion batteries in the ultrafine particle size range, *Journal of Aerosol Science*, 15, 673–682.

Sinclair, D. and Hoopes, G.S. (1975) A novel form of diffusion battery, *American Industrial Hygiene Association Journal*, 36, 39–42.

Soderholm, S.C. (1979) Analysis of diffusion battery data, *Journal of Aerosol Science*, 10, 163–175.

Stern, S.C., Zeller, H.W. and Schekman, A.I. (1962) Collection efficiency of jet impactors at reduced pressures, *Industrial and Engineering Chemistry*, 1, 273–277.

Vincent, J.H., Aitken, R.J. and Mark, D. (1993) Porous plastic foam filtration media penetration characteristics and applications in particle size-selective sampling, *Journal of Aerosol Science*, 24, 929–944

Wake, D. and Brown, R.C. (1991) Filtration of monodisperse aerosols and polydisperse dusts by porous foam filters, *Journal of Aerosol Science*, 22, 693–706.

Walton, W.H. (1954) Theory of size classification of airborne dust clouds by elutriation, *British Journal of Applied Physics*, 5 (Suppl. 3), S29–S40.

White, H.J. (1963) *Industrial Electrostatic Precipitation*, Addison-Wesley, Reading, MA, USA.

Wright, B.M. (1954) A size-selecting sampler for airborne dust, *British Journal of Industrial Medicine*, 11, 284–288.

Yeh, H.C. (1993) Electrical techniques. In: *Aerosol Measurement* (Eds K. Willeke and P.A. Baron), Van Nostrand Reinhold, New York, pp. 410–426.

Part B

Standards for Aerosols

9

Framework for Aerosol Sampling in Working, Living and Ambient Environments

9.1 Introduction

The science of aerosol sampling is closely interwoven with its applications in the practical world. That includes not only sampling technology in the form of devices and methods, all based on the scientific and technical aspects discussed in the preceding chapters, but also the establishment of frameworks of policies and standards within which sampling takes place. Much of the latter is concerned with regulating the environment for the protection of the health and well-being of people.

Aerosols are an important aspect of the natural environment. In the widest sense they may represent challenges to the earth's ecology, both globally and locally. At the global level, one impact of long-lived atmospheric aerosols is to reduce the total amount of sunlight reaching the surface of the earth, with the potential to influence climate accordingly. Locally, aerosols emitted into the atmosphere may influence ecosystems and people close to – and sometimes not so close to – sources. Some may enter the environment as aerosols of one type but may be transformed into other species and/or be transported into other environmental media. Some chemicals emitted into the atmosphere as gases and vapors may undergo physical and chemical transformation and ultimately arrive as aerosols. Through these various processes of transformation and transportation, bioaccumulation of toxic species arising from some such atmospheric aerosols may occur in living systems. At the yet more local level, aerosols released into workplace atmospheres may be very specific to the industrial processes taking place there, and may in turn be associated with correspondingly specific health outcomes. In summary, for nearly all scenarios like those described, aerosols are generally thought of as having the potential to be harmful. Hence, responsible modern societies seek to regulate them.

In order to discuss a rational regulatory framework for aerosols, it is important first to establish a conceptual basis and a working context. It begins with an environmental policy in the form of an overarching bureaucratic instrument designed to address the overall concern. In democratic systems such a policy emerges from discussions between interested parties, often with conflicting interests, by some process leading to consensus. The specific processes – and indeed the ultimate emphasis – vary

Aerosol Sampling: Science, Standards, Instrumentation and Applications   James Vincent
© 2007 John Wiley & Sons, Ltd

between societies, depending on cultural, legal, constitutional and a wide range of other factors. In an ideal world the principles should be consistent. A given policy should first clarify the need, and then identify and prioritise the contributing factors, prescribing actions to address the nature and extent of the problem and providing guidelines for remediation. It should also refer to mechanisms or structures by which management systems may be employed to ensure that the appropriate actions are carried out and in the correct sequence, and to confirm that those actions are in fact working with respect to the original stated goals. Moreover, it should prescribe actions that may be taken if the requirements of the policy are not satisfied after it has been implemented. These may include sanctions. Finally, the policy should also indicate the resources that will be made available to operate the policy realistically.

A *standard* is an important component of an environmental policy. In general, it defines a measurable reference point, consisting of specific criteria and guidelines, by which progress towards satisfying the desired policy objective can be quantified and measured. In the context of this book, we might say that it is environmental policy that drives the justification for aerosol sampling. It is the standard that provides the underlying rationale for aerosol sampling in the real world, and points the way towards how it should be applied.

9.2   Exposure to aerosols

In occupational and environmental health, *exposure* is the central ingredient in a paradigm that links humans with aerosols in the environment, addresses the questions of dose, biological response and adverse health outcome, and points to technical and management options for controlling the adverse outcome. This paradigm is summarised in Figure 9.1. In this form it may be applied to most potentially harmful environmental agents, including aerosols. It provides a good basis for thinking about the issues that need to be addressed during the development of a standard. In the first instance, almost any type of aerosol may be viewed as a *hazard*. The associated *risk* to human health is realised only when interaction between the hazard and humans occurs such that there is exposure.

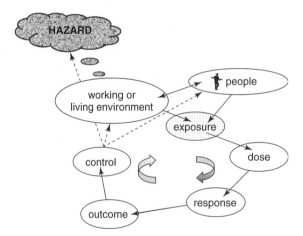

Figure 9.1   *General overview to indicate the linkages between hazards in the environment, human exposure, dose, response, health outcome and control options*

9.2.1 The human respiratory tract

The route of aerosol exposure that is by far the most important, and is therefore reflected in nearly all standards, is *inhalation*. This may take place through either the nose or the mouth, or both. Here, the inhalation of an aerosol is directly analogous to the aspiration of aerosols by aerosol samplers, where the nose and/or mouth are equivalent to the inlet that has been the feature of much discussion in earlier chapters. What happens to an aerosol after inhalation immediately after it has entered the body is similarly analogous to what happens in a sampler after aspiration and before it reaches the filter or other collecting substrate. Initial discussion of the human respiratory tract may be based on the rudimentary diagram shown in Figure 9.2, indicating three primary regions – the *extrathoracic region* (including the *nasopharynx* and *mouth* in the head), the *tracheobronchial region* (including the trachea, main bronchi, bronchi and terminal bronchioles) and the *alveolar region* (including the respiratory bronchioles, alveolar ducts and alveolar sacs). More detailed reviews of the anatomy and physiology of the human respiratory tract as they relate to aerosol inhalation have been given by the International Commission of Radiological Protection (ICRP, 1994), Phalen (1999) and others. In summary, the lung has two parts, the left being divided into two lobes and the right into three lobes. These contain a complex system of airways, starting off with the trachea and developing through successive generations of branching, followed by several million alveoli that have a total surface area of about 100 m², approximately equivalent to the area of a tennis court. The total air capacity of the lung includes the inspiratory and expiratory reserve volumes, along with the tidal and residual volumes. The latter is the volume left in the lung after maximal exhalation. The total lung capacity depends on an individual's sex, age, weight and height, and varies from about 4 to 6 L. The tidal air volume is the volume inspired during each breath, and is typically about 0.5 L, but can be as high as 1 L. A typical adult breathing pattern has a breathing rate of from about 7 to 30 breaths min⁻¹, depending on the level of activity undertaken. The lung comprises the tracheobronchial conducting airways and the alveolar alveolar region. The conducting airways serve as a conduit by which inspired air is conveyed to the alveolar region. It is reinforced by strong cartilage and smooth muscle that provide variable mechanical flow resistance. In addition to providing the conduit for the inspired air, this region also humidifies the air to saturation and warms it to about 37 °C before it passes to the alveolar region. This is where the gas exchange takes place – oxygen into the blood from the inspired air and carbon dioxide from the blood into the expired air. A summary of the structure of

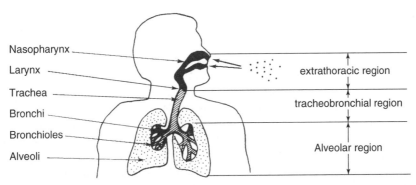

Figure 9.2 *Rudimentary picture of the human respiratory tract on which to base discussion of health-related aerosol exposures. Reproduced with permission from Vincent, Aerosol Sampling: Science and Practice. Copyright (1989) John Wiley & Sons, Ltd*

Table 9.1 *Summary of morphological and physiological data for the typical human lung, describing branching, dimensions of successive branchings, and air velocities and residence times at each generation (from Lippmann, 1977)*

Region	Airway	Generation, G	Number per generation, N_G	Diameter at generation, D_G (mm)	Length at generation, L_G (mm)	Area at generation, X_G (cm^2)	V_G (m s^{-1})	Residence time at generation, t_G (ms)
Tracheobronchial	Trachea	0	1	18	120	2.5	3.9	30
	Bronchi	1–13	up to 2×10^3	down to 1	down to 3.9	up to 20	down to 0.52	down to 7.4
	Bronchioles	14–16	up to 66×10^3	down to 0.6	down to 1.6	up to 180	down to 0.05	16–31
Alveolar	Respiratory bronchioles	17–18	up to 26×10^4	down to 0.5	down to 1.2	up to 530	down to 0.03	60
	Alveolar ducts	19–21	up to 2×10^6	down to 0.43	down to 0.7	up to 3×10^3	down to 0.003	210
	Alveolar sacs	22–23	up to 8×10^6	down to 0.41	down to 0.5	up to 10^4	down to 0.001	550

the human lung, indicating the number of branching generations and the morphological dimensions at each level, along with the corresponding air velocities and residence times, is given in Table 9.1.

It is the nasopharyngeal region that the particles first encounter immediately they have been aspirated through the portals of the nose and/or mouth. Nasally inhaled particles entering the complex system of nasal passages encounter highly distorted and turbulent flows. Larger particles may be deposited mainly by inertial and gravitational forces, smaller ones by diffusion and (possibly) electrostatic forces. Particles deposited in the *anterior* part of the nasopharyngeal region may be eliminated almost completely by external nose blowing and those in the *posterior* part by the mechanical action of cilia on some of the internal passages, by which means they may eventually reach the epiglottis and be swallowed, and so enter the gastrointestinal (GI) system. They then remain in the body until they are either absorbed or excreted. Particles entering through the mouth meet similarly distorted and turbulence air motions, and particle deposition in the head takes place by the same range of mechanisms. Mouth-inhaled particles deposited in the mouth, throat and larynx may be removed directly by expectoration, or may be swallowed and enter the GI system.

Particles that are not deposited in the head pass through the larynx and enter the lung. This contains the region of airways through which the air is conducted to the alveolar region, and includes the trachea, and successive branches of bronchi, bronchioles and – finally – the terminal bronchioles. Particles are deposited in this tracheobronchial region mainly by inertial and gravitational forces. Deposited particles may be cleared by the efficient mechanical action of the cilia, by which particles are conveyed back upwards towards the larynx and are eventually swallowed, and so go to the GI system. Particles that pass through the tracheobronchial region arrive in the alveolar region defined by the respiratory bronchioles, alveolar ducts and alveolar sacs. They may be deposited mainly by the influence of gravitational forces and diffusion, possibly aided by electrostatic forces. Particles deposited in this region may be cleared mainly by the action of the scavenging *macrophage* cells that are present on the alveolar walls. Such cells may engulf the particles and then carry them back up towards the conducting airway region, after which they are transported onwards by the ciliary action there and so are eventually swallowed. In terms of the fate of deposited particles, it is an important distinction between the tracheobronchial and

Table 9.2 *Values for respiratory anatomical and physiological parameters for defining a reference worker (Phalen, 1999)*

Parameter	Value
Weight (kg)	70
Height (cm)	175
Age (years)	20–30
Body surface area (m^2)	1.8
Lung weight (kg)	1
Lung surface area (m^2)	80
Trachea weight (g)	10
Trachea length (cm)	12
Total lung capacity (L)	5.6
Functional residual capacity (L)	2.2
Vital capacity (L)	4.3
Residual volume (L)	1.3
Respiratory dead space (mL)	160
Breathing rate (breaths min^{-1})	15
Tidal volume (L)	1.45
Minute volume (L)	21.75
Inspiratory period (s)	2
Expiratory period (s)	2

alveolar regions that particles deposited in the conducting airways are cleared relatively quickly by the action of the cilia. By contrast, the clearance of particles deposited in the alveolar region by the action of the macrophages is much slower.

It is clear, even from the brief summary given above, that the human respiratory tract – in terms of both its anatomy and its physiological performance – will vary greatly from one individual to another, depending on age, body weight, gender, lifestyle and many other factors. In order to rationalise the knowledge that has been gained from experiments to characterise factors that may be influential in exposure to aerosols so that it might be applied most effectively in aerosol standards, Phalen summarised the parameters for describing a 'reference worker'. This is shown in Table 9.2. Here the *minute volume* is the volume of external air that is inspired per minute, equivalent to the *tidal volume* inspired per breath times the number of breaths per minute. The *breathing flow rate* is the total volumetric flow rate of air passing through the respiratory tract, including both inhalation and exhalation.

9.2.2 Definitions of exposure

In general, *exposure* may be defined in the first instance as the instantaneous *intensity* of an agent that is relevant to a particular adverse health outcome at an appropriate interface between the environment and the exposed individual. In this definition, the intensity is reflected in the metric that is chosen to measure the exposure. For aerosols it is expressed in terms of the concentration, expressing the amount of particulate material per uniform volume of air. Measurement of exposure intensity strictly according to the definition might be achieved by assessing the instantaneous aerosol concentration in real-time using an appropriate direct-reading sampling instrument. However, in practice it is usually more appropriate to measure it in terms of a value that is averaged over an appropriate time interval. This might range from a

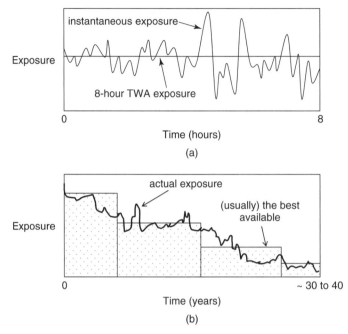

Figure 9.3 *Schematic illustration of the relationship between daily time-weighted exposures and lifetime exposure history for an individual exposed to a given type of aerosol exposure*

few minutes to as much as many hours. In occupational aerosol exposures, for example, time-averaging typically takes place over a working shift of up to 8 h. For ambient atmospheric exposures, a period of 24 h is more common. Figure 9.3(a) illustrates how the instantaneous and time-averaged exposures are linked. In terms of toxicological effects, the extent to which the time-averaged intensity is relevant to adverse aerosol-related health effects is determined by the timescale of the processes in the human body by which the substance of the particles lead to biological responses and in turn to actual health effects. For most aerosols found in working and living environments, periods of time-averaging like those indicated are short compared with the latency periods of the long-term health effects that might arise.

In general, for most aerosol-related health effects, neither the instantaneous nor even the time-averaged intensity of exposure are particular helpful on their own for the estimation of risk. For this, an assessment of the *exposure history* is needed, taking into account the trends of aerosol exposures over much longer periods, possibly over the whole lifetime of an exposed individual and for each type of aerosol. Figure 9.3(b) illustrates how this relates to the individual measures of time-weighted exposure that have been acquired over a long period of time. It is this type of exposure history which provides the most appropriate information upon which to base an estimate of risk associated with exposure to aerosol of a given type, taking into account the nature of the dosimetric processes after the particles have arrived in the body, including both the fate and subsequent transport of the particles and the biological responses that are result of the interactions between particles and cells in the body. Aerosol sampling strategies need to be developed with all this in mind. Generally, as indicated, it is possible in practice only to obtain historical exposure information based on relatively infrequent periodic sampling campaigns.

9.2.3 Variability of exposure

Experience has shown that, in working and ambient environments, human exposures to airborne contaminants, including aerosols, are highly variable. Part of that variability is due to uncertainty in the measurement method itself. A major part of this book deals with the factors that are associated with biases in the sampling procedure associated with factors external and internal to the sampling instrument in question. In addition, there is random variability associated with stochastic departures of any of these factors from assigned or expected values. However, in practice, with careful attention to the choice of sampling instruments and procedures, such errors are generally small. By far the greatest uncertainty in the measurement of exposure is the variability of the exposure itself. Such variability arises as the result of random changes in contaminant emissions, external factors (e.g. changing wind speed or direction) and human activity, all of which are external to the process of sampling itself. In the overall context of aerosol sampling for human exposure assessment, although it is not part of the physical process of sampling itself, it is essential to be conversant with the nature of such variability. This is particularly important in considerations of sampling strategy by which to determine where, when, for how long and how often to take samples.

9.3 Framework for health-related aerosol sampling

There are three primary rationales that may be identified for health-related aerosol sampling in working and living environments. The first is aerosol sampling for exposure assessment as part of epidemiology enquiry or risk estimation, where the aim is to provide a set of relevant exposure data which may be folded into an analysis, along with health effects and other data, to examine associations between exposure and aerosol-related ill-health. The second is the sampling that is carried out in a specific exposure situation, usually at the initiative of a professional occupational or environmental hygienist, to characterise exposure in order to enable identification of options for control action. The third is the sampling that is carried out in the context of a standard, in order to determine whether air quality meets or exceeds policy requirements. In general, an ideal framework for conducting practical aerosol sampling may be viewed as comprising five primary ingredients; criteria, sampling instrumentation, sampling strategy, analytical instrumentation and – finally – a *limit value*. The epidemiology rationale requires the first four of these and has the aim of providing information leading to the fifth. For the other two rationales, all five ingredients form an essential framework.

9.3.1 Criteria

Health-related aerosol sampling should begin with the establishment of a clear scientific basis from which to choose the most appropriate measurement procedure. This defines the metric by which the measurement is expressed. It includes reference to the health effect – or range of health effects – that might result from the exposure in question. In the first instance, the single most important route of exposure by which aerosols may enter the body is by inhalation. The definition of exposure given above refers to the interface where the particles are separated from the air and arrive at the body. So, at the outset, a criterion should include consideration of the particle size-dependent processes that govern how the inhaled particles penetrate through, and are deposited in, the different parts of the respiratory tract. This is where Figure 9.2 becomes useful. Details of experimental studies to quantify the processes of inhalation and deposition in this system will be given in subsequent chapters. For the present, it is sufficient to note that, for a polydisperse aerosol, larger particles are the most likely to be removed during inhalation and by deposition higher up in the respiratory tract. Smaller particles are more likely to penetrate to the lower regions, and may be deposited there.

In addition to the particle size-selectivity of the human respiratory tract, consideration is needed of the composition (or species) of the particles, and how this relates to the health effect of interest. This is important in practical situations in working and living environments, where the aerosols to which people are exposed usually take the form of mixtures, with some species being more harmful than others or are more or less relevant to the given health effect. So, for example, for mine dust where one health effect of concern is silicosis, the relevant chemical species is the free crystalline silica (quartz) fraction; for airborne dust in a nickel smelter, where the health effect of significant concern is lung cancer, sulfidic nickel species are thought to be relevant. Furthermore, criteria for a standard should also include reference to the metric that is used to define exposure – the aerosol concentration. This may be defined in a number of ways. Particle *number concentration* and *mass concentration* are metrics that have been widely used over the years, but particle *surface area concentration* has been noted in some instances. Strictly speaking, for the purpose of a standard the choice of exposure metric should relate to the toxic action of the particle. While this has indeed driven consideration of exposure limits over the years, a major factor in the choice of metric for the purpose of standards has been the availability of practical, workable sampling instrumentation. As such instrumentation has advanced over the years, there have been periodic shifts in the metrics chosen for specific exposure limits. In order to maintain consistency, this has required the development of correlation procedures – conversion factors or algorithms – that have customarily been obtained through experimental studies in field situations of the type described in later chapters.

The three criteria aspects described above provide the basis for choosing a measurement procedure that will provide information by which to determine, for exposed humans, how much of the most appropriate species is deposited at the appropriate part of the respiratory tract. The particle size-selective criterion dictates the physical performance required of the sampling instrument that will be used. The species criterion dictates the analytical method that will be used to analyze samples. The two criteria, taken together, provide the essential starting point for means to determine the most relevant measure of the health-related dose or uptake.

9.3.2 Sampling instrumentation

As described in the preceding chapters, aerosol sampling involves physical processes that include the mechanics of both the air and particle motions. Aerosol inhalation and the transport of particles through the human respiratory tract may be viewed in the same way, so that humans themselves may be viewed as 'aerosol samplers'. It follows therefore that the ideal sampler should reflect the same physical processes, both qualitatively and quantitatively, as the human respiratory tract. In relation to a specific situation, where there is a specific aerosol-related health risk, the most appropriate particle size-selective criterion should be identified. In turn, an instrument should be chosen whose physical particle size-selective performance most closely matches that defined in the criterion.

9.3.3 Analytical methods

For most practical purposes, the aerosol collected at a filter or a substrate in a sampling instrument must be converted into an aerosol concentration. Usually this is expressed in terms of particulate mass per unit volume of air sampled, in which case the mass concentration is given by:

$$c = \frac{M}{Qt} \qquad (9.1)$$

where M is the mass of particulate matter collected, Q is the sampling flow rate and t the sampling time. In this expression, Q and t are known from the sampling conditions, and M is the subject of interest for analysis. It is here where analytical methods are needed. In many practical situations, M is measured in terms of the overall mass that has been collected, including all species of particulate material. In that case, it may be obtained gravimetrically by using an appropriate analytical balance, where it is usual to weigh the filter or substrate before and after sampling, with the difference providing M and in turn c from Equation (9.1). In many other practical situations, it is desired to measure the concentration of a chemical, or sometimes biological, fraction of the overall aerosol that has been collected. This requires the use of analytical methods, either by 'wet chemistry' procedures involving the use of reagents or indicators or – more commonly nowadays – by the application of sophisticated instrumentation developed for the measurement of individual classes of species or elements. A wide range of such instrumentation is commercially available today.

9.3.4 Sampling strategies

For aerosol sampling, attention needs to be given to the question of sampling strategy. Whatever the rationale for sampling, it is important to be able to acquire exposure data of sufficient quality and quantity to enable satisfactory completion of the task at hand. However, differing objectives may point to different sampling strategies. For health-based aerosol sampling, a decision needs to be made about whether sampling should be performed on an *area* or a *personal* basis. *Area sampling* involves the measurement of aerosols in the environment, working or otherwise, by the use of one or more sampling devices that are placed in fixed locations, with the aim of providing results for aerosol concentration that are representative of those locations and, hopefully, people nearby. *Personal sampling* involves the measurement of aerosols close to individual human subjects by the use of small personal samplers that are mounted on the body, usually in the so-called 'breathing zone'. This has been made possible by the emergence during the past four decades, since the pioneering work of Sherwood and Greenhalgh (1960), of miniature sampling pumps that can provide the desired air flow rate for sufficient duration, and yet be small and light enough not to greatly inconvenience the wearer. The breathing zone is generally poorly defined, reflecting the philosophy that any particles sufficiently within range of the face of the wearer are available for inhalation. Several versions have been suggested. The United States Occupational Safety and Health Administration, in OSHA Standard 29 CFR, suggests only that it is defined by '... air that would most nearly represent that inhaled by the employee' (Occupational Safety and Health Administration, 1999). The United Kingdom Health and Safety Executive, in its methods document MDHS 14/3, defines it as '... the space around then worker's face from where the breath is taken, and is generally accepted to extend no more than 30 cm from the mouth. Personal sampling instruments are normally mounted therefore on the upper chest, close to the collar-bone' (Health and Safety Executive, 2000). The American Industrial Hygiene Association (AIHA), in the latest edition of its compendium widely known as 'The White Book' (DiNardi, 2003), defines it as 'The volume surrounding a worker's nose and mouth from which he or she draws breathing air over the course of a work period. This zone can be pictured by inscribing a sphere with a radius of about 10 inches centered at the worker's nose'. The United States Environmental Protection Agency (EPA) defines it very broadly as 'a zone of air in the vicinity of an organism from which respired air is drawn'.[1] Variations on these definitions are used around the world, none of which has a particularly strong underlying scientific basis. As becomes apparent from a grasp of physical pictures like those presented in preceding chapters, simply placing a sampler in the general proximity of the nose and/or mouth portals of aerosol entry into the body does

---

[1] See http://www.epa.gov/ttn/atw/hapsec1.html.

not, in itself, guarantee that a truly representative sample is obtained. That said, however, it is fair to say that the breathing zone idea is a useful starting point because it at least ensures that the measurement is made in the most appropriate environmental location and follows the subject as he/she goes about their daily activities.

Comparisons between area and personal sampling measurements of aerosols in workplaces have consistently shown that the area-derived concentrations are consistently lower than personal-derived ones, even by as much as factors of 3 to 5 (e.g. Sherwood, 1965, and many others since). The same has been found for nonoccupational situations. One possible explanation is that, by personal sampling, the sampling instrument is able to reach locations which are generally closer to aerosol sources. Another is the possibility that each worker generates his or her own 'personal dust cloud' (e.g. Ozkaynak *et al.*, 1996), although the scientific foundation for this suggestion has never been clear. Nowadays, however, it is customary in most workplace exposure situations in most countries for professional occupational hygienists to adopt the personal sampling approach, on the basis of the argument that this is the only way to assess the exposures of actual people. Indeed, it is now – more often than not – a specified requirement in occupational exposure standards for airborne contaminants. For ambient environment exposures, the logistics of this approach are more difficult because exposure takes place over 24 h, day after day, and not just for limited periods corresponding to working shifts. In addition the population of interest may include children and elderly or infirm people. In such situations, the wearing of a personal sampler may be a considerable imposition for many people. However, even there, personal sampling is seen as a desirable ultimate goal for exposure assessment.

Beyond the philosophy of whether to use area or personal sampling, practical aerosol sampling for the purpose of characterising the aerosol exposures of population groups or sub-groups requires strategic considerations that are based on an understanding not only of the nature of aerosols, aerosol samplers and human exposures themselves, but also of the intrinsic large – and uncontrollable – variability described earlier and which we have come to accept as inevitable in the real world. The statistical nature of exposure is therefore an important feature in any discussion about sampling strategy. For workplace aerosols it has long been known that, for many separate measurements of exposures in a given situation where workplace environmental conditions are generally the same from day to day, the statistical distributions of those exposures tends to be log-normal. The same is true of other airborne contaminants. A typical cumulative distribution for a *similarly exposed group* (SEG) of workers is shown graphically in Figure 9.4, referring to a group of people working in the same general area and exposed to the same aerosol sources. In practice an SEG is customarily defined through the expert judgement of the professional occupational hygienist involved. In Figure 9.4 the log-normal tendency is clear from the linear appearance of the data when plotted on log-probability axes. Also noted is the value of the geometric standard deviation for the exposure, given here as $\sigma_{g,exp}$. Generally speaking, for workplace aerosols, $\sigma_{g,exp} = 1.5$ indicates low variability, 2.5 moderate variability, 3.5 high variability and 5.0 very high variability.

In relatively simple statistical considerations of exposure, the distribution of exposures is assumed to be same for all workers within the SEG. This is sufficient for most purposes, and is the model used in the field by most occupational hygienists (Mulhausen and Damiano, 1998). However, as pointed out in the body of work of Rappaport and his colleagues (much of it summarised in Rappaport, 1991), there are considerable differences in the statistical properties of exposures *within* individual workers (for repeat exposure measurements) and *between* different workers, and there are some situations where these differences might be important, especially in the identification of individuals more prone to higher exposures than others.

In exposure assessment for the purpose of epidemiology, the aim is to obtain a complete and reliable description of the statistical properties of the exposure in question. This means that the sampling strategy

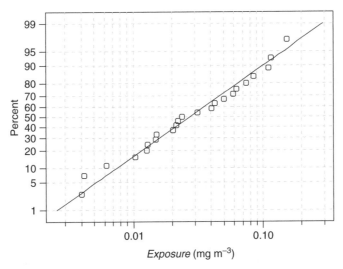

Figure 9.4 *Example of a typical cumulative aerosol exposure, indicating the log-normality that is characteristic of most aerosol exposures*

should be designed to achieve enough single personal exposure data records for workers that are representative of the whole SEG, chosen randomly and over a representative number of shifts to provide the exposure distribution without bias. In this case, all cases are represented uniformly, including both 'best-case' and 'worst-case' examples. Experience has shown that at least 10 samples are needed to reasonably define a distribution of aerosol exposure concentrations, and ideally should be considerably more.

In principle the same approach should be adopted when sampling is carried out with respect to compliance with standards. In particular, for a set of unbiased samples sufficient in number to accurately define the exposure distribution, it is possible to estimate the probability that a limit value (see below) will be exceeded, from whence a decision about control actions can be made. However, this can be very time-consuming and labor-intensive. So in many practical situations, it is a reality that too few samples are taken such that actions are taken on the basis of weak scientific evidence. In other scenarios, a pragmatic decision is made to focus the sampling effort on what appear to be 'worst-case' conditions, with a view to narrowing down where the actions to control the environment might in turn be directed.

Another aspect of sampling strategy relates to the frequency with which an exposure should be carried out in order to define the exposure history for a particular cohort of workers. Again this is especially important for aerosol sampling based on the epidemiology rationale. Here the frequency of the sampling campaigns over time depends on the duration and level of detail required for the exposure history to be relevant to the health effect in question. It needs to take into account the nature and dynamics of the sequence of exposure, dose and response in the population of interest, allowing for the update of the contaminant into the body via the inhalation route, its redistribution, storage, metabolisation and elimination, and its toxic effects. A good example of a long-running series of sampling campaigns is the large, long-running project conducted by the British coal industry in relation to coalworkers' pneumoconiosis, known as the 'Pneumoconiosis Field Research' (PFR) (e.g. see Attfield and Kuempel, 2003). In setting up this large project, it was recognised that dust-related lung disease in mineworkers may develop over decades of exposure, and that the probability of disease relates to the whole exposure history. So a sampling strategy was set up from the outset where aerosol sampling of an appropriate particle size fraction would be conducted on a regular basis for many years, accompanied by surveys of the health status of the working population.

9.3.5 Exposure limits

In a given situation, an *exposure limit* (EL) relates to the aerosol concentration – usually time-averaged – that defines the upper end of the aerosol exposure concentration to an appropriate aerosol particle size fraction and species that should not be exceeded. This is the quantitative part of the standard, and is the metric by which the population in question is protected by the policy. Generically, an EL for a given aerosol type is determined on the basis of known toxicology and epidemiology, reflecting the maximum level of exposure that can be accepted, according to whatever definition of 'acceptable' is applied. The development of an actual numerical value for an EL is a difficult process. For a single, relatively well-defined adverse health outcome, the principle of the EL is usually discussed formally by reference to the *dose–response* curve which relates the probability of the outcome arising from a given exposure. Usually this is accompanied by the assumption that the exposure in question occurs day after day for the life of the subject. Ideally, the EL may be defined by the point in the curve below which the outcome becomes unobservable, which toxicologists call the 'no observed adverse effect level' (or NOAEL). Further, for a whole population of such subjects, the curve becomes bounded by uncertainties arising from inter-subject variability in the population of interest. In addition, there are additional uncertainties which derive from the quality and quantity of the available experimental or observational data. In reality, therefore, expert judgement is required in order to interpret the overall uncertainty. This is usually achieved in practice by the application of certain language in the standard that provides the desired flexibility. For example, in relation to the occupational ELs that are set by the United Kingdom Health and Safety Executive, Carter (1989) referred to '... the ability to identify, *with reasonable certainty*, a concentration averaged over a reference period, at which there is no indication that the substance is likely to be injurious to employees if they are exposed by inhalation day after day to that concentration...'. Similarly, the American Conference of Governmental Industrial Hygienists states that its own ELs, known as 'threshold limit values' (or TLVs), represent levels of exposure to which '... nearly all workers may be repeatedly exposed, day after day, over a working lifetime, without adverse health effects' (American Conference of Governmental Industrial Hygienists, 2004). In the USA, National Ambient Air Quality Standards (NAAQS) are developed for the ambient atmosphere by the EPA to '... protect public health with an adequate margin of safety...', also reflecting a similar respect for the inevitable uncertainty.[2] Fairhurst (1995) noted the need for the application of an uncertainty factor – as a safety margin – to bridge the gap between an actual EL which standards setting bodies are obliged to assign in the real world and the hypothetical NOAEL. The uncertainty factor that needs to be applied will be lower the greater the quality and quantity of the available data, and will be lower the more actual human data are available.

There are other unknowns – even *unknowables* – that further complicate things. If there is more than one health effect associated with a given aerosol exposure, which one should take priority in the setting of the EL? For animal studies, what is known about interspecies differences and extrapolation to humans? For epidemiological studies, what are the quality and quantity of the exposure data? How do we deal with the mixed exposures which prevail in most practical situations, perhaps involving different species? For such mixtures, there may be biological interactions which are additive, synergistic or antagonistic in ways which are not at all well understood.

---

[2] See http://www.epa.gov/ttn/naaqs/standards/pm/s_pm_index.html.

From the preceding it is clear that, in the discussion about actual numerical values for ELs, there is scope for wide disagreement even among scientists within the same discipline. The situation is compounded still further by the fact that many of the scientific issues involved in the discussion cut across scientific disciplinary boundaries, where the terminologies and basic underlying philosophies can be quite different. Further, the preceding description relates to ELs which are strictly health-based, representing the ideal. The occupational ELs (TLVs) developed by the American Conference of Government Industrial Hygienists and some other standards setting bodies come into this category. The intention of these is primarily to provide guidance to professional occupational and environmental hygienists in their work to protect the populations to whom they are responsible. But, although they are widely quoted around the world, they have no regulatory force. On the other hand, in the real world, there are ELs that are set by government agencies, and are embodied in standards as part of public policy and so are enforceable by law. For these, it is common to include not only the scientific argument about how much exposure leads to how much ill-health but also considerations of technical feasibility and socioeconomic and sociopolitical factors. The result then is a set of 'pragmatic' ELs which, hopefully, represents fair compromises between all the competing factors and interests. For many substances it is inevitable that such ELs will be set at a higher level than the corresponding health-based ELs.

9.3.6   Overview

From the preceding discussion, it becomes obvious that a limit value *cannot* be assigned until consideration has been given to the first four ingredients in the framework described. That is, the EL-value must conform to criteria that have already been assigned, the measurements of aerosol exposure must have been made by methods that have been shown to accurately collect and analyze the appropriate fractions, and those measurements must have been acquired in a way such that the results are demonstrably representative of the population that is being protected. If these requirements are not satisfied, a situation may arise where different practitioners – perhaps in different industries or even in different countries – may carry out sampling according to other guidelines leading to different results, even though the actual exposure situation may in fact be the same. In that event, the quantitative EL-value, and in turn the standard itself, has only limited meaning. This is clearly unsatisfactory. The framework described above provides a way in which the standard may be harmonised.

9.4   Nonhealth-related aerosol standards

There is a wide variety of adverse aerosol-related effects that do not relate to human health. In general they are associated with aerosols that are found in the ambient atmospheric environment, and are of more general environmental concern, some at short and longer range, and others globally. Stationary sources may include industrial operations, including coal-burning power stations, municipal incinerators, cement plants, etc., and outdoor operations such as landfill, agricultural burning, etc. Mobile sources include trucks, cars, ships and aircraft.

Local effects include nuisance or aesthetic effects arising from deposition of large dust or grit particles on surfaces leading to adverse impacts on plants and buildings, and with finer particles of black smoke that impact on atmospheric visibility. The 1993 United Kingdom Clean Air Act refers to the '. . . 'statutory nuisance' (that) can arise from dust, odour, smoke, fumes or gases which is prejudicial to health or a nuisance. For something to be serious enough to be a statutory nuisance, it must be a serious and persistent problem interfering with reasonable living.' But it has proved very difficult to arrive at a scientifically defensible rationale by which such 'nuisance' may be defined. Perhaps this is because – according to Vallack and Shillito (1998) – there is in the UK no actual legal definition of the term 'nuisance'! For large dust or grit particles, sampling is conventionally carried out using samplers in the form of passive 'deposit gauges', of which there are many types, having widely differing performance characteristics (see Chapter 17), and such samplers provide dust/grit levels in terms of the mass collected per unit surface area per unit time (e.g. $mg\,m^{-2}\,day^{-1}$). With this in mind, Vallack and Shillito developed a framework for coarse dust and grit particles based on the interesting concept of 'likelihood of complaint'. Surveys had already indicated that dust/grit levels of $100-150$ $mg\,m^{-2}\,day^{-1}$ were 'noticeable with occasional complaints', while levels of $200-300$ $mg\,m^{-2}\,day^{-1}$ were 'very noticeable producing regular complaints'. Noting that typical annual average levels were of the order of 50 $mg\,m^{-2}\,day^{-1}$ for open country, 100 $mg\,m^{-2}\,day^{-1}$ for town centres, and 150 $mg\,m^{-2}\,day^{-1}$ for industrial areas, Vallack and Shillito suggested a rationale based on excursions of daily averages from such annual averages. It was suggested that daily local averages of 2.5 times the annual local average would lead to 'complaints possible', and 3.5 times the annual local average would lead to 'complaints likely'. A similar logical sequence was proposed for the daily local averages in relation to nationwide 90th and 95th percentile values. The Vallack and Shillito framework, combining these statistics, is summarised in Figure 9.5. It was suggested that such a framework could be used to define 'action thresholds'.

Longer range effects include the wet and dry deposition of finer particles potentially harmful to ecological environments and their inhabitants, including lakes and rivers, sometimes after chemical transformations. Global effects include climatic changes. Large bodies of aerosol sampling activity take place in all these contexts. One has been in the determination of aerosol emissions from specific stationary or mobile sources. Another has been in the ground-based sampling of aerosols, another in airborne sampling at high altitudes from balloons or aircraft.

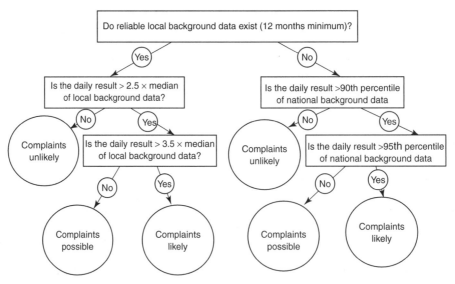

Figure 9.5 *Framework for determining the 'likelihood of complaint' for coarse dust/grit particles for daily levels measured using a dust deposit gauge (see Chapter 17), referenced to local and national averages (Vallack and Shillito, 1998). Adapted from Atmospheric Environment, 32, Vallack and Shillito, 2737–2744. Copyright (1998), with permission from Elsevier*

References

American Conference of Governmental Industrial Hygienists (2004) Threshold limit values for chemical substances and physical agents, and biological exposure indices (2004–2005), ACGIH, Cincinnati, OH.

Attfield, M.D. and Kuempel, E.D. (2003) Commentary: pneumoconiosis, coalmine dust and the PFR, *Annals of Occupational Hygiene*, 47, 525–529.

Carter, J.T. (1989) Indicative criteria for the new occupational exposure limits under COSHH, *Annals of Occupational Hygiene*, 33, 651–652.

DiNardi, S.R. (Ed.) (2003) *The Occupational Environment: Its Evaluation, Control and Management*, 2nd Edn, American Industrial Hygiene Association, Fairfax, VA.

Fairhurst, S. (1995) The uncertainty factor in the setting of occupational exposure standards, *Annals of Occupational Hygiene*, 39, 375–385.

Health and Safety Executive (2000) General methods for sampling and gravimetric analysis of respirable and inhalable dust, Methods for the Determination of Hazardous Substances, MDHS 14/3, HSE, London.

International Commission on Radiological Protection (1994) Human respiratory tract model for radiological protection, ICRP Publication 66, *Annals of the ICRP*, 24, 1–482.

Lippmann, M. (1977) Regional deposition of particles in the human respiratory tract. In: *Handbook of Physiology, Section 9, Reactions to Environmental Agents* (Eds D.H.K. Lee, H.L. Falk and S.D. Murphy), American Physiological Society, Bethesda, MD, pp. 213–232.

Mulhausen, J.R. and Damiano, J. (Eds) (1998) *A Strategy for Assessing and Managing Occupational Exposures*, 2nd Edn, American Industrial Hygiene Association, Fairfax, VA.

Occupational Safety and Health Administration (1999) Sampling and Analytical Methods for MDA Monitoring and Measurement Procedures, OSHA Regulations (Standards - 29 CFR), 1910.1050 App D, Washington, DC.

Ozkaynak, H., Xue, J., Spengler, J., Wallace, L., Pellizzari, E. and Jenkins, P. (1996) Personal exposure to airborne particles and metals: results from the particle team study in Riverside, California. *Journal of Exposure Analysis and Environmental Epidemiology*, 6, 57–78.

Phalen, R.F. (1999) Airway anatomy and physiology. In: *Particle Size-Selective Sampling for Particulate Air Contaminants* (Ed. J.H. Vincent), American Conference of Governmental Industrial Hygienists, Cincinnati, OH.

Rappaport, S.M. (1991) Assessment of long-term exposures to toxic substances in air, *Annals of Occupational Hygiene*, 35, 61–121.

Sherwood, R.J. (1965) Representativeness of air sampling for particulates. In: *Radiological Monitoring of the Environment*, Proceedings of the Conference held by the Joint Health Physics Community and the Central Electricity Generating Board at Berkeley, UK, October 1963, Pergamon Press, Oxford, pp. 69–83.

Sherwood, R.J. and Greenhalgh, D.M.S. (1960) A personal air sampler, *Annals of Occupational Hygiene*, 2, 58–63.

Vallack, H.W. and Shillito, D.E. (1998) Suggested guidelines for deposited ambient dust, *Atmospheric Environment*, 32, 2737–2744.

Vincent, J.H. (1989) *Aerosol Sampling: Science and Practice*, John Wiley & Sons, Ltd, Chichester.

10

Particle Size-Selective Criteria for Coarse Aerosol Fractions

10.1 Introduction

In the early days of aerosol sampling, it was a primary objective to collect a sample of airborne particles that selected particles uniformly across the whole particle size range. That is, the aspiration efficiency of the sampler should be 100 % for all particle sizes, defining what has long been referred to as 'total aerosol'. This, for example, is what would be sampled by a thin-walled tube sampler facing into the wind and operated isokinetically. But isokinetic sampling is an ideal situation, however, pertaining to well-defined air flow conditions in stacks and ducts, and one that cannot be achieved for samplers in working and living spaces with varying conditions of wind speed and direction. Despite this difficulty, some aerosol standards are still written in terms of 'total' aerosol, sometimes referred to as 'total suspended particulate' (or TSP').

It was first proposed in the 1970s that, for health-related aerosol sampling, coarse aerosol should better be defined in terms of what is inhaled by people – that is, by what is inhaled (or aspirated) through the nose and/or mouth during breathing. Thus, it was acknowledged for the first time that the human head may be treated as an aerosol sampler (Ogden and Birkett, 1977). This was an important conceptual step towards a rational new set of criteria for health-related aerosol sampling, in which it was recognised that aerosol sampling should reflect the nature of human exposure. The concept of aerosol 'inhalability' emerged out of this new thinking.

10.2 Experimental studies of inhalability

10.2.1 Early experimental measurements of inhalability

Once it was decided that inhalation was an important part of the exposure process, experiments began to be conducted to obtain quantitative data for the aspiration efficiency of the human head over representative ranges of breathing, aerosol and external wind conditions. Although the original driving force was the practical need to provide information on which to base the development of criteria for sampling health-related fractions in workplaces, it was recognised from the outset that the case of the

Aerosol Sampling: Science, Standards, Instrumentation and Applications   James Vincent
© 2007 John Wiley & Sons, Ltd

aerosol exposure of humans in the ambient atmosphere was also important. Here an extended range of environmental conditions applies. In particular, whereas wind speeds in workplaces are usually less than 1 m s^{-1} and rarely exceed 4 m s^{-1}, they could reach 10 m s^{-1} – or even higher – at ground level in the outdoors environment.

The first experimental studies were performed in Britain and Germany in the late 1970s and early 1980s, and were continued into the 1990s. Most were carried out using life-sized human models based on the forms of tailors' mannequins. The experiments themselves were performed in aerosol wind tunnels, generating comprehensive sets of data for the aspiration efficiency of the human head. Such experiments were distinct from those concerned with determining the regional deposition of particles in different parts of the respiratory tract (which will be discussed in a separate chapter) since they involved the human anatomy only minimally and so did not require the use of actual human subjects. Since the aerosol and fluid mechanical phenomena which govern inhalability are all external to the human body, it was widely agreed that these could be reliably represented by inert experimental systems, with appropriate simulation of breathing through the nose and/or mouth as desired. The principle behind most of the work that has been reported in this area involved measures of the concentration of particles upstream of, and of those inhaled by, the mannequin, the latter collected on filters installed inside the head just behind the nose and/or mouth. From these, the aspiration efficiency (A) of the mannequin was obtained by the indirect (or comparison) method using the earlier Equation (1.20). As before in previous chapters, the quantity A is a fundamental property describing the entry of particles into aerosol sampling devices, and was defined in the experiments for each particle size, wind speed and inhalation flow rate conditions, along with the shape of the body of the sampling system and its orientation with respect to the wind. Ogden and Birkett (1977) and Ogden *et al.* (1977) reported the pioneering first experimental data, noting the importance of the aspiration efficiency of the human head as an index of human exposure, and the correspondence of A with *inhalability*.

The earliest experiments of Ogden and his colleagues were carried out using a life-size head-and-shoulders mannequin in a small wind tunnel. They used monodisperse aerosols of di-2-ethyl-hexyl sebacate labelled with a fluorescent dye, generated using a spinning disk aerosol generator in the range of d_{ae} up to about $30 \mu\text{m}$, and for moving air with wind speed (U) ranging from 0.75 to 2.75 m s^{-1}. Steady sampling flow rates over the range corresponding to the various parts of the human inhalation cycle for someone carrying out 'normal work' were applied using an external pump. In successive experiments the mannequin was placed at $45°$ intervals with respect to forwards-facing, covering the range from 0 to $360°$. Aspiration efficiency (A) for both mouth and nasal breathing was measured by obtaining the sampled masses from fluorimetric analysis of collected particles for each combination of conditions. The raw data were combined to construct 'data' for uniformly orientation-averaged aspiration efficiency for more realistic cyclical breathing. The results for orientation-averaged A as a function of d_{ae} are summarised in Figure 10.1(a), where the hatched area encloses nearly all the actual data points that were reported in the original publications. The results are portrayed in this way here, and for other such data presented below and in the following chapter, since they are used as a basis for identifying broad trends that can be expressed in terms of conventions for aerosol sampling. Figure 10.1(a) shows a band of A that starts at values close to unity for small d_{ae} and falls steadily to about 0.5 at d_{ae} about $30 \mu\text{m}$. Although it was not evident in this simplistic form of presentation, Ogden and his colleagues noted from the raw data that dependencies on sampling flow rate and wind speed were small in relation to the main trend; nor were there any significant differences between nose and mouth breathing. Somewhat later, Armbruster and Breuer (1982) reported experimental studies for a model head (without shoulders), wind speeds from 1 to 8 m s^{-1}, and cyclical breathing conditions approximating to 'at rest', 'normal work' and 'hard work' achieved by means of a purpose-built, reciprocating, breathing machine. Orientation-averaging was achieved by averaging the individual sets of results obtained for angles with respect

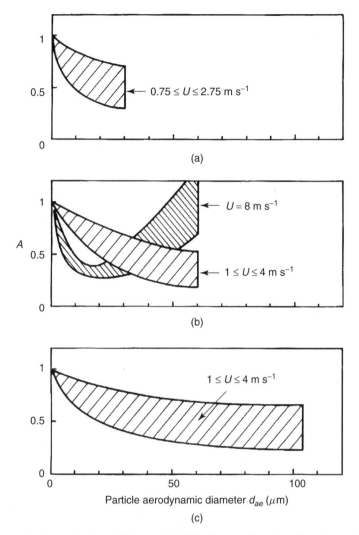

Figure 10.1 *Earlier results for aspiration efficiency (A) of the human head as a function of particle aerodynamic diameter (d$_{ae}$), for various wind speeds (U) in the range from about 0.75 to 8 m s^{-1}: (a) Ogden and Birkett (1977) and Ogden et al. (1977); (b) Armbruster and Breuer (1982); (c) Vincent and Mark (1982). All are for simulated breathing broadly representative of people at work. The hatched areas contain nearly all the experimental data. Reproduced with permission from Vincent, Aerosol Sampling: Science and Practice. Copyright (1989) John Wiley & Sons, Ltd*

to forwards-facing (θ) of 0°, 90° and 180°. Polydisperse test aerosols were generated mechanically from coal dust. Collected particulate material was dispersed in an electrolytic solution and analyzed as a function of particle equivalent volume diameter (d_V) using a Coulter counter,[1] after which the

---

[1] This instrument, used in some aerosol sampling research, measures particle size distribution for aerosols that have been collected and redispersed into a weakly conducting solution, where they are counted and sized on the basis of their equivalent volume diameter.

particle sizes were converted to corresponding particle aerodynamic diameters (d_{ae}) using estimates of particle density and dynamic shape factor. In this way, A was measured as a function of d_{ae} up to about 60 μm, for both mouth and nasal breathing. The results are summarised in Figure 10.1(b), again indicated by hatched areas that enclose most of the actual data. This time there are two hatched areas, reflecting two distinct families that were apparent in the original data, one for the range of wind speed $1 \leq U \leq 4$ m s^{-1} and the other for $U = 8$ m s^{-1}, which showed a markedly different trend. It is seen that the results for the lower range of wind speed and for the two higher breathing rates exhibited a trend very similar to that reported by Ogden *et al.*, except that the range of particle size was extended. Again, for both ranges of wind speed indicated, the main trend of A was with d_{ae}, with differences associated with mouth versus nasal breathing and wind speed being relatively small. In Figure 10.1(b) it is interesting to note for the lower range of wind speed that the values of A levelled off at around 0.5, consistent with what had been observed by Ogden and his colleagues, again with relatively small dependence on wind speed and breathing rate. For the higher wind speed and for the 'at rest' breathing rate, however, A dipped quite sharply as d_{ae} increased from small values, but then rose sharply, reaching values of the order of 1 for d_{ae} about 60 μm.

At about the same time, Vincent and Mark (1982) reported similar experiments for a life-size model head with a full torso, which was placed in the large wind tunnel shown earlier in Chapter 3 (see Figure 3.4). Wind Speeds ranged from 1 to 4 m s^{-1}, and the mannequin was rotated incrementally during each experiment to achieve orientation-averaging, spending equal times at angle $\theta = 0$, 45, 90, 135 and 180°, respectively. Cyclical breathing conditions approximating to 'normal work', with about 20 breaths min^{-1} and 1 L per breath, were applied by means of a purpose-built breathing machine, and for mouth breathing only. Relatively monodisperse test aerosols were generated using the narrowly graded powders of fused alumina that had been used in some of the studies reported in earlier chapters, providing d_{ae} in the range from about 6 to 90 μm. The results are summarised in Figure 10.1(c). It is seen that the trend was very consistent with what had been observed by Ogden *et al.* and Armbruster and Breuer for the same ranges of wind speed and breathing rate. In particular the tendency of A to first fall and then level out at around 0.5 continued right up to $d_{ae} = 90$ μm.

The results from these three early sets of experiments were notable for the consistency which they showed in terms of the broad tendencies exhibited by the data. The main trend for wind speeds below 4 m s^{-1} and for 'at work' breathing, was found to be the one between A and d_{ae}. Detailed inspection of the actual data revealed that there were some other internal trends with wind speed and breathing conditions (not shown), but that these were relatively weak. Such trends were clearly stronger for wind speeds higher than about 4 m s^{-1} or for low breathing rates. The overall broad consistency between all the data sets, especially for conditions thought to be most representative of workplaces and workers, provided the first hint that a single curve to represent inhalability might be achievable.

Most of the major trends found in the earlier studies were repeated in the results of the later work reported by Vincent *et al.* (1990), summarised more recently by Vincent (2005). These too were for orientation-averaged aspiration efficiency, for mouth breathing conditions corresponding to 'normal work', but now for wind speeds as high as 9 m s^{-1}. They are shown in Figure 10.2 in the same summary form as for the earlier results. Here again the same clear relationship between A and d_{ae} is seen that had been observed in the earlier experiments. Again, for the lower wind speeds, A first decreases from unity as d_{ae} increases from zero, then tends to level off at around 0.5 for larger particle sizes. Importantly, there is no evidence of any 'cut-off' anywhere in the particle size range indicated. Within these latest results, however, Vincent *et al.* noted a clear dependency on wind speed, most obviously for higher wind speeds. They also noted the sharp upturn in A for larger particle sizes that had previously been noted by Armbruster and Breuer.

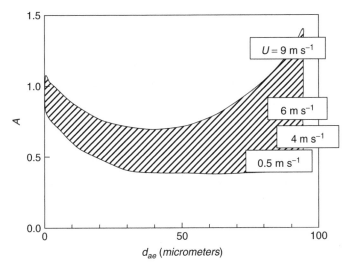

Figure 10.2 *Later experimental results for aspiration efficiency (A) of the human head as a function of particle aerodynamic diameter (d$_{ae}$), for various wind speeds (U) in the range from about 0.75 to 9 m s^{-1} (Vincent et al., 1990), again for simulated breathing broadly representative of people at work, as summarised by Vincent (2005). The hatched area contains nearly all the experimental data. Reproduced with permission from Vincent, Journal of Environmental Monitoring, 7, 1037–1053. Copyright (2005) Royal Society of Chemistry*

All the experiments described so far related to full-grown adults and for a range of levels of physical exertion relevant mostly to people at work. This body of work was supplemented by the study of Phalen *et al.* (1992) which involved wind tunnel experiments to also investigate human aspiration efficiency as a function of body mass, extending down to the range applicable to small children. In this way, Phalen *et al.* invoked the *exposure rate parameter*, the product of inhalability, inhaled minute volume and aerosol mass concentration divided by the body mass. They recommended that this type of reasoning might be applied in the wider context of risk estimation.

10.2.2 Physical basis of inhalability

It is desirable to find a physical explanation for the observed trends in order that the data can be applied with confidence. It is well-known from the physical theories for describing the performances of aerosol samplers outlined in previous chapters that aspiration efficiency may be described broadly as a function of particle aerodynamic diameter, wind speed, sampling flow rate, sampler body and orifice shapes and sizes and sampler orientation. However, at this stage, quantitative predictive modeling of aspiration efficiency has been achieved only for very simple sampler configurations facing the wind. A rigorous theory that fully explains the experimental inhalability results seems a long way off. Some progress has been made by assuming highly simplistic flow models, including, for example, two-dimensional geometry, perfectly spherical head, unidirectional sampling flow rate, etc. (Dunnett and Ingham, 1988; Erdal and Esmen, 1995). But, while such mathematical models are instructive, they are currently of limited practical value. A dimensional approach might be more useful. For this, the human head problem may be simplified somewhat by assuming fixed shape and dimensions (i.e. a 'standard' human head) and by imposing fixed cyclical breathing conditions (i.e. 'normal work'). Then, the semi-empirical so-called impaction-model approach for moving air described in Chapter 5, with negligible effect of gravity, permits the functional statement:

$$A \equiv \frac{c_{inh}}{c_0} = f\{K, U\} \tag{10.1}$$

where c_{inh} is the inhaled aerosol concentration and, as before, c_0 is the concentration in the undisturbed freestream, and where:

$$K = d_{ae}^2 U \propto St \tag{10.2}$$

Here, K is an inertial parameter which reflects the ability of particles to follow the streamlines of a distorted air flow just outside the nose and mouth. As expressed in Equation (10.2), K is not dimensionless. But it is directly proportional to the dimensionless sampler Stokes number, St, that has featured prominently in previous chapters. In drawing the comparison between A for the human head and the version that has been discussed extensively in earlier chapters for blunt aerosol samplers more generally, it is of course important to remember that the sampling flow rate in this case is cyclical, and so varies widely during a single breathing cycle.

As originally portrayed by Vincent *et al.*, the actual aspiration efficiency results from their 1990 study were plotted in the form suggested by Equations (10.1) and (10.2). This is reproduced here in Figure 10.3(a), and the proposed relations appear to have been confirmed. The data were then inspected by reference to the empirical form:

$$A \exp[a(1 - U)] \equiv A^* = [\exp(bK^c) + dK] \tag{10.3}$$

in which a, b, c and d are constant coefficients. Equation (10.3) is consistent with Equation (10.1) and has physical meaning in that A is a function of both K and U. The fact that the breathing flow rate was constant throughout these experiments meant that the effective sampler inlet velocity, U_s, as defined in earlier chapters, was also constant. This, together with the preceding discussion about K, leads to the sensible statement that A is a function of the same St and R (= U/U_s) as previously discussed. Nonlinear regression of the data with respect to Equation (10.3) produced a = 0.0267, b = −0.0393, c = 0.370 and d = 1.04 × 10⁻⁵, when d_{ae} was expressed in μm and U in m s⁻¹. The new correlation is shown here in Figure 10.3(b), revealing the marked upturn in aspiration efficiency for larger particles and higher wind speeds.

The observed trends for mouth breathing are plausible within the framework of what is known about the physics of the sampling process for aspiration in moving air in the absence of significant gravity effects. This plausibility includes not only the tendency for there to be no cut-off in A for the particle size range indicated but also the tendency for A to increase as the wind speed increases and/or the aspiration flow rate decreases. Both derive from the fact that, as wind speed increases, a greater fraction of larger particles impact – by virtue of their inertia – onto the mouth of the mannequin when it is oriented in forwards-facing directions.

It is important to note that the physical scenario described above assumes that particle inertia is the dominant physical mechanisms governing aspiration and that the effect of gravity is negligible. The latter assumption may not be true for very large particles or for very low wind speeds. So the wider interpretation of aspiration efficiency results like those shown needs to be treated with some caution.

10.2.3   Inhalability for very large particles

As already noted, the earlier data for the aspiration efficiencies of mannequins in wind tunnels were remarkable for their consistency, providing a major stimulus towards their application in developing

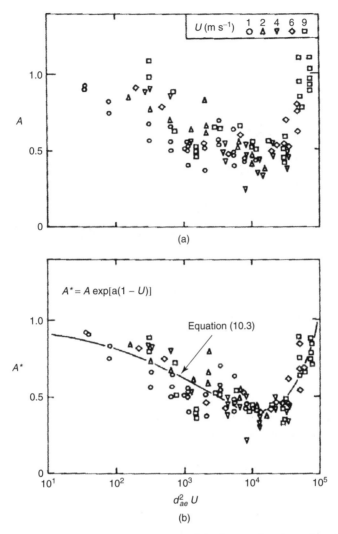

Figure 10.3 *The later results for aspiration efficiency (A) of the human head, re-plotted as a function of $d_{ae}^2 U$ as suggested by Equation (10.1), and the modified aspiration efficiency (A*) as a function of $d_{ae}^2 U$ in the form suggested by Equation (10.3), for wind speeds in the range indicated (Vincent et al., 1990). Reprinted from Journal of Aerosol Science, 21, Vincent et al., 577–586. Copyright (1990), with permission form Elsevier*

a particle size-selective criterion for the inhalable fraction (see below). However, it has always been acknowledged that those experiments did not cover the full range of conditions pertaining to all possible workplaces. In recent studies of particle size distributions in workplaces using sampling instrumentation capable of covering a very wide range of d_{ae}, Spear *et al.* (1998) showed that *very* large airborne particles may be present, even beyond the range of particle sizes covered in the wind tunnel experiments. Hinds and his colleagues set out to investigate A of a breathing mannequin in their large wind tunnel (see Figure 3.5) for nearly monodisperse test aerosols mechanically generated from narrowly graded fused fused alumina powders with d_{ae} up to 140 μm (Kennedy and Hinds, 2002). This involved solving

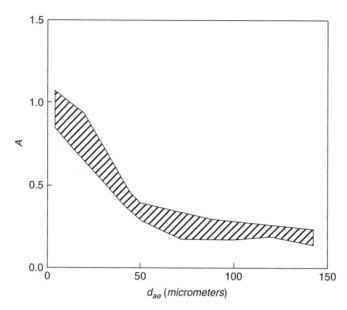

Figure 10.4 *Aspiration efficiency (A) of the human head as a function of particle aerodynamic diameter (d_{ae}) for an extended range of particle size and for wind speeds (U) in the range from 0.4 to 1.6 m s⁻¹. The hatched area contains nearly all the experimental data reported by Kennedy and Hinds (2002)*

some very difficult technical challenges, but these were successfully overcome and Hinds and his colleagues were ultimately able to achieve uniform particle concentration in the working section of their large wind tunnel. Unlike the previous studies cited above, the test aerosols were partially neutralised before introduction into the wind tunnel. Also, orientation-averaging was achieved in these experiments by rotating the mannequin steadily and continuously. Their results for orientation-averaged aspiration efficiency as a function of d_{ae} for a range of wind speeds from 0.4 to 1.6 m s⁻¹ and for breathing rates between about 14 Lpm and 37 Lpm showed that, as in the earlier studies cited, *A* was relatively independent of both wind speed and breathing rate. The combined results for mouth breathing are summarised in Figure 10.4, embodying the data for the various wind speeds and breathing rates. These results show the same characteristic steady decline of *A* with increasing d_{ae} that was shown for the earlier studies. However the rate of decline was greater. For d_{ae} about 100 μm, *A* fell to about 0.3, towards the low end of the bounded region in Figure 10.1(c) enclosing the corresponding results of Vincent and Mark. A number of reasons were argued for this slight difference, the most plausible one being – in the light of all that is now known about aspiration efficiency for blunt samplers – that the inhaled aerosol was influenced by the way in which the air was inspired and then expired by the mannequin in their studies. They speculated in particular the possibility that the exhalation of clean air in the immediate vicinity of the mouth may have momentarily diluted the aerosol there, hence leading to a reduced concentration for the aspirated aerosol.

10.2.4 Inhalability at very low wind speeds

The technical difficulties associated with investigating sampler aspiration efficiency and inhalability for larger particles extend to experiments with smaller particles at very low wind speeds. Yet low wind speeds are, as has become increasingly evident in recent years, of considerable relevance to many aerosol sampling situations, especially indoors in workplaces which have been shown to be characterised by wind speeds mostly lower than $0.5 \, \mathrm{m\,s^{-1}}$, often much lower (Berry and Froude, 1989; Baldwin and Maynard, 1997).

In their early study, in addition to their main body of work for moving air, Ogden *et al.* also reported some results from experiments with their breathing mannequin in a calm air chamber. Few technical details were reported, and the work was not followed up because, according to Ogden and Birkett (1977), the results would be of '... less (practical) importance ...'. Now, however, in the light of what has since been learned about workplace wind speeds, those early results are of considerable interest. They are summarised in Figure 10.5. They show that the rate of fall in A with increasing d_{ae} was less than had been observed in moving air [Figure 10.1(a)], with A falling only to between about 0.6 and 0.8 for the upper end of the particle size range studied. In addition (not shown), Ogden and Birkett noted that there was a clear distinction between the results for mouth and nasal breathing, respectively, with A-values for nasal breathing tending to be generally higher.

Later, Aitken *et al.* (1999) reported a collaborative study of aspiration efficiency for the human head in this interesting regime, involving two research teams that performed experiments in their respective laboratories, working with mannequins in nearly calm air chambers where the peak local air velocities were reported as being no greater than $0.2 \, \mathrm{m\,s^{-1}}$. For the purpose of measuring aspiration efficiency, the reference concentration in each chamber was measured ingeniously by means of a 'pseudo-isokinetic' thin-walled sampler mounted on a slowly rotating arm, where the angular velocity of rotation was chosen so that the effective relative velocity between the samplers and the main air was the same as the sampler inlet velocity. In each laboratory the mannequin too was rotated slowly. Nearly monodisperse test aerosols were generated from narrowly graded fused alumina powders and d_{ae} ranged from 6 to

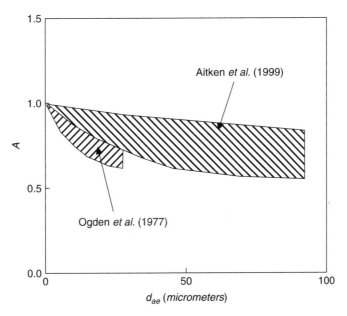

Figure 10.5 *Aspiration efficiency (A) of the human head as a function of particle aerodynamic diameter (d$_{ae}$) for calm air for breathing corresponding to people at work. The hatched areas contain nearly all the experimental data reported by Ogden et al. (1977) and Aitken et al. (1999)*

90 μm. At one laboratory, the aspiration efficiency of the mannequin (A) was measured as a function of d_{ae} for a range of cyclical breathing flow rates from 6 Lpm (minute volume, corresponding to 'at rest') to 20 Lpm (corresponding to 'normal work'). The second laboratory performed experiments only for 20 Lpm. The combined results from the two laboratories are summarised in Figure 10.5 alongside the data from Ogden and his colleagues. They include results for both mouth and nose breathing. The results were quite consistent between the two laboratories. Most striking is the fact that the measured A-values fell markedly above the trends exhibited by the data for the wind tunnel studies described earlier. This tendency was found to be greatest at the highest ('normal work') breathing flow rate. Importantly, this general tendency confirms what was reported by Ogden *et al.* in 1977. However, unlike Ogden *et al.*, Aitken and his colleagues were not able to detect any significant dependency on whether breathing was through the nose or the mouth. Nor – interestingly – did the results depend on whether or not the

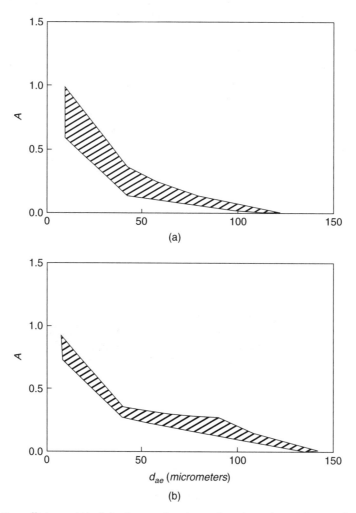

Figure 10.6   *Aspiration efficiency (A) of the human head as a function of particle aerodynamic diameter (d_{ae}) for calm air: (a) nose breathing; (b) combined nose and mouth (oronasal) breathing. The hatched areas contain nearly all the experimental data reported by Hsu and Swift (1999)*

mannequin was heated, even though it has been suggested many times over the years that the air flow around a living human subject might be influenced by the thermal air currents arising from the elevated temperature of the human body, especially for low external air movement.

Elsewhere, Hsu and Swift (1999) performed their own experiments to examine the aspiration efficiencies of breathing mannequins in a calm air chamber. The mannequins were chosen to represent both adults and children, and simulated cyclical breathing corresponded to 'at rest' and 'moderate exercise', respectively. The results are shown in Figure 10.6, and it is seen in general that A fell steadily as d_{ae} increased, becoming close to zero for d_{ae} exceeding about 80 μm. Figure 10.6(a) for nasal breathing contains results for both adult and child mannequins, for which Hsu and Swift reported no significant differences between any of the combinations studied. A good fit with the experimental data was obtained using:

$$A = 3.01 + 0.64(\ln d_{ae})^2 - 2.78 \ln d_{ae} \qquad (10.4)$$

Figure 10.6(b) shows some results for the important combined mouth-and-nasal ('oronasal') breathing condition for just the adult mannequin, undergoing 'heavy exercise'. Here, it was reported that A for oronasal breathing tended to lie above that for mouth breathing alone, especially for smaller particles. For these results, a good fit with the data was found using:

$$A = 1.44 - 0.66 \ln d_{ae} \qquad (10.5)$$

In general, the results of Hsu and Swift were very consistent in themselves, and seemed sensible in the light of the trends in A predicted from work carried out for simple sampling systems in calm air (see Chapter 6). However, they were at odds with the experimental results of both Ogden *et al.* and Aitken *et al.* which, too, were very internally consistent. From what has been published, there is no obvious explanation for the discrepancy.

The studies that have been cited above all described conditions where the air was either moving (as in a wind tunnel) or is essentially still (as in a calm air chamber). So far there has been no work to explore inhalability in the transition regime between moving and purely calm air, where both convection and gravity may play significant, competing roles. Yet it is this regime which is likely to be of considerable practical significance, and so remains an area for future work.

10.3 Particle size-selective criteria for the inhalable fraction

The criteria that will be presented below as options for the inhalable fraction are shown as curves describing the probability of particle inhalation as functions of particle aerodynamic diameter (d_{ae}). Such curves are intended for use as 'yardsticks' against which the performances of various candidate aerosol sampling instruments may be compared.

For some types of aerosol, inhaled particles constitute a risk to health, regardless of where they are eventually deposited inside the respiratory tract. This is considered to be the case for all substances that may be carcinogenic, or that may be soluble in body fluids so that molecules can enter the bloodstream regardless of where they are actually deposited. For other types of aerosol, inhaled particles may lead to adverse health effects if they are deposited soon after entering the body – in the mouth or nasal passages – even if they do not penetrate any further into the respiratory tract. So, in the first place, inhalable aerosol is a fraction which is an objective for measurement in itself in many occupational and environmental situations. With this in mind, the search for a criterion for inhalable aerosol began with what had been learned about the aspiration efficiency of the human head.

For many years the recommendations for the health-related sampling of coarse particles in most countries have been based on the concept of so-called 'total' aerosol. This therefore was a primary basis of aerosol standards, and still is to this day in many countries and for many standards-setting bodies. For example, in the USA, the permissible exposure limits (PELs) that form the basis of regulation by the Occupational Safety and Health Administration (OSHA) still retain the 'total aerosol' rationale. For ambient atmospheric aerosol, the rationale of 'total suspended particulate' (TSP) persisted in Environmental Protection Agency (EPA) standards until as recently as 1987. As seen from the data presented above, where the aspiration efficiency of the human head varied markedly with d_{ae}, it becomes apparent that the scientific justification for continued use of 'total aerosol' in standards and sampling is flawed. Meanwhile, however, practical sampling instruments for 'total' aerosol continue to be sold commercially and are extensively deployed for health-related aerosol sampling in occupational and environmental hygiene contexts. Most such instruments were originally developed without particular regard to specific quantitative criteria or performance indices, and now – on closer inspection – it is becoming apparent that their performance characteristics have varied greatly from one to the other. It follows, therefore, that what we have referred to as 'total aerosol' has been effectively defined in each particular situation by the particular sampling instrument chosen to do the job. This means that, in a given practical situation, switching from one instrument to another might well produce different measurements of exposure. With the preceding in mind, the concept of inhalability has emerged as important in unifying aerosol exposure in a scientific manner which is strongly related to how aerosols enter the bodies of people during breathing, and hence to health. Data from experiments like those described above form the basis of recommendations for replacing the old 'total' aerosol concept with a quantitative sampling convention based on human inhalability.

10.3.1   Early recommendations

In their search for a scientific basis for the sampling of coarse aerosol, the German occupational health standards body, the Maximale Arbeitsplatzkonzentration Commission proposed the purely physical criterion that the mean velocity of the air entering the sampling device (i.e. U_s in earlier chapters in this book) should be $1.25 \mathrm{~m\,s}^{-1} \pm 10\%$ (Maximale Arbeitsplatzkonzentration, 1981). According to sampler theories available at the time, it was expected that this simple criterion should provide aspiration efficiency close to unity, at least for particles with aerodynamic diameter (d_{ae}) up to about 30 μm. Later, Holländer (1990) proposed another physical criterion based on considerations of the stop distances of particles of given size moving near a sampling inlet placed in a moving air stream of known wind speed. However, neither of these physical criteria became widely accepted for health-related aerosol sampling by the international occupational and environmental hygiene community.

The first recommended formal criterion for the *inhalable* fraction was the one proposed by the International Standards Organisation (ISO) in 1983 (International Standards Organisation, 1983).[2] During the meetings of its ad hoc working group in the late 1970s, the only human head aspiration efficiency data available for consideration were those from the experiments of Ogden and his colleagues for d_{ae} up to only about 30 μm. In the absence then of data for larger particle sizes, those results were extrapolated

---

[2] With respect to terminology, the original form of expression for the aspiration of the human head was indeed the 'inhalable' fraction. But at the time of the publication of the 1983 ISO recommendations, the alternative term 'inspirable' was adopted in order to avoid confusion with the use of the term 'inhalable' elsewhere, notably in research publications sponsored by the United States EPA for describing the fraction of inhaled aerosol that penetrates deeper into the human respiratory tract. Later, however, after the EPA had adopted other terminology in its standards documentation, the international aerosols standards community reverted back to the term 'inhalable', and so it remains to this day.

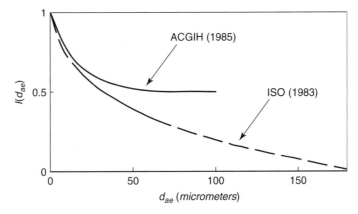

Figure 10.7 *Early inhalability curves, I(d_ae), as a function of particle aerodynamic diameter (d_ae), showing the original ISO curve (International Standards Organisation, 1983) and the original ACGIH curve (Phalen, 1985)*

to larger particle sizes and used as a the basis of a curve described by the purely empirical expression for *inhalability*, $I(d_{ae})$, given by:

$$I(d_{ae}) = 1 - 0.15[\log_{10}(1 + d_{ae})]^2 - 0.10 \log_{10}(1 + d_{ae}) \tag{10.6}$$

where d_{ae} is expressed in μm and the aspiration efficiency for the human head (*A*) has been replaced by inhalability (*I*) to reflect that the curve is now a convention for aerosol sampling. The curve calculated using this expression is shown in Figure 10.7, where the dashed part of the line indicates where the curve has been extrapolated from the then-available actual experimental data. One interesting feature of this early inhalability convention is that *I* continued to fall with increasing d_{ae} and eventually reached zero at about $d_{ae} = 185$ μm. That is, there was a 'cut-off'.

10.3.2 Modern criteria for the inhalable fraction

The later, more comprehensive data set published in 1982 by Armbruster and Breuer and by Vincent and Mark, respectively, for particles with d_{ae} up to about 100 μm, provided the starting point for a curve more representative of actual human inhalation over a wider range of particle sizes. Vincent and Armbruster (1981) reviewed the data for the aspiration efficiency of the human head (*A*) that were available at that point in time, and used them to propose a new definition that was expressed in terms of a table of values of *I* for various d_{ae} up to 100 μm that could be plotted as a curve of *I* versus d_{ae}. The curve reflected the main trend that was shown in all the data sets for *A* in moving air situations, in particular that *I* fell steadily as d_{ae} in the range from 0 to about 30 μm, but then leveled of at $I = 0.5$ for d_{ae} up to about 100 μm. This trend was notable for the fact that there was now no 'cut-off', a proposal that surprised some commentators but one supported by what we now know from blunt sampler theory.

In its 1985 report, the Air Sampling Procedures Committee of the American Conference of Governmental Industrial Hygienists (ACGIH) took the new proposal into account (Phalen, 1985). The empirical expression:

$$I(d_{ae}) = 0.5[1 + \exp(-0.06 d_{ae})] \tag{10.7}$$

for d_{ae} (again expressed in μm) up to and including 100 μm. Beyond 100 μm it was acknowledged that there was no information on which to base a firm recommendation. It should be noted, however, that this did *not* imply a 'cut-off', as the 1985 (ACGIH) report is sometimes misunderstood as saying. The curve calculated from Equation (10.7) is shown in Figure 10.7 alongside the earlier ISO version. When the ACGIH definition of the inhalable fraction emerged in 1985, it immediately superceded the earlier one from the ISO.

In its original 1985 version of the inhalability convention, the ACGIH also recommended that, as a performance band for practical sampling instruments for the inhalable fraction, I may vary by ± 10 percentage points for each value of d_{ae}. That is, there is a working envelope that is representative of a high proportion of the original data and so is a reasonable basis for assessing the performances of practical devices. Later, ACGIH modified its tolerance bands slightly (Vincent, 1999). Specifically: (a) the majority of the performance data – in terms of sampling efficiency for a given particle size and wind speed – for the instrument in question should fall within ± 10 percentage points of the I-value indicated for each particle size; (b) the total sampled mass, as measured using the instrument in question, should – under the conditions of practical use – fall within $\pm 20\%$ of that which would be measured using a hypothetical ideal instrument whose performance perfectly matched the inhalability curve.

In its earlier, as well as its later, recommendations, ACGIH focused primarily on particle size-selective criteria for sampling in indoor workplaces where wind speeds even as high as 4 m s^{-1} are uncommon. Therefore the 1985 inhalability curve shown in Figure 10.7 was considered to be adequate for most practical workplace situations. However, it has long been recognised that there are some outdoor workplaces where conditions might lie outside these wind conditions. For the ISO, sampling in the ambient atmosphere for the purpose of evaluating the risk to the community at large was an important part of its general charge. From the same the earlier experimental data [e.g. Armbruster and Breuer, 1982, see Figure 10.1(b)] as well as some of the later ones (Vincent *et al.*, 1990, see Figure 10.2), it was clear that the ACGIH convention did not properly represent what happens at those higher wind speeds. In particular, the use of samplers with performance based just on this curve could lead to a significant underestimation of the exposure of humans to large particles. This could be important in situations where there are large particles containing potentially hazardous substances such as radioactive nuclides, heavy metals, polycyclic aromatic hydrocarbons, etc. It was therefore considered necessary to extend the formal definition of the convention for the inhalable fraction. The earlier Equation (10.3) represents the experimental data for the aspiration efficiency of the human head quite well over the range of conditions examined, including the higher wind speeds. However, it is substantially more complicated than the relatively simple 1985 ACGIH convention, so it is not really suitable for direct application as a practical criterion for the performance of sampling instruments. But it provides a good starting point. With this in mind, a new convention based on a simple empirical modification was proposed, thus (Vincent *et al.*, 1990):

$$I(d_{ae}) = 0.5[1 + \exp(-0.06d_{ae})] + \text{f}(d_{ae}, U) \tag{10.8}$$

where the new, additional term on the right-hand side was offered as an empirical function of d_{ae} and U given empirically by:

$$f(d_{ae}, U) \equiv pU^q \exp(rd_{ae}) \tag{10.9}$$

in which d_{ae} is again in μm and the wind speed U is in m s^{-1}. Best fit with the experimental data was achieved for $p = 1 \times 10^{-5}$, $q = 2.75$ and $r = 0.055$ for wind speeds in the range $1 \leq U \leq 9$ m s^{-1}. For low wind speeds ($U < 4$ m s^{-1}), the new Equation (10.8) reverts closely back to the original ACGIH curve in Equation (10.7). The generalised expression in Equations (10.8) and (10.9) was subsequently incorporated into a revised ISO set of proposals (International Standards Organisation, 1992). Meanwhile, the version applying to lower wind speeds, corresponding to Equation (10.7), was embodied in the

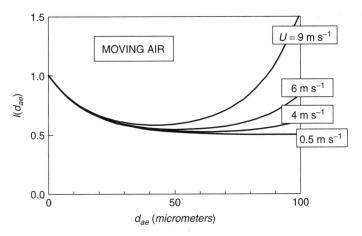

Figure 10.8 *Modern inhalability curves, I(d_{ae}), as a function of particle aerodynamic diameter (d_{ae}) for various wind speeds (International Standards Organisation, 1992), calculated from Equation (10.8) with Equation (10.9)*

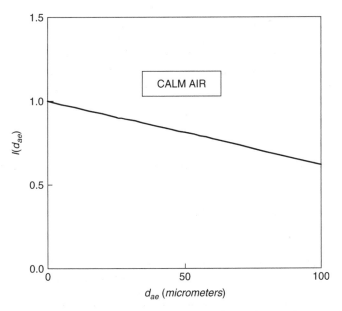

Figure 10.9 *Suggested inhalability curve, I(d_{ae}), as a function of particle aerodynamic diameter (d_{ae}) for calm air (Aitken et al., 1999), calculated from Equation (10.10)*

criteria adopted by the Comité Européen de Normalisation (CEN, 1992). This extended modern criterion for the inhalable fraction is shown as a set of continuous curves in Figure 10.8, all calculated from Equation (10.8), each one for a given wind speed as indicated.

10.3.3   Further recommendations

The conventions described above were firmly based on the tacit assumption that the air is moving to the extent that particle motion is governed entirely by inertial forces. The shapes of the curves in question can all be justified by reference to blunt sampler theory for moving air situations. At very low wind speeds, however, gravity may play a much stronger role, as is evident from the basic studies described in Chapter 6. So the same trends may not be followed. In fact, the small number of mannequin studies in very slowly moving air that have been reported showed clearly that the aspiration efficiency of the human head (A) differed markedly from the curves for I that have been proposed and accepted as criteria for the inhalable fraction based on data from experiments in moving air. Based on their 'calm air' experimental results shown earlier, Aitken *et al.* proposed a possible revised convention of the form:

$$I(d_{ae}) = 1 - 0.0038d_{ae} \qquad (10.10)$$

for the inhalable fraction in very slowly moving air. This was a plausible suggestion, supported broadly by the earlier experimental study for such conditions reported by Ogden *et al.* It is shown as a continuous curve – a straight line – in Figure 10.9. Although Hsu and Swift presented a pair of empirical expressions that were fitted to their own experimental data, they stopped short of proposing either of them as a possible convention for the inhalable fraction in slowly moving air situations.

   Ideally, a revised convention for the inhalable fraction should contain explicit reference to the actual wind speeds that are present, covering the transition regime between pure calm air and the moving air situation where inertial effects are dominant. Further, for it to be a realistic practical contribution, it should be capable of being combined with the existing inhalable aerosol criterion that has already been widely accepted, in the same way as was done for the high wind speed situations discussed earlier. As already noted by Aitken *et al.*, further work in this important regime is needed before standards setting bodies are likely to adopt an extended criterion along the lines suggested.

10.4   Overview

Modern criteria for the inhalable fraction are summarised in Table 10.1. For moving air, the general convention for the inhalable fraction is given by:

$$I(d_{ae}) = 0.5[1 + \exp(-0.06d_{ae})] + 10^{-5}U^{2.75}\exp(0.055d_{ae}) \qquad (10.11)$$

Table 10.1   *Summary of particle size-selective sampling criteria; shaded area represents a version that has been discussed but which has not yet been adopted by any occupational or environmental standards bodies*

Rationale	Function	Variables
Penetration into the nose and/or mouth in moving air environments	$0.5[1 + \exp(-0.06d_{ae})]$ $+10^{-5}U^{2.75}\exp(0.055d_{ae})$ $\rightarrow 0.5[1 + \exp(-0.06d_{ae})]$	$d_{ae} \leq 100 \ \mu m$ $0.25 < U < 9 \ m \ s^{-1}$ for $U < 4 \ m \ s^{-1}$
Penetration into the nose and/or mouth in calm air environments	$1 - 0.0038d_{ae}$	$d_{ae} \leq 100 \ \mu m$ $U < 0.25 \ m \ s^{-1}$

for wind speeds from about 0.5 to 9 m s^{-1} and for particles with d_{ae} up to and including 100 μm. This was the form adopted by the ISO. It becomes:

$$I(d_{ae}) = 0.5[1 + \exp(-0.06d_{ae})] \qquad (10.12)$$

for wind speeds (U) below about 4 m s^{-1}, as adopted by the ACGIH and the CEN, as well as many occupational exposure standards-setting bodies around the world.

It has been suggested, for calm air or very low wind speeds, that:

$$I(d_{ae}) = 1 - 0.0038d_{ae} \qquad (10.13)$$

may be used, although this has not yet been formally adopted by any of the appropriate standards-setting bodies. That said, however, it is now widely recognised that a comprehensive inhalability convention needs to account for aspiration efficiency of the human head at very low wind speeds; Equation (10.13) is included in Table 10.1 (shaded area).

References

Aitken, R.J., Baldwin, P.E.J., Beaumont, G.C., Kenny, L.C. and Maynard, A.D. (1999) Aerosol inhalability in low air movement environments, *Journal of Aerosol Science*, 30, 613–626.

Armbruster, L. and Breuer, H. (1982) Investigations into defining inhalable dust. In: *Inhaled Particles V* (Ed. W.H. Walton), Pergamon Press, Oxford, pp. 21–32.

Baldwin, P.E.J. and Maynard, A.D. (1997) Measurement of wind speeds in indoor workplaces, *Proceedings of the 11th Annual Aerosol Society Conference*, The Aerosol Society, Bristol.

Berry, R.D. and Froude, S. (1989) An investigation of wind conditions in the workplace to asses their effect on the quantity of dust inhaled, UK Health and Safety Executive Report IR/L/DS/89/3, Health and Safety Executive, London.

Comité Européen de Normalisation (1992) *Workplace atmospheres: size fraction definitions for measurement of airborne particles in the workplace*, CEN Standard EN 481.

Dunnett, S.J. and Ingham, D.B. (1988) An empirical model for the aspiration efficiencies of blunt aerosol samplers oriented at an angle to the oncoming flow, *Aerosol Science and Technology*, 8, 245–264.

Erdal, S. and Esmen, N.A. (1995) Human head model as an aerosol sampler: calculations of aspiration efficiencies for coarse particles using an idealised human head model, *Journal of Aerosol Science*, 26, 253–272.

Holländer, W. (1990) Proposed performance criteria for samplers of total suspended particulate matter, *Atmospheric Environment*, 24, 173–177,

Hsu, D.-J. and Swift, D.L. (1999) The in vitro measurements of human inhalability of ultra-large aerosols in calm air conditions, *Journal of Aerosol Science*, 30, 1331–1343.

International Standards Organisation (1983) Air quality-particle size fraction definitions for health-related sampling, Technical Report ISO/TR/7708-1983 (E), ISO, Geneva.

International Standards Organisation (1992) Air quality-particle size fraction definitions for health-related sampling, Technical Report ISO/TR/7708-1983 (E), Revised version, ISO, Geneva.

Kennedy, N.J. and Hinds, W.C. (2002) Inhalability of large particles, *Journal of Aerosol Science*, 33, 237–255.

Maximale Arbeitsplatzkonzentration (1981) *Mitteilungen der Senatskommission zur Prüfung gesundheitsschädlicher Arbeitsstoffe*. In: Maximale Arbeitsplatzkonzentration, Harald Boldt, Boppard.

Ogden, T.L. and Birkett, J.L. (1977) The human head as a dust sampler. In: *Inhaled Particles IV* (Ed. W.H. Walton), Pergamon Press, Oxford, pp. 93–105.

Ogden, T.L., Birkett, J.L. and Gibson, H. (1977) Improvements to dust measuring techniques, IOM Report No. TM/77/11, Institute of Occupational Medicine, Edinburgh.

Phalen, R.F. (Ed.) (1985) Particle size-selective sampling in the workplace, Report of the ACGIH Air Sampling Procedures Committee, American Conference of Governmental Industrial Hygienists, Cincinnati, OH.

Phalen, R.F., Oldham, M.J. and Dunn-Rankin, D. (1992) Inhaled particle mass per unit body mass per unit time, *Applied Occupational and Environmental Hygiene*, 7, 246–252.

Spear, T.M., Werner, W.A., Bootland, J., Murray, E., Ramachandran, G. and Vincent, J.H. (1998) Assessment of particle size distributions of health-related aerosol exposures of primary lead smelter workers, *Annals of Occupational Hygiene*, 42, 73–80.

Vincent, J.H. (1989) *Aerosol Sampling: Science and Practice*, John Wiley & Sons, Ltd, Chichester.

Vincent, J.H. (1995) *Aerosol Science for Industrial Hygienists*, Pergamon Press, Elsevier Science, Oxford.

Vincent, J.H and Armbruster, L. (1981) On the quantitative definition of the inhalability of airborne dust, *Annals of Occupational Hygiene*, 24, 245–248.

Vincent, J.H. and Mark, D. (1982) Application of blunt sampler theory to the definition and measurement of inhalable dust. In: *Inhaled Particles V* (Ed. W.H. Walton), Pergamon Press, Oxford, pp. 3–19.

Vincent, J.H., Mark, D., Miller, B.G., Armbruster, L. and Ogden, T.L. (1990) Aerosol inhalability at higher wind speeds. *Journal of Aerosol Science*, 21, 577–586.

Vincent, J.H. (Ed.) (1999) *Particle Size-selective Sampling for Particulate Air Contaminants*, American Conference of Government Industrial Hygienists, Cincinnati, OH.

Vincent, J.H. (2005) Health-related aerosol measurement: a review of existing sampling criteria and proposals for new ones, *Journal of Environmental Monitoring*, 7, 1037–1053.

11

Particle Size-selective Criteria for Fine Aerosol Fractions

11.1 Introduction

The behavior of inhaled airborne particles *after* they have entered the human respiratory tract and their fate by regional deposition during the inspiration and expiration parts of the breathing cycle are governed by the same range of physical processes that were discussed in Chapter 8 in the context of the behavior of particles after aspiration by an aerosol sampler. The distribution of the regional deposition of particles in the respiratory tract may be linked causally with a wide range of adverse health effects arising from aerosol inhalation. This therefore provides an important basis of criteria for aerosol exposure standards.

The processes of aerosol inhalation described in the previous chapter involve events that occur entirely outside the human body. So it has been easy to justify conducting experiments to examine the dependency of the efficiency of inhalation of aerosols by the use of inert mannequins to investigate inhalation and quantify its dependency on particle size, wind speed, breathing parameters, body orientation, and external shape and dimensions. The same cannot be said, however, for the investigation of what happens to particles after they have been inhaled. Here, although some useful work has been reported over the years from experiments using plaster cast models of the human respiratory tract, it is generally felt that the anatomical and physiological factors that influence particle transport after inhalation are so complex that only experiments with actual human subjects can provide the information that is needed to underpin the development of particle size-selective criteria for the finer aerosol fractions.

11.2 Studies of regional deposition of inhaled aerosols

11.2.1 Framework

In Chapter 9 a rudimentary picture of the human respiratory tract was presented as a basis for a general discussion of the nature of human exposure to aerosols. The process of inhalation itself is directly analogous to the aspiration of aerosols by aerosol sampling devices. The penetration of particles into the respiratory tract, however, is more complex, involving successive stages where one set of physical fluid and aerosol mechanical processes gives way to another. A clear rational framework is needed by

Aerosol Sampling: Science, Standards, Instrumentation and Applications   James Vincent
© 2007 John Wiley & Sons, Ltd

which to guide experimental studies in order to provide data that may be used in the development of particle size selective criteria that may in turn be used in health-related aerosol exposure standards.

In its 1994 report, the International Commission on Radiological Protection (ICRP) described a morphological model in which the dose of inhaled radioactive particles to relevant regions of the respiratory tract was represented, allowing for whether breathing took place through the nose or mouth and accounting for particle deposition during both the inhalation and exhalation phases of the breathing cycle. The version shown in Figure 11.1 is simplified somewhat from the one described by ICRP, but it embodies the most important features. It shows that both inspiration and expiration may be partitioned between nose and mouth breathing, not necessarily in the same proportions. For nose breathing, there are two nasal extrathoracic regions, the anterior and the posterior nasal passages (ET_1 and ET_2 respectively). Particles inhaled through the nose that are large enough for their motions to be dominated by inertial and gravitational forces tend to be deposited primarily towards the rear of ET_1 (James *et al.*, 1994). For mouth breathing, particles deposited in the extrathoracic head region (ET_3) are collected primarily by impaction in the larynx (Emmett and Aitken, 1982). Not surprisingly, therefore, extrathoracic deposition differs between nose and mouth breathing, both qualitatively *and* quantitatively. In turn, so too does penetration below the larynx and into the lung. Particles entering the tracheobronchial region are deposited mainly by inertial forces. The smaller ones not deposited and penetrating down into the alveolar region may be deposited there mainly by gravitational forces or diffusion. Particles not deposited in the inspiration half of the breathing cycle may then be deposited in the alveolar, tracheobronchial and extrathoracic regions during the expiration half. Those which are not deposited during either inspiration *or* expiration are exhaled.

Figure 11.1 shows the regions that have been identified as well as the various particle fluxes through the respiratory tract and to its surfaces (from Vincent 2005). In the first instance the aspiration efficiency (A) of the human head is given by:

$$A = \frac{N_{inh}}{N_0} \longrightarrow \frac{c_{inh}}{c_0} \tag{11.1}$$

where N_0 is the number of particles carried per unit time in the original undisturbed inspired air volume and N_{inh} the number inhaled into the nose and/or mouth, while c_0 and c_{inh} are the corresponding particle concentrations. Again it is noted that the inspiration flow rate is cyclical. For inhaled particles the efficiency with which particles are deposited in the *extrathoracic* (including both *nasopharyngeal* and *mouth*) region is given by:

$$E_{ETin} = \frac{D_{ETin}}{N_{inh}} = 1 - \frac{N_{TBin}}{N_{inh}} \tag{11.2}$$

where D_{ETin} is the flux of particles to nasal and mouth surfaces inside the extrathoracic region and N_{TBin} is the flux of particles that progress further and so enter the lung below the larynx. Conceptually, E_{ETin} may be viewed as the collection efficiency of the extrathoracic region acting as a filter during the inspiration part of the breathing cycle. The converse of Equation (11.2) is the efficiency with which inhaled particles penetrate further and so enter the *thoracic* region, given by:

$$P_{thor} = \frac{N_{TBin}}{N_{inh}} = 1 - E_{ETin} \tag{11.3}$$

Next, the local efficiency with which particles are deposited in the tracheobronchial region is given by:

$$E_{TBin} = \frac{D_{TBin}}{N_{TBin}} = 1 - \frac{N_{alvin}}{N_{TBin}} \tag{11.4}$$

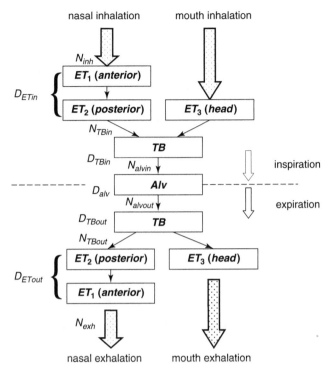

Figure 11.1 *Schematic diagram to identify the regions of the human respiratory tract and the deposition of inhalation particles that takes place there during inspiration and expiration (modified from the 1994 ICRP respiratory tract model). TB, tracheobronchial; Alv, alveolar. Reproduced with permission from Vincent, Journal of Environmental Monitoring, 7, 1037–1053. Copyright (2005) Royal Society of Chemistry*

where D_{TBin} is the flux of particles to the surfaces of the tracheobronchial region and N_{alvin} is the flux of particles that progress further and so enter the alveolar region of the lung. Equation (11.4) describes the filtration efficiency of the tracheobronchial region alone. It is especially useful in the context of standards to express it alternatively as a fraction of the aerosol that was *inhaled* initially. That is:

$$E^*_{TBin} = \frac{D_{TBin}}{N_{inh}} = \frac{N_{TBin} - N_{alvin}}{N_{inh}} \longrightarrow P_{thor} E_{TBin} \equiv (1 - E_{ETin}) E_{TBin} \qquad (11.5)$$

which now involves combining the filtration efficiencies of both the extrathoracic and the tracheo-bronchial regions. Now, if D_{alv} is the flux of particles to the surfaces of the alveolar region, the local efficiency of deposition there is:

$$E_{alv} = \frac{D_{alvin}}{N_{alvin}} = 1 - \frac{N_{alvout}}{N_{alvin}} \qquad (11.6)$$

where N_{alvout} is the flux of particles that, having reached the alveolar region, remain airborne and so leave there during the exhalation part of the breathing cycle. Again it is useful to express the efficiency of alveolar region deposition as a fraction of the aerosol that was inhaled initially. That is:

$$E^*_{alv} = \frac{D_{alvin}}{N_{inh}} \longrightarrow P_{alv} E_{alv} = (1 - E_{ETin})(1 - E_{TBin}) E_{alv} \qquad (11.7)$$

where

$$P_{alv} = \frac{N_{alvin}}{N_{inh}} \longrightarrow (1 - E_{ETin})(1 - E_{TBin}) \tag{11.8}$$

is the efficiency with which particles penetrate down to the alveolar region.

Those particles remaining airborne at the end of the inspiration part of the breathing cycle are transported in the opposite direction inside the respiratory tract during the expiration part. This provides for further opportunities for deposition of the inhaled particles. It follows from the preceding that:

$$E^*_{TBout} = (1 - E_{ETin})(1 - E_{TBin})(1 - E_{alv})E_{TBout} \tag{11.9}$$

and

$$E^*_{ETout} = (1 - E_{ETin})(1 - E_{TBin})(1 - E_{alv})(1 - E_{TBout})E_{ETout} \tag{11.10}$$

In these, it is a fair approximation that $E_{ETin} = E_{ETout}$ and $E_{TBin} = E_{TBout}$.

Combining all of the above, we have the total efficiencies for each region, including deposition during both inspiration and expiration, as follows:

$$_{total}E^*_{TB} = E^*_{TBin} + E^*_{TBout} \text{ and } _{total}E^*_{ET} = E^*_{ETin} + E^*_{ETout} \tag{11.11}$$

Finally, the efficiency of the grand total of aerosol deposited throughout the respiratory tract is given by:

$$E_{total} = 1 - (1 - E_{ETin})(1 - E_{TBin})(1 - E_{alv})(1 - E_{TBout})(1 - E_{ETout}) \tag{11.12}$$

The preceding equations provide a set of relationships that enables calculation of the fluxes penetrating to and depositing at the various parts of the respiratory tract. They also provide a basis for designing experiments to investigate regional lung deposition and, also, for defining practical criteria for aerosol fractions that might relate to specific types of health effect. In what follows, attention will be focused firstly on the range of experimental data for which particle motion will have been governed primarily by aerodynamic effects – that is, where the relevant metric of particle size is aerodynamic diameter (d_{ae}). This emphasis reflects the way in which particle size has been – and is being – expressed in the criteria that have been defined so far for various aerosol fractions of health-related interest. Attention will then be turned to the deposition of smaller particles whose motion is dominated by diffusion, and where the more relevant metric of particle size is the equivalent volume diameter (d_V).

11.2.2   Theories, simulations and models

Experimental studies of the regional deposition of inhaled aerosols in actual human subjects are difficult and fraught with constraints, and provide data that are very variable. So they have been complemented from the beginning by studies based on mathematical and numerical modeling. These have been driven by the need for a dosimetric framework that can be applied, along with toxicology, in the processes of quantitative risk estimation. Here, the ultimate goal of achieving an entirely computer-based simulation of the dosimetric process, one that works across all aerosol and physiological conditions, is highly desirable. Then, experiments with human subjects will no longer be needed. The problem has been that the passage of air – and in turn particles – into and through the lung is extremely complicated. Not only are the geometries themselves of the airways and other lung spaces very diverse and complex, but the nature of the air flow changes greatly from the earlier airway generations down to the higher-order ones. However, in recent years, the evolution of powerful new computer technologies, characterised by fast processing and very large storage capacity, have made the goal of a comprehensive simulation of the lung, and the fluid and aerosol mechanical processes taking place in it, a more realistic one.

Early models assumed relatively simple lung morphometries. For example, Taulbee and Yu (1975) treated the airway system as analogous to a single 'horn-shaped' channel of increasing cross-sectional area, using the lung model of Weibel (1963) to provide the functional form needed to link the channel properties with the actual lung and allowing for physical expansion and contraction during the breathing cycle. Over the years, other approaches applied other analogs, one being the granular bed model for deposition in the deep lung by Zhou *et al.* (1996). The mainstream approach, however, has involved explicit and realistic anatomical description of the human respiratory tract, taking account of the roles of inertia, gravity, diffusion and electrostatic forces in particle deposition at all generations, and considering particle deposition both in the airways, ducts and sacs and at the bifurcations where branching occurs between generations. As computing capability has increased over the years, such models have become progressively more sophisticated, some of which have been deterministic and others stochastic. A large number of lung deposition modeling studies have by now been reported and the range of conceptual approaches, from experimentally based empirical models to more sophisticated numerical simulations, was reviewed recently by Hofmann (1996). Special cases have been variously addressed over the years, including deposition of fibrous aerosols (e.g. Asgharian and Yu, 1989; Podgorski and Gradan, 1990; Zhou *et al.* 1996), ultrafine particles (e.g., Gradon and Podgorski, 1999; Zhang *et al.*, 2005), the role of intersubject variability (Hofmann *et al.*, 2002) and the role of electrostatic forces (e.g. Yu, 1985).

In its earlier report in 1966, the Task Group on Lung Dynamics of ICRP used the then available experimental data on aerosol inhalation and regional deposition to create a set of physiologically based empirical models, aimed at providing a framework for the calculation of deposition, and in turn dose, of radioactive aerosols to all parts of the human respiratory tract and beyond (Bates *et al.*, 1966). This was exhaustively revised in the later ICRP report in 1994, taking into account the large body of new research that had been conducted during the intervening years. It contained an annex in the form of a paper describing the development of a comprehensive system of models for regional aerosol deposition at the various parts of the respiratory tract (James *et al.*, 1994). This paper now provides an important contemporary basis for discussing the experimental data from studies of the regional deposition of inhaled particles and, in turn, their application in particle size-selective aerosol exposure criteria.

11.2.3   Experiments for studying regional deposition

The literature on experimental studies of regional deposition in humans is more extensive than for studies of inhalability using mannequins. Also, as seen from the preceding outline, the issues of interest are diverse and more complex. This means that a considerable range of experimental methods has been needed to meet the challenges of measuring the desired range of parameters. Inhalation experiments with human volunteer subjects began in the 1950s with the pioneering work of Hatch and colleagues (e.g. Hatch and Gross, 1964) and continued in the bodies of research led by Lippmann, Stahlhofen, Heyder and others. In all of them, human subjects were asked to inhale idealised monodisperse, spherical test aerosols of a nontoxic material (e.g. polystyrene latex) with known particle size under controlled breathing conditions. In most of them, inhalation was achieved through a tube inserted into the mouth of the subject, whose nose was clipped in order to avoid additional air entering or leaving through that portal. Other experiments were designed specifically to investigate inhalation through the nose. In most of the reported experiments the aerosols were neutralised to Boltzmann equilibrium in order to eliminate effects associated with electrostatic forces, hence removing one source of bias or variability.

In one experimental design, the concentration of aerosols delivered from a continuous, well-defined source and entering during the inspiration part of the breathing cycle was measured, along with the concentration of particles exiting during the exhalation part. This was most easily achieved by the application of light scattering apparatus to count individual particles in the air close to the mouth during each part of the breathing cycle. In this way, the efficiency of total aerosol deposition (E_{total}) was immediately obtained in real-time for each particle size tested. In an extension of this approach, the aerosol supply was delivered to the inspired air as a *bolus*, in the form of a short pulse of test aerosol injected at a specific, predetermined part of the breathing cycle. The concentration of particles in the exhaled air now also took the form of a time-varying trace whose properties – concentration as a function of time – in relation to the original bolus reflected the fate of the particles in the original bolus, including how much had been deposited in the respiratory tract, and where. By appropriate interpretation of the results, it was therefore possible to obtain information about the efficiency of aerosol deposition in the various regions of the respiratory tract.

Another experimental approach that has been widely applied involved the use of radioactively labeled test aerosols that could be detected in the respiratory tract using methods derived from applied nuclear physics. For example, if the label was a gamma emitter, the particles deposited in the respiratory tract could be measured noninvasively – regionally or overall, instantaneously or over time – by means of gamma camera techniques. By observing the clearance over time of deposited particles from the respiratory tract, the results could be partitioned in order to obtain the original deposition in the various parts of the respiratory tract characterised by faster or slower clearance mechanisms – for example, ciliary clearance in the conducting airways and macrophage-mediate clearance in the alveolar region, respectively. Other approaches applied sensitive detection methods, for example superconductive quantum interference devices (or SQUIDS), to noninvasively measure the regional deposition of inhaled aerosols containing magnetic materials. Such methods have been described in detail in the literature and will not be further elaborated here.

An important feature of such experiments is that, because they have involved studies with real people, the results have reflected the large biological variability between individuals, including differences in physiology and health status.

11.2.4   Results for total deposition

Experimental results for the efficiency of total aerosol deposition (E_{total}) have been published by many workers, including George and Breslin (1967), Giacomelli-Maltoni *et al.* (1972), Shanty (1974), Heyder *et al.* (1975) and others for particle sizes from about 0.1 to 10 μm. In these it is noted that the metric of particle size shifted between particle aerodynamic diameter (d_{ae}) for larger particles, reflecting the role of inertia and gravity in particle deposition, and particle equivalent volume diameter (d_V) for smaller particles reflecting the role of diffusion. The results are summarised in Figure 11.2(a) and (b) for nose and mouth breathing, respectively. The summary form of presentation shown there follows the style that was adopted in Chapter 10 for coarser particles. Within the bands shown, enclosing nearly all the actual data, the individual experimental data sets were generally consistent with one another. As already mentioned, in these figures and others like them described below, the wide scatter reflects mainly the considerable variability observed between experimental methods and among human subjects. In general the results show that, on average, E_{total} was unity for large particles with d_{ae} greater than about 5 μm for nasal breathing and greater than about 10 μm for mouth breathing, indicating that all particles that were large enough were deposited *somewhere* in the respiratory tract. However, as particle size decreased in the range of d_{ae} down to about 0.5 μm, E_{total} fell steadily reaching about 0.2 for both nasal and mouth breathing. However, as the particle size decreased further, all the way down to $d_V \approx 0.1$ μm, there was no further decline in E_{total}. The indicated trend in E_{total} with particle size reflects the fact that particles

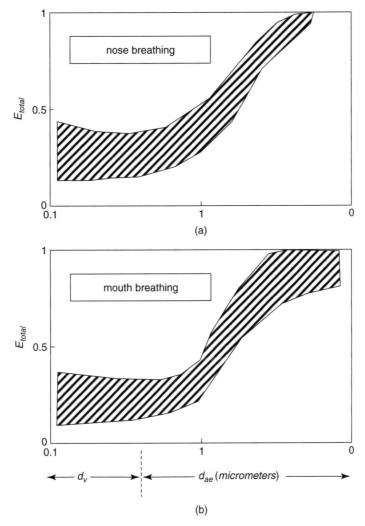

Figure 11.2   *Summary of the measured efficiency of total deposition (E_{total}) in the human respiratory tract as a function of particle diameter (d_{ae} for larger particles and d_v for smaller ones): (a) nose breathing; (b) mouth breathing. Experimental data include those from George and Breslin (1967), Giacomelli-Maltoni et al. (1972), Shanty (1974) and Heyder et al. (1975), and the hatched areas include nearly all the data*

of different sizes were not deposited with the same efficiency, and that for particles below about 10 μm, there was a significant proportion that was exhaled.

11.2.5   Results for extrathoracic deposition

Experimental results for deposition in the head (E_{ET}) have been published by several groups, including Hounam *et al.* (1971), Lippmann (1977) and Heyder and Rudolf (1977) for nasal deposition in ET_1 and ET_2 following nasal breathing, and Lippmann (1977), Stahlhofen *et al.* (1980) and Chan and Lippman (1980) for mouth deposition in ET_3 following mouth breathing.

In the first instance, James *et al.* considered extrathoracic deposition in terms of d_{ae}, breathing flow rate Q (the sum of both inspiration and expiration) and the tidal volume V_T. As mentioned earlier, the deposition of larger particles during nasal breathing takes place mainly towards the rear of the anterior part (ET_1). During mouth breathing it takes place mainly in the larynx part of ET_3. From their inspection of all the experimental data, James *et al.* proposed (after Rudolf *et al.*, 1986; Stahlhofen *et al.*, 1989) the empirical expression:

$$_{nose}E_{ET} \approx E_{ET1} = 1 - \left(\frac{1}{1 + 0.005 d_{ae}^2 Q} \right) \tag{11.13}$$

for nose breathing and deposition primarily in ET_1, along with:

$$_{mouth}E_{ET} = E_{ET3} = 1 - \left(\frac{1}{1 + 0.00017(d_{ae}^2 Q^{0.6} V_T^{-0.2})^{1.4}} \right) \tag{11.14}$$

for mouth breathing and deposition in ET_3.[1] In these expressions, d_{ae} is given in μm, Q in Lpm and V_T in L. Guided by Equations (11.13) and (11.14), the experimental data for E_{ET} are summarised for nose breathing in Figure 11.3(a), shown as a function of $d_{ae}^2 Q$, and for mouth breathing in Figure 11.3(b) as a function of $d_{ae}^2 Q^{0.6} V_T^{-0.2}$. Again, the shaded areas represent nearly all the experimental data. Also shown are the corresponding curves calculated using the above equations, with breathing parameters chosen to corresponding to the 'reference worker' defined in Chapter 9. They take into account deposition during both inspiration and expiration. Overall, it is seen that the calculated curves are quite representative of the experimental data, with E_{ET} increasing steadily with particle size in both cases, eventually reaching unity for larger particles. In addition, for any given particle size, E_{ET} is significantly greater for nasal than for mouth breathing. That is, deposition in the extrathoracic region is more efficient during nose breathing than during mouth breathing. It follows that, for mouth breathing, particles are better able to enter the tracheobronchial region and beyond.

11.2.6   Results for tracheobronchial deposition

Aerosol deposition in the conducting airways was studied experimentally by Lippmann and Albert (1969), Foord *et al.* (1976), Chan and Lippmann (1980), Stahlhofen *et al.* (1980) and others, in all of which the efficiency of deposition for mouth breathing was expressed as a fraction of the aerosol entering that part of the respiratory tract. This is the intrinsic local filtration efficiency of the tracheobronchial region, E_{TB}, and so is – in itself – independent of whether breathing is through the nose and/or mouth. Again for particles large enough for their motion to be governed primarily by aerodynamic effects, James *et al.* inspected the available experimental data and proposed an empirical expression for deposition in the trachea and bronchi – generations 0 to 8 in Table 9.1 – of the form:

$$E_{BB} = 1 - \exp[-0.0001(d_{ae}^2 Q)^{1.152}] \tag{11.15}$$

where, as before, d_{ae} is given in μm and Q in Lpm. James *et al.* also suggested an additional expression for deposition in the terminal bronchiolar airways – generations 9 to 19 in Table 9.1 – given by:

$$E_{bb} = 1 - \exp\{-0.1147[(0.056 + t_{bb}^{1.5})d_{ae}^{t_{bb}^{-0.25}}]^{1.173}\} \tag{11.16}$$

---

[1] Note that, in Equations (11.12) and (11.13), and elsewhere in this chapter Q is given in Lpm and V_T in L, as opposed to cm^3 s^{-1} and cm^3, respectively, in the original publication of James *et al.* (1994). This accounts for the differences in the coefficients shown here in comparison with those in the original publication.

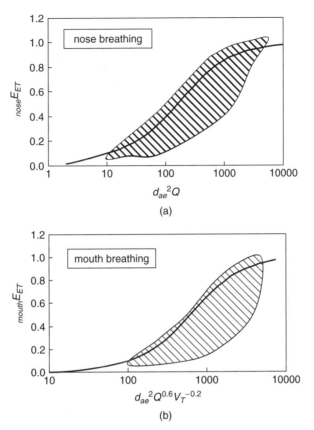

Figure 11.3 *Summary of the measured efficiency of extrathoracic deposition (E_{ET}): (a) for nasal deposition (mainly in the anterior region) following nose breathing (Hounam et al., 1971; Heyder and Rudolf, 1977; Lippmann, 1977): (b) for head deposition (mainly in the larynx) after mouth breathing (Lippmann, 1977; Chan and Lippman, 1980; Stahlhofen et al., 1980). The data are plotted as functions of combinations of variables suggested by James et al. (1994). The shaded areas enclose nearly all the experimental data and the solid lines are calculated using equations proposed by James et al. Reproduced with permission from Vincent, Journal of Environmental Monitoring, 7, 1037–1053. Copyright (2005) Royal Society of Chemistry*

where t_{bb} is the residence time of aerosols as they pass through the terminal bronchiolar region which, according to Lippmann (1977), lies between about 16 and 30 ms (see Table 9.1). Equations (11.15) and (11.16) may be combined to give the total local efficiency of tracheobronchial deposition, thus:

$$E_{TB} = 1 - (1 - E_{BB})(1 - E_{bb}) \qquad (11.17)$$

The available experimental data for tracheobronchial deposition are summarised in the form of E_{TB} versus $d_{ae}^2 Q$ in Figure 11.4(a), again portrayed as a shaded area enclosing most of the measurements. Within the limits of experimental variability, it is seen that the curve calculated from the preceding equations, again for a 'reference worker', is a fair reflection of what was observed. In general, the trend is for E_{TB} to increase steadily with particle size, rising from zero to close to unity.

The model for E_{TB} may usefully be combined with those for E_{ET}–in the manner described by Equation (11.5) – to calculate the efficiency of the deposition in the conducting airways region as a

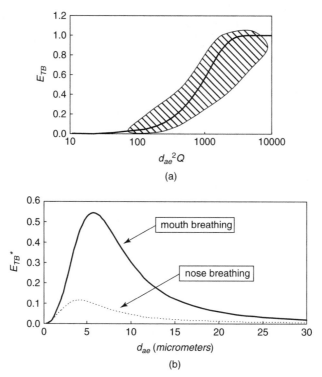

Figure 11.4   *Summary of the measured efficiency of tracheobronchial deposition: (a) shown as intrinsic efficiency (E_{TB}), from the data of Lippmann and Albert (1969), Foord et al. (1976), Chan and Lippmann (1980) and Stahlhofen et al. (1980): and (b) efficiency relative to what was inhaled (E_{TB}^*), both plotted as functions of the combinations of variables suggested by James et al. (1994). The shaded area encloses nearly all the experimental data, and the continuous lines in both graphs are calculated from the equations proposed by James et al. Adapted with permission from Vincent, Journal of Environmental Monitoring, 7, 1037–1053. Copyright (2005) Royal Society of Chemistry***

proportion of what was inhaled, E_{TB}^*. The results are shown for both nose and mouth deposition in Figure 11.4(b). Here the results are expressed in the form of E_{TB}^* versus d_{ae}, again for a 'reference worker'. The difference in tracheobronchial deposition for nose and mouth breathing is now very clear, most notably in the higher deposition in this region for mouth breathing. Also shown is the peak in tracheobronchial deposition for d_{ae} about 5 μm, decreasing for both smaller and larger particles.

11.2.7   Results for deposition in the alveolar region

Experimental results for aerosol deposition in the alveolar region have been reported by Altshuler *et al.* (1957), George and Breslin (1967), Lippmann and Albert (1969), Shanty (1974) and Foord *et al.* (1976) and others, all for mouth breathing. This time the efficiency of particle deposition in the alveolar region was presented as a fraction of the aerosol originally inhaled; that is E_{alv}^*. The results are summarised in Figure 11.5, with E_{alv}^* plotted as a function of d_{ae}, showing a peak at between 0.2 and 0.5 for d_{ae}

---

* Here and in the figures that follow, the calculated curves have been updated from the original published ones by the inclusion of the terminal bronchiolar (bb) deposition contribution, which had earlier been neglected

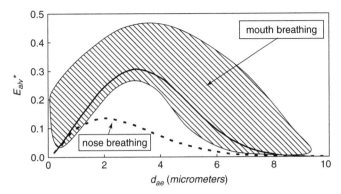

Figure 11.5 *Summary of the measured efficiency of alveolar deposition as a function of particle aerodynamic diameter (d_{ae}), shown relative to what was inhaled (E^*_{alv}). The shaded area encloses nearly all the experimental data for mouth breathing. (Altshuler et al., 1957; George and Breslin, 1967; Lippmann and Albert, 1969; Shanty, 1974; Foord et al., 1976). The continuous curves – for both mouth and nose breathing – are calculated from equations proposed by James et al. (1994) Adapted with permission from Vincent, Journal of Environmental Monitoring, 7, 1037–1053. Copyright (2005) Royal Society of Chemistry*

about 4 μm, falling off on either side, and indicating virtually no deposition for particles greater than about $d_{ae} = 10$ μm. This observed trend reflected the fact that all particles larger than this had been deposited in the higher regions of the respiratory tract and so had not reached the alveolar region.

Again, James *et al.* inspected the available experimental data and developed an empirical equation for the local efficiency of alveolar deposition, this time as:

$$E_{alv} = 1 - \exp[-0.146(d_{ae}t_{alv})^{0.65}] \tag{11.18}$$

where t_{alv} is the residence time for particles in the alveolar region which, according to Lippmann (1977), is about 800 ms (see Table 9.1). Equation (11.18) may now be combined with Equation (11.7), involving the individual modeled filtration efficiencies of the extrathoracic and tracheobronchial regions (E_{ET} and E_{TB}, respectively), in order to calculate E^*_{alv}. This was carried out for mouth breathing by the now-familiar 'reference worker', and the results are shown in Figure 11.5 alongside the summary of the experimental data. Again the model provided a quite good match with the experimental data. Also as indicated, calculations of the corresponding efficiency for nose breathing show values much less than for mouth breathing.

11.2.8 Results for the deposition of fibrous aerosols

As already mentioned in Chapter 9, interest in fibers as a subject for aerosol sampling began many years ago with the emergence of greater awareness about the significant health risks associated with asbestos exposures. However, there appear to be no experimental data on regional deposition in the human respiratory tract for fibers. But this is not surprising since the geometry and dimensions of fibers – asbestos or otherwise – raise serious concerns about the risk associated with *any* exposure, no matter how small, even under controlled clinical research conditions. Although experimental data do exist for deposition in the alveolar regions of rat lungs, the link with human exposure is tenuous without human data. The best approach, therefore, for discussing the regional discussion of fibers lies in the availability of large amounts of human data for *nonfibrous* particles. One of the links is the set of equations for fibrous particle aerodynamic developed by Cox (1970) and Stöber (1971), as described in the earlier Equations (2.46)–(2.48). For fibers in the range where the equivalent value of d_{ae} is greater than about 0.5 μm, it may be assumed that inertia and/or gravity are predominant influences on deposition in the confined spaces of the respiratory tract. Here, therefore, the available experimental data for deposition in the extrathoracic, conducting airways and alveolar regions can be reasonably translated to fibers from the results for nonfibrous aerosols. However, for fibers where the calculated d_{ae} is less than 0.1 μm, where it must be assumed that diffusion becomes the primary mechanism for deposition, the most appropriate metric for particle diameter for the purpose of translating the experimental data for nonfibrous particles to fibrous ones will be a function of fiber length and width, l_f and d_f, respectively. As described in Chapter 2, Gentry *et al.* (1991) showed the relationship between diffusion coefficient (D_B) and fiber dimensions to be a complex one. Their results suggested that, to a fair approximation, an appropriate metric for translating regional respiratory tract deposition from the results for spherical particles to fibers in the diffusion regime is the fiber width (d_f).

11.2.9   Results for the deposition of very fine and ultrafine aerosols

Specific concerns about the very fine and ultrafine aerosol fractions are more recent than those about the other fractions discussed above. The earliest studies of the regional deposition of ultrafine particles in humans appear to be the ones reported by Heyder *et al.* (1986) and Schiller *et al.* (1988). In these, the efficiency of total respiratory tract deposition (E_{total}) was measured in human subjects for radioactively labeled spherical particles in the range of geometric diameter (d) from 0.005 to 0.5 μm, for both nasal and mouth breathing and a range of breathing flow rates. From observing the rates of clearance from the respiratory tract, E_{ET} and its nasopharyngeal and head components, E_{TB} and

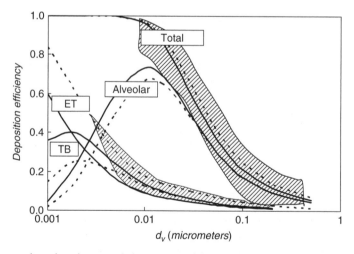

Figure 11.6 *Efficiency of total and regional deposition of fine and ultrafine particles as a function of particle equivalent volume diameter (d_v) expressed as a proportion of what was inhaled, all fractions calculated from equations proposed by James et al. (1994) within the framework outlined at the beginning of this chapter. The dashed and solid lines represent nose and mouth breathing, respectively. Also shown are summaries of the measured data for total respiratory tract deposition for ultrafine particles during mouth breathing by human subjects (from Heyder et al., 1986; Schiller et al., 1988) and nose breathing for a human nasal cast for a range of steady inspiratory flow rates from 4 to 30 Lpm (from Cheng et al., 1988). The shaded areas enclose nearly all the experimental data referred to*

E_{alv} were obtained experimentally. The results for E_{total} as a function of d in the nominal ultra-fine range are summarised in Figure 11.6, with the results enclosed within the shaded area containing data for both nose and mouth breathing. The indicated trend shows E_{total} rising steeply as the particles decreased in size, reaching close to unity for $d = 0.005$ μm. Closer inspection of the actual data revealed only slight differences between nasal and mouth breathing and for the different breathing flow rates investigated.

Around the same time, Y.-S. Cheng *et al.* (1988) reported results for the efficiency of nasal deposition, $_{nose}E_{ET}$, in an inert cast of a human nasal tract for a range of steady inspiratory flow rates from 4 to 30 Lpm. Some typical results in the form of $_{nose}E_{ET}$ versus d are shown also in Figure 11.6. They showed nasal deposition increasing quite sharply as particle size decreased below about 0.02 μm. Cheng *et al.* reported relatively small dependence on the inspiratory flow rate. Such results drew particular attention to the role of nasal deposition in the overall uptake of ultrafine particles. Swift and Strong (1996) also conducted experiments to investigate the nasal deposition of ultrafine particles in human subjects, and obtained very similar results. At about the same time, K.-H. Cheng *et al.* (1996) reported experiments that also investigated the role of differences in nasal airway dimensions and noted the existence of significant variability in $_{nose}E_{ET}$ from subject to subject.

In addition to their equations for the regional deposition of particles in the aerodynamic regime, James *et al.* used experimental data like those described to develop a set of corresponding empirical equations for the local regional deposition of fine particles whose motions are governed by diffusion, as follows:

$$E_{ET1} = 1 - \exp[-1270(D_B Q^{-0.25})^{0.500}] \text{ for nasal deposition in } ET_1$$

$$E_{ET2} = 1 - \exp[-1460(D_B Q^{-0.25})^{0.535}] \text{ for nasal deposition in } ET_2$$

$$E_{ET} = 1 - (1 - E_{ET1})(1 - E_{ET2}) \text{ for nasal deposition (in } ET_1 \text{ and} ET_2)$$

$$E_{ET3} = 1 - \exp[-630(D_B Q^{-0.25})^{0.500}] \text{ for head deposition (mouth breathing) in } ET_3$$

$$E_{BB} = 1 - \exp[-7930(D_B t_{TB})^{0.639}] \text{ for bronchial deposition (in } BB)$$

$$E_{bb} = 1 - \exp[-7930(D_B t_{TB})^{0.639}] \text{ for bronchiolar deposition (in } bb)$$

$$E_{TB} = 1 - (1 - E_{BB})(1 - E_{bb}) \text{ for tracheobronchial deposition (in } BB \text{ and } bb)$$

$$E_{alv} = 1 - \exp[-75260(D_B t_{alv})^{0.610}] \text{ for alveolar deposition}$$

(11.19)

In all these, D_B is the coefficient of Brownian diffusion in $m^2 s^{-1}$ given by the earlier Equation (2.62), while t_{BB}, t_{bb} and t_{alv} are the residence times of particles (in s) in the bronchial, bronchiolar and alveolar regions, respectively, all of which may be estimated from the right-hand column in Table 9.1. For nasal deposition it is notable that the form of the equation listed is very similar to the $D_B^{-0.5} Q^{-0.125}$ dependency suggested by Y.-S. Cheng *et al.*

The calculated deposition efficiencies for the fractions indicated, along with total deposition (E_{total}), are all plotted together in Figure 11.6, alongside the available experimental data summaries for nasal and total deposition already mentioned. Results are shown for both nose breathing (dashed curves) and mouth breathing (solid curves). All are expressed as proportions of what was inhaled and are shown for values of d_V all the way down to 0.001 μm (1 nm). Figure 11.6 reveals some important trends. Firstly, in the range below 0.1 μm, the efficiencies of deposition for all fractions increase steadily. For total aerosol, it continues to rise, reaching unity for d_V close to 0.01 μm. For extrathoracic deposition, deposition efficiency also continues to rise, becoming predominant for extremely small particles below about 0.005 μm. This tendency is especially marked for nasal deposition during nose breathing. Tracheobronchial deposition efficiency also rises steadily as d_V decreases, but there is an indication of a downturn for d_V close to 0.002 μm. Generally speaking, however, tracheobronchial deposition in this range of particle size is less than for the other regions. Finally, the efficiency of alveolar deposition rises steeply as d_V decreases, but then peaks at around 0.7 for d_V close to 0.01 μm, and thereafter falls steeply as extrathoracic deposition increasingly takes over as the predominant component.

In view of the current increased interest in health effects associated with exposure by inhalation to ultrafine particles, the relationships discussed above take on special importance. In the first instance, the general high efficiency of alveolar deposition is not surprising and is very important. But, in addition, the role of nasal deposition is also very important, identifying the nose as a significant target organ, especially for very small particles. Overall, it is also important to further note that, after deposition, very small, ultrafine particles may penetrate readily into the body beyond the initial site of deposition, even if they are of intrinsically insoluble material (e.g. Nemmar *et al.*, 2002).

11.3   Criteria for fine aerosol fractions

Data and models like those described provide a basis for developing criteria for health-related fine particle size fractions that can be used in occupational and environmental health standards, providing guidelines for scientifically rational, health-related aerosol sampling. Most of the criteria that will be presented below appear as curves describing the probability of particle penetration or deposition as functions of particle size. Such curves become the 'yardsticks' against which the performances of aerosol sampling instruments should be compared when chosen for specific purposes.

11.3.1   Historical overview

Even in the earliest days of occupational and environmental hygiene, it was recognised that aerosol-related health effects were broadly linked with one or other of two classes of aerosol. The first consisted of coarse particles which may, depending on their chemical composition, lead to adverse health effects regardless of where in the respiratory tract they are deposited. These were discussed in the preceding chapter. The second consisted of finer particles which may penetrate into the lung and contribute to aerosol-related lung disease.

As early as 1913, McCrea noted from his microscope studies of the lungs post mortem of South African mineworkers that particles in the alveoli of the deep lung did not exceed about 7 μm in physical diameter. This led him to identify the need to selectively measure fine particles as an appropriate index of exposure relevant to certain types of lung disease, in particular pneumoconiosis. This would appear to be the first reference to what is now widely referred to as a 'particle size-selective sampling criterion'. Later McCrea's observation was confirmed by the earliest controlled inhalation experiments with human volunteer subjects where it became apparent that, of the aerosol entering through the nose and mouth during breathing, only the finest particles penetrated down to the deep lung (e.g. Davies, 1952). The effect of the physical filtration of coarser particles by combinations of impaction and gravitational settling in the higher, conducting airways of the lung, and – notably – the role of particle aerodynamic diameter, was recognised. For many years, however, and in some industries in some countries in particular, the idea of a criterion based on the visual observation of particles and their discrimination on the basis of their physical dimensions, persisted as the basis of standards, even until as late as the 1980s. Most notably, in the extraction industries of Canada, Australia and South Africa, a criterion was applied for many years that required the counting of particles that, when viewed under the microscope at appropriate magnification, were seen to be less than 10 μm in diameter. This approach was accompanied by occupational exposure limits that expressed aerosol concentrations in terms of particle number per unit volume of air sampled.

From the late 1950s onwards, the concept of particle size selection based on particle aerodynamic diameter began to emerge, and a number of quantitative definitions for the aerodynamically defined fine

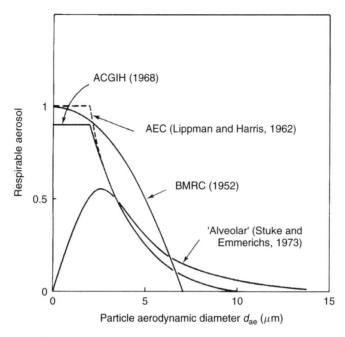

Figure 11.7 *Particle size selection curves for a number of historically important, earlier definitions for the respirable fraction, as indicated. Reproduced with permission from Vincent, Aerosol Sampling: Science and Practice. Copyright (1989) John Wiley & Sons, Ltd*

respirable fraction followed. In 1952, the British Medical Research Council (BMRC) recommended the definition of respirable airborne dust in terms of the *penetration* of particles to – as opposed to *deposition* in – the alveolar region as a function of d_{ae}. At the time, the thinking behind this was driven by concerns about the high prevalence of pneumoconiosis among workers in the mining industries. The resultant, now familiar, BMRC version of the respirable fraction took the form of a curve where penetration was unity for $d_{ae} = 0$ and fell to zero for $d_{ae} = 7.1$ μm, passing through 0.5 at $d_{ae} = 5$ μm (see Figure 11.7). It was noted at the time that, whilst this curve was reasonably representative of the relevant fraction, based on the experimental information then available, it also described the penetration characteristics for particles moving through a horizontal rectangular channel in which deposition took place – on the lower surface – by gravitational settling (Walton, 1954). Hence the availability of appropriate instrumentation for sampling respirable aerosol was built explicitly into the formal criterion. Later, during the 1959 International Pneumoconiosis Conference in Johannesburg, South Africa, the BMRC version was adopted by much of the international occupational hygiene community (Orenstein, 1960), and thereafter became widely known as the 'Johannesburg curve'.

At about the same time, in the United States, a group meeting under the auspices of the Atomic Energy Commission (AEC) proposed its own definition for respirable aerosol, also based on the available alveolar deposition data while at the same time capable of being matched by suitable instrumentation – in this case, cyclones (Lippmann and Harris, 1962). Later, the ICRP Task Group on Lung Dynamics (Bates *et al.* 1966) proposed a representative model for alveolar deposition in typical subjects under typical breathing conditions, based on which the American Conference of Governmental Industrial Hygienists (ACGIH) defined its own version of respirable aerosol (American Conference of Governmental Industrial

Hygienists, 1968). The AEC and ACGIH curves are both shown in Figure 11.7 alongside the BMRC curve. The AEC and ACGIH curves differed from one another only in the region $d_{ae} < 2$ μm where, while the AEC curve assumed a penetration of unity, the ACGIH definition allowed it to flatten out at 0.90. Both curves passed through 0.5 for $d_{ae} = 3.5$ μm. In the years that followed, all three definitions for respirable aerosol – BMRC, AEC and ACGIH – were cornerstones of aerosol health-related research and standards-setting throughout the world.

In addition to these primary conventions, a variety of others were proposed for describing the fine dust fraction associated with health effects in the deep lung. One suggested in Germany for the sampling of coalmine dust in a manner relevant to the pneumoconiosis health risk was especially interesting because it took account of the fact that, of the aerosol which penetrated to the deep lung, not all the particles were deposited there, but some were exhaled (Stuke and Emmerichs, 1973). This curve too is shown in Figure 11.7. In principle this, like the BMRC, AEC and ACGIH curves, was linked explicitly with sampling instrumentation, this time a pair of cyclones placed in series.

In the early 1980s, in response to the ever-increasing public awareness in many countries of the risks associated with the inhalation of airborne particles, in both occupational and general environmental situations, fresh deliberations were undertaken. These represented the integration of thinking about coarse and fine aerosols, respectively, within a single scientific framework. A new approach emerged in which three aerosol fractions were identified, defining coarse particles (the *inhalable fraction*, representing aerosol that passed through the nose and/or mouth during breathing), intermediate-sized particles (the *thoracic fraction*, representing that part of the inhaled aerosol that penetrated past the larynx and into the lung) and fine particles (the *respirable fraction*, representing that part of the inhaled aerosol that penetrated down to the alveolar region of the lung). It was felt that defining these fractions would provide the basis for the scientifically based sampling of aerosols relevant to the majority of aerosol-related health effects. The emergence of this new framework was led initially by expert committees of the International Standards Organisation (ISO, 1981 and 1983) and ACGIH (Phalen, 1985), and represented a substantial widening of the scope of particle size-selective criteria for aerosol sampling in relation to health. Many other standards setting bodies have since followed their lead.

In this emerging new framework, the approach to the respirable fraction was already well developed from previous efforts, driven again by the high prevalences of occupational lung diseases such as pneumoconiosis. The new ingredients were the coarser inhalable and intermediate thoracic fractions. Development of the inhalable fraction represented a rationalisation of the old 'total' aerosol approach. But prior to the 1980s there was little discussion about an intermediate fraction. There *was* one notable exception. In the 1970s, the United States National Institute for Occupational Safety and Health (NIOSH) proposed a criterion for the sampling of cotton dust in relation to the occupational disease of byssinosis, an airways disease prevalent among cotton mill workers (NIOSH, 1975). For this, NIOSH recommended that sampling should be carried out with a sampling device that had a pre-selector which would provide a collection efficiency of 50 % at $d_{ae} = 15$ μm.

In the late 1970s interest was stirring in the United States Environmental Protection Agency (EPA) in a standard that reflected concerns about aerosols in the ambient atmospheric environment that were most likely associated with a wide range of respiratory tract illnesses in the general population. In this context, in 1979 Miller *et al.* used the then-available experimental lung deposition data to propose a definition for thoracic aerosol – although they referred to it at the time as 'inhalable' (see footnote 1 in Chapter 10). They suggested a selection characteristic that exhibited a 'cut-off' at $d_{ae} = 15$ μm. Later, however, EPA (Environmental Protection Agency, 1984) decided to opt for a formal definition closer to the ISO definition that was emerging at about the same time, specifying a selection characteristic with a 'cut-off' at d_{ae} equal to the lower value of 10 μm. In the field of ambient atmospheric aerosol monitoring, this has since become widely known as the 'PM$_{10}$' standard. Meanwhile, however, the use of the 'total suspended particulate' (TSP) rationale remained as the primary basis of standards until it was eventually eliminated in 1987.

The ISO establishes criteria relating to air quality with respect both to occupational and general environmental exposures. In its 1983 recommendations for the thoracic fraction, ISO took into account the available experimental lung deposition data, following in particular the 1979 review of Miller *et al.* It acknowledged that the division between extrathoracic and thoracic aerosols would be different depending on whether breathing took place through the nose or the mouth, as shown earlier in Figures 11.3 and 11.4. It took into account that, for given particle size and breathing flow rate, extrathoracic deposition was less for mouth than for nasal breathing. It followed that larger particles, and hence greater overall particulate mass, penetrated to the lung during mouth breathing. Miller *et al.* had suggested that, in order to establish a practical, working aerosol sampling criterion relevant to health effects that could be used in a standard, this 'worst-case' scenario was the more appropriate. The ISO philosophy on this was basically the same, and a curve was proposed that described the penetration of inhaled particles beyond the larynx as a function of d_{ae} having the form of a cumulative log-normal function with its median at $d_{ae} = 10$ μm and geometric standard deviation 1.5. In this definition, unlike EPA's PM$_{10}$, it was explicitly stated that the thoracic fraction was a sub-fraction of the inhalable fraction. In addition, in its 1983 report, ISO recommended the continued use of existing criteria for the respirable fraction like those already described. The rationale for this was to provide continuity between the new and the older criteria, smoothing the path towards their implementation in actual standards.

The ACGIH published its first major review of particle size-selective aerosol sampling criteria in 1985 (Phalen, 1985). The thoracic fraction was addressed by examining the experimental data for particle deposition in the head during mouth inhalation, notably those reported by Lippmann (1977), Stahlhofen *et al.* and Chan and Lippmann. These experimental results were adjusted to a 'reference worker' with an average minute volume of 21.75 Lpm (i.e. 1.45 L inspired per breath, 15 breaths min^{-1}, hence $Q = 43.5$ Lpm, see Table 9.2) and then plotted in the form of P_{thor} versus d_{ae}. They are summarised in this form by the shaded area in Figure 11.8. For sampling purposes, however, it was decided – as a matter of policy–to recommend a selection curve for thoracic aerosol which erred on the side of conservatism, providing a greater level of protection when used as a criterion for a standard. The selection curve recommended by ACGIH in 1985 was therefore a cumulative log-normal function with its median at $d_{ae} = 10$ μm and geometric standard deviation 1.5, and this too is shown in Figure 11.8. It is seen to lie substantially to the right of the actual experimental data. The corresponding curve for respirable aerosol took the same mathematical form, but with its median at $d_{ae} = 3.5$ μm and geometric standard deviation 1.5, and this is shown in Figure 11.9, together with a summary of the measured deposition efficiency data. Here, again, a conservative approach was taken, where the loss of the smaller particles exhaled during the expiratory part of the breathing cycle was neglected. For both the thoracic and respirable fractions, there was at that time no acknowledgement of the role of the particle size-dependent inhalable fraction. This came later.

Meanwhile, it is relevant to mention that, as part of the historical evolution of the criteria for fine particles leading up to the present day, there have been shifts in the metrics by which particles are assessed *after* they have been sampled. In the early days of health-related sampling, it was the common convention that the concentration of airborne particles should be obtained in terms of *number* per unit volume of air in the size fraction of interest. Since then, however, epidemiological studies confirmed that, for most aerosol-related illnesses, *mass* concentration was a more appropriate index. This is the philosophy underlying most of the criteria that will be discussed below. However there do remain some cases where particle number concentration remains the preferred index. Further, there are some situations where particle *surface area* concentration might be given serious consideration.

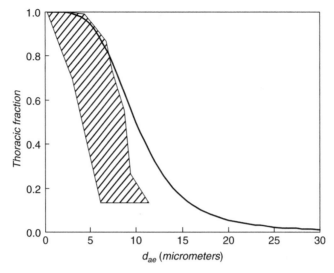

Figure 11.8 *1985 ACGIH curve for the thoracic fraction, expressed as a function of particle aerodynamic diameter (d_{ae}). The shaded area encloses most of the experimental data for P_{thor} from experiments with human subjects. Note that the chosen curve lies well outside the actual human data, a conscious decision based on the intention to err on the side of over-protection when the curve is used as the basis of a standard*

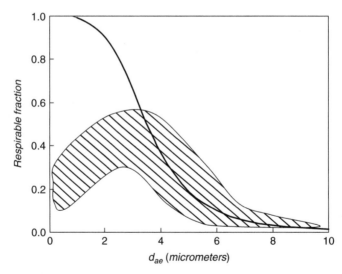

Figure 11.9 *1985 ACGIH curve for the respirable fraction, expressed as a function of particle aerodynamic diameter (d_{ae}). The shaded area encloses most of the experimental data for alveolar deposition (E^*_{alv}) from experiments with human subjects. Note that the chosen curve oversamples for the finer particles, making the assumption – valid for some, but not all, exposures – that most of the aerosol mass is contained in the large particles*

11.3.2 Modern criteria for the thoracic and respirable aerosol fractions

The criteria for the inhalable, thoracic and respirable fractions continued to evolve during the 1980s and 1990s. In its more recent comprehensive review, ACGIH presented its clarified recommendations where, as before, the finer fractions continued to be defined in terms of the penetration of particles into the lung and, in turn, down to the alveolar region (Vincent, 1999). In these later recommendations, the intermediate *thoracic fraction*, $T(d_{ae})$, was defined by the curve:

$$T(d_{ae}) = I(d_{ae})[1 - F(x)] \tag{11.20}$$

where $I(d_{ae})$ was as defined in the previous chapter – see Equation (10.12) – and $F(x)$ was described formally as a cumulative log-normal function with median $\Gamma_x = 11.64\ \mu\text{m}$ and geometric standard deviation $\Sigma_x = 1.5$. The net result was that the function $T(d_{ae})$ reached 0.5 at $d_{ae} = 10\ \mu\text{m}$. Here it was acknowledged that the ideal mathematical form was unwieldy and so not convenient for routine use. But it had been noted, usefully, that the simple analytical form:

$$T(d_{ae}) = I(d_{ae}) \left\{ \frac{\exp[a + b \cdot \ln(d_{ae})]}{1 + \exp[a + b \cdot \ln(d_{ae})]} \right\} \tag{11.21}$$

provided results indistinguishably close to the formal definition. Here:

$$b = 1.658 \ln(\Sigma) \text{ and } a = -b \ln(\Gamma) \tag{11.22}$$

The finer *respirable fraction*, $R(d_{ae})$ was defined by the:

$$R(d_{ae}) = I(d_{ae})[1 - F(x)] \tag{11.23}$$

where $F(x)$ was again the cumulative log-normal distribution function, this time with $\Gamma_x = 4.25\ \mu\text{m}$ and $\Sigma_x = 1.5$. The net result was that $R(d_{ae})$ reached 0.5 at $d_{ae} = 4.0\ \mu\text{m}$. The same simpler analytical formulae as those given above for the thoracic fraction may also be used. A significant difference from the equivalent definition in the original 1985 (Phalen, 1985) ACGIH report was the increase in the value of d_{ae} for $R(d_{ae}) = 0.5$ from 3.5 to 4.0 μm.

These new ACGIH criteria now became consistent with the later ones proposed by the ISO (International Standards Organisation, 1995), with two important exceptions. Firstly, ISO explicitly retained the wind speed dependency for the inhalable fraction, as given earlier in Equation (10.11), in recognition of its possible application in outdoor aerosol exposure situations. Secondly, ISO proposed an alternative definition for the respirable fraction, acknowledging the importance of a vulnerable target population, including children and sick and infirm people. Again, this reflected the scope of the ISO criteria beyond occupational exposures and to the population at large. For this target population, ISO defined the respirable fraction as in Equation (11.23) but with $\Gamma_x = 2.5\ \mu\text{m}$ and $\Sigma_x = 1.5$.

The criteria as stated in the above equations are also consistent with those of the Comité Européen de Normalisation (CEN, 1992). Now, therefore, during their evolution over many years, a considerable – and highly welcome – degree of international harmonisation has been achieved (Soderholm, 1989). The thoracic and respirable fractions as defined according to the primary criteria described above are shown as continuous curves in Figure 11.10. Discrete values for those same fractions are listed in Table 11.1. In the figure, the thoracic and respirable fractions are shown as sub-fractions of the inhalable fraction. For the purpose of a sampling instrument, they define the full performance that is required

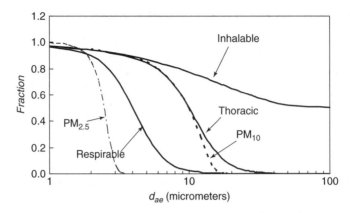

Figure 11.10 *Summary of the latest, currently accepted particle size-selective criteria for health-related aerosol sampling. Reproduced with permission from Vincent, Journal of Environmental Monitoring, 7, 1037–1053. Copyright (2005) Royal Society of Chemistry*

Table 11.1 *Numerical values for the inhalable, thoracic, PM$_{10}$, respirable and PM$_{2.5}$ aerosol fractions (Comité Européen de Normalisation, 1992; International Standards Organisation, 1995; Environmental Protection Agency, 1996a; Vincent, 1999)*

d_{ae} (μm)	$I(d_{ae})$	$T(d_{ae})$	PM$_{10}$	$R(d_{ae})$	PM$_{2.5}$
1	0.97	0.97	0.98	0.97	0.995
2	0.94	0.94	0.95	0.90	0.854
3	0.92	0.91	0.91	0.74	0.067
4	0.89	0.88	0.88	0.50	0.001
5	0.87	0.84	0.84	0.30	
6	0.85	0.80	0.80	0.17	
7	0.83	0.73	0.74	0.10	
8	0.81	0.67	0.66	0.06	
9	0.79	0.59	0.58	0.04	
10	0.77	0.50	0.50	0.02	
15	0.70	0.18	0.04	0	
20	0.65	0.06	0		
25	0.61	003			
30	0.58	0.01			
100	0.50	0			

from the inlet down, hence including both aspiration efficiency and subsequent particle selection inside the instrument.

Elsewhere, in the United States, EPA continued to promulgate its PM$_{10}$ criterion for particulate matter in the ambient atmospheric environment within its National Ambient Air Quality Standards (NAAQS), based on a similar rationale to that for the ISO thoracic convention. In its 1996 criteria document (Environmental Protection Agency, 1996a), it was stated that, considering – among other things – the similar convention on particles penetrating down to the thoracic region adopted by ISO, '. . . EPA staff recommended that the size-specific indicator include particles of diameter less or equal to a nominal 10 μm "cut point" generally referred to as "PM_{10}". In terms of collection efficiency, this represents a 50 % cut point, the aerodynamic particle diameter for which particle collection is 50 %. With such a cut point, larger particles are not excluded entirely but are collected with substantially decreasing efficiency, and smaller particles are collected with increasing (up to 100 %) efficiency.' In a subsequent staff paper (Environmental Protection Agency, 1996b), it was noted more explicitly that PM_{10} '. . . is an indicator for thoracic particles (i.e. particles that penetrate to the tracheobronchial and the gas-exchange regions of the lung)'. The curve for PM_{10} appears in the 1996a criteria document and is shown here alongside the other fractions in Figure 11.10. Discrete values for PM_{10} are listed also in Table 11.1

11.3.3   Criteria for the extrathoracic aerosol fraction

Of the various standards setting bodies, neither ISO, ACGIH, CEN nor EPA have yet proposed criteria – or developed standards – for a specific extrathoracic fraction. So, if a new criterion is to be proposed, it is necessary to first identify a need. ICRP has already acknowledged the importance of extrathoracic deposition in general in relation to dosimetry associated with inhaled radioactive aerosols. More generally, however, it is reasonable to argue that deposition in the head during mouth breathing is not of great interest since most particles deposited in this way, mainly in the larynx, are rapidly ingested. So, for the purpose of aerosol sampling standards, such particles are already covered by the inhalable fraction. On the other hand, there *are* some exposure situations where the *nasal* deposition of particles may be of significant specific interest. For example, there are some well-documented relationships between aerosol exposure and a range of nasal health effects, including nasal cancers associated with the inhalation into the nose of some radioactive aerosols, some wood dusts and some nickel species. For aerosol sampling to be carried out in relation to aerosol exposure assessment specifically in relation to such diseases, there is a clear need for a particle size-based sampling criterion that is not met by one of the aerosol fractions that are currently defined. We therefore seek a nasal deposition fraction, say $N(d_{ae})$.

It is recalled that the thoracic fraction was developed from data for the extrathoracic deposition of particles in the nasopharynx or head regions for *mouth* breathing. This is because it is for mouth breathing that the lung itself receives the greatest exposure. So the choice of mouth breathing in that case was based from the outset on 'worst-case' considerations. Further, in using the data for mouth breathing to arrive at a working practical criterion for the thoracic fraction, an additional safety margin was applied, as is seen in the comparison between the data and the 1985 version of the proposed thoracic fraction curve in Figure 11.8. In view of these compromises, it is out of the question to arrive at a relevant curve for $N(d_{ae})$ simply by referring to the conventionalised thoracic fraction. Instead, therefore, it is necessary to go back to the original nasal deposition data, and this is most plausibly achieved by reference to the empirical equations suggested by James *et al.* In particular it is noted again that the deposition of particles large enough for their motion to be governed by inertial forces is confined mainly to the rear of the anterior region of the nasal passages, identified earlier as ET_1, for which the efficiency of particle deposition is given by the earlier Equation (11.10). This is an appropriate starting point for the development of $N(d_{ae})$.

In proceeding, the problem in relation to the possible application of a nasal deposition criterion is now complicated by the fact that, for most people, inspiration is partitioned between nose and mouth breathing. It will be mostly through the nose for low levels of activity (lower Q) but mostly through the mouth for higher levels of activity (higher Q). It is at low levels of activity, therefore, where the greatest exposure of the nasal passages is likely to occur. With this in mind, the discussion is focused on

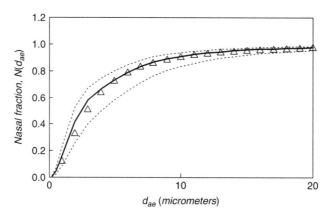

Figure 11.11   *Calculated results for the efficiency of nasal deposition as a function of particle aerodynamic diameter (d_{ae}) for relatively low values of 10, 20 and 30 Lpm (continuous and dashed lines), using the equations suggested by James et al. (1994). Also shown is the curve describing a possible new criterion for the nasal fraction (open triangles) Adapted with permission from Vincent, Journal of Environmental Monitoring, 7, 1037–1053. Copyright (2005) Royal Society of Chemistry*

relatively light activity. Calculated results for the efficiency of nasal deposition as a function of d_{ae} and for relatively low Q-values of 10, 20 and 30 Lpm (corresponding to minute volumes 5, 10 and 15 L), respectively, are shown in Figure 11.11. The calculated curves reveal that nasal deposition efficiency does not vary excessively over the range of Q indicated. So there is encouragement to seek a single function that may serve as a criterion for sampling. One suggestion, expressed in the same form as for the other criteria recommended by ACGIH, is:

$$N(d_{ae}) = I(d_{ae})F(x) \qquad (11.24)$$

where $F(x)$ is again the cumulative log-normal function, but now with $\Gamma_x = 3.0$ μm and $\Sigma_x = 2.5$. The term $F(x)$ in Equation (11.24) corresponds to the efficiency of extrathoracic deposition, and this is shown in Figure 11.11 alongside the curves for the three Q-values calculated from the James *et al.* model. It closely matches the calculated nasal deposition for $Q = 20$ Lpm.

The difficulty with writing this – or any similar expression – as a criterion for a formal health-based standard is that it is rarely possible to assume that inhalation takes place exclusively through the nose. This cannot readily be accommodated within a simple criterion like the one shown. It may be possible to account for it within an overall standard for nasal exposure. For example, in a specific sampling situation with respect to such a standard, it might be possible to weight measured concentrations by the application of a 'partition factor', a number between 0 and 1, to be decided by the professional hygienist conducting the sampling survey.

11.3.4   Criteria for the tracheobronchial and alveolar aerosol fractions

The criteria for the thoracic and respirable fractions discussed earlier are based on the penetration of particles to the lung and to the alveolar region, respectively. They do not refer to what is actually deposited, and so do not strictly relate to the health-related dose. For most applications in standards, however, they are sufficient, adequately reflecting dose. They have the advantage that they are simple in form, and in turn suggest relatively simple options for sampling instrumentation. However, there are some exposure assessment situations where exposure information more closely associated with dose is desirable, leading to the need for criteria that more closely reflect actual particle deposition.

It is a default that most aerosol-related diseases associated specifically with deposition in the tracheobronchial region are covered by the thoracic fraction as already defined. For a more representative measure of aerosol dose to that region, a criterion that more closely follows tracheobronchial deposition deserves consideration, say $B(d_{ae})$. One suggestion is:

$$B(d_{ae}) = I(d_{ae})[1 - F(x)]F(y) \tag{11.25}$$

where $F(x)$ and $F(y)$ are both cumulative log-normal distribution functions, one with $\Gamma_x = 7$ and $\Sigma_x = 2.0$ and the other with $\Gamma_y = 3.5$ and $\Sigma_y = 1.50$. The term $[1 - F(x)]F(y)$ corresponds to the efficiency of tracheobronchial deposition. It is shown in Figure 11.12 together with the curve calculated on the basis of the James *et al.* equations applied to the 'reference worker'. There is a good match.

A similar approach may be taken towards a criterion for the aerosol fraction for particles that deposit in the alveolar region. McCawley (1999) had already noted that the respirable fraction as defined above may be adequate for applications like those that drove the earliest developments of criteria for the fine aerosol fraction, specifically dust exposures leading to pneumoconiosis. He argued that it was less satisfactory for aerosols containing higher proportions of fine aerosols for which a significant fraction is exhaled. For small submicrometer particles, yet large enough to continue for their motion to be governed by aerodynamic processes, McCawley suggested:

$$A(d_{ae}) = 1 - \left\{ 1.03 \exp \left[-\frac{(\log_{10} d_{ae} + 0.49)^2}{1.77} \right] - 0.18 \right\} \tag{11.26}$$

only for $d_{ae} \leq 1$ μm. No lower limit was specified, even though diffusion would influence particle motion increasingly for smaller particles, and be dominant for $d_V \leq 0.1$ μm. Interestingly, McCawley noted that the curve described by Equation (11.26) could be matched by a commercially available

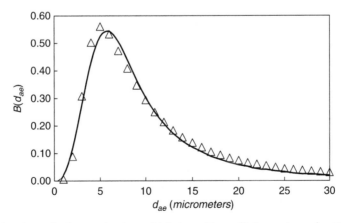

Figure 11.12 *Calculated results for tracheobronchial deposition efficiency (as a fraction of what is inhaled) as a function of particle aerodynamic diameter (d_{ae}) for mouth breathing by a 'reference worker', using the equations suggested by James et al. (1994). Also shown is the curve describing a possible new criterion for the tracheobronchial fraction (open triangles) Adapted with permission from Vincent, Journal of Environmental Monitoring, 7, 1037–1053. Copyright (2005) Royal Society of Chemistry*

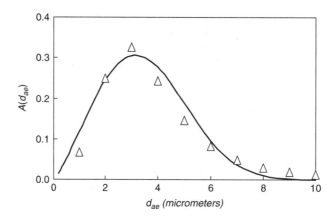

Figure 11.13 *Calculated results for alveolar deposition efficiency (as a fraction of what is inhaled) as a function of particle aerodynamic diameter (d_{ae}) for mouth breathing by a 'reference worker', using the equations suggested by James et al. (1994). Also shown is the curve describing a possible new criterion for the alveolar deposition fraction (open triangles) Adapted with permission from Vincent, Journal of Environmental Monitoring, 7, 1037–1053. Copyright (2005) Royal Society of Chemistry*

sampling instrument in the form of a multi-screen diffusion cell. For more general use, however, a form of expression consistent in form with the preceding criteria may be obtained using

$$A(d_{ae}) = I(d_{ae})[1 - F(x)]F(y) \qquad (11.27)$$

where $F(x)$ and $F(y)$ are again cumulative log-normal functions, this time with $\Gamma_x = 3.5$ μm and $\Sigma_x = 3.0$, and $\Gamma_x = 1.5$ μm and $\Sigma_y = 2.0$, respectively. The term $[1 - F(x)]F(y)$ now corresponds to the efficiency of alveolar deposition. It is shown in Figure 11.13 together with the curve calculated on the basis of the James *et al.* equations, again as applied to the 'reference worker'. The match is seen to be good.

Finally, if the suggested criteria for the tracheobronchial and alveolar fractions for the forms shown were to be considered for applications in a standard, previous practical experience for the thoracic and respirable fractions suggests that sampler instrumentation options will be available.

11.3.5   Criteria for very fine aerosol fractions

In the late 1990s it emerged that many of the health effects attributable to aerosols in the ambient atmospheric environment were closely associated with a fine particle fraction within PM_{10}. This fraction comprised particles having origins very distinctive from those of the coarse mode that make up the bulk of aerosol concentrations usually contained in the PM_{10} fraction. In particular, particles in the range of d_{ae} below about 2 μm are derived in large measure from combustion processes. The choice of parameters for a new fine-aerosol fraction for use in extended EPA NAAQS for atmospheric aerosol was based on the following considerations (Lippmann, 1999):

- Fine particles produce adverse health effects primarily on account of their chemical composition, and need to be regulated separately.
- Any separation between the accumulation and coarse modes, respectively, must be approximate because the modes overlap.

- The perceived separation point between the two modes will vary with aerosol composition and climate, and so will be regionally biased.
- The prevention of the intrusion of coarse mode into the desired finer fraction may be best achieved by specifying a relatively sharp cut-characteristic.

For regulatory purposes, EPA defined a $PM_{2.5}$ fraction, expressed in terms of the physical performance of a specific practical sampling reference sampler, the well impactor ninety-six (WINS) impactor-based instrument (e.g., Peters *et al.*, 2001). It was also specified that this impactor should be placed in a sampling line immediately downstream of an inlet that accurately samples the PM_{10} fraction accurately over a defined range of wind speeds. The corresponding particle size-selection curve is shown in Figure 11.10 alongside the other fractions already mentioned. Discrete values for $PM_{2.5}$ are likewise tabulated in Table 11.1. To be consistent with the way in which other sampling criteria have been expressed, a suitable definition satisfying these considerations would be:

$$PM_{2.5}(d_{ae}) = I(d_{ae})[1 - F(x)] \tag{11.28}$$

where $F(x)$ has $\Gamma_x = 2.5$ μm and $\Sigma_x = 1.15$. Here it is noted that the 'cut' for the $PM_{2.5}$ fraction is much sharper than for the thoracic and respirable fractions.

11.3.6 Criteria for fibrous aerosols

The inhalation of fine fibrous aerosols poses especially serious risks to health, especially for materials such as asbestos. Because of their distinctive morphological properties described elsewhere in this book, such particles are not specifically covered by any of the aerodynamically defined particle size-selective criteria described above. Particle size selection on the basis of particle aerodynamic diameter alone is not sufficient for such particles. Instead a new set of criteria is needed that takes into account both the aerodynamic properties of fibers that govern their regional deposition in the respiratory tract after inhalation *and* the biological effects and responses that follow deposition. As background to the development of criteria for fibers, it is also important to note that concentrations of fibers of materials such as asbestos are usually very low in practical situations. In addition, the fibers are invariably accompanied by much larger relative concentrations of nonfibrous particles. So gravimetric sampling, or any other assessment of aerosol mass concentration, for such particles is rarely a viable option.

Taking the preceding into account, it has become an almost-universal convention that fibers should be assessed in terms of their number concentration; also that they should be selected according to criteria that relate to shape and dimensions when fibers collected onto a filter are viewed – post-sampling – under an optical microscope. It is therefore important to note, therefore, that the concept of a 'respirable fiber' diverges sharply from the aerodynamic definition that has been described above for respirable aerosols more generally.

Particle size-selective criteria for the routine assessment of fibers depend inextricably on the microscopic methods that are used to visualise them. This is because differences in the techniques for sample preparation and microscope set-up, including magnification and type of light used (e.g. wavelength, state of polarization, etc.), can greatly influence the way in which micrometer-sized objects appear to the observer. For these reasons, therefore, the evolution of fiber selection criteria has taken place alongside the development of microscopy procedures. The phase contrast microscopy (PCM) method was developed to optimise the visibility of thin fibers close to the limits of observability under visible light. The PCM method involves the use of cellulose ester membrane filters (typically 0.45–1.2 μm pore size), mounting them on a glass slide, and 'clearing' them in an atmosphere of acetone vapor. The cleared filter is observed under the microscope under phase contrast illumination at a magnification of ×400 and with an appropriate graticule that permits accurate evaluation of individual fibers (e.g. Walton and Beckett, 1977).

The development of criteria and laboratory methods for the visual selection and counting of 'respirable' fibers by the PCM method has been led by the United States National Institute for Occupational Safety and Health (NIOSH) and the European-based Asbestos International Association (AIA), starting in the late 1970s and continuing into the 1990s (e.g. NIOSH, 1977; Asbestos International Association, 1979). There are slight differences between the two approaches as they have evolved. The criteria in the widely applied NIOSH 7400 method specified that a 'respirable' fiber should be taken to be one with length (l_f) greater than 5 μm and aspect ratio (l_f/d_f) greater than or equal to 3, where d_f is fiber width. The AIA criteria specified that, in addition to $l_f > 5$ and $l_f/d_f \geq 3$, d_f should be less than 3 μm. Other differences included how to deal with fibers that appeared to be in contact with nonfibrous particles, or with fibers that lay partially outside the graticule area, or with split fibers or fiber bundles. Such issues have been treated in successive refinements of the NIOSH and AIA methods. Similar methods and criteria have also been published by other bodies (e.g. the Australian National Health and Medical Research Council, the World Health Organisation, etc.).

The particle size-based criteria that are summarised above were intended initially for asbestos fibers. However, to complete what is needed for a standard, criteria are also needed to define the material of the fiber. The PCM method is not ideally suited to identifying asbestos fibers in the presence of other nonasbestiform fibers. Therefore in the application of the PCM method, it has become customary to assume that all fibers meeting the geometrical criteria *are*, in fact, asbestos. In any case, in recent years the health-related concerns about fine asbestos fibers have been extended to all fine fibers that are similarly durable in the lung after inhalation. Over the years, scanning and transmission electron microscope techniques have been developed that allow not only the observation of fibrous particles much smaller than is possible by optical microscopy but also their identification.

Finally, it is apparent that the criteria for fibers that have been described have less to do with the process of sampling (including aspiration) and more with the assessment of particles after they have been collected.

11.3.7   Criteria for ultrafine aerosols

There are at present no health-related particle size-selective criteria for ultrafine particles in the range of d_V below 0.1 μm. But, because of the interest stimulated by concerns about health effects in populations exposed to ultrafine particles in working and ambient atmospheric environments, current research is striving to provide the information needed in order to develop a standard. There are several issues that need to be addressed and integrated. These include the particle size-dependent deposition of ultrafine particles in the various regions of the respiratory tract, the particle size-dependent properties of the particles that make smaller particles apparently intrinsically more toxic than larger particles of the same material, the particle size-dependent rate at which deposited particles are dispersed around the body after exposure, and the relationship between these factors and health effects. The picture with respect to these is still far from complete. However, earlier in this chapter, the regional deposition of ultrafine particles – dominated by Brownian diffusion – was described. Most importantly, Figure 11.6 shows clearly that there are two significant target regions of the human respiratory tract for particles in this size range, the nasal passages (for nose breathing) and the alveolar region (for mouth breathing). As described later in Chapter 12, there are distinctly different, potentially serious, health effects thought to arise from deposition of ultrafine particles in these two regions. This knowledge provides a useful starting rationale, suggesting the need for *two* criteria for ultrafine particles, one relating to each primary target region.

For ultrafine particles, it is not possible to separate the criteria from the methods by which the sampled particles are assessed. Although, as described earlier, there are data for the particle size-dependent deposition efficiency of very small particles in the nose and alveolar region, we do not yet know the manner in which particle size influences the nature and level of the toxic response. Nor do we yet know the most appropriate metric for aerosol concentration. From inhalation toxicology (e.g. Oberdörster, 2000), mass would appear to be inappropriate. But, in any case, mass is not directly measurable in most practical situations because mass concentrations are usually likely to be very small for the particle size range of interest. A promising option might involve application of direct-reading instrumentation of the type that has already been applied for the evaluation of ultrafine aerosols in the field. For example, the differential mobility analyzer (DMA) has been used extensively in both ambient and workplace atmospheric environments to provide particle number size distributions for ultrafine particles. In the first instance, such information may be used, along with appropriate particle size-selective criteria for the nasal and alveolar deposition for such particles, together with assumptions about the relationship between particle number and other indices of exposure, to make realistic estimates of dose by whatever metric is deemed appropriate.

The experimental data for the regional deposition of very small particles whose motions are dominated by diffusion were embodied in the equations proposed by James *et al.*, as described in the earlier set of Equations (11.19) and in their application in calculating the various deposition fractions summarised in Figure 11.6. These may be used as the basis of criteria for the ultrafine nasal fraction, $UN(d_v)$, and ultrafine alveolar fraction, $UA(d_v)$, respectively. Expressed similarly to the previous ones, criteria for these ultrafine fractions are suggested in the forms:

$$UN(d_v) = 1.4[(1 - F(x)] \tag{11.29}$$

where $F(x)$ has $\Gamma_x = 0.002$ μm and $\Sigma_x = 7.0$, and

$$UA(d_v) = [1 - F(x)]F(y) \tag{11.30}$$

where $F(x)$ has $\Gamma_x = 0.002$ μm and $\Sigma_x = 2.5$, and $F(y)$ has $\Gamma_y = 0.03$ μm and $\Sigma_y = 4.0$, respectively. These functions are shown alongside the model calculations in Figure 11.14, and the matches are again seen to be good all the way down to $d_V = 0.001$ μm.

Following the rationale suggested, the preceding equations may be combined with the measured particle size distribution, $f(d_V)$, as obtained for example using a DMA, for the particle size range of interest to provide the particle size distributions of the respective doses, thus:

$$_{nasal}\text{Dose}\ (d_v) = f(d_v)UN(d_v) \text{ and } _{alveolar}\text{Dose}(d_v) = f(d_v)UA(d_v) \tag{11.29}$$

The way in which these criteria are described is notable in that there is no mention of their relation to the inhalable fraction. This is acceptable because the aspiration efficiency of the human head for such small particles will inevitably be close to unity and so be independent of particle size. Two problems remain, however. The first is that it is still not possible to define the particle size range for the application of these models. At present, 'ultrafine' is described notionally as comprising particles with diameter less than 0.1 μm. There is at present no toxicological or physical basis for this. The second applies specifically to the ultrafine nasal criterion. As described earlier for larger particles in the aerodynamic regime, it is necessary to identify the extent to which breathing takes place through the nose in any given practical situation. How might this be written into a standard in order to protect people from nasal exposures that might relate to specific serious health effects?

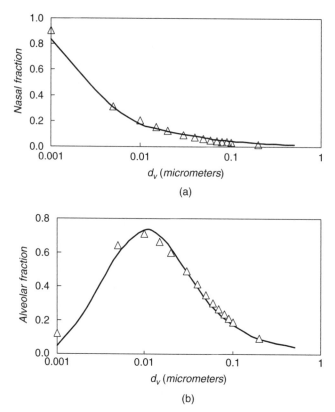

Figure 11.14   *Calculated results for deposition efficiency for ultrafine particles as a function of particle equivalent volume diameter (d_V), using the equations proposed by James et al. (1994): (a) nasal deposition; (b) alveolar deposition (mouth breathing). Also shown are the corresponding curves for the proposed new criteria for the nasal and alveolar deposition fractions (open triangles) Adapted with permission from Vincent, Journal of Environmental Monitoring, 7, 1037–1053. Copyright (2005) Royal Society of Chemistry*

11.4   Overview

11.4.1   Summary

A set of criteria for fine aerosol fractions relevant to penetration to and deposition in the various regions of the human respiratory tract after inhalation has been described in preceding sections. Taken together these form an overall framework for health-related aerosol sampling for which the philosophy for each particle size-selective criterion has been: (a) it should embody recognition of the various processes by which particles eventually come into contact with the human respiratory system; (b) it reflects the wide range of people and environmental situations that pertain to the standard in which it will be used, taking into account worst-case scenarios where appropriate; (c) it provides a realistic basis for practical sampling instrumentation; and (d) it is defined in terms of a consistent mathematical form where it is known that options for sampling instrumentation can be identified. The criteria that have been adopted by the occupational and environmental hygiene community and many regulatory bodies around the world for providing the modern basis for aerosol exposure standards are summarised in Table 11.2.

Table 11.2 *Summary of complete adopted particle size-selective sampling criteria. Function $F(x)$ is a cumulative log-normal function with median Γ_x and geometric standard deviation Σ_x*

Fraction	Term	Rationale	Function	Variables
Inhalable (see Chapter 10)	$I(d_{ae})$	Penetration into the nose and/or mouth in moving air environments	$0.5[1 + \exp(-0.06d_{ae})]$ $+10^{-5}U^{2.75}\exp(0.055d_{ae})$	$d_{ae} \leq 100\ \mu m$ $0.25 < U < 9\ m\,s^{-1}$
Thoracic	$T(d_{ae})$	Penetration into the lung	$I(d_{ae})[1 - F(x)]$	$\Gamma x = 11.64\ \mu m$ $\Sigma_x = 1.5$
PM_{10}	PM_{10}	Penetration into the lung	See curve in Figure 11.12	Curve passes through 0.5 at $d_{ae} = 10\ \mu m$
Respirable	$R(d_{ae})$	Penetration into the alveolar region	$I(d_{ae})[1 - F(x)]$	$\Gamma x = 4.25\ \mu m$ $\Sigma_x = 1.5$
$PM_{2.5}$	$PM_{2.5}$	Particles generated during combustion or hot processes	$[1 - F(x)]$	$\Gamma x = 2.5\ \mu m$ $\Sigma_x = 1.15$
Fine fibers	'Respirable' fibers	Particles fine enough to penetrate to the alveolar region and exhibit toxic behavior there	NIOSH $l_f > 5\ \mu m$ $l_f/d_f \geq 3$ AIA $l_f > 5\ \mu m$ $d_f < 3\ \mu m$ $l_f/d_f \geq 3$	Particle sizes as determined in terms of geometrical dimensions by optical microscopy

Table 11.3 contains some of the suggested new criteria that have not yet been widely discussed, but which might point towards the development of criteria and standards for aerosol fractions that might in the future be applied in certain special situations.

11.4.2 Precision and tolerance bands

Differences between conventions aimed at sampling for essentially the same fraction can produce significant differences when they are applied in the real world. For example, the thoracic fraction and the PM_{10} criteria are intended to represent essentially the same fraction in relation to the same range of health effects. But, as seen in Figure 11.10, the curves differ slightly, in particular for larger particle sizes. Here, the thoracic fraction allows for the sampling of particles with finite collection efficiency for d_{ae} up to about 25 μm, whereas the PM_{10} curve cuts off at around 15 μm. It therefore clear that, for a hypothetical pair of samplers that exactly follow the respective conventions, the one designed for the thoracic fraction will collect more than the one designed for PM_{10}. Further, the magnitude of the difference will be greater for coarser aerosol.

More generally, similar differences will occur for: (a) a sampler designed to match a single given criterion but does not exactly follow the target performance curve, in which case the collected sample will be consistently biased with respect to the ideal; and (b) a pair of samplers designed to match a single given criterion but where their performances are not identical, in which case they will collect different amounts of aerosol. The latter was first discussed by Bartley and Doemeny (1986), with particular

Table 11.3 *Summary of some suggested additional particle size-selective sampling criteria. Function F(x) is a cumulative log-normal function with median Γ_x and geometric standard deviation Σ_x, F(y) a cumulative log-normal function with median Γ_y and geometric standard deviation Σ_y*

Fraction	Term	Rationale	Function	Variables	
Inhalable (see Chapter 10)	$I(d_{ae})$	Penetration into the nose and/or mouth in calm air environments	$1 - 0.0038 d_{ae}$	$d_{ae} \leq 100\ \mu m$ $U < 0.25\ ms^{-1}$	
Nasal	$N(d_{ae})$	Deposition in the nasal passages	$I(d_{ae})F(x)$	$\Gamma_x = 3.0\ \mu m$ $\Sigma_x = 2.5$	
Tracheobronchial	$B(d_{ae})$	Deposition in the conducting airways	$I(d_{ae})[1 - F(x)]F(y)$	$\Gamma_x = 7.0\ \mu m$ $\Sigma_x = 2.0$	$\Gamma_y = 3.5\ \mu m$ $\Sigma_y = 1.5$
Alveolar	$A(d_{ae})$	Deposition in the alveolar region	$I(d_{ae})[1 - F(x)]F(y)$	$\Gamma_x = 3.5\ \mu m$ $\Sigma_x = 3.0$	$\Gamma_x = 1.5\ \mu m$ $\Sigma_x = 2.0$
Ultrafine nasal	$UN(d_V)$	Ultrafine particles deposited in the nasal passages	$1.4[1 - F(x)]$	$\Gamma_x = 0.002\ \mu m$ $\Sigma_x = 7.0$	
Ultrafine alveolar	$UA(d_V)$	Ultrafine particles deposited in the alveolar region	$[1 - F(x)]F(y)$	$\Gamma_x = 0.002\ \mu m$ $\Sigma_x = 2.5$	$\Gamma_y = 0.03\ \mu m$ $\Sigma_y = 4.0$

reference to the earlier ACGIH criterion for the respirable fraction. They analyzed numerically the possible excursions in respirable aerosol concentration as might be measured by hypothetical samplers with performances falling within defined tolerance bands for aerosols with particle size distributions typical of many practical occupational situations. They demonstrated that so-called 'acceptable' samplers may yield mass concentrations that vary – depending on the particle size distribution – by as much as a factor of 6!

Such biases are recognized in the implementation of criteria like those described at the point where a specific sampler is chosen to do the job. For this, test methods are being developed by which sampler performance may be assessed relative to given sampling criteria. In its proposed sampler performance protocol, CEN describes procedures in which not only is the collection efficiency of a sampler as close as possible to the curve identified in the target criterion, but also the collected particulate mass must fall within $\pm 10\%$ of that which would be collected by a hypothetical ideal sampler that follows the curve perfectly (Comité Européen de Normalisation, 2001).

11.4.3 International harmonisation of sampling criteria

Initially there were wide differences in the criteria proposed by different bodies. This was especially true for the respirable fraction which attracted so much attention during the early years of occupational and environmental hygiene. Some differences still remain. However, considerable progress has been made towards the international harmonisation of the main set of criteria for the inhalable, thoracic and respirable fractions for applications in occupational health standards, involving ISO, CEN and ACGIH (Soderholm, 1989).

References

American Conference of Governmental Industrial Hygienists (1968) Threshold limit values of airborne contaminants, ACGIH, Cincinnati, OH.

Altshuler, B, Yarmus, L., Palmes, E.D. and Nelson, N. (1957) Aerosol deposition in the human respiratory tract, I. Experimental procedures and deposition, *American Medical Association Archives of Industrial Health*, 15, 293–303.

Asbestos International Association (1979) *Airborne asbestos fiber concentrations at workplaces by light microscopy (membrane filter method)*, AIA Health and Safety Publication RTM1, AIA, Paris.

Asgharian, B. and Yu, C.P. (1989) Deposition of fibers in the rat lung, *Journal of Aerosol Science*, 20, 355–366.

Bartley, D.L. and Doemeny, L.J. (1986) Critique of the 1985 ACGIH report on particle size-selective sampling in the workplace, *American Industrial Hygiene Association Journal*, 47, 443–447.

Bates, D.V., Fish, B.R., Hatch, T.F., Mercer, T.T. and Morrow, P.E. (1966) Deposition and retention models for internal dosimetry of the human respiratory tract, Report of the International Commission on Radiological Protection, ICRP, Task Group on Lung Dynamics, *Health Physics*, 12, 173–207.

British Medical Research Council (1952) Recommendations of the BMRC panels relating to selective sampling, from the minutes of a joint meeting of Panels 1,2 and 3 held on March 4, 1952.

Chan, T.L. and Lippmann, M. (1980) Experimental measurements and empirical modelling of the regional deposition of inhaled particles in humans, *American Industrial Hygiene Association Journal*, 41, 399–409.

Cheng, K.-H, Cheng, Y.-S., Yeh, H.-C., Guilmette, R.A., Simpson, S.Q., Yang, Y.-H. and Swift, D.L. (1996) In vivo measurements of nasal airway dimensions and ultrafine aerosol deposition in the human nasal and oral airways, *Journal of Aerosol Science*, 27, 785–801.

Cheng, Y.-S., Yamada, Y., Yeh, H.-C. and Swift, D.L. (1988) Diffusional deposition of ultrafine aerosols in a human nasal cast, *Journal of Aerosol Science*, 19, 741–751.

Comité Européen de Normalisation (1992) *Workplace atmospheres: size fraction definitions for measurement of airborne particles in the workplace*, European Standard EN 481, CEN, Brussels, Belgium.

Comité Européen de Normalisation (2001) *Workplace atmospheres – assessment of performance of instruments for measurement of airborne particle concentrations*, European Standard EN 13205, CEN, Brussels, Belgium.

Cox, R.G. (1970) The motion of long slender bodies in a viscous fluid: Part 1, General theory, *Journal of Fluid Mechanics*, 44, 791–810.

Davies, C.N. (1952) Dust sampling and lung disease, *British Journal of Industrial Medicine*, 9, 120–126.

Emmett, P.C. and Aitken, R.J. (1982) Measurements of the total and regional deposition of inhaled particles in the human respiratory tract, *Journal of Aerosol Science*, 13, 549–560.

Environmental Protection Agency (1984) National ambient air quality standard, proposed rule, *Federal Register*, 49(55), 10408–10462.

Environmental Protection Agency (1996a) *Air Quality Criteria for Particulate Matter*, EPA/600/P-95/001, Washington, DC.

Environmental Protection Agency (1996b) *Review of the National Ambient Air Quality Standards for Particulate Matter*, Office of Air Quality Planning and Standards (OAQPS) Staff Paper, EPA-452/R-96-013, Research Triangle Park, NC.

Foord, N., Black, A. and Walsh, M. (1976) Regional deposition of 2.5–7.5 μm diameter inhaled particles in healthy male non-smokers, ML. 76/2892. Atomic Energy Research Establishment, Harwell.

Gentry, J.W., Spurny, K.R. and Schörmann, J. (1991) The diffusion coefficients for ultrathin chrysotile fibers, *Journal of Aerosol Science*, 22, 869–880.

George, A.C. and Breslin, A.J. (1967) Deposition of natural radon daughters in human subjects, *Health Physics*, 13, 375–378.

Giacomelli-Maltoni, G., Melandri, C., Prodi, V. and Tarroni, G. (1972) Deposition efficiency of monodisperse particles in the human respiratory tract, *American Industrial Hygiene Association Journal*, 33, 603–610.

Gradon, L. and Podgorski, A. (1999) Deposition and retention of fine and ultrafine particles in the pulmonary region of human lungs: normal and pathological cases, *Journal of Aerosol Science*, 30, S801–S802.

Hatch, T.E. and Gross, P. (1964) *Pulmonary Deposition and Retention of Inhaled Aerosols*, Academic Press, New York.

Heyder, J. and Rudolf, G. (1977) Deposition of aerosol particles in the human nose, *Inhaled Particles IV* (Ed. W.H. Walton), Pergamon Press, Oxford pp. 107–125.

Heyder, J., Armbruster, L., Gebhart, J., Grein, E. and Stahlhofen, W. (1975) Total deposition of aerosol particles in the human respiratory tract for nose and mouth breathing, *Journal of Aerosol Science*, 6,311–328.

Heyder, J., Gebhart, J., Rudolf, G., Schiller, C.F. and Stahlhofen, W. (1986) Deposition of particles in the human respiratory tract in the size range 0.005–15 μm, *Journal of Aerosol Science*, 17, 811–825.

Hofmann, W. (1996) Modeling techniques for inhaled particle deposition: the state of the art, *Journal of Aerosols in Medicine*, 9, 369–388.

Hofmann, W., Asgharian, B. and Winkler-Heil, R. (2002) Modeling intersubject variability of particle deposition in human lungs, *Journal of Aerosol Science*, 33, 219–235.

Hounam, R.F., Black, A. and Walsh, M. (1971) The deposition of aerosol particles in the nasopharyngeal region of the human respiratory tract, In: *Inhaled Particles III* (Ed. W.H. Walton), Unwin Brothers Ltd, Surrey, pp. 71–80.

International Commission on Radiological Protection (1994) *Human Respiratory Tract Model for Radiological Protection*, ICRP Publication 66, Pergamon Press, Elmsford, NY.

International Standards Organization (1981) Recommendation on size definitions for particle sampling, *American Industrial Hygiene Association Journal*, 42, A64–A68.

International Standards Organisation (1983) Air quality – particle size fraction definitions for health-related sampling, Technical Report ISO/TR/7708-1983 (E), ISO, Geneva.

International Standards Organisation (1995) *Air quality – particle size fraction definitions for health-related sampling*, ISO Standard ISO 7708, ISO, Geneva.

James A.C., Stahlhofen, W., Rudolf, G., Köbrich, R., Briant, J.K., Egan, M.J., Nixon, W. and Birchall, A. (1994) Deposition of inhaled particles. In: *Human Respiratory Tract Model for Radiological Protection*, Annex D, *Annals of the ICRP*, Pergamon Press, Oxford, pp. 231–299.

Lippmann, M. (1977) Regional deposition of particles in the human respiratory tract. In: *Handbook of Physiology, Section 9, Reactions to Environmental Agents* (Eds D.H.K. Lee, H.L. Falk and S.D. Murphy), American Physiological Society, Bethesda, MD, pp. 213–232.

Lippmann, M. (1999) Sampling criteria for fine fractions of ambient air, In: *Particle Size-selective Sampling for Particulate Air Contaminants* (Ed. J.H. Vincent), American Conference of Governmental Industrial Hygienists, Cincinnati, OH, pp. 97–118.

Lippmann, M. and Albert, R. (1969) The effect of particle size on the regional deposition of inhaled aerosols in the human respiratory tract, *American Industrial Hygiene Association Journal*, 30, 257–275.

Lippmann, M. and Harris, W.B. (1962) Size-selective samplers for estimating 'respirable' dust concentrations, *Health Physics*, 8, 155–163.

McCawley, M.A. (1999) Particle size-selective criteria for deposited submicrometer particles, In: *Particle Size-Selective Sampling for Particulate Air Contaminants* (Ed. J.H. Vincent), American Conference of Governmental Industrial Hygienists, Cincinnati, OH.

McCrea, J. (1913) *The Ash of Silicotic Lungs*, Publication of the South African Institute of Medical Research, No. 1, Johannesburg, p. 6.

Miller, F.J., Gardner, J.A., Graham, J.A., Lee, R.E., Willson, W.E. and Bachmann, J.D. (1979) Size considerations for establishing a standard for inhalable particles, *Journal of the Air Pollution Control Association*, 29, 610–615.

National Institute for Occupational Safety and Health (1975) *Criteria for a Recommended Standard – Occupational Exposure To Cotton Dust*, NIOSH Publication 75–118, Washington, DC.

National Institute for Occupational Safety and Health (1977) *Manual of Analytical Methods*, 2nd Edn, Vol. 1, P&CAM 239, US Department of Health, Education and Welfare, NIOSH Publication 77-157-A, Washington DC.

Nemmar, A., Hoet, P.H.M., Vanquickenborne, B., Dinsdale, D., Thomeer, M., Hoylaerts, M.F., Vanbilloen, H., Mortelmans, L. and Nemery, B. (2002) Passage of inhaled particles into the blood circulation in humans, *Circulation*, 105, 411.

Oberdörster, G. (2000) Toxicology of ultrafine particles: in vivo studies, *Philosophical Transactions of the Royal Society of London, Series A*, 358, 2719–2740.

Orenstein, A.J. (Ed.) (1960) Recommendations adopted by the Pneumoconiosis Conference. In: *Proceedings of the Pneumoconiosis Conference*, Churchill, London, pp. 619–621.

Peters, T.M., Vanderpool, R.W. and Wiener, R.W. (2001) Design and calibration of the EPA PM$_{2.5}$ Well Impactor Ninety-Six (WINS), *Aerosol Science and Technology*, 34, 389–397.

Phalen, R.F. (Ed.) (1985) Particle size-selective sampling in the workplace, Report of the ACGIH Air Sampling Procedures Committee, American Conference of Governmental Industrial Hygienists, Cincinnati, OH.

Podgorski, A. and Gradon, L. (1990) Motion and deposition of fibrous flexible particles in laminar gas flow through a pipe, *Journal of Aerosol Science*, 21, 957–968.

Rudolf, G., Gebhart, J., Heyder, J., Scheuch, G. and Stahlhofen, W. (1986) An empirical formula describing aerosol deposition in man for any particle size, *Journal of Aerosol Science*, 17, 350–355.

Schiller, C.F., Gebhart. J., Heyder, J., Rudolf, G. and Stahlhofen, W. (1988) Deposition of monodisperse insoluble particles in the 0.005 to 0.2 μm size range within the human respiratory tract, *Annals of Occupational Hygiene*, 32 (Supp. 1), 41–49.

Shanty, F. (1974) Deposition of ultrafine aerosols in the respiratory tract of human volunteers, PhD Thesis, School of Hygiene and Public Health, The Johns Hopkins University, Baltimore, MD.

Soderholm, S.C. (1989) Proposed international conventions for particle size-selective sampling, *Annals of Occupational Hygiene*, 33, 301–320.

Stahlhofen, W., Gebhart, J. and Heyder, J. (1980) Experimental determination of the regional deposition of aerosol particles in the human respiratory tract, *American Industrial Hygiene Association Journal*, 41, 385–398.

Stahlhofen, W., Rudolph, G. and James, A.C. (1989) Intercomparison of experimental regional aerosol deposition data, *Journal of Aerosols in Medicine*, 2, 285–308.

Stöber, W. (1971) A note on the aerodynamic diameter and mobility of non-spherical aerosol particles, *Journal of Aerosol Science*, 2, 453–456.

Swift, D.L. and Strong, J.C. (1996) Nasal deposition of ultrafine [218]Po aerosols in human subjects, *Journal of Aerosol Science*, 27, 1125–1132.

Stuke, J., and Emmerichs, M. (1973) Das gravimetrische Staubprobennahmegerat TBF50, *Silikoebericht Nordrhein-Westfalen*, 9, 47–51.

Taulbee, D.B. and Yu, C.P. (1975) A theory of aerosol deposition in the human respiratory tract, *Journal of Applied Physiology*, 38, 77–85.

Vincent, J.H. (1989) *Aerosol Sampling: Science and Practice*, John Wiley & Sons, Ltd,, Chichester.

Vincent, J.H. (Ed.) (1999) *Particle Size-selective Sampling for Particulate Air Contaminants*, American Conference of Government Industrial Hygienists, Cincinnati, OH.

Vincent, J.H. (2005) Health-related aerosol measurement: a review of existing sampling criteria and proposals for new ones, *Journal of Environmental Monitoring*, 7, 1037–1053.

Walton, W.H. (1954) Theory of size classification of airborne dust clouds by elutriation, *British Journal of Applied Physics*, 5 (Suppl. 3), S29–S40.

Walton, W.H. and Beckett, S. T. (1977) A microscope eyepiece graticule for the evaluation of fibrous dusts, *Annals of Occupational Hygiene*, 20, 19–23.

Weibel, E.R. (1963) *Morphometry of the Human Lung*, Springer, Berlin.

Yu, C.P. (1985) Theories of electrostatic lung deposition of inhaled particles, *Annals of Occupational Hygiene*, 29, 219–227.

Zhang, Z., Kleinstreuer, C., Donohoe, J.F. and Kim, C.S. (2005) Comparison of micro- and nano-size particle deposition in a human upper airway model, *Journal of Aerosol Science*, 36, 211–233.

Zhou, Y., Podgorski, A., Marijnissen, J.C.M., Lemkowtz, S.M. and Bibo, H.B. (1996) Deposition model for fibers in the deep parts of the lung, *Journal of Aerosol Science*, 27, (Suppl. 1), S491–S492.

12

Health Effects and Limit Values

12.1 Introduction

To reiterate, particle size-selective, health-related sampling has two primary aims: the first to provide a scientific basis for biologically relevant aerosol measurement for epidemiological research leading to the setting of meaningful exposure limits, and the second to underpin routine aerosol measurement for the purpose of monitoring and controlling environments so that people may be protected from the effects of exposure. In the first, one endpoint is the development of the exposure limit itself. In the second, the exposure limit is the quantitative reference point for exposure assessment and monitoring. Here a summary is given of the development of exposure limits, both historically and then in the context of the modern framework of the particle size-selective criteria described in the preceding chapters.

12.2 Aerosol-related health effects

As described earlier in this book, the human respiratory tract may be broken down into three main regions: the *extrathoracic region*, including the *nasopharynx* and *mouth* in the head; the *tracheobronchial region*, including the trachea, main bronchi and bronchi of the conducting airways; and the *alveolar region*, including the bronchioles, alveolar ducts and alveolar sacs. The penetration of particles into the respiratory tract, and hence the aerosol exposures of these regions by particle deposition, has been shown to be a strong function of particle size. Once deposited, particles may be cleared by various mechanisms at rates that are characteristic of the site of deposition. Accordingly, after inhaled particles have deposited on a surface of the respiratory tract, they may either express their toxicity at the site of deposition, or may be translocated to other parts of the body and be influential there. Either way, consideration of the original site of deposition is important to the establishment of a limit value based on particle size. This identifies the interface between the environment and the subject by which exposure is defined, and in turn points to the most appropriate criterion for exposure assessment.

The first step in the process of setting a standard requires identification of the disease that is associated with particle inhalation and the anatomical location where the pathological process is initiated. Diseases are usually diagnosed on the basis of symptoms reported by patients to medical professionals or through clinical and pathological examinations. The anatomical locations of pathological lesions can frequently

Aerosol Sampling: Science, Standards, Instrumentation and Applications   James Vincent
© 2007 John Wiley & Sons, Ltd

be determined from signs, symptoms and tests, but in some cases only biopsy observations can actually locate specific sites. In many parts of the world, clinical diagnoses are coded in hospital records and on death certificates in accordance with the International Classification of Diseases (ICD) of the World Health Organisation (WHO, 2003), and such records provide an important source of information on which to base standards development.

12.2.1 Diseases of the respiratory tract

Table 12.1 lists some noninfectious, noncancerous diseases that are listed for regions of the respiratory tract that are known to be associated with aerosol exposure. The table also suggests the most appropriate

Table 12.1 *Some health conditions associated with aerosol exposure (see also World Health Organisation, 2003), listed together with the part of the respiratory tract affected for each and the particle size-selective (PSS) criteria that might be most appropriate to each. Note: Criteria in the right-hand column marked with an asterisk are suggested criteria, not yet adopted by any of the standards-setting bodies referred to in this book*

Disease	Region of deposition	PSS criterion
Acute rhinitis Chronic rhinitis Chronic pharyngitis Chronic sinusitis Allergic rhinitis	Extrathoracic – nasal breathing	Inhalable, $I(d_{ae})$ Nasal*, $N(d_{ae})$
Chronic laryngitis Chronic larnygotracheitis	Extrathoracic – mouth breathing	Inhalable, $I(d_{ae})$
Bronchitis, not specific as acute or chronic Chronic bronchitis Asthma Chronic airway obstruction, not elsewhere classified	Tracheobronchial	Thoracic, $T(d_{ae})$ PM_{10} Tracheobronchial*, $B(d_{ae})$
Extrinsic allergic alveolitis Coalworkers' pneumoconiosis Pneumoconiosis due to other silica and silicates Pneumoconiosis due to other inorganic dust Pneumopathy due to inhalation of other dust Emphysema Chemical pulmonary edema Extrinsic allergic alveolitis	Alveolar	Respirable, $R(d_{ae})$ Alveolar*, $A(d_{ae})$
Asbestosis Mesothelioma	Alveolar	'Respirable' fibers as defined by the PCM method

sampling criteria for the most relevant exposure assessment. The original ICD classifications identify the general anatomical regions of the respiratory tract associated with the listed diagnoses, although these do not always correspond to the anatomical regions that have been defined in this book for the purpose of aerosol sampling. They are therefore modified here to relate to the specific particle size-selective criteria described, including not only the ones that are already on the books but also some of the additional ones introduced in Chapters 10 and 11 that have not yet been widely discussed or adopted. Asbestosis, a serious form of pneumoconiosis associated with exposure to fine, durable asbestiform fibers is also listed.

In addition to the diseases listed in Table 12.1, cancers of the respiratory tract are a significant class of aerosol-related disease and relate quite specifically to the regions of the respiratory tract where particles are deposited or accumulate during clearance. For these it is an option to indicate criteria that reflect this reality. However, for substances that are carcinogenic, it is generally considered that all inhaled particles have the potential to be very hazardous *wherever* they are deposited. Therefore the inhalable fraction is usually adopted for such substances.

In the simplest view of aerosol-related disease it is assumed that, when there is a sufficient dose of toxic particles at a given type of surface in the respiratory tract, it may initiate a range of biological responses leading to a disease outcome specifically at that surface. This is true for most forms of the pneumoconioses, and for many aerosol-related diseases of the extrathoracic and tracheobronchial regions. But it is not always the case. For example, insoluble particles which are deposited in the alveolar region may be captured by alveolar macrophages which in turn migrate into the bronchioles and other proximal parts of the pulmonary system where disease processes can develop in the respiratory tract remote from the original site of deposition.

12.2.2   Diseases beyond the respiratory tract

For soluble particles, dissolution in pulmonary fluids may occur through physico-chemical reactions so that toxic molecules may enter the systemic circulation. One well-known example comes from exposure to aerosols containing soluble species of lead, where inhalation may lead to serious diseases of the central nervous system. Such exposures are still widely experienced in both occupational and ambient atmospheric environments. Another example is exposure to particles containing soluble cadmium species, leading to kidney damage and prostate cancer; another to soluble nickel species, leading to effects in the cardiovascular system, kidneys and central nervous system. Here, in the first instance, since the particles are available for dissolution after deposition at any point in the respiratory tract, albeit at different regional rates, the inhalable fraction represents the most appropriate criterion for a standard. However, for soluble substances like the examples mentioned, additional exposure assessment metrics may also be available in the form of biological indices – for example, lead in blood, protein products in urine directly associated with cadmium aerosol inhalation, and nickel in urine. For such substances, biological exposure indices (BEIs) represent important complementary tools, to be applied alongside air sampling for overall exposure assessment and hazard evaluation.

Aerosols of biological origin, including bacteria, viruses, spores and other infectious or antigenic proteins, usually act through immunological responses, which too can be remote from the site of deposition. Micro-organisms such as bacteria, many of which are as small as $1-2$ μm, and viruses, which are even smaller at $0.01-1$ μm, can penetrate quite readily to the alveolar region, and may either act directly at the site of deposition or elicit antibodies that enter the systemic circulation, leading to responses elsewhere in the body. Fungal spores are generally larger, typically from 3 to 30 μm, and most of those inhaled are captured largely in the extrathoracic region where they are commonly linked with allergic rhinitis. Even some of these have the potential to be associated with conditions elsewhere (e.g. asthma) through systemic responses.

In short, from the above, it should not necessarily *always* be assumed that health effects are experienced in the same region of the respiratory tract where the particles are deposited. This means that the choice of a sampling criterion for particle size-selective sampling, and ultimately a limit value in relation to a given aerosol-related health effect, needs to be carried after due consideration of all the available physical, chemical, toxicological and medical information.

12.3   The processes of standards setting

The primary core of scientific knowledge for all environmental standards setting is that which is available in the open literature, containing information linking exposure with health effects as assessed in toxicological and epidemiological research. This is a common thread linking most of the world's standards setting bodies in the modern era, where there is general mutual access to information and awareness of each other's activities, priorities and thought processes, along with considerable interchange of information about the data and their application. The published exposure and health data provide a common starting point for the discussions within the various expert working groups involved in setting exposure limits. The basic process requires drawing together and distilling the relevant exposure and health-related data in a way that enables construction of dose–response information by which to determine *how much* exposure leads to *what prevalence* of a particular health outcome. As already discussed in Chapter 9, this is a multi- and interdisciplinary problem, and many scientific and nonscientific facets come into play: the quantity, quality and sources of the exposure and health data; the choice of health endpoint; how to deal with mixed exposures; the consideration of what prevalence of disease may be acceptable; national legal and cultural frameworks; the role of scientists and other experts; the dynamics of the interactions between scientists from different disciplines; moral and ethical issues; the international dimension; priorities for choice of substances – or types of aerosols – to be considered for inclusion, and so on. The difficulties are compounded when limit values obtained based purely on health considerations are adjusted to take into account considerations such as feasibility and cost benefit, as is the case for many regulatory standards setting bodies. It is therefore inevitable that simplifications and compromises are made along the way, simply to enable progress to be made. The process is highly complicated and interesting, but more detailed discussion of the dynamics of standards setting belongs in a different type of book (e.g. Funtowicz and Ravetz, 1990).

12.4   Occupational exposure limits (OELs)

Over the years, many tables have been published of airborne concentrations of toxic substances in workplaces that were considered safe. One of the earliest was the list published by Rudolf Kobert in 1912 that provided information on 20 substances falling into four categories: (a) rapidly fatal to man and animals; (b) dangerous in 0.5–1 h; (c) exposure may be experienced for 0.5–1 h without serious harm; and (d) only minimal symptoms observed after several hours. Sayers and Dallavalle (1935) later published a list of recommended exposure limits for 38 substances based on reports in the international toxicological literature. An expanded list was developed in 1937 from recommendations by the Massachusetts Division of Occupational Hygiene and faculty members at the Harvard School of Public Health and Yale University, respectively (Bowditch *et al.*, 1940; Bowditch, 1944). At the fifth annual meeting of the American Conference of Governmental Industrial Hygienists (ACGIH)[1]

---

[1] At that time, ACGIH was referred to as the National Conference of Governmental Industrial Hygienists.

in 1942, a subcommittee on limit values presented a table of 'maximum permissible concentrations of atmospheric contaminants' in the form of a list compiled by state industrial hygienists. In 1946 the same subcommittee presented an enlarged list (American Conference of Governmental Industrial Hygienists, 1946) which included values from the 1942 report, together with new limits suggested by Cook (1945) and values published by the American Standards Association.[2] The list has been republished – and regularly expanded and updated – each year ever since. In 1948 these numbers became known for the first time as 'threshold limit values' (or TLVs). Currently, in 2006, the list of ACGIH TLVs extends to over 600 chemical compounds, of which about 300 are experienced in workplaces in the form of aerosols. With the exception of fibrous aerosols (e.g. asbestos), such limits are currently almost universally expressed in terms of the mass of particulate matter per unit volume of air (e.g. mg m^{-3}).

12.4.1   Health-based exposure limits

The ACGIH TLVs are health-based OELs defined as airborne concentrations of substances to which workers may be repeatedly exposed day after day without adverse health effects that might shorten life expectancy, damage physiological function, impair the capability for resisting other toxic substances or disease processes, adversely affect reproductive function or development processes, or – more generally – cause irritation, narcosis, nuisance, or other forms of stress. As discussed earlier in Chapter 9, the aim is that such limit values should protect '... nearly all workers', acknowledging that workplace exposures and peoples' responses to them are very variable, so much so that the definition in any given exposure situation of a specific concentration level below which no harm will occur, and which might *never* be exceeded, is not achievable and so has little meaning. The ACGIH TLVs may be spoken of as 'scientific OELs', developed on the basis solely of scientific information from studies of human populations and/or animals, and without regard to technical or economic feasibility that such levels of exposure can be achieved in practice.

The preceding short historical overview focused on developments that took place mainly in the USA. This was indeed very significant because the original development of the ACGIH TLVs, as well as corresponding regulatory limit values, coincided with the evolution in the USA of the (then) new professional field of *industrial hygiene*, later to become known around the world as 'occupational hygiene'. Significant developments took place even earlier in some other countries. In Germany, for example, the first proposals for OELs were made by Lehmann as early as 1886. The list published by the Maximale Arbeitsplatzkonzentration (MAK) Commission first came into existence in the 1940s and currently lists over 500 substances. For these, the underlying philosophy is similar to that of ACGIH, in particular that they are set without specific regard to technical or economic feasibility. In the former Russian Federation, OELs have a similarly long history. Here, however, limit values were developed from the strict standpoint that *no* effects should arise from exposure, and so were generally significantly lower than OELs elsewhere for similar substances.

12.4.2   Regulatory exposure limits

Scientific OELs like those described have been highly influential in the development of limit values applied within regulatory standards. In the case of the ACGIH TLVs, their initial intention was – and still is – to provide guidance to professional occupational hygienists in their day-to-day mission of protecting workers. At various times, parts of the TLV and German MAK lists have found their way into the lists of limit values in regulatory standards in countries and jurisdictions around the world.

---

[2] Now the American National Standards Institute (ANSI).

For example, in the USA, the ACGIH TLVs were incorporated in full into the Occupational Health and Safety Act when it was first launched by the US Congress in 1970. The OELs published by the United States Occupational Health and Safety Administration (OSHA) are referred to as 'permissible exposure limits' (PELs). Since their original introduction it is fair to say that they have not kept up with subsequent changes in the ACGIH TLVs.

In many countries, government agencies promote worker health by promulgating legally binding OELs. Depending on the laws of the particular country, some may be treated as absolute limits not to be exceeded on any single day. Government inspectors may evaluate exposures in a given work-place, and if excessive exposures are found – by whatever statistical criterion is deemed appropriate for defining *exceedance* – the employer may be subject to legal sanctions varying from monetary fines to plant closure. Regulatory OELs are usually incorporated into more inclusive laws and guidelines that include protective measures besides controlling airborne exposure levels, including medical surveillance, personal protective equipment, work practices, and routine exposure monitoring (Ratney, 1999).

12.4.3   OELs for aerosols

Nowadays, mass concentration is the most widely used exposure metric for OELs for aerosols encoun-tered occupationally. In the OEL development process, since most of the standards-setting bodies around the world have access to essentially the same such information, the OEL values themselves tend not to vary greatly from one body to the next, unless there is an overarching policy directive that drives the processes to greater or less stringency. By way of illustration, it is instructive to compare OELs for two contrasting standards-setting systems, those of ACGIH and the UK Health and Safety Executive (HSE), respectively. ACGIH specifies 8 h (workday) time-weighted average (TWA) exposure concen-tration levels in the form of TLVs, and these are not intended for use in regulatory standards. The British exposure limits published by the UK HSE *are* intended for use in regulatory standards, although they are, in the first instance, philosophically similar to the ACGIH TLVs. For a given substance, HSE first establishes an *occupational exposure standard* (OES) which, like the corresponding ACGIH TLV, is strictly health based. The HSE OES is applied in a regulatory framework provided that it can reasonably be complied with by industry. Otherwise HSE proceeds to the development of a higher *maximum expo-sure limit* (MEL) that embodies considerations of technical feasibility and socioeconomic impact. In the application of this limit value it is expressly stated that industry should strive to reduce its workers' exposures as far as is reasonably practicable below the MEL. In addition to 8 h TWA limit values, both ACGIH and HSE recommend 15 min TWA *short-term exposure limits* (STELs) for some substances which should not be exceeded at any time even if the 8 h TWA exposure is within the prescribed limit. Table 12.2 shows some 8 h TWA OELs published by ACGIH and HSE, respectively, for substances that are found in workplace atmospheres as aerosols (in mg m^{-3}). The list is not exhaustive, but serves to illustrate the similarities and the differences between the two listings. The full lists are available in publications released by the two organisations (Health and Safety Executive 2002; American Conference of Governmental Industrial Hygienists 2005), and it should be noted that they are continuously under-going changes as individual substances come under review. An important difference is that the HSE explicitly requires – indeed, has done so since 1989 – that the inhalability criterion should be applied to the coarse aerosol fraction for all relevant substances, replacing 'total' aerosol. So too now, incidentally, does the German MAK Commission. On the other hand, ACGIH recommends the inhalable fraction only for some substances that have been reviewed relatively recently and where where it is felt that sufficient information exists to 'safely' make that recommendation in the light of full knowledge of the possible impact on industry. For substances not yet reviewed, the old 'total' aerosol definition remains in effect.

Table 12.2 Examples of 8 h TWA ACGIH TLVs for selected substances occurring in workplaces as aerosols for 'total' aerosol and the inhalable (I), thoracic (T) and respirable (R) fractions, shown in comparison with the corresponding UK HSE OESs/MELs

Substance	ACGIH TLV (mg m^{-3})				HSE OES (mg m^{-3})			
	'Total'	I	T	R	'Total'	I	T	R
Aluminum (as Al)								
Metal dust	10	–	–	–	–	10	–	4
Oxides	–	–	–	–	–	10	–	4
Soluble	2	–	–	–	–	2	–	–
Cadmium and compounds (as Cd)	0.01	–	–	0.002	–	0.025 (MEL)	–	–
Oxide fume (as Cd)	–	–	–	–	–	0.025 (MEL)	–	–
Sulfide (as Cd)	–	–	–	–	–	–	–	0.03 (MEL)
Calcium carbonate	10	–	–	–	–	10	–	4
Carbon black	3.5	–	–	–	–	3.5	–	–
Coal dust	–	–	–	–	–	–	–	2
Anthracite	–	–	–	0.4	–	–	–	–
Bituminous	–	–	–	0.9	–	–	–	–
Cotton dust	–	–	0.2^a	–	–	2.5 (MEL)^b	–	–
Grain dust	4	–	–	–	–	10 (MEL)	–	–
Lead and compounds (as Pb)	–	–	–	–	0.15^c	–	–	–
Dust	–	–	–	–	–	–	–	–
Fume	–	–	–	–	–	–	–	–
Lead inorganic compounds	0.05	–	–	–	–	–	–	–
Manganese and compounds (as Mn)	–	–	–	–	–	5	–	–
Inorganic compounds	0.2	–	–	–	–	–	–	–
Dust	–	–	–	–	–	–	–	–
Fume (as Mn)	–	–	–	–	–	1	–	–
Nickel and compounds (as Ni)	–	1.5	–	–	–	1	–	–
Water-soluble inorganic	–	0.1	–	–	–	0.1 (MEL)	–	–
Water-insoluble inorganic	–	0.2	–	–	–	0.5 (MEL)	–	–
Subsulfide	–	0.1	–	–	–	–	–	–

(*continued overleaf*)

Table 12.2   (continued)

Substance	ACGIH TLV (mg m^{-3})				HSE OES (mg m^{-3})			
	'Total'	I	T	R	'Total'	I	T	R
Silica								
Amorphous	–	10	–	3	–	6	–	2.4
Cristobalite	–	–	–	0.05	–	–	–	0.3 (MEL)
Quartz	–	–	–	0.05	–	–	–	0.3 (MEL)
Tridymite	–	–	–	0.05	–	–	–	0.3 (MEL)
Fused								0.08
Sulfuric acid mist	–	–	0.2	–	–	1	–	–
Titanium dioxide	10	–	–	–	–	10	–	4
Wood dust								
Hardwoods	1	–	–	–	–	5 (MEL)	–	–
Softwoods	5	–	–	–	–	5 (MEL)	–	–
Asbestos (all forms)	–	–	–	0.1 f cm^{-3}	–	–	–	–
Chrysotile	–	–	–	–	–	–	–	0.5 f cm^{-3d}
Other forms	–	–	–	–	–	–	–	0.2 f cm^{-3d}

TLV, ACGIH threshold limit value; OES, HSE occupational exposure limit; MEL, HSE maximum exposure limit.
[a] TLV for cotton dust is based on the use of a vertical elutriator sampler with a 50% 'cut' at $d_{ae} = 15$ μm.
[b] MEL for cotton dust applies during handling of raw and waste cotton.
[c] As regulated separately in the UK by the Control of Lead at Work Regulations 1998, and measured according to the Approved Code of Practice 'Control of Lead at Work'.
[d] As regulated separately in the UK by the Control of Asbestos at Work Regulations 1999, limit values averaged over 4 h.

It is seen from Table 12.2 that, for many of the substances listed, the TLVs and OESs are quite similar, not surprisingly for the reasons already mentioned. Such differences as do exist reflect to some extent the differing dynamics of the individual standards setting bodies in question (Vincent, 1998). The classification of substances with respect to the particle size-selective sampling criteria shown provides the occupational hygienist with firm guidance about the aerosol sampling options that should be considered for exposure assessment with respect to each substance. Further, identification of individual chemical species for many substances points to specific analytical methods for the evaluation of collected samples. ACGIH provides guidance about the choice of sampling and analytical methods in other documents, including its regularly updated volume on air sampling instrumentation (American Conference of Governmental Industrial Hygienists, 2001) and other publications (e.g. Vincent, 1999). Similarly HSE provides such guidance through its own accompanying documentation (Health and Safety Executive, 2000). Further, in the UK, some substances – specifically lead and asbestos – are controlled by different regulations, as indicated in Table 12.2. Some substances, as already mentioned, are also controlled by reference to biological indices (e.g. blood lead levels), where the expectation is that the airborne exposures and biological indices should be applied together for the overall protection of workers.

Table 12.2 illustrates that OELs are listed extensively for the inhalable and respirable fractions. So far there has been little interest either in the ACGIH or HSE, or in other occupational standards-setting bodies, in developing OELs based on the thoracic fraction. The ACGIH TLV for cotton dust is a long-standing exception, going back to 1975 (National Institute for Occupational Safety and Health, 1975), although this is defined with respect to the aerosol penetration characteristic of a vertical elutriator-type sampler which has a value of 50 % at $d_{ae} = 15$ μm. A more recent exception is the ACGIH TLV for sulfuric acid mist, where the criterion to be applied is the thoracic fraction. It is expected that others will follow as more substances come under review and the TLVs are updated.

12.5   Ambient atmospheric aerosol limits

OELs like those referred to above relate only to exposures in the workplace and cannot be used as guides to exposures in the general environment or in residences. The first difference arises out of the fact that the human population at large is much more diverse than the working sub-population. In addition to working people, it also includes children, the elderly and persons – many of working age – whose health may have been impaired by infections and other disease processes. Of these, children undergo rapid growth and cell turnover, and have undeveloped homeostatic defense systems, while the elderly may have defenses which have been compromised by degenerative diseases or earlier toxic exposures. The second difference arises because workplace exposures normally occur only during working hours, where there is the opportunity for detoxification and elimination occurring during nights, holidays and weekends when there are no occupational exposures. Ambient environmental exposures continue during these times, such that not only is the overall period of potential exposure longer but there is no respite.

12.5.1   Black smoke and fine particles

The first efforts to deal legislatively with air pollution were made during the late 1800s, directed at visible black smoke (soot) emissions from steam boilers. In the USA, the cities of Cincinnati and Chicago passed smoke regulations in 1881, and Ohio passed a law to limit these emissions in 1897. The effects of particulate air pollution on visibility were initially the primary concern, although health effects soon became the main driving force once it became recognised that aerosols in the ambient atmospheric environment were associated with excess mortality and morbidity. The strength of the association was known to be quite strong even early on where epidemiological studies depended on crude aerosol exposure indices based on *black smoke* (BS) and *total suspended particulate* (TSP). More recent epidemiological studies, many of them based on exposure data obtained in terms of thoracic particulate matter, as measured according to the PM_{10} criterion, have generally produced stronger associations between ambient aerosol and mortality and morbidity in the general population. Even more recent studies have shown that those associations are stronger still for indices of exposure based on the finer $PM_{2.5}$ criterion (Lippmann and Thurston, 1996). Nonetheless, standards and limits for BS have continued to play a prominent role in air pollution regulation in some countries. In Europe, for example, BS has been regulated since 1980 under the European Community (EC) Directive 80/779/EEC, and this regulation remains active even to this day. For this standard, BS levels in the ambient atmosphere are assessed simply by sampling aerosol directly onto a filter and assessing the 'blackness' (e.g. by an optical scanning device) from which to estimate – by empirical calibration – an equivalent gravimetric smoke concentration. Limit values for BS are therefore expressed in terms of mass per unit volume of air (μg m^{-3}). Current EC Directive limit and guide values for BS are summarised in Table 12.3, where the former are mandatory and the latter are intended to serve as long-term precautionary levels.

Sulfate ion (SO_4^{2-}), most of which is formed in the atmosphere from the oxidation of the SO_2 emitted from fossil fuel combustion sources, is known to be a major mass constituent of the $PM_{2.5}$ fraction in North America (Lippmann, 1999). Other major constituents include ammonium nitrate, along with fixed and organic carbon. Much of the fixed carbon is attributable to diesel engine exhaust, and most of the organic carbon is formed in the atmosphere in photochemical reactions leading to ozone. The sulfate,

Table 12.3   *European Community (EC) Directive Limit and Guide Values for BS*

Reference period	Limit value (μg m^{-3})
EC Directive Limits	
Year (median of daily values)	68
Year (98th percentile of daily values)	213
Winter (median of daily values, October–March)	111
EC Directive Guide Values	
Year (arithmetic mean of daily values)	68
24 h (daily mean value)	111

nitrate and fixed carbon components of the PM$_{2.5}$ fraction are now thought to be causal factors for the excess mortality and morbidity that have been observed, but the actual responsible constituents have not yet been identified. The United States Environmental Protection Agency (EPA) chose the PM$_{2.5}$ mass fraction as the criterion for a new fine atmospheric aerosol standard (Environmental Protection Agency, 1996a).

Epidemiological evidence for health effects attributable to exposures to the PM$_{2.5}$ fraction is more limited than for PM$_{10}$, but some important investigations have been reported. One major study in which fine particles corresponding to PM$_{2.5}$ were measured, along with the PM$_{10}$ and TSP fractions, has become known as the 'Harvard Six-Cities Study', named after a group of cities chosen to be representative of the majority of exposures in the USA. Results of daily time-series mortality analyses (Schwartz *et al.*, 1996; Environmental Protection Agency, 1996b) showed not only that the PM$_{2.5}$ fraction correlated better with daily mortality than PM$_{10}$ but also that there was an especially good correlation with the sulfate chemical subfraction. Further, it was shown that the excess deaths reflected in these results were accounted for by cardiopulmonary effects and lung cancer (Dockery *et al.*, 1993; Pope *et al.*, 1995). In another study in 22 US and Canadian cities, adverse effects on the lung capacity and bronchitis in children were also shown to correlate better with the finer aerosol fraction (Dockery *et al.*, 1996; Raizenne *et al.*, 1996).

12.5.2   Establishment of the EPA PM NAAQS limit values

Interest and actions with respect to standards for aerosols in the ambient atmospheric environment around the world have been greatly influenced by what has transpired in the USA during the past two decades. Until relatively recently, EPA aerosol standards were based on the TSP rationale. This was eliminated in 1987 in favor of the more scientific approach derived from knowledge of particle inhalation and penetration into the lungs that led to the PM$_{10}$ philosophy (and subsequently PM$_{2.5}$). Nonetheless, TSP sampling is still carried out in some states, even to this day, especially for lead-containing particles.

A major responsibility of EPA under the Clean Air Act is to periodically review the National Ambient Air Quality Standards (NAAQS) and, if necessary, to propose and promulgate new or revised standards (Lippmann, 1987). The primary NAAQS are concentration limits for specified averaging times that are intended to prevent adverse acute and chronic health effects associated with exposures to pollutants having numerous and widespread sources. Furthermore, they are intended to protect sensitive or susceptible segments of the general population with an adequate margin of safety. There are currently six categories of ambient air pollutants having specified NAAQS, and atmospheric aerosol – commonly referred to as 'particulate matter' (or PM) – is one of them. The others are ozone, sulfur dioxide, nitrogen dioxide, carbon monoxide and lead.

In 1997, after consideration of health effects in the general population, as well as in vulnerable sub-groups, EPA established new daily and annual fine particle NAAQS in the form of limit values based on the $PM_{2.5}$ particle size-selective criterion. The pre-existing PM_{10} limit values were retained (Environmental Protection Agency, 1996b). Both were expressed in terms of daily and annual time-weighted average concentrations. It was argued that NAAQS based on both the $PM_{2.5}$ and PM_{10} criteria would serve to protect people from the differing adverse health effects identified as being associated with the two fractions, and hence for the majority of aerosol-related ill-health. It also provided continuity with the earlier approach based solely on the use of the PM_{10} criterion.

The philosophy underlying the EPA standards for ambient atmospheric aerosol is that sampling should take place using fixed point samplers placed at locations considered to be representative of human populations – or, as described by the EPA, 'community-oriented'. On the basis of the available health effects and exposure data, and taking into account the impacts on vulnerable populations, the NAAQS limit values currently applied by the EPA are as summarised in Table 12.4.[3] As seen, this table refers to both *primary* and *secondary* standards. In the EPA framework, the primary standard is the one that is aimed at protecting human health, while the secondary standard is aimed at protecting the general environment and property. It is only for the first of these that the scientific link exists between the measurement (via the definitions of PM_{10} or $PM_{2.5}$) and the objective of the standard. However, EPA deemed it convenient to include the rationale underlying the secondary standard into the practical implementation of the primary standard, not least for convenience because it removes the need to conduct separate aerosol sampling campaigns.

The EPA standards are very influential in other countries. In particular, the PM_{10} criterion has been widely adopted for many years. Limit values for the UK are shown in Table 12.5.[4] In Europe, an

Table 12.4 *EPA NAAQS 2005 limit values for particle size-selective aerosol fractions*

Fraction	Primary standard (μg m^{-3})	Averaging procedure	Secondary standard (μg m^{-3})
Particulate matter PM_{10}	50	Annual arithmetic mean for a given monitor must not exceed stated value	
	150	24 h TWA must not exceed stated value more than once per year	
Particulate matter $PM_{2.5}$	15	3 year average of the annual arithmetic mean for a given monitor must not exceed stated value	Same as primary
	65	3 year average of the 98th percentile of 24 h concentrations must not exceed stated value	

---

[3] See EPA website at http://epa.gov/air/criteria.html.
[4] See website at http://www.airquality.co.uk/archive/standards#std.

Table 12.5   *Air quality limit values for the UK for particle size-selective aerosol fractions*

Fraction	Limit value (μg m^{-3})	Averaging procedure	To be achieved by
Particulate matter PM_{10} (England and Wales)	50	24 h arithmetic mean, not to be exceeded more than 35 times per year	December 31, 2004
	40	Annual mean	
Particulate matter PM_{10} (Scotland)	50	24 h arithmetic mean, not to be exceeded more than 7 times per year	December 31, 2010
	18	Annual mean	

Expert Panel of Air Quality Standards (EPAQS) reports to the European Commission's Clean Air for Europe (CAFE) program. In 2003 it conducted a review of the then-current directive for air quality that was based on the PM_{10} philosophy, and recommended that monitoring the $PM_{2.5}$ fraction should be introduced, while continuing measurement of the PM_{10} fraction.[5] EPAQS noted that the average ratio of $PM_{2.5}$ and PM_{10} mass concentrations in Europe ranged from about 0.3 to 0.6, and recommended that an annual-average limit value for the $PM_{2.5}$ fraction should be assigned in the range from 12 to 20 μg m^{-3}, with a 24 h value around 35 μg m^{-3} not to be exceeded on more than 10 % of the days in any given year. These figures correspond approximately to those of the EPA.

12.5.3   Limits for nonhealth-related aerosols

Nonhealth-related aerosol standards cannot be related to exposure in the sense that has been discussed in relation to human health – that is, through linking the physics of aerosol behavior during the process of exposure with the physiology of the human respiratory system. However, as mentioned at the end of Chapter 9, there are significant impacts on the environment beyond those relating just to human health. In the USA, these are addressed by EPA in the first instance through the secondary standards within the NAAQS. However, they are also addressed within the framework of the 1990 US Clean Air Act that requires limits on emissions from stationary and mobile sources.

Dust deposition (or *dustfall*) is one important, nonhealth aspect of atmospheric aerosol. This relates to ecological and aesthetic concerns about (mostly coarse) dust, grit and smuts, emitted for example from coal-fired electricity generating stations and other industrial plants, open-cast mining, waste dumps, etc. Although there are currently no particle size-selective criteria that correspond to those that have been developed for aerosols that can impact on health after inhalation, some standards have been proposed in the form of limit values expressed in terms of the rate of mass deposited per unit area. Some of these, originally summarised by Vallack and Shillito (1998), are shown in Table 12.6, where it is noted that these are generally long-term TWA values and that no reference is made to any specific sampling instrumentation, let alone criteria. In addition to those listed, other countries have also proposed standards, including New Zealand. It is notable that the UK is absent from the list. Nominally the 1993 UK Clean Air Act addresses all aspects of air pollution and, in addition to the regulatory limits on

---

[5] See website at http://www.defra.gov.uk/environment/airquality/aqs/meetings/sep03/epaqs0315.pdf.

Table 12.6 *Summary of dust deposition limit values (from Vallack and Shillito, 1998)*

Country/region	Basis	Limit ($mg\ m^{-2}\ day^{-1}$)
USA		
Kentucky	Annual average	196
Louisiana	Annual average	262
Maryland	Annual average	183
Mississippi	Monthly average	175
Montana	Annual average[a]	196
New York		100–130[b]
North Dakota	3 monthly average	196
Pennsylvania	Annual average	267/500
Washington	Annual average	183
Wyoming	Monthly average	170
Canada		
Alberta	Annual average	180
Manitoba	Annual average[c]	153
Newfoundland	Annual/monthly average	153/233
Ontario	Annual/monthly average	170/200
Finland	Annual average	333
Germany	Long-term/short-term	350/650
Spain	Annual average	200
Western Australia		
		133[d]
		133[e]

[a]Residential areas.
[b]During any 12 months no more than 5 % of 30 day values to exceed 100 and 84 % to be below 130.
[c]Maximum acceptable 266 and desirable 200.
[d]Loss of amenity first perceived.
[e]Unacceptable reduction in air quality.

health-related aerosol fractions, specifies limits on industrial emissions, these relate only loosely to dust deposition. However, a mass deposition rate of $200\ mg\ m^{-2}\ day^{-1}$ is widely used in the UK in environmental assessments as a 'custom-and-practice' limit for coarse 'nuisance' aerosols.

In the context of this book, it is important to note that, in addition to all the above, most countries have regulations that require aerosol sampling in stacks and ducts in order to estimate emissions from stationary sources such as industrial plants.

12.6 Special cases

The framework outlined above describes what might be regarded as a 'classical' approach to aerosol sampling and standards, based on knowledge about particle penetration into the respiratory tract (underpinning particle size-selective sampling) and the chemical properties of the particles (underpinning considerations of biological response and toxicology). However, some 'hot-button' issues have emerged over the years that require adjustments to this basic approach.

12.6.1   Fibrous aerosols

Interest in sampling fibrous aerosols began in earnest when it became clear, especially during the post World War II years, that the inhalation by humans of fine fibers of asbestos leads to serious lung disease, including asbestosis, lung cancer and mesothelioma. Considerable research has been conducted into exposure assessment (including sampling), toxicology and health effects. So this is not a new, and has not been a neglected, area. Out of this effort has come the development of a set of distinctive criteria, and in turn methods, for aerosol sampling. The subject remains highly topical to this day.

Asbestos in its bulk form is a fibrous mineral that is mined from the ground in South Africa, Canada and a few other countries. Early on it was found capable of being processed to form bulk media that have outstanding thermal and mechanical properties. So it became a primary ingredient around the world in thermal and electrical insulation and other construction materials. It exists in a number of mineralogical forms, classified as amphiboles (including most commonly the 'brown' *amosite* and 'blue' *crocidolite*) and serpentine (including 'white' *chrysotile*). It is also present in other minerals that have been used in construction, including *vermiculite*. Health effects associated with inhaling the dust from asbestos-containing materials and products started to become evident around the early 1900s. The health effects in question were soon to be found to be very specific to asbestos, notably *asbestosis*, a form of pneumoconiosis reflected in progressive disabling inflammation and scarring of the lung tissue of the alveolar region. Later, in the 1940s the link with *lung cancer* of the conducting airways was established, followed in the 1960s by the link with *mesothelioma*, a cancer of the outside lining of the lungs (the pleura). Initial associations between exposure and disease related to occupational settings, including workplaces in asbestos mining, processing and using industries. By the 1970s concerns were increasingly being expressed about exposures in general ambient and living environments. The asbestos-related diseases in question are now known to be very slow in their development, with the time between the onset of exposure and the appearance of clinically or pathologically observable signs typically being as long as from 15 to 30 years. For this reason, interest in the past exposures of people that might have been exposed by inhalation to asbestos fibers continues to this day.

Since the problem first surfaced, a great deal of toxicological research has revealed that the health effects in question are directly associated with the long, thin asbestos fibers that are released into the air when asbestos-containing materials are processed or used. Many reviews have been written (e.g. Meldrum, 1996; Agency for Toxic Substances and Disease Registry, 2001). Three primary features have emerged in the toxicological profile. In the first, the risk is greater for some types of asbestos than for others (e.g. crocidolite is generally more harmful than chrysotile), and in the second the risk may be associated with exposures even at very low levels. The third feature is that long, thin, durable fibers are the ones most likely to be associated with diseases like those mentioned. All three have important implications to aerosol sampling and measurement, including both the physical act of sampling and the identification and analysis of sampled fibers. That is, it is necessary to sample the fibers efficiently and then to identify and quantify those that qualify – on the basis of an appropriate set of criteria – as being harmful.

An important feature of human exposure to asbestos is that, when it occurs, it nearly always involves exposure to the asbestos fibers of concern along with other, nonasbestiform particles. Indeed, experience has shown that the concentrations of asbestos fibers in typical occupational or environmental exposure settings are typically very small in relation to the overall aerosol exposure. Further, the concentration of the asbestos fiber component is so low that physical methods to analyze them on the basis of a bulk metric such as mass are not feasible. This in turn has pointed the way to alternative indices of the concentrations of fibers meeting appropriate health criteria in terms of fiber type and dimensions. Mainly for this reason, the evolution of sampling and analytical procedures for assessing exposures to asbestos fibers has taken its own singular path within the history of aerosol measurement, notably involving optical and electron microscopy.

Finally, although asbestos continues to be identified as a major health hazard, other materials containing durable fine fibers, including some introduced to replace asbestos, have subsequently emerged as being of similar concern. According to some prominent experts, the very properties of asbestos that govern their harmfulness are present in other materials that can produce fibrous aerosols. Issues of fiber shape, dimensions and durability have again dictated those concerns.

12.6.2   Bioaerosols

Bioaerosols are formed by the aerosolisation of particles of biological origin. Interest in them from the point of view of aerosol sampling and characterisation stems from their biological properties, and not their chemical properties as is the case for most other aerosols of concern to occupational and environmental hygienists. They take a wide variety of broad classifications, including viruses and bacteria, actinomycete, fungal and other spores, algal and plant cells, fragments of insects and mites, protein fragments from plant and animal sources, antibiotic materials from certain pharmaceutical processes, endotoxin from Gram-negative bacteria, and mycotoxins and glucans from fungi (Lacey and Dutkiewicz, 1994). Within each individual classification there is a very wide variety of species. So the scope of aerosol science relating to bioaerosols is vast. In terms of sampling and analysis, this provides some stern scientific and technical challenges. Many of these are discussed in the extensive review carried out by the Bioaerosols Committee of ACGIH (Macher, 1999). For many years, the study of bioaerosols fell within the area of *aerobiology*, where research and practice were dominated by biology and biologists. In more recent years, however, the field has expanded into a much wider interdisciplinary field located firmly within the mainstream of aerosol science (Ho and Griffiths, 1994).

Lacey and Dutkiewicz reviewed the range of lung diseases associated with exposure by inhalation to bioaerosols. Their primary attention was focused on occupational settings, but most of what they reviewed is also relevant to other environmental exposures. Infection is an important relevant class of disease, and arises from exposures to viruses, bacteria and fungi. Since that review was written, this area has become of great topical concern as bioterrorism has emerged as a significant threat. Noninfectious diseases arise from the allergenic properties and/or immunotoxic properties of other bioaerosols. Some are associated with, for example, occupational exposures to cotton dust (*byssinosis*) and mouldy hay (*Farmer's lung*), and in nonoccupational exposures to house mite dusts (*asthma*).

The challenges in sampling for bioaerosols are closely linked with the methods that are required to assay collected samples. Griffiths and DeCosemo (1994) reviewed the options for bioaerosol assessment, and identified a number of problem areas where further research was needed. A major one was the collection of the organisms in a manner that minimises stress that might affect their viability, especially since continued exposure of some organisms to sampled air while present on the filter or substrate of a sampler can lead to desiccation. The matter of criteria for sampling bioaerosols in a way that is relevant to the range of possible health effects is closely tied up with the ability to sample and conserve viable samples. At the time of writing this book, the ACGIH Bioaerosols Committee has been given the charge to '... compile and disseminate information on biologically derived airborne contaminants, to develop recommendations for assessment, control, remediation and prevention of such hazards' and '... to establish criteria for bioaerosol exposure limits'. The only attempt to define an actual limit value for this very broad aerosol category appears to be that recommended in Germany by Technische Regeln für Biologische Arbeitsstoffe (TRBA 430) (see Engelhart and Exner, 2002). Other standards-setting bodies have not yet felt ready to follow suit.

12.6.3   Ultrafine aerosols

There was a time when the finest particles of interest in occupational and environment health were considered to be those which, after inhalation, could penetrate down to the alveolar region of the lung. However, interest has been drawn more recently towards still finer particle fractions, not least because of their possible association with serious health effects specific to those sizes of particle. The term 'ultrafine' is used widely nowadays to refer broadly to particles whose diameters are less than 0.1 μm. Attention is now being directed towards particles deep within this size range. Previously, recognition of the potential problems associated with particles in the ultrafine size range did not surface because the mass concentrations of this aerosol fraction in practical situations were very small in relation to those for the coarser fractions that had been the primary subjects of interest. This lack of awareness was compounded by the fact that, until even now, and for most types of aerosol, aerosol standards have usually been expressed in terms of particulate *mass* concentration. The question of an appropriate metric for the measurement of exposure to ultrafine aerosol fractions remains one of the key issues for discussion in the context of standards development.

A 0.1 μm ultrafine particle measures 100 nm on the nanometer scale. Some particles in environmental aerosols have been detected down to well below 10 nm. Particles in this range are usually formed by nucleation in thermal and chemical processes involving the combustion of fuels (e.g. as in the engines of cars, trucks or planes) or the extreme heating of other materials (e.g. as in smelting or welding). To obtain a sense of scale for such particles, it is noted that a typical atom or molecule has dimensions of about 0.3–0.5 nm, depending on the atom or molecule in question and on its electronic state. The latter arises from the fact that, at the atomic or molecular level, the size of the entity is determined by the probability densities of the orbiting electrons. In turn, therefore, a molecule which is excited will have a larger characteristic size by virtue of the larger radius of those orbiting electrons. What is important in relation to particles is that, as they become smaller and smaller, approaching the order of magnitude of the molecules themselves (say, a few nm), their physical state becomes distinctly different from that of larger particles (Preining, 1998). For a particle that is so small, the number of molecules or atoms the particle contains becomes so small that a high proportion of them lie at the surface of the particle. Preining estimated, for example, that a 20 nm particle has 12 % of its molecules at the surface, and a 10 nm particle has 25 % at its surface. At this level, therefore, even the concept of a 'surface' has limited meaning because the material structure of the particle can no longer be regarded as a continuum. In turn, the material itself can no longer be thought of simply as 'solid' or 'liquid', and we enter the world of 'cluster physics'. Here, the magnitude and spatial distribution of the surface reactivity of a particle become strongly influenced by the way in which the individual molecules combine in the overall particulate entity. This is what has stimulated the great current interest in the study of nanometer-sized aerosol particles in reactive systems and their applications in the synthesis of new materials. It also provides much food for thought in relation to how such particles might interact with biological organisms or cells, and hence on their potential toxicological effects.

Some basic biological research has been conducted to ascertain the toxicological responses to very small particles in the size range from about 10 nm up to about 500 nm, in relation to what makes very fine particles more toxic to biological systems than larger particles of the same material. Oberdörster *et al.* (1992, 1995) and Donaldson *et al.* (1998) carried out animal (rat) studies for aerosols of Teflon and titanium dioxide, respectively, and showed that particles of diameter less than about 50 nm produced much stronger inflammatory responses than 500 nm particles of the same substances at the same mass concentrations on the basis of mass concentration exposures. The question was raised about what might be the most appropriate metric for health-related exposure to ultrafine aerosol fractions. Many researchers have speculated that particle number concentration might be better than particle mass concentration for ultrafine particles. However, Oberdörster (2000) reported important results from animal experiments suggesting that the metric of exposure that best correlated with the inflammatory response was particle surface area concentration. Other workers have addressed the role of particle chemical properties for ultrafine aerosols. For example, Chen *et al.* (1992) showed in guinea pigs an association between hyperresponsiveness and exposures to acid-coated ultrafine particles.

Coupled with concerns about their toxicological impact, evidence has emerged in recent years about the significant – perhaps increasing – levels of ultrafine particles in living and working environments. For example, Kittelson (1998) noted that, although the later generations of diesel and spark engines have improved in terms of the *mass* of particulate emissions, the relative *number* of nano-sized particles emitted has increased sharply. This may also be true for emissions from aircraft engines. Measurements of the particle size distribution of atmospheric aerosols have confirmed that an increase in the number concentrations of particles in the low ultrafine range has occurred in recent years, especially in urban areas (e.g. Kulmala *et al.*, 2004). Similar evidence is now starting to appear to support the view that the ultrafine component in some workplace aerosols may be significant (e.g. Möhlmann, 2004; Peters *et al.*, 2006). Studies are continuing in Europe and North America in both occupational and environmental situations.

General interest in health effects of ultrafine particles has come from growing awareness during the past decade or so of health effects associated with exposures to particles finer than those contained just within the aerosol fractions which underpin current air quality regulations. Such interest was initially driven by concerns that such particles in ambient air may be linked with the observed increases in mortality linked to cardiopulmonary disease in vulnerable populations, and the possibility of the involvement of particles in the population much finer even than described by $PM_{2.5}$. Specifically with respect to the observed mortality, Seaton and colleagues (1995; 1996) considered the possible causative factors. They proposed a hypothesis in which exposure to ultrafine particles in the size range around 50 nm '... characteristic of air pollution (may) provoke alveolar inflammation leading to acute changes in blood coagulability and release of mediators able to provoke attacks of acute respiratory illness in susceptible individuals. The blood changes result in an increase in the exposed population's susceptibility to acute episodes of cardiovascular disease; the most susceptible suffer the most. This hypothesis, being based on the number, composition and size – rather than on the mass – of particles accounts for the observed epidemiological relations.'

The Seaton hypothesis relates to ultrafine particles that deposit in the alveolar region of the lung, and the suggestion that such small particles are readily translocated into the blood. A plausible particle size-selective criterion for sampling relative to this health effect might be the $UA(d_V)$ fraction described in the preceding chapter. Meanwhile, there has been more recent discussion about the deposition of inhaled ultrafine particles in the nasal passages, and their direction translocation through the olfactory nerve into the olfactory bulb and thence to the brain and central nervous system (Oberdörster *et al.*, 2005). The potential importance of this specifically to ultrafine particles of manganese, and hence to manganese-related health effects in some welders, has been raised by Dorman *et al.* (2002). Here a plausible criterion for health-related sampling might be the $UN(d_V)$ fraction also described in Chapter 11. It is clear that there is a need for further research to confirm the plausibility of such hypotheses or to provide a basis for new ones. But, for the time being, there are no exposure limits for ultrafine aerosols.

References

Agency for Toxic Substances and Disease Registry (2001) Toxicological profile for asbestos, US Department of Health and Human Services, Public Health Service, Atlanta, GA.

American Conference of Governmental Hygienists (1946) Proceedings of the Eighth Annual Meeting of the American Conference of Governmental Hygienists, April 7–13, Chicago, IL, pp. 54–55.

American Conference of Governmental Hygienists (2001) *Air Sampling Instruments*, 9th Edn, ACGIH, Cincinnati, OH.

American Conference of Governmental Hygienists (2005) *Threshold Limit Values for Chemical Substances and Physical Agents and Biological Exposure Indices*, ACGIH, Cincinnati, OH.

Bowditch, M. (1944) In setting threshold limits, Transactions of the Seventh Annual Meeting of the National Conference of Governmental Industrial Hygienists, May 9, St Louis, MO, pp. 29–32.

Bowditch, M., Drinker, C.K., Drinker, P., Haggard, H.H. and Hamilton, A. (1940) Code for safe concentrations of certain common toxic substances used in industry, *Journal of Industrial Hygiene and Toxicology*, 22, 251.

Chen, L.C., Miller, P.D., Amdur, M.O. and Gordon, T. (1992) Airway hyperresponsiveness in guinea pigs exposed to acid-coated ultrafine particles, *Journal of Toxicology and Environmental Health*, 35, 165–174.

Cook, W.A. (1945) Maximum allowable concentrations of industrial atmospheric contaminants, *Industrial Medicine*, 14, 936–946.

Dockery, D.W., Cunningham, J., Damokosh, A.I., Neas, L.M., Spengler, J.D., Koutrakis, P., Ware, J.H., Raizenne, M. and Speizer, F.E. (1996) Health effects of acid aerosols on North American children: respiratory symptoms, *Environmental Health Perspectives*, 104, 500–505.

Dockery, D.W., Pope, C.A. III, Xu, X., Spengler, J.D., Ware, J.H., Fay, M.E., Ferris, B.G. and Speizer, F.E. (1993) An association between air pollution and mortality in six US cities, *New England Journal of Medicine*, 329, 1753–1759.

Donaldson, K., Li, X.-Y. and MacNee, W. (1998) Ultrafine (nanometer) particle mediated lung injury, *Journal of Aerosol Science,* 29, 553–560.

Dorman, D.C., Brenneman, K.A., McElveen, A.M., Lynch. S.E., Roberts, K.C. and Wong, B.A. (2002) Olfactory transport: a direct route of delivery of inhaled manganese phosphate to the rat brain, *Journal of Toxicology and Environmental Health*, 65, 1493–1511.

Engelhart, S. and Exner, M. (2002) Fungi in indoor air – determination of mesophilic and thermo-tolerant fungi in indoor air using occupational standards on bioaerosols, *Gefahrstoffe Reinhaltung der Luft*, 62, 79–82.

Environmental Protection Agency (1996a) *Air Quality Criteria for Particulate Matter*, EPA/600/P-95/001, Washington, DC.

Environmental Protection Agency (1996b) *Review of the National Ambient Air Quality Standards for Particulate Matter*, Office of Air Quality Planning and Standards (OAQPS) Staff Paper, EPA-452/R-96-013, Research Triangle Park, NC.

Funtowicz, S.O. and Ravetz, J.R. (1990) *Uncertainty and Quality in Science for Policy*, Kluwer Academic, Dordrecht.

Griffiths, W.D. and DeCosemo, G.A.L. (1994) The assessment of bioaerosols: a critical review, *Journal of Aerosol Science*, 25, 1425–1458.

Health and Safety Executive (2000) *General Methods for the Gravimetric Determination of Respirable and Inhalable Dust*, HSE, MDHS14/3 (revised), HMSO, London.

Health and Safety Executive (2002) *Occupational Exposure Limits 2002*, HSE, EH40/2002, HMSO, London.

Ho, J. and Griffiths, W.D. (1994) Guest Editorial to Special Issue on Bioaerosols, *Journal of Aerosol Science*, 25, 1369–1370.

Kittelson, D.B. (1998) Engines and nanoparticles: a review, *Journal of Aerosol Science,* 29, 575–588.

Kobert, R. (1912) *Kompendium der Praktischen Toxicologie zum Gebrauche für Ärzte, Studierende um Medizinalbeamte*, Ferdinand Enke, Stuttgart.

Kulmala, M., Vehkamäki, H., Petäjä, T., Dal Maso, M., Lauri, A., Kerminen, V.-M., Birmili, W. and McMurry, P.H. (2004) Formation and growth rates of ultrafine atmospheric particles: a review of observations, *Journal of Aerosol Science*, 35, 143–176.

Lacey, J. and Dutkiewicz, J. (1994) Bioaerosols and occupational lung disease, *Journal of Aerosol Science*, 25, 1371–1404.

Lehmann, K.B. (1886) Experimentelle Studien über den Einfluss Technisch und Hygienisch Wichtiger Gase und Dampfe auf Organismus: Ammoniak und Salzsauregas, *Archives of Hygiene*, 5, 1–12.

Lippmann, M. (1987) Role of science advisory groups in establishing standards for ambient air pollutants, *Aerosol Science and Technology*, 6, 93–114.

Lippmann, M. (1999) Sampling criteria for fine fractions of ambient air. In: *Particle Size-selective Sampling for Particulate Air Contaminants* (Ed. J.H. Vincent), American Conference of Governmental Industrial Hygienists, Cincinnati, OH, pp. 97–118.

Lippmann, M. and Thurston, G.D. (1996) Sulfate concentrations as an indicator of ambient particulate matter air pollution for health risk evaluations, *Journal of Exposure Analysis and Environmental Epidemiology*, 6, 123–146.

Macher, J. (1999) *Bioaerosols: Assessment and Control*, American Conference of Governmental Industrial Hygienists, Cincinnati, OH.

Meldrum, M. (1996) Review of fibre toxicology, HSE Report, Health and Safety Executive, London.

Möhlmann, C. (Ed.) (2004) Ultrafine aerosols in workplaces, BIA Report 7/2003E, Berufsgenossenschaftliches Institute für Arbeitsschutz (BIA), Proceedings of the Workshop held at the BIA, Sankt Augustin, Germany, August 2002.

National Institute for Occupational Safety and Health (1975) *Criteria for a recommended standard – occupational exposure to cotton dust*, NIOSH Publication No. 75–118, Washington, DC.

Oberdörster, G. (2000) Toxicology of ultrafine particles: in vivo studies, *Philosophical Transactions of the Royal Society of London, Series A*, 358, 2719–2740.

Oberdörster, G., Ferin, J., and Gelein, R. (1992) Role of the alveolar macrophage in lung injury: studies with ultrafine particles, *Environmental Health Perspectives*, 97, 193–199.

Oberdörster, G., Gelein, R. and Ferin, J. (1995) Association of particulate air pollution and acute mortality: involvement of ultrafine particles, *Inhalation Toxicology*, 71, 111–124.

Oberdörster, G., Sharp, Z., Atudorei, V., Elder, A., Gelein, R., Kreyling, W. and Cox. C. (2005) Translocation of inhaled ultrafine particles to the brain, *Inhalation Toxicology*, 16, 437–445.

Peters, T.M., Heitbrink, W.A., Evans, D.E., Slavin, T.J. and Maynard, A.M. (2006) The mapping of fine and ultrafine particle concentrations in an engine machining and assembly facility, *Annals of Occupational Hygiene*, 50, 249–257.

Pope, C.A. III, Thun, M.J., Namboodiri, M., Dockery, D.W., Evans, J.S., Speizer, F.E. and Heath, C.W. (1995) Particulate air pollution is a predictor of mortality in a prospective study of US adults, *American Journal of Respiratory and Critical Care Medicine*, 151, 669–674.

Preining, O. (1998) The physical nature of very, very small particles and its impact on their behavior, *Journal of Aerosol Science*, 29, 481–495.

Raizenne, M., Neas, L.M., Damokosh, A.I., Dockery, D.W., Spengler, J.D., Koutrakis, P., Ware, J.H. and Speizer, F.E. (1996) Health effects of acid aerosols on North American children: pulmonary function, *Environment Health Perspectives*, 104, 506–514.

Ratney, R.S. (1999) Application of particle size-selective sampling criteria in establishing TLVs. In: *Particle Size-selective Sampling for Particulate Air Contaminants* (Ed. J.H. Vincent), American Conference of Governmental Industrial Hygienists, Cincinnati, OH, pp. 179–207.

Sayers, R.R. and Dallavalle, J.M. (1935) Prevention of occupational diseases other than those that are caused by toxic dust. *Mechanical Engineer*, April, 13–17.

Schwartz, J., Dockery, D.W. and Neas, L.M. (1996) Is daily mortality associated significantly with fine particles? *Journal of the Air and Waste Management Association*, 46, 927–939.

Seaton, A. (1996) Particles in the air: the enigma of urban air pollution, *Journal of the Royal Society of Medicine*, 89, 604–607.

Seaton, A., MacNee, W. and Donaldson, K. (1995) Particulate air pollution and acute health effects, *The Lancet*, 345, 176–178.

Vallack, H.W. and Shillito, D.E. (1998) Suggested guidelines for deposited ambient dust, *Atmospheric Environment*, 32, 2737–2744.

Vincent, J.H. (1989) *Aerosol Sampling: Science and Practice*, John Wiley & Sons, Ltd, Chichester.

Vincent, J.H. (1998) International occupational exposure standards: a review and commentary, *American Industrial Hygiene Association Journal*, 59, 729–742.

Vincent, J.H. (Ed.) (1999) *Particle Size-selective Sampling for Particulate Air Contaminants*, American Conference of Governmental Industrial Hygienists, Cincinnati, OH.

World Health Organisation (2003) *International Classification of Diseases, 10th Revision, Version for 2003*, WHO, Geneva.

Part C
Aerosol Sampling Instrumentation

13

Historical Milestones in Practical Aerosol Sampling

13.1 Introduction

This chapter points out the main highlights of the historical development of practical aerosol sampling instrumentation, from its origins in the 1800s to the sophisticated approaches which are regarded as routine today. It will refer to criteria and strategies described in greater detail in previous chapters, and will link them up with some of the instruments that have played important roles during the developments that have taken place, and which will be described in upcoming chapters. So what follows provides a bridge between the past and the future, and between the theory and the practice of aerosol sampling. It focuses on the separate, but generally interconnected, developments that have taken place in aerosol sampling in workplaces and the ambient atmosphere, respectively, both nationally and internationally. To illustrate these historical time lines, a number of specific instruments are mentioned as important milestones. Further details on these and many other aerosol samplers will appear later in greater detail.

The historical perspective is important because the older sampling methods are of more than just passing interest. This is because exposures from as far back as 50 years ago remain relevant to current epidemiology and hence to standards setting. Indeed, there are people still alive today who may yet suffer the consequences of exposures that occurred even that many years ago. Yet, modern occupational and environmental epidemiology, primary sources of information for the purpose of the setting of reliable exposure limits, frequently remain encumbered by the lack of knowledge or understanding about what was really being measured in the distant past. It therefore remains important to be able to elucidate the nature and intensity of those early exposures and to be able to relate them to modern concepts of what constitutes 'health-related' exposure. *Retrospective exposure assessment* represents a challenging area for occupational and environmental hygiene scientists, requiring full appreciation of the rationales and physical principles underpinning the sampling instrumentation used in the past, and how these – and the instruments themselves – have evolved as scientific knowledge has advanced.

Aerosol Sampling: Science, Standards, Instrumentation and Applications   James Vincent
© 2007 John Wiley & Sons, Ltd

13.2   Occupational aerosol sampling

13.2.1   Sampling strategies and philosophies

Workplace aerosol exposures take place over limited periods defined by working shifts and patterns. In modern developed societies these are typically 8 h a day, 5 days a week, although in other eras, or even today in many other countries, these patterns differ, often tending towards greater proportions of life spent at work. In reality, occupational exposures are superimposed on the exposures which members of the general population experience in the ambient environment. However, they are highly specific to the type of organisation and type of work done, changing as old technologies are phased out and new ones emerge. Occupational hygiene addresses only those exposures associated with work. For specific agents characteristic of the work done, these tend to be significantly higher there than might be found elsewhere, and this argument has usually been sufficient to justify the paucity of effort in combining the considerations of ambient and occupational exposures.

As described in earlier chapters, there have been essentially two strategies or methods of approach driving the development and application of aerosol instrumentation in occupational hygiene: area (or static) sampling, involving monitoring the environment in specific work areas or associated with specific processes, and personal sampling, involving the determination of the health-related exposures of individual workers. Historically, the approach adopted has been dictated as much by the practical and economic constraints imposed by the availability of suitable instruments (and personnel to operate them), and by requirements – if any – to measure the chemical or mineralogical composition of the sampled aerosol, as by the overarching philosophy of hazard evaluation. So instruments and strategies of their deployment have evolved in parallel with improved understanding of hazards and their associated risks to workers. Such technical progress has gone through a number of important transitions over the years. During the early 1900s to 1930s, relatively short-period gravimetric samples (for nominally 'total dust') were usually taken with area samplers placed at breathing height in the general vicinities of representative workers during active operations, primarily to locate sources and to test the effectiveness of dust prevention measures. This invariably required the participation of an attendant operator. During the 1920s to 1950s, following realisation that the mass of the nonrespirable large particles in the total dust sampled over-estimated the extent of the risk, short-period or 'snap' sampling for microscopic counts of particles considered to be small enough to penetrate deep into the lung was widely carried out, using instruments such as the konimeter (described later in Chapter 15). Such sampling was carried out by an attendant operator who usually ensured that the samples were taken as close as possible to the workers' breathing zones. In this way, such samples could be regarded as 'personal'. But then, from the 1950s onwards, increasing emphasis was given to full-shift *time-weighted average* (TWA) particle-number sampling, by area sampling either near representative workers as a measure of health-related exposure or at strategic sampling positions to monitor dust control, again (usually) under the supervision of an attendant operator. This changed in the 1960s, with the emergence of long-running, full-shift area or static samplers that employed the aerodynamic selection of the respirable fraction of airborne particles, enabling them to operate unattended and providing samples of sufficient quantity that could be assessed in terms of mass and/or chemical (or mineralogical) composition. Eventually, however, from the 1970s onwards, increasing emphasis was given to TWA personal sampling, especially in countries with highly developed occupational hygiene cultures. This ability arrived with the advent of samplers and (particularly) pumps that could be miniaturised to the point where they could conveniently be carried by – or worn on the person of – the worker. As related in the historical account by Sherwood (1997), the possibility of small sampling pumps first became apparent in the late 1950s when health physicists at the United Kingdom Atomic Energy Research Establishment at Harwell came to realise that small, constant-speed DC electric motors like those developed for the early phonographs might be applied to air pumps, much smaller than anything that had previously been conceived. The first practical sampling device based on this new concept was reported in 1960 (Sherwood and Greenhalgh, 1960). The first commercial personal sampling pumps subsequently appeared around 1962, and by the early 1970s were becoming routine occupational hygiene tools. Now, 30 or so years further on, personal sampling is widely regarded in most countries as the *only* satisfactory way to assess the exposures of a workforce.

Until the 1960s considerable attention was directed at the peak levels of dust concentration during working shifts, often confined to periods of aerosol-producing activity. The relevance of full-shift sampling to health was not clearly understood at the time, and so the focus of much sampling was placed on the needs to identify sources in order to guide control measures. However, the eventual introduction of full-shift samplers coincided with acceptance that the TWA aerosol concentration over this period was the more appropriate index of risk to health (Wright, 1953; Orenstein, 1960). But another, increasingly important, practical consideration was the desirability of reducing the burden on occupational hygienists associated with evaluating multiple short-period samples.

13.2.2 Indices of aerosol exposure

Again, an important issue in aerosol exposure assessment is to identify the most appropriate *concentration* of aerosol particles. Even as far back as the early 1900s, particles of certain specific composition (e.g. containing crystalline silica) were already acknowledged to be dangerous. Indeed, at that time, many authorities considered crystalline silica (or quartz) to be the only significantly dangerous component of some mineral dusts. The mass proportion of such chemical species in high-volume gravimetric samples was initially inferred from the overall dust collected on filter media, or – when the sample mass was sufficient (usually of the order of grams) – quantitated using wet chemical or X-ray diffraction methods. Often, however, the amount of particulate matter collected was insufficient for the use of such methods, and so compositional analysis was often performed on settled dust collected from floors or ledges or taken from industrial filter units. In some cases, samples taken for microscope-based particle counting were heated to burn-off carbonaceous material and acid-washed to remove the soluble component, allowing limited mineralogical discrimination of the residue when viewed under the microscope. Later, however, more advanced analytical instrumentation became available for analyzing samples of particulate matter *in situ* on the collection filters. During the 1930s to the 1960s, some authorities considered that, for free crystalline silica, the surface area of the respirable particles might be a better hazard index than the mass (Orenstein, 1960). Here it was suggested that surface area concentrations could be assessed by means of light scattering or absorption measurements on liquid suspensions of sampled particles (after large particles had been removed by sedimentation) or on particles sampled onto glass slides.

13.2.3 Early gravimetric samplers for 'total' aerosol

Gravimetric sampling involves determination of the overall mass of particulate matter collected, from which, knowing the sampling time and the sampling flow rate, the airborne mass concentration of the sampled aerosol may be determined. South Africa led the way in the early days of gravimetric sampling, followed by the UK and the USA. From about 1900 onwards, *cotton-wool filters* were used in mine dust sampling, where the mass of sampled aerosol was determined from weighing of the cotton-wool filter media before and after sampling. In the method described by Thomas and McQueen (1904), the cotton plug was incinerated and the residue weighed. Around the same time, from about 1902 to 1903, some early dust determinations were made using the *sugar tube*, in which a packed bed of sugar granules was used as the filter medium, and the mass of collected insoluble material was determined after the sugar granules had been dissolved in water (e.g. Boyd, 1930). Here, incidentally, it is of interest to note that this method first emerged much earlier as a bioaerosol sampler (Frankland, 1886). The sugar tube was used routinely in South African mines from 1914 until 1936. Indeed, records show that over 35 000 such samples were taken and analyzed in South African underground mines during 1919 alone. From 1923 onwards, *paper thimble-type collectors* became popular. In these, porous paper was used as the filter media, and concentration was again obtained from determination of the change in the filter mass after sampling.

The disadvantages of these earlier devices soon became apparent. Firstly, in industries such as mining and elsewhere, the observed prevalence of lung disease did not fall consistently with reductions in dust levels as measured by such sampling methods. Secondly, the earlier instrumentation was cumbersome, a special disadvantage in many confined working environments like those found in mines. Thirdly, it later emerged also that very variable results were obtained because particles of different sizes were not collected with the same efficiency, depending on the aerodynamics governing how particles were aspirated into and collected within sampling instruments. This problem was particularly acute for the sampling of coarser aerosols. To overcome this particular problem in the case of the early sugar tube, crude attempts were made to separate the over-large particles from collected samples. But, in due course, the cumulative impact of disadvantages like those mentioned became overwhelming, leading to a switch of emphasis to aerosol measurement based on the counting of individual particles.

13.2.4 Particle count samplers

According to Drinker and Hatch (1934), the earliest mention of particle sampling and counting by microscope was by Cunningham (1873). Somewhat later, as mentioned in Chapter 11, McCrea had noted in 1913 that particles in the lungs of miners observed post mortem were between 1 μm and 7 μm in diameter, thus identifying the need to selectively measure the finer particles as an appropriate index of exposure for certain types of dust-related lung disease. This appears to be the first reference to what we now refer to as 'particle size-selective' exposure assessment for aerosols.

The basic problem for particle-count sampling is to collect particles in a form that can be examined microscopically, classified according to their size and counted. In the early days this raised some technical difficulties because filters were not available that would permit collected particles to be examined directly. The membrane filters which later became the primary means for the routine collection and counting of asbestos fibers by optical microscopy did not become available until the 1960s. So, for the sampling instruments that were used pre-1960, particles were deposited onto glass slides by combinations of impaction, impingement or thermal precipitation. Such slides were very convenient for examination by conventional optical microscopy.

One such instrument which found very wide usage is the portable *konimeter* first described by Kotzé (1916), which allowed short snap samples to be taken close to a worker followed by microscopic analysis of particles deposited inside the instrument by impaction (more details are given in Chapter 15). Various types of konimeter were widely used from the 1930s until as recently as the 1980s, particularly in the extraction and metals industries in Canada, Australia and South Africa. Before the advent of gravimetric standards, the konimeter was the basis of dust exposure standards in those and many other countries. Le Roux (1970) reported that, in South Africa, mine officials took about half a million konimeter samples per annum while government inspectors took about 30 000 samples during routine inspections. In Western Australian underground mining operations, of the order of 40 000 konimeter samples were recorded for the period from 1925 to 1977 (Hewson, 1996).

Other, related aerosol sampling instruments were fashionable during the same period, including the *impinger* proposed by Greenberg and Smith (1922), followed by a miniaturised version, the *midget impinger*. These were popular for many years among occupational hygienists in North America in the period before gravimetric sampling of fine particles was introduced, providing the basis of occupational exposure limits for many years. Also in the 1920s, the *Owens jet* appeared (Owens, 1922), and was used in early investigations of airborne asbestos dusts in the UK, the USA and Germany.

The *thermal precipitator* that first appeared during the 1930s was attractive in that it provided a 'gentle' means of depositing the particles such that particles that were subsequently observed under the microscope would not be damaged by fluid mechanical shear forces or impact fracture during deposition. It was felt by aerosol scientists of the day that particles would be more representative of those inhaled by the exposed workers, in particular in the sense that the individual airborne particulate entities would not have undergone any significant changes during the act of sampling. The *standard thermal precipitator* (STP) developed by Green and Watson (1935) operated at the low continuous flow rate of 8 mLpm, and thermophoretic particle deposition took place onto glass slides placed either side of a heated wire. It was used as the standard instrument in British coal mines during 1949 to 1965. Later a modified device, the *long-running thermal precipitator* (LRTP) (Hamilton, 1956), was built which excluded the collection of large particles by virtue of a laminar-flow horizontal elutriator placed ahead of the thermal collection section. This instrument became the standard instrument in British coal mines during 1965 to 1970, and was also used elsewhere and for collecting asbestos fibers. A similar instrument was used in South Africa (Kitto and Beadle, 1952). In most versions of the thermal precipitator, collected particles were usually assessed by optical microscopy.

Particle counting approaches had become popular when it became clear that the earliest 'total dust' gravimetric methods were fundamentally flawed, based on emerging new knowledge derived from aerosol science. But it later transpired that particle counting too had its disadvantages. Firstly, the effort to visually count the collected particles was very labor intensive and there were considerable inter- and intra-observer variabilities. Secondly, for instruments like those described, aspiration and collection efficiencies were poorly defined, so that biases associated with particle size-dependent entry and deposition effects were unknown. Thirdly, measured concentrations based on particle count were found not to correlate particularly well with health effects. Rather, it was shown quite early on that the mass concentrations of fine particles provided much better correlations, in particular for certain types of lung disease such as pneumoconiosis (e.g. Bedford and Warner, 1943). So, from the 1950s onwards, new knowledge of the physics of how particles are transported in the air, are inhaled, and are deposited in the respiratory tract, facilitated a return to gravimetric sampling, now much improved (compared with earlier efforts) by the identification of health-relevant aerosol fractions defined in terms of their aerodynamic properties and the development of instruments that could measure them. As a result, particle counting-based sampling devices were gradually superseded by gravimetric samplers.

Meanwhile, however, interest in particle counting approaches have continued to the present day in certain key areas. For fine fibrous particles, the interest was stimulated originally by concerns about the very serious health effects arising from the inhalation of asbestos fibers, but has since been extended to related concerns about all fine fibrous particulate matter. Although primary asbestos manufacturing or using industries virtually disappeared during the 1990s, at least in developed countries, occupational exposures still occur during maintenance or demolition work involving older facilities or equipment. Occupational hygienists have therefore needed to remain conversant with the appropriate sampling and analytical methods. From the 1960s onwards, the membrane filter method has been critical in this application, providing means for the efficient collection of particles which are aerodynamically very fine and then – after sampling has been completed – the ability to make the filter media itself transparent (i.e. 'cleared') so that the collected particles can be viewed by optical microscopy. Here, phase contrast microscopy is needed because of the small difference in refractive index between the fibers and the membrane filter material. In Britain, Holmes (1965) described the application of the membrane filter method towards the assessment of the concentration of 'respirable' airborne asbestos fibers, defined as particles which, when viewed under the microscope, had length greater than 5 μm and aspect ratio greater than 3 to 1, later adapted in some quarters to refer additionally to fibers with diameter less than 3 μm. These definitions have continued to be the basis of regulatory standards for airborne asbestos fibers almost world-wide.

From the beginning, bioaerosols have been assessed in terms of the number of organisms collected, expressed in terms of colony forming units (CFUs). Interest in bioaerosols in working environments has risen quite sharply in recent years, as better understanding has been gained into the nature and causes of lung disease such as asthma.

13.2.5   Emergence of gravimetric samplers for the respirable fraction

In the early 1950s, there was sufficient information, experimental and theoretical, to enable consideration of the sizes of inhaled particles which were capable of penetrating to the alveolar region of the lung (see Chapter 11). In particular, this highlighted the significance of particle aerodynamic diameter (d_{ae}), as distinct from the geometrical measure of particle size that is associated with microscope counting. Driven at the time largely by concerns about the high incidence of pneumoconiosis among coal miners, this led to a succession of conventions for what has been referred to ever since as the 'respirable fraction'. Most sampling for the respirable aerosol fraction has subsequently been based on the premise that the index of interest is the mass of overall respirable aerosol or of particular health-related species. However, in view of the continuing interest in some quarters in particle surface area concentration as a relevant metric (e.g. for crystalline silica in relation to silicosis), Talbot (1966) devised an instrument, based on the analysis of optical diffraction patterns, for assessing respirable surface area concentrations for samples collected by thermal precipitator. This instrument was used for a while in South Africa.

The first respirable sampling instruments took the form of horizontal elutriators, in which the particles of interest corresponding to the British Medical Research Council (BMRC) curve were those which penetrated in laminar flow through the narrow rectangular channel of an elutriator pre-separator and were collected on a filter. The very first commercial sampler was the British *HEXHLET* (Wright, 1954), a large device with a flow rate of 100 Lpm. This was used widely in British foundries in the 1950s and 1960s. Later, a smaller 2.5 Lpm version, the portable *MRE Type 113A* (Dunmore *et al.*, 1964), was developed specifically for use in British coal mines, where it was used as the basis of regulatory standards from 1970 until quite recently. This device, although portable, was intended for use as a static (or area) sampler. Other horizontal elutriator-based devices were used in European coal mines and other workplaces during the same period.

Initially in the USA, and later elsewhere, respirable aerosol samplers emerged with pre-selectors based on cyclones. Indeed, the 1968 American Conference of Governmental Industrial Hygienists (ACGIH) respirable aerosol curve was originally derived from consideration of the penetration characteristics of miniature cyclones. The great advantage of the cyclone over the horizontal elutriator was that it could be miniaturised to the point where the sampler could be worn on the lapel of the worker, and so be deployed as a personal sampler along with a miniaturised sampling pump worn on the worker's belt. Many commercial versions of cyclone samplers have since appeared, most with sampling flow rates between 1.5 and 2.5 Lpm, some with performance characteristics matching the ACGIH curve, some matching the BMRC curve, and an increasing number matching the contemporary respirable fraction definitions that have come along in recent years.

Finally, it should be noted that, during the transition from particle counting to gravimetric sampling that took place during the 1960s and 1970s, there was the need – at least from the point of view of occupational epidemiology, and hence standards setting – to be able to convert particle count exposure data to equivalent respirable mass concentration data. This set difficult problems for aerosol science and occupational hygiene researchers, because each index was determined by local factors such as aerosol type and particle size distribution such that any single conversion could not be generalised for all situations. So such conversions were carried out as needed on the basis of field studies in specific industrial situations (see Chapter 22).

13.2.6 Emergence of gravimetric samplers for 'total' and inhalable aerosol

As described in Chapter 10, it has long been recognised that ill-health arising from aerosol exposure is not necessarily confined to the lung. There are substances which can produce health effects after deposition anywhere in the body, including not only the lung but also other parts of the respiratory tract, as well as elsewhere in the body if the aerosol material is soluble. During the 1970s, occupational hygiene scientists began to question 'total airborne particulate' as a valid, health-related metric of aerosol exposure. In addition, by the late 1970s, the science of aerosol sampling had advanced to the point where it became clear that simply drawing a known volume of air through a sampling orifice of arbitrary dimensions and onto a filter did not – in itself – provide a measure of true total aerosol. So, in the late 1970s, a new approach was first proposed by Ogden and Birkett (1977), in which 'total' aerosol would be represented by what is actually inhaled, based on knowledge of the particle size-dependent efficiency with which particles were aspirated into the nose and/or mouth during breathing. Out of this the *inhalable fraction* emerged. Wind tunnel studies during the 1980s and 1990s, and field studies during the 1990s, confirmed that the performances of most of the samplers previously used for collecting 'total' aerosol did not adequately match this definition (e.g. Mark and Vincent, 1986; Kenny *et al.*, 1997). So a new generation of sampling instruments began to emerge to fill this instrumentation gap. The first was the *ORB sampler* suggested by Ogden and Birkett (1978). It was originally suggested as a sampler that could be applied in both the area and personal sampling modes. Although the ORB was developed only as a prototype, was never applied in field studies and was never made available commercially, it represented an important first milestone in the development of instruments that, from the outset, were intended for collecting specifically the inhalable fraction. This prompted other aerosol scientists towards the development inhalable aerosol samplers that could realistically be commercialised and be applied in the real world of occupational hygiene. Out of this, the Institute of Occupational Medicine (IOM) personal inhalable aerosol sampler first appeared during the late 1980s, and its commercial availability was a significant factor in decisions by some leading standards-setting bodies to begin the establishment of occupational exposure limits based on this new fraction. Subsequently other inhalable aerosol samplers have been proposed, and some of these too are commercially available.

13.2.7 Other aerosol fractions

As described in Chapter 11, the past two decades has seen a widening of the scope of particle size-selective criteria for aerosol exposure assessment. In the early 1980s, ISO produced its first set of criteria, encompassing the coarser inhalable, the intermediate thoracic and the finer respirable fractions. These were followed by a somewhat similar, but quantitatively different, set of criteria from ACGIH, and another from the Comité Européen de Normalisation (CEN). In the late 1980s, however, these combinations were harmonised into a single set of criteria for the three fractions indicated, and this was widely accepted by many of the world's standards-setting bodies (Soderholm, 1989).

13.2.8 Sampling to measure aerosol particle size distribution

For investigational purposes, occupational hygienists have frequently sought aerosol information beyond knowledge of the airborne concentration alone, including particle size distribution. In the occupational hygiene context, such information can be especially valuable in (a) enabling estimation of the dose of particulate matter deposited in – or penetrating to – particular regions of the respiratory tract, and (b) the choice of control strategy or technology. In the early years of aerosol sampling, the most common approach to estimating particle size distribution was by the microscopic analysis of samples collected on suitable substrates, usually glass slides. For many years, for example, samples obtained using the thermal precipitator were analyzed to provide quite accurate particle size number distribution information over the particle size range from about 0.2 to 10 μm, enabling other parameters – such as the particle mass distribution – to be estimated based on assumptions about particle shape and density. By the use of the thermal precipitator, it was possible to argue that the particle size distribution was relatively unbiased with respect to the collection efficiency characteristic of the sampler. In such an approach, particles collected on glass slides were counted and sized visually under the microscope by comparison with circular profiles depicted on an appropriate eyepiece graticule. The *Patterson-graticule* used by Green and Watson had 10 such circles with apparent diameters ranging from 0.2 to 5 μm. The procedure was tedious and subject to error for 'overcrowded' samples. It was from exposure data obtained in this way, coupled with corresponding health effects data, that Bedford and Warner were able to make their highly influential conclusion that it was the mass of coal dust particles smaller than 5 μm that provided the index of exposure most relevant to the risk of pneumoconiosis for workers in underground coal mines.

The term 'aerosol spectrometer' has been used to refer to sampling instruments that segregate particles directly into a continuous size spectrum or, more correctly, into well-defined size bands, from which the continuous particle size distribution can be inferred or calculated. A wide range of such instruments has emerged over the years, utilising a variety of aerosol mechanical principles. These include inertial, gravimetric and centrifugal spectrometers, some of which will be described later in Chapter 18. Many such devices had sampling flow rates that were too small to enable the determination of particle size distribution according to mass, and so were of limited value in relation to occupational hygiene applications. They nevertheless have provided valuable information about the relationships between 'microscopical' and aerodynamic particle size. One class of aerosol spectrometer has risen above all the others and found very widespread use by occupational hygienists in the workplace environment. This is the *cascade impactor* which first emerged in the 1940s (May, 1945). The general principle of the cascade impactor is that sampled aerosol is passed through a succession of impactor stages with progressively smaller particle 'cut' sizes. The amount of particulate matter collected masses on the various stages of the instrument define the cumulative particle size distribution (by mass) to an accuracy determined – indeed limited – by the sharpness of the cuts of the impactor stages. The earliest cascade impactors were relatively large devices, and so could be used in the workplace only as area samplers. When small personal sampler versions appeared during the late 1980s, interest in their potential as investigational exposure assessment tools for occupational hygienists rose sharply. In particular the eight-stage personal cascade impactor first described by Marple and his colleagues (Rubow *et al.*, 1987) has been commercially available for several years and has been widely used by occupational hygienists.

13.2.9   Direct-reading instruments

In parallel with the direct-reading sampling instruments that have found use for the measurement of ambient atmospheric aerosols in the past three decades, similar instrumentation has been of increasing interest to occupational hygienists. Today it is seen as an approach to aerosol sampling that, after the initial capital investment, can provide fast results and yet be economic in terms of the time and effort on the part of occupational hygienists. The earliest practical direct-reading aerosol instrument for workplace applications appears to have been the '*hazard*' device described by Drinker and Hatch as early as 1934, since when a large variety of instruments has appeared, many of which were portable but were used

Table 13.1 Summary of time lines of approaches to aerosol sampling in occupational settings

Strategy/*Index*	1900	1910	1920	1930	1940	1950	1960	1970	1980	1990	2000
'Snap' in workers' breathing zones *Number concentration*				▓	▓	▓	▓	▓	▓		
Full shift TWA fixed point, static (or area) *Number concentration*					▓	▓	▓				
Full shift TWA fixed point, static (or area) *Mass concentration ('total')*	▓	▓	▓	▓	▓	▓	▓	▓	▓	▓	▓
Full shift TWA fixed point, static (or area) *Mass concentration (respirable)*							▓	▓	▓	▓	
Full shift TWA personal *Mass concentration ('total')*								▓	▓	▓	▓
Full shift TWA personal *Mass concentration (respirable)*								▓	▓	▓	▓
Full shift TWA personal *Mass concentration (inhalable)*										▓	▓
Fixed point and portable direct-reading instruments							▓	▓	▓	▓	▓

as fixed point static samplers. The development of direct-reading instruments in the 1990s that were sufficiently miniaturised that they could be deployed as personal monitors opened up the possibility of some very interesting and useful control strategies for workplace aerosols. For example, real-time measurements of worker exposure using an appropriate personal monitor have been correlated – through the use of video techniques – directly with worker activity (e.g. Rosen, 1993). Such an approach is proving especially helpful in identifying what types of activity leads to the greatest exposure and, in turn, has helped educate workers on how to minimise such exposures.

13.2.10 Overview

Table 13.1 provides a concise summary of the history of aerosol sampling in occupational settings, drawing together the main threads and time line of what has been reviewed above. It shows the considerable overlap between philosophies and approaches over the years. It reveals the primary milestones, most notably the transitions between particle number and particle mass concentration measurement, between static and personal sampling, and the emergence of the different phases of particle size-selective sampling. It hides the differences between countries or between industry sectors, and so paints only a very broad picture.

13.3 Ambient atmospheric aerosol sampling

13.3.1 Sampling strategies and philosophies

The exposures of people to aerosols in the ambient atmospheric environment takes place over 24 h a day, 7 days a week, and for every week of the year. The nature of such exposures changes as individuals move around during their daily lives and spend differing proportions of their time indoors (at home or elsewhere), at work, or outdoors. Specific such exposures differ widely, both qualitatively and quantitatively. So an aerosol sampling strategy that is truly representative should ideally involve a method that allows all those various contributions to exposure to be identified, measured and combined in a way that reflects the overall risk to health, and so is representative of each exposed individual. However, this presents a very difficult technical and strategic challenge. Sampling approaches have evolved, therefore, that address aerosol levels in the general ambient environment to which most people are exposed most of the time. These are specified mainly by geographical location. Most air pollution measurement for ambient aerosols has so far been carried out using such fixed point, static (or area) sampling instrumentation.

As discussed above, personal aerosol sampling has been the preferred mode of aerosol exposure assessment in occupational settings in many countries since the 1970s. It is also desirable in relation to ambient exposures. Recently, acknowledging that static sampling at fixed point outdoor locations provides less-than-adequate assessment of the actual exposures of individuals or groups of people, efforts have been made to introduce personal sampling to the ambient atmospheric environment. But, so far, such measurements have been confined to specific research enquiries and there are as yet no air quality standards based on personal exposures.

13.3.2   Indices of health-related aerosol exposure

The primary index of aerosol exposure is the *concentration* of aerosol particles, since this is what may be related to the dose received by exposed people. However it has long been an important question as to how concentration should be defined. Particle size fraction? Species? Mass or some other index of 'amount'? In these, as for occupational aerosol exposures, there has been a long history of changing viewpoint.

In Britain, Cohen and Rushton (1912) described air pollution data records, taken mainly in the City of Leeds in the north of England, that went back as far as the late 1800s. Even then, the primary focus of attention was atmospheric pollution in the form of smoke arising from combustion, from both industrial and domestic sources. For the latter, much of it was the result of the long culture of home heating by the burning of coal. Although very variable, the smokes in question were later universally characterised by the presence of carcinogenic hydrocarbon species (e.g. 3, 4-benzpyrene), thus confirming a clear link with public health. Eventually, as a result of the 1956 Clean Air Act, a National Survey was introduced in Britain in 1961 that required the measurement of particulate air pollution expressly in terms of 'black smoke'. Somewhat later this was formalised still further by the prescription by the British Standards Institution (BSI, 1969) of the 'smoke stain' method, involving optical assessment of the 'blackness' of particulate matter deposited on a filter (see Chapter 17). The black smoke approach has persisted in Britain and elsewhere in Europe to this day, where aerosol sampling and analysis along the lines of the BSI method still features in atmospheric environmental regulations (e.g. as in European Community Directive 80/779/EEC). The APHEIS (Air Pollution and Health: a European Information System) program, coordinated by the Institut de Veille Sanitaire (InVS) in Saint-Maurice, France and the Institut Municipal de Salut Publica de Barcelona (IMSPB) in Spain, recently reported measurements of particulate air pollution in 26 cities in 12 European countries taken during 2001, expressed in terms of two indices, one of which was black smoke (the other was PM_{10}, see below) (Medina *et al.*, 2002). The World Health Organisation (WHO) also continues to list guidelines based on the black smoke index. But the United States Environmental Protection Agency (EPA) lists no such requirement.

The term 'smog' was first used by Des Voeux in 1905 to describe the aerosol that he observed when fog occurred in the presence of smoke. Smog episodes were most likely to occur during meteorological conditions associated with calm air and fog, along with high concentrations of smoke. The health implications of the effects of such combinations were later dramatically illustrated during the 1952 smog incident in London where as many as 4000 people were thought to have died; another 700 or so were thought to have died in the subsequent 1962 episode. In general, it is a reasonable assumption that smog levels would have been quite well reflected in the measurements of black smoke, and this is likely to have spurred the continuing use of the black smoke index in Britain and elsewhere in Europe. More recently, however, use of the term 'smog' has been extended to describe the complex form of aerosol pollution that occurs in the presence of photochemical reactions.

More recently, the term 'total suspended particulate' (TSP) has referred in principle to the mass concentration of all particles considered to be airborne, although in reality it has always been defined by whatever sampling instrument was chosen to collect it. In Britain and elsewhere in Europe, this fraction was measured using sampling instruments whose design varied over time, depending on the technology currently available at the time. In the early 1900s, filter media were quite rudimentary, and cotton wool filters were commonly used. Later, however, improved filter technology increased the options for sampling. Many samplers for TSP were proposed, and some of them are described later in Chapter 17. But, as will be seen there, they are notable for the general lack of consistency in performance from one to the other. This is partly because TSP has never been defined in terms of any 'target' sampler performance criterion.

TSP, regardless of how poorly defined, formed the basis of primary standards in the USA for many years, indeed up until 1987. Then, however, in a flurry of activity, EPA published revised particulate standards to account for the deeper penetration of particles into the respiratory tract, and hence to a finer aerosol fraction. The aerosol mass concentration for particles with aerodynamic diameter nominally less that 10 μm was chosen for defining the new index, providing in turn a new criterion for sampling. This index is what has become widely known as 'PM$_{10}$', and it became the primary atmospheric aerosol standard in the USA. As such it was the first particle size-selective index (or criterion) for ambient atmospheric aerosol sampling that was linked directly to the physical nature of human exposure to aerosols, representing the penetration of particles which, after inhalation, may penetrate beyond the larynx and into the lower respiratory tract. Many new samplers appeared to meet the new need, and the old TSP standard was relegated to the status of a secondary standard. The PM$_{10}$ criterion subsequently became widely accepted in other countries, as reflected in the APHEIS report mentioned above. Then in 1997 EPA proposed a new, additional fine particulate criterion for those particles that, nominally, have aerodynamic diameter less than 2.5 μm, aimed at providing measures of combustion-related aerosol. This new criterion was referred to as 'PM$_{2.5}$' and it became a second standard (on a par with PM$_{10}$), while the TSP standard was dropped. With the cementing of EPA's philosophy on particulate matter (PM$_{10}$ and PM$_{2.5}$), and its wide acceptance in other countries, a whole new generation of sampling instruments, with particle size-selective performance curves matching the definitions in question, emerged.

Meanwhile, in 1981 the ISO had first proposed a quantitatively different but similar definition for the same intermediate fraction, with a view to applications both in ambient atmospheric and occupational aerosol exposure assessment. ISO referred to this as the 'thoracic' fraction. In addition to the 'thoracic' fraction, ISO also recommended criteria for the coarser, inhalable fraction that can be inhaled through the nose and/or mouth during breathing and also for a finer, respirable fraction. But neither of the ISO criteria were adopted for routine ambient aerosol monitoring and harmonisation of criteria and methods now appear to have coalesced around the EPA PM$_{10}$ and PM$_{2.5}$ criteria.

Interest in airborne fibers grew from the 1960s onwards as long, thin particles of asbestos fibers were acknowledged to be especially hazardous. Apart from potential exposures in some workplaces, there was special concern about exposures to such particles in indoor living spaces, particularly in locations inhabited by vulnerable populations such as children (e.g. schools). For these, mass concentration was found to be difficult to measure at low concentrations, especially in the presence of other, nonfibrous particulate material. So the particle number concentration was determined by the manual counting of particles of appropriate shape and dimensions after they had been collected on filters when viewed under the light or electron microscope. Later, in the 1990s, although they had long been of interest to aerobiologists, concerns about biological organisms in indoor living environments rose sharply. Such organisms were sampled and assayed, for example, by collecting them onto, and culturing them in, nutrient media and observing the colonies visually. The concentrations of collected organisms were then expressed in terms of the numbers of colony forming units (CFUs), following basic methodology that went back even as far as the 19th century.

13.3.3   Indices for coarse 'nuisance' aerosols

Interest in sampling atmospheric aerosol has extended in some cases beyond the adverse health effects experienced by people after inhalation. A significant economic burden may be placed on individuals or businesses impacted by the unwanted deposition of coarse, 'nuisance' particles. Such effects were felt in the near vicinities of certain industrial operations, including quarrying, electricity generating stations, landfill sites, etc. and have been manifested, for example, in psychological stress and lowering of property values (Ridker, 1970). Such concerns surfaced in Britain and a number of other countries during the 1970s and were serious enough to require action, beginning with monitoring by aerosol sampling.

13.3.4   Direct-reading instruments

Development and applications of direct-reading instrumentation for the measurement of atmospheric aerosols began to appear during the second half of the 20th century. These derived from the range of potential physical interactions involving airborne particles, or by particles after they have been deposited onto surfaces. In all such devices it was a general feature that information about aerosol properties was converted into electrical signals that could be processed and recorded. Some of the earliest automatic instruments for the direct measurement of airborne particles were reviewed by Green and Lane (1964), including photometers and particle counters based on light scattering by ensembles of particles or just single particles, and detectors operating on the basis of the electrical mobility of charged particles. In later decades other devices appeared, operating on the basis of particle growth by condensation in supersaturated vapors (condensation particle counter) or time-of-flight in a changing flow field (aerodynamic particle sizer). Others were based on mechanical properties associated with the changing mass of particulate material on substrates as particles were collected there, including piezoelectric and tapered element mass balances. The commercial development and availability of such instruments has increased sharply in subsequent years as more sophisticated and cost effective electronic systems have become available, accelerating still further with the arrival of enhanced computer-based capability. Direct-reading aerosol instrumentation is now seen as an approach to aerosol sampling that, after the initial capital investment, can provide fast results that are also economic in terms of the time and effort on the part of the environmental hygienist. More details are given later in Chapter 20.

13.3.5   Overview

Some of the main points of what has been described are summarised in Table 13.2, providing a short overview of the history of aerosol sampling in the ambient atmospheric environment. It shows the considerable overlap between philosophies and approaches over the years. It reveals the primary milestones,

Table 13.2   *Summary of time lines of approaches to aerosol sampling in the ambient atmospheric environment*

	1900	1910	1920	1930	1940	1950	1960	1970	1980	1990	2000
Strategy/*Index*											
Fixed point *'Nuisance' dust*								▓	▓	▓	▓
Fixed point *Mass concentration (TSP)*		▓	▓	▓	▓	▓	▓	▓	▓	▓	▓
Fixed point *Black smoke*						▓	▓	▓	▓	▓	▓
Fixed point *Mass concentration (PM$_{10}$)*								▓	▓	▓	▓
Fixed point *Mass concentration (PM$_{2.5}$)*									▓	▓	▓
Fixed point and portable direct-reading instruments									▓	▓	▓
Personal sampling *PM$_{10}$* and *PM$_{2.5}$*										▓	▓

most notably the introduction of the black smoke index in some countries, the emergence of PM$_{10}$ and PM$_{2.5}$, and the beginnings of personal sampling for members of the general population.

References

Bedford, T. and Warner, C. (1943) Physical studies of the dust hazard and thermal environment in certain coalmines. In: *Chronic Pulmonary Disease in South Wales Coalminers. II. Environmental Studies*, Special Report Series No. 244, British Medical Research Council, HMSO, London, p. xi and pp. 1–78.

Boyd, J. (1930) Methods for determining the dust in mine air, as practised on the Witwatersrand, *Silicosis* (Supplement, Records of the International Conference held in Johannesburg), Studies and Reports F13, International Labour Office (ILO), Geneva, pp. 141–150.

British Standards Institution (1969) Non-reflective (dark) particulate matter, associated with the smoke stain measurement method, BS 1747 Part 2, London.

Cohen, J.B. and Rushton, A.G. (1912) *Smoke*, Edward Arnold, London.

Cunningham, D.D. (1873) *Microscopic Examination of the Air*, Superintendent of Government Printing, Calcutta.

Des Voeux, H.A. (1905) Fog and Smoke, Remarks during a presentation to the Public Health Congress of Great Britain.

Drinker, P. and Hatch, T. (1934) *Industrial Dust*, McGraw-Hill, New York.

Dunmore, J.H., Hamilton, R.J. and Smith, D.S.G. (1964) An instrument for the sampling of respirable dust for subsequent gravimetric assessment, *Journal of Scientific Instruments*, 41, 669–672.

Frankland, L.F. (1886) A new method for the quantitative estimation of the micro-organisms present in the atmosphere, *Philosophical Transactions of the Royal Society (London)*, 178, 113–152.

Green H.L. and Lane, W.R. (1964) *Particulate Clouds: Dust, Smokes and Mists*, 2nd Edn, Van Nostrand, London.

Green, H.L. and Watson, H.H. (1935) Physical methods for the estimation of the dust hazard in industry, with special reference to the occupation of the stone mason, Medical Council of the Privy Council, Special Report Series 199, HMSO, London.

Greenberg, L. and Smith, G.W. (1922) A new instrument for sampling aerial dust, United States Bureau of Mines Report of Investigations 2392.

Hamilton, R.J. (1956) A portable instrument for respirable dust sampling, *Journal of Scientific Instruments*, 33, 395–399.

Hewson, G.S. (1996) Estimates of silica exposure among metalliferous miners in Western Australia, *Applied Occupational and Environmental Hygiene*, 11, 868–877.

Holmes, S. (1965) Developments in dust sampling and counting techniques in the asbestos industry, *Annals of the New York Academy of Sciences*, 132, 288–297.

Kenny, L.C., Aitken, R.J., Chalmers, C., Fabries, J.F., Gonzalez-Fernandez, E., Kromhout, H., Lidén, G., Mark, D., Riediger, G. and Prodi, V. (1997) A collaborative European study of personal inhalable aerosol sampler performance, *Annals of Occupational Hygiene*, 41, 135–153.

Kitto, P.H. and Beadle, D.G. (1952) A modified form of thermal precipitator, *Journal of the Chemical Metallurgical and Mining Society of South Africa*, 52, 284–287.

Kotzé, Sir R. (1916) Final Report of the Miners' Phthisis Prevention Committee, Union of South Africa, Johannesburg (January 10, 1916).

Le Roux, W.L. (1970) Recorded dust conditions and possible new sampling strategies in South African gold mines. In: *Proceedings of the Johannesburg International Pneumoconiosis Conference* (Ed. H.A. Shapiro), Oxford University Press, Cape Town, pp. 467–469.

Mark, D. and Vincent, J.H. (1986) A new personal sampler for airborne total dust in workplaces, *Annals of Occupational Hygiene*, 30, 89–102.

May, K.R. (1945) The cascade impactor: an instrument for sampling coarse aerosols, *Journal of Scientific Instruments*, 22, 187–195.

McCrea, J. (1913) *The ash of silicotic lungs*, Publication of the South African Institute of Medical Research, Johannesburg.

Medina S., Plasència A., Artazcoz L. Quénel P., Katsouyanni K., Mücke H.-G., De Saeger E., Krzyzanowsky M. and Schwartz J. (2002) APHEIS Health Impact Assessment of Air Pollution in 26 European Cities, Second year report, 2000–2001. Institut de Veille Sanitaire, Saint-Maurice, France.

Ogden, T.L. and Birkett, J.L. (1977) The human head as a dust sampler. In: *Inhaled Particles IV* (Ed. W.H. Walton), Pergamon Press, Oxford, pp. 93–105.

Ogden, T.L., and Birkett, J.L. (1978) An inhalable dust sampler, for measuring the hazard from total airborne particulate, *Annals of Occupational Hygiene*, 21, 41–50.

Orenstein, A.J. (1960) Recommendations adopted by the Pneumoconiosis Conference. In: *Proceedings of the Johannesburg Pneumoconiosis Conference* (Ed. A.J. Orenstein), Churchill, London, pp. 619–621.

Owens, J.S. (1922) Suspended impurities in air, *Proceedings of the Royal Society (London)* A, 101, 18–28.

Ridker, R. (1970) *Economic Costs of Air Pollution: Studies in Measurement*, Praeger Publishers, New York.

Rosen, G. (1993) PIMEX–combined use of air sampling instruments and video filming: experience and results during six years of use, *Applied Occupational and Environmental Hygiene*, 8, 344–347.

Rubow, K.L., Marple, V.A., Olin, J. and McCawley, M.A. (1987) A personal cascade impactor: design, evaluation and calibration, *American Industrial Hygiene Association Journal*, 48, 532–538.

Sherwood, R.J. (1997) Realization, development and first applications of the personal air sampler, *Applied Occupational and Environment Hygiene*, 12, 229–234.

Sherwood, R.J. and Greenhalgh, D.M.S. (1960) A personal air sampler, *Annals of Occupational Hygiene*, 2, 127–132.

Soderholm, S.C. (1989) Proposed international conventions for particle size-selective sampling, *Annals of Occupational Hygiene*, 33, 301–321.

Talbot, J.H. (1966) A diffraction size frequency analyser with automatic recording of size frequency distribution and total and respirable surface areas, *Journal of Scientific Instruments*, 43, 747–749.

Thomas, R.A. and McQueen, W.P.O. (1904) The dust in the air and the gases from explosives in a Cornish mine (Dolcoath), and the efficacy of methods of dealing with them, Paper given at the Institution of Mining and

Metallurgy, December 15, 1904 (reported in Report of the Royal Commission on the Ventilation and Sanitation of Mines, Government Printer, Perth).

Vincent, J.H. (1989) *Aerosol Sampling: Science and Practice*, John Wiley & Sons, Ltd, Chichester.

Wright, B.M. (1953) The importance of the time factor in the measurement of dust exposure, *British Journal of Industrial Medicine*, 10, 235–244.

Wright, B.M. (1954) A size-selective sampler for airborne dust, *British Journal of Industrial Medicine*, 11, 284–288.

14

Sampling for Coarse Aerosols in Workplaces

14.1 Introduction

Chapter 10 laid out the scientific background to particle size-selective criteria for inhalable aerosol and how these form the basis of modern health-related aerosol exposure assessment for particles entering the body during inspiration through the nose and/or mouth. For workplaces the types of aerosols that are present are usually quite well defined, relating to identifiable processes taking place and materials used. A strong rationale has long existed for exposure assessment for aerosols where coarse fractions are known to be associated with health effects, and the traditional approach involved sampling for 'total' aerosol, a fraction defined not so much in terms of a conventional 'target' performance curve, but rather – and solely – by the sampling instrument chosen to do the job. A large number of such samplers have been available commercially over the years. The redefinition of 'total' aerosol in terms of the more scientifically relevant inhalability concept that emerged during the 1980s has led to the development of sampling instruments designed specifically for that fraction. Meanwhile, however, samplers originally intended for collecting 'total' aerosol have continued to be available and used, and many have been reevaluated in terms of their ability to collect the inhalable fraction. As will be seen below, some have passed this test but others have not.

In this chapter, samplers intended for the collection of the coarser aerosol fraction are reviewed, in particular in terms of their abilities to sample the inhalable fraction. In each case, sampler performance is assessed in terms of the recommended use of the instrument in question; that is, either on the basis of aspiration efficiency (A) where everything that is aspirated is assessed, or overall sampling efficiency (A^*) where it is only the aerosol that is collected on a filter (or other substrate) inside the body of the sampler that is assessed (see Chapter 1).

14.2 Static (or area) samplers for coarse aerosol fractions

14.2.1 'Total' aerosol

In general, fixed point, static (or area) aerosol samplers are intended to provide measurements of appropriate aerosol fractions in workplace air in specific locations, often under the assumption that this may be related to the actual aerosol exposures of people present in those locations. Long before the concept

Aerosol Sampling: Science, Standards, Instrumentation and Applications   James Vincent
© 2007 John Wiley & Sons, Ltd

of inhalability, and before the emergence of personal sampling, fixed point sampling for 'total' aerosol was carried out routinely in many workplaces, and was the strategy widely adopted by occupational hygienists. In turn it was tacitly acknowledged as such in occupational health standards, and many samplers were built for this purpose. The performances of many of these as functions of particle size have been studied and reported in the literature. Such information allows assessment of the performances of these instruments against the inhalability criterion.

One type of static sampler used routinely at one time for 'total' aerosol used in the UK in a number of industries consisted simply of a 60 mm diameter *open-face filter holder* through which air was drawn at flow rates in the range from 30 to 100 Lpm by means of a portable vacuum pump. There were two modes of operation. When used to sample the general workplace environment, the filter holder was fixed onto the sidewall of the outer casing of the pump. In the other mode, the filter holder was separated from the pump and mounted independently on an adjustable extension arm so that it could be positioned as close as possible to a worker or a particular operation. The device was used in this way in British nuclear industry workplaces for many years. Wind tunnel experiments reported by Mark *et al.* (1986) showed that the sampling efficiency was close to unity (100 %) for low wind speeds up to 1 m s^{-1} and for d_{ae} up to about 30 μm, reflecting weak inertial effects during aspiration from slowly moving air into the large entry of this instrument. In Germany in the 1970s, the omnidirectional, high flow rate *Gravicon* was proposed as an area sampler for use in workplaces (Coenen, 1973). Armbruster *et al.* (1983) tested a 400 Lpm version and showed that performance, as reflected in the sampling efficiency by which particles arrived at the filter inside the instrument, was close to the inhalability curve for d_{ae} up to about 50 μm when averaged over wind speeds from 1.25 to 8 m s^{-1}. However, less satisfactorily, a strong dependence on wind speed was observed.

There were several samplers that, although intended primarily as selective samplers for a finer sub-fraction (e.g. respirable), were used to provide information about 'total' aerosol. One was the horizontal elutriator-based British 2.5 Lpm MRE Type 113A area sampler for respirable aerosol that was used extensively for many years from the 1970s onwards – most notably in the British coal mining industry – in field research that also required estimation of 'total' aerosol (Dodgson *et al.*, 1971). Another was the cyclone-based German TBF50 static sampler that was intended primarily for the measurement of airborne alveolar dust but had also been used for sampling 'total' aerosol (Stuke and Emmerichs, 1973). In general, from all the available information on static aerosol samplers proposed for collecting 'total' aerosol like those described, it became apparent that most of them did not come close to matching the inhalability criterion. This stirred interest in the development of new samplers specifically designed to match the inhalability criterion.

14.2.2   Inhalable aerosol

The first static sampling instrument developed specifically for collecting the inhalable fraction was the prototype sampler proposed by Ogden and Birkett (1978). Referred to as the '*ORB sampler*' (an acronym for overall respirable burden), the device is shown in Figure 14.1. The body of the sampler was formed from a pair of hemispheres pressed out from thin aluminum or brass sheet, and the resultant spherical sampling head had 32 circular entry ports of diameter 2.5 mm located at about 10 mm above the equator (where the hemispheres joined). The top hemisphere also carried a thin brass 'halo'. Inside the head, sampled aerosol was deposited onto a 47 mm filter located in the plane of the equator and supported on a base of wire gauze. The prescribed sampling flow rate was 2 Lpm. In developing the ORB, Ogden and Birkett began from the point of view of trying to simulate the general aerodynamics of what happens in the case of human exposure and to relate performance to their own experimental human head data (Ogden and Birkett, 1977). Since the inhalability definition from the beginning had been based on

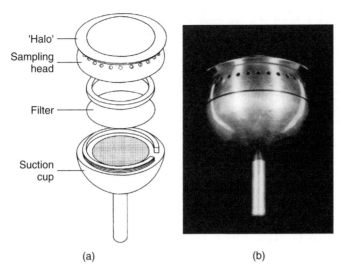

'Halo'
Sampling head
Filter
Suction cup

(a)   (b)

Figure 14.1   *The ORB static inhalable aerosol sampler: (a) exploded diagram to show external and internal features; (b) photograph of the actual instrument. Reproduced with permission from Ogden and Birkett, Annals of Occupational Hygiene, 21, 41–50. Copyright (1978) British Occupational Hygiene Society*

orientation-averaged aspiration efficiency for the human head, the first objective in the new instrument was to achieve a sampler which had no preferred orientation with respect to the wind. The uniform, multi-orifice arrangement of the ORB achieved this objective. Beyond this, progress towards the final instrument shown in Figure 14.1 was based largely on the age-old process of trial-and-modification. Although the flow pattern around the instrument was not studied in detail, the role of the halo was seen by Ogden and Birkett as helping to '... *prevent the air near the entry moving too fast...*', producing an effect similar to the idea behind stagnation point sampling as described by May (1967). Ogden and Birkett evaluated the performance of the ORB in both a wind tunnel and a calm air chamber covering wind speeds (U) ranging from close to zero up to 2.75 m s^{-1}, using monodisperse liquid-droplet test aerosols in the range of d_{ae} from 6 to 23 μm of di2-ethyl-hexyl sebacate produced by means of a spinning-disk generator. Performance was assessed in terms of the aerosol reaching the filter inside the sampler, and hence in terms of A^*. The results as a function of d_{ae} are summarised in Figure 14.2. Also shown are some later results reported by Vincent *et al.* (1983) in a wind tunnel for test aerosols generated from narrowly-graded fused alumina powders. For moving air and for smaller particles in the range of d_{ae} up to about 20 μm, although A^* approximately matched the general shape of the inhalability curve, it fell progressively below it for larger particles. For calm air, on the other hand, while the ORB oversampled with respect to the inhalability convention, its performance came close to what has since been proposed as an inhalability criterion for calm air (Aitken *et al.*, 1999; see Chapter 10). For larger particles, a significant difference was observed in the Vincent *et al.* results between cases where the sampler surface was greased and ungreased, respectively. These latter data provided a clear indication that there was oversampling associated with particle blow-off of the type described earlier in Chapter 7.

The ORB sampler has an important place in the historical development of aerosol samplers that set out to collect aerosol in a particle size-selective manner closely representing how particles are inhaled by aerosol-exposed humans. However, it was tested over only a limited particle size range compared with what we now know about human inhalability. Moreover, it was never taken beyond the prototype stage and was never made available commercially. In turn it was never used under practical conditions

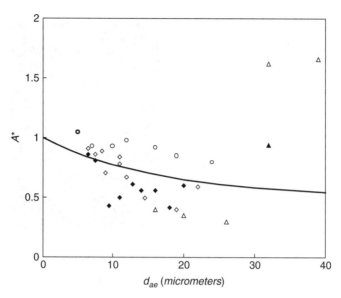

Figure 14.2 *Performance of the 2 Lpm ORB static inhalable aerosol sampler, shown in the form of sampling efficiency (A*) as a function of particle aerodynamic diameter (d_{ae}) for various windspeeds and types of aerosol (Ogden and Birkett, 1978: ○ calm air, ◇ 1 m s⁻¹, ◆ 3 m s⁻¹; Vincent et al., 1983: △ 2.4 m s⁻¹ for ungreased sampler, ▲ 2.4 m s⁻¹ for greased sampler, ___ inhalability curve). Reproduced with permission from Vincent, Aerosol Sampling: Science and Practice. Copyright (1989) John Wiley & Sons, Ltd*

in the field. However it was the forerunner to subsequent developments that *did* eventually lead to samplers of practical importance. In the next stage of searching for a sampler for inhalable aerosol, the principles underlying the inhalability definition itself were examined. On this basis, it was initially postulated that the *ideal* inhalable aerosol sampler should: (a) consist of a full-scale model of the human head with a sampling flow rate corresponding to simulated breathing; and (b) be rotated (stepwise or continuously) through 360° about a vertical axis in order to provide the required orientation-averaging. For practical sampling, however, a device was needed that would be smaller and altogether simpler. In work carried out at the Institute of Occupational Medicine (IOM), Mark *et al.* (1985) investigated the feasibility of achieving this. They began with experiments with a number of sampling heads in the form of short metal cylinders each containing a single short horizontal slot, providing a range of values for the critical dimensions. These were tested experimentally in a wind tunnel for ranges of wind speed and sampling flow rate. During each test, the sampling head in question was mounted with its axis vertical and rotated incrementally through 360°, thus providing the desired orientation-averaging. In evaluating these sampling heads, all the aerosol entering each sampling head was assessed, not only particles collected on the internal filter but also those deposited on the internal walls, thus providing performance in terms of aspiration efficiency (A) as a function of d_{ae}. From the results of this exhaustive study, Mark *et al.* were able to estimate by inspection the combination of variables that provided the best match with the inhalability curve. This was achieved for sampling head diameter $D = 32$ mm and sampling flow rate $Q = 3$ Lpm.

The knowledge gained from these experiments led to the development of what became known as the 3 Lpm *IOM static inhalable aerosol sampler*. The sampling head for the prototype version of this instrument is shown dismantled in Figure 14.3(a), and the complete assembled instrument in Figure 14.3(b).

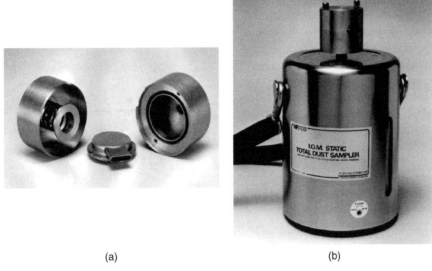

(a)   (b)

Figure 14.3   *The IOM 3 Lpm static inhalable aerosol sampler: (a) sampling head, shown disassembled to reveal the aerosol-collecting cassette; (b) the complete assembled instrument (Mark et al., 1985). Reproduced with permission from Vincent, Aerosol Sampling: Science and Practice. Copyright (1989) John Wiley & Sons, Ltd*

The sampling head had a cylindrical body of diameter 50 mm and length 52 mm, and featured a transverse entry slot measuring 3 mm × 15 mm with metal lips protruding by about 2 mm from the body of the sampler. In the assembled sampler, the lips and entry slot formed integral parts of an aerosol-collecting capsule (or cassette) which was located mainly inside the body of the sampling head. This capsule in turn incorporated a 37 mm filter, making a leak-proof seal with the body of the sampling head by means of a spring-loaded cap which pressed the capsule down onto an O-ring seal. The parts of the capsule were stamped out from aluminum sheet, giving an overall tare weight (including the filter) of about 7.5 g. In use, the whole capsule was weighed before and after sampling, so that evaluation of aerosol concentration was based on all that entered through the plane of the entry slot. Thus the combination of lips and capsule ensured that the performance of the sampler was not complicated either by external wall effects (i.e. particle blow-off) or by internal wall effects (i.e. unwanted deposition) of the type described in Chapter 7. Thus the performance of the device was based purely on aerodynamic factors and hence could be described in terms of aspiration efficiency (A) so that it could reasonably be expected to be consistent from one type of aerosol to another. There were two particular problems involved in weighing the whole capsule. The first was that an appropriate analytical balance was needed that was capable of accurately weighing small differences in relatively large tare weights. The second was that the mass stability of the capsule under varying environmental conditions could become a significant factor in the overall accuracy of sampling. However, it was shown that, provided that the capsule was allowed to stabilise overnight in the balance room, variability in weight was usually less than ±0.1 mg when measured using an electronic balance.

In the complete instrument shown in Figure 14.3(b), the sampling head was mounted on a combined pump and drive system located within its outer casing. The electrically driven, single-acting diaphragm pump drew air through the filter at the prescribed 3 Lpm. Rotation of the sampling head was provided by an electric motor geared down to give slow but steady motion at about 2 rpm. Leak-free connection

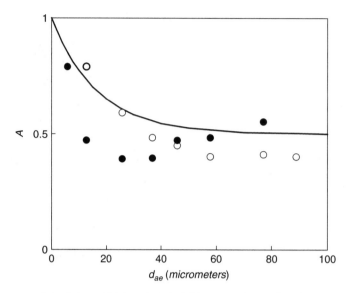

Figure 14.4   *Performance of the 3 Lpm IOM static inhalable aerosol sampler shown in the form of aspiration efficiency (A) as a function of particle aerodynamic diameter (d_{ae}) for two windspeeds (Mark et al., 1985: ○ 1 m s^{-1}, ● 3 m s^{-1}, ___ inhalability curve). Reproduced with permission from Mark et al., American Industrial Hygiene Association Journal, 46, 127–133. Copyright (1985) American Industrial Hygiene Association*

between the sampling head and the pump was achieved by means of a spring-loaded PTFE cup that sealed onto the tapered end of a rotating brass pipe carrying the sampling head itself. With overall dimensions of about 150 mm in diameter and 250 mm in height, and weighing 2.5 kg, the complete instrument was readily portable and suitable for use in the factory environment. Its performance was evaluated in wind tunnel experiments using test aerosols generated from narrowly graded powders of fused alumina, and the results are presented in Figure 14.4. These show that, for wind speeds of 1 and 3 m s^{-1}, agreement between aspiration efficiency (A) and the inhalability curve was quite good for values of d_{ae} up to 90 μm. The IOM static inhalable aerosol sampler was used in research projects to assess the exposures of workers to inhalable aerosol in a number of industries, including coal mining, textiles, cement, etc. It was available commercially briefly during the earlier 1990s, but disappeared as interest switched to the more relevant personal sampling for inhalable aerosol.

Elsewhere, researchers at the Institut National de Recherche de Sécurité (INRS) developed a static sampler for collecting the inhalable fraction and health-related sub-fractions (Fabries *et al.*, 1998). This is the CATHIA (French acronym for thoracic, inhalable and respirable aerosol sampler), a variant on the smaller CIP10 personal sampler that has been widely used in France for the respirable fraction during the past two decades (Courbon *et al.*, 1983) (see Chapter 15). The key feature of this instrument is an annular, downwards-facing slot inlet that was designed to provide aspiration efficiency that followed the inhalability convention. Görner *et al.* (1996) showed that aspiration efficiency for this instrument lay between about 0.9 and 0.4 as d_{ae} increased from about 10 to 60 μm.

The developments described above are interesting as part of the overall history of the search for sampling instrumentation in recent years matching the emerging new particle size-selective sampling criteria. However, progress did not extend much beyond the instruments described as occupational hygienists – driven by the requirements of modern standards – have turned increasingly towards the monitoring of personal aerosol exposures by means of instruments that could be carried by individual workers, thus providing results that directly relate to their own exposures.

14.3   Personal samplers for coarse aerosol fractions

14.3.1   'Total' aerosol

Occupational hygienists have long advocated that personal sampling is the most appropriate way to obtain truly representative workers' exposures to airborne contaminants. In turn, many personal aerosol sampling heads emerged and have since been used extensively by occupational hygienists in workplace aerosol exposure assessments. A typical personal sampling set-up is shown in Figure 14.5, indicating the sampler – or sampling head – itself (which is usually worn on the lapel or the breast so that it is well within the breathing zone), the sampling pump (which is usually worn on the belt) and a flexible plastic connecting hose. The sampling head contains the filter for the collection of sampled particulate matter.

A number of the most commonly used personal aerosol samplers originally intended for sampling 'total' aerosol have been studied experimentally in the laboratory to characterise their performances.

Figure 14.5   *Photograph of a typical personal sampling set-up*

Most of those experiments – the ones considered to be the most reliable – were carried out in wind tunnels or test chambers large enough that the samplers of interest could be mounted on the torsos of life-sized mannequins. In this way, the results reflected the presence of the bluff human body on sampler performance, the importance of which was generally assumed in view of its effect on the shape of the air flow in the vicinity of the sampler attached to it. Several research groups have contributed to the large body of information that has subsequently been built up over the years for a wide range of personal aerosol samplers considered to be candidates for collecting the inhalable fraction. Some of the ones most widely used in the field and most extensively studied in relation to the inhalability criterion include:

- The *closed-face 37 mm plastic cassette* [see Figure 14.6(a)]. This is a cheap, disposable sampling head that is widely used to this day by occupational hygienists in the USA and some other countries when sampling for 'total' aerosol. It is a three-part system, molded from clear nonconducting polystyrene. One part contains a circular, 4 mm diameter sampling orifice that is raised about 2 mm from the flat face of the sampler. The second part has a nipple that enables connection with the hose that leads to the pump and contains a 37 mm diameter filter for particle collection. The third part is inserted between the two other parts, and ensures firm location of the filter. After the filter is installed, the sampler is assembled by bringing the three parts together in a tight friction fit, which – in some practical applications – occupational hygienists seal with masking tape. The sampling flow rate for this sampler is usually 2 Lpm. When it is worn, the sampling orifice points generally downwards. It is commercially available from several manufacturers. In practical use, only the particulate material collected on the filter is assessed, so that the index of performance relevant to this instrument is overall sampling efficiency, A^*.
- The *open-face 37 mm plastic cassette* [see Figure 14.6(b)]. This is a simpler version of the closed-face sampler described above. Now the first part of the cassette takes the form of a ring that retains the 37 mm filter, such that the effective sampling orifice is 35 mm. Again, the sampling flow rate is usually 2 Lpm. This too is commercially available from several manufacturers. A 25 mm version of the open filter sampler has been used in some European countries, also operated at 2 Lpm. Again, in practical use, only the particulate material collected on the filter is assessed, so that the index of

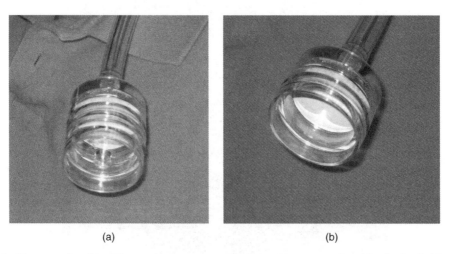

(a)   (b)

Figure 14.6 *Photographs of the 37 mm plastic cassette widely used by occupational hygienists in North America and elsewhere as a personal sampler for 'total' aerosol: (a) closed-face 37 mm version; (b) open-face version*

performance relevant to this instrument is A^*. Both the open- and closed-face 37 mm samplers are available from a number of sources.

- The *single-hole sampler* [see Figure 14.7(a)]. This closed face sampler was originally developed for use in the context of the United Kingdom Health and Safety Executive standard for airborne lead (Health and Safety Executive, 1981). Although the most recent version of that standard calls for sampling with reference to the inhalable aerosol fraction, the single-hole sampler remains a permitted option for occupational hygienists in Britain. It is a closed-face sampler where the aerosol enters through a 4 mm circular orifice in a flat aluminum cover and is collected on a 25 mm filter that is mounted in a recessed chamber behind the cover. When this sampler is worn, the sampling orifice faces directly outwards. The prescribed sampling flow rate is 2 Lpm. Again, in practical use, only the particulate material collected on the filter is assessed, so that the index of performance relevant to this instrument is A^*. The version shown in Figure 14.7(a) is in aluminum (Casella CEL, Bedford, UK).

- The *seven-hole sampler* [see Figure 14.7(b)]. This sampler originally evolved out of the single-hole sampler during the early 1980s. Supposedly, the original rationale was that a version of the sampler was needed that would provide a more uniform deposit of particulate matter on the filter. It rapidly became popular and the seven-hole sampler later became a recommended sampler for use in the sampling for the inhalable fraction in the earlier versions of the UK standard method for workplace aerosol evaluation (Health and Safety Executive, 1986). Indeed, it still remains an option in the latest version (Health and Safety Executive, 2000). The instrument is identical to the single-hole sampler described above, except that six additional holes, each of 4 mm, are drilled in the face plate. One version of this sampler is manufactured in nonconducting plastic (SKC Ltd, Blandford Forum, UK). Another is the aluminum version shown in Figure 14.7(b) (Casella CEL, Bedford, UK). Again, the

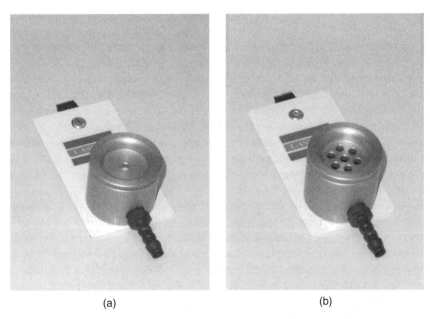

(a)   (b)

Figure 14.7 *Two personal aerosol samplers for 'total' aerosol widely used over the years by occupational hygienists in the UK: (a) single hole; (b) seven-hole. Photographs courtesy of Gary Noakes, Casella CEL, Bedford, UK*

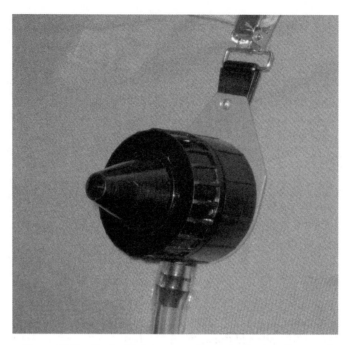

Figure 14.8 *Conical inlet personal sampler for 'total' aerosol*

prescribed sampling flow rate is 2 Lpm and, in practical use, only the particulate material collected on the filter is assessed, so that the index of performance relevant to this instrument is A^*.

- The *conical inlet sampler* (CIS) (see Figure 14.8). This sampler comprises a conical inlet section connecting with a cassette that contains a 37 mm filter. The diameter of the sampling orifice at the apex of the cone is 8 mm. When worn, the sampling orifice faces directly outwards. The sampling flow rate for this instrument is 3.5 Lpm, higher than for the other samplers described here. This sampler was originally developed in Germany and is available there as the GSP sampler (Ströhlein GmbH and Co., Kaarst, Germany). Other versions of the CIS have been available at various times, for different orifice sizes and flow rates, including the Dutch PAS-6 sampler and an earlier Italian instrument. The conical inlet configuration therefore appears to be a popular attractive option, perhaps on account of the more streamlined internal geometry that may reduce internal wall losses of aspirated particles. Again, in practical use, only the particulate material collected on the filter is assessed, so that the relevant index of performance is A^*.
- The *CIP10* (see Figure 14.9). The CIP10 personal sampler was developed in France originally as a sampler for the respirable aerosol, and will be discussed further as such in Chapter 15. However, over the years, interest has been maintained in this instrument as a sampler for the inhalable fraction, notably in Europe. This unusual sampler aspirates air at the very high flow rate of 10 Lpm by virtue of the pumping action of a rapidly rotated porous plastic foam plug. The total mass of particulate material collected in both this rotating plug plus that in the stationary foam plug placed just above it, intended – as will be seen in Chapter 15 – for selecting the respirable fraction, provides a measure of 'total' aerosol, such that the relevant index of performance is overall sampling efficiency, A^*. It is currently available commercially from ARELCO ARC (Fontenay sous Bois, France).

Figure 14.9   *Photograph of a recent commercial version of the CIP10 personal aerosol sampler, where – as indicated – starting and stopping the pump is achieved by passing a special magnet over the sampler's case. Photograph courtesy of Christian Champion, ARELCO ARC, Fontenay sous Bois, France*

- The *PERSPEC* (see Figure 14.10). This instrument was originally intended for use as a versatile personal sampler for providing simultaneous measurements of several fractions (see Chapter 15). It was first described by Prodi *et al.* (1986). As seen in the figure, aerosol enters through a pair of crescent-shaped openings, whereafter it undergoes inertial separation into the desired sub-fractions. The aspirated flow rate is 2 Lpm and the entire catch of particulate material inside the instrument is used to provide a measure of 'total' aerosol. This means that, in addition to material collected at the sites desired for the determination of the finer sub-fractions, it is necessary to recover material deposited on the inner surfaces – an inconvenient and laborious process. In this way, however, the most relevant index of performance is the aspiration efficiency, *A*. Although this instrument appears to be no longer available commercially, it is of historic interest.

Experimental data for the sampling efficiencies of these instruments in moving air as functions of particle aerodynamic diameter (d_{ae}) and wind speed (U) from experiments in wind tunnels have been reported by many authors (Buchan *et al.*, 1986; Mark and Vincent, 1986; Chung *et al.*, 1987; Vincent and Mark, 1990; Vincent, 1991; Kenny *et al.*, 1997; Aizenberg *et al.*, 2001). In all of these studies, the

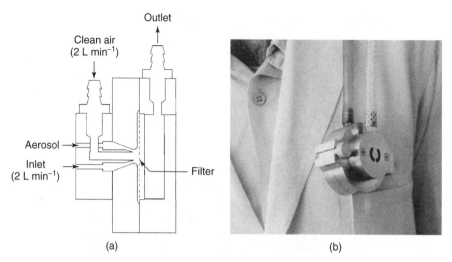

Figure 14.10   *PERSPEC personal aerosol sampler: (a) schematic to indicate its principle of operation; (b) photograph of the sampler itself. (a) Reprinted from Journal of Aerosol Science, 17, Prodi et al., 576–581. Copyright (1986), with permission from Elsevier*

samplers were mounted on life-sized mannequins and results were obtained where the orientation of the mannequin–sampler system was averaged over all possible orientations in the horizontal plane. Overall, d_{ae} ranged from about 5 to close to 100 μm, and U from 0.5 to 4 m s^{-1}. For most of them, relatively monodisperse test aerosols were generated from narrowly graded dry powders of various materials, for which fused alumina was the most common choice. The method adopted by Chung *et al.* was somewhat different in that polydisperse test aerosols were generated from silicon carbide powders and the catches of particles in the test and reference samplers were analyzed for both number and particle size by Coulter counter. For each sampler, performance was expressed in the form of overall sampling efficiency, A^*, or in the case of the PERSPEC aspiration efficiency, A, reflecting their respective modes of operation.

The published results from the various sources are collected together and summarised in Figures 14.11–14.18. Most of the data points shown for each instrument for most of the studies cited represent averages from multiple runs. In the first instance it is encouraging that the data from the various sources cited are generally quite consistent with one another for the majority of the samplers studied. The Chung *et al.* results are, perhaps, the least useful because, although they follow the same general trends as for the other experiments for the same samplers, the particle size range was limited to d_{ae} no greater than about 20 μm. Overall, the primary object of interest in the figures presented here is the extent to which any of the samplers studied exhibited performance characteristics that matched the inhalability convention. The answer is clear. For the moving air conditions studied, the performances of nearly all of the samplers studied fell well below the inhalability curve most of the time. Furthermore, they were generally more wind speed-dependent than the human head during the experiments of human head aspiration efficiency that led to the inhalability convention. In turn, therefore, it is reasonable to expect that all of them would undersample with respect to a hypothetical ideal sampler that perfectly follows the desired curve. More specifically, the closed-face 37 mm plastic cassette undersampled very significantly in relation to the inhalability curve across the whole range of conditions studied, which is of particular concern in view

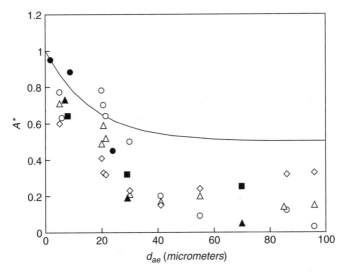

Figure 14.11　*Performance of the 2 Lpm personal closed-face 37 mm plastic cassette in terms of sampling efficiency (A*) as a function of particle aerodynamic diameter (d_{ae}) for various windspeeds, results from wind tunnel studies in a number of laboratories (Buchan et al., 1986: ● 1 m s^{-1}; Kenny et al., 1997: ○ 0.5 m s^{-1}, △ 1 m s^{-1}, ◇ 4 m s^{-1}; Aizenberg et al., 2000: ▲ 0.5 m s^{-1}, ■ 2 m s^{-1}; __ inhalability curve)*

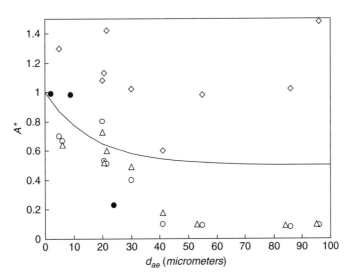

Figure 14.12　*Performance of the 2 Lpm personal open-face 37 mm plastic cassette in terms of sampling efficiency (A*) as a function of particle aerodynamic diameter (d_{ae}) for various windspeeds, results from wind tunnel studies in a number of laboratories (Buchan et al., 1986: ● 1 m s^{-1}; Kenny et al., 1997; ○ 0.5 m s^{-1}; __ inhalability curve)*

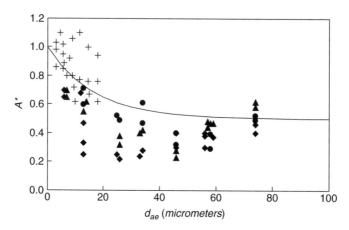

Figure 14.13   *Performance of the 2 Lpm open-face personal 25 mm plastic cassette in terms of sampling efficiency (A*) as a function of particle aerodynamic diameter (d_{ae}) for various windspeeds, results from wind tunnel studies in a number of laboratories (Mark and Vincent, 1986: ● 0.5 m s^{-1}, ♦ 1 m s, ▲ 2.6 m s^{-1}; Chung et al., 1987: +0.6–3.5 m s^{-1}; ___ inhalability curve). Reproduced with permission from Vincent, Aerosol Sampling: Science and Practice. Copyright (1989) John Wiley & Sons, Ltd*

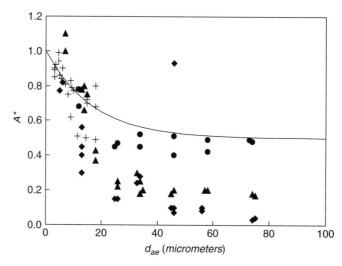

Figure 14.14   *Performance of the 2 Lpm single-hole personal sampler in terms of sampling efficiency (A*) as a function of particle aerodynamic diameter (d_{ae}) for various windspeeds, results from wind tunnel studies in a number of laboratories (Mark and Vincent, 1986: ● 0.5 m s^{-1}, ♦ 1 m s^{-1}, ▲ 2.6 m s^{-1}; Chung et al., 1987: +0.6–3.5 m s^{-1}; ___ inhalability curve)*

of the very widespread use of this sampler. The corresponding open-face sampler also undersampled markedly at the lower wind speeds, as too did the 25 mm version. But it suddenly oversampled at the highest wind speed studied ($U = 4$ m s^{-1}). The single-hole sampler consistently undersampled across the whole range of conditions, as did its seven-hole cousin. The latter is of concern because, long

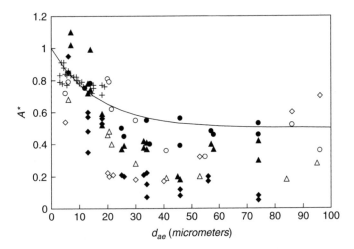

Figure 14.15 *Performance of the 2 Lpm seven-hole personal sampler in terms of sampling efficiency (A*) as a function of particle aerodynamic diameter (d_{ae}) for various windspeeds, results from wind tunnel studies in a number of laboratories (Mark and Vincent, 1986:* ● *0.5 m s⁻¹,* ◆ *1 m s⁻¹,* ▲ *2.6 m s⁻¹; Chung et al., 1987: +0.6–3.5 m s⁻¹; Kenny et al., 1997:* ○ *0.5 m s⁻¹,* △ *1 m s⁻¹,* ◊ *4 m s⁻¹; ___ inhalability curve)*

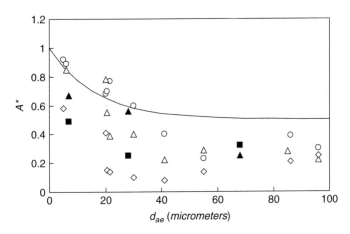

Figure 14.16 *Performance of the 2 Lpm conical inlet personal sampler (CIS) in terms of sampling efficiency (A) as a function of particle aerodynamic diameter (d_{ae}) for various windspeeds, results from wind tunnel studies in a number of laboratories (Kenny et al., 1997:* ○ *0.5 m s⁻¹,* △ *1 m s⁻¹,* △ *4 m s⁻¹; Aizenberg et al., 2000:* ▲ *0.5 m s⁻¹,* ■ *2 m s⁻¹; ___ inhalability curve)*

after the experimental data were published, and even though they are unequivocal, the seven-hole is still recommended in the UK by the HSE as a suitable sampler for the inhalable fraction. In short, it is fair to say that most of the samplers tested did not come sufficiently close to being regarded as successful candidates for use as a sampler for the inhalable fraction. One exception, based on the available experimental data, is the idiosyncratic CIP10, which exhibits quite a good match with the

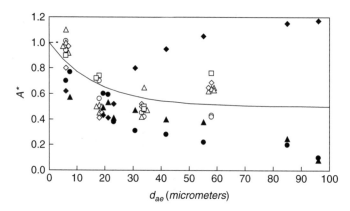

Figure 14.17 *Performance of the 2 Lpm CIP10 used as a personal sampler in terms of sampling efficiency (A*) as a function of particle aerodynamic diameter (d_{ae}) for various windspeeds, results from wind tunnel studies (Vincent and Mark, 1990: O 0.5 m s^{-1}, △ 1 m s^{-1}, ◇ 2.6 m s^{-1}, □ 3 m s^{-1}; Kenny et al., 1997: ● 0.5 m s^{-1}, ▲ 1 m s^{-1}, ◆ 4 m s^{-1}; ___ inhalability curve)*

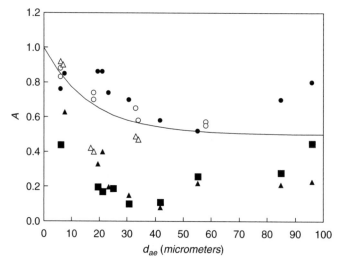

Figure 14.18 *Performance of the 2 Lpm PERSPEC personal sampler in terms of aspiration efficiency (A*) – see main text – as a function of particle aerodynamic diameter (d_{ae}) for various windspeeds, results from wind tunnel studies (Vincent and Mark, 1990: O 1 m s^{-1}, △ 3 m s^{-1}; Kenny et al., 1997: ● 0.5 m s^{-1}, ▲ 1 m s^{-1}, ■ 4 m s^{-1}; ___ inhalability curve)*

inhalability curve for a wide range of relevant conditions. To a somewhat lesser extent so does the PERSPEC, although – again – this instrument is no longer available commercially.

Elsewhere, a separate study of some of the same samplers reported by Li and Lundgren (2000) described their performances while mounted in isolation, simulating how they might be used in practice as static (or area) samplers. As expected, the results indicated the presence of strong orientation dependencies. They were not presented in a form that would allow the performances of the samplers as area samplers to be compared with the corresponding results – described above – for orientation-averaged conditions.

All the experiments discussed so far were for air that was moving at wind speeds no less than 0.5 m s^{-1}. As already discussed in Chapter 3, studies of air movements in workplace environments had revealed that local average air velocities – including relative velocities between the air and people who themselves are in motion – rarely exceeded 0.2 to 0.3 m s^{-1} (e.g. Berry and Froude, 1989; Baldwin and Maynard, 1998). Such wind speeds lie significantly below the range of conditions for which the samplers described above were studied in the cited works. So there is a need for those samplers – and other candidates for collecting the inhalable fraction – to be evaluated under such low wind speed conditions. Further, their performances need to be considered in the light of what is know about inhalability at correspondingly low wind speeds (e.g. Aitken *et al.*, 1999). Here, however, much less experimental data are available and for a smaller range of sampling devices. A major problem in conducting such research has been the difficulty in generating and dispersing test aerosols into low wind speed test environments whilst achieving uniform air flow and aerosol concentration distributions in relevant parts of the test apparatus; also in making reliable measurements of the reference aerosol concentration. One important study is the one reported by Kenny *et al.* (1999) in work complementary to that for calm-air inhalability reported by Aitken *et al.* The experimental methods were essentially the same. Overall-averaged results for two of the samplers already mentioned – the 37 mm closed-face plastic cassette and the seven-hole sampler – are shown together in Figure 14.19 again for the samplers in question mounted on the torso of a mannequin to simulate actual wearing by a human subject. Here the results were averaged for different positions of the sampler on the body, including on the left, center and

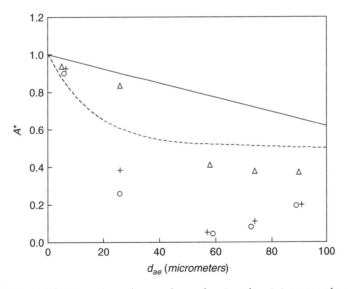

Figure 14.19 *Performance of the two personal aerosol samplers in calm air in terms of sampling efficiency (A*) as a function of particle aerodynamic diameter (d$_{ae}$); 2 Lpm closed-face plastic cassette and 2 Lpm seven-hole sampler used as personal samplers, and 2 Lpm 37 mm closed-face plastic cassette used as a static sampler (Kenny et al., 1999: ○ 37 mm closed-face plastic cassette as a personal sampler, △ seven-hole sampler as a personal sampler, +37 mm plastic cassette as a static sampler; - - - inhalability curve for moving air; —— suggested inhalability curve for calm air)*

right sides of the front of the mannequin's torso, for which differences were found to be small. They are presented in the form of A^* versus d_{ae}, alongside the current inhalability conventional curve, as originally determined from experiments in moving air, and also for the curve suggested by Aitken *et al.* for human exposure in calm or slowly moving air (see Chapter 10). It is seen that the sampling efficiency of the 37 mm sampler fell well below both curves, so that this instrument would clearly undersample with respect to either, the more so for the current convention. The data for the seven-hole sampler also fell below both curves, albeit modestly so in relation to the current convention. Overall, however, it is again apparent that neither device can be regarded as a suitable sampler for the inhalable fraction. Also shown in Figure 14.19 are some results obtained for the 37 mm plastic cassette used as an isolated static sampler, and these were not significantly different from those used as personal samplers.

14.3.2   Inhalable aerosol

A small number of personal aerosol samplers have so far been developed with the specific aim, from the outset, of collecting the inhalable fraction. The first emerged from work carried out at the Institute of Occupational Medicine (IOM) in the mid-1980s (Mark and Vincent, 1986). This involved an experimental study of a range of idealised single-hole sampling heads of simple geometry whose aspiration efficiencies when mounted on the torso of an orientation-averaged life-size mannequin – as for the personal samplers already described above – were determined in wind tunnel studies for ranges of orifice dimensions, sampling flow rates and wind speeds. From the results of these experiments, it emerged that performance – in terms of aspiration efficiency (A) as a function of d_{ae} – closely matching the inhalability curve, relatively independently of wind speed, could be achieved for a version with a single, outwards-facing inlet of diameter 15 mm operated at a sampling flow rate of 2 Lpm. It was also shown that performance was not significantly dependent on the exact location of the sampler on the general breast area within the breathing zone.

   These studies led to the development of a new practical device, for which commercial versions have been available since the early 1990s. A recent version of the 2 Lpm *IOM personal inhalable aerosol sampler* is shown in Figure 14.20 (SKC Ltd, Blandford Forum, UK and SKC Inc., Eighty Four, PA, USA). It comprises a cylindrical body 37 mm in diameter and 27 mm long. Onto one end of the cylinder is screwed a flat circular cap incorporating a 15 mm circular orifice having a thin lip that protrudes 1.5 mm outwards. The purpose of the lip, as in the case of the static inhalable aerosol sampler described earlier, is to minimise unpredictability in performance arising from the blow-off of particles from the outer surfaces of the sampler. Again, this lip and orifice form integral parts of a dust-collecting cassette which is located mainly inside the body of the sampler. The cassette incorporates a 25-mm filter and backing grid, and makes a leak-proof seal with the body of the sampler, aided by pressure from the screwed top cap which is transmitted to the cassette by a sealing ring which in turn, presses the cassette down onto an O-ring seal. A further washer is positioned inside the cassette between the filter and the cassette top to prevent the filter from tearing or buckling when the sampler is assembled. In one modern version of the instrument, the cassette shell is molded in conducting plastic, providing a tare weight of about 0.8 g. In another version, the cassette is fabricated from stainless steel so that, although the tare weight is significantly greater, the mass stability of the cassette associated with moisture uptake is much improved (e.g. Smith *et al.*, 1998; Paik and Vincent, 2002). In the practical use of the instrument for gravimetric assessment of the sampled aerosol, the whole catch of particulate material aspirated through the entry, and hence collected within the cassette, is weighed, including not only what is on the filter but also everything that is deposited on other internal surfaces within the cassette. It follows, therefore, that an appropriate metric of sampler performance is the aspiration efficiency, A, so that performance is dominated entirely by the aerodynamic process taking place outside the sampler, irrespective of interactions between particles and sampling walls.

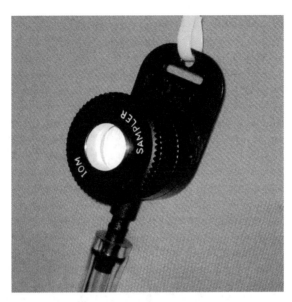

Figure 14.20   *The 2 Lpm IOM personal inhalable aerosol sampler, shown with conducting plastic body and a stainless steel cassette (SKC Inc., Eighty Four, PA, USA)*

The performance of this instrument has been characterised in several laboratory studies over the years (Mark and Vincent, 1986; Kenny *et al.*, 1997, 1999; Aizenberg *et al.*, 2000, 2001). Some of the results from moving air experiments in wind tunnels are shown in Figure 14.21, presented in the form of aspiration efficiency (A) as a function of d_{ae} for a range of wind speeds. It is seen that, for most conditions, performance of this sampler was quite close to the inhalability curve over the whole range of d_{ae} up to close to 100 μm. A notable exception is the trend for the results of Kenny *et al.* for the highest wind speed of 4 m s^{-1}. However, this wind speed lies outside the range of most workplaces, even those with environments characterised as being 'moving air'. This condition apart, there was the inevitable – and considerable – variability. There were no obvious biases from one wind speed to the next. The corresponding results of Kenny *et al.* (1999) for calm air are summarised in Figure 14.22, and these show that the IOM sampler oversampled markedly with respect to the inhalability conventional curve across the whole range of particle size. However performance was respectably close to the new inhalability curve for calm air suggested by Aitken *et al.* Results for the same sampler used as an isolated static sampler were not significantly different from those obtained for the sampler used in the personal mode.

Further experimental data for the IOM sampler in moving air but at very low wind speeds ($U = 0.05$ and 0.1 m s^{-1}, respectively) were reported by Aizenberg *et al.* (2001). These results are notable also for the extended range of particle size that was studied, achieved by the use of a novel experimental set-up, involving a closed-loop, open cross-section wind tunnel and careful placement of the aerosol source with respect to the location of the mannequin that carried the sampler. In addition, to avoid unnecessary disturbance associated with the injection of test aerosol, coarse particles in the size range of interest were delivered by means of a 'spinning-tray particle spreader'. In this way, it was possible to continue to maintain uniform air velocity in the working section of the tunnel, with tolerable uniformity of aerosol concentration, especially when averaged over time. The results for the IOM sampler are shown, for the two wind speeds, in Figure 14.23. Also shown are some corresponding results for the

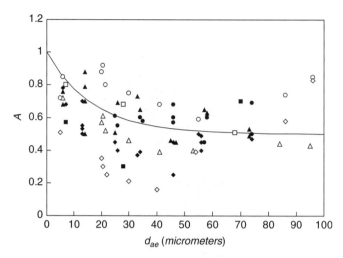

Figure 14.21 *Performance of the 2 Lpm IOM personal inhalable aerosol sampler in terms of aspiration efficiency (A) as a function of particle aerodynamic diameter (d_{ae}) for various windspeeds, results from wind tunnel studies in a number of laboratories (Mark and Vincent, 1986: ● 0.5 m s^{-1}, ▲ 1 m s^{-1}, ◆ 2.6 m s^{-1}; Kenny et al., 1997: ○ 0.5 m s^{-1}, △ 1 m s^{-1}, ◇ 4 m s^{-1}; Aizenberg et al., 2000: □ 0.5 m s^{-1}, ■ 2 m s^{-1})*

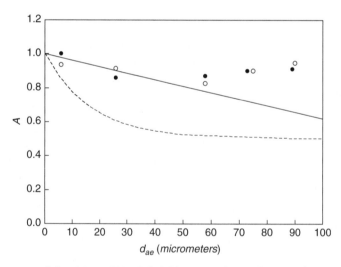

Figure 14.22 *Performance of the 2 Lpm IOM inhalable aerosol sampler in calm air in terms of aspiration efficiency (A) as a function of particle aerodynamic diameter (d_{ae}), used both as a personal sampler and as a static sampler (Kenny et al., 1999: ○ IOM sampler as a personal sampler, ● IOM sampler as a static sampler; - - - inhalability curve; —— suggested inhalability curve for calm air)*

CIS sampler discussed earlier. In addition, the conventional inhalability curve is shown, along with the proposed Aitken *et al.* version for calm air, both for d_{ae} up to 100 μm. The trends exhibited in Figure 14.23 are altogether more complex than the ones that were reflected in the results for moving air and calm air described above. Here, the experimental data for the aspiration efficiency (*A*) of the

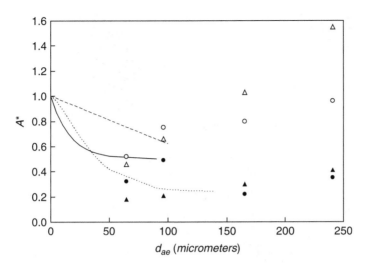

Figure 14.23   *Performance of the 2 Lpm IOM and CIS personal samplers in terms of A (for the IOM) and A* (for the CIS) as functions of particle aerodynamic diameter (d_{ae}) for slowly moving air and very large range of particle sizes (Aizenberg et al., 2001:* ○ *IOM sampler 0.05 m s⁻¹,* △ *IOM sampler 0.1 m s⁻¹,* ● *CIS sampler 0.05 m s⁻¹,* ▲ *CIS sampler 0.1 m s⁻¹; ___ inhalability curve; - - - suggested inhalability curve for calm air; . . . trend from Kennedy and Hinds, 2002)*

IOM sampler rose well above either of the trends suggested by the two inhalability curves, while the sampling efficiency (A^*) of the CIS sampler fell substantially below. For the latter, however, they were closer to the trend for large particle sizes suggested by Kennedy and Hinds (2002) for d_{ae} up to about 140 μm, albeit for larger wind speeds. Unfortunately, the lack of experimental inhalability data for both very low wind speed and very large particles means that it is not yet possible to resolve the apparent inconsistency between these results for the IOM sampler and the earlier ones for moving air in terms of its performance relative to inhalability for such conditions.

For several years, the IOM sampler was the only instrument designed specifically to collect the inhalable fraction and available commercially as such. However, Aizenberg *et al.* (2000) have described a new personal sampler for the inhalable fraction. Known as the *Button personal inhalable aerosol sampler*, it is shown schematically in Figure 14.24 in the version that is commercially available (SKC Inc., Eighty Four, PA, USA). This sampler is quite different from the IOM sampler and from any of the other samplers described so far. The inlet comprises a curved porous surface in the form of a section of a spherical stainless steel shell containing a large number of individual circular orifices of diameter 381 μm, equi-spaced such that the porosity of the surface is 21 %. The section of the sphere has a subtended angle of 160° such that the overall surface area of the screen itself is about 20 cm². The aspirated aerosol is collected on a 25-mm filter located directly behind the screen. The prescribed sampling flow rate is high at 4 Lpm. There are several advantages to this design. Firstly, the pressure drop across the screen inlet is such that particles are deposited very uniformly over the filter surface. In addition, the streamlined profile of the inlet surface reduces the level of flow distortion in the immediate proximity of the sampler, while the close placement of the filter to the screen greatly reduces internal wall losses. In the practical use of this instrument, only the particulate material collected on the filter is assessed. So, in the terminology adopted in this book, the appropriate metric of performance is overall sampling efficiency, A^*.

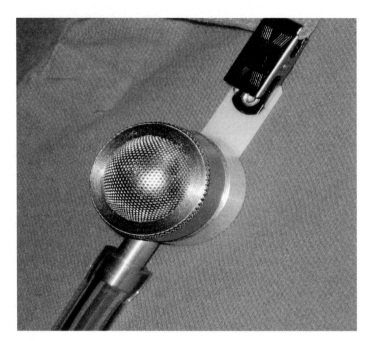

Figure 14.24   *The 4 Lpm Button inhalable aerosol sampler*

The Button sampler is a relatively recent development so that there are fewer reports of its experimental characterisation in the laboratory. The most comprehensive data are those by Aizenberg *et al.* (2000) for orientation-averaged conditions when the sampler is mounted on a life-size mannequin, for d_{ae} up to about 70 μm and for wind speeds (U) of 0.5 and 2 m s^{-1}, and by Aizenberg *et al.* (2001) for d_{ae} up to about 240 μm and much lower U of 0.05 and 0.1 m s^{-1}. The complete data set from Aizenberg and colleagues is summarised in Figure 14.25, alongside the inhalability conventional curve, the version for calm air suggested by Aitken *et al.* and the version for large particles suggested by Kennedy and Hinds. In general, the performance of the Button sampler fell somewhat below the inhalability convention that is currently the target performance criterion for inhalable aerosol samplers.

More recently still, interest has emerged in samplers for the inhalable fraction that would operate at lower flow rates such that smaller sampling pumps may be used. At significantly lower sampling flow rates, and at the aerosol levels nowadays found in most workplaces, even those previously considered to be 'dusty', gravimetric analysis of samplers is no longer feasible. However, for many applications, the component of the collected aerosol that is relevant to the sampling exercise in question may be one that may be measured at extremely low levels using appropriate analytical instrumentation. For example, metals can be quantitated to nanogram levels using modern versions of inductively coupled plasma mass spectrometry. All that is required, therefore, is that the aerosol is collected on surfaces from which it may be recovered and so made available for analysis. With this in mind, Professor Yngvar Thomassen of the Norwegian National Institute for Occupational Health (personal communication) suggested a modified version of the IOM sampler, based on the scaling laws that had been described for aerosol sampling (see Chapter 3). For this, a modification was made that involved incorporation of an orifice adapter directly into the orifice of the original IOM sampler. As shown in Figure 14.26, the adapter was

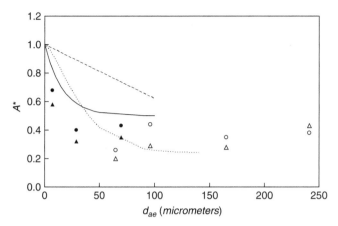

Figure 14.25 *Performance of the 4 Lpm Button personal inhalable aerosol sampler in terms of sampling efficiency (A*) as a function of particle aerodynamic diameter (d_{ae}) for various windspeeds and for an extended range of particle sizes (Aizenberg et al., 2001: ○ IOM sampler 0.05 m s⁻¹, △ IOM sampler 0.1 m s⁻¹, ● CIS sampler 0.05 m s⁻¹, ▲ CIS sampler 0.1 m s⁻¹; ── inhalability curve; - - - suggested inhalability curve for calm air; . . . trend from Kennedy and Hinds, 2002)*

Figure 14.26 *The prototype 0.3 Lpm Baby-IOM inhalable aerosol sampler, with stainless steel cassette and insert, developed on the basis of aerosol sampler scaling laws*

very simple, and allowed for an orifice size that was smaller than the original 15 mm. Table 14.1 shows how scaling may be achieved in order to achieve a sampler that should have the same performance of an equivalent full-size IOM sampler. Here, the simplifying assumption was made that the performance of the sampler should not depend significantly on the size of the body on which it was mounted. In

Table 14.1 *Scaling relationships for the modified, low flow rate version of the IOM personal inhalable aerosol sampler shown in Figure 14.26. Here, as in earlier chapters, δ is the diameter of the sampling orifice (achieved by the use of an adapter) and D is the size of the body on which the sampler is mounted. The full-scale situation matches some of the experimental ones of Mark and Vincent (1986) and Kenny et al. (1997). The variables in the column to the right were determined by the application of the scaling laws outlined in Chapter 3*

	Full-scale (Mark and Vincent, 1986; Kenny et al., 1997)	Small-scale (Vincent et al., 2003)
d_{ae} (μm)	8–96	6–89.5
δ (mm)	15	6
D (mm)	300	120
U (m s^{-1})	1	1
Flow rate, Q (L min^{-1})	2	0.32

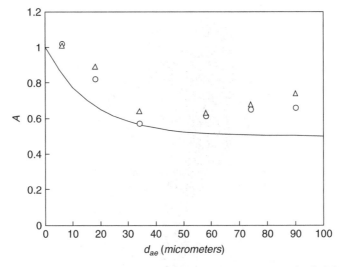

Figure 14.27 *Comparison between the performance of the 2 Lpm IOM personal inhalable aerosol sampler and that of the corresponding 0.3 Lpm Baby-IOM personal inhalable aerosol sampler, shown in terms of aspiration efficiency (A) as a function of particle aerodynamic diameter (d_{ae}) for one windspeed (Vincent et al., 2003: ○ Baby-IOM, △ original IOM, ⎯ inhalability convention)*

this way, the new sampler could be tested in a small wind tunnel mounted on a smaller and simpler bluff body than had been the case for the experiments to characterise personal sampler performance described above. According to the scaling laws, equivalent performance should exist for a modified IOM sampler whose orifice diameter is 6 mm (as opposed to the original 15 mm) and with sampling flow rate 0.3 Lpm (as opposed to the original 2 Lpm). This is the version shown in Figure 14.26. Averaged experimental results are shown in Figure 14.27 for two systems: the first with the new, modified IOM sampler (the Baby-IOM) and the second with the original IOM sampler, both mounted on a simple rectangular bluff body of width 120 mm (compared with about 300 mm for the life-size mannequin). The experiments were carried out for a wind speed of 1 m s^{-1}. The results for the original IOM sampler and the Baby-IOM were in almost exact agreement, confirming the validity of the scaling laws. The data for both samplers were quite close to the conventional inhalability curve (Vincent *et al.*, 2003).

14.3.3 Other samplers

A small number of other samplers that have been developed for use as personal samplers for 'total' aerosol are worthy of mention by virtue of their interesting features. The *full-shift true breathing zone sampler* was described by Allen *et al.* (1981). As shown in Figure 14.28(a) this sampler was based on a small filter head which, instead of being worn on the lapel, was located within a few centimeters of the nose and mouth by mounting it on a stalk attached to an adjustable headband. It was intended that, by placing the sampler so close – closer than any other of the samplers mentioned so far – to where aerosol was inhaled into the body, a representative sample of true aerosol exposure would be obtained. Although a prototype version of this device was used in practical studies of the exposures of lead workers in a battery factory, it was never tested with respect to the inhalability curve and does not appear to have been widely used or made commercially available. In another instrument, the *helmet-mounted sampler* shown in Figure 14.28(b) (Ter Kuile, 1982) set out to achieve the same end by similar means. This too does not appear to have been developed beyond the prototype stage.

A final personal sampler worthy of mention by virtue of its unusual features is the one developed in Poland and described by Kucharski (1980). Here the sampling flow rate was varied in a manner simulating the pattern of breathing of the wearer, by means of an electronic system that monitored the pulse rate of the wearer which in turn was used to control the pump speed. This was a most interesting idea if what was required was to measure the actual aerosol uptake of a worker. However, there is no evidence that this idea was carried beyond the laboratory prototype stage.

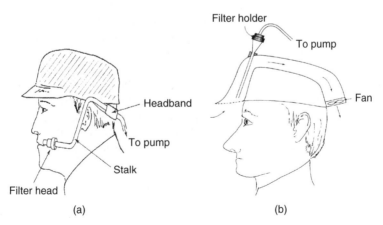

Figure 14.28 *Examples of two alternative types of personal sampler: (a) the full-shift true breathing zone sampler (Allen et al., 1981); (b) the helmet sampler (Ter Kuile, 1982). Reproduced with permission from Vincent, Aerosol Sampling: Science and Practice. Copyright (1989) John Wiley & Sons, Ltd*

14.4   Analysis of performance data for inhalable aerosol samplers

14.4.1   Statistics

As described above, a number of personal samplers have been suggested for the coarser aerosol that is now known – for sampling in relation to health effects – to be best represented by the inhalable fraction. A large amount of experimental test data has been generated. Several of the samplers mentioned above have been toted as 'personal inhalable aerosol samplers', either by the inventors, the manufacturers or national standards-setting bodies, some with more scientific justification than others. The fact is, although there are certain broad trends in the overall body of data that has been assembled, there are sufficient differences and inconsistencies to permit latitude in many such claims.

The most comprehensive single laboratory study of personal samplers in moving air was the large test program described in 1997 by Kenny *et al.*, aiming specifically to generate sufficient information from which to base rational choices of aerosol samplers for specific situations. In an earlier paper, Kenny and Bartley (1995) had described mathematical procedures by which to critically examine: (a) the mass bias that might exist between the aerosol collected by a given sampler in relation to that collected by a hypothetical sampler that perfectly matches the desired performance criterion; and (b) the accuracy of any given sampler from one experiment to the next using the same sampler unit or between separate units of the same sampler. This framework was summarised in Chapter 3, and provides the analytical foundation for the sampler test procedures described in the Comité Européen de Normalisation (CEN) standard EN 13205 (Comité Européen de Normalisation, 2002). Importantly, for any given sampler, EN 13205 provides the option to apply a correction factor to the collected mass, so long as the bias can be shown to be similar for all aerosol particle size distributions of interest and for all influencing variables. The Kenny *et al.* study is unique among the other works cited in that its results have been subjected to the rigorous statistical analysis along the lines indicated. In the first instance, estimates were made of sampler bias and accuracy in the earlier Kenny and Bartley paper. Bartley (1998) subsequently conducted an important extended analysis of the results, focusing specifically on the data for the lower wind speeds of 0.5 and 1 m s^{-1}, regarding these – at the time – as being of greatest relevance to most workplaces. He considered a range of hypothetical aerosol particle size distributions with mass median particle aerodynamic diameter (MMAD) up to 50 μm and geometric standard deviation (σ_g) greater than 1.50, covering almost all realistic practical eventualities. From his bias estimates, he calculated correction factors for each of the eight samplers that Kenny *et al.* had studied experimentally. Table 14.2 summarises the results for seven of the samplers studied, showing both correction factors and the estimated 95 % level confidence limits on those factors. Bartley noted that these correction factors differed somewhat from earlier ones generated by Kenny *et al.*, but explained the differences in terms of the wider range of particle size distributions he had chosen, and hence wider applicability of the results of his analysis. It was concluded that the IOM, CIS and CIP10 samplers were the most appropriate choices for general applications in the sampling of the inhalable fraction. The Button sampler, which – as mentioned – was developed later specifically to match the inhalability convention, was not examined in this analysis, and there remain to this day insufficient data for this sampler to be subjected to such analysis.

14.4.2   Modeling

Earlier chapters provided details of the physical processes of aspiration under various conditions of particle size, sampler physical shape and dimensions, sampling flow rate, external wind speed and orientation with respect to the wind. It is a large step from the models presented there to the more complicated systems that are found in most real-world sampling. Some progress has been made. As

Table 14.2 *Correction factors for the samplers for their performances in relation to the inhalation criterion, also showing estimated confidence limits at the 95 % level (from Bartley, 1998)*

Sampler	Correction factor	Confidence limit (95 % level)
IOM	0.95	0.44
CIS	1.21	0.49
CIP10	1.47	0.50
Seven-hole	1.33	0.57
37 mm closed-face	1.49	0.60
37 mm open-face	1.57	0.67
PERSPEC	1.23	0.73

described in Chapter 5, Tsai and Vincent (1993) had developed semi-empirical models for blunt aerosol samplers of simple shape for aspiration efficiency at orientations with respect to the wind (θ) of 90° and 180°, adding to what was already known for $\theta = 0°$. Later, Tsai *et al.* (1995) applied these towards the development of a model for aspiration efficiency of relatively simple *actual* samplers under orientation-averaged conditions. They began by stating that aspiration efficiency averaged over all possible such orientations is given by:

$$A = \frac{1}{2\pi} \int_0^{2\pi} A(\theta)\mathrm{d}\theta \qquad (14.1)$$

where $A(\theta)$ is aspiration efficiency at angle θ. The dependence of A on θ is very complex and detailed information is not available. The models for A_0, A_{90} and A_{180} provide useful starting points, along with knowledge that A does not vary uniformly at intermediate angles, but – rather – favours collection for forwards-facing angles. Tsai *et al.* proposed empirical expressions of the form:

$$A(\theta) = A_0 - (A_0 - A_{90}) \left(\frac{2\theta}{\pi}\right)^{\mu_f} \text{ for } 0° \leq \theta \leq 90° \qquad (14.2)$$

and

$$A(\theta) = A_{90} - (A_{90} - A_{180}) \left[\frac{2(\theta - 90)}{\pi}\right]^{\mu_b} \text{ for } 90° \leq \theta \leq 180° \qquad (14.3)$$

In these, the terms μ_f and μ_b embody the effects of St, R and r (see Chapter 5) on the distribution of $A(\theta)$, expressed in the form $\mu_f = k_1 St^{k2} R^{k3} r^{k4}$ and $\mu_b = k_5 St^{k6} R^{k7} r^{k8}$, where the k-values are constant coefficients. It was found that a unique set of these coefficients on the basis of experimental data for aspiration efficiency were available for a range of sampling systems, including the human head during the inhalability experiments (see Chapter 10), the 30 Lpm IOM sampler for inhalable atmospheric aerosol (see later Chapter 17), and the 3 Lpm IOM static inhalable aerosol sampler for workplaces described earlier in this chapter. With appropriate choice of these coefficients, agreement between experiment and theory was very good across the whole range of systems studied. The particular results for the 3 Lpm IOM static inhalable aerosol sampler and versions tested during its development – with different dimensions and operated at different sampling flow rates (Mark *et al.*, 1985) – are shown

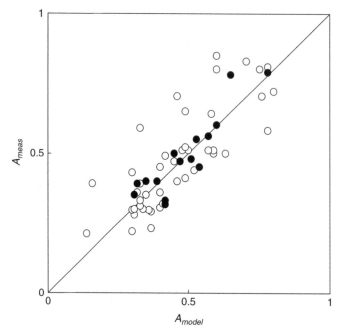

Figure 14.29 *Comparison between theory (A_{model}) and experiment (A_{meas}) for the 3 Lpm IOM static inhalable aerosol sampler and other versions tested during its development (Tsai et al., 1995: ○ 1 m s^{-1}, ● 3 m s^{-1}, —— 1:1 line)*

in Figure 14.29. The slope of the relationship was found to be 1.01 and the corresponding R^2-value was 0.96.

Encouraged by this, Tsai *et al.* (1996) proceeded to apply similar reasoning in an effort to develop a model for personal samplers of the type described in this chapter. This proved to be an even more elaborate exercise, requiring considerations beyond those of the simpler model just described, including the fact that personal samplers are usually worn asymmetrically on the body. Although fair agreement was obtained between this new model and the available experimental data for the aspiration efficiencies of a number of personal samplers, this exercise was ultimately rendered less satisfying due to the even greater number of coefficients that needed to be fitted.

14.5 Passive aerosol samplers

Passive, nonaspirating aerosol samplers are of considerable interest because, if they can be designed appropriately and shown to perform according to desired specifications, they may allow sampling to take place without the need for pumping and air flow rate monitoring systems, along with the associated tubing and associated paraphernalia. Although there is quite a long history of such sampling for the collection of large 'nuisance' particles in the ambient atmosphere (see Chapter 17), relatively little of the same thinking has yet made its way into the sampling of workplace aerosol. There have been some stirrings of late, and a small number of important contributions have been reported of this alternative mode of aerosol sampling for workplace applications.

Workplace aerosols are usually charged to levels exceeding those corresponding to Boltzmann equilibrium (Johnston *et al.*, 1985). In 1994, Brown and his colleagues exploited this in a prototype aerosol sampling system by which such charged particles were deposited locally in the electric field generated in the vicinity of a rectangular plate of electret material (Brown *et al.*, 1994a, b). The principle of operation of the system described is shown in Figure 14.30(a). From the original basic work, a practical personal sampler emerged, comprising a 25 mm diameter electret disk inset into a compact housing. This is shown in Figure 14.30(b). It was not possible to test this sampler in the laboratory in the same way as all the other samplers described above. From consideration of the estimated distribution of particle charge on particles of different sizes (from Johnston *et al.*, 1985), Brown *et al.* (1994b) noted that the efficiency of collection may not be expected to be strongly dependent on particle size. So there would be no prospect of being able to design such an instrument directly matching any of the health-related particle size-selective criteria that have been described. Nonetheless, Brown *et al.* (1995) found in field trials that weighable deposits were collected by the electret sampler and that they were consistently linearly correlated with the aerosol concentrations as measured using the IOM personal inhalable sampler in workplace trials.

In Denmark, Vinzents (1996) proposed a small passive sampler that operated on the basis of particle deposition by gravitational and by turbulent and Brownian diffusion. As shown in Figure 14.31(a), the sampler comprised three sticky transparent foils facing upwards (for gravitational deposition), downwards and horizontally, respectively (for deposition by turbulent and Brownian diffusion). The prototype personal sampling instrument is shown in Figure 14.31(b). One method of analysis of collected material involved application of the principle of optical extinction for estimating the concentration of the deposit on each foil. Another involved optical microscopy with automated image analysis, whereby the rapid counting and sizing of particles was converted into a deposited mass. Schneider *et al.* (2002) showed that a fair linear correlation could be achieved with inhalable aerosol as measured simultaneously using the IOM personal sampler. Somewhat later, Wagner and Leith (2001a, b) described a passive sampler in which particles were deposited onto a horizontal collection surface located about 1 mm below

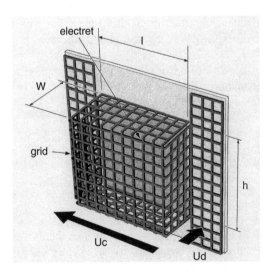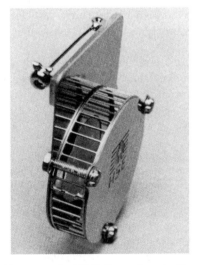

Figure 14.30 *Electret-based passive personal aerosol sampler (Brown et al., 1995). Reproduced with permission from Brown et al., Annals of Occupational Hygiene, 39, 603–622. Copyright (1995) British Occupational Hygiene Society*

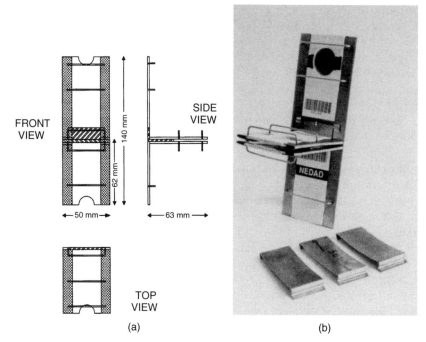

Figure 14.31 *Danish personal passive sampler (Vinzents, 1996): (a) diagram to shown basic lay-out; (b) photograph of prototype sampler. Reproduced with permission from Vinzents, Annals of Occupational Hygiene, 40, 261–280. Copyright (1996) British Occupational Hygiene Society*

a stainless steel mesh. This was intended for use as a static sampler, aimed at potential applications in clean rooms, indoor air sampling, etc. The physical deposition mechanisms involved gravitational and inertial forces (for the larger particles) and diffusion (for the smaller ones). For this instrument too, it was proposed that collected samples should be analyzed by automated microscopy and image analysis.

None of the three passive samplers described appears to have yet been widely applied in practical aerosol sampling situations, or to have become commercially available.

References

Aitken, R.J., Baldwin, P.E.J., Beaumont, G.C., Kenny, L.C. and Maynard, A.D. (1999) Aerosol inhalability in low air movement environments, *Journal of Aerosol Science*, 30, 613–626.

Aizenberg, V., Choe, K., Grinshpun, S.A., Willeke, K. and Baron. P.A. (2001) Evaluation of personal aerosol samplers challenged with large particles, *Journal of Aerosol Science*, 32, 779–793.

Aizenberg, V., Grinshpun, S.A., Willeke, K., Smith, J. and Baron. P.A. (2000) Performance characteristics of the Button personal inhalable aerosol sampler, *American Industrial Hygiene Association Journal*, 61, 398–404.

Allen, J., Bellinger, E.G. and Higgins, R.I. (1981) A full-shift true breathing zone air sampler and its application to lead workers, *Proceedings of the Institution of Mechanical Engineering*, 195, 325–328.

Armbruster, L., Breuer, H., Vincent, J.H. and Mark, D. (1983) The definition and measurement of inhalable dust. In: *Aerosols in the Mining and Industrial Work Environment* (Eds V.A. Marple and B.Y.H. Liu), Ann Arbor Science, Ann Arbor, MI, pp. 205–217.

Baldwin, P.E.J. and Maynard, A.D. (1998) A survey of wind speeds in indoor workplaces, *Annals of Occupational Hygiene*, 20, 303–313.

Bartley, D.L. (1998) Inhalable aerosol samplers, *Applied Occupational and Environmental Hygiene*, 13, 274–278.

Berry, R.D. and Froude, S. (1989) An investigation of wind conditions in the workplace to assess their effect on the quantity of dust inhaled, Internal Report IR/L/DS/89/3, Health and Safety Executive, London.

Brown, R.C., Hemingway, M.A., Wake, D. and Thompson, J. (1995) Field trials of an electret-based passive dust sampler in metal-processing industries, *Annals of Occupational Hygiene*, 39, 603–622.

Brown, R.C., Wake, D., Thorpe, A., Hemingway, M.A. and Roff, M.W. (1994a) Theory and measurement of the capture of charged particles by electrets, *Journal of Aerosol Science*, 25, 149–163.

Brown, R.C., Wake, D., Thorpe, A., Hemingway, M.A. and Roff, M.W. (1994b) Preliminary assessment of a device for passive sampling of airborne particulate, *Annals of Occupational Hygiene*, 38, 303–318.

Buchan, R.M., Soderholm, S.C. and Tillery, M.I. (1986) Aerosol sampling efficiency of 37-mm filter cassettes, *American Industrial Hygiene Association Journal*, 47, 825–831.

Chung, K.Y.K., Ogden, T.L. and Vaughan, N.P. (1987) Wind effects on personal dust samplers, *Journal of Aerosol Science*, 18, 159–174.

Coenen, W. (1973) A new procedure for evaluation of fibrogenic dusts in the working place, *Staub Reinhaltung der Luft* (English translation), 33, 97–102.

Comité Européen de Normalisation (2002) *Workplace atmospheres – assessment of performance instruments for measurement of airborne particle concentrations*, European Standard EN 13205, Brussels, Belgium.

Courbon, P., Froger, C. and Le Bouffant, L. (1983) Personal dust sampler CIP10, Presented at the 20th International Conference of Safety in Mines Research Institutes, Sheffield, Paper L2.

Dodgson, J., Hadden, G.G., Jones, C.O. and Walton, W.H. (1971) Characteristics of the airborne dust in British coal mines. In: *Inhaled Particles III* (Ed. W.H. Walton), Unwin, Old Woking, pp. 757–781.

Fabries, J.F., Görner, P., Kauffer, E., Wrobel, R. and Vigneron, J.C. (1998) Personal thoracic CIP10-T sampler and its static version CATHIA-T, *Annals of Occupational Hygiene*, 42, 453–465.

Görner, P., Witscher, O. and Fabries, J.F. (1996) Annular aspiration slot entry efficiency of the CIP10 aerosol sampler, *The Analyst*, 121, 1257–1260.

Health and Safety Executive (1981) Control of lead: air sampling techniques and strategies, Guidance Note EH28, HSE, London.

Health and Safety Executive (1986) General methods for the gravimetric determination of respirable and total inhalable dust, *Methods for the determination of hazardous substances*, MDHS 14, HSE, London.

Health and Safety Executive (2000) General methods for sampling and gravimetric analysis of respirable and inhalable dust, *Methods for the determination of hazardous substances*, MDHS 14/3, HSE, London.

Johnston, A.M., Vincent, J.H. and Jones, A.D. (1985) Measurements of electric charge for workplace aerosols, *Annals of Occupational Hygiene*, 29, 271–284.

Kennedy, N.J. and Hinds, W.C. (2002) Inhalability of large particles, *Journal of Aerosol Science*, 33, 237–255.

Kenny, L.C. and Bartley, D.L. (1995) The performance evaluation of aerosol samplers tested with monodisperse aerosols, *Journal of Aerosol Science*, 26, 109–126.

Kenny, L.C., Aitken, R.J., Baldwin, P.E.J., Beaumont, G.C. and Maynard, A.D. (1999) The sampling efficiency of personal inhalable aerosol samplers in low air movement environments, *Journal of Aerosol Science*, 30, 627–638.

Kenny, L.C., Aitken, R.J., Chalmers, C., Fabries, J.F., Gonzalez-Fernandez, E., Kromhout, H., Lidén, G., Mark, D., Riediger, G. and Prodi, V. (1997) A collaborative European study of personal inhalable aerosol sampler performance, *Annals of Occupational Hygiene*, 41, 135–153.

Kucharski, R. (1980) A personal dust sampler simulating variable human lung function, *British Journal of Industrial Medicine*, 37, 194–196.

Li, S.-N. and Lundgren, D.A. (2000) Evaluation of six inhalable samplers, *American Industrial Hygiene Association Journal*, 61, 506–516.

Mark, D. and Vincent, J.H. (1986) A new personal sampler for airborne total dust in workplaces, *Annals of Occupational Hygiene*, 30, 89–102.

Mark, D., Vincent, J.H., Gibson, H. and Lynch, G. (1985) A new static sampler for airborne total dust in workplaces, *American Industrial Hygiene Association Journal*, 46, 127–133.

Mark, D., Vincent, J.H., Stevens, D.C. and Marshall M. (1986) Investigation of the entry characteristics of dust samplers of the type used in the British nuclear industry, *Atmospheric Environment*, 20, 2389–2396.

May, K.R. (1967) Physical aspects of sampling airborne microbes, *Symposium of the Society of General Microbiology*, 17, 60–80.

Ogden, T.L. and Birkett, J.L. (1977) The human head as a dust sampler. In: *Inhaled Particles IV* (Ed. W.H. Walton), Pergamon Press, Oxford, pp. 93–105.

Ogden, T.L. and Birkett, J.L. (1978) An inhalable dust sampler, for measuring the hazard from total airborne particulate, *Annals of Occupational Hygiene*, 21, 41–50.

Ogden, T.L., Birkett, J.L. and Gibson, H. (1978) Large particle entry efficiencies of the MRE 113A gravimetric dust sampler, *Annals of Occupational Hygiene*, 21, 251–263.

Paik, S. and Vincent, J.H. (2002) Filter and cassette mass instability in ascertaining the limit of detection of inhalable airborne particulates, *American Industrial Hygiene Association Journal*, 63, 698–702.

Prodi, V., Belosi, F. and Mularoni, A. (1986) A personal sampler following ISO recommendations on particle size definitions, *Journal of Aerosol Science*, 17, 576–581.

Schneider, T., Schlünssen, V., Vinzents, P.S. and Kildesø, J. (2002) Passive sampler used for simultaneous measurement of breathing zone size distribution, inhalable dust concentration and other size fractions involving large particles, *Annals of Occupational Hygiene*, 46, 187–195.

Smith, J.P., Bartley, D.L. and Kennedy, E.R (1998) Laboratory investigation of the mass stability of sampling cassettes from inhalable aerosol samplers, *American Industrial Hygiene Association Journal*, 59, 582–585.

Stuke, J. and Emmerichs, M. (1973) Das gravimetrische Staubprobennahmegerat TBF50, *Silikosebericht Nordrhein-Westfalen*, 9, 47–51 (in German).

Ter Kuile, W.M. (1982) Ontwikkeling van de totaalstof helm, Report No. F1893, IMG-TNO, Delft (in Dutch).

Tsai, P.J. and Vincent, J.H. (1993) Impaction model for the aspiration efficiencies of aerosol samplers at large angles with respect to the wind, *Journal of Aerosol Science*, 24, 919–928.

Tsai, P.J., Vincent, J.H., Mark, D. and Maldonado, G. (1995) Impaction model for the aspiration efficiencies of aerosol samplers in moving air under orientation-averaged conditions, *Aerosol Science and Technology*, 22, 271–286.

Tsai, P.J., Vincent, J.H. and Mark, D. (1996) Semi-empirical model for the aspiration efficiencies of personal aerosol samplers of the type widely used in occupational hygiene, *Annals of Occupational Hygiene*, 40, 93–114.

Vincent, J.H. (1989) *Aerosol Sampling: Science and Practice*, John Wiley & Sons, Ltd, Chichester.

Vincent, J.H. (1991) Joint investigations of new generations of dust sampling instrument, Synthesis Report of Five-Nations Project, Commission of European Communities, EUR 13414 EN, Office for Official Publications of the European Communities, Luxembourg.

Vincent, J.H. and Mark, D. (1990) Entry characteristics of practical workplace aerosol samplers in relation to the ISO philosophy, *Annals of Occupational Hygiene*, 34, 249–262.

Vincent, J.H., Mark, D., Gibson, H., Botham, R.A., Emmett, P.C., Witherspoon, W.A., Aitken, R.J., Jones, C.O. and Miller, B. (1983) Measurement of inhalable dust in wind conditions pertaining to mines, IOM Report No. TM/83/7, Institute of Occupational Medicine, Edinburgh.

Vincent, J.H., Paik, S.Y. and Evans, D.E. (2003) Development of new personal aerosol samplers, Final Report on NIOSH Project RO1-OH03687-03, National Institute for Occupational Safety and Health, Atlanta, GA.

Vinzents, P. (1996) A passive personal dust monitor, *Annals of Occupational Hygiene*, 40, 261–280.

Wagner, J. and Leith, D. (2001a) Passive aerosol sampler. Part I: Principle of operation, *Aerosol Science and Technology*, 34, 186–192.

Wagner, J. and Leith, D. (2001b), Passive aerosol sampler. Part II: Wind tunnel experiments, *Aerosol Science and Technology*, 34, 193–201.

15

Sampling for Fine Aerosol Fractions in Workplaces

15.1 Introduction

In the case of sampling for 'total' aerosol, the performances of sampling instruments are expressed largely in terms of their aspiration efficiencies, or closely related quantities, and hence depend largely on aerodynamic processes outside the sampler. Here, therefore, for the most part, particle selection processes inside the body of the instrument are regarded as interfering factors. By contrast, in sampling instruments for collecting the finer fractions, performance is expressed mainly in terms of particle size-selection processes which take place after aspiration. Now the geometry and dimensions of the transmission section between the entry and filter are specifically designed to provide penetration characteristics matching the selection curve corresponding to the desired fine fraction, applying options likes like those outlined in Chapter 8. However, the harmonised new criteria agreed by the Comité Européen de Normalisation (CEN, 1993), the International Standards Organisation (ISO, 1995) and the American Conference of Governmental Industrial Hygienists (ACGIH) (Vincent, 1999) make specific mention of the role of the inhalable fraction in the measurement of fine fractions, acknowledging that the finer fractions representing penetration down into the lung are in fact sub-fractions of what was inhaled (see Chapter 11). In this chapter and subsequently, and widely elsewhere, these criteria are now referred to as the 'CEN/ISO/ACGIH criteria'. For these, it is useful to reiterate that the most appropriate performance of a sampler intended to measure a specific lung fraction should follow the general relationship:

$$\text{Sampling efficiency } (d_{ae}) = I(d_{ae}) \cdot \text{lung fraction } (d_{ae}) \qquad (15.1)$$

in which the inhalable fraction is explicitly included.

15.2 Samplers for the respirable fraction

Respirable aerosol refers to the particle size fraction of inhaled aerosol that penetrates down to the alveolar region of the lung. Over the years, this has been the most widely applied to the fine aerosol

Aerosol Sampling: Science, Standards, Instrumentation and Applications   James Vincent
© 2007 John Wiley & Sons, Ltd

fraction for health-related aerosol exposure assessment, driven in the early days of occupational hygiene by concerns about lung disease in the mining industries and continuing to the present day. Samplers for this fraction therefore need to select particles according to their aerodynamic diameter in a manner that reflects the shape of an appropriate respirable aerosol curve. The quantitative definition of respirable aerosol has shifted somewhat over the years since it was first introduced in 1952, but now appears to have settled down to the version described in the harmonised set of CEN/ISO/ACGIH conventions. Respirable aerosol samplers were once referred to as 'two-stage' samplers (e.g. Lippmann, 1983), describing a first (pre-selector) stage inside the instrument where particles were separated with efficiency varying appropriately with particle aerodynamic size and a second (removal) stage where the respirable particles that had penetrated through the first stage were collected with high efficiency. Although to be strictly consistent with the CEN/ISO/ACGIH criterion, samplers should first aspirate the inhalable fraction and then select the respirable sub-fraction, it is implicitly assumed that – for most respirable aerosol samplers under most conditions – the particles of interest are small enough that the effects of inlet geometry, wind speed and orientation may be neglected. With this in mind, the performance data for respirable aerosol samplers that are described below are presented in terms of the efficiency (P) by which particles penetrate down to the collection substrate, but they are compared directly with desired overall performance in the form described generally by Equation (15.1).

15.2.1   Early samplers

As mentioned in earlier chapters, it had been noted in the early 1900s that only small particles were able to reach the alveolar region of the lung. So very early on the concept of a health-related fine fraction was influential in occupational aerosol exposure assessment. One instrument that emerged at that time to fill the need to measure this fraction was the *konimeter* (Kotzé, 1916). Several versions were sold commercially and were widely used for many decades by occupational hygienists throughout the world. The one shown in Figure 15.1 was built in Germany by the Carl Zeiss company. In general, konimeters were portable so that the occupational hygiene investigator could carry them to the workplace, and use them to take samples directly from the breathing zone of a worker, or at a strategic point within arm's reach. The basic principle of the instrument was to take a 'snap' sample by rapidly aspirating a small air volume of about 5 to 15 ml, and to be able to take repeated samples such that exposure assessment could be carried out for a number of workers throughout a working shift. In practice, the pre-set air volume was drawn by means of a spring-loaded aspirator, and passed through a circular-jet impactor arrangement by which the particles were collected by impaction onto the predetermined part of a greased glass slide. The circular slide was designed so that up to 40 different parts of its surface could be exposed (and hence up to 40 individual samples obtained) and later identified. After sampling, the dust spots on the slide could be viewed under the microscope, using a graticule to identify the field of view and allow the collected particles to be sized. Fine particles within the field of view were counted and, from knowledge of the size of the field viewed and the air volume sampled, the airborne concentration was calculated in terms of *particle counts per cubic centimeter (ppcc)*. In the operating instructions for the Zeiss version of the instrument, fine particles were defined as particles with diameter less than 10 μm. Viewing was by light or dark field, usually at a magnification of 150×. Some konimeter models, including the Zeiss version shown, had a microscope built directly into the sampler itself, providing the opportunity for on-site sample assessment. This *in situ* mode was found to be particularly useful if the investigator wished to provide immediate feedback to the worker about exposure levels or to initiate a rapid response in terms of dust control. More usually, however, samples were analyzed back in the comfort and quiet of the laboratory. Various types of konimeter were widely used from the 1930s until as recently as the 1980s, particularly in the extraction and metals industries in Canada, Australia and South Africa where

Figure 15.1 *Photograph of the Carl Zeiss konimeter*

the instrument was an important basis of dust exposure standards. Sebestyen *et al.* (1987) compared the particle size-dependent sampling efficiencies of a number of models in the laboratory, and Verma *et al.* (1987) carried out a side-by-side field comparison in hard rock mines of their performances in terms of particle counts. These studies revealed that there were significant differences in both sampling efficiency and counts obtained from one model to another. The field studies showed in particular that the differences were dependent on the type of dust.

The *Owens jet sampler* (Owens, 1922) had some similar features (see Figure 15.2). Here, the jet orifice in the impactor took the form of a narrow slit rather than a circular hole. In this device, the sampled air was pre-saturated with water vapor in a wet-paper-lined entry tube before passing through the jet where adiabatic expansion caused condensation of water molecules onto the particles, aiding not only the deposition onto the glass slide by impaction but also their adhesion to the collection surface. The deposited particles were subsequently examined in the laboratory by optical microscopy. Like the konimeter, the Owens-jet device involved taking a 'snap' sample by one or more strokes of a 50 ml, hand-operated, piston pump.

Other, related, aerosol sampling instruments were fashionable during the same period. In the *impinger* of Greenberg and Smith (1922), the sampled air jet was directed onto a surface immersed in water or alcohol, and the particles were separated into the liquid and subsequently plated out so that they could be size-classified and counted by optical microscopy, typically by light-field at $100\times$ after airborne 'clumps' had been dispersed into their constituent particles. The earliest version was a high flow rate

Figure 15.2   *Photograph of the Owens jet sampler*

device, sampling at 28 Lpm. In 1937 a miniaturised version, the 2.8 Lpm midget impinger, appeared which, when personal sampling pumps later became available, was then sometimes used for personal sampling. The sampling time was much longer than for the konimeter, typically about 20 min, but its use was more flexible since samples could be diluted to adjust the on-slide density for optimal particle counting. This instrument was quite widely used by occupational hygienists in North America in the period before gravimetric sampling of fine particles was introduced, providing the basis of occupational exposure limits for many years. Impingers are still used for some types of aerosol sampling, most notably for bioaerosols (see Chapter 19), and so remain commercially available from several sources. One modern version of the Greenberg–Smith impinger is shown in Figure 15.3.

The *thermal precipitator* was attractive in that it provided a 'gentle' means of depositing the particles with the intention that they would not be damaged by fluid mechanical shear forces or impact fracture during deposition. In that way, the particles that were subsequently observed under the microscope would be more representative of those inhaled by the exposed workers, in particular in the sense that the individual airborne particulate entities would not have undergone any significant changes during sampling. In the thermal precipitator, particles were deposited by the action of thermophoretic forces in the strong temperature gradient applied between heated and cooled surfaces in a manner not strongly dependent on particle size for particles smaller than about 10 μm (see Chapter 8). The standard thermal precipitator (STP) developed by Green and Watson (1935) is shown schematically in Figure 15.4. This sampler was operated at the very low continuous flow rate of 8 mLpm, with air movement achieved ingeniously by means of a water-filled aspirator. The aerosol was aspirated through a slit, entering a narrow chamber in which thermophoretic particle deposition took place onto a pair of glass slide cover slips placed on brass heat sinks and located about 0.5 mm apart on either side of 0.25 mm diameter heated nickel–chromium resistance wire held at 120° K. The version shown in the figure was used as the standard instrument in British coal mines from 1949 to 1965. Later a modified device, the long-running thermal precipitator (LRTP) (Hamilton, 1956) was built which excluded the collection of large particles by virtue of a laminar-flow horizontal elutriator placed ahead of the thermal collection section. This

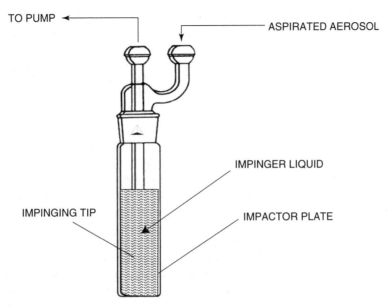

Figure 15.3 *Diagram of the generic Greenberg–Smith impinger (typical capacity 500 mL, typical sampling flow rate 28.3 Lpm)*

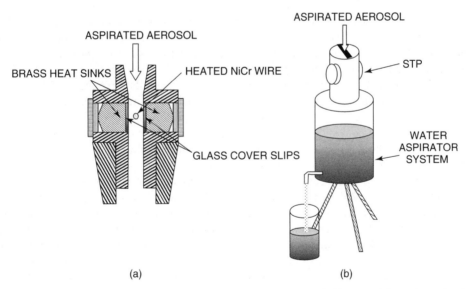

Figure 15.4 *Standard thermal precipitator (STP): (a) sketch of the STP itself, viewed from the side; (b) diagram of sampling system, including the water aspirator*

latter instrument became the standard instrument in British coal mines during 1965 to 1970, and was also used elsewhere and for collecting asbestos fibers. A similar instrument was used in South Africa (Kitto and Beadle, 1952). As with the other earlier samplers described here, in most versions of the thermal precipitator the collected particles were assessed by optical microscopy. The lower end of the particle size range was later extended by some workers by the application of electron microscopy.

15.2.2   Horizontal elutriators

Awareness of the role of the aerodynamic selection of particles in the penetration and deposition of inhaled particles in the respiratory tract began in the 1950s. This emerged soon after it had become apparent that particulate mass was a more relevant index to many of the health effects of interest – including the pneumoconioses – than the particle count index that had been used for so long and had underpinned the use of so many of the earlier sampling instruments (e.g. Bedford and Warner, 1943). Walton (1954) described a complete deterministic model for gravitational elutriation in both vertical and horizontal systems. Of these, the horizontal elutriator was subsequently by far the most widely applied for respirable aerosol sampling. Early on, Walton recognised that the penetration characteristics of the horizontal elutriator were governed by broadly the same physical factors as, and so were similar in form to, the penetration of inhaled particles into the alveolar regions of the lung. Even before publication of his 1954 paper, his ideas had been influential in the 1952 deliberations of the British Medical Research Council (BMRC) towards defining respirable aerosol. As a result, the original BMRC curve for respirable aerosol was firmly rooted in the principles of horizontal elutriation. In this way, the criterion for respirable aerosol sampling pointed directly to the horizontal elutriator option for a sampling instrument.

One of the first practical such instruments was the *Hexhlet sampler* described by Wright (1954) (see Figure 15.5). In the original version the elutriator part comprised 118 channels of length 255 mm and cross-section 0.81×35.5 mm. Particles penetrating through the elutriator were collected in the glass-fiber thimble of the 'soxhlet' type described by Griffiths and Jones (1940). Operated at the high flow rate of 100 Lpm, elutriator penetration (P) as a function of particle aerodynamic diameter (d_{ae}) should, in theory, match the BMRC curve perfectly. Wright's experimental data for the Hexhlet are shown in the earlier Figure 8.3 and are seen to be in good agreement with the predicted curve, and hence – since the dimensions of the device had been chosen to produce performance exactly matching the BMRC curve – with the respirable convention prevailing at the time. This instrument was the first to be used for the gravimetric sampling of respirable airborne dust in British coalmines (Fay, 1960). Later, a lower flow rate 50 Lpm version was developed, with the aim of reducing problems associated with the re-entrainment of deposited dust from the elutriator plates. Although the Hexhlet was later superceded in mining applications by the smaller, lower flow rate MRE sampler – which will be discussed in detail shortly – it continued for many years to find occasional applications elsewhere. An application was reported as recently as 1998 in connection with an epidemiological study in the Croatian paper recycling industry (Zuskin *et al.*, 1998). However, this instrument is now no longer available commercially.

A range of other elutriator-based samplers emerged during the same period, including the British 3 Lpm *SIMGARD sampler* described by Critchlow and Proctor (1955) and the German 46 Lpm *MPG II*. Most, like the Hexhlet, were originally intended for use in investigating the fine airborne dust in coal industry workplaces. The development of the portable, four-channel, 2.5 Lpm *MRE Type 113A* static sampler represented the response of the British mining industry to the need for a sampler for both routine exposure assessment for standards compliance purposes and also for epidemiological research. As shown in Figure 15.6, the MRE was a refinement of the Hexhlet and other previous such instruments. It was first described in detail by Dunmore *et al.* (1964), and changed little over the years that followed, except for improvements that included, for example, the implementation of an improved inlet to reduce

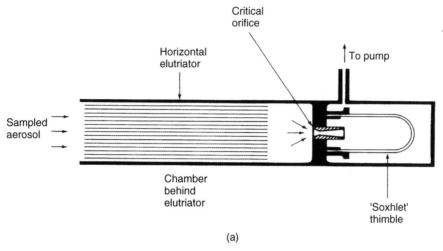

(a)

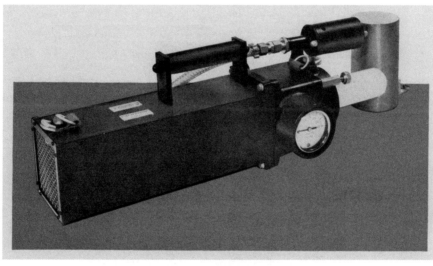

(b)

Figure 15.5 *(a) Schematic of the Hexhlet horizontal elutriator-based static respirable aerosol sampler (not to scale); (b) photograph of the (then) commercially available instrument. Photograph courtesy of Gary Noakes, Casella CEL, Bedford, UK*

wind speed effects. In the version that was most widely used, the rectangular entry slits for the four channels measured 40 mm × 0.4 mm, leading into channels of cross-section 40 mm × 2.38 mm and length 171.9 mm. In the previous chapter, the efficiency with which particles were aspirated into the sampler through the inlet slits was discussed, since some workers had wanted to use the MRE as a sampler for 'total' aerosol. Now, in this chapter, consideration is given to the primary role of this instrument – as a sampler for the respirable fraction. With the stated choice of channel dimensions and with overall sampling flow rate 2.5 Lpm, and neglecting inlet effects, P as a function of d_{ae} should match the BMRC curve perfectly. The experimental data of Dunmore *et al.* are plotted in Figure 15.7,

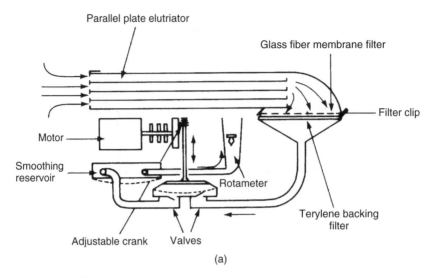

Parallel plate elutriator

Glass fiber membrane filter

Filter clip

Motor

Smoothing reservoir

Rotameter

Terylene backing filter

Adjustable crank   Valves

(a)

(b)

Figure 15.6   *MRE Type 113A horizontal elutriator-based static respirable aerosol sampler: (a) schematic of layout of the sampler (not to scale); (b) photograph of the (then) commercially available instrument. Photograph courtesy of Gary Noakes, Casella CEL, Bedford, UK*

together with more data from Hamilton *et al.* (1967), Ettinger *et al.* (1970) and Fabries and Wrobel (1987). Also shown are Wright's corresponding data for the Hexhlet. The various data sets are seen to be in general good agreement and are consistent with Walton's horizontal elutriator theory, and hence with the original BMRC curve. In Figure 15.7 they are shown alongside the new, harmonised CEN/ISO/ACHGIH respirable aerosol curve where it is seen that both the MRE and the Hexhlet tend to consistently oversample with respect to the current convention.

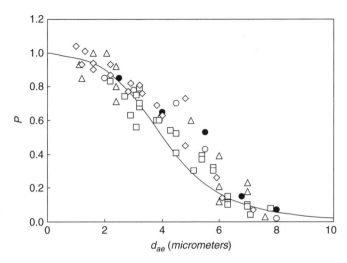

Figure 15.7 *Penetration (P) as a function of particle aerodynamic diameter (d_{ae}) for the horizontal elutriator-based, 2.5 Lpm MRE static respirable aerosol sampler (○ Dunmore et al., 1964; △ Hamilton et al., 1967; □ Ettinger et al., 1970; ◊ Fabries and Wrobel, 1987; ── CEN/ISO/ACGIH respirable aerosol curve); also shown are corresponding data for the 100 Lpm Hexhlet sampler (● Wright, 1954)*

The MRE was usually operated in underground mine environments with its entry facing directly into the wind.[1] Effects of wind speed and orientation on the entry of coarse particles into the elutriator through the narrow entry slots may be expected to be quite marked, on the basis of what is now known about aspiration efficiency described in the first half of this book. However, there has also been some evidence to suggest that such effects may extend down even to fine particles in the respirable range, especially for the higher wind speeds that were prevalent in some British mines (Ford, 1971; Ogden *et al.*, 1977). Ford speculated that such effects may have been associated with the penetration of freestream turbulence into the elutriator ducts, and this prompted efforts to make the modifications to the 'narrow-slit' entry that featured in later versions of the instrument.

Since the introduction of the MRE, various related devices subsequently appeared. One was the *SIMSLIN* described by Leck and Harris (1978), in which the respirable particles penetrating through the elutriator were sensed in real-time by a light-scattering method. The *optical scattering instantaneous respirable (dust) indication system (OSIRIS)* later described by Leck (1983) represented an extension of the same basic idea. In the *gravimetric dust monitor (GDM)* described by King (1984), the penetrating particles were collected onto a filter, as in the MRE, and were assessed *in situ* by means of a beta-attenuation mass balance technique in order to provide a continuously updated measure of respirable dust concentration over extended, multishift sampling durations.

Of the various horizontal elutriator-based devices that have appeared and been used in countries around the world, all were used as fixed-point static samplers. The main advantage of all such instruments was that their performances were predictable from first principles on the basis of Walton's elutriation theory. However, the main disadvantage common to all of them lay in the difficulty of achieving miniaturisation sufficient to allow the development of versions suitable for use as personal samplers. Eventually,

---

[1] In British coal mines, the most common mode of mining is that known as 'longwall mining' in which the coal face and its associated roadways are characterised by generally unidirectional air flow driven by the mine ventilation.

the move by the occupational aerosol standards-setting bodies – and the occupational hygiene community at large – to acceptance of personal sampling as the only truly reliable way to conduct assessment of workers' exposures meant that such instruments fell out of use. So few if any remain commercially available today and hence it is rather moot that instruments like the Hexhlet and the MRE samplers do not accurately follow the modern CEN/ISO/ACGIH criterion.

15.2.3   Cyclones

Horizontal elutriators were extensively used as static aerosol samplers for the respirable fraction for many years in Europe. They were never common in the USA. There, development of the early criteria for respirable aerosol was based directly on application of the cyclone concept (Lippmann and Harris, 1962), an important feature of which lies in its ability to be realised in versions sufficiently miniature for personal sampling, thus presenting a significant advantage over horizontal elutriators. In the long run, as occupational hygiene practices around the world pointed increasingly to aerosol exposure assessment based on personal sampling, the cyclone gained in prominence and eventually surpassed the horizontal elutriator as the respirable aerosol sampler 'of choice'.

As discussed earlier in Chapter 8, the particle size-selectivity of the cyclone, and hence its penetration (P), is – like that of the horizontal elutriator – a function of d_{ae}. Unlike the horizontal elutriator, however, that relationship cannot be predicted exactly. But, by careful empirical design, experience has shown that it is possible to find a cyclone configuration which provides performance characteristics closely matching a desired selection curve.

In Europe, the German 50 Lpm *TBF50 static sampler* was based on a pair of cyclones operating in series. It was actually intended primarily as a sampler for the fine fraction, where the aerosol of interest was aerodynamically selected by the first cyclone and the desired fraction was taken to be that which collected in the pot of the second cyclone. Since the second cyclone was not a perfect collector, then the overall selectivity of the TBF50 was determined by the penetration characteristics of both cyclones. An important feature, therefore, was the fact that some of the finer particles penetrated through the second cyclone, and so were not collected. This resembled somewhat the exhalation of some fine particles during actual human inhalation (Stuke and Emmerichs, 1973), and so the instrument was never intended to match any of the prevailing respirable aerosol conventions. In France, the 50 Lpm *CPM3 static sampler* operated on somewhat similar principles, with the pre-selector stage derived from the cyclone stage of the TBF50 but with the collector stage based on the particle collecting properties of a rapidly rotating plug of porous plastic foam which also provided the pumping action of the device (Courbon, 1972). Fabries and Wrobel later reported (Fabries and Wrobel, 1987) that the particle size selectivity of the CPM3 was quite close to what is now recognised as the CEN/ISO/ACGIH curve for the respirable fraction, although – like the TBF50 – there was a noticeable progressive loss of finer particles as d_{ae} fell below about 2 μm. Both the TBF50 and CPM3 samplers were used extensively for many years in the German and French mining industries, but were not applied much elsewhere. In addition to these instruments, many other cyclone-based static samplers appeared, some of them brought over from applications in ambient air sampling. However, in recent decades, interest has turned increasingly to personal sampling where the cyclone concept has provided the ability to realise instruments that could be miniaturised sufficiently that they could be used as personal samplers.

One version of the cyclone that has been widely used for many years was the American *10 mm nylon (Dorr-Oliver), single-inlet cyclone*, comprising a cyclone and lightweight aluminum frame which located a 37 mm plastic filter cassette (see Figure 15.8). Its performance, in terms of the particles collected on the filter, was evaluated experimentally by several workers, including Ettinger *et al.* (1970) Blachman and Lippman (1974) and Caplan *et al.* (1977), all using monodisperse test aerosols produced

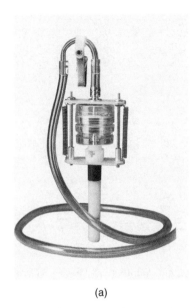

(a)   (b)

Figure 15.8   *10 mm nylon (Dorr–Oliver) single-inlet cyclone personal sampler for respirable aerosol: (a) basic assembly, showing the locations of the cyclone and the 37 mm plastic cassette collector; (b) in the process of being assembled with the spring-loaded holder arrangement for the 37 mm plastic cassette. Photograph courtesy of Zefon International Inc., Ocala, FL, USA*

by means of spinning disk generators and for sampling flow rates from 1.7 to 2.1 Lpm. The results for P as a function of d_{ae} are summarised in Figure 15.9 (Phalen, 1985). They are compared there with the CEN/ISO/ACGIH criterion. The results for this sampler dipped somewhat below the modern conventional curve. In an important later experimental study of the same sampler, Bartley *et al.* (1994) applied polydisperse test aerosols along with a rapid testing method to analyze sampling efficiency across the whole particle size range of interest using the aerodynamic particle sizer (APS). Their results, shown in Figure 15.10(a), indicated more clearly the strong dependency of performance of the 10 mm nylon cyclone sampler on the sampling flow rate. They suggested that the flow rate of 1.7 Lpm would provide the most plausible operating condition for the use of this instrument, broadly in agreement with the results from the earlier studies reflected in Figure 15.9. The mass bias map in Figure 15.10(b) confirms this to be so.

As for the horizontal elutriator, the possibility of wind speed effects can not be ruled out. Cecala *et al.* (1983) experimentally investigated the wind speed dependency of the 10 mm nylon cyclone in a wind tunnel using polydisperse test aerosols generated from coal dust. They showed that significant effects were present for wind speeds above about 2 m s^{-1}. For most workplaces, therefore, where prevailing wind speeds are lower than this, there is not likely to be a problem. But for some situations – including in underground mining workplaces requiring high ventilation rates – local velocities might lie outside this range. For these, Cecala *et al.* showed that the introduction of an inlet shield could significantly reduce the wind speed dependency. The rationale – and the result – is similar to that for the modified inlet suggested for the MRE horizontal elutriator sampler by Ford. Later, Gautam and Sreenath (1997) investigated not only wind speed dependencies in the performance of the 10 mm nylon cyclone but also directional effects. They carried out experiments in a wind tunnel for wind speeds in the range

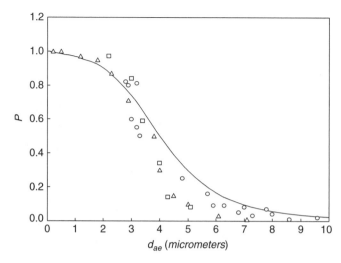

Figure 15.9 *Penetration (P) as a function of particle aerodynamic diameter (d$_{ae}$) for the 10-mm nylon cyclone for flow rates ranging from 1.7 to 2.1 Lpm (○ Ettinger et al., 1970, for 1.7 Lpm; △ Blachman and Lippmann, 1974, for 2.1 Lpm; □ Caplan et al., 1977, for 2.0 Lpm; ── CEN/ISO/ACGIH respirable aerosol curve). Adapted from Report of the ACGIH Air Sampling Procedures Committee. Copyright 1985. ACGIH®, Cincinnati, OH. Reprinted with permission*

up to 4 m s^{-1} for polydisperse test aerosol generated from coal dust using a fluidised bed generator and for a range of sampling flow rates. In these experiments, the sampled aerosol was analyzed across the whole particle size range of interest using the APS in the mode mentioned previously. From the large amount of data rapidly generated in this way for the sampling efficiency of the sampler, Gautam and Sreenath developed not only curves for *P* versus *d$_{ae}$*, as had previously been done for the original 10 mm nylon cyclone, but also – like Bartley *et al.* – constructed a set of maps reflecting the mass bias of the instrument in relation to a hypothetical sampler that perfectly matched the CEN/ISO/ACGIH curve for the respirable fraction across wide ranges of particle size distributions. Results like these showed that the 10 mm cyclone consistently undersampled particulate mass with respect to the desired criterion by an average of about 30 % for aerosol particle size distributions typical of those expected to be found in workplaces. This tendency was broadly consistent with the earlier experimental data shown in Figure 15.9, except that the mass bias was significantly greater than had been observed by Bartley *et al.* In addition, strong dependencies on wind speed and sampler orientation to the wind were observed. The overall picture of the performance of the 10 mm nylon cyclone from this work led Gautam and Sreenath to propose a modified version of the sample aimed at reducing the trends that had been noted. The result was a new sampler having the same external geometry and dimensions as the 10 mm nylon cyclone but containing a new multi-orifice inlet. This new inlet had three separate inlets, as opposed to the original single one, each designed so that aspirated air entered the cyclone tangentially. The new sampler is shown in Figure 15.11. Experiments with this instrument showed that the performance was indeed much less dependent on wind speed and orientation. Further, as shown in Figure 15.12, penetration for a sampling flow rate of 2.6 Lpm was found to lie close to the CEN/ISO/ACGIH curve. Importantly, the average mass bias with respect to a hypothetical ideal sampler for a representative range of particle size distributions was shown to be less than 10 %, much lower than they had shown for the original 10 mm nylon cyclone. This sampler is now commercially available as the *GS-3 respirable dust cyclone*

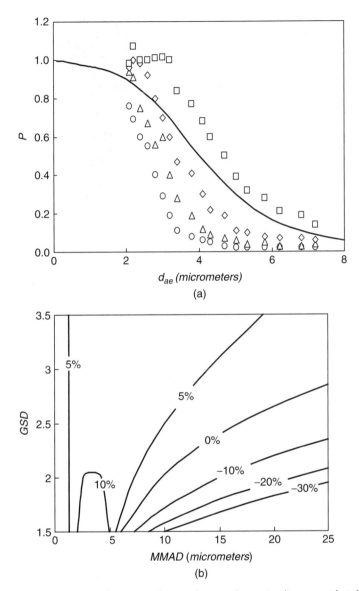

Figure 15.10 (a) Penetration (P) as a function of particle aerodynamic diameter (d$_{ae}$) for the 10 mm nylon cyclone for various flow rates (experimental data from Bartley et al., 1994 using the APS-based rapid data acquisition method: □ sampling flow rate 1.5 Lpm, ◊ 2.0 Lpm, △ 2.5 Lpm, ○ 3.0 Lpm; —— CEN/ISO/ACGIH respirable aerosol curve); (b) bias map representation of the sampler's ability to accurately measure respirable mass for a flow rate of 1.7 Lpm, showing bias contours for particles with various particle size distributions described in terms of mass median particle aerodynamic diameter (MMAD) and geometric standard deviation (GSD). Adapted with permission from Bartley et al., American Industrial Hygiene Association Journal, 55, 1036–1046. Copyright (1994) American Industrial Hygiene Association

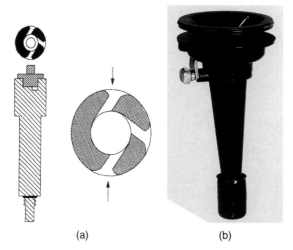

(a)   (b)

Figure 15.11 *The multi-inlet respirable aerosol sampler proposed by Gautam and Sreenath (1997): (a) diagram of the original multi-inlet modification to the generic 10 mm nylon cyclone; (b) photograph of the commercially available version, the 2.75 Lpm GS-3 respirable dust cyclone-based personal sampler. (a) Reprinted from Journal of Aerosol Science, 28, Gautam and Sreenath, 1265–1281. Copyright (1997), with permission from Elsevier. (b) Photograph courtesy of SKC Inc., Eighty Four, PA, USA*

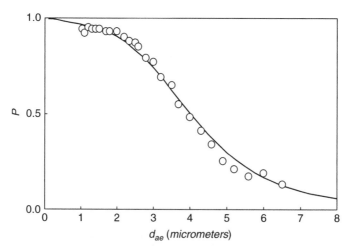

Figure 15.12 *Penetration (P) as a function of particle aerodynamic diameter (d_{ae}) for the GS-3 respirable dust cyclone when operated at a sampling flow rate of 2.6 Lpm (O Gautam and Sreenath, 1997; ___ CEN/ISO/ACGIH respirable aerosol curve). (Note: Subsequent unpublished calibrations indicated that an even better match with the CEN/ISO/ACGIH criterion could be achieved at a sampling flow rate of 2.75 Lpm, and this is the flow rate recommended by the manufacturer). Reprinted from Journal of Aerosol Science, 28, Gautam and Sreenath, 1265–1281. Copyright (1997), with permission from Elsevier*

(SKC Inc., Eighty Four, PA, USA) where its recommended sampling flow rate is 2.75 Lpm. Based on subsequent unpublished tests, it is said to provide an even better match with the CEN/ISO/ACGIH curve. As far as respirable aerosol is concerned, however, the original 10 mm nylon cyclone remains to this day the personal sampler recommended for regulatory compliance purposes in the USA by the Occupational Safety and Health Administration. So it is still widely available commercially (e.g. Zefon International Inc., Ocala, FL, USA).

Beyond the 10 mm nylon cyclone, there has been a long history of development of alternative cyclone-based personal samplers for the respirable fraction, thus providing many alternative options for occupational hygienists. In Britain, a family of cyclone-based instruments began with the 1.9 Lpm instrument first proposed by Higgins and Dewell (1967, 1968) in the form originally known as the BCIRA sampler and now generally referred to as the *Higgins–Dewell sampler*. Its performance was assessed using monodisperse aerosols of fluorescein produced by means of a spinning-disk generator. A modified version (the *SIMPEDS*) was developed for applications in the coalmining industry and its performance was investigated experimentally by Harris and Maguire (1968) and Maguire *et al.* (1973), also using fluorescein test aerosols. Although it was used for research purposes, the routine use of SIMPEDS in British coalmines was limited for a long time by the decision in that industry to opt for the fixed-point static sampling approach – using the MRE – as the basis of dust monitoring and control. Another version, shown in Figure 15.13, was developed for use in surface industry applications (Gwatkin

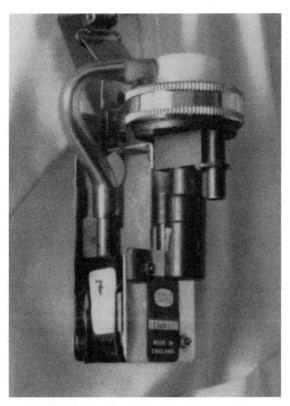

Figure 15.13 *A version of the British Higgins–Dewell 1.9 Lpm cyclone-based personal sampler, in a form developed for surface (non-mining) applications. Reproduced with permission from Vincent, Aerosol Sampling: Science and Practice. Copyright (1989) John Wiley & Sons, Ltd*

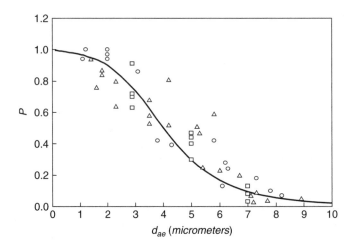

Figure 15.14   *Penetration (P) as a function of particle aerodynamic diameter (d$_{ae}$) for the 1.9 Lpm cyclone shown in Figure 15.13 (O Higgins and Dewell; △ Maguire et al., 1973; □ Ogden et al., 1983; ___ CEN/ISO/ACGIH respirable aerosol curve). Reproduced with permission from Vincent, Aerosol Sampling: Science and Practice. Copyright (1989) John Wiley & Sons, Ltd*

and Ogden, 1979). It was studied experimentally by Ogden *et al.* (1983) using both monodisperse test aerosols of dioctyl phthalate. Experimental data for this family of instruments, again in the form of *P* as a function of *d$_{ae}$*, are shown in Figure 15.14. Although the results were quite widely scattered, and were originally intended to be compared with the BMRC respirable dust curve, they are seen to lie quite close to the CEN/ISO/ACGIH curve.

From the European and American experience, a whole new generation of cyclone-based personal samplers evolved, drawing on – and combining – the strengths of the earlier ones, and featuring different internal geometries and inlet designs, different materials and different flow rates. One issue for the generic plastic-bodied 10 mm nylon cyclone noted by some users was its tendency for it to be influenced by electrostatic interferences. This has led to the development of a generation of cyclone-based samplers that would not suffer in this regard. Examples are shown in Figure 15.15, including the 2.5 Lpm *respirable dust aluminum cyclone* and the 2 Lpm *single-inlet GS-1 conductive plastic cyclone* from SKC Inc. (Eighty Four, PA, USA), and the 2.2 Lpm *respirable dust cyclone* from BGI Inc. (Waltham, MA, USA). All three of the examples shown were designed to interface with the 37 mm plastic cassette system widely used by occupational hygienists in North America and elsewhere. The GS-1 was designed to be equivalent to the 10 mm nylon cyclone. For the sampling flow rates indicated, all are described as meeting the requirements of the CEN/ISO/ACGIH convention (Maynard and Kenny, 1995; Harper *et al.*, 1998). Kenny and Gussman (1997) carried out an experimental study of a range of such samplers, both personal and static, using a rapid, APS-based testing method similar to that already mentioned. Here, as in most modern work on particle size-selective samplers, the bias map approach was applied as an important analytical tool in a procedure that assessed sampler performance not only in terms of its proximity to an appropriate 'target' conventional curve but also in terms of the sampler's ability to accurately collect the mass contained in the fraction of interest. From the Kenny and Gussman study, one sampler in particular emerged of particular interest for the personal sampling of fine aerosol fractions, the 4.2 Lpm *GK2.69* (BGI Inc., Waltham, MA, USA). This sampler is shown in Figure 15.16(a), where – like the

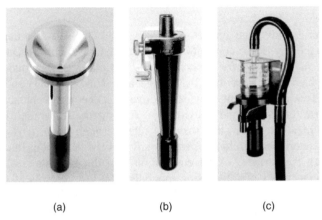

(a)   (b)   (c)

Figure 15.15 *Some examples of cyclone-based personal respirable aerosol samplers currently available: (a) the 2.5 Lpm respirable dust aluminum cyclone; (b) the 2 Lpm conductive plastic cyclone; (c) the 2.2 Lpm respirable dust cyclone. All three of these samplers were designed to be used together with the 37 mm plastic cassette. In the figure, only (c) is shown with the cassette incorporated. (a) and (b) Photographs courtesy of SKC Inc., Eighty Four, PA, USA (c) Photograph courtesy of Robert Gussman, BGI Inc., Waltham, MA, USA*

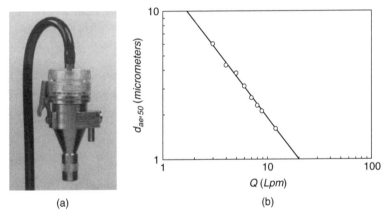

(a)   (b)

Figure 15.16 *The GK2.69 cyclone-based personal sampler: (a) the sampler itself; (b) experimental data showing the d_{ae}-value at which penetration (P) is 50 % as a function of sampling flow rate (from Kenny and Gussmann, 1997). (a) Photograph courtesy of Robert Gussman, BGI Inc., Waltham, MA, USA. (b) Adapted from Journal of Aerosol Science, 28, Kenny and Gussman, 677–688. Copyright (1997), with permission from Elsevier*

10 mm nylon cyclone – it is seen to be integrated into a personal sampling system involving the 37 mm plastic cassette. Also shown [see Figure 15.16(b)] are data that Kenny and Gussman reported for the d_{ae}-values at which penetration fell to 50 % (i.e. $_{50}d_{ae}$). Shown as a function of sampling flow rate, it is seen that the 'cut-size' for the cyclone rose steadily as the sampling flow rate decreased. From these results together with bias-map analysis, the GK2.69 sampler was shown to provide a good match with the CEN/ISO/ACGIH curve for respirable aerosol.

15.2.4 Impactors

As described in Chapter 8, individual impactors have particle size-selective deposition characteristics that are too sharp to match any of the conventional respirable aerosol curves. One solution was embodied in the *single-stage, multi-orifice-size impactor-based sampler system* proposed by Marple (1978), working versions of which were described by Marple and Rubow (1983) and are shown schematically in Figure 15.17. This arrangement provides an overall penetration characteristic that is built up by superposition of those for each of the individual jet sizes acting in parallel. In principle, by careful 'tweaking' of the individual impactors and the flow rates through them, it is possible to achieve an overall penetration characteristic that closely matches any desired definition. In one design, Marple and Rubow set out to match the earlier ACGIH convention in a three-nozzle-size, 2 Lpm personal sampler. This impactor system was calibrated using monodisperse test aerosols of oleic acid produced by means of a vibrating-orifice generator. The results in Figure 15.18 show that the performance of the system tested was in quite good agreement with the modern CEN/ISO/ACGIH respirable aerosol curve. Similarly impressive agreement was also achieved for other versions aimed at matching the BMRC curve. In order to reduce the effects of particle bounce, oil-soaked, porous impaction surfaces, like those described by Reischl and John (1978), were employed. Commercial versions of this sampler are currently available, for example, the 2 Lpm *parallel particle impactor* (PPI) (SKC Inc., Eighty Four, PA, USA) (see Figure 15.19).

To overcome some of the difficulties associated with impactors, including particle bounce and other types of loss, especially at high particle loadings on the impaction surface, the related concept of virtual impaction has been considered. Here, particles 'impact' not onto a solid surface – as in conventional impactors – but onto a quiescent region of air where they may enter and so effectively be removed from the flow. An early practical respirable aerosol sampler based on such principles was the one proposed by Zurlo (1987). Others have appeared. More recently, as described later in this chapter, the principles of virtual impaction were applied for the selection of the fine particle fractions in an instrument (the Respicon) designed to simultaneously collect the inhalable, thoracic and respirable fractions.

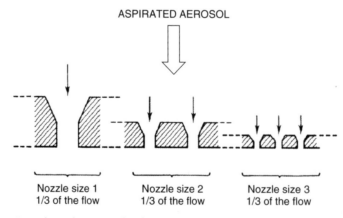

ASPIRATED AEROSOL

Nozzle size 1
1/3 of the flow

Nozzle size 2
1/3 of the flow

Nozzle size 3
1/3 of the flow

Figure 15.17 *Schematic to show the principle of operation of the multi-orifice-size parallel impactor, indicating how the aspirated aerosol is divided between the impactor nozzles of different sizes. Reproduced with permission from Vincent, Aerosol Sampling: Science and Practice. Copyright (1989) John Wiley & Sons, Ltd*

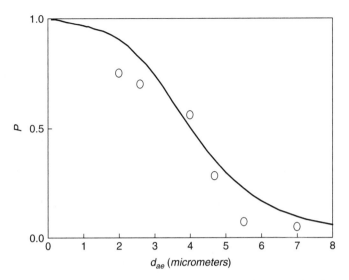

Figure 15.18 *Penetration (P) as a function of particle aerodynamic diameter (d_{ae}) for a 2 Lpm version of the multi-orifice-size respirable aerosol impactor (O based on the experimental data of Marple and Rubow, 1983; ___ CEN/ISO/ACGIH respirable aerosol curve)*

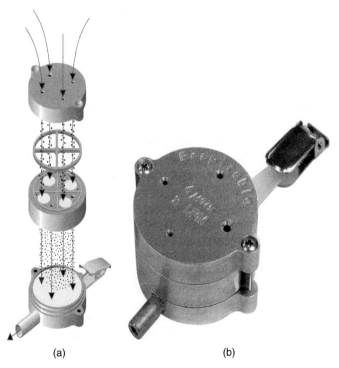

(a)   (b)

Figure 15.19 *Commercial version of the 2 Lpm multi-orifice-size, respirable aerosol, parallel impactor: (a) schematic of the disassembled sampler; (b) photograph of the assembled sampler. Photograph and drawing courtesy of SKC Inc., Eighty Four, PA, USA*

15.2.5   Porous plastic foam filter samplers

As already noted earlier in this book, foam media are relatively inefficient as filters for particulate matter, and so are not particularly useful for mainstream applications in, for example, aerosol collection and air cleaning. But this very feature, manifested in the form of a penetration that changes relatively slowly with particle size, makes foam media potentially very useful for particle size-selective sampling, as in the case of the respirable fraction. Practical applications of this idea first emerged in the late 1970s at the Institute of Occupational Medicine (IOM) in Edinburgh, with the suggestion of a sampling respirator that would be worn by, and powered by, an aerosol-exposed worker. Here the respirator in question would incorporate porous foam media as a pre-selector that, during the normal breathing action of the wearer, would allow the respirable fraction to pass to an absolute filter. Gibson and Vincent (1981) developed a prototype sampling respirator that met the desired rationale. At the time, in the early 1980s, there were no models for foam penetration, so Gibson and Vincent arrived at an appropriate arrangement based empirically on inspection of their own experimental data for P as a function of d_{ae} for various grades of foam and ranges of face velocity. This process suggested a system comprising a pre-filter which consisted of two porous plastic foam disks of diameter 11.3 cm placed together in series, the first one of 60 ppi and thickness 1.25 cm and the second of 80 ppi and thickness 0.6 cm.[2] The resultant pre-filter was backed by an 'absolute' filter for collection of the desired respirable fraction. In practice it was intended that this package of filtration media would be incorporated into the cartridge of the respirator worn by the worker. In the experiments the package was contained within an open-faced sampling probe, where the aspiration efficiency was assumed to be unity. Air flow through the system was provided by means of a cam-driven breathing machine set up to deliver 21 breaths min^{-1} and breathing minute volumes of 18 and 34 Lpm, respectively. Here the 'exhaled' air was made to bypass the filter by means of a non-return valve. The experiments themselves were carried out in a calm air chamber using polydisperse test aerosols mechanically generated from coal dust, and collected particulate matter was analyzed – counted and sized – by means of a Coulter counter. Reference aerosol measurements in the chamber were made using an inverted open filter. In addition to the test system of interest, an MRE respirable aerosol sampler (of the type described earlier in this chapter) was also placed in the chamber to provide a reference sample of the respirable fraction, as defined by the BMRC curve which was then the target convention. The results in Figure 15.20 show that those for the two sampling flow rates for the same sampling respirator were in good agreement with one another. They were also in similar agreement with the BMRC curve. It is seen that agreement with the CEN/ISO/ACGIH convention was less good for the system that was tested. However, that is not to say that some other combination of porous foam disks might not be found that would provide better agreement.

Although the original idea of a sampling respirator was never translated into a practical device, it provided encouragement for the subsequent research that has since led to a variety of more conventional particle size-selective samplers, some of which are now available commercially. An important step was the development of a validated empirical mathematical model (Vincent *et al.*, 1993; Kenny *et al.*, 2001) that could be used to make a good first estimate of the penetration for any choice of foam media, geometrical configuration and dimensions, and flow parameters. Subsequent developments – and, eventually, commercially available instruments – have been based on extensions of existing instruments for coarser aerosol fractions. In particular, it has been a relatively simple matter to insert foam plugs into the inlets of some existing samplers for the inhalable fraction, where the foam plug is chosen – by use of the empirical model – to provide as closely as possible the desired respirable pre-selection curve. In this way, the resultant samplers for the respirable fraction have met the first important requirement

---

[2] ppi ≡ pores per inch (see Chapter 8).

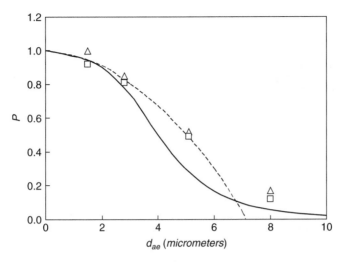

Figure 15.20 *Penetration (P) as a function of particle aerodynamic diameter (d_{ae}) for a prototype sampling respirator for cyclical simulated breathing (Gibson and Vincent, 1981: △ simulated minute volume of 18 Lpm; □ 34 Lpm; ___ CEN/ISO/ACGIH respirable aerosol curve; - - - BMRC respirable aerosol curve). Adapted with permission from Gibson and Vincent, Annals of Occupational Hygiene, 24, 205–215. Copyright (1981) British Occupational Hygiene Society*

of the CEN/ISO/ACGIH respirable aerosol criterion – namely, that the sampler should first aspirate the inhalable fraction and then select the desired subfraction, thus reflecting what happens in actual human aerosol exposure.

The IOM personal inhalable aerosol sampler described in the preceding chapter has been of special interest in this regard, firstly because it is a sampler that had already been demonstrated as following the inhalable aerosol criterion, and so has become perhaps the most widely accepted in that regard, and secondly because the entry allows the insertion of a simple cylindrical foam plug with minimum change to other features of the sampler. Prototype versions were first proposed as long ago as 1988 in an unpublished conference paper by Mark *et al.* (1988) and subsequently by Vincent *et al.* and Aitken *et al.* (1993). Later, other prototypes emerged that had been developed for specific applications. A good example is the one reported by Chung *et al.* (1997). This was a prototype sampler for collecting respirable welding fume aerosol. It is shown assembled in intended mode of use in Figure 15.21(a) and disassembled in Figure 15.21(b). This instrument was essentially similar to the original IOM personal inhalable aerosol sampler, but with an extended nose-piece that contained a snug-fitting 90 ppi foam of diameter 30 mm and length 25 mm. The penetration characteristics of this plug had previously been determined experimentally by Aitken *et al.* in a calm air chamber, using monodisperse polystyrene latex test aerosols for which collected samplers were analyzed by optical microscopy. The results are shown in Figure 15.22. It is important to note that penetration is now the penetration of just the foam plug so that the results do not include the contribution to overall performance of the aspiration efficiency of the inlet (which, for the IOM sampler, was assumed to closely follow the inhalability criterion). Figure 15.22 shows quite good agreement between the measured performance of the prototype sampler and the CEN/ISO/ACGIH criterion. A commercial respirable aerosol sampler along the lines described above is now available (SKC Inc., Eighty Four, PA, USA and SKC Ltd, Blandford Forum, UK). Another sampler, similarly based on the CIS sampler, described in the preceding chapter, is also available (BGI

(a)   (b)

Figure 15.21   *A prototype, porous plastic foam-based 2 Lpm personal sampler designed for applications in sampling respirable welding fume (as described by Chung et al., 1997): (a) assembled sampler, worn in the intended mode of use; (b) disassembled to show the porous plastic foam respirable aerosol pre-selector. Reproduced with permission from Chung et al., Annals of Occupational Hygiene, 41, 355–372. Copyright (1997) British Occupational Hygiene Society*

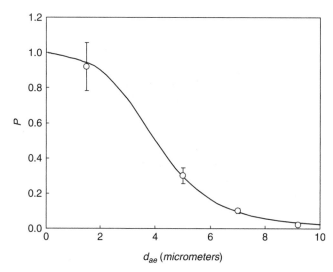

Figure 15.22   *Penetration (P) as a function of particle aerodynamic diameter (d_{ae}) for the prototype porous plastic foam-based 2 Lpm respirable aerosol sampler shown in Figure 15.21 (O experimental data from Aitken et al., 1993; ___ CEN/ISO/ACGIH respirable aerosol curve). Adapted from Applied Occupational and Environmental Hygiene, Application of Porous Foams as Size Selections for Biologically Relevant Samplers, 8(4), pages 363–369. Copyright 1993 ACGIH®, Cincinnati, OH. Reprinted with permission*

Inc., Waltham, MA, USA and Casella CEL Ltd, Kempston, UK). In recent years, both have been deployed in practical occupational hygiene situations.

The preceding approach – making simple modifications to existing samplers by the insertion of porous plastic foam plugs – has been applied elsewhere, in particular in the modification of some direct-reading aerosol measuring instruments for the respirable fraction. For example, the *tapered element oscillating microbalance* (*TEOM*) *ambient particulate monitor* (from Thermo Electron Corporation, Boston, USA) has been for several years a popular direct-reading instrument for aerosol measurement in a range of occupational and environmental hygiene situations. It operates on the principle that particles aspirated through a nozzle are collected on a filter that is mounted on a tapered oscillating element whose mass-dependent natural frequency may be determined accurately. Recently, a respirable particle size-selective inlet has been developed for the TEOM incorporating a cylindrical foam plug (see Figure 15.23), chosen on the same scientific basis as for the IOM and CIS adaptations described above.

Elsewhere, Page *et al.* (2005) developed a low flow rate respirable aerosol sampler comprising a cylindrical pre-selector of 90 ppi porous foam media with overall diameter 4 mm and length 25 mm and operated at a flow rate of 0.25 Lpm. In chamber experiments with polydisperse liquid droplet test aerosols, with particles counted and size using an APS, they showed that the results for penetration as a function of d_{ae} were very close to corresponding results obtained for the ubiquitous 1.7 Lpm 10 mm nylon cyclone. The possibility of using a sampler at such a low flow rate was considered to provide the opportunity for using much smaller and lighter personal sampling pumps, much to the pleasure of the wearers!

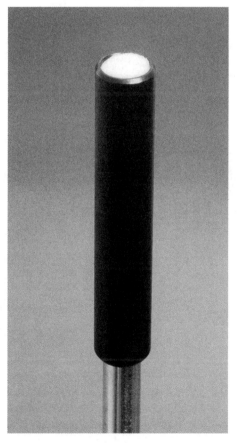

Figure 15.23 *Porous plastic foam-based respirable inlet for the TEOM ambient particulate monitor also widely used in workplace sampling applications. Photograph courtesy of Thermo Electron Corporation, Waltham, MA, USA*

What has been described so far represents the outcome of a recent sequence of research into the nature of particle penetration through porous plastic foam media. Earlier, however, and quite independently, porous foam media had already been applied for pre-selection purposes in the ingenious French respirable aerosol sampler, the 10 Lpm *CIP10*. It was introduced in the preceding chapter as a candidate sampler for the inhalable fraction. In the form first described by Courbon *et al.* (1983), it was devised as a sampler for the finer respirable fraction. The inside of the sampling head of the instrument shown in Figure 14.9 is now shown schematically in Figure 15.24. As shown in Figure 15.24(a), the aspirated air passed through a coarse pre-selector foam to remove the largest particles and then a second pre-selector to separate the fine particles of interest. The relatively high airflow was generated by the centrifugal pumping action of a rapidly rotating second foam plug [Figure 15.24(b)], located just below the pre-selector foams, which also served as the collector for the desired aerosol fraction. There was no backing filter, so the overall flow resistance was small, a necessary prerequisite for the successful delivery of such a high flow rate. In this instrument, the penetration of the upper, coarser foam plug was similar to that of the foam-based samplers described above. The overall performance of the sampler was also strongly influenced by the aerosol collection properties of the lower foam plug, notably because this was far from being an 'absolute' filter of the type used in most sampling instruments. However, the fact that the second foam plug allowed finer particles to penetrate – and so not be collected – provided a collection characteristic closer to the actual lung deposition of finer particles in the lungs of exposed human subjects. As described, the performance of the CIP10 therefore may be expected to have some features similar to those of the German TBF50 and the French CPM3 static samplers described earlier. Courbon *et al.* (1988) measured its performance using test aerosols generated from fused alumina and coal dust, respectively, and the results are shown in Figure 15.25 for the standard version where the static selecting foam was specified as 45 ppi. Here it is important to note that, based on how the

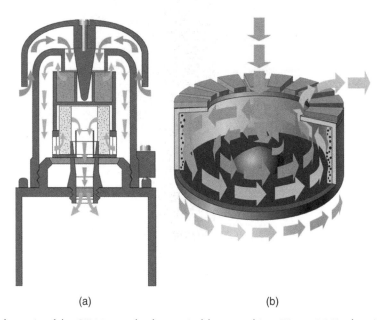

(a)   (b)

Figure 15.24 *Schematic of the CIP10 sampler for respirable aerosol (see Figure 14.9), showing the porous foam pre-selector and the rotating foam pumping/collection stage: (a) entry and pre-selection stage; (b) rotating foam collecting stage. Diagrams courtesy of Christian Champion, ARELCO ARC, Fontenay sous Bois, France*

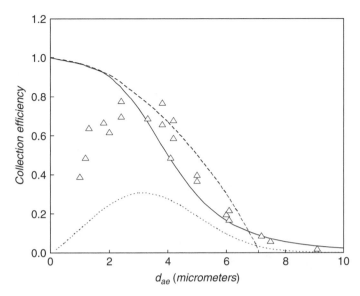

Figure 15.25 *Collection efficiency as a function of particle aerodynamic diameter (d_{ae}) for the 10 Lpm CIP10 respirable aerosol sampler shown in Figure 14.9) ($\triangle$ experimental data from Courbon et al., 1988; ___ CEN/ISO/ACGIH curve; - - - suggested curve based on human alveolar deposition, see Chapter 11)*

experiments were performed, the results reflected the overall sampling efficiency of the instrument, involving contributions not only from the penetration characteristics of the foam pre-selector but also from aspiration efficiency, as well as the collection efficiency of the second porous foam plug. So the results are plotted in terms of overall collection efficiency. Also shown in the figure is the modern CEN/ISO/ACGIH criterion. Overall, the experimental results showed the expected trends. For larger particle sizes ($d_{ae} > 3$ μm) the trend is seen to be similar to that of both the CEN/ISO/ACGIH and the BMRC curves. Quantitative agreement, however, is only fair. Also shown on the graph is the curve representing the suggested new alveolar deposition criterion introduced in Chapter 11. Here, although the results reflect the overall trend for human lung deposition, quantitative agreement is rather poor. The CIP10 is currently available commercially from ARELCO ARC (Fontenay sous Bois, France) and has been widely used in Europe, particularly in France, both as a static and a personal sampler.

In the broader sense, it is a significant attraction to potential manufacturers that the engineering challenges in the construction of porous foam-based particle size-selective samplers for fine aerosol sub-fractions samplers are much less than those involved in the manufacture of, for example, elutriators, cyclones and impactors. Manufacturing tolerances are lower and structural requirements are simpler, leading to potential lower cost to users.

15.2.6 Other samplers

The devices that have been described above represent the vast majority of respirable aerosol samplers that have been used by professional occupational hygienists. However, a number of other approaches have been tried and are worthy of mention. Operating on the principle of centrifugal separation, the *Conicycle* was developed primarily for applications in British coalmines (Wolff and Roach, 1961). The sampling head of this instrument is shown schematically in Figure 15.26. Rotation of the head at

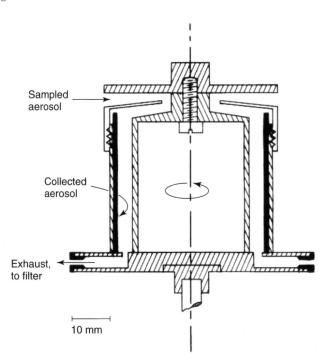

Sampled aerosol

Collected aerosol

Exhaust, to filter

10 mm

Figure 15.26 *The 8 Lpm Conicycle centrifuge-based respirable aerosol sampler (Wolff and Roach, 1961). Adapted from Wolff and Roach, Inhaled Particles and Vapours, pp. 460–464. Copyright (1961) British Occupational Hygiene Society*

8000 rpm about the vertical axis provided centrifugal pumping which aspirated air at 8 Lpm. Coarser particles were deposited under the influence of centrifugal forces on the inner wall of the outer cylinder. Finer respirable particles exited through the base of the instrument. For an instrument like this, a theoretical description of particle motion inside the annular space between the inner and outer cylinders is similar to that for the horizontal elutriator, the only difference being that – in the Conicycle – the gravitational force is replaced in the equations of particle motion by the centrifugal one. In principle, therefore, it should be possible to devise a system whose penetration characteristics come very close to the BMRC curve. In practice, however, it was found that some empirical adjustment was needed. Although the Conicycle underwent field trials in mines during the 1960s, it never enjoyed widespread use.

Filter media are usually employed for the highly efficient separation of particles from gases. Relatively inefficient filters other than the porous plastic foam media already mentioned also have the potential for use as pre-selectors for a particle size-selective sampler. For example, Parker *et al.* (1977) and Cahill *et al.* (1977) proposed using large-pore polycarbonate membrane filters for selecting the respirable fraction. However, Lippmann (1983) recommended against this approach on two counts. He argued firstly that, for particle sizes, pore sizes and face velocities in the ranges of interest, the main mechanism of particle collection is interception, so that penetration is dependent on geometric – and not aerodynamic – particle size. Secondly, he pointed out that particle bounce or blow-off is difficult to eliminate in such a pre-selector.

15.2.7   Sampling for 'respirable' fibers

Fine fibrous aerosols present special sampling problems. To begin with, the practical definition of what is 'respirable' for such particles is based not only on the aerodynamic factors that govern the deposition of fibers in the lung after inhalation but also on their known dimension-associated health risk, especially as far as asbestos is concerned. Selection of 'respirable' particles in this context is conventionally carried out by visual sizing and counting under the microscope. This means that, in sampling, the main priority is to achieve deposition of aerosol onto a membrane filter which can then be mounted and cleared for subsequent microscopic evaluation. Unlike all the respirable aerosol samplers described so far, there is now no specific need for aerodynamic selection inside the sampler.

Practical methods for the sampling and analysis of 'respirable' fibrous aerosols, usually with particular reference to asbestos, have been described by a number of bodies, including the Asbestos International Association (AIA, 1979), the US National Institute of Occupational Safety and Health (NIOSH, 1979) and others. The details of the sample preparation and analysis are not repeated here, but an excellent summary was provided by Baron (1993). The samplers themselves that are recommended for this purpose are very simple, mainly because the aerodynamic diameters of the fibers of interest are usually small enough that inertial effects leading to inlet (i.e. aspiration efficiency) bias are usually negligible. This aspect was investigated in experiments by Johnston *et al.* (1982) into the effects of both sampling flow rate and external wind speed. For 25 mm diameter open-faced samplers of the type used for personal sampling by professional occupational hygienists, aspiration efficiency for asbestos fibers was found to change negligibly over a wide range of sampling flow rates and wind speeds. One version of this sampler that is favored nowadays features a 50 mm long cylindrical conductive plastic cowl, one purpose of which is to reduce the possibility of contamination of the filter by inadvertent contact by personnel. Another is to reduce the possibility of electrostatic charge effects that might influence both the amount and spatial distribution of fibers collected on the filter.

One feature of most of the published methods is the prescription of a particular flow rate (or range of flow rates). In some practical situations, this has been found to be quite restrictive, especially where the airborne concentration of fibers is such that, for typical sampling times, the on-filter density of fibers is so low that visual counting and sizing under the microscope by human operatives becomes unreliable or subject to bias (Beckett, 1980; Cherrie *et al.*, 1986). One way to overcome this problem is to increase the flow rate so that, for a given sampling period, the on-filter fiber density is raised. As indicated by Johnston *et al.*, this can be achieved over a wide range of sampling flow rate without significant bias associated with inertial effects during particle aspiration.

15.3   Samplers for the thoracic fraction

For sampling thoracic aerosol, the pre-selector options are the same as the ones that have just been discussed in relation to respirable aerosol sampling, but with appropriate adjustment of primary dimensions and sampling flow rate. Now, however, because the aerosol fraction in question is coarser, the role of sampler aspiration efficiency may be expected to be significantly greater and so should not be neglected in the way that was justified for the finer respirable fraction. In addition, for thoracic aerosol sampling, the situation is quite different from that for respirable aerosol because there are currently very few occupational exposure standards expressed in terms of this fraction. Consequently the demand for thoracic samplers for occupational settings has so far been low. That said, sampling for aerosols in the ambient atmosphere has been dominated in recent years by the PM_{10} standards, which is qualitatively the same as – and quantitatively similar to – the CEN/ISO/ACGIH thoracic aerosol convention. There are many samplers for PM$_{10}$ aerosol intended for applications in ambient atmospheric sampling that have been thoroughly tried and tested. So there is a good opportunity for the transfer of knowledge from one branch of aerosol sampling application to another. In the meantime, an increasing number of thoracic aerosol samplers are emerging specifically for applications in occupational aerosol exposure assessment.

15.3.1   Vertical elutriators

Although the idea of thoracic aerosol sampling in workplaces has not yet been widely applied, the US cotton industry provides one notable exception. Here, for some years, in recognition of inhaled particle deposition in the upper airways of the respiratory tract and its possible role in cotton workers' byssinosis, a sampling criterion based on a pre-selector with 50 % penetration at $d_{ae} = 15$ μm has been prescribed since the 1970s (National Institute of Occupational Safety and Health 1974). The recommended sampling method employs the concept of 'vertical elutriation'. In this application, air is aspirated into a downwards-facing vertical column in which the upward velocity of the air flow is equal and opposite to the falling speed of a particle with $d_{ae} = 15$ μm. The device that has been widely used in the US cotton industry is shown in Figure 15.27. Its basic performance characteristics were investigated by a number of workers, notably at the Southern Regional Research Center of the US Department of Agriculture in New Orleans. This work was reviewed in a paper by Robert and Baril (1984). The basic performance of the device was defined in terms of the particle size selectivity of the vertical elutriator itself, allowing for the effects of flow separation and the effects of crosswinds and updrafts (see also Neefus *et al.*, 1977; Fairchild *et al.*, 1978). The accumulation of cotton lint fragments were also found to influence performance. Robert (1980) suggested that such effects, singly or in combination, could lead to errors of up to a factor of ×2 in the mass sampling of the desired aerosol fraction, most of it a direct consequence of the excursions in aspiration efficiency. Once again, from the theories of aspiration efficiency described earlier in this book, such biases are not surprising. Perhaps the greatest concern with this sampler is that such effects cannot be controlled or accounted for under practical conditions. Therefore, based on what is now known, the vertical elutriator does not look promising for more general applications in thoracic aerosol sampling.

15.3.2   Cyclones

In the 1980s, an alternative static sampler was proposed for applications in the US cotton industry that employed a cyclone pre-selector designed to achieve a match with the PM$_{10}$ curve (McFarland *et al.*, 1987). In contrast to the vertical elutriators described above, wind effects were found to be less significant for this instrument. Later, as the new particle size-selective criteria emerged, more cyclone-based samplers were proposed for the thoracic fraction, in particular versions that could be used as personal samplers.

In principle, all the cyclones described earlier for the respirable fraction have the potential for modification to meet the requirements of the thoracic fraction, most obviously by a change – specifically a decrease – in the sampling flow rate. This was demonstrated most clearly by Kenny and Gussman who developed the GK2.69 personal cyclone sampler that accurately collected the respirable fraction when the sampling rate was set at 4.2 Lpm. The same sampler collected the thoracic fraction with equal accuracy when the flow rate was reduced to 1.6 Lpm [see Figure 15.16(b)]. As already mentioned, this sampler is commercially available (BGI Inc., Waltham, MA, USA). Although it has been well characterised in

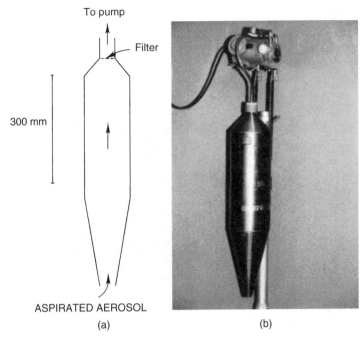

To pump

Filter

300 mm

ASPIRATED AEROSOL

(a)   (b)

Figure 15.27 *The 7.4 Lpm vertical elutriator used widely for sampling cotton dust aerosols in the US cotton industry (for which penetration is nominally 50 % for $d_{ae} = 15$ μm): (a) schematic diagram; (b) photograph of the instrument. (a) Reproduced with permission from Vincent, Aerosol Sampling: Science and Practice. Copyright (1989) John Wiley & Sons, Ltd. (b) Photograph courtesy of John N. Zey, Department of Safety Sciences, Central Missouri State University, Warrensburg, MO, USA*

terms of the penetration characteristics of the cyclone separator, nothing has been reported about its aspiration efficiency. So the true extent of how it truly matches the strict CEN/ISO/ACGIH criterion is not yet known. Another example of a sampler originally designed for the respirable fraction but adapted for the thoracic fraction is the SIMPEDS sampler originally developed for sampling respirable aerosol but now operated at the lower sampling flow rate of 0.8 Lpm (Jones *et al.*, 2005).

15.3.3 Impactors

Perhaps the first impactor-based sampler for thoracic aerosol in occupational settings was the device introduced in the late 1970s for applications in the US cotton industry, intended as another alternative to the vertical elutriator (Batra *et al.*, 1980). Since then, impactors and virtual impactors have been widely considered for the thoracic aerosol fraction, many of them initially intended for the sampling of PM_{10} aerosol in the ambient atmosphere. Examples of these are discussed in Chapter 17. Many of them are not suitable for occupational hygiene applications, however, because they are generally too large to be adapted for use as personal samplers. One instrument originally developed for use in PM_{10} sampling in the ambient atmosphere, but which has already found applications in occupational aerosol exposure assessment, is the small *personal environmental monitor* (*PEM*) (Buckley *et al.*, 1991). It is available commercially (SKC Inc., Eighty Four, PA, USA; MSP Corporation, Minneapolis, MN, USA) in

(a)   (b)

Figure 15.28 *The personal environmental monitor (PEM), available in several versions: (a) schematic of the disassembled sampler in the 2 Lpm version intended for the collection of PM$_{10}$; (b) photograph of the assembled PM$_{10}$ sampler. Pictures courtesy of SKC Inc., Eighty Four, PA, USA*

a number of versions with flow rates 2, 4 or 10 Lpm. The version shown in Figure 15.28 is particularly appropriate for personal sampling. It features a single five-orifice impactor stage that provides a sharp cut at $d_{ae} = 10$ μm when the sampler is operated at 2 Lpm. Particles larger than this are collected on an annular impaction ring and smaller ones pass through to a filter. In the use of this system the collecting surface is greased in order to reduce the effects of particle bounce and blow-off.

15.3.4   Porous plastic foam filter samplers

In recent years, porous foam media have emerged as excellent candidates for the pre-selectors of samplers for the thoracic fraction. The first attempt to develop a thoracic aerosol pre-selector was reported in 1988 by Mark *et al.* in their description of an instrument that would simultaneously provide the inhalable, thoracic and respirable fractions (see below). The original Mark *et al.* sampler explicitly embodied the principle underlying the CEN/ISO/ACGIH set of criteria – namely that the sampler should first aspirate the inhalable fraction and then select the thoracic sub-fraction. As already mentioned, while strict adherence to this principle might not be a significant issue for the finer respirable fraction, it is likely to be important for the coarser thoracic fraction.

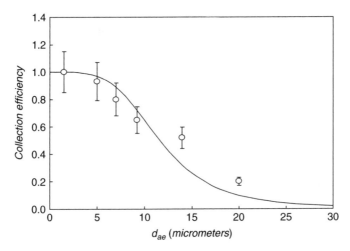

Figure 15.29 *Collection efficiency as a function of particle aerodynamic diameter (d_{ae}) for the porous plastic foam-based modified 2 Lpm IOM sampler for thoracic aerosol sampler (experimental data from Aitken et al., 1993 for a 30 ppi foam plug of diameter 29 mm and length 24 mm), shown in comparison with the CEN/ISO/ACGIH thoracic aerosol convention without the contribution due to inhalability. Adapted from Applied Occupational and Environmental Hygiene, Application of porous foams as size selectors for biologically relevant samplers, 8(4), pp. 363–369. Copyright 1993 ACGIH®, Cincinnati, OH. Reprinted with permission*

Again the IOM personal inhalable aerosol sampler has been the starting point for some of the versions that have been proposed. Aitken *et al.* designed a version of the IOM sampler that contained a cylindrical plug of 30 ppi porous foam media 29 mm in diameter and 24 mm long, and operated at a flow rate of 2 Lpm. The efficiency of penetration (P) of particles through this plug was measured as a function of d_{ae} using the same method described earlier, and Figure 15.29 shows some of the experimental results. Here the performance of the sampler system was obtained in terms of the efficiency of aerosol penetration through the foam pre-selector. The results are compared in the figure with the version of the CEN/ISO/ACGIH thoracic penetration aerosol curve that does *not* include the contribution of inhalability. In this sampler, this important feature is accounted for directly by the fact that its aspiration efficiency already closely matches the inhalability criterion by virtue of its inlet design. A commercial version of this instrument is now available (SKC Inc., Eighty Four, PA, USA). So too is a similarly modified version of the CIS inhalable aerosol sampler (BGI Inc., Waltham, MA, USA).

As described above, the French CIP10 sampler, in the form originally developed in the early 1980s as a respirable aerosol sampler, employed porous plastic foam media to provide the respirable fraction. More recently, efforts have been made to realise new versions of the CIP10 aimed at the thoracic fraction. In 1998, Fabries *et al.* described a model intended for personal sampling, the 7 Lpm *CIP10-T*, where the porous plastic foam pre-selector was replaced by an impactor arrangement. Its use in the cotton industry had previously been described earlier by Görner *et al.* (1994). As shown in Figure 15.30, the pre-selector comprised a stainless steel cone with eight 1.6 mm orifices through which the aerosol passed so that larger particles were deposited initially in the cone and the finer ones, corresponding to the desired fraction, passed through to the lower rotating cup containing the pumping/collecting foam stage, as in the original CIP10. The performance of this sampler was evaluated using polydisperse tests

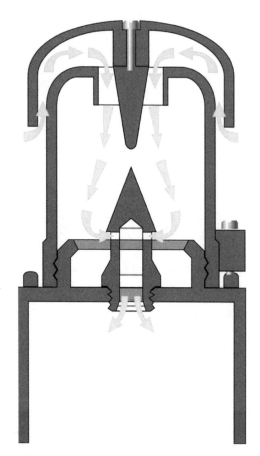

Figure 15.30 *Diagram to show the design the inlet for the French 7 Lpm CIP10-T personal sampler for the thoracic fraction (Fabries et al., 1998). Diagram courtesy of Christian Champion, ARELCO ARC, Fontenay sous Bois, France*

aerosols where the penetrating particles were sized and counted using the rapid APS-based method. The results are shown in Figure 15.31(a) in terms of the efficiency with which the particles were aspirated and then were collected in the rotating cup. Now therefore they are compared with the version of the CEN/ISO/ACGIH thoracic aerosol curve that *does* include the contribution due to inhalability. The observed trends were very similar to those exhibited by the CIP10 in the version intended for collecting the respirable fraction – that is, although they agreed well with the desired target curve for the larger particles, there was a clear tendency towards undersampling the finer particles for d_{ae} below about 2 μm. Importantly, however, when the results were subjected to bias-map analysis, the instrument was found to sample thoracic aerosol mass accurately – to within ±10 % – over wide ranges of particle size distributions corresponding to ones expected in practical workplace situations. Fabries *et al.* also proposed a static version of the instrument, the 10 Lpm *CATHIA-T*, which featured an external pump and where the particles of interest were collected on a conventional filter. As shown in Figure 15.31(b), the experimental results for this version were in good agreement with the

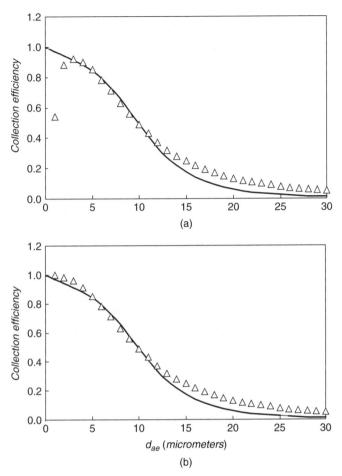

Figure 15.31 *Collection efficiency as a function of particle aerodynamic diameter (d_{ae}) for: (a) the CIP10-T personal thoracic aerosol sampler; (b) the CATHIA-T static thoracic aerosol sampler (△ model of Fabries et al., 1998, based on their experimental data; ___ CEN/ISO/ACGIH thoracic aerosol curve). Adapted with permission from Fabries et al., Annals of Occupational Hygiene, 42, 453–465. Copyright (1998) British Occupational Hygiene Society*

CEN/ISO/ACGIH criterion and, not surprisingly, did not reflect the down-turn at small particle sizes exhibited by both the original CIP10 and the CIP10-T. Again, this agreement was well supported by bias-map analysis.

15.4 Samplers for PM$_{2.5}$

The PM$_{2.5}$ convention emerged from considerations of the particle size distribution of atmospheric aerosol, and so was not originally intended for application in relation to workplace environments.

However, some occupational hygienists have considered it useful to be able to make measurements of workers' personal exposures to this fine aerosol fraction. A version of the PEM previously described above for the thoracic fraction has also been developed for sampling $PM_{2.5}$. The version shown in Figure 15.32 is very similar to the one shown earlier in Figure 15.28, but contains a larger number of inlet orifices and collects $PM_{2.5}$ when it is operated at 10 Lpm. In this instrument the sharp cut associated with the impaction process is particularly appropriate to the definition of the $PM_{2.5}$ fraction.

Figure 15.32   *The personal environmental monitor (PEM), shown in the 10 Lpm version intended for the collection of $PM_{2.5}$. Photograph courtesy of SKC Inc., Eighty Four, PA, USA*

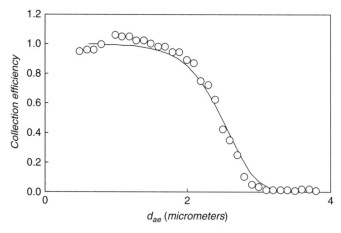

Figure 15.33   *Collection efficiency as a function of particle aerodynamic diameter (d_{ae}) for a porous plastic foam-based modified CIS sampler for $PM_{2.5}$ aerosol (O experimental data from Kenny and Stancliffe, 1997; ___ EPA $PM_{2.5}$ convention)*

Porous plastic foam-based samplers have also been proposed for $PM_{2.5}$. A high flow, 16 Lpm personal sampler for this fraction was described by Adams *et al.* (2001). This flow rate is far higher than is usually possible for personal sampling, but was considered possible for short periods using currently available portable sampling pumps. This instrument performed well in chamber tests when compared with a co-located EPA-recommended reference sampler. Kenny and Stancliffe (1997)[3] developed a version of the CIS sampler with a porous plastic foam insert chosen to select the $PM_{2.5}$ fraction. The results, obtained in a calm air chamber with a polydisperse test aerosol and the sampled particles counted and sized by an APS, are shown in Figure 15.33. It is seen that agreement with the $PM_{2.5}$ criterion was excellent. For this fine fraction, there are no concerns about biases associated with the sampler aspiration efficiency. This instrument is available commercially (BGI Inc., Waltham, MA, USA).

15.5   Thoracic particle size selection for fibrous aerosols

Maynard (2002) noted that the sampling of respirable fibrous aerosols in the membrane filter method, as described for the measurement of asbestos and other fibrous substances, is likely to lead to the collection of large nonfibrous particles, agglomerates or fiber clumps. These are not of interest. However, their presence on the filter may significantly interfere with the microscope-based analysis of the filter deposit which is used to determine the concentration of 'respirable' fibers, leading to reduced accuracy and precision in a procedure that is already fraught with uncertainty. Earlier, Lippmann (1994) had suggested that the exclusion of such particles may lead to improved overall performance. Maynard proposed that pre-selection according to the thoracic aerosol criterion would provide the desired improvement. He also argued that, in addition to the immediate technical improvements, this would bring sampling for fibrous aerosols into line with the rest of health-related environmental aerosol sampling. He performed an experimental study to investigate the dependence of penetration as a function of fiber length for five existing thoracic aerosol samplers, including the GK2.69 cyclone, a modified SIMPEDS cyclone, the CATHIA-T, and two samplers based on the insertion of porous plastic foam plugs into the nozzle of an IOM personal inhalable aerosol sampler. For the latter, one foam plug was made from 30 ppi foam media and the other from 45 ppi foam. The results for the cyclone and impactor based pre-selectors (GK2.69, modified SIMPEDS and CATHIA-T) showed that they performed well in the sense that the penetration of fibers through the thoracic pre-selection was independent of fiber length. For the foam-based pre-selectors, however, caution was recommended since the dimensions of the flow path through the media in question was of the same order of magnitude as the physical dimensions of the longer fibers of interest. Here, therefore, particle collection mechanisms could well be modified by interception.

Jones *et al.* (2005) went further to suggest that the aerodynamic separation of thoracic fibers by the use of samplers like those studied by Maynard might eliminate the need to discriminate fibers on the basis of their diameter during counting, since – as shown in Chapter 2 – particle aerodynamic diameter for fibers is primarily a function of fiber width and is only weakly dependent on fiber length. The uniformity of sample deposit on the filter samples, which is important when microscope counts are carried out for randomly selected fields, was examined for compact and fibrous particles with aerodynamic diameter spanning the thoracic range. For three of the samplers tested – the GK2.69, CATHIA-T and IOM thoracic aerosol sampler – the spatial distribution of fibers collected on the filter were as good as for the cowled open filter that is routinely used by professional occupational hygienists for fiber sampling. The modified SIMPEDS exhibited some undesirable 'hot-spots' on the filter. However, the results for penetration as a function of d_{ae} suggested that all the tested samplers followed the thoracic sampling convention for fibers

---

[3] See http://www.bgiusa.com/ihi/cisis.htm.

quite well. Overall, the results of both the Maynard and the Jones *et al.* studies supported the proposal that thoracic pre-selection of aerosols during sampling of fibers would be effective in improving the quality and reliability of samples taken when there was likely to be significant coarse background dust.

15.6 Sampling for very fine aerosols

15.6.1 Ultrafine aerosols

The recent surge in interest in the effects associated with human exposures to ultrafine aerosols were driven initially by concerns about such exposures to ambient particles in the atmospheric environment. There have more recently been increasing concerns about some workplaces (Brown *et al.*, 2000). For the sampling and particle size-selection of particles in this range, gravitational deposition is not a feasible option. Inertial deposition is also problematical, except under special conditions such as those found in micro-orifice, low pressure, hypersonic and focused beam impactors. Diffusive and electrostatic deposition provide potentially more useful options. Progress towards the development of samplers for ultrafine particles has been modest since, at the time of writing, formal criteria for particle size-selective sampling for particles in the ultrafine range have not yet been established. Although, as mentioned in Chapter 11, models for alveolar and nasal deposition of ultrafine aerosols have been suggested (Vincent, 2005), neither these – nor alternative versions – are as yet even close to formal adoption. Furthermore, studies of how the dose of such fine particles may lead to health effects have not identified sufficiently the nature of their toxicity, so we do not yet know the extent to which particle size-selectivity needs to go beyond simply the matter of the particle size-dependent deposition in the two respiratory tract regions of interest.

In the meantime, however, some progress towards the development of samplers for ultrafine aerosols has been made. For example, the deposition velocities of ultrafine particles due to thermophoresis may be large enough in feasible sampling systems to allow efficient collection, and a prototype of a thermal precipitator designed to sample ultrafine aerosols in order to allow subsequent electron microscopical analysis of individual nano-sized particles was described by Maynard and Brown (1991). It was noted that this instrument could be useful for the routine sampling of ultrafine particles for microscopical investigations, but attention was drawn to the technical challenge involved in maintaining a stable and adjustable temperature gradient inside the body of the sampling instrument. In one recent commercially available option, the *nanometer aerosol sampler* (NAS), charged particles are collected by electrostatic deposition onto sample substrates for further analysis (TSI Inc., St Paul, MN, USA). This instrument efficiently collects particles in the particle size range from 0.1 μm down to about 2 nm and is intended to be used in tandem with a *differential mobility analyzer* (DMA) that pre-selects particles in the size range of interest. In view of the continued interest, it is expected that there will be a sharp increase in the availability of such samplers in the years ahead.

15.6.2 Combustion-related aerosols

As noted earlier in this book, there are some aerosol exposure situations where there may be distinct particle size modes comprising aerosols from totally differing sources and associated with totally different health effects. One example, illustrated in Figure 1.3, is typical of aerosols found in many underground mining environments where, in addition to a dust component associated with rock cutting and other extraction procedures, there is a fine component associated with the exhausts of diesel vehicles operating in the same environment. The fine diesel component is of considerable interest to many occupational hygienists, so much so that sampling instrumentation that can select that specific fraction has been

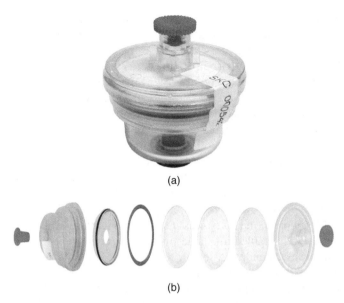

(a)

(b)

Figure 15.34   *Sampling cassette for the separation of diesel particulate matter (DPM), in a version (shown capped) intended for use with the 2 Lpm conductive plastic cyclone shown earlier in Figure 15.15: (a) shown assembled; (b) exploded view. Pictures courtesy of SKC Inc., Eighty Four, PA, USA*

developed. In 1992, Cantrell and Rubow proposed an impactor-based pre-selector that could be used in conjunction with a personal respirable cyclone, providing a sharp 'cut' at $d_{ae} = 0.9$ μm such that, based on extensive studies of such bimodal aerosols in mine environments, the dust and diesel-associated fractions would be clearly separated. The sampling cassette is shown in Figure 15.34, in the version designed to operate with the conductive-plastic respirable cyclone shown earlier in Figure 15.15 (SKC Inc., Eight Four, PA, USA).

15.7   Simultaneous sampling for more than one aerosol fraction

Again, the CEN/ISO/ACGIH recommendations embody the important concept that not only is the thoracic fraction a subfraction of the inhalable fraction but also the respirable fraction is a subfraction of the thoracic fraction. The framework is therefore provided for a sampler that simultaneously yields the mass concentrations of inhalable, thoracic and respirable aerosols respectively. In principle, this – and more – may be achieved using versatile aerosol spectrometers, and these will be discussed later in Chapter 18. Here discussion will be confined to instruments dedicated to specific fractions.

The first instrument to be proposed along these lines was the 2 Lpm *PERSPEC sampler* described by Prodi *et al.* (1986) and shown earlier in Figure 14.10. In this instrument, air entered the device through its pair of crescent-shaped orifices, forming a narrow (almost) ring-shaped entry. In addition to the 2 Lpm of aspirated air, 2 Lpm of particle-free winnowing air was also circulated inside the instrument so that the total flow rate of air passing through the filter was 4 Lpm. The trajectories of particles in the cylindrical layer entering through the ring-shaped entry were governed by inertial forces as the air flow diverged to pass through the 47 mm diameter circular filter. Particles with larger aerodynamic diameter were deposited inertially closer to the center of the filter, finer particles further out. Prodi *et al.* proposed that, by cutting the filter along perimeters at radii chosen to match cut-off values of d_{ae} consistent with the then-available recommendations for inhalable, thoracic and respirable aerosols, respectively, the masses of aerosol sampled within those fractions could be obtained. In assessing the performance of PERSPEC, it is noted that, although the aspiration efficiency of this instrument was quite close to the inhalability criterion (see Figure 14.18), particle losses to internal walls between the entry and the filter were considerable, especially for larger particles. In practical use of the instrument, it was therefore recommended that such aerosol should be recovered and added to the overall filter mass in order to correct for the shortfall in inhalable mass. Similar corrections were recommended to the other fractions. However, the PERSPEC is no longer available commercially.

The second instrument derived from the IOM personal inhalable aerosol sampler and its thoracic and respirable modifications described earlier in this chapter. The idea is that, with a 15 mm diameter circular entry orifice and sampling flow rate of 2 Lpm, the inhalable fraction should be aspirated. The first new instrument of this type is shown in the prototype form described by Mark *et al.* in Figure 15.35. As with the original IOM samplers, the entry was integral to an aerosol-collecting capsule which acted as the receptacle for the overall sampled inhalable fraction. Now, however, compared with the original IOM sampler, the capsule was extended in length to house the multiple porous foam selectors which were the special features of this instrument. The first foam plug was chosen to provide penetration matching the thoracic sub-fraction and the second matching the respirable sub-fraction. In practical use of the instrument, the whole capsule was weighed before and after sampling in order to provide the inhalable mass fraction. Then the fine foam plug and backing filter were removed and weighed

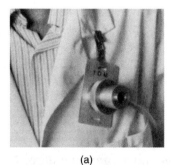

(a)

(b)

Figure 15.35   *An early prototype porous plastic foam-based 2 Lpm sampler for the simultaneous collection of inhalable, thoracic and respirable aerosol (Mark et al., 1988): (a) assembled; (b) disassembled to show the two particle size-selective foam media. Reproduced with permission from Vincent, Aerosol Sampling: Science and Practice. Copyright (1989) John Wiley & Sons, Ltd*

separately. The sum of the resultant two aerosol masses provided the thoracic fraction. The mass on the backing filter provided the corresponding respirable fraction. This instrument has not yet been made available commercially. However, versatile versions of the IOM personal inhalable aerosol sampler are available in which porous foam plugs may be inserted as pre-selectors for either the thoracic and respirable fraction, as desired (SKC Inc., Eighty Four, PA, USA).

An important recent addition to this category of instrument was the 3.11 Lpm *Respicon personal sampler* for the simultaneous sampling of the inhalable, thoracic and respirable aerosol fractions (Koch *et al.*, 1997) (see Figure 15.36). The principle of operation is shown in Figure 15.36(b), where the separation of particles into the desired particle size fractions was achieved by three successive virtual impactor stages. Particles were first aspirated through an annular inlet and the coarser ones passed through the first virtual impactor and so passed on to the next stage. The finer ones with d_{ae} below about 4 μm were deposited onto a filter. At the second virtual impactor stage, the remaining larger particles passed through while the finer ones with d_{ae} below about 10 μm, and yet were not collected at the first stage, were available for deposition onto a second filter. Finally, those particles not collected at the second stage were collected by a third filter. In this way, the aerosol deposited on the first filter was intended to represent the respirable fraction, the combined aerosol on the first and second filters the thoracic fraction, and the combined aerosol on all three filters the inhalable fraction. This sampler was tested in the laboratory under calm air conditions for relatively monodisperse test aerosols with a range of particle size up to about 60 μm (Koch *et al.*, 1999). As shown in Figure 15.37, the results were in generally quite good agreement with CEN/ISO/ACGIH curves (see Figure 11.12) for all three

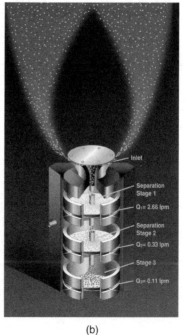

(a)   (b)

Figure 15.36 *The Respicon personal sampler for the simultaneous sampling of the inhalable, thoracic and respirable fractions: (a) photograph of the sampler; (b) schematic, showing the principle of operation. Pictures courtesy of Gilmore Sem, TSI Inc., St Paul, MN, USA*

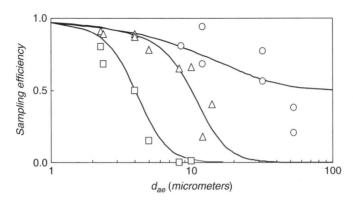

Figure 15.37 *Sampling efficiency as a function of particle aerodynamic diameter (d_{ae}) for the three aerosol fractions obtained using the Respicon sampler as measured under calm air conditions (Koch et al., 1999: O based on the sum of all three stages, corresponding to the inhalable fraction; △ based on the sum of just the top two stages, corresponding to the thoracic fraction; □ just the top stage, corresponding to the respirable fraction; solid curves are the CEN/ISO/ACGIH inhalable, thoracic and respirable fractions as defined in Figure 11.12). Adapted from Aerosol Science and Technology, 31, 231–246. Copyright (1999). Mount Laurel, NJ. Reprinted with permission*

fractions of interest. Experiments were also reported for moving air conditions in a wind tunnel where the instrument was challenged with polydisperse test aerosols and the total mass collected on the three stages was compared with the inhalable fraction as calculated from measurements of the particle size distribution using a cascade impactor. These results showed that the Respicon provided quite accurate measurements of the inhalable fraction for wind speeds up to 2 m s^{-1}, above which the sampler indicated a tendency towards oversampling. Altogether, it was shown that the Respicon was a very promising sampling for the simultaneous collection of all three of the primary health-related aerosol fractions listed by CEN/ISO/ACGIH. This sampler is available commercially (TSI Inc., St Paul, MN, USA). Koch *et al.* also described a prototype direct-reading version of the same sampler in which the aerosol at each stage was detected and recorded optically by a system of miniaturised photometers.

References

Adams, H.S., Kenny, L.C., Nieuwenhuijsen, M.J., Colvile, R.N. and Gussman, R.A. (2001) Design and validation of a high-flow personal sampler for PM$_{2.5}$, *Journal of Exposure Analysis and Environmental Epidemiology*, 11, 5–11.

Aitken, R.J., Vincent, J.H. and Mark, D. (1993) Application of porous foams as size selectors for biologically relevant samplers, *Applied Occupational and Environmental Hygiene*, 8, 363–369.

Asbestos International Association (1979) Reference method for determination of airborne asbestos fiber concentration by light microscopy, Recommended Technical Method No. 1 (RTM 1), AIA Health and Safety Publication.

Baron, P.A. (1993) Measurement of asbestos and other fibers. In: *Aerosol Measurement* (Eds K. Willeke and P.A. Baron), Van Nostrand Reinhold, New York, pp. 560–590.

Bartley, D.L., Chen, C.C., Song, R. and Fischbach, T.J. (1994) Respirable aerosol sampler performance testing, *American Industrial Hygiene Association Journal*, 55, 1036–1046.

Batra, S.K., Shang, P.P., Hersh, S.P. and Robert, K.O. (1980) The SRCC cotton dust sampler as a substitute for the vertical elutriator, *Proceedings of the ASME Symposium on Cotton Dust*, 23–31.

Beckett, S.T. (1980) The effects of sampling practice on the measured concentration of airborne asbestos, *Annals of Occupational Hygiene*, 23, 259–272.

Bedford, T. and Warner, C. (1943) Physical studies of the dust hazard and thermal environment in certain coalmines. In: *Chronic Pulmonary Disease in South Wales Coalminers. II. Environmental Studies*, Special Report Series No. 244, British Medical Research Council, HMSO, London, p. xi and pp. 1–78.

Blachman, M.W. and Lippmann, M. (1974) Performance characteristics of the multicyclone aerosol sampler, *American Industrial Hygiene Association Journal*, 35, 311–326.

Brown, L.M., Collings, N., Harrison, R.M., Maynard, A.D. and Maynard, R.L. (2000) Ultrafine particles in the atmosphere, *Philosophical Transactions of the Royal Society (London) A*, 358, 2563–2565.

Buckley, T.J., Waldman, J.M., Freeman, N.G., Lioy, P.J., Marple, V.A. and Turner, W.A. (1991) Calibration, inter-sampler comparison and field application of a new PM_{10} personal air-sampling impactor, *Aerosol Science and Technology*, 14, 380–387.

Cahill, T.A., Ashbaugh, L.L. and Barone, J.B. (1977) Analysis of respirable fractions in atmospheric particulate via sequential filtration, *Journal of the Air Pollution Control Association*, 27, 675–678.

Cantrell, B.K. and Rubow, K.L. (1992) Measurement of diesel exhaust aerosol in underground coal mines. In: *Diesels in Underground Mines: Measurement and Control of Particulate emissions*, Proceedings of the Bureau of Mines Information and Technology Transfer Seminar, Minneapolis, M N, September 29–30, pp. 11–17.

Caplan, K.J., Doemeny, L.J. and Sorenson, S.D. (1977) Performance characteristics of the 10 mm cyclone respirable mass sampler. Part 1. Monodisperse studies, *American Industrial Hygiene Association Journal*, 38, 83–95.

Cecala, A.B., Volkwein, J.C., Timko, R.J. and Williams, K.L. (1983) Velocity and orientation effects on the 10 mm Dorr-Oliver cyclone, US Bureau of Mines Report of Investigations RI 8764.

Cherrie, J.W., Jones, A.D. and Johnston, A.M. (1986) The influence of fiber density on the assessment of fiber concentration using the membrane filter method, *American Industrial Hygiene Association Journal*, 47, 465–474.

Chung, K.Y.K., Aitken, R.J. and Bradley, D.R. (1997) Development and testing of a new sampler for welding fume, *Annals of Occupational Hygiene*, 41, 355–372.

Comité Européen de Normalisation (1993) *Workplace atmospheres: size fraction definitions for measurement of airborne particles in the workplace*, European Standard EN 481, CEN, Brussels, Belgium.

Courbon, P. (1972) Mesure ponderale des empoussierages, *Proceedings of the Conference on Dust Prevention in Mines,* 179–194.

Courbon, P., Froger, C. and Le Bouffant, L. (1983) Personal dust sampler CIP10, Presented at the 20th International Conference of Safety in Mines Research Institutes, Sheffield, Paper L2.

Courbon, P., Wrobel, R. and Fabries, J.-F. (1988) A new individual respirable dust sampler: the CIP10, *Annals of Occupational Hygiene*, 32, 129–143.

Critchlow, A. and Proctor, T.D. (1955) SIMGARD: a new portable instrument for gravimetric estimation of respirable airborne dust, *Colliery Guardian*, 13 August, 208–209.

Dunmore, J.H., Hamilton, R.J. and Smith, D.S.G. (1964) An instrument for the sampling of respirable dust for subsequent gravimetric assessment, *Journal of Scientific Instruments*, 41, 669–672.

Ettinger, H.J., Partridge, J.E. and Royer, G.W. (1970) Calibration of two-stage air samplers, *American Industrial Hygiene Association Journal*, 31, 537–545.

Fabries, J.-F. and Wrobel, R. (1987) A compact high-flow rate respirable dust sampler: the CPM 3, *Annals of Occupational Hygiene*, 31, 195–209.

Fabries, J.-F., Görner, P., Kauffer, E., Wrobel, R. and Vigneron, J.C. (1998) Personal thoracic CIP10-T and its static version CATHIA-T, *Annals of Occupational Hygiene*, 42, 453–465.

Fairchild, C.I., Ortiz, L.W., Tillery, M.I. and Ettinger, H.J. (1978) Aerosol research and development related to health hazard analysis, Los Alamos Scientific Laboratory Progress Report LA-7380-PR.

Fay, J.W.J. (1960) Experiments with the Hexhlet dust sampler in British coal mines, *Annals of Occupational Hygiene*, 1, 314–323.

Ford, V.H.W. (1971) Experimental investigations into the dispersion and transport of respirable dust in mechanised coal mining, PhD Thesis, University of Newcastle-upon-Tyne, UK

Gautam, M. and Sreenath, A. (1997) Performance of a respirable multi-inlet cyclone sampler, *Journal of Aerosol Science*, 28, 1265–1281.

Gibson, H. and Vincent, J.H. (1981) The penetration of dust through porous foam filter media, *Annals of Occupational Hygiene*, 24, 205–215.

Görner, P, Fabries, J.-F. and Wrobel, R. (1994) Thoracic fraction measurement of cotton dust, *Journal of Aerosol Science*, 25, S487–S488.

Green, H.L. and Watson, H.H. (1935) Physical methods for the estimation of the dust hazard in industry, with special reference to the occupation of the stone mason, Medical Council of the Privy Council, Special Report Series 199, HMSO, London.

Greenberg, L. and Smith, G.W. (1922) A new instrument for sampling aerial dust, US Bureau of Mines Report of Investigations 2392.

Griffiths, J. and Jones, T.D. (1940) The determination of dust concentrations in mine atmospheres, *Transactions of the Institution of Mining Engineers*, 99, 150–180.

Gwatkin, G. and Ogden, T.L. (1979) The SIMPEDS respirable dust sampler: side-by-side comparisons with the 113A, *Colliery Guardian*, 227, 326–331.

Hamilton, R.J. (1956) A portable instrument for respirable dust sampling, *Journal of Scientific Instruments*, 33, 395–399.

Hamilton, R.J., Morgan, G.D. and Walton, W.H. (1967) Measurement of dust by mass and by number. In: *Inhaled Particles and Vapours II* (Ed. C.N. Davies), Pergamon Press, Oxford, pp. 533–549.

Harper, M., Fang. C.P., Bartley, D.L. and Cohen, B.S. (1998) Calibration of the SKC Inc. aluminum cyclone for operation in accordance with ISO/CEN/ACGIH respirable aerosol sampling criteria, *Journal of Aerosol Science*, 29, S347–S348.

Harris, G.W. and Maguire, B.A. (1968) A gravimetric dust sampling instrument (SIMPEDS): preliminary results, *Annals of Occupational Hygiene*, 11, 195–201.

Higgins, R.I. and Dewell, P. (1967) A gravimetric size-selecting personal dust sampler. In: *Inhaled Particles and Vapours, Vol. II* (Ed. C.N. Davies), Pergamon Press, Oxford, pp. 575–586.

Higgins, R.I. and Dewell, P. (1968) A gravimetric size-selecting personal dust sampler, *British Cast Iron Research Association Reports*, 908, 112–119.

International Standards Organisation (1995) *Air quality–particle size fraction definitions for health-related sampling*, ISO Standard ISO 7708.

Johnston, A.M., Jones, A.D. and Vincent, J.H. (1982) The influence of external aerodynamic factors on the measurement of the airborne concentration of asbestos fibers by the membrane filter method, *Annals of Occupational Hygiene*, 25, 306–316.

Jones, A.D., Aitken, R.J., Fabries, J.-F., Kauffer, E., Lidén, G., Maynard, A., Riediger, G. and Sahle, W. (2005) Thoracic size-selective sampling of four types of thoracic sampler in laboratory tests, *Annals of Occupational Hygiene*, 49, 481–492.

Kenny, L.C. and Gussman, R.A. (1997) Characterization and modelling of a family of cyclone aerosol separators, *Journal of Aerosol Science*, 28, 677–688.

Kenny, L.C. and Stancliffe, J. (1997) Unpublished report from the Health and Safety Laboratory, Sheffield.

Kenny, L.C., Aitken, R.J., Beaumont, G. and Görner, P. (2001) Investigation and application of a model for porous foam aerosol penetration, *Journal of Aerosol Science*, 32, 271–285.

King, A.M. (1984) The gravimetric dust monitor – a new instrument which provides integral assessment of multiple shift-length respirable dust samples, *Annals of Occupational Hygiene*, 28, 107–115.

Kitto, P.H. and Beadle, D.G. (1952) A modified form of thermal precipitator, *Journal of Chemical and Metallic Mining Society of South Africa*, 52, 284–287.

Koch, W., Dunkhorst, W. and Lödding, H. (1997) RESPICON TM-3: a new personal personal measuring system for size-segregated dust measurement at workplaces, *Staub Reinhaltung der Luft*, 57, 177–184

Koch, W., Dunkhorst, W. and Lödding, H. (1999) Design and performance of a new personal aerosol monitor, *Aerosol Science and Technology*, 31, 231–246.

Kotzé, Sir R. (1916) Final Report of the Miners' Phthisis Prevention Committee, Union of South Africa, Johannesburg (January 10, 1916).

Leck, M.J. (1983) Optical scattering instantaneous respirable dust indication system. In: *Aerosols in the Mining and Industrial Work Environments* (Eds V.A. Marple and B.Y.H. Liu), Ann Arbor Science, Ann Arbor, MI, pp. 701–717.

Leck, M.J. and Harris, G.W. (1978) SIMSLIN II: an airborne dust-measuring instrument that uses a light scattering method, *Colliery Guardian*, 226, 676–677.

Lippmann, M. (1983) Size-selective health hazard sampling. In: *Air Sampling Instruments* (Eds P.J. Lioy and M.J.Y. Lioy), American Conference of Governmental Industrial Hygienists, Cincinnati, OH, pp. H1–H22.

Lippmann, M. (1994) Workshop on the health risks associated with chrysotile asbestos, *Annals of Occupational Hygiene*, 38, 459–467.

Lippmann, M. and Harris, W.B. (1962) Size-selective samplers for estimating 'respirable' dust concentrations. *Health Physics*, 8, 155–163.

Maguire, B.A., Barker, D. and Wake, D. (1973) Size-selection characteristics of the cyclone used in the SIMPEDS 70 Mk 2 gravimetric dust sampler, *Staub Reinhaltung der Luft*, 33, 95–9.

Mark, D., Borzucki, G., Lynch, G. and Vincent, J.H. (1988) The development of personal sampler for inspirable, thoracic and respirable aerosol, Presented at the Annual Conference of the Aerosol Society, Bournemouth, UK.

Marple, V.A. (1978) Simulation of respirable penetration characteristics by inertial impaction, *Journal of Aerosol Science*, 9, 125–34.

Marple, V.A. and Rubow, K.L. (1983) Impactors for respirable dust sampling. In: *Aerosols in the Mining and Industrial Work Environments* (Eds V.A. Marple and B.Y.H. Liu), Ann Arbor Science, Ann Arbor, MI, pp. 847–860.

Maynard, A. (2002) Thoracic size-selection of fibers: dependence of penetration on fiber length for five thoracic sampler types, *Annals of Occupational Hygiene*, 46, 511–522.

Maynard, A.M. and Brown, L.M. (1991) The collection of ultrafine aerosol particles for analysis by transmission electron microscopy, using a new thermophoretic precipitator, *Journal of Aerosol Science*, 22, S379–S382.

Maynard, A.M. and Kenny, L.C. (1995) Performance assessment of three personal cyclone models, using an aerodynamic particle sizer, *Journal of Aerosol Science*, 26, 671–684.

McFarland, AR., Hickman, P.D. and Parnell, C.B. (1987) A new cotton dust sampler for PM_{10} aerosol, *American Industrial Hygiene Association Journal*, 48, 293–297.

National Institute of Occupational Safety and Health (1974) Criteria for a recommended standard:occupational exposure to cotton dust, DREW (NIOSH) Publication No. 75–118, USGO, Washington, DC.

National Institute of Occupational Safety and Health (1979) ASPHS/NIOSH membrane filter method for evaluating airborne asbestos fibers, NIOSH Technical Report.

Neefus, J.D., Lumsden, J.C. and Jones, M.T. (1977) Cotton dust sampling, II. Vertical elutriation, *American Industrial Hygiene Association Journal*, 38, 394–400.

Ogden, T.L., Barker, D. and Clayton, M.P. (1983) Flow-dependence of the Casella respirable-dust cyclone, *Annals of Occupational Hygiene*, 27, 261–271.

Ogden, T.L., Birkett, J.L. and Gibson, H. (1977) Improvements to dust measurement techniques, IOM Report No. TM/77/11, Institute of Occupational Medicine, Edinburgh.

Owens, J.S. (1922) Suspended impurities in air, *Proceedings of the Royal Society (London) A*, 101, 18–28.

Page, S.J., Volkwein, J.C., Baron, P.A. and Deye, G. (2005) Particulate penetration of porous foam used as a low flow rate respirable dust size classifier, *Applied Occupational and Environmental Hygiene*, 15, 561–568.

Parker, R.D., Buzzard, G.H., Dzubay, T.G. and Bell, J.P. (1977) A two-stage respirable aerosol sampler using Nuclepore filters in series, *Atmospheric Environment*, 11, 617–621.

Phalen, R.F. (Ed.) (1985) Particle size-selective sampling in the workplace, Report of the ACGIH Air Sampling Procedures Committee, American Conference of Governmental Industrial Hygienists, Cincinnati, OH.

Prodi, V., Belosi, F. and Mularoni, A. (1986) A personal sampler following ISO recommendations on particle size definitions, *Journal of Aerosol Science*, 17, 576–581.

Reischl, G.P. and John, W. (1978) The collection efficiency of impaction surfaces: a new impaction surface, *Staub Reinhaltung der Luft*, 38, 55.

Robert, K.Q. (1980) Relationships between crossflow, isokinetic and calm-air sampling with a vertical elutriator, *Proceedings of the ASME Symposium on Cotton Dust*, 45–53.

Robert, K.Q. and Baril, A. (1984) Sampling for respirable cotton dust, *Journal of the American Oil Chemists' Society*, 10, 1553–1558.

Sebestyen, A., Verma, D.K. and Muir, D.C.F. (1987) An evaluation of konimeter performance – I. Determination of the particle acceptance curves of continuously aspirated konimeters using latex spheres, *Annals of Occupational Hygiene*, 31, 441–449.

Stuke, J. and Emmerichs, M. (1973) Das gravimetrische Staubprobennahmegerat TBF50, *Silikosebericht Nordrhein-Westfalen*, 9, 47–51 (in German).

Verma, D.K., Sebestyen, A. and Muir, D.C.F. (1987) An evaluation of konimeter performance – II. A field comparison of konimeters, *Annals of Occupational Hygiene*, 31, 451–461.

Vincent, J.H. (1989) *Aerosol Sampling: Science and Practice*, John Wiley & Sons, Ltd, Chichester.

Vincent, J.H. (1999) *Particle Size-selective Sampling for Particulate Air Contaminants*, American Conference of Government Industrial Hygienists, Cincinnati, OH.

Vincent, J.H. (2005) Health-related aerosol measurement: a review of existing sampling criteria and proposals for new ones, *Journal of Environmental Monitoring*, 7, 1037–1053.

Vincent, J.H., Aitken, R.J. and Mark, D. (1993) Porous plastic foam filtration media penetration characteristics and applications in particle size-selective sampling, *Journal of Aerosol Science*, 24, 929–944.

Walton, W.H. (1954) Theory of size classification of airborne dust clouds by elutriation, *British Journal of Applied Physics*, 5 (Suppl. 3) S29–40.

Wolff, H.S. and Roach, S.A. (1961) The conicycle selective sampling system. In: *Inhaled Particles and Vapours* (Ed. C.N. Davies), Pergamon Press, Oxford, pp. 460–464.

Wright, B.M. (1954) A size-selecting sampler for airborne dust, *British Journal of Industrial Medicine*, 11, 284–288.

Zurlo, N (1987) Nouveau pneumo-classificateur pour poussieres fines, Presented at the Fourth Colloquium on Dust, Paris (in French).

Zuskin, E., Mustajbegovic, J., Schachter, E.N., Kanceljak, B., Kern, J., Macan, J. and Ebling, Z. (1998) Respiratory function and immunological status in paper-recycling workers, *Journal of Occupational and Environmental Medicine*, 40, 986–993.

16

Sampling in Stacks and Ducts

16.1 Introduction

As discussed in earlier chapters, sampling with cylindrical thin-walled sampling tubes is the area that provoked most of the initial research into aerosol sampling science, and one which produced the initial base of knowledge from which our understanding of all other sampling devices and systems has flowed. So the scientific study of the aspiration efficiencies of cylindrical thin-walled tubes has been a cornerstone of most of what is contained in earlier parts of this book. But, in addition, it has been a subject of considerable practical importance in its own right, and remains so to this day.

Sampling in stacks and ducts is usually carried out for the purpose of determining the emission of aerosols to the atmosphere. It therefore forms the basis of assessment of the performances of air pollution control equipment installed on the exhausts of industrial processes, and of compliance with emission regulations. Unlike most of those found in other aerosol sampling applications, the gas flows in stacks and ducts may usually be reasonably well defined in terms of direction and velocity. Therefore aerosol sampling with thin-walled probes provides the main methodology for practical stack and duct sampling.

16.2 Basic considerations

Unlike for most other practical aerosol sampling scenarios, the sampling of particles in stacks and ducts does not necessarily take place from air – or at least not what is commonly regarded as the atmospheric air that humans breathe (and which is the subject of most of the sampling scenarios described elsewhere in this book). Many industrial processes discharge gas mixtures which, depending on the process in question, may have constituents that make it very different from ordinary atmospheric air in terms of many of the physical properties that are important to the mechanics of aerosol sampling, notably density and viscosity. In addition, industrial discharges may be rich in gaseous constituents such that, during the act of sampling, phase changes may take place, leading to the presence of particulate matter which had not previously been present. Such considerations need to be borne in mind when it comes to applications of the science described in Chapter 5 to the real world of the sampling of industrial gas emissions. In turn, what might seem initially to be a relatively simple application of aerosol sampling science may be seriously complicated by such factors.

Aerosol Sampling: Science, Standards, Instrumentation and Applications   James Vincent
© 2007 John Wiley & Sons, Ltd

For the aerosol component, the basic limitations of sampling with cylindrical thin-walled probes have been studied in detail in the research cited in earlier chapters of this book. In particular, these relate to (a) the factors that may cause departures from isokineticity during aspiration, and (b) losses inside the sampling line (or *train*) after aspiration. For the former, the effects of excursions in flow conditions from isokinetic, the angular relationship between the probe and gas stream, and particle blow-off from external walls, are all by now quite well understood. Although most of the gas flows of interest are highly turbulent, it is fortunate that the fluctuating nature of the flow under such conditions appears to have little (if any) effect on aspiration efficiency (Vincent *et al.*, 1985). For internal losses, however, the effects of the role of external freestream turbulence on particle motion just inside the probe entry (Wiener *et al.*, 1988), the line losses associated with impaction, gravitation settling and diffusion, together with condensation and electrostatic effects, greatly complicate the sample extraction process, often in ways that are not predictable. In practice, the sampling location in a stack or duct needs to be chosen to minimise possible such interfering factors. In particular, close to bends, sudden changes in cross-section or some other form of flow disturbance, the nature of the gas flow – both upstream and downstream – may be characterised by nonuniform and unstable distribution of velocity and direction (including reverse flow), and by enhanced turbulence. Whenever possible, it has therefore been long considered to be good practice to avoid proximity of the sampling station to such regions. Having chosen the sampling station, practical stack sampling then becomes a matter of choosing an appropriate probe, sampling train and strategy. The latter involves sampling at a number of points in the chosen cross-section, usually by successive deployment of the same probe in traverses across the duct. Such sampling points are chosen so that, by combining all the samples at the end of a complete run, a representative measure of the overall mass flux of particulate material in the duct may be calculated. The actual number of sampling points in the traverse desirable for achieving representative measurement may be allowed to be smaller the more uniform the velocity distribution in the duct; and vice versa if the flow is less uniform. The minimum numbers of sampling points for given distance of the disturbance away from the sampling station may be prescribed based on empirical information obtained in the laboratory or in the field. From fluid mechanical considerations like those outlined in Chapter 2, the flow disturbances upstream of the sampling point may be expected to have much greater influence on the flow uniformity at the plane of the sampling station than those downstream. Hence, for a given magnitude of the chosen distance from the disturbance, more sampling points are required for sampling downstream than for upstream of a disturbance. This is reflected in the prescribed methods described below.

In addition to the number of samples to be taken, the choice of individual sampling locations over the chosen duct cross-section needs to be made so that the overall sampling exercise can provide results that truly reflect the total flux of particles through the duct. To achieve this, sampling points in the cross-section should be representative of segments of the cross-section that are equal in area, and sampling should be carried out in them isokinetically with respect to the air velocity at the same points, respectively. These locations are indicated in Figure 16.1 for ducts of both circular and rectangular cross-section. Here, in principle, and as close as practically possible, each sampling location is placed at the geometric centroid of the area segment in question. This figure points clearly to the need, in order to ensure isokineticity for aerosol sampling, to be able to measure the air velocity at the same point where the aerosol is to be aspirated, and in turn to be able to control the sampling flow rate there.

16.3   Stack sampling methods

The purpose of making measurements of aerosol concentrations in gas streams emitted from stationary sources is to provide data that will enable the determination of the total rate (e.g. in $kg\,day^{-1}$) of the

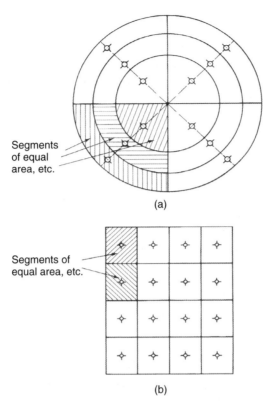

Segments
of equal
area, etc.

(a)

Segments of
equal area, etc.

(b)

Figure 16.1 *Location of sampling points over the cross-sections of ducts: (a) circular; (b) rectangular. The individual sampling points are located at the centroid of each equal-area segment. The same principles apply to ducts of other cross-section. Reproduced with permission from Vincent, Aerosol Sampling: Science and Practice. Copyright (1989) John Wiley & Sons, Ltd*

emission of particulate emissions. Such data may then be used in order to assess compliance with emission control regulations and/or for environmental management. For evaluating such emissions, detailed test methods have been prescribed in many countries so that control criteria and emission regulations may be standardised and the quality of the data can be assured and maintained. Some are described below. It is interesting to note that much of what is described in the methods outlined was described in the 1972 review paper by Morrow *et al.*, reflecting that the science of stack aerosol sampling has not moved on significantly during the past 30 years, perhaps because the earlier science is sufficient for most practical purposes.

16.3.1 United States of America

The primary source of guidelines for stack sampling in the USA is contained in the documentation for a series of methods contained within the Code of Federal Regulations 'Protection of the Environment', listed as 40 CFR. These are developed and applied by the Environmental Protection Agency (EPA) (e.g. Environmental Protection Agency, 1971). They comprise many parts, the most relevant for present purposes being Part 60 (*Standards of performance of new stationary sources*), Part 61 (*National emission standards for hazardous air pollutants*) and Part 63 (*National emission standards for hazardous air pollutants for source categories*). These are supported by a long list of individual methods (*'CFR Promulgated Test Methods'*) for the various measurement procedures that are needed for stack sampling of both particulate and gaseous emissions. They are aimed at identifying the test methods to be used as reference methods to the facility subject to regulation in question, and describing any special instructions or conditions to be followed. Full details are given in the cited documents.[1] For the ones most relevant to aerosol sampling, the main points may be summarised as follows:

- Method 1 (*Sample and velocity traverses for stationary sources*) A method is described that is applicable to gas streams in ducts, stacks and flues provided that the flow is reasonably straight and without swirl, and that the duct is greater than 0.3 m in hydraulic diameter. Any directional velocity-sensing probe may be used that is capable of measuring both pitch and yaw angles in the flow. Depending on the sampling location with respect to any disturbance – in the form of a bend, an expansion or a contraction, or from a visible flame – the minimum number of sampling locations is specified for both velocity measurement and aerosol sampling, respectively. Ideally, sampling and velocity measurements should be carried out at a location whose distance X is at least eight duct diameters downstream, and at least two diameters upstream, of any bend, expansion or contraction, or from a visible flame. Under these conditions, the minimum number of traverse points for measurement is twelve for circular and rectangular ducts of hydraulic diameter greater than 0.61 m, eight for circular ducts with hydraulic diameter between 0.3 and 0.61 m, and nine for rectangular ducts with hydraulic diameter between 0.3 and 0.61 m. Here hydraulic diameter, D_h, is given by:

$$D_h = \frac{4 \cdot \text{cross - section}}{\text{perimeter}} \tag{16.1}$$

which becomes *actual* geometric diameter for a circular duct. When the 'eight-and-two-diameter' condition cannot be met, then the number of traverse points for sampling and velocity measurement, respectively, needs to be greater. The main features of the recommendations for these numbers are summarised graphically in Figures 16.2 (for aerosol sampling) and 16.3 (for velocity measurement), indicating the recommended minimum number of incremental cross-sections, N, vs. the ratio X/D_h.

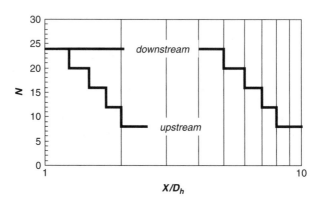

Figure 16.2 *Minimum number of traverse points (N) for aerosol sampling in stacks and ducts as a function of X/D_h, where X is the distance upstream or downstream from a disturbance (as indicated) and D_h is the duct hydraulic diameter, for both circular and rectangular duct cross-sections, dashed lines as indicated reflecting recommended high N-values for rectangular ducts (from EPA 40 CPR, Method 1)*

[1] See also http://www.epa.gov/ttn/emc/promgate.html.

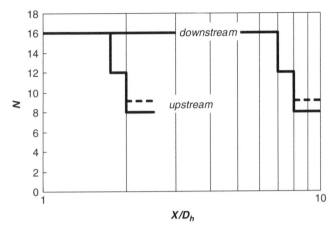

Figure 16.3 *Minimum number of traverse points (N) for air velocity measurement in stacks and ducts as a function of X/D_h, where X is the distance upstream or downstream from a disturbance (as indicated) and D_h is the duct hydraulic diameter, for both circular and rectangular duct cross-sections, dashed lines as indicated reflecting high recommended N-values for rectangular ducts (from EPA 40 CPR, Method 1)*

- Method 1A (*Sample and velocity traverses for stationary sources with small stacks or ducts*) Here a method is described for ducts of hydraulic diameter less than 0.3 m. Again, it is not applicable to swirling flows. Now, because the duct width is so small, the sampling probe may cause significant relative blockage to the flow, leading to possible errors in gas velocity measurement. To overcome this, the gas velocity is measured at a distance downstream of the actual sampling location. The location for sampling should be chosen at least eight duct hydraulic diameters downstream of, and at least 10 diameters upstream of, any flow disturbances of the type already mentioned. The velocity measuring site should be located eight hydraulic diameters downstream of the aerosol sampling location. This location is therefore no less than two diameters upstream of any flow disturbance. When all the preceding conditions are met, the minimum number of traverse points to be chosen is eight for circular ducts and nine for rectangular ducts, again representing equal-area segments of the duct cross-section.
- Methods 5 and 17 (*Determination of particulate matter emissions from stationary sources*) These methods describe the actual aerosol sampling part of the overall stack sampling process. The basic requirement is that aerosol is extracted from the duct gas flow using a cylindrical tube inlet operated isokinetically while facing into the gas stream, and particles thus aspirated are collected onto a glass fiber filter that is maintained at the same temperature as the gas in the stack. The collected particulate material is usually evaluated gravimetrically after the removal of adsorbed moisture. In principle, therefore, it is seen that the recommended method for the practical stack sampling of aerosols derives directly from thin-walled probe research like that described earlier in Chapter 5. On the face of it, therefore, the sampling head is relatively simple and the conditions for accurate performance, at least in terms of aspiration efficiency, are straightforward. In practice, however, a complex system – a *sampling train* – is required. The EPA-recommended sampling train is shown in Figure 16.4. It incorporates all the elements needed to enable sampling in a manner that is accurate and reliable in terms of both the basic aerosol sampling science and the local practical requirements. It includes a cylindrical probe nozzle (the thin-walled probe itself), a probe extension that supports the probe and pitot-static tube (for the air velocity measurement required to ensure isokinetic sampling

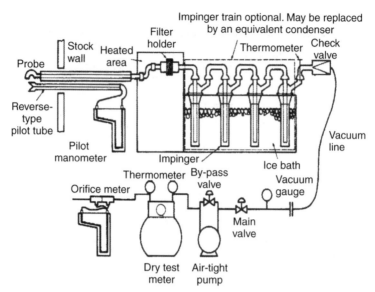

Figure 16.4   *Diagram of the sampling train described in the EPA stack sampling method. Reproduced with permission from Perry and Young, Handbook of Air Pollution Analysis. Copyright (1977) Chapman and Hall*

conditions), a borosilicate or quartz glass or stainless steel filter holder, and a condenser in the form of an impinger train (in order to allow determination of the moisture content or other condensible components in the sampled gas). As shown, in this method the filter holder is usually located outside the stack. The method also calls for apparatus by which to recover particulate matter collected on the inside walls of the probe assembly between the inlet and the filter, including a probe-liner and probe-nozzle brushes. The method provides details about how to prepare the sampling train for a sampling run, including leak-check procedures and flow calibration, the actual operation of the sampling train during a sampling run, sample recovery and sample analysis.

- Method 201 (*Determination of* PM_{10} *emissions*) The preceding methods are intended for the assessment of all particulate emissions from stacks, including particles of all sizes with equal efficiency. The PM_{10} aerosol fraction is a particular target for air quality standards, and there are instances where it might be considered appropriate to measure the emission of particles in that size fraction. In addition to in-stack PM_{10}, EPA recognises that there may also be condensible emissions that, once in the atmosphere, may contribute to the overall PM_{10} level. With these considerations in mind, therefore, the sampling train for PM_{10} contains an in-stack cyclone with penetration characteristics matching the PM_{10} curve shown earlier in Figure 11.10. The condensible fraction is obtained from analysis of the catch of the impinger train.

Elsewhere, the American Society for Testing and Materials (ASTM) (American Society for Testing and Materials, 1973) also provides its own guidelines for standardised procedures for the measurement of stack emissions, notably in the form of its D3685/D3685M-98 (*Test methods for sampling and determination of particulate matter in stack gases*) and D6331-98 [*Standard test method for determination of mass concentration of particulate matter from stationary sources at low concentrations (manual gravimetric methods)*]. The essential elements are very similar to those described in the EPA methods. One difference, as seen in Figure 16.5, is that the ASTM method calls specifically for a filter holder that is located

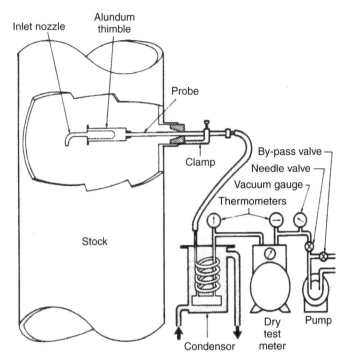

Figure 16.5 *Diagram of the sampling train described in the ASTM method. Reproduced with permission from Perry and Young, Handbook of Air Pollution Analysis. Copyright (1977) Chapman and Hall*

'in-stack' so that it is much closer to the inlet nozzle. This is advantageous (Hemeon and Black, 1972) because it is then much easier to recover everything that has passed isokinetically through the sampling inlet. However it has been suggested that, overall, the documentation for the ASTM method contains insufficient quality control requirements compared with the EPA method (McIntyre and Moore, 2004).

16.3.2 United Kingdom and elsewhere

In the UK, the primary source of recommended methods for stack sampling is the British Standards Institution (BSI) (British Standards Institution, 1971). Its technical sub-committee BSI EH/2/1 is specifically charged with stationary source measurement standardisation and promulgates a wide range of relevant methods, including BS 1756 pt4 (*Sampling and analysis of flue gases*). Other countries in Europe have similar standards institutions that advise on and recommend such methods (e.g. the German Engineering Society, or VDI). Importantly, BSI and many of the other such national bodies are formally linked with – and coordinate their efforts with – the leading international standards institutions, notably the Comité Européen de Normalisation (CEN) through its technical committee CEN TC 264, and the International Standards Organisation (ISO) through ISO TC 146 SC1. Through these bodies, there has emerged over the years a comprehensive system of methods documents for the sampling of stack emissions. Again, the actual technical procedures are science-based and so are quite similar to those promulgated by EPA.

16.4   Sampling probes for stack sampling

From the above, it is clear that effective stack sampling requires a quite complex system of apparatus to enable measurements to be carried out consistently and to deal with the wide-ranging gas and flow conditions that are encountered in practice. Aerosol sampling per se is only a small part of that system. However, reliable extraction of the desired aerosol fraction is the central requirement of the system. So the choice of the sampling probe itself remains of great importance. In what follows, some variants on the simple thin-walled cylindrical sampling probe are described, reflecting some of the thinking that has gone into what is, on the surface, a simple matter but is, in reality, fertile ground for innovation.

16.4.1   Standard probes

In the most common approach to the sampling of aerosols in stacks, the first step is to measure the distribution of gas velocity over the cross-section of the duct at the chosen sampling station. This is usually achieved using a pitot-static tube, in the form of a snub-nosed cylindrical tube which is pointed directly into the freestream. The difference between the stagnation pressure at the leading edge of the tube and the freestream static pressure, as measured using a manometer connected between the ports at the leading edge and sidewall, respectively, provides gas velocity by the application of Bernoulli's expression as given by the earlier Equation (2.9). The velocity at each point of interest in the duct may thus be defined so that, when the time comes to sample, the aspiration flow rate required in order to achieve isokinetic conditions at each point may be prescribed. As described in the EPA method, the pitot-static tube is integrated explicitly into the whole sampling train system.

The design of the sampling probe itself is governed by the sort of considerations outlined in earlier chapters. Possible effects due to departures from ideal isokinetic flow conditions, for example associated with finite probe blockage or probe orientation, are minimised by careful probe design. So too are the effects of particle blow-off from external surfaces. In addition, there is the problem of particle losses inside the probe entrance. This is where the so-called '*standard probe*' is preferable to other types, since the possibility exists of placing the particulate-collecting filter very close to the probe entrance. This is sometimes referred to as '*in-stack*' particle collection. A typical such probe is shown in Figure 16.6 for the case where the filter takes the form of a 'thimble' or cartridge. In standard probes where the collecting filter is located further away from the probe entry, it is necessary that, after sampling, the particulate material collected in the sampling line in front of the filter is recovered separately and

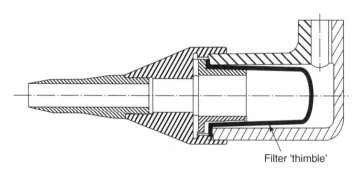

Filter 'thimble'

Figure 16.6   *Typical in-stack sampling probe, showing the filter in the form of a 'thimble' so that it can be located as close as possible to the inlet. Reproduced with permission from Vincent, Aerosol Sampling: Science and Practice. Copyright (1989) John Wiley & Sons, Ltd*

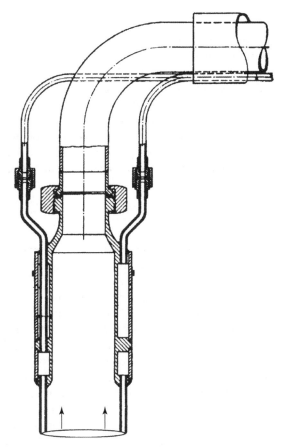

Figure 16.7 *Typical velocity-sensing in-stack sampling probe. Reproduced with permission from Vincent, Aerosol Sampling: Science and Practice. Copyright (1989) John Wiley & Sons, Ltd*

its weight added to that on the filter. The standard-probe approach is used in both the BSI and the ASTM methods already mentioned. Its main advantage lies in its simplicity. A disadvantage is that, before sampling itself can begin, the velocity traverse must be carried out and flow conditions must be assumed to subsequently remain constant.

16.4.2 Velocity-sensing probes

In some probes, the thin-walled sampling head also incorporates a gas velocity sensor. This enables simultaneous measurement of the local velocity of the freestream during the sampling run, thus eliminating the need for a separate velocity traverse. Sampling flow rate can then be manually adjusted during the run to accommodate changes in gas velocity. This is basically the approach described in the EPA method indicated in Figure 16.4. A similar system was described by Boubel (1971). A more sophisticated arrangement, in which a pitot-static velocity sensor system was actually built in to the walls of the sampling tube itself, was reported by Bohnet (1978) and is illustrated in Figure 16.7. Boothroyd (1967) proposed an approach based on hot-wire anemometry.

16.4.3   Null-type probes

The '*null-type*' *sampling probe* is an extension of the combined probe idea described in the preceding paragraph, and – likewise – aims to eliminate the need for pre-sampling velocity traversing and to provide more sensitive control of the isokinetic conditions during sampling itself. It is based on the principle that, for an idealised thin-walled probe, if the gas velocity is the same inside the probe as that outside it, then the static pressure at the inside and outside wall, respectively, should be the same. However, for less-idealised practical sampling probes with finite blockage ratios, errors may arise due to flow distortions like those described in earlier chapters. Nevertheless, workable systems can be achieved by careful design to minimise such distortions. The practical probe described by Dennis *et al.* (1957) is shown in Figure 16.8. A similar probe design that has been used in pulverised power plant stacks in Germany was proposed by Narjes (1965).

16.4.4   Self-compensating probes

A particularly interesting variation of the null-type probe is the novel device proposed by Steen and his colleagues (Steen, 1977; Steen *et al.*, 1981). Its basic idea is illustrated in Figure 16.9 where it is seen that there was no external pump or other mechanical source of gas movement through the sampling tube. Instead, the driving force was the negative static pressure formed in the near wake of the conical deflector plate located coaxially with the tube at several diameters downstream of the tube entrance. In principle, the dimensions and shape of this plate and of the tube itself were chosen such that, for the probe facing into the wind, gas always entered the tube isokinetically, regardless of the freestream gas velocity. Changes in gas velocity were automatically compensated for by the corresponding changes taking place in the static pressure behind the plate. Such a device would only be possible if any resistance to the flow was negligible, precluding the use of a filter or any such collecting medium. In the Steen device, particles were therefore collected by electrostatic precipitation in the corona discharge which was sustained between the high-voltage, axial-rod electrode and the inside wall of the tube, and particle deposition took place onto an aluminum foil tube liner which could be removed for analysis after sampling had taken place. In principle, this type of device required no measurements of gas velocity or

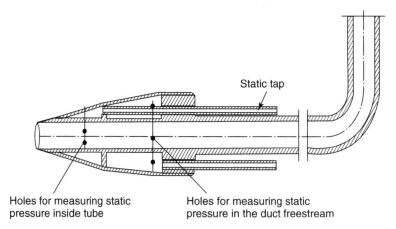

Figure 16.8 *Typical null-type in-stack sampling probe. Reproduced with permission from Vincent, Aerosol Sampling: Science and Practice. Copyright (1989) John Wiley & Sons, Ltd*

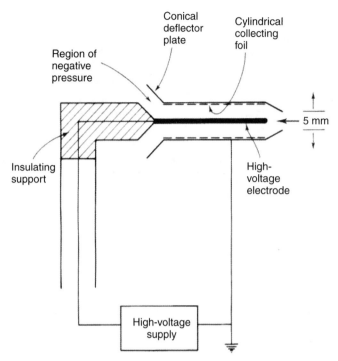

Figure 16.9 *Schematic of a self-compensating in-stack sampling probe of the type suggested by Steen et al. (Steen, 1977; Steen et al., 1981). Reproduced with permission from Vincent, Aerosol Sampling: Science and Practice. Copyright (1989) John Wiley & Sons, Ltd*

static pressure to enable the setting of isokinetic flow conditions. Despite its originality, the Steen *et al.* inlet has not been widely used and does not appear to be commercially available. A few alternative, albeit less elegant, approaches to self-compensating sampling have been tried. All involved automated feedback-control systems that provided means to sense the local instantaneous air velocity at the sampling location in the stack and thence to continuously control the sampling air flow accordingly by means of appropriate control circuitry. A similar approach was developed by Lombardi *et al.* (1984) and Delprato and Lombardi (1990). Here, both the gas velocity and temperature very close to the sampler inlet were sensed by hot-wire anemometry.

16.4.5   Dilution

In stack sampling for the evaluation of industrial emissions, aerosol concentrations tend to be higher than in most other areas of interest to occupational and environmental hygienists. In some such situations it is desirable to dilute the aerosol with particle-free gas, for example to reduce the adsorption and condensation of volatile compounds on particle surfaces and to suppress particle coagulation (that would change the particle size distribution) and chemical reactions (that would change the aerosol composition). It is an important requirement of any such dilution system that the aerosol is physically and chemically unchanged from that which was originally aspirated. A number of *diluters* for stack sampling applications have been described over the years, and several were reviewed in the early 1980s by Felix *et al.* (1981), some for in-stack applications and some for out-of-stack applications. All have

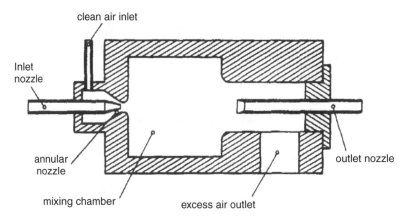

Figure 16.10   *Ejector (or Venturi)-type dilution system of the type originally described by Koch et al. (1988), shown here in the form described by Helsper et al. (1990). Reprinted from Journal of Aerosol Science, 21, Helsper et al., S637–S640. Copyright (1990), with permission from Elsevier*

applied a variety of air ejector configurations by which the air aspirated through the sampling inlet was mixed with clean air under controlled flow conditions. The devices described provided dilution ratios in the range from $\times 10$ to $\times 1500$. Later, Koch *et al.* (1988) described a dilution system that operated on the basis of *ejector* (or *Venturi-type*) dilution, providing a dilution ratio of $\times 10$. It is shown schematically in Figure 16.10. Helsper *et al.* (1990) tested this system for aerosol with particles in the range of aerodynamic diameter from about 0.1 to 10 μm and showed that dilution factors as high as $\times 10\,000$ could be achieved by cascading several such individual dilutors in series. More recently still, Hueglin *et al.* (1997) developed an alternative dilution system in which the dilution gas was admitted through a sequence of cavities in a rotatable disk that allowed for continuously variable dilution ratios in the range from $\times 10$ to $\times 10\,000$ for particles in the range below 1 μm. This system was constructed so that it could be placed close to the sampling inlet and entirely within the stack. A variety of dilution systems suitable for use in aerosol sampling applications are available commercially.

16.5   Sampling for determining particle size distribution in stacks

There are situations where it is desired to determine particle size in aerosols in stacks and ducts, for example when there is the need to investigate properties or effects that vary with particle size – for example, the collection efficiency of air pollution control equipment. The EPA Method 201 described earlier identifies a procedure for measuring in-stack PM_{10} concentrations that employs a cyclone with penetration characteristics matching the PM_{10} curve.

A review of sampling for the measurement of aerosol particle size distribution is provided later in this book in Chapter 18. Whilst the measurement of the particle size distribution of stack aerosols is secondary to the primary purpose of stack sampling, driven as it is by emission control regulation, it is part of the armoury of the environmental hygienist in areas where investigative or exploratory research is needed to underpin technical or engineering decisions. For example, the choice of the type of air pollution control equipment and its detailed specification will be highly dependent on particle size distribution. For such applications, a variety of types and models of particle size measuring instruments are available. For these, cascade impactors have been favoured. A range of so-called '*University of*

Washington in-stack impactors' developed by Professor M. J. Pilat and his colleagues at the University of Washington have been prominent, the first of which – the Mark I – appeared as long ago as 1968, intended originally for measuring the particle size distribution in stacks as the basis of assessing the collection efficiencies of air pollution control devices and the relationship of plume opacity to particle properties (see below) (Pilat *et al.*, 1970, 1971; Ensor and Pilat, 1971). Many versions have since emerged and have been used successfully in the evaluation of many emission sources including coal fired power boilers, pulp mill recovery boilers, aluminum reduction plants, and oil fired turbines.

In addition to cascade impactors, *in-stack cascade cyclones* have also been used, and a range of such devices are commercially available from several sources. In a separate development, Campbell *et al.* (1986) described the modification of an electrostatic analyzer to be used for the measurement of particle size distributions in condensing aerosols in hot, moist stack gases. All such devices may be incorporated into sampling trains like those described in the EPA Method 5.

16.6   Direct-reading stack-monitoring instruments

A number of real-time, direct-reading monitoring instruments have been developed for assessing aerosols in stacks and ducts. Most operate on the principle of the scattering of electromagnetic radiation by airborne particles of solid or liquid matter, as described later in greater detail in Chapter 20. Since it is frequently the appearance of aerosol emission from a stack that is of concern, assessment on the basis of *opacity* is then appropriate, assessed on a scale from 0 to 100 %. *Transmissometer-type* opacity meters for this purpose, involving measurement of the attenuation of a beam of light as it traverses the stack, have been available commercially from several sources for many years. The simplest involves a single pass of the light, and more complicated ones – with the help of mirrors – involve multiple passes that allow for longer optical path length, and hence greater sensitivity. Others involve beam chopping in the process of electronic signal processing in order to achieve still greater sensitivity of detection.

Such instruments are not samplers in the sense as discussed in this book. Moreover, the index of opacity cannot be consistently related to the measures of actual aerosol concentration (e.g. number, surface area, volume or mass per unit volume of air) that come from direct sampling as mentioned elsewhere. But it is relevant to touch on them here because such measurements are directly relevant to the visual appearance of the stack emission, complementing the sampling methods that are the main priority in stack emissions evaluation. In the USA, EPA sets out requirements for continuous in-stack opacity monitoring for stationary sources in Part 60 of 40 CFR, specifically under *Performance Specification 1* (PS1). The EPA requirements incorporate in their entirety the guidelines first set out in ASTM D6216-98, including any ongoing changes currently being undertaken. This documentation covers the procedure for certifying continuous opacity monitors, including design and performance specifications, test procedures, and quality assurance requirements to ensure that continuous opacity monitors meet minimum design and calibration requirements. It is stated that following such guidelines will permit accurate opacity monitoring requirements in regulatory environmental opacity monitoring applications subject to 10 % or higher locally applied opacity standards.

References

American Society for Testing and Materials (1973) Standard method of test for sampling stacks for particulate matter. In: *ASTM Annual Book of Standards,* D2928-71, Part 23., ASTM, West Conshohocken, PA.

Bohnet, M. (1978) Particulate sampling. In: *Air Pollution Control*, Part III: *Measuring and Monitoring Air Pollutants,* (Ed. W. Strauss), John Wiley & Sons, Inc., New York, pp. 79–119.

Boothroyd, R.G. (1967) An anemometric isokinetic sampling probe for aerosols, *Journal of Scientific Instruments*, 44, 249–253.

Boubel, R.W. (1971) A high volume stack sampler, *Journal of the Air Pollution Control Association*, 21, 783–787.

British Standards Institution (1971) *Simplified methods for measurement of grit and dust emission*, British Standard 3405: 1971 (revised 1983).

Campbell, M.J., Cronin, D.R. and Bamesberger, W.L. (1986) A modified commercial electrostatic aerosol analyzer for size distribution measurements of condensing aerosols in hot, wet stack gases, *Journal of Aerosol Science*, 17, 179–190.

Delprato, U. and Lombardi, V. (1990) An automatic system for the isokinetic sampling and PM_{10} concentration monitoring of industrial stack emissions, *Journal of Aerosol Science*, 21 (Suppl. 1), S629–S632.

Dennis, R., Samples, W.R., Anderson, D.M. and Silverman, L. (1957) Isokinetic sampling probes, *Industrial and Engineering Chemistry*, 49, 294–302.

Ensor, D.S. and Pilat, M.J. (1971) Calculation of smoke plume opacity from particulate air pollutant properties, *Journal of the Air Pollution Control Association*, 21, 496–501.

Environmental Protection Agency (1971) Standards for performance for new stationary sources, *Federal Register*, Part II, 36 (247), 24876–24895.

Felix, L.G., Merritt, R.L., McCain, J.D. and Ragland, J.W. (1981) Sampling and dilution system design for measurement of submicron particle size and concentration in stack emissions aerosols, *TSI Quarterly*, 7, 3–12 (available from TSI Inc., St Paul, MN)

German Engineering Society (1974) Standard for efficiency measurement on dust collectors, VDI 2066.

Helsper, C., Mölter, W. and Haller, P. (1990) Representative dilution of aerosols by a factor of 10,000, *Journal of Aerosol Science*, 21 (Suppl. 1), S637–S640.

Hemeon, W.C.L. and Black, A.W. (1972) Stack dust sampling: in-stack filter or EPA train, *Journal of the Air Pollution Control Association*, 22, 516–518.

Hueglin, Ch., Scherrer, L. and Burtscher, H. (1997) An accurate, continuously adjustable dilution system (1:10 to $1:10^4$) for submicron aerosols, *Journal of Aerosol Science*, 28, 1049–1955.

Koch, W., Lödding, H., Mölter, W. and Munzinger, F. (1988) Verdünnungssystem für die Messung hochkonzentrierter Aerosole mit optischen Partikelzählern, *Staub Reinhaltung der Luft*, 48, 341–344.

Lombardi, V., Salmi, M., Allegrini, I. and Febo, A. (1984) The sampling of radioactive particulate matter in ducts and stacks by means of an active isokinetic probe, *Journal of Aerosol Science*, 15, 385–386.

McIntyre, K.L. and Moore, M.B. (2004) Seventh Annual Report on Federal Agency Use of Voluntary Consensus Standards and Conformity Assessment, United States National Institute of Standards and Technology, NISTIR 7118.

Morrow, N.L., Brief, R.S. and Bertrand, R.R. (1972) Sampling and analyzing air pollution sources, *Chemical Engineering*, 79, 84–98.

Narjes, L. (1965) Anwendung neuartiger Nulldrucksonden zur quasiisokinetischen Staubprobenahme in Dampfkraftanlagen, *Staub Reinhaltung der Luft*, 25, 148–153.

Perry, R. and Young, R.J. (1977) *Handbook of Air Pollution Analysis*, John Wiley & Sons, Ltd, New York.

Pilat, M.J., Ensor, D.S. and Bosch, J.C. (1970) Source test cascade impactor, *Atmospheric Environment*, 4, 671–679.

Pilat, M.J., Ensor, D.S. and Bosch, J.C. (1971) Cascade impactor for sizing particulates in emission sources, *American Industrial Hygiene Association Journal*, 32, 508–511.

Steen, B. (1977) A new simple isokinetic sampler for the determination of particle flux, *Atmospheric Environment*, 11, 623–627.

Steen, B., Keady, P.B. and Sem, G.J. (1981) A sampler for direct measurement of particle flux, *TSI Quarterly*, 7, 3–9.

Vincent, J.H. (1989) *Aerosol Sampling: Science and Practice*, John Wiley & Sons, Ltd, Chichester.

Vincent, J.H., Emmett, P.C. and Mark, D. (1985) The effects of turbulence on the entry of airborne particles into a blunt dust sampler, *Aerosol Science and Technology*, 4, 17–29.

Wiener, R.W., Okazaki, K. and Willeke, K. (1988) Influence of turbulence on aerosol sampling efficiency, *Atmospheric Environment*, 22, 917–928.

17

Sampling for Aerosols in the Ambient Atmosphere

17.1 Introduction

The criteria described in earlier chapters for coarse and fine aerosol fractions are intended for application in health-based standards, covering scenarios where human exposure takes place through inhalation. This theme is continued in this chapter, and health-related sampling for ambient atmospheric aerosols will be discussed in detail below. In addition, however, there are important practical situations where aerosol sampling may be driven by other rationales. For aerosols in the ambient atmospheric environment, for example, the criteria for sampling are more diverse than those for workplaces. In the first place, the justifications for health-related sampling now involve not only the inhalation route but also the *ecological* route – for example, by which particulate matter in the atmosphere can be transferred into other media and, ultimately, present a risk via the food chain. In addition, there are nuisance and ecological justifications associated with the deposition of large particles onto surfaces, along with climatic justifications arising from the extinction of light associated with fine suspended particles. The resultant widening of the overall rationale for aerosol sampling now presents a picture that goes beyond the one that relates just to inhalation.

17.2 Sampling for coarse 'nuisance' aerosols

In addition to possible direct human health effects, coarse particles emitted by large-scale industrial processes, such as power stations, waste tips, quarries and mines, etc., pose a significant risk of damage to nearby ecosystems and – not trivially – considerable nuisance to human populations. For the latter, soiling of buildings and windows, and deposition on plants and crops, are serious concerns. Indeed, such deposition of particulate matter is one of the main causes of complaints about air pollution (Hall *et al.*, 1993) – the other is odor. However, as summarised by Vallack and Shillito (1998), no consistent approach to standards appears to exist that is in any way commensurate with the elaborate framework that has evolved for health-related aerosols. Nonetheless, the local problem of the precipitation of dust, grit and 'smuts' has been acknowledged for many years, and has long been a concern of local

Aerosol Sampling: Science, Standards, Instrumentation and Applications   James Vincent
© 2007 John Wiley & Sons, Ltd

populations, and in turn of industries themselves. According to Brimblecombe (1987), measurements of deposited dust were among the earliest air pollution studies. This interest has continued into recent decades as concerns about deposited dust – or *dustfall* – have persisted. In the UK, for example, the (formerly) Central Electricity Generating Board and others invested significantly from the 1960s onwards in research and development of sampling technology for collecting particles relevant to this problem.

For sampling in a manner thought to be broadly relevant to the problem, the concept of the *dust deposit gauge* was introduced, intended for the collection of airborne particles of aerodynamic diameter (d_{ae}) up to and beyond $100 \, \mu m$, depending on the proximity of sampling to the emission source. The principle of operation of one type of deposit gauge in common usage is shown schematically in Figure 17.1. This comes into the general *horizontal deposit gauge* category, and takes the simple form of an upwards-facing bowl. Particles moving under the influence of both the external wind and gravity follow trajectories some of which may enter the bowl. In the first instance, performance is determined by the external wind speed (U), particle relaxation time (τ), acceleration due to gravity (g) and dimensional scale of the bowl. The mechanism of collection involves a combination of gravitational settling and impaction. It is also strongly dependent on the geometrical shape of the system, external freestream turbulence and secondary flows inside the bowl, all influencing both deposition and particle blow-out of particles actually deposited inside the bowl – re-entrainment post-deposition occurs especially in strong wind gusts. The performance of such a device in all these respects reflects quite well the corresponding aerosol mechanical processes that occur during the deposition of particles that constitute the nuisance in question, at least qualitatively. Since such devices operate at zero sampling flow rate, performance can hardly be described in terms of aspiration efficiency or, for that matter, any of the other parameters introduced for aspirating samplers described in preceding chapters, because there is no net air flow into the sampler. Rather, the device may be regarded as a 'passive' sampler, so that the definition of collection efficiency is different than for the 'active' samplers that are the primary subjects of this book. Here, it is more appropriate to define *collection efficiency*, C, as:

$$C = \left(\frac{1}{c_0 g \tau} \right) \text{(mass collected per unit area per unit time)} \qquad (17.1)$$

where c_0 as usual is the aerosol concentration in the undisturbed freestream upwind of the sampler and the product $g\tau$ is the particle settling velocity. In Equation (17.1), the collected aerosol is typically expressed in terms of mg m^{-2} day^{-1}.

Horizontal deposit gauges along the lines indicated have been around for a long time. The type known as the *British standard gauge* (British Standard Institution, 1969a), shown in Figure 17.2, has been in use for up to 60 years, not only in Britain but – in related forms – also in other countries. It is very simple. A metal stand supports an upwards-facing bowl at a height of 1.2 m above the ground. The wire guard over the top of the collecting bowl is intended to prevent interference from feathered and other wildlife. The primary subject of interest with this instrument is the particulate matter collected in the bowl. However a hole in the bottom of the bowl allows collected water to drain into a receptacle whose contents, if desired, can also be analyzed. Various other dust gauge versions emerged over the years, including one from the Norwegian Institute for Air Research (NILU), another from the International Standards Organisation (ISO), and another suggested by the American Society for Testing and Materials (ASTM).

Ralph and Barrett (1984) carried out a wind tunnel investigation of the performances of a number of such gauges, including the ones suggested by the British Standards Institution (BSI), NILU and ISO, respectively. The experiments were carried out using scale models in experiments where, by careful application of the laws of similarity, experimental results for C, as defined in Equation (17.1),

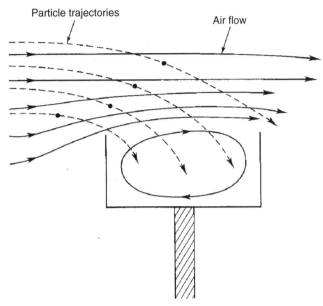

Figure 17.1 *Diagram to indicate the principle of operation of the dust deposit gauge. Reproduced with permission from Vincent, Aerosol Sampling: Science and Practice. Copyright (1989) John Wiley & Sons, Ltd*

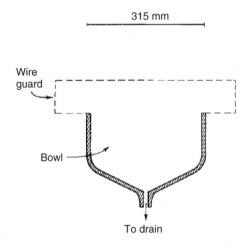

Figure 17.2 *Diagram of the British standard horizontal dust deposit gauge (British Standards Institution, 1969a). Reproduced with permission from Vincent, Aerosol Sampling: Science and Practice. Copyright (1989) John Wiley & Sons, Ltd*

were obtained appropriate to full-scale systems. Test aerosols consisted of glass spheres with d_{ae} up to 400 μm. Freestream aerosol concentration was determined using isokinetic thin-walled probes. The results for the gauges placed 1 m above the ground in a freestream where $U = 3$ m s^{-1}, considered to be the nominal average wind speed for British locations, are summarised in Figure 17.3. Here, the

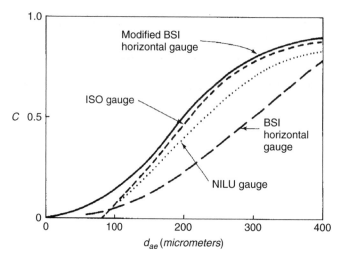

Figure 17.3   *Summary of experimental results for the performance of a number of dust deposit gauges in terms of collection efficiency (C) as a function of particle aerodynamic diameter (d_{ae}) (Barrett et al., 1984). Reproduced with permission from Vincent, Aerosol Sampling: Science and Practice. Copyright (1989) John Wiley & Sons, Ltd*

smooth curves represent best fits with the actual experimental data. They show that, for all three gauges, C increased steadily with d_{ae}. This tendency was entirely as to be expected since gravitational and inertial effects were known to be the predominant influences for such devices. Of the gauges tested, the BSI gauge was markedly the least efficient, and this was thought to be associated with the blow-out of deposited particles from the collecting bowl. In an attempt to reduce this effect, Ralph and Barrett fitted a cross-shaped baffle inside the bowl of the BSI gauge to reduce the recirculation of the air enclosed there. As shown in Figure 17.3, the performance of this modified BSI gauge was much improved, indeed to the extent where it actually exceeded those for the other gauges tested. Overall, however, the general finding from this study was that the performances of all the gauges investigated fell substantially below unity for particles with d_{ae} less than about 100 μm.

Hall and his colleagues (Hall and Waters, 1986, Hall and Upton, 1988) commented on the poor performances of these horizontal dust deposit gauges, especially the very low values of C for smaller values of d_{ae}. This observation took no account of the extent to which any such instrument might match a hypothetical performance curve reflecting the nature of the dust nuisance in question, and it was therefore considered a worthy goal to achieve a sampler that exhibited values of C that were as high as possible over as wide a range of d_{ae} as possible. Hall *et al.* noted that, in addition to the problem of particle blow-out already referred to for the dust gauges studied, the aerodynamic bluffness of the gauge body would cause the air flow to be strongly accelerated and deflected upwards over the opening of the bowl, enhancing the possibility of weak collection there or of re-entrainment. So they sought an alternative sampler configuration. The more aerodynamic '*inverted Frisbee*' configuration shown in Figure 17.4 was introduced. This arrangement was assessed in wind tunnel experiments similar to those reported by Ralph and Barrett. In the first instance, on the question of blow-out, it was shown that the threshold wind speed at which removal of collected particles took place was much higher than for the BSI gauge (Hall and Waters, 1986). Secondly, as summarised in Figure 17.5, the results for C as a function of d_{ae} showed that performance was much improved compared with the modified BSI gauge, and hence to all the other gauges that had been proposed up until then, particularly in the range of d_{ae}

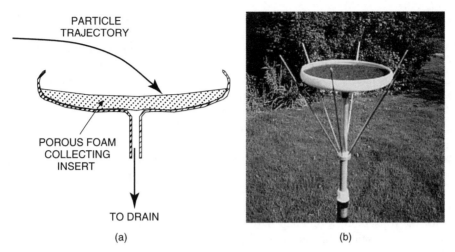

Figure 17.4 *Inverted frisbee deposit gauge: (a) general layout (Hall and Waters, 1986; Hal and Upton, 1988) also showing the porous plastic foam collecting insert (Vallack, 1995); (b) photograph of actual instrument, also showing the supporting structure and bird strike preventor. Photograph courtesy of Dr. H.W. Vallack, Stockholm Environment Institute, University of York, UK*

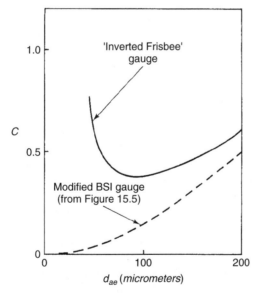

Figure 17.5 *Summary of experimental results for the performance of the inverted frisbee dust gauge in terms of collection efficiency (C) as a function of particle aerodynamic diameter (d_{ae}), shown in comparison with the modified BSI gauge (the best of these represented in Figure 17.3) (Hall and Waters, 1986; Hall and Upton, 1988). Reproduced with permission from Vincent, Aerosol Sampling: Science and Practice. Copyright (1989) John Wiley & Sons, Ltd*

below 100 μm. One additional feature of particular interest, absent for the other gauges, was the sharp upturn in C at smaller particle sizes in the range $d_{ae} < 70$ μm. It was suggested that this might be associated with the enhanced deposition of small particles by turbulent diffusion across the separated shear layer over the mouth of the shallow bowl of the instrument. In the use of this gauge, and indeed the other ones described so far, the importance of maintaining the mouth of the collecting bowl accurately in the horizontal plane was stressed. So too was the positioning of the instrument away from sources of major flow disturbance (e.g. buildings, trees, etc.) which would influence the nature of the approaching flow. Quite recently, Vallack (1995) described modified versions of the inverted frisbee sampler aimed at reducing re-entrainment or blow-off losses from the inner surface of the bowl. The modifications included the inclusion of a polyester foam insert into the base of the bowl to provide a collecting surface less susceptible to re-entrainment. In addition, a 'bird-strike preventor' was incorporated, in the form of a ring of fine nylon fishing line supported at 5 cm above the plane of the top of the inverted frisbee by six thin struts. Finally, in order to catch the inevitable rainwater, an opening was located at the center of the frisbee bowl, draining into a plastic bottle located below the sampler itself. These improvements are embodied in the version of the sampler shown in Figure 17.4.

Despite the obvious simplicity of these passive samplers, practical experience has shown that the recovery of the collected dust presents some significant technical challenges. With this in mind, Vallack[1] created a detailed protocol that applied specifically to the inverted frisbee dust deposit gauge but which could also be applicable to dust deposit gauges like the other ones mentioned. This protocol addressed all the challenges that had been listed, allowing for consistent, reliable use of the instrument in question. It described how the collected particulate matter, both in the dry state in the bowl of the gauge and in the rainwater collected in the drainage bottle, should be combined together in suspension and recovered by filtration.

Instruments like those described are, by their design, omni-directional. Earlier, Lucas and Moore (1964) of the UK Central Electricity Research Laboratories (CERL) had noted the limitations of early versions, including their inability to collect nuisance particles in the size range relevant to the emissions from the chimneys of modern plants and – in particular – their inability to distinguish between sources. A new device was proposed, which became known as the '*CERL vertical dust gauge*' (Lucas and Snowsill, 1967). As shown in Figure 17.6, this sampler was made up of four cylindrical collecting elements, mounted together symmetrically in a square array on a vertical pillar, each element having a vertical slot cut into the outwards-facing side of the cylinder. The cylinder was closed at the bottom to provide a receptacle for the dust which entered the cylinder through the slot. Bush *et al.* (1976) described a simple mathematical model for particle transport in this system in terms of impaction onto a single cylinder with a slit-shaped opening, identifying the relative roles of particle characteristics and wind speed magnitude and direction on particle transport into the slit. Their model supported the desired directionality of particle collection.

The mechanism of collection in the CERL sampler mainly involved the inertial deposition of particles through the plane of the slot, with a bottle located at the bottom of each cylindrical element, both for the collection of rainwater and also to trap dry-deposited particles transferred there during wash-off after the completion of sampling. Clearly, based on the aerosol mechanical principles discussed earlier in this book, collection efficiency for this instrument would be greater for higher wind speeds and for particles with larger d_{ae}. The overall effect of the four collecting elements was to provide information about airborne coarse particles arriving simultaneously from all directions, and hence to provide directional information that might enable identification and assessment of the contributions of particular emission sources. Like the nondirectional BSI horizontal dust deposit gauge shown in Figure 17.2, the CERL

---

[1] See also website at www.york.ac.uk/inst/sei/dust/frisbeeprotocol.pdf.

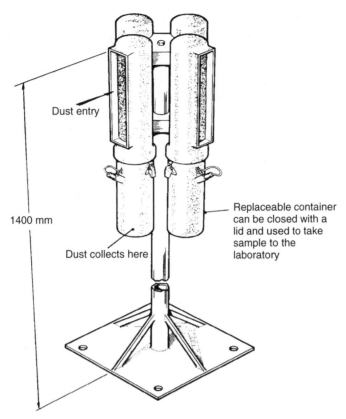

Figure 17.6   *Diagram of the CERL vertical directional dust deposit gauge (Lucas and Snowsill, 1967; British Standards Institution, 1972). Reproduced with permission from Vincent, Aerosol Sampling: Science and Practice. Copyright (1989) John Wiley & Sons, Ltd*

instrument also became the subject of a British standard (British Standards Institution, 1972). It was widely used in Britain for assessing coarse airborne 'nuisance' particles near industrial sources, and is still occasionally used to this day.

17.3   Sampling for 'black smoke'

In the USA, the EPA 40 CFR provides for the assessment of atmospheric visibility through its Method 9 (*Visual determination of the opacity of emissions from stationary sources*). This, however, is directed at the evaluation of the appearance of a smoke plume (e.g. from an industrial plant) by a remote observer that has been trained to make a semi-quantitative – albeit subjective – assessment of plume opacity relative to the background. This has formed an important part of air pollution surveillance in the USA. However, it does not involve aerosol sampling as it is referred to in this book.

In Europe, aerosol sampling to assess black smoke by direct evaluation of what is collected onto a filter has long featured in atmospheric environmental regulation, as in EC Directive 80/779/EEC (see Chapter 13). Indeed, until the early 1990s, almost all measurements of air pollution in the UK involved this approach. The methodology behind this regulation was based on the *British smoke shade sampler* (BSS) adopted as a standard method by the British Standards Institution (BSI, 1969b). It relied on the drawing of air through an inverted funnel, collecting particles on a filter, followed (post-sampling) by assessment of the 'blackness' of the stain by measuring the light reflectance from the surface of the filter, using an appropriate light scattering instrument whose output had previously been calibrated in terms of the mass of collected black smoke. Properly set up, this method should be sensitive only to the light-absorbing components in the deposited aerosol, and so to the carbonaceous components associated with incomplete combustion that were predominant over other components. The calibration of the optical procedure was the one that was developed back in the 1960s when smoke from coal combustion – mainly domestic – still represented a significant proportion of the atmospheric aerosol in the UK (Quality of Urban Air Review Group, 1993). A similar method was described in 1964 by the Organisation for Economic Cooperation and Development (OECD) (Organisation for Economic

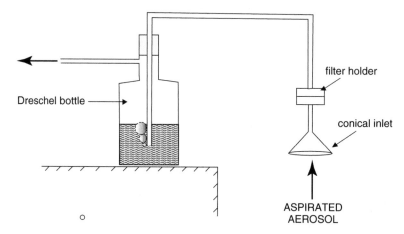

Figure 17.7 *Diagram of the apparatus used in the UK for the sampling of black smoke, known as the British smoke shade sampler (British Standards Institution, 1969b) (not to scale)*

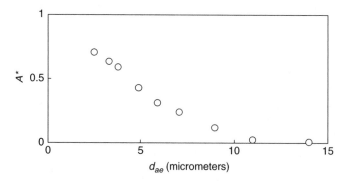

Figure 17.8 *Overall sampling efficiency (A*) as a function of particle aerodynamic diameter (d_{ae}) for the British smoke shade sampler (McFarland et al., 1982). Adapted from Atmospheric Environment, 16, McFarland et al., 325–328. Copyright (1982), with permission from Elsevier*

Cooperation and Development, 1964). Direct-reading instruments for black smoke operating on the same basis principle have also appeared, including the *aethalometer* (Hansen *et al.*, 1984).

One familiar sampling system used for black smoke is shown in Figure 17.7, and this was designed for sampling flow rates in the range from 1 to 2 Lpm. It shows the funnel-shaped entry attached directly by a plastic tube to a separate filter holder, and thence to a sampling train comprising a Dreschel bottle placed there to permit simultaneous collection of sulfur dioxide. The performance of this system in terms of the efficiency of collection of aerosol at the filter was assessed by McFarland *et al.* (1982), and the results are shown in Figure 17.8. Here it is seen that the overall collection efficiency (A^*) fell off sharply for particles of increasing aerodynamic diameter, reaching 0.5 for d_{ae} about 4.5 μm. However, since the aerosol fraction of interest was the one containing carbonaceous particles in the atmosphere, derived from incomplete combustion processes, performance of the sampler could reasonably be expected to be biased towards small particles where sampling efficiency would be quite close to unity.

17.4 Sampling for total suspended particulate in the ambient atmosphere

17.4.1 'Total' aerosol

When sampling is required for health-related purposes, we now know that such sampling needs to be carried out with respect to specific particle size fractions. This a relatively recent consideration. Earlier, it was considered sufficient to attempt to measure all particles that may be considered to be airborne since these were all potentially available for inhalation. For many years, this was a major basis of aerosol pollution measurement, complementing black smoke in many countries. Over the years a large number of instruments were developed for the sampling of aerosol in the ambient atmosphere in the form of what has been known variously as 'suspended particulate matter' (SPM) or 'total suspended particulate' (TSP). Samplers for 'total aerosol' first emerged during the time when it was widely thought that it was sufficient to simply draw an aerosol sample through an inlet and to collect the particles on a filter. A wide variety of such instruments appeared, with widely differing configurations and, as was later discovered, having widely differing performance characteristics.

In 1984, Barrett *et al.* reported wind tunnel experiments carried out at the UK Department of Trade and Industry's Warren Spring Laboratory (now closed) to investigate the performances of a selection of devices already widely in use in Europe for the general sampling of total aerosol. These included: the Italian 20 Lpm *ISTISAN* (1981) featuring a single, downwards-facing open-faced filter; the German 242 (minimum) Lpm *LIS/P-Filtergerat* (Verein Deutscher Ingenieure, 1981a) featuring a downwards-facing, cowled filter holder; the 45 Lpm *Kleinfiltergerat* GS050/3 (Verein Deutscher Ingenieure, 1981b) also featuring a downwards-facing, cowled filter holder; and the American 28.3 Lpm *Andersen Mark II nonviable sampler* (Andersen, 1965) featuring a single, upwards-facing circular 25 mm sampling orifice. These samplers had the feature in common that there was no preferred orientation with respect to the wind. Barrett *et al.* assessed their performances in terms of overall sampling efficiency (A^*), using monodisperse uranine aerosols generated by a spinning disk generator for d_{ae} up to about 25 μm, and for wind speeds from 1 to 9 m s^{-1}. In 1985, Barrett and his colleagues (Barrett *et al.*, 1985) reported a similar study into the performances of a further pair of devices, this time with a view to their possible applications for the sampling of airborne lead. These were: the French 25 Lpm *Type PPA 60 sampler* (from Environment SA, Poissy), featuring a mushroom-shaped head into which aerosol was sampled through its downwards-facing annular orifice; and the British 4.4 Lpm *M-type sampler* (from Warren Spring Laboratory), featuring a single, downwards-facing open-faced filter. Like the four instruments described in the earlier 1984 study, both sampled omnidirectionally. The results for A^* as a function of d_{ae} for various wind speeds (U) up to 9 m s^{-1} for all six samplers tested are shown in

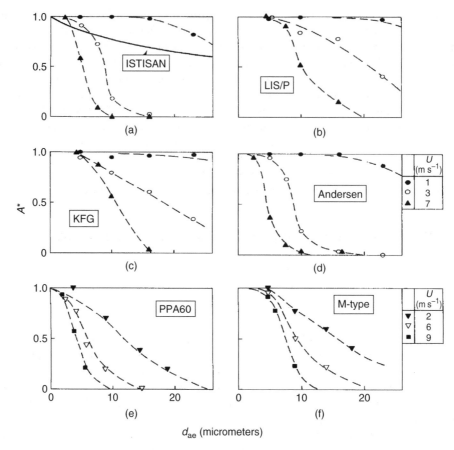

Figure 17.9 *Experimental results for overall sampling efficiency (A*) as a function of particle aerodynamic diameter (d_{ae}) for a range of samplers suggested – and used – for the collection of TSP in Europe, for the ranges of wind speeds (U) indicated (Barrett et al., 1984, 1985). Reproduced with permission from Vincent, Aerosol Sampling: Science and Practice. Copyright (1989) John Wiley & Sons, Ltd*

Figure 17.9. Qualitatively, they show very consistent patterns. A^* varied strongly with both d_{ae} and U falling progressively with increasing d_{ae} and with the rate of fall being more rapid for larger values of U. These trends were in general agreement with those for similar instruments proposed elsewhere for sampling the ambient atmosphere (e.g. Pattenden and Wiffen, 1977) as well as similar ones intended for applications in workplaces (see Chapter 14). As far as the downwards-facing samplers are concerned (five out of the six), the observed trends were broadly consistent with expectations based on the theory of vertical elutriation. Although there were substantial quantitative variations in performance from one sampler to another, it is very clear than none of them truly collected total aerosol, except – in some cases – for the lowest wind speed tested. Rather, and to varying degrees, they all undersampled with respect to true total aerosol, and to an extent that decreased as d_{ae} increased. In addition, all showed strong wind speed dependencies, and their performances were a long way from matching the curve for the inhalable fraction. In summary, therefore, none of these instruments performed in a satisfactory manner with respect to any known scientific, health-related criterion. In turn, therefore, results obtained using these instruments could only be regarded as of limited value in relation to modern standards requirements.

In the USA, standards for atmospheric aerosol were based on TSP for many years up to and including 1987. The EPA definition of TSP was never more than a loosely defined one, and was never actually intended to reflect true total aerosol. Rather, it was to be regarded more as a nominal (or 'statutory') measure of airborne particulate material identified for regulatory purposes in the absence at that time of anything more scientific, and was based on the performances of specific approved reference instruments. It was only eliminated when EPA implemented the PM_{10} standards based on new scientific information about the particle size dependency of the lung deposition of inhaled aerosols. Nonetheless, the measurement of TSP persists to this day in some American states and in some countries, and sampling instruments aimed at that fraction – albeit poorly defined – are still commercially available. A working protocol still appears in EPA's 40 CFR Part 50 in Appendix B (*Reference method for the determination of suspended particulate matter in the atmosphere*). The sampling system described there is illustrated in Figure 17.10. This figure shows a 20.3 cm × 25.4 cm (8 in. × 10 in.) rectangular filter housed in a shelter that also contains the pump and flow control apparatus that can provide high flow rates to as high as 1.7 $m^3 min^{-1}$ – hence the term '*High Volume*' or '*Hi-Vol*'. The performance of this apparatus depends on the efficiency by which particles are first aspirated and then transferred to the filter, and so is expressed in terms of A^*. There are a few published data suggesting a decline in A^* with d_{ae}, reaching 50 % in the range from about 30 to 60 μm, depending on the wind speed, decreasing at higher wind speeds and increasing at lower ones (Wedding *et al.*, 1977; McFarland *et al.*, 1979; Lodge, 1989). Such instruments are widely available commercially (e.g., Tisch Environmental, Cleves, OH, USA).

Another family of TSP samplers has emerged based on a technical criterion that the air velocity at the inlet should be equal to the settling velocity of a particle with $d_{ae} = 100$ μm (0.25 $m s^{-1}$) and the underlying assumption that all particles smaller than this would be collected with 100 % efficiency. Two

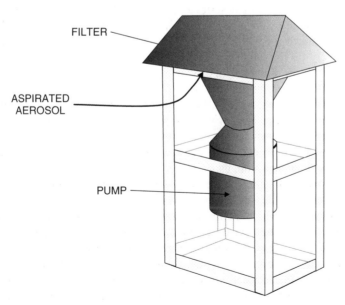

Figure 17.10 *Sketch of a sampler widely used for TSP in the USA, recommended sampling flow rates from 1.1 to 1.7 $m^3 min^{-1}$ (not to scale)*

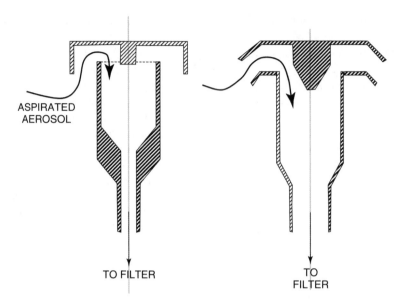

Figure 17.11   *Diagrams of the TSP and louvered TSP sampler inlets tested by Kenny et al. (2005) (not to scale)*

of them are shown in Figure 17.11, and were tested by Kenny *et al.* (2005), who referred to them as the '*TSP inlet*' and '*TSP louvered inlet*', respectively. Both featured omnidirectional annular inlets whose planes faced generally downwards. Versions operating at different sampling flow rates are quite widely available commercially. Those shown were both operated at 16.7 Lpm, and Kenny *et al.* reported the results of experiments with them in wind tunnels at two laboratories, using both gravimetric and rapid data acquisition methods of the types described earlier in Chapter 3. Their results for the sampling efficiencies of the two inlets are shown in Figure 17.12, showing that A^* fell steadily as d_{ae} increased for all the conditions examined for both samplers. The general observed trend was for both samplers to undersample progressively for the larger particle sizes, similarly to the other samplers described above. The sampling efficiency of the louvered inlet was significantly greater than for the TSP inlet. Wind Speed effects were noted but were secondary.

From the preceding it is clear that none of the samplers described was able to collect a representative sample of true total aerosol. That in itself would not be particularly awkward if the performances of the samplers described were at least compatible with one another. However, as seen from the selection of experimental data that has been presented, this was certainly not the case. So the application of any one of these instruments in a given situation, especially in the context of compliance with standards, would be problematical. Over the years, there have been a number of alternative approaches. The *Rotorod* (May *et al.*, 1976) is a nonaspirating instrument that operates on the principle of particle impaction onto a narrow, square-sectioned rod which is translated rapidly with respect to the ambient air. This was achieved in practical versions of the device by bending the rod into a U-shape having two vertical arms, the centre of which was attached to the vertical shaft of an electric motor. It is shown schematically in Figure 17.13. In the version described by May and his colleagues, the rod width was 1.6 mm, the arm lengths 60 mm and the radial location of each rod 40 mm from the axis. Thus, by rotating the shaft at 2500 rpm, the two vertical arms were swept through the air to produce an effective sampling rate – obtained in terms of the rate at which air volume was geometrically incident on the collecting arms as they rotated – of about 120 Lpm. Collection efficiency, by impaction, of the Rotorod was high

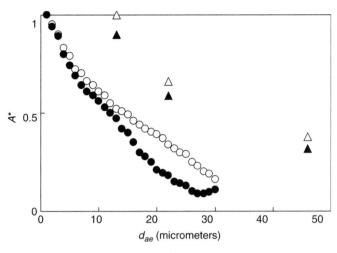

Figure 17.12 *Experimental results for the sampling efficiency (A*) of the TSP and louvered inlets (see Figure 17.11) as functions of particle aerodynamic diameter (d_{ae}) for various wind speeds; for the TSP inlet using a rapid data acquisition method involving the use of the aerodynamic particle sizer (APS), and for the louvered inlet using a gravimetric method involving the weighing of collected samples of aerosols generated from narrowly graded powders (Kenny et al., 2005: ○ TSP inlet 1 m s^{-1}, ● TSP inlet 2 m s^{-1}, ▲ louvered inlet 0.5 m s^{-1}, △ louvered inlet 1 m s^{-1}). Adapted with permission from Kenny et al., Journal of Environmental Monitoring, 7, 481–487. Copyright (2005) Royal Society of Chemistry*

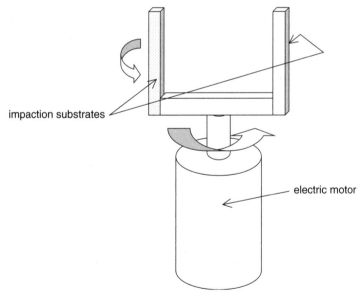

Figure 17.13 *Diagram to illustrate the principle of the Rotorod sampler for collecting true total aerosol (not to scale)*

by virtue of the high value of Stokes' number for particle sizes in the range of interest, derived from the high relative velocity between the collecting surface and the air. Particles collected on the greased leading surfaces of the rods could subsequently be recovered for analysis. This type of device has been widely used as an atmospheric aerosol sampler in its own right. In recent studies of the performances of other, aspirating samplers under outdoor conditions, it has also been used as a means of providing reference measurements of the ambient aerosol concentration. Its own performance has been studied extensively over the years (e.g. Vrins and Hofschreuder, 1982). One area where this instrument appears to have been of special interest has been in the sampling of bioaerosols (see Chapter 19). It is still commercially available today (Multidata Sampling Technologies, St Louis Park, MN, USA).

A significant separate development in the sampling of true total aerosol in the ambient atmosphere was first described in the same paper by May *et al*. Subsequently referred to as the '*aerosol tunnel sampler*', the idea was subsequently extended by Hofschreuder, Vrins and their Dutch colleagues (Hofschreuder *et al.*, 1983; Vrins *et al.*, 1984; Hofschreuder and Vrins, 1986). As shown in Figure 17.14, their version of the device consisted of a 150 mm diameter tube through which air was drawn by means of a fan. The purpose of the honeycomb was to straighten the flow in the tube and to prevent the transmission of flow distortions back upstream from the fan. The air velocity in the tube was maintained at 9 m s^{-1}. Just in front of the honeycomb, a 10 mm diameter thin-walled secondary sampler was located axially, with its flow rate of 42 Lpm chosen to ensure isokinetic sampling conditions. Thus it was reasonable to assume that the axially located probe collected aerosol that was representative of the concentration inside the main tube. The main tube was thus seen to take the form of a small wind tunnel. The whole arrangement was mounted on a pivot with a wind vane so that, under actual sampling conditions, it was always oriented with the tube mouth facing into the wind. At the chosen tube diameter and entry velocity, the Stokes number for the aspiration of particles through its entry plane was seen to be small for most particle sizes of practical interest. Therefore aspiration efficiency for the tube entry should be close to unity. The net effect, therefore, was that the aerosol collected by the axial thin-walled secondary probe inside the tube should be representative of that in the ambient air outside. Hofschreuder, Vrins and their colleagues confirmed this to be the case. This ingenious instrument does not appear to have been widely used in practical ambient air monitoring.

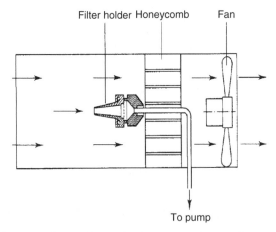

Figure 17.14   *Diagram of the aerosol tunnel sampler for true total aerosol (not to scale) (Hofschreuder et al., 1983). Reprinted from Journal of Aerosol Science, 14, Hofschreuder et al., 65–68. Copyright (1983), with permission from Elsevier*

17.4.2 Inhalable aerosol

Sampling for inhalable aerosol – as defined in modern particle size-selective criteria – has not yet been widely applied in the sampling of atmospheric aerosols. As far as is known, just one such instrument has been built, the prototype 30 Lpm sampler developed at the Institute of Occupational Medicine (IOM) (Aitken *et al.*, 1987). Shown in Figure 17.15, it evolved from the smaller 3 Lpm *inhalable aerosol sampler* developed earlier for applications in workplaces (Mark *et al.*, 1985), with which it had many features in common. The new instrument was intended originally for application in the collection of polycyclic aromatic hydrocarbon (PAH)-containing aerosols in the atmospheric environment near ground level close to British coking plants. Air was drawn through a single, 20 mm diameter circular orifice located in the side of a 50 mm diameter cylindrical head and sampling took place whilst the head was rotated slowly, at about 2 rpm, about a vertical axis to achieve orientation-averaging. The lipped circumference of the entry was intended to reduce oversampling associated with particle blow-off from external surfaces of the sampler. Like the other instruments discussed above, the performance of this sampler was assessed in terms of the overall sampling efficiency for particles reaching the filter (A^*). Some experimental performance data are shown in Figure 17.16, obtained in a wind tunnel for a range

Figure 17.15 *Prototype sampler developed for collecting inhalable aerosol in the ambient atmospheric environment, the 30 Lpm IOM sampler (not to scale) (Aitken et al., 1987). Reproduced with permission from Vincent, Aerosol Sampling: Science and Practice. Copyright (1989) John Wiley & Sons, Ltd*

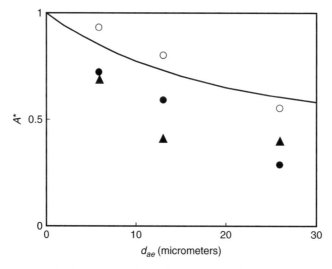

Figure 17.16 *Experimental results for the sampling efficiency (A*) of the prototype 30 Lpm inhalable aerosol sampler (see Figure 17.15) as a function of particle aerodynamic diameter (d_{ae}) for various wind speeds (from Aitken et al., 1987: O 1 m s^{-1}, ● 3 m s^{-1}, ▲ 7 m s^{-1}; ___ inhalability curve). Reproduced with permission from Vincent, Aerosol Sampling: Science and Practice. Copyright (1989) John Wiley & Sons, Ltd*

of relevant wind speeds using a range of narrowly graded test aerosols of fused alumina reflecting relevant particle sizes with d_{ae} up to about 30 μm and for wind speeds up to 7 m s^{-1}. These show that A^* fell somewhat below the inhalability curve, but less so than for the other samplers described above.

As far as coarser aerosol fractions are concerned, sampling for TSP has no scientific basis in relation to health, and so has quite rightly either been eliminated or, at least, relegated to secondary consideration. A more appropriate, health-based approach would involve sampling for the inhalable fraction, as defined for atmospheric aerosol by the ISO criterion described earlier in Chapter 10. However, in more recent years, concerns have turned away from coarser aerosol fractions in the atmosphere, and towards the finer fractions. So interest in pursuing new versions of the inhalable aerosol sampler described in the preceding paragraph has receded, and no new instruments beyond the prototype described have appeared.

17.5   Sampling for fine aerosol fractions in the ambient atmosphere

17.5.1   PM₁₀

In the latter part of the 20th century there emerged the widespread view that, if just one aerosol fraction is to be collected relevant to health for a wide range of types of aerosol in the ambient atmosphere, then that fraction should be relevant to the deposition of particles in the lung, associated with a wide range of aerosol-related respiratory illness. In July of 1987, EPA published revised particulate standards along these lines. After much deliberation, a sampler performance curve was chosen based on what penetrates into the lung below the larynx. This curve, passing through 50 % at $d_{ae} = 10$ μm, defines what is now referred to universally as PM₁₀ (see Chapter 11). Limit values have been assigned accordingly.

Promulgation of the new physiologically-based standard was dependent on the emergence of a new family of aerosol sampling instruments. Such instruments started to appear during the late 1970s and early 1980s, initially aimed at an anticipated PM_{15} standard that then appeared likely. In all of them, selection of the desired fraction from the aspirated aerosol was achieved variously by impaction, virtual impaction, vertical elutriation and/or cyclone separation. Many of them were originally intended as inlet systems for so-called '*dichotomous*' samplers, aimed at providing two aerosol fractions simultaneously. In these, the sampled aerosol – selected according to PM_{15} – was further classified by virtual impaction to provide a still finer fraction corresponding to the fine atmospheric mode with d_{ae} below 2.5 μm. The choice back then of a 2.5 μm cut for the finer sub-fraction was later to prove critical in the decision about the criterion for the finer, combustion-related fraction – $PM_{2.5}$ – that emerged later during the 1990s.

Typical early impactor-based inlets designed for use with 16.7 Lpm dichotomous samplers included, for example, the Liu–Pui inlet (Liu and Pui, 1981) and the *TAMU inlet* (McFarland *et al.*, 1977). Both are shown in Figure 17.17. A range of cyclone-based inlets was developed by Wedding and his colleagues, and one version is shown in Figure 17.18. Experimental results for A^* as a function of d_{ae} for various wind speeds (U) are summarised in Figure 17.19 for all three of these instruments, as reported by Liu and Pui (1986), Wedding (1982) and Wedding *et al.* (1980), all using monodisperse test aerosols generated by means of a vibrating orifice generator. In the graphs shown the results are compared with the performance envelope suggested for the original 15 μm standard suggested – pre-PM_{10} – by Ranade and Kashden (1979). It is seen that performance was fair for all three sampling inlets. It was particularly good for the Liu–Pui and Wedding inlets. For the TAMU inlet, however, there were quite strong variations with wind speed that took some of the experimental data outside the performance envelope. In addition to these inlets, McFarland *et al.* (1984) proposed a PM_{10} inlet for use with a Hi-Vol sampler like that shown in Figure 17.10.

Regulations based on PM_{10} eventually appeared in 1987. On the basis of aerosol sampling science it was a relatively simple matter to modify inlets like those just described – by combinations of adjustments of geometrical and flow rate variables – to derive families of samplers matching what was by now the definitive sampling criterion. Liu and Pui (1986) had already tested modified versions of their inlet,

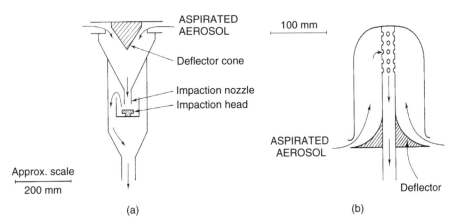

Figure 17.17 *Two impactor-based sampling heads for collecting fine atmospheric aerosol fractions: (a) the 16.7 Lpm Liu–Pui inlet (Liu and Pui, 1981); (b) the 16.7 Lpm TAMU inlet (McFarland et al., 1977) (not to scale). Reproduced with permission from Vincent, Aerosol Sampling: Science and Practice. Copyright (1989) John Wiley & Sons, Ltd*

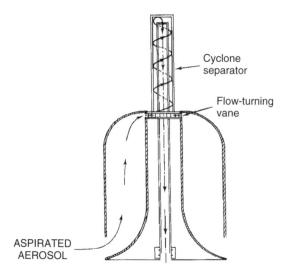

Cyclone
separator

Flow-turning
vane

ASPIRATED
AEROSOL

Figure 17.18   *The 16.7 Lpm cyclone-based sampling head for the fine atmospheric aerosol fraction (Wedding, 1982) (not to scale). Reproduced with permission from Vincent, Aerosol Sampling: Science and Practice. Copyright (1989) John Wiley & Sons, Ltd*

and the results in Figure 17.20 are seen to be in good agreement with the PM_{10} curve. The same applies for the many other devices that have since been approved as reference methods and come onto the market commercially. In 1984, EPA (Environmental Protection Agency, 1984) introduced the first *Federal Reference Method* (*FRM*) by which new sampling instruments for ambient atmospheric aerosol may be evaluated in relation to the performance criteria, requiring both laboratory characterisations and field comparative studies. Although there have been several updates, the basic framework has remained essentially the same (see later version, Environmental Protection Agency, 1997), and has formed the basis of the development and commercialisation of a large number of samplers. Most of these embody various combinations of the various inlet design features and particle size-selection mechanisms contained in the earlier instruments described above. A typical instrument, a modification of the Hi-Vol TSP incorporating an impactor-based pre-selector stage, is shown in Figure 17.21 (Tisch Environmental, Cleves, OH, USA). Many such samplers have come and gone, or have been improved.

In Europe, interest at one time was shown in developing an approach that was related to EPA's PM_{10} standard but whose principles were more firmly rooted in what were viewed by some as the somewhat more rigorous ISO approach. In this, sampling should reflect particle size-selection processes taking place both outside and inside the human body. The ideal sampler conforming to this rationale should first aspirate the inhalable fraction, then select the thoracic subfraction. It is the first stage in this process that was missing from the PM_{10} standard. To demonstrate this approach, the IOM 30 Lpm inhalable aerosol sampler described earlier (see Figure 17.15) was extended whereby the aspirated inspirable aerosol fraction was delivered to a selection region comprising a 123 mm diameter, 25 mm thick cylindrical plug of 30 ppi porous plastic foam. Penetration (*P*) of particles through this foam plug at the prescribed flow rate was assessed using monodisperse test aerosols of polystyrene latex spheres (Aitken *et al.*, 1987). As shown in Figure 17.22, the results were in good general agreement with the curve for thoracic aerosol. A prototype version of the instrument was tested in outdoor field trials in Britain and the Netherlands. But this sampler was never widely used beyond those early campaigns.

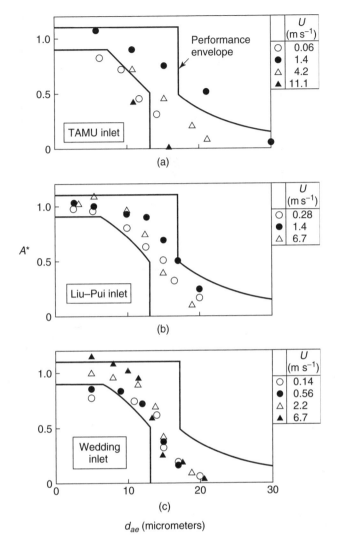

Figure 17.19 *Performance of the three sampling heads shown in Figures 17.17 and 17.18 in terms of sampling efficiency (A*) as a function of particle aerodynamic diameter (d_{ae}) and for a range of wind speeds (Wedding et al., 1980; Wedding, 1982; Liu and Pui, 1986). Also shown is the proposed performance envelope for the PM_{15} fraction that was being considered as a possible standard at the time. Reproduced with permission from Vincent, Aerosol Sampling: Science and Practice. Copyright (1989) John Wiley & Sons, Ltd*

17.5.2   PM$_{2.5}$

As interest in still finer aerosol fractions in the ambient atmosphere rose more recently, leading to the concept of PM$_{2.5}$, so too has the search for appropriate sampling instrumentation. Indeed, in the USA, the formal definition of PM$_{2.5}$ was linked directly with a specific sampler developed by EPA scientists, the '*well impactor ninety-six*' (*WINS*) (Peters and Vanderpool, 1996; Peters *et al.*, 2001). The WINS is shown schematically in Figure 17.23. In its application, the EPA defines a Federal Reference

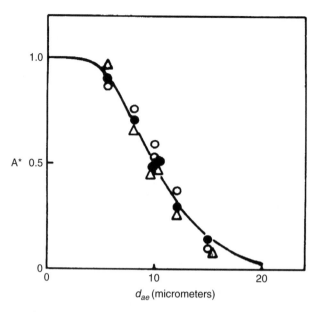

Figure 17.20 *Performance of a version of the 16.7 Lpm Liu–Pui inlet [see Figure 17.17(a)] modified to meet the PM_{10} criterion that eventually became the basis of standards for atmospheric aerosol, shown in terms of sampling efficiency (A^*) as a function of particle aerodynamic diameter (d_{ae}) and for a range of wind speeds (Liu and Pui, 1986: ○ 0.56 m s^{-1}, ● 2.2 m s^{-1}, △ 6.7 m s^{-1}). Also shown is the PM_{10} curve. Reproduced with permission from Vincent, Aerosol Sampling: Science and Practice. Copyright (1989) John Wiley & Sons, Ltd*

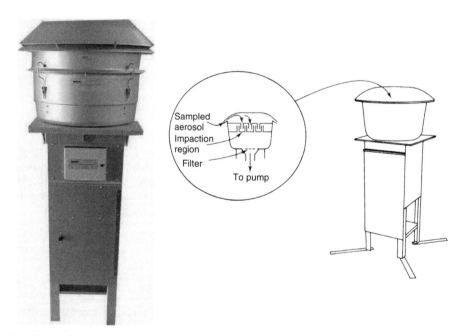

Figure 17.21 *The Hi-Vol sampler for PM_{10}, incorporating a louvre-type inlet and an impactor-based pre-selector for PM_{10}. Photograph courtesy of Tisch Environmental, Cleves, OH, USA*

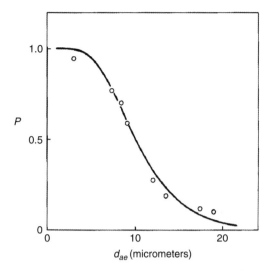

Figure 17.22 *Performance of the porous foam pre-selector stage of the prototype 30 Lpm IOM thoracic aerosol sampler, shown in terms of penetration of the foam (P) as a function of particle aerodynamic diameter (d_{ae}) (Aitken et al., 1987). Also shown is the CEN/ISO/ACGIH thoracic curve. Reproduced with permission from Vincent, Aerosol Sampling: Science and Practice. Copyright (1989) John Wiley & Sons, Ltd*

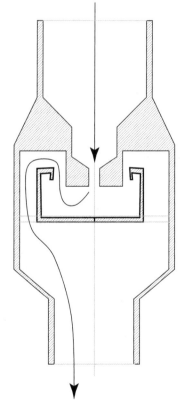

Figure 17.23 *Schematic of the 16.7 Lpm EPA well impactor ninety-six (WINS) for selecting the PM$_{2.5}$ fraction (not to scale) (Peters et al., 2001)*

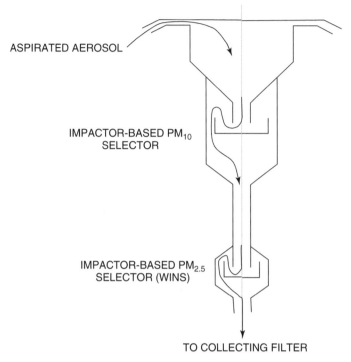

ASPIRATED AEROSOL

IMPACTOR-BASED PM$_{10}$
SELECTOR

IMPACTOR-BASED PM$_{2.5}$
SELECTOR (WINS)

TO COLLECTING FILTER

Figure 17.24 *Schematic of a 16.7 Lpm EPA generic reference sampling system for PM$_{2.5}$ (Environmental Protection Agency, 1997b)*

Method (FRM) in which the WINS must be incorporated into a 16.7 Lpm sampling system that also features a specific inlet and impactor system that first selects the PM$_{10}$ fraction (Environmental Protection Agency, 1997). Here the chosen generic PM$_{10}$ inlet is the same as the louvred inlet shown earlier in Figure 17.11. The complete sampling system for the defined reference sampler with this inlet is shown schematically in Figure 17.24. In addition to the sampling instrumentation, the FRM also provides a detailed protocol defining the ranges of conditions over which the system should operate accurately, and how measurements should be made and analyzed. For practical aerosol sampling for PM$_{2.5}$ in the ambient atmosphere, the EPA regulations require either the use of the FRM or a corresponding Federal Equivalent Method (FEM). It is required that any equivalent PM$_{2.5}$ sampler must have a sampling efficiency of 0.5 at a d_{ae}-value of 2.5 ± 0.2 μm and a mass sampling bias for PM$_{2.5}$ concentrations (with respect to the ideal) of less than 5 % as calculated for particle size distributions designated in the regulations as 'fine', 'typical' and 'coarse'. Further tests under field conditions are then required. Many samplers have been approved in this way. Following this process, for example, Kenny *et al.* (2000) carried out a study of three cyclone-based candidate samplers, including laboratory tests to determine the penetration curves of the instruments in question, both for clean and for 'loaded' samplers, and also conducted field tests in a suburban garden and a car park. This work identified the *sharp-cut cyclone* (*SCC*) as one sampler in particular successfully meeting the formal requirements for PM$_{2.5}$ sampling. It is shown in Figure 17.25 (BGI Inc., Waltham, MA, USA). An improved version, the *very sharp-cut cyclone* (*VSCC*), has subsequently appeared.

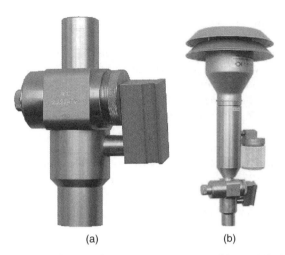

(a)   (b)

Figure 17.25 *The sharp cut cyclone for sampling PM$_{2.5}$ (BGI Inc, Waltham, MA, U.S.A.): (a) the sampler itself; (b) the sampler incorporated into a generic PM$_{10}$ sampling system like that shown in Figure 17.24. Photographs courtesy of Robert Gussman, BGI Inc., Waltham, MA, USA*

17.5.3 Ultrafine aerosols

As noted elsewhere in this book, interest in ultrafine particles in the range below 0.1 μm in the ambient atmospheric environment has risen in recent years. Measurement options for this fraction have been available for many years, in particular in the form of the direct-reading *differential mobility analyzer* (*DMA*) that enables the determination of the particle size distribution deep down into the nanometer range (i.e. below 10 nm). Such instruments will be described later in Chapter 20. For the present,

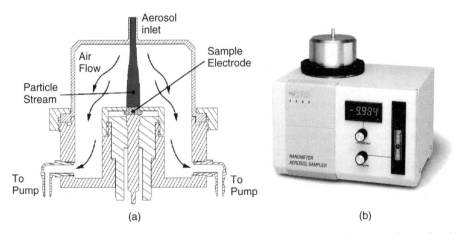

(a)   (b)

Figure 17.26 *The Nanometer Aerosol Sampler (NAS), TSI Model 3089: (a) schematic of sampling head to show the principle of operation; (a) photograph of the packaged instrument. Pictures courtesy of Gilmore Sem, TSI Inc., St Paul, MN, USA*

however, it is relevant to discuss options by which such small particles can be sampled and collected in a form that would be available for subsequent analysis – physical, chemical or biological. In one recent commercially-available option, the *nanometer aerosol sampler* (NAS), charged ultrafine particles are collected by electrostatic deposition onto sample substrates (TSI Inc., St Paul, MN, USA) [see Figure 17.26(a)]. The charged particles entering this sampler may be those, for example, emerging from the outlet of a DMA used in the mode of selecting particles according to their size in the range of interest. As shown in Figure 17.26(b), the charged particles enter the NAS axially into a cylindrical chamber and are attracted towards the sampler electrode under the influence of a strong externally applied electric field. The collection substrate, in the form of a transmission electron microscope grid or a glass slide, is placed on the sampler electrode. It is possible with this sampler to control the size of the deposition 'spot', in order to optimise the uniformity of the deposit for subsequent electron microscopical analysis, by the deployment of deposition electrodes of different sizes. This instrument efficiently collects particles in the particle size range from 0.1 μm down to about 2 nm.

17.6   Meteorological sampling

Sampling of aerosols for the purposes of meteorological research is usually carried out at high altitude where the suspended particles may be assumed to be small. Atmospheric aerosols high up in the troposphere and stratosphere, where there are much wider ranges of humidity, temperature and pressure than is found at ground level, are important in that they influence the radiative balance of the atmosphere. Also the very fine particles found at such high altitudes act as condensation nuclei for the production of larger particles. These are influential in the mechanics and chemistry of cloud formation and precipitation, as well as in the wider nature of climate, including climate change.

As in most other aerosol measurement situations, the most reliable method of determining the desired physical and chemical properties of tropospheric and stratospheric aerosols involves the collection of a representative sample onto a filter and subsequent analysis. Here, although many of the aerosol sampling challenges are generally similar to those encountered on the ground, there are some special additional considerations that need to be taken into account, associated with the extreme conditions that are encountered at such high altitude. Some aerosol sampling high in the atmosphere – at altitudes of the order of thousands of meters – has employed tethered and untethered helium-filled balloons (e.g. Rankin and Wolff, 1978, and others). However, most sampling for this branch of atmospheric aerosol sampling takes place from aircraft, where the most commonly used approach is the deployment of versions of the thin-walled cylindrical sampling probe mounted on the fuselage of the aircraft and directed to face the approaching airflow. It is therefore seen to be analogous to the application of isokinetically operated thin-walled probes for aerosol sampling in stacks and ducts, as discussed in an earlier chapter. Some of the features of typical airborne sampling systems are shown schematically in Figure 17.27 where the first thing to note is that the sampling head is supported on a stem such that it can be held parallel to the approaching air flow and be located so that it is positioned outside the boundary layer of the aircraft's fuselage and distant from locally generated turbulence.

In their report of the 1991 Airborne Aerosol Inlet Workshop in Boulder, Colorado, Baumgardner and Huebert (1993) discussed the ranges of physical conditions that pertain to aerosol sampling at high altitude from aircraft. These are summarised in Table 17.1, and it is seen that, in many respects, they extend conditions greatly beyond those pertaining to sampling at – or close to – ground level, even in the stacks of industrial processes. Sampling under such extreme conditions poses many technical

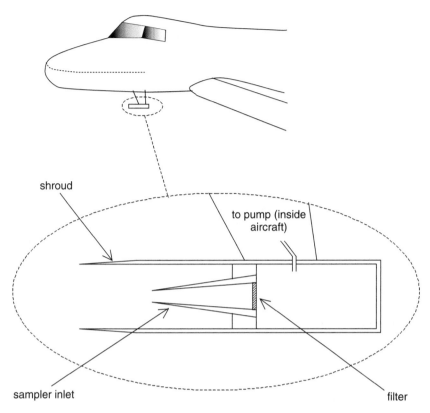

Figure 17.27 *Diagram to show the lay-out of a typical sampling system for airborne aerosol sampling, showing (inset) a typical shrouded inlet*

Table 17.1 *Summary of conditions for meteorological sampling at high altitude from aircraft (from Baumgardner and Hubert, 1993)*

Variable	Range
Airspeed (m s^{-1})	40–200
Pressure (m bar)	40–1000
Temperature (°C)	−80 to +30
Relative humidity (%)	0–100
Sampling inlet flow rate (Lpm)	1–10 000
Attitude	±10° pitch, ±5° yaw, ±5° roll
Particle size (μm)	0.001–10
Particle concentration (number) (m^{-3})	10^6–10^{11}
Particle shape	Spherical to complex structures
Particle phase	Mixed solid and liquid
Particle composition	Any combination of species

difficulties. One notable factor is the very high relative velocity between the sampler and the aerosol to be sampled, and this poses an immediate – and unavoidable – problem. For the isokinetic sampling that is required for unbiased aerosol aspiration a very high inlet velocity is required. Yet the air entering the sampler needs to be rapidly decelerated to much lower velocities in order to meet the needs of filtration and other sampling line apparatus. As indicated in Figure 17.27, this is achieved by means of an expanding taper – or diffuser – inside the sampling tube. The resultant dynamic heating may be considerable, with increases in air temperature that may even be as high as 20 °C, leading in turn to the possibility of significant changes in the aerosol particle size distribution, particle number concentration and even chemical composition (e.g. Porter *et al.*, 1992; Strapp *et al.*, 1992; Hermann *et al.*, 2001). Another factor that may be significant during airborne sampling is the role of electrostatic forces, especially during aerosol sampling from certain types of clouds where high electric fields exist and the water droplets may be highly charged. Sharp entries, which are desirable from a mechanical point of view under 'normal' terrestrial conditions, may then result in very high localised electric fields, leading to uncontrollable modifications to aspiration efficiency (Romay *et al.*, 1996). A third factor is the extent to which sampling conditions may or may not be truly isokinetic – or, more to the point, the extent to which it is *anisoaxial*. Although it is usual to locate the sampler far enough away from the aircraft fuselage to avoid flow distortions associated with the presence of the aircraft fuselage, it cannot always be ensured that the inlet is always aligned with the local airflow. As articulated by Hermann *et al.*, it is in the nature of flight that an aircraft rides on the air such that the 'angle of attack' depends on flight conditions and so is not constant. A partial solution to this is obtained by the use of the 'shrouded inlet'. This too is shown in Figure 17.27, and appears as a cylindrical shroud that is axial with the sampling probe itself but extends out beyond the inlet, acting to both partially decelerate the air flow and also align the flow with the sampler (e.g. Ram *et al.*, 1995).

While many ingenious versions of the basic system illustrated in Figure 17.27 have been offered over the years, there remain many problems in airborne inlet sampling for atmospheric aerosols. More than a decade after the Boulder meeting in 1991, a follow-up meeting was held in Leipzig, Germany in 2002 (Wendisch *et al.*, 2004), where there was further discussion about how to get aerosol particles into aircraft-borne sampling instruments without changing their ambient properties. Improved numerical modeling has allowed significant advances in recent years in the design of inlets in order to minimise problems like those outlined above, including the effects of the blunt-body housing that is used in some cases (Dhaniyala *et al.*, 2004). However it was acknowledged that there remain many unresolved problems. So there is scope for much continued research.

References

Aitken, R.J., Gibson, H., Lynch, G., Vincent, J.H. and Mark, D. (1987) Development of a static sampler for the measurement of suspended particulate matter in the ambient atmosphere, IOM Report No. TM/87/02, Institute of Occupational Medicine, Edinburgh.

Andersen, A.A. (1965) A sampler for respiratory health hazard assessment, *American Industrial Hygiene Association Journal*, 27, 1605.

Barrett, C.F., Carroll, J.D., Irwin, J.G., Ralph, M.O. and Upton, S.L. (1985) A wind tunnel study and field comparison of three samplers of suspended particulate matter, Report No. LR 544 (AP), Warren Spring Laboratory, Stevenage.

Barrett, C.F., Ralph, M.O. and Upton, S.L. (1984) Wind tunnel measurements of the inlet efficiency of four samplers of suspended particulate matter, Final Report on CEC Contract No. 6612/10/2, Report EUR 9378 EN, Commission of European Communities, Luxembourg.

Baumgardner, D. and Huebert, B. (1993) The Airborne Aerosol Inlet Workshop: meeting report, *Journal of Aerosol Science*, 24, 835–846.

Brimblecombe, P. (1987) *The Big Smoke: A History of Air Pollution in London since Mediaeval Times*, Methuen, London (Routledge, London, since 1989).

British Standards Institution (1969a) *Methods for the measurement of air pollution, 1. Deposit gauges*, British Standard 1747: Part 1.

British Standards Institution (1969b) *Methods for the measurement of air pollution, 2. Determination of concentration of suspended particulate matter*, British Standard 1747: Part 2.

British Standards Institution (1972) *Methods for the measurement of air pollution, 5. Directional dust gauges*, British Standard 1747: Part 5.

Bush, A.W., Cross, M., Gibson, R.D. and Owst, A.P. (1976) The collection efficiency of directional dust gauges, *Atmospheric Environment*, 10, 997–1000.

Dhaniyala, S., Wennberg, P.O., Flagan, R.C., Fahey, D.W., Northway, M.J., Gao, R.S. and Bui, T.P. (2004) Stratospheric aerosol sampling: effect of a blunt-body housing on inlet sampling characteristics, *Aerosol Science and Technology*, 38, 1080–1090.

Environmental Protection Agency (1984) National ambient air quality standard, proposed rule, *Federal Register*, 49(55), 10408–10462.

Environmental Protection Agency (1997) Ambient air monitoring reference and equivalent methods, *United States Federal Register*, 40 CFR Parts 50, 53 and 58.

Hall, D.J. and Upton, S.L. (1988) A wind tunnel study of the particle collection efficiency of an inverted Frisbee used as dust deposition gauge, *Atmospheric Environment*, 22, 1383–1394.

Hall, D.J. and Waters, R.A. (1986) An improved, readily available dustfall gauge, *Atmospheric Environment*, 20, 219–222.

Hall, D.J., Upton, S.L. and Marsland, G.W. (1993) Improvements in dust gauge design. In: *Measurements of Airborne Pollutants* (Ed. S. Couling) Butterworth Heinemann, London

Hansen, A.D.A., Rosen, H. and Novakov, T. (1984) The aethalometer – an instrument for the real-time measurement of optical absorption by aerosol particles, *Science of the Total Environment*, 36, 191–196.

Hermann, M., Stratmann, F., Wilck, M. and Wiedensohler, A. (2001) Sampling characteristics of an aircraft-borne aerosol inlet system, *Journal of Atmospheric and Oceanic Technology*, 18, 7–19.

Hofschreuder, P. and Vrins, E. (1986) The aerosol tunnel sampler: a total airborne dust sampler. In *Aerosols: Formation and Reactivity*, Pergamon, Oxford, pp. 491–494.

Hofschreuder, P., Vrins, E. and van Boxel, J. (1983) Sampling efficiency of aerosol samplers for large wind-borne particles – a preliminary report, *Journal of Aerosol Science*, 14, 65–68.

ISTISAN (1981) Metodi di prelievo a di analisi degli inquinanti del' aria: Appendice 2-Determinazione del materiale particellare in sospensione nell' aria, Instituto Superiore di Sanita, Rome.

Kenny, L., Beaumont, G., Gudmundsson, A., Thorpe, A. and Koch, W. (2005) Aspiration and sampling efficiencies of the TSP and louvered particulate matter inlets, *Journal of Environmental Monitoring*, 7, 481–487.

Kenny, L. C., Gussmann, R. and Meyer, M., (2000) Development of a sharp-cut cyclone for ambient aerosol monitoring applications, *Aerosol Science and Technology*, 32, 338–358.

Liu, B.Y.H. and Pui, D.Y.H. (1986) Aerosol sampling and sampling inlets. In: *Aerosols: Research, Risk Assessment and Control Strategies* (Eds S.D. Lee, T. Schneider, L.D. Grant and P.J. Verkerk), Lewis Publishers Inc., Chelsea, MI,, pp. 175–183.

Liu, B.Y.H. and Pui, D.Y.H. (1981) Aerosol sampling inlets and inhalable particles, *Atmospheric Environment*, 15, 589–600.

Lodge, J.P. (Ed.) (1989) *Methods of Air Sampling and Analysis*, 3rd Edn, Lewis Publishers, Inc., Chelsea, MI.

Lucas, D.H. and Moore, D.J. (1964) The measurement in the field of pollution by dust, *International Journal of Air and Water Pollution*, 8, 441–453.

Lucas, D.H. and Snowsill, W.L. (1967) Some developments in dust pollution measurement, *Atmospheric Environment*, 1, 619–636.

Mark, D., Vincent, J.H. and Gibson, H. (1985) A new static sampler for airborne total dust in workplaces. *American Industrial Hygiene Association Journal*, 46, 127–133.

May, K.R., Pomeroy, N.P. and Hibbs, S. (1976) Sampling techniques for large windborne particles, *Journal of Aerosol Science*, 7, 53–62.

McFarland, A.R., Ortiz, C.A. and Bertch, R.W. (1984) A 10 µm cut-point size selective inlet for Hi-Vol samplers, *Journal of the Air Pollution Control Association*, 34, 544–547.

McFarland, A.R., Ortiz, C.A. and Rodes, C.E. (1979) Characteristics of aerosol samplers used in ambient air monitoring, Paper presented at the 86th National Meeting of the American Institute of Chemical Engineers, Houston, TX.

McFarland, A.R., Ortiz, C.A. and Rodes, C.E. (1982) Wind tunnel evaluation of the British smoke shade sampler, *Atmospheric Environment*, 16, 325–328.

McFarland, A.R., Wedding, J.B. and Cermak, J.E. (1977) Wind tunnel evaluation of a modified Andersen impactor and an all-weather sampler inlet, *Atmospheric Environment*, 11, 535–539.

Organisation for Economic Cooperation and Development (1964) Methods of measuring air pollution, Report of the Working Group on Methods of Measuring Air Pollution and Survey Techniques, OECD, Paris.

Pattenden, N.J. and Wiffen, R.D. (1977) The particle size dependence of the collection efficiency of an environmental aerosol sampler, *Atmospheric Environment*, 11, 677–681.

Peters, T.M. and Vanderpool, R.W. (1996) Modification and evaluation of the WINS impactor, RTI Report No. 6360-011, Research Triangle Institute, NC.

Peters, T.M., Vanderpool, R.W. and Wiener, R.W. (2001) Design and calibration of the EPA $PM_{2.5}$ well impactor ninety-six (WINS), *Aerosol Science and Technology*, 34, 389–397.

Porter, J., Clarke, A. and Pueschl, R. (1992) Aircraft studies of size-dependent aerosol sampling through inlets, *Journal of Geophysical Research*, 97, 3815–3824.

Quality of Urban Air Review Group (1993) *Urban Air Quality in the United Kingdom*, Department of the Environment, London.

Ralph, M.O. and Barrett, C.F. (1984) A wind tunnel study of the efficiency of three deposit gauges, Report No. LR 499 (AP), Warren Spring Laboratory, Stevenage.

Ram, M., Cain S.A. and Taulbee, D.B. (1995) Design of shrouded probe for airborne aerosol sampling in a high velocity airstream, *Journal of Aerosol Science*, 26, 945–962.

Ranade, M.B. and Kashden, E.R. (1979) Critical parameters for the federal reference method for the inhalable particulate standard, Final Report Technical Directive 222, EPA Contract 68-02-2720, Research Triangle Institute, Research Triangle Park, NC.

Rankin, A.M. and Wolff, E.W. (1978) Aerosol profiling using a tethered balloon in Coastal Antarctica, *Journal of Atmospheric and Oceanic Technology*, 19, 1978–1985.

Romay, F.J., Pui, D.Y.H., Smith, T.J., Ngo, N.D. and Vincent, J.H. (1996) Corona discharge effects on aerosol sampling efficiency, *Atmospheric Environment*, 30, 2607–2613.

Strapp, J.W., Leaitch, W.R. and Liu, P.S.K. (1992) Hydrated and dried aerosol-size distribution measurements from the particle measuring system FSSP-300 probe and the de-iced PCASP-100X probe, *Journal of Atmospheric and Oceanic Technology*, 9, 548–555.

Vallack, H.W. (1995) Protocol for using the dry Frisbee (with foam insert) dust deposit gauge, Stockholm Environment Institute at York, February 1995 (see also www.york.ac.uk/inst/sei/dust/frisbeeprotocol.pdf).

Vallack, H.W. and Shillito, D.E. (1998) Suggested guidelines for deposited ambient dust, *Atmospheric Environment*, 32, 2737–2744.

Verein Deutscher Ingenieure (1981a) Determination of mass concentration of particulates in ambient air: filter method: LIS/P filter device, VDI Report No. 2463, Blatt 9.

Verein Deutscher Ingenieure (1981b) Determination of mass concentration of particulates in ambient air: filter method: small filter device GS050/3, VDI Report No. 2463, Blatt 7.

Vincent, J.H. (1989) *Aerosol Sampling: Science and Practice*, John Wiley & Sons, Ltd, Chichester.

Vrins, E. and Hofschreuder, P. (1982) Sampling total suspended particulate matter, *Journal of Aerosol Science*, 14, 318–332.

Vrins, E., Hofschreuder, P., Ter Kuile, W.M., van Nieuwland, R. and Oeseburg, F. (1984) Sampling efficiency of aerosol samplers for large windborne particles, In *Aerosols* (Eds B.Y.H. Liu, D.Y.H. Pui and H. Fissan), Elsevier, New York, pp. 154–157.

Wedding, J.B. (1982) Ambient aerosol sampling: history, present thinking and a proposed inlet for inhalable particulate matter, *Environmental Science and Technology*, 16, 154–161.

Wedding, J.B., McFarland, A.R. and Cermak, J.E. (1977) Large particle collection characteristics of ambient aerosol samplers, *Environmental Science and Technology*, 11, 387–390.

Wedding, J.B., Weigand, M., John, W. and Wall, S. (1980) Sampling effectiveness of the inlet to the dichotomous sampler, *Environmental Science and Technology*, 14, 1367–1370.

Wendisch, M., Coe., H., Baumgardner, D., *et al.* (2004) Aircraft particle inlets: state-of-the-art and future needs, *Bulletin of the American Meteorological Society*, 85, 89–91.

18

Sampling for the Determination of Particle Size Distribution

18.1 Introduction

Most of the aerosol samplers described in preceding chapters were dedicated to collecting one aerosol fraction only, and so their performances were expected to be matched to specific selection criteria. Such instruments feature in the vast majority of aerosol sampler applications, both for routine measurement and for research. However, there are many practical circumstances where it is desirable to obtain more information and therefore a more flexible approach is called for. Detailed particle classification into narrow ranges of diameter is sometimes needed. Sampling instruments capable of achieving this are often referred to as 'aerosol spectrometers'. They have existed in one form or another for many years and are well represented in the literature. A review of just some of the more important developments is given here.

18.2 Rationale

In principle, if the particle size distribution of an aerosol is known, then the particle size distribution may be defined for any fraction that can be defined numerically. This rationale is illustrated in Figure 18.1. Here, the original particle size distribution 1 is multiplied point-by-point by the selection curve 2 to yield the modified particle size distribution 3. If the initial and resultant distributions, respectively, are expressed in terms of particulate mass, the mass contained within the fraction of interest is represented by the shaded area under the resultant curve 3. This process may be applied for any selection curve, whether it is one of the current health-related particle size-selective criteria or suggested new ones described in Chapter 11, or some other curve not yet defined or contemplated.

This rationale may be extended so that, in practical situations, the particle size distribution may be given for each species, chemical or biological, within an overall aerosol sample. This may be achieved, for example, by appropriate quantitative analysis of the collected sample, and may be of interest where there is more than one species present that could lead to health effects and/or where there might be separate applicable standards. By thus describing an aerosol sample, effectively in three dimensions,

Aerosol Sampling: Science, Standards, Instrumentation and Applications   James Vincent
© 2007 John Wiley & Sons, Ltd

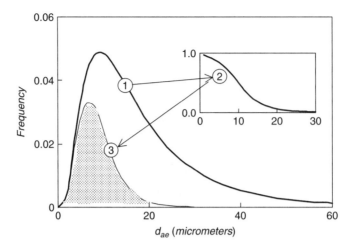

Figure 18.1 *Diagram to illustrate the principle of particle size-selective sampling using the measured particle size distribution: 1, the measured aerosol particle size distribution; 2, the particle size selection curve of interest; and 3, the particle size distribution of the resultant aerosol fraction*

in terms of the distributions of both particle size and species, all the information is available for dose evaluation, and in turn risk. This might be referred to as a *'fingerprint'* for the aerosol in the environment of interest.

18.3   Aerosol spectrometers

Aerosol spectrometers have been developed for applications in both the ambient atmospheric and working environments. The general scientific issues are essentially the same for the two application areas so that, even though there are some differing challenges in implementation, they are described here together.

18.3.1   Horizontal elutriators

Horizontal elutriators, in which particles are classified according to their aerodynamic sizes, are effective by virtue of gravitational forces acting on particles moving through a two-dimensional rectangular channel in which the air flow is laminar. In earlier chapters devices have been described that were designed as pre-selectors for sampling instruments aimed at specific fine aerosol fractions (e.g. respirable). In applying the same basic concept of horizontal elutriation to aerosol spectrometers, the physics is basically the same. However, in the pre-selectors already referred to, information about particle size distribution is not immediately accessible from their penetration characteristics, or even from the inspection of the spatial distribution of material deposited along the length of the elutriator duct. Now, in order to achieve the desired information, modifications are required, notably in the manner in which the aerosol is introduced into the elutriator channel.

One well-known such instrument is the aerosol spectrometer first described by Timbrell (1972). Its principle of particle classification is illustrated in Figure 18.2. In this sampler, particles were first aspirated and then introduced into the air stream through a narrow tube located just in front of the

ASPIRATED AEROSOL

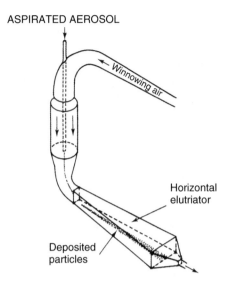

Winnowing air

Horizontal
elutriator

Deposited
particles

Figure 18.2 *Diagram to illustrate the principle of aerosol particle size classification by horizontal elutriation, as featured in the aerosol spectrometer proposed by Timbrell (1972). Particles move towards the bottom surface of the elutriator under the influence of gravity so that the larger particles are deposited closer to the inlet to the channel and the finer particles penetrate further into the sampler. Reproduced with permission from Timbrell, Assessment of Airborne Particles. Copyright (1972) Charles C. Thomas Publisher*

elutriator channel. The particles thus entering were winnowed in a stream of clean air. Particles of given aerodynamic diameter followed well-defined trajectories as they fell under gravity and were collected on the floor of the channel on a succession of glass slides spaced out along its length. Since all particles had to fall the same distance in order to be deposited, there was a unique relationship between the aerodynamic diameters of particles found on a given slide and the location of the slide with respect to the channel entrance. So Timbrell was able to define specific particle size ranges for particles collected on each of the slides.

The advantage of this type of spectrometer over most other types is that the theory of elutriation is simple and deterministic so that calibration may be carried out from first principles. Thus it is seen that the sensitivity of classification varies inversely with the square of d_{ae}. This means that, for very coarse particles, differences in location on the channel floor close to the channel entrance of particles with slightly different d_{ae} is not easily resolved. On the other hand, for very fine particles, excessively long channel lengths are required to permit detection of even quite small changes in d_{ae}. These constraints impose limits on what can be achieved by horizontal elutriation. But as far as its relevance to health effects is concerned, the process of gentle gravitational elutriation in the manner indicated admirably represents what happens to particles inside the human respiratory tract. Another attraction is that, because the forces involved are relatively weak, sampled aerosols – some of which may be agglomerates – are less likely to be damaged or changed physically during the classification process itself. The same may not necessarily be assumed for all other spectrometers described below.

In its practical realisation, the length of the elutriator channel of the *Timbrell spectrometer* was 40 cm, giving acceptably sensitive classification of particles in the range $1 < d_{ae} < 15$ μm. In order to avoid flow disturbance during admittance of the sampled aerosol to the elutriator channel, the sampling flow rate of the instrument was only 1 mLpm. Herein lay some of the disadvantages of the horizontal elutriator-type aerosol spectrometer as a practical sampler. Firstly, the physical size of the elutriator channel needs to be quite large in order to achieve the desired performance. Secondly, the sampling flow rate is such that the amount of particulate material collected at any given part of the instrument is too small to be assessed gravimetrically. Thirdly, at such a low sampling flow rate, substantial biases to the particle aerodynamic size distribution may be expected from particle losses occurring not only during aspiration but also during subsequent transmission to the entrance of the elutriator channel. The Timbrell spectrometer found some applications as a research tool, but was never made available commercially.

18.3.2   Centrifuges

The physical basis of centrifuge-type aerosol spectrometers is essentially similar to that for horizontal elutriators. The primary difference is that gravitational settling forces are replaced by the potentially greater centrifugal ones arising from rapid rotation of the body of air containing the particles of interest. Meanwhile, however, the penetration of particles through the channel has the same general functional dependence on d_{ae} as for the horizontal elutriator.

An early instrument built along these lines was the *Conifuge* described by Sawyer and Walton (1950). Its principle of operation is shown schematically in Figure 18.3. The sampled aerosol entered the duct through a narrow tube near the center of rotation and, as in the Timbrell spectrometer described above, was winnowed in a stream of clean air. The coarser particles were deposited by the relatively weak centrifugal forces as they passed along the first part of the conical duct at small distances from the axis. Finer particles were deposited further along the duct under the influence of the stronger forces acting at greater radii. The Conifuge was capable of classifying particles within the range $0.5 < d_{ae} < 30$ μm at a sampling flow rate of 25 mLpm. Later, *spiral-duct centrifuges* were introduced (e.g. Stöber and Flaschbart, 1969; Kotrappa and Light, 1972). For these, operating under the same general physical principles as the Conifuge, overall particle classification took place within the range $0.01 < d_{ae} < 5$ μm but at sampling flow rates up to about 1 Lpm. One advantage of a spiral configuration was that a long duct, and hence a wide range of particle classification, could be achieved within a system that remained physically compact. Whilst there has been considerable disagreement about the details of duct design, especially on the question of long versus short spirals (Stöber, 1976), most of this has related to the ability of centrifuges to give good classification for very fine – particularly submicron – particles. This follows from the fact that the development of such instruments was directed largely towards applications in characterising the fine aerosol mode in the ambient atmosphere. Like horizontal elutriators, many centrifuge-type aerosol spectrometers have similar disadvantages arising from the low sampling flow rate. They have not been widely applied in practical environmental and occupational hygiene.

18.3.3   Inertial spectrometers

Inertial spectrometers represent a large family of devices that separate particles by virtue of their relative inabilities, dependent on particle aerodynamic diameter and air velocity, to follow rapid changes in the motion (velocity and direction) of the sampled air. One instrument that was of interest for a while was the *cascade centipeter* described by Hounam and Sherwood (1965). It was a compact multi-stage device in which particle separation on the basis of particle size took place by virtual impaction, and therefore had some features similar to the Respicon sampler mentioned in the previous chapter. Sampling at the high flow rate of 30 Lpm, it could classify particles in the size range $1 < d_{ae} < 15$ μm and could collect sufficient classified material at each stage that – unlike the other spectrometers mentioned above – allowed gravimetric assessment of collected, classified material. One significant disadvantage, however, was that there was substantial deposition of particles between the intended collection stages,

ASPIRATED AEROSOL

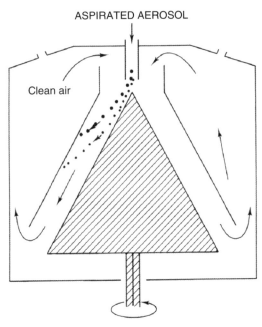

Clean air

Figure 18.3 *Diagram to illustrate the principle of aerosol particle size classification by centrifugation, as featured in the Conifuge proposed by Sawyer and Walton (1950). Particles move by centrifugal forces towards the inside wall of the outer cone – the coarser particles are collected near the entry while the finer particles penetration further down into the sampler. Reproduced with permission from Vincent, Aerosol Sampling: Science and Practice. Copyright (1989) John Wiley & Sons, Ltd*

with the result that there was a relatively large proportion of the sampled aerosol that was not accounted for. The cascade centipeter is of historic interest only today.

An interesting subsequent development was the *inertial spectrometer (INSPEC)* described by Prodi *et al.* (1979). As shown schematically in Figure 18.4, the sampled aerosol – as in some of the other samplers referred to above – entered the instrument at a low flow rate (0.1 Lpm) and joined a laminar stream of winnowing air. Here, separation took place as particles were projected inertially onto a rectangular filter when that winnowing air was deflected sharply through 90°. Particle aerodynamic diameter was related to location along the length of the filter and evaluation of particle aerodynamic size distribution itself was achieved through the visual inspection of the spatial distribution of the deposit of particulate matter along the filter. The instrument was shown to provide good aerodynamic particle size classification for the aerosol which arrived at the sensing zone. But, because of its low sampling flow rate, its disadvantages were the same as for the horizontal elutriator and centrifuge types. Prodi *et al.* (1986) subsequently developed a smaller-scale device suitable for use as a personal sampler. Operating under similar physical principles, but at the higher sampling flow rate of 2 Lpm and in an axially symmetric flow system, this was the *PERSPEC* first referred to in Chapter 14. This meant that collected particles could be assessed gravimetrically. This, however, was achieved at some expense to the sensitivity of particle size selectivity. At the time, the greatest potential for this instrument appeared to be in the simultaneous selection of a number of specific aerosol fractions (e.g. inhalable, thoracic and respirable, as described in Chapter 17). However, ineither the INSPEC nor PERSPEC are available commercially today.

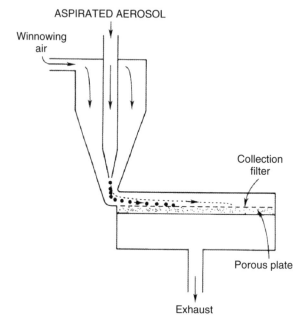

Figure 18.4 *Diagram to illustrate the principle of particle size classification in the INSPEC inertial aerosol spectrometer (Prodi et al., 1979). Particles are deposited by inertial forces as the flow changes direction at the entrance to the collecting section. Larger particles are collected on the filter closer to the entrance (left), and the finer ones penetrate further along (right). Reproduced with permission from Vincent, Aerosol Sampling: Science and Practice. Copyright (1989) John Wiley & Sons, Ltd*

18.4 Cascade impactors

18.4.1 Outline

Cascade impactors first appeared in the 1940s (May, 1945). Now they represent by far the largest single family of aerosol spectrometers and a large variety of instruments has appeared, many of which are available commercially and are widely used in occupational and environmental hygiene. Historically, this is closely linked with the advances that have been made in the evolution of our understanding of the science of impactors (May, 1982; Marple, 2004). The principle of operation of the cascade impactor is shown in Figure 18.5. Sampled aerosol passes through a succession of stages, at each of which the aerosol-containing jet is directed through a nozzle onto a solid surface. Particle deposition takes place by impaction, the efficiency of which (E) is a function of Stokes number (St), involving not only d_{ae} but also the jet dimensions and air velocity in the jet. For each stage, it is useful in the first instance to define the 'cut-size' as the value of $d_{ae} = {}_{50}d_{ae}$ at which $E = 0.5$. The narrower the jet and the higher the air velocity, the finer the particles that are collected. In a succession of such impactors placed together in series, it is arranged for progressively finer particles to be collected as the particle-laden air passes through the device. Fine particles passing through the final impactor stage are usually collected onto a backing filter. The distribution of collected particulate matter between the stages – including the filter – therefore provides a means for assessing the particle size distribution of the aerosol that enters initially. For example, for a hypothetical n-stage device, if the masses on the impactor stages are $m_1, m_2, \ldots, m_n$, and that on the backing filter is m_F, then the cumulative particle size distribution

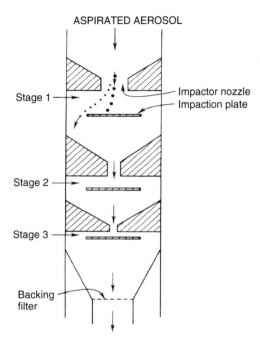

ASPIRATED AEROSOL

Stage 1 — Impactor nozzle
— Impaction plate

Stage 2 —

Stage 3 —

Backing filter

Figure 18.5   *Diagram to illustrate the principle of operation of a cascade impactor. Reproduced with permission from Vincent, Aerosol Sampling: Science and Practice. Copyright (1989) John Wiley & Sons, Ltd*

is represented in the first instance by the combined masses m_F, $m_F + m_n$, $m_F + m_n + m_{n-1}$, and so on. That is, m_F is the mass of particles that are, in principle at least, smaller than the $_{50}d_{ae}$ for the nth impactor stage; $m_F + m_n$ is the mass of particles smaller than the $_{50}d_{ae}$ for the n-1th impactor stage; and so on. For an eight-stage cascade impactor, this whole process may be summarised as follows:

d_{ae}	Cumulative mass, CUM(d_{ae})
$_{50}d_{ae}$ for Stage 8	m_F less than $_{50}d_{ae}$
$_{50}d_{ae}$ for Stage 7	$m_F + m_8$
$_{50}d_{ae}$ for Stage 6	$m_F + m_8 + m_7$
$_{50}d_{ae}$ for Stage 5	$m_F + m_8 + m_7 + m_6$
$_{50}d_{ae}$ for Stage 4	$m_F + m_8 + m_7 + m_6 + m_5$
$_{50}d_{ae}$ for Stage 3	$m_F + m_8 + m_7 + m_6 + m_5 + m_4$
$_{50}d_{ae}$ for Stage 2	$m_F + m_8 + m_7 + m_6 + m_5 + m_4 + m_3$
$_{50}d_{ae}$ for Stage 1	$m_F + m_8 + m_7 + m_6 + m_5 + m_4 + m_3 + m_2$
Estimated d_{ae} for largest particle seen on Stage 1	$m_F + m_8 + m_7 + m_6 + m_5 + m_4 + m_3 + m_2 + m_1$

All the $_{50}d_{ae}$-values for the stages of any given cascade impactor would have been determined experimentally as part of the calibration process. The final line of the process outlined above requires an estimate of the maximum sampled particle size, and this is usually made on the basis of expert knowledge of the sampled aerosol on the part of the operator. From the preceding, in a practical application of the hypothetical cascade impactor described, the cumulative masses indicated might be plotted against the corresponding $_{50}d_{ae}$-values shown. The resultant plot of CUM(d_{ae}) versus $_{50}d_{ae}$ may then be used graphically to estimate important parameters of the particle size distribution, including for example the mass median particle aerodynamic diameter and the geometric standard deviation.

Ideally, collection at each stage should be such that all particles with d_{ae} above $_{50}d_{ae}$ are removed and all those below penetrate through to the next stage. This would mean that the impactor stage has a 'perfect cut' and, if this could be achieved, then the collected masses on the various stages, when combined as indicated and plotted as a function of $_{50}d_{ae}$, would accurately represent the cumulative distribution of the sampled particulate mass. In reality, however, the collection characteristics of impactors are less idealised and perfect cuts are not achievable, despite efforts to optimise them (e.g. Marple, 1970 and many others). Instead, E versus d_{ae} for each stage follows the familiar 'S-shaped' curve. Now, depending on the degree of lack-of-sharpness, there is a strong likelihood that particles of the same size might be found on more than one stage. This means that any analysis based on the simple plotting of cumulative particulate mass versus $_{50}d_{ae}$ is only approximate. More accurate determination of particle size distribution requires mathematical procedures that take account of the actual E versus d_{ae} characteristics for the various stages. Known as 'inversion procedures', these will be discussed later.

18.4.2   Earlier cascade impactors

In 1945, K.R. May described his original cascade impactor, a four-stage sampler in which particles were impacted from progressively narrower cylindrical nozzles onto glass slides. Figure 18.6 shows a photograph of an early prototype built at Harvard University from engineering drawings supplied by May himself, together with a photograph of the first commercially available instrument built along these lines (the *Casella Mk 1*). The usefulness of such an instrument was immediately recognised (e.g. Laskin, 1949) and several instruments soon appeared operating on the same physical principles, albeit with different impactor geometries and in different configurations. Later, May (1975) acknowledged the presence of unaccounted for particle losses to the inner walls of cascade impactors, and so proposed an *ultimate cascade impactor*, designed so that such wall losses would be negligible. Improvements continued over the years and many versions of the cascade impactor subsequently appeared. In 1986 these were reviewed by Cohen, who also provided a long table of the specific instruments that were built during the 40 years following May's original description, together with a list of some of the prominent manufacturers. Cohen therefore provided an excellent impression of the huge impact of May's invention on the sampling and characterisation of environmental aerosols.

The early May cascade impactor and its derivatives employed a single air jet at each impactor stage. The first multi-circular-jet system was the sampler proposed by Andersen (1958), with the goal of providing a deposit at each stage that was more uniformly distributed over a relatively large collection surface, thus facilitating optical evaluation for particle identification and counting. It was aimed in the first instance at the sampling and classification of aerosols of biological origin, but a version soon appeared specifically for nonbacteriological aerosols. Several versions of these instruments have appeared over the years, many of which are still commercially available.

18.4.3   Static cascade impactor-based samplers

Of the cascade impactors that emerged over the years intended for use as fixed-point, static samplers, some were developed for the measurement of ambient atmospheric aerosols, others for workplace aerosols, some for both.

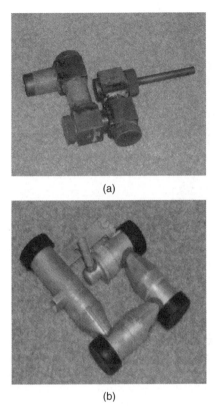

(a)

(b)

Figure 18.6 *Early cascade impactor (May, 1945): (a) version made at Harvard University to plans supplied by K.R. May; (b) early commercial version, the Casella Mk I cascade impactor. Photographs courtesy of Robert Gussman, BGI Inc. Waltham, MA, USA*

The eight-stage *Andersen Mk-II nonviable sampler* emerged as an important instrument for the measurement of particle size distributions for wet or dry particles, having both low particle bounce and wall loss characteristics. This instrument is shown in Figure 18.7(a), and is still commercially available (Thermo Electron Corporation, Waltham, MA, USA) and remains widely used to this day. The high sampling flow rate of 28.3 Lpm, aspirating through an upwards-facing conical inlet of diameter 2.54 cm, allows the collection of large amounts of particulate material at each stage, in turn facilitating both gravimetric and chemical analysis. Each impactor stage is made up of from 100 to 400 separate nozzles, with jets all impacting onto a 82 mm diameter stainless steel collection plate The detailed aerosol collection characteristics of each stage, in the form of E versus d_{ae}, were determined definitively in experiments carried out by Mitchell *et al.* (1988). Kerr *et al.* (2001) showed that it was reasonable to represent the experimental data to a good approximation by the logistic function:

$$F(d_{ae}) = \frac{\exp(\alpha + \beta \cdot d_{ae})}{1 + \exp(\alpha + \beta \cdot d_{ae})} \qquad (18.1)$$

in which the coefficients α and β were obtained by nonlinear least-squares regression of the Mitchell *et al.* raw data. The resultant fitted curves for the top seven stages are shown in Figure 18.7(b). The

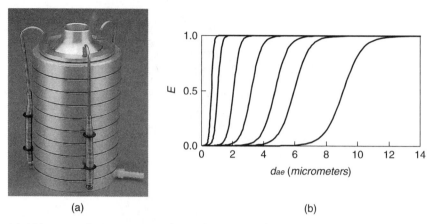

(a)   (b)

Figure 18.7   *Multi-jet cascade impactor: (a) the eight-stage Andersen Mk-II nonviable sampler; (b) smoothed experimental data for the top seven impactor stages (data from Mitchell et al., 1988, modeled by Kerr et al., 2001). Photograph courtesy of Thermo Electron Corporation, Waltham, MA, USA*

$_{50}d_{ae}$-values for all eight stages were determined to be 9.0, 5.8, 4.7, 3.3, 2.1, 1.1, 0.7 and 0.4 μm, respectively. Particles penetrating through the bottom stage are collected on a 81 mm diameter filter. It is seen from Figure 18.7(b) that the particle size range of this instrument is quite small, with no useful particle size-selectivity above about $d_{ae} = 12$ μm. In terms of the potential of this instrument to characterise aerosol fractions like those described by the CEN/ISO/ACGIH criteria, this is rather limiting. However the instrument has been successfully used for the sampling and characterisation of relatively fine aerosols.

Kerr *et al.* described a prototype modified version of the Andersen Mk-II sampler. It featured a new omnidirectional inlet aimed at providing more predictable aspiration efficiency characteristics. This was achieved by choosing the dimensions of the inlet such that, for the known flow rate of the instrument, and based on calm air theory like that outlined earlier in Chapter 6, aspiration may be expected to be close to unity for a wide range of particle sizes. Secondly it included a new top stage that featured a cylindrical plug of porous plastic foam filter media with properties chosen such that, at the 28.3 Lpm working flow rate, it provided a E versus d_{ae} curve that would provide information about particle deposition for much higher particle sizes than was possible with the original sampler. In the use of the modified Andersen, the contents of the entry containing the foam would be weighed or otherwise quantitated, along with the deposits on the impactor stages. Figure 18.8(a) shows a photograph of the new sampler with an inset showing the disassembled modified inlet and porous plastic foam top stage. The entry itself contained a horizontal circular cap, intended to prevent the unwanted settling of very large, nonairborne particles directly into the sampler. Figure 18.8(b) shows the curve of E versus d_{ae} for the new top stage alongside the curves – from Figure 18.7(b) – for the top seven impactor stages (the bottom stage with $_{50}d_{ae} = 0.4$ μm was omitted in this version). This shows that there was particle size-dependent collection efficiency for d_{ae} all the way out to greater than 50 μm. Thus, by comparison with Figure 18.7(b), this version provided clear potential for the retrieval of particle size information for much large particle sizes than for the original Andersen Mk-II sampler. However, it is important to note that the shape of the E versus d_{ae} curve for the new top stage was much less sharp than for the impactor stages. Indeed the curve in Figure 18.8(b) is seen to stretch all the way from 0.1 at about $d_{ae} = 10$ μm up to 0.9 at about $d_{ae} = 50$ μm. Although $_{50}d_{ae} \approx 35$ μm, it was not realistic to apply the

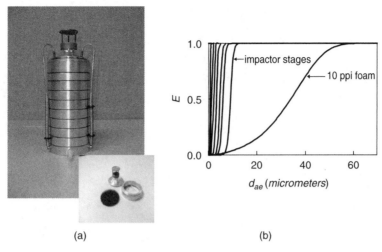

(a)　　　　　　　　　　　　　　(b)

Figure 18.8　*Modified Andersen Mk-II nonviable sampler: (a) photograph of the new instrument (new inlet shown disassembled in the inset); (b) smoothed experimental data for the top seven impactor stages (see Figure 18.7) and for the new top, porous foam-based collection stage (Kerr et al., 2001). Reproduced with permission from Kerr et al., Annals of Occupational Hygiene, 45, 555 – 568. Copyright (2001) British Occupational Hygiene Society*

simple graphical method described earlier for obtaining the particle size distribution, pointing instead to the need for an inversion procedure. The modified Andersen Mk-II sampler described here was used in a small set of aerosol-related field studies in occupational settings (e.g. Vincent *et al.*, 2001). But it has never appeared in a commercially available form.

The family of Andersen cascade impactor-based samplers has been prominent in aerosol characterisation for many years. Another family is based on the *micro-orifice uniform deposit impactor* (MOUDI), the principle of which was introduced earlier in Chapter 8. MOUDI precision cascade impactors are now available in multiple versions covering a very wide range of particle size-range options (MSP Corporation, Shoreview, MN, USA). They include a range of flow rates from 10 to 100 Lpm and diverse combinations of impactor stages with $_{50}d_{ae}$-values 18, 5.6, 3.2, 1.8, 1.0, 0.56, 0.32, 0.18, 0.1, 0.056, 0.032, 0.018 and 0.01μm. There are traditional eight-stage models and also models with rotating impaction plates in order to provide even more uniform particle deposits, also reducing problems associated with particle bounce or the evaporation of semi-volatile material [see Figure 18.9(a) and (b)]. A specialised new version has recently been introduced – the *Nano-MOUDI* – providing three cut points of 32, 18 and 10 nm, respectively [see Figure 18.9(c)].

The principle of low pressure impaction, also described in Chapter 8, has also been applied to the development of cascade impactors with enhanced capability at very small particle sizes. This was achieved, for example, by the placement of a high pressure-drop orifice plate at an appropriate location between high and lower stages of the cascade impactor. As early as 1963, McFarland and Zeller reported how, by adding three low-pressure stages to one of the Andersen samplers available at that time, the particle size range could be extended significantly at the lower end. Since then, several such samplers have appeared, including the *Berner impactor* (Berner *et al.*, 1979) and the *Hering low pressure impactor* (Hering *et al.*, 1979). One instrument, the *electrical low pressure impactor* (*ELPI*), was developed that allowed the determination of particle size distribution in close to real-time. In this, the particles were electrically charged as they were aspirated and were then detected at each impactor stage by highly

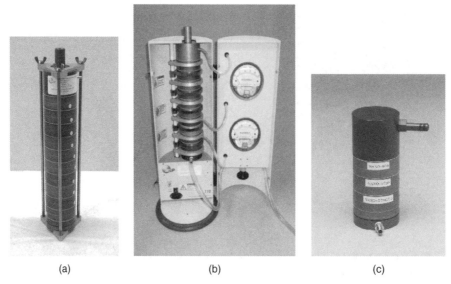

(a)   (b)   (c)

Figure 18.9   *A selection of versions of the micro-orifice uniform deposit impactor (MOUDI): (a) eight-stage nonrotating version; (b) eight-stage rotating version; (c) Nano-MOUDI (MSP Corporation, Shoreview, MN, USA). Photographs courtesy of Virgil A. Marple and Francisco Romay, University of Minnesota and MSP Corporation*

sensitive electrometers as they impacted onto the collection plate. In another real-time instrument, the collected masses of particulate material on the impactor stages were detected and measured by the application of quartz crystal microbalances built into the collection substrates (Fairchild and Wheat, 1984). Various versions of these – and others not mentioned – are still commercially available today.

Several static cascade impactor-based samplers were developed out of research conducted during the 1980s aimed at providing means to better evaluate workers' dust exposures in the European mining industries. One cascade impactor arising out of this body of work was the 10 Lpm *Institute of Occupational Medicine (IOM) static inhalable dust spectrometer* (SIDS).[1] This featured a slowly rotating single circular inlet designed, on the basis of what was learned from research like that leading to samplers described in earlier chapters for aspirating the CEN/ISO/ACGIH inhalable fraction (Mark *et al.*, 1984). Particulate matter collected in the entry stage – in front of the first impactor stage – was recovered and assessed for use in the determination of the particle size distribution of the aspirated, inhalable aerosol. For the impactor stages themselves, impaction at each took place from narrow rectangular slots onto slowly rotating drums. This feature was very similar to a cascade impactor first suggested by Lundgren in 1967. Here, as for the Lundgren sampler, by virtue of the synchronous rotation of the collecting surfaces for the various stages, and by analyzing the deposits at specific locations on the circumference of each drum, the particle size distribution of the aspirated aerosol could be determined for intervals within an overall sampling shift. In the IOM instrument, particle deposits were collected onto polycarbonate membrane films that had been stretched over the cylindrical collection drums, and were subsequently removed for analysis using an automated beta-attenuation mass balance procedure. The SIDS subsequently featured in some field research, but was never developed beyond the prototype stage and was never available commercially. Another instrument deriving out of the same overall body of European

---

[1] At the time it was referred to as the 'static inspirable dust spectrometer'.

coal industry-related work was the 40 Lpm *precision cascade impactor* (*PCI*, sometimes known as the 'Retsch impactor') described by Emmerichs and Armbruster (1981). The latter was interesting because it aspirated true total aerosol approximately isokinetically by virtue of the multiplicity of inlet nozzles, the number of which could be chosen in order to match mean entry velocity to the prevailing wind speed (that could be identified in most long-wall coalmining environments). This instrument had a wide range of particle size capability, with the $_{50}d_{ae}$ for top impactor stage being as high as 32 μm, higher than for most cascade impactors described in the literature. Since all the aspirated aerosol was assessed in the seven-stage cascade impactor of this instrument, the particle size distribution of both the inhalable fraction and other subfractions could be readily obtained numerically. Its high flow rate ensured that relatively large masses were collected at each impactor stage.

Quite recently, Marple and his colleagues at the University of Minnesota have produced a range of versatile new cascade impactors aimed at a wide range of aerosol sampling and aerosol characterisation applications in the pharmaceuticals industry, in particular for the testing of inhalers. It was introduced as the '*next generation pharmaceutical impactor*' (*NGI*) (e.g. Marple *et al.*, 2003), a seven-stage cascade impactor featuring collection cups that were held in a tray that was located horizontally in a user-friendly 'flat-pack' configuration in which each tray could be removed from the instrument easily and as a single unit. In order to facilitate drug recovery, each tray could contain up to 40 ml of an appropriate liquid. The NGI was designed to operate at various sampling flow rates from 30 to 100 Lpm, providing $_{50}d_{ae}$-values in the range from 0.54 to 11.7 μm for 30 Lpm and from 0.24 to 6.12 μm at 100 Lpm. The instrument is shown open in Figure 18.10(a) and closed in Figure 18.10(b), and is available commercially (MSP Corporation, Shoreview, MN, USA).

18.4.4 Personal cascade impactors

Cascade impactors like those described above are too large to be considered for use as personal samplers. However, in view of considerable interest in the characterisation of occupational aerosol exposure assessment, personal sampling instruments that can provide the same sort of information as that from those static aerosol spectrometers are of considerable interest. A small number of such samplers have appeared in response to this demand.

The 2 Lpm *IOM personal inhalable dust spectrometer* (*PIDS*) was one of the first such instruments (Gibson *et al.*, 1987). It was developed so that its performance would be consistent with the (then) emerging new particle size-selective sampling criteria. That is, the sampler should first aspirate the inhalable fraction and then provide information leading to the particle size distribution of that well-defined fraction. A prototype version of the PIDS is shown in Figure 18.11. To achieve the desired aspiration efficiency, the 15 mm circular, lipped entry was incorporated into the top stage of the sampler, similar to that of the IOM personal inhalable aerosol sampler described in Chapter 14. The cascade impactor part of the instrument was an eight-stage device plus backing filter. The stages were formed from circular aluminum disks of overall diameter 37 mm which were stacked one on top of the other. Thus each disk served both as an impaction jet plate and a collection plate (for the preceding stage) at the same time. At each stage, the impactor jets were made up of circular orifices with diameters chosen to provide eight $_{50}d_{ae}$-values in the range from about 1 to 20 μm, spaced closely enough to provide sufficient detail in the particle size distribution yet not so close that the E versus d_{ae} curves overlapped significantly. The disk of each impactor stage was lipped at its outer edge so that, when the system was assembled, the flow boundaries were defined entirely by the disks themselves. This meant that, by weighing each stage before and after sampling, material that was deposited other than in the immediate regions of the impactor jets was specifically included. Thus there was no particulate material anywhere in the system that was unaccounted for. In this respect, the device was similar to May's 'ultimate'

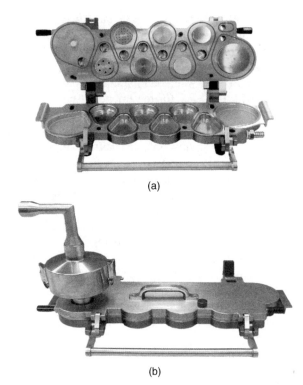

(a)

(b)

Figure 18.10   *The new generation pharmaceutical cascade impactor (NGI) designed for applications in the pharmaceuticals industry (Marple et al., 2003): (a) open to reveal the collecting trays; (b) closed ready for sampling, also showing the pre-selector which may be used with the instrument. Photographs courtesy of Virgil A. Marple and Francisco Romay, University of Minnesota and MSP Corporation*

cascade impactor. Degradation of performance due to particle bounce or blow-off was minimised by the application of grease to each stage prior to assembling the instrument. Based on calibration of the PIDS using monodisperse test aerosols, the $_{50}d_{ae}$-values for the eight impactor stages were 18.2, 14.4, 10.6, 6.0, 4.8, 3.3,1.7 and 0.9 μm, respectively. In addition, the E versus d_{ae} curve was obtained experimentally for the particle size-dependent deposition of particles in the entry stage (between the inlet and the first impactor stage), and this information was used – along with the impactor calibrations – in the final determination of the particle size distribution across the whole of the inhalable range up to as high as $d_{ae} = 100$ μm. The PIDS was used in a number of field-based occupational aerosol exposure assessment research studies and was commercially available briefly during the 1990s. It is no longer obtainable.

A more successful personal cascade impactor was the eight-stage sampler first described at about the same time by Marple and his colleagues at the University of Minnesota (Rubow *et al.*, 1987). Now widely known among occupational hygienists as the '*Marple personal cascade impactor*', it has been commercially available for several years (Thermo Electron Corporation, Waltham, MA, USA) and has featured in many reported studies of occupational aerosol exposures. The sampler is shown in Figure 18.12(a) where it is seen to have eight impactor stages, and an entry stage that incorporates

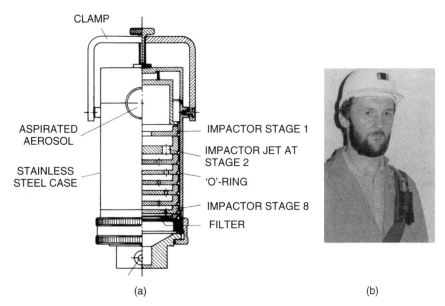

CLAMP

ASPIRATED
AEROSOL

STAINLESS
STEEL CASE

IMPACTOR STAGE 1

IMPACTOR JET AT
STAGE 2

'O'-RING

IMPACTOR STAGE 8

FILTER

(a)   (b)

Figure 18.11 *The prototype 2 Lpm personal inhalable dust spectrometer (PIDS): (a) diagram to show lay-out and main features; (b) photograph of the assembled instrument when worn (Gibson et al., 1987). Reproduced with permission from Gibson et al., Annals of Occupational Hygiene, 31, 463–479. Copyright (1987) British Occupational Hygiene Society*

a large-diameter inlet, often used with a shield to prevent the aspiration of unwanted very large particles. Particles collected at each stage are collected onto polycarbonate membrane or stainless steel impaction substrates that can be removed after sampling for separate analysis. The impactor stages were calibrated using monodisperse test aerosols, yielding $_{50}d_{ae}$-values of 21.3, 14.8, 9.8, 6.0, 3.5, 1.55, 0.93 and 0.52 μm, respectively. Fitted logistic curves – see Equation (18.1) – to the full calibration data of Rubow *et al.* are shown in Figure 18.12(b). Those original calibration data did not take account of internal wall losses and Rader *et al.* (1991) later re-calibrated the sampler in order to obtain correction factors for the deposits at the various stages. But for practical aerosol sampling these are relatively small and may usually be neglected. It is widely acknowledged that the Marple sampler is satisfactory for relatively fine aerosols. However, for the many workplaces where coarse aerosols are present, the limitations already identified above for the Andersen static sampler likewise apply to the Marple sampler. With this in mind, Wu (2005) described a version modified in a manner similar to that applied by Kerr *et al.* to the Andersen Mk-II sampler, featuring in particular a new entry system and top stage incorporating porous plastic foam filtration media.

A small number of other personal cascade impactors have been developed recently and are available commercially. One is the 2 Lpm personal sampling version of the MOUDI (MSP Corporation, Shoreview, MN, USA). Another is the 9 Lpm *Sioutas personal cascade impactor sampler (PCIS)*. First described by Misra *et al.* (2002), this features four impaction stages with $_{50}d_{ae}$-values at 2.5, 1.0, 0.50 and 0.25 μm, respectively. Its performance is described as separating out 'coarse, fine and ultrafine' particles, providing aerosol fractions nominally in the ranges 10 to 2.5, 2.5 to 1.0, 1.0 to 0.50, 0.50 to 0.25 and less than 0.25 μm, respectively. Thus portrayed, this sampler is most appropriate for the sampling

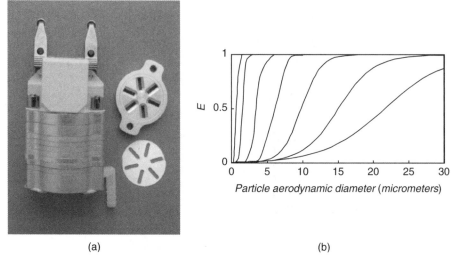

(a)   (b)

Figure 18.12   *2 Lpm personal cascade ('Marple') cascade impactor (Rubow et al., 1987): (a) photograph of the sampler, also showing one of the impactor stages and a corresponding impactor substrate; (b) fitted curves to the experimental data for the collection efficiencies of the top seven stages of the sampler (bottom stage curve not shown). Photograph courtesy of Thermo Electron Corporation, Waltham, MA, USA*

of health-related fractions in ambient aerosol. Although it has a relatively small number of impactor stages as compared with the other cascade impactors mentioned so far, it is still possible in principle to use this sampler to provide actual particle size distributions. The version shown in Figure 18.13 is available commercially (SKC Inc., Eighty Four, PA, USA).

18.4.5   Cascade impactors for stack sampling

Cascade impactors have been used extensively for many years for the sampling of aerosols from stacks and ducts, where one important rationale is to assess the particle size-selective performance of particulate air pollution control equipment. Pilat and his colleagues at the University of Washington developed a number of such instruments, beginning in the late 1960s (Pilat *et al.*, 1970) with what became known as the *Pilat source test cascade impactor Mark I*, followed successively by new versions, including more recently the seven-stage *Mark III* and the eleven-stage *Mark V*. For each of these samplers, the thin-walled inlet nozzle provides the first impaction stage. The inlet nozzle is fixed, and so sampling is arranged to be isokinetic by varying the sampling flow rate in the range from 2.8 to 28 Lpm. The aspirated stack gas then passes immediately to the first of a series of axially located impactor stages, the first comprising a single circular jet and subsequent ones containing multiple jets. The sampling arrangement for this sampler is shown schematically in Figure 18.14(a) and the Mark III is shown disassembled in Figure 18.14(b). The Mark II and Mark V are both currently available commercially (Pollution Control Systems Corporation, Seattle, WA, USA). These instruments are unusual in that, in their use, it is necessary to determine the $_{50}d_{ae}$-values for the cascade impactor stages for each given sampling flow rate. The manufacturer provides data reduction software that performs the necessary calculations.

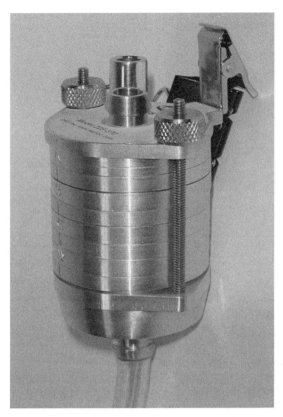

Figure 18.13 *The 9 Lpm Sioutas personal cascade impactor. Photograph courtesy of SKC Inc., Eighty Four, PA, USA*

18.4.6 Inversion procedures for cascade impactors

The cascade impactor performance ambiguities associated with 'imperfect' stage collection alluded to earlier suggest the need for mathematical models capable of generating acceptable *continuous* particle size distribution information from the small number of discrete values of the particulate amounts – usually masses – collected on the impactor stages. It seems sensible that such models should combine data for those masses with knowledge of the actual shapes of the individual E versus d_{ae} curves. Whilst it is easy to see that, by starting off with a known particle size distribution and known deposition curves, the mass of material collected on each stage may be calculated numerically, the problem is less straightforward in reverse. Fuchs (1978) observed that '... by no device can more information ... be squeezed out of a cascade impactor than that given by one histogram ...' and that '... the experimental errors in calibrating cascade impactors make at present illusory all methods of processing the experimental data obtained ... based on the use of (stage) characteristics'. The first is a clear statement that the problem of retrieval of the desired particle size distribution is mathematically 'ill-posed'. But the second seems to argue against placing too much faith in the prospect of developing workable retrieval methods that are any more accurate than the simple graphical one described earlier. This view is now generally regarded as being overly pessimistic.

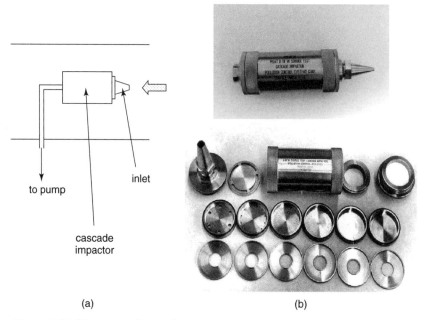

(a)   (b)

Figure 18.14   *The variable flow rate Pilat Mark III source test cascade impactor: (a) schematic showing mode of operation; (b) sampler shown assembled and disassembled. Photographs courtesy of John W. Paul, Pollution Control Systems Corporation, Seattle, WA, USA*

Several early attempts were reported in the 1960s and 1970s to develop mathematical inversion procedures for cascade impactors (e.g. Sundelof, 1967; Cooper and Davis, 1972; Picknett, 1972; Cooper and Spielman, 1976; and others). The ill-posed problem described here is not confined just to cascade impactors. It is also relevant not only elsewhere in aerosol science, including other systems for measuring particle size distributions, but also to many aspects of science further afield. So there has been a great deal of interest in developing reliable and optimal mathematical solutions, and Twomey (1977) and Tikhonov and Arsenin (1977) provided key milestones on the path towards acceptable inversion procedures. In 1999, Kandlikar and Ramachandran published a useful review of the options that might be applicable to aerosol spectrometer measurements. A short summary suffices for present purposes. In general, the fraction of particles entering the instrument that is collected on the ith stage may be described as the kernel function, K_i which for the ith stage is given by:

$$K_i(d_{ae}) = E_i(d_{ae})[1 - E_{i-1}(d_{ae})][1 - E_{i-2}(d_{ae})] \cdots [1 - E_1(d_{ae})] \qquad (18.2)$$

The mass collected on the ith stage, m_i, is now given by:

$$m_i = \int_a^b K_i(d_{ae}) \cdot f(d_{ae}) \cdot \mathrm{d}d_{ae} + \varepsilon_i \qquad (18.3)$$

where $f(d_{ae})$ is the frequency distribution of mass for the aspirated aerosol, and ε_i is the measurement error for the ith stage, while a and b are the limits within which the particle size distribution lies. Tikhonov and Arsenin proposed a 'regularisation' procedure in which the intrinsic ill-posedness of a particular problem, such as the analysis of a set of cascade impactor data, is overcome by substituting a nearby 'well-posed' problem whose solution may be easily obtained and approximates reasonably well to what is required. In their model, the problem to be solved is contained in:

$$\sum_{i=1}^{n} \left[\frac{m_i - \int_a^b K_i(d_{ae}) \cdot f(d_{ae}) \cdot dd_{ae}}{\varepsilon_i} \right] = R(\lambda) + \lambda \cdot J(\lambda) \tag{18.4}$$

where λ is a regularization parameter. Here, the term R defines the residual reflecting the agreement between the solution and the measured data and J reflects the smoothness of the solution. Both R and J are functions of λ. The solution remains complex, in particular in deciding how to deal with J. However, as noted by Lekhtmakher and Shapiro (2000) in a response to the Kandlikar and Ramachandran paper, the assumption of an a priori functional form for the particle size distribution means that the parameters for the distribution may be determined by straightforward minimisation of the residuals between the measured and estimated collected masses on the impactor stages, thus eliminating specific consideration of J. The observation that particle size distributions in many environments are log-normal – or are of combinations of relatively small numbers of separate log-normal distributions – means that this simpler approach may be appropriate in many applications. For a single log-normal distribution, the parameters to be fitted are simply the median and geometric standard deviation. For a more complex particle size distribution that may be described in terms of a larger number of individual such distributions then, in addition to the median and geometric standard deviation for each component, the weighting factors for each of them also need to be determined. Such a large number of coefficients to be fitted in the minimisation analysis significantly extends the computing power required for the analysis. But this approach is entirely accessible with modern spreadsheets.

18.5 Other spectrometers

18.5.1 Parallel impactors

Cascade impactor-based samplers have been dominant for many years in the determination of particle size distributions in workplace and environmental aerosols. They are basically quite simple and the science of impactors is well understood. However, some interesting alternative methods have been proposed over the years. One is the *parallel impactor* system, where the aspirated aerosol is directed to a number of impactors placed in parallel to one another, each one separating out a different particle size fraction. Samplers based on this concept have been used in both the USA and Europe to determine the concentrations and size distributions of atmospheric suspended particles with d_{ae} up to as large as 200 μm. One such sampler was the *wide-ranging aerosol classifier* (WRAC) developed by Lundgren and his colleagues (e.g. Burton and Lundgren, 1987). This was a very large four-stage device, mounted on a trailer that could be readily transported from one outdoor sampling location to another. It is shown in Figure 18.15. Air was aspirated into this sampler at the very high flow rate of 40 000 Lpm through an upwards-facing circular opening shrouded by an outer cylindrical shield in order to reduce the effects of cross-winds. In this way, it was argued that aspiration efficiency for sampling the ambient air would be close to unity for most of the particle size range of interest. The aerosol thus aspirated was delivered – inside the instrument – to four parallel, upwards-facing secondary entries, each leading to a single-stage impactor where classification according to particle aerodynamic diameter took place. The design of each impactor entry was arranged so that sampling from the main aspirated stream inside the shroud was isokinetic at the individual impactor flow rate of 1600 Lpm. Each impactor took the form

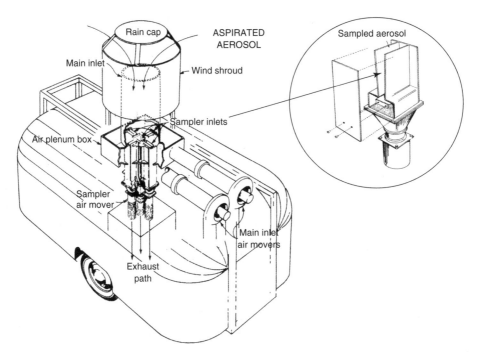

Figure 18.15   *The wide-ranging aerosol classifier (WRAC), a very high flow rate device based on parallel impactors (Burton and Lundgren, 1987). Reprinted from Aerosol Science and Technology, 6, Burton and Lundgren, 289–301. Copyright (1987), with permission from Elsevier*

of a rectangular jet directed onto a flat plate, suitably greased in order to reduce losses due to particle bounce. By suitable choice of impactor dimensions and flow rate, $_{50}d_{ae}$-values of about 9, 19, 34 and 47 μm were achieved in the practical version of the instrument described by Burton and Lundgren. In addition to the four impactors, a fifth sample was extracted isokinetically from the main aspirated air, this time directly onto a filter, to provide a measure of true total aerosol. In the practical use of the instrument, the desired particle size distribution was obtained in terms of a histogram derived from the masses collected at each impactor stage. The range of particle size encompassed by the WRAC was very large for the very high sampling flow rate indicated. However, compared with most of the other aerosol spectrometers referred to above, particle size resolution was limited. The WRAC was a very interesting variant on the impactor-based aerosol spectrometer idea, but has not been widely used since its initial appearance and is not commercially available today. A *Mini-WRAC* was built at the Fraunhofer Institute for Toxicology and Aerosol Research in Hannover, Germany, and featured in some European trials to identify samplers for health-related aerosol fractions in the ambient atmosphere. However, this sampler featured cascade – rather than parallel – impactors (Dr Werner Holländer, personal communication).

18.5.2   Cascade cyclones

A *cascade cyclone* has been developed and is sometimes used for the assessment of particle size distributions of stack aerosol emissions. The overall principle is similar to that for the cascade impactor, but now aerosol at each stage is separated on the basis of particle aerodynamic diameter by a cyclone and collected on a filter. Here, however, the sharpness-of-cut for the separate cyclone stages is much less than for impactors. But this sampler has the advantage of being able to collect large amounts of particulate material, and so is capable of operation for long periods in higher concentration aerosol streams.

18.5.3 Diffusion batteries

The diffusion battery was first suggested in 1900 by J.S.E. Townsend for the purpose of measuring the distributions of the diffusion coefficients of ions in gases (Knudson, 1999). However its application to aerosols came much later. Chapter 8 contained descriptions of the mechanisms by which fine particles are deposited by diffusion from an aerosol that is passed through a narrow capillary tube or a fine wire-mesh screen. It was shown how collection efficiency is a function of the geometrical properties of the tube or screen, the flow rate and particle size, expressed in terms of equivalent volume diameter, d_V. The earlier Figure 8.16 showed how the penetration of particles through a single screen (P) increased steadily as d_V increased. In general, for a given screen and flow rate, a curve of collection efficiency ($E = 1 - P$) versus d_V may be defined, along with a $_{50}d_V$-value. By analogy with the cascade impactor, the clear opportunity exists for an instrument based on a cascaded system of capillaries or screens, with stages providing different values $_{50}d_V$, by which to obtain information leading to the particle size distribution, this time in the sub-micrometer particle size range. Such an instrument – a *diffusion battery* – would therefore have particularly useful applications in the characterisation of ultrafine aerosols.

Practical sampling instruments of both capillary and screen types have been developed. Screen-type diffusion batteries, first described by Sinclair and Hoopes (1975), are especially compact and convenient in use. Scheibel and Porstendörfer (1984) carried out experiments to determine P as a function of d_V for combinations of individual screens stacked together. Figure 18.16 shows curves of E versus d_V estimated from their experimental data for P for combinations of screens stacked together in numbers ranging from one to 64. Such combinations are seen to provide a wide range of $_{50}d_V$-values. The 4 Lpm 10-stage *Model 3040 diffusion battery* (TSI Inc., St Paul, MN, USA) operates on this principle and was derived from the original prototype instrument described by Sinclair and Hoopes. As shown in Figure 18.17, it has 10 stages, the first containing one screen such that particles passing through it have encountered just one screen, the second containing two screens such that particles passing through

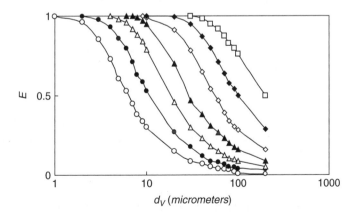

Figure 18.16 *Collection efficiency (E) as a function of particle equivalent volume diameter (d_V) for stacks of screens as indicated (based on experimental data from Scheibel and Porstendörfer, 1984: ○ 1 screen, ● 2 screens, △ 4 screens, ▲ 8 screens, ◇ 16 screens, ◆ 32 screens, □ 64 screens)*

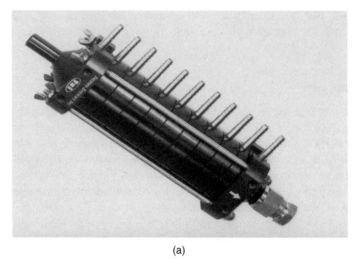

(a)

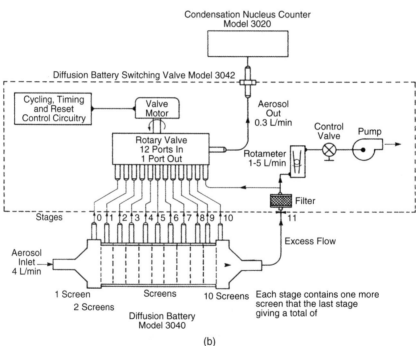

(b)

Figure 18.17 *Typical 10-stage diffusion battery: (a) photograph of the instrument; (b) schematic diagram showing the layout of the instrument in use. Pictures courtesy of Gilmore Sem, TSI Inc., St Paul, MN, USA*

it have encountered three screens, and so on until the particles passing through the tenth screen have encountered all 55 screens. The $_{50}d_V$-values for penetration through Stage 1 is 8.6 nm, through Stage 2 is 19.8 nm, and successively 34.0, 51.6, 72.0, 96.0, 123, 156, 192 and 234 nm, respectively, for succeeding stage penetrations. An 11-stage version is also available (Model 3041), containing a modified screen arrangement yielding improved resolution at smaller particle sizes. In the use of both versions of this sampler, a condensation particle counter (CPC, see Chapter 20), that can accurately detect and count particles in the size range of interest, is first connected to the sampling port for Stage 1 and a reading taken of particles generally less than 8.6 nm, after which the CPC is connected to the port at Stage 2 providing counts for particles generally less than 34 nm; then on to Stage 3, and beyond.

The numbers of particles passing beyond each stage provide basic information that may be used to construct the particle size distribution over the range of interest. This may be done graphically in much the same as described earlier cascade impactor. However, inspection of Figure 18.16 reveals that the E versus d_V curves for diffusion battery screen combinations are not sharp – indeed, significantly less so than for inertial impactors. So the problem of particle size distribution recovery from diffusion battery data like those indicated is even more 'ill-posed' than for the cascade impactor. This too therefore indicates the need for appropriate data inversion methodology. Similar procedures to those outlined for cascade impactors have been developed specifically for diffusion batteries (e.g. Lesnic *et al.*, 1995).

18.6   Particle size distribution analysis by microscopy

The earliest methods for assessing particle size involved collecting particles onto a substrate or a filter that could be prepared for visualisation under an optical microscope. Such methods are still used occasionally today, although visualisation methods have been extended in recent years to include also electron microscope techniques. One advantage of this approach to particle size analysis is that it also allows inspection of particle features such as shape and – in the case of electron microscopy – composition. However, the particle size obtained in this way is expressed in terms of particle geometrical size, so that it needs adjustment in order to make it applicable to the sampling conventions outlined in Chapter 11. Since most particles found in aerosols in working and ambient environments are irregular in shape, this requires assumptions about the relationship between geometrical and equivalent spherical diameter and about particle density. Since particles observed under an optical microscope are seen in two dimensions, the counting of particles on-filter is usually obtained in the first instance in terms of the equivalent projected area diameter (d_{PA}).

Aerosol sampling for particle size analysis by microscopy generally involves the collection of particles directly onto a filter or substrate that allows deposition in a way such that the particles may be visible, sometimes, as in the case of some filters, after preparation to enhance visibility. Alternatively, collected samples may be suspended into appropriate liquid media and re-deposited onto a preferred substrate in order to achieve – by appropriate dilution – manageable on-filter concentration. The latter is useful if the original deposit is too dense for individual particles to be observed and discriminated. Once the sample has undergone the desired preparation, it is then ready for microscopy. In optical microscopy, particle counting and sizing takes place with the aid of a suitable graticule containing graduated markings against which individual particles can be compared and hence their size estimated. The most common is the *Porton graticule* first described by May (1965). It is available in several versions available from several sources (e.g. Electron Microscopy Sciences, Hatfield, PA, USA; Canemco Inc. & Marivac Inc., Lakefield, Quebec, Canada). One known as the '*New Porton NG-12*' is shown in Figure 18.18. The graticule is a glass disk on which are etched markings like those shown, including clear and dark circles with diameters increasing progressively by a factor of $(2)^{1/2}$ (i.e. doubling of their areas) and vertical straight lines that provide corresponding linear dimensions (e.g. such that the distance between the reference line and the line marked '4' is the same as the diameter of circle #4). In use, the graticule is calibrated for a given microscope setting against a stage micrometer. Each individual particle observed in the microscope field of view may be compared with the circles or linear distances, and thus be assigned the most appropriate diameter on the basis of projected area or some linear dimension. By the examination

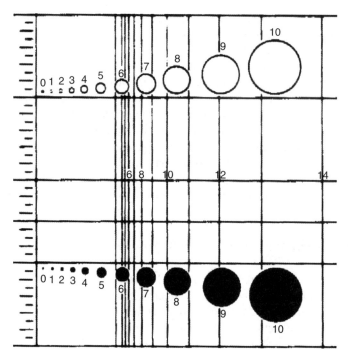

Figure 18.18   *Porton graticule for the counting and sizing of particles collected onto a substrate or filter. The version shown is known as the 'New Porton NG-12' (e.g. Electron Microscopy Sciences, Hatfield, PA, USA; Canemco Inc. & Marivac Inc., Lakefield, Quebec, Canada). Reproduced with permission from May, Journal of Scientific Instruments, 42, 500–501. Copyright (1965) The Institute of Physics*

of an ensemble of particles, individual particles may be placed into size classifications. Multiple fields of view may be examined until sufficient particles have been counted and classified to identify the desired particle size distribution. Since most environmental aerosols are distributed log-normally, or are otherwise skewed, the procedure as outlined will involve the counting of very large numbers of small particles and smaller numbers of larger ones. In order to reduce the effort and obtain statistically viable particle counts across the whole range of particle sizes, it is commonplace to conduct *stratified counting* (e.g. see Hinds, 1999). Here, the counting of the smaller particles is stopped when enough particles have been observed, while the counting of large ones continues. The results are balanced by normalising the results in terms of counts per field.

References

Andersen, A. (1958) New sampler for the collection, sizing and enumeration of viable airborne particles, *Journal of Bacteriology*, 76, 471–484.

Berner, A., Lurzer, C.H., Pohl, F., Preining, O. and Wagner, P. (1979) The size distributions of urban aerosols in Vienna, *Science of the Total Environment*, 13, 245–261.

Burton, R.M. and Lundgren, D.A. (1987) Wide range aerosol classifier: a size selective sampler for large particles, *Aerosol Science and Technology*, 6, 289–301.

Cohen, B.S. (1986) Introduction: the first 40 years. In: *Cascade Impactor: Sampling and Data Analysis* (Eds J.P. Lodge and T.L. Chan), American Industrial Hygiene Association, Akron, OH, pp. 1–22.

Cooper, D.W. and Davis, J.W. (1972) Cascade impactors for aerosols: improved data analysis, *American Industrial Hygiene Association Journal*, 33, 79–89.

Cooper, D.W. and Spielman, L.A. (1976) Data inversion using non-linear programming with physical constraints: aerosol size distribution measurement by impactor, *Atmospheric Environment*, 10, 723–729.

Emmerichs, M. and Armbruster, L. (1981) Improvement of a multi-stage impactor for determining the particle size distribution of airborne dusts, *Silikosebericht Nordrhein-Westfalen*, 13, 111–115.

Fairchild, C.I. and Wheat, L.D. (1984) Calibration and evaluation of a real-time cascade impactor, *American Industrial Hygiene Association Journal*, 45, 205–211.

Fuchs, N.A. (1978) Aerosol impactors: a review. In: *Fundamentals of Aerosol Science* (Ed. D.T. Shaw), John Wiley & Sons, Ltd, New York, pp. 1–83.

Gibson, H., Vincent, J.H. and Mark, D. (1987) A personal inspirable aerosol spectrometer for applications in occupational hygiene research, *Annals of Occupational Hygiene*, 31, 463–479.

Hering, S.V., Friedlander, S.K., Collins, J.J. and Richards, L.W. (1979) Design and evaluation of new low-pressure impactor, *Environmental Science and Technology*, 13, 184–188.

Hinds, W.C. (1999) *Aerosol Technology*, 2nd Edn, John Wiley & Sons, Inc, New York.

Hounam, R.F. and Sherwood, R.J. (1965) The cascade centipeter: a device for determining the concentration and size distribution of aerosols, *American Industrial Hygiene Association Journal*, 26, 122–131.

Kandlikar, M. and Ramachandran, G. (1999) Inverse methods for analyzing aerosol spectrometer measurements: a critical review, *Journal of Aerosol Science*, 30, 413–437.

Kerr, S.M., Vincent, J.H. and Ramachandran, G. (2001) A new approach to sampling for particle size distribution and chemical species 'fingerprinting' of workplace aerosols, *Annals of Occupational Hygiene*, 45, 555–568.

Knudson, E.O. (1999) History of diffusion batteries in aerosol measurements, *Aerosol Science and Technology*, 31, 83–128.

Kotrappa, P. and Light, M.E. (1972) Design and performance of the Lovelace Aerosol Particle Separator, *Review of Scientific Instruments*, 43, 1106–1112.

Laskin, S. (1949) Measurement of particle size. In: *Pharmacology and Toxicology of Uranium Compounds* (Eds C. Voegtlin and H.C. Hodge), McGraw-Hill, New York, pp. 463–505.

Lekhtmakher, S. and Shapiro, M. (2000) On the paper 'Inverse methods for analyzing aerosol spectrometer measurements: a critical review', *Journal of Aerosol Science*, 31, 867–873.

Lesnic, D., Elliott, L. and Ingham, D.B. (1995) An inversion method for the determination of the particle size distribution from diffusion battery measurements, *Journal of Aerosol Science*, 26, 797–812.

Lundgren, D.A. (1967) An aerosol sampler for determination of particle concentration as a function of size and time, *Journal of the Air Pollution Control Association*, 17, 225–229.

Mark, D., Vincent, J.H., Gibson, H., Aitken, R.J. and Lynch G. (1984) The development of an inhalable dust spectrometer, *Annals of Occupational Hygiene*, 28, 125–143.

Marple, V.A. (1970) A fundamental study of inertial impactors, PhD Thesis, University of Minnesota, Minneapolis, MN.

Marple, V.A. (2004) History of impactors – the first 110 years, *Aerosol Science and Technology*, 38, 247–292.

Marple, V.A., Roberts, D.L., Romay, F.J., Miller, N.C., Truman, K.G., Van Oort, M., Olsson, B., Holroyd, M., Mitchell, J.P. and Hochrainer, D. (2003) Next generation pharmaceutical impactor (a new impactor for pharmaceutical inhaler testing) – Part 1: Design, *Journal of Aerosols in Medicine*, 16, 283–299.

May, K.R. (1945) The cascade impactor: an instrument for sampling coarse aerosols, *Journal of Scientific Instruments*, 22, 187–195.

May, K.R. (1965) A new graticule for particle counting and sizing, *Journal of Scientific Instruments*, 42, 500–501.

May, K.R. (1975) An 'ultimate' cascade impactor for aerosol assessment, *Journal of Aerosol Science*, 6, 413–419.

May, K.R. (1982) A personal note on the history of the cascade impactor, *Journal of Aerosol Science*, 13, 37–47.

McFarland, A.R. and Zeller, H.W. (1963) Study of a large volume impactor for high-altitude aerosol collection, United States Atomic Energy Commission Report No. TID-18624, Washington, DC.

Misra, C., Singh, M., Shen, S., Sioutas, C. and Hall, P.M. (2002) Development and evaluation of a personal cascade impactor sampler (PCIS), *Journal of Aerosol Science*, 33, 1027–1047.

Mitchell, J.P., Costa, P.A. and Waters, S. (1988) An assessment of an Andersen Mark-II cascade impactor, *Journal of Aerosol Science*, 19, 213–221.

Picknett, R.G. (1972) A new method of determining aerosol size distributions from multistage sampler data, *Journal of Aerosol Science*, 3, 185–198.

Pilat, M.J., Ensor, D.S. and Bosch, J.C. (1970) Source test cascade impactor, *Atmospheric Environment*, 4, 671–679.

Prodi, V., Belosi, F. and Mularoni, A. (1986) A personal sampler following ISO recommendations on particle size definitions, *Journal of Aerosol Science*, 17, 576–581.

Prodi, V., Melandri, C., Tarroni, G., De Zaiacomo, T., Formignani, M. and Hochrainer, D. (1979) An inertial spectrometer for aerosol particles, *Journal of Aerosol Science*, 10, 411–419.

Puttock, J.S. (1981) Data inversion for cascade impactors: fitting sums of lognormal distributions, *Atmospheric Environment*, 15, 1709–1716.

Rader, D.J., Mondy, L.A., Brockmann, J.E., Lucero, D.A. and Rubow, K.L. (1991) Stage response calibration of the Mark III and Marple personal cascade impactors, *Aerosol Science and Technology*, 14, 365–379.

Rubow, K.L., Marple, V.A., Olin, J. and McCawley, M.A. (1987) A personal cascade impactor – design, evaluation and calibration, *American Industrial Hygiene Association Journal*, 48, 532–538.

Sawyer, K.F. and Walton, W.H. (1950) The 'Conifuge' – a size-separating device for airborne particles, *Journal of Scientific Instruments*, 27, 272–276.

Scheibel, H.G. and Porstendörfer, J. (1984) Penetration measurements for tube and screen-type diffusion batteries in the ultrafine particle size range, *Journal of Aerosol Science*, 15, 673–682.

Sinclair, D. and Hoopes, G.S. (1975) A novel form of diffusion battery, *American Industrial Hygiene Association Journal*, 36, 39–42.

Stöber, W. (1976) Design, performance and applications of spiral duct aerosol centrifuges, In: *Fine Particles* (Ed. B.Y.H. Liu), Academic Press, New York, pp. 352–397.

Stöber, W. and Flaschbart, H. (1969) Size-separating precipitation of aerosols in a spinning spiral duct, *Environmental Science and Technology*, 3, 1280–1296.

Sundelof, L. (1967) On the accurate calculation of particle size distributions in aerosols from impactor data, *Staub Reinhaltung des Luft* (English translation), 27, 22–28.

Tikhonov, A.N. and Arsenin, V.Y. (1977) *Solutions of Ill-posed Problems*, John Wiley & Sons, Ltd, New York.

Timbrell, V. (1972) An aerosol spectrometer and its applications, In *Assessment of Airborne Particles: Fundamentals, Applications and Implications to Inhalation Toxicology* (Eds T.T. Mercer, P.E. Morrow and W. Stöber), Charles C. Thomas, Springfield, IL, pp. 290–330.

Twomey, S. (1977) *Introduction to the Mathematics of Inversion in Remote Sensing and Indirect Measurements*, Elsevier, Amsterdam.

Vincent, J.H. (1989) *Aerosol Sampling: Science and Practice*, John Wiley & Sons, Ltd, Chichester.

Vincent, J.H., Ramachandran, G. and Kerr, S.M. (2001) 'Fingerprinting' of aerosol exposures of nickel industry workers by particle size and chemical species distributions, *Journal of Environmental Monitoring*, 3, 565–574.

Wu, Y.H. (2005) Application of particle size distribution measurement to the characterization of workplace aerosols, PhD Dissertation, University of Michigan, Ann Arbor, MI.

19

Sampling for Bioaerosols

19.1 Introduction

Aerobiology defines the scientific discipline that addresses the transport of organisms and biologically significant materials through the atmosphere, including the source of organisms or materials, release into the atmosphere, dispersion and deposition, and their impact on animal, plant or human systems (e.g. Spieksma, 1991). One of the traditional practical aims of aerobiology has been to provide measurements of airborne pollen concentrations as a service to allergy sufferers (Larsson, 1993). The study of *bioaerosols* is a very distinct branch of aerosol science, embracing all aspects of airborne particles of biological origin which may affect living organisms through not only allergenicity but also infectivity and related toxicology. Bioaerosol particles are ubiquitous in all environments and are extremely diverse, including viruses, bacteria, proteins, fungal spores, pollens, and plant or animal detritus, as well as fragments and products derived from them. They are particularly prevalent in certain specific environmental and occupational settings, including for example those involving – or proximal to – waste recycling and composting, agriculture, food processing, etc. Particle size for bioaerosols ranges typically from less than 1 μm up to greater than 100 μm. Although some individual bioaerosol particles may be very small indeed – for example, viruses, protein and microbial products such as endotoxins – they are rarely found as individual separate particles but rather as components of larger organisms or contained within liquid droplets. These properties, as well as descriptions of sampling for bioaerosols, are reviewed in detail in the volumes edited by Cox and Wathes (1995) and Macher (1999).

Interest in bioaerosols has risen sharply in recent years as their relevance to occupational and environmental health has become better understood. Their general importance in infectious disease, acute toxic effects, allergies and – even – cancer is now well recognised, and their specific importance in relation to respiratory symptoms and lung function impairment has become a particular public health priority (Douwes *et al.*, 2003). As a result, the study of bioaerosols has been brought fully into the mainstream of aerosol science, in particular the science and practice of aerosol sampling and other aspects of measurement. This growing interest is reflected, for example, in the special issues of the *Journal of Aerosol Science* that appeared in 1994, 1997 and 2005, and of *Aerosol Science and Technology* in 1999, as well as the books already mentioned.

Aerosol Sampling: Science, Standards, Instrumentation and Applications   James Vincent
© 2007 John Wiley & Sons, Ltd

19.2   Standards for bioaerosols

In earlier chapters, the currently accepted scientific framework for standards and limit values was presented. This is now seen as applicable to most types of aerosol. The American Conference of Governmental Industrial Hygienists (ACGIH) has quite recently considered how it might be applied to bioaerosols (Macher, 1999). It was concluded that insufficient information is available about any of the five components – that is, the criteria (or scientific basis), sampling methods, analytical methods, sampling strategy and toxicological dose–response information – that are needed to define a standard. So ACGIH has not yet been able to suggest threshold limit values (TLVs), except for the few substances of biological origin that have been treated on an individual basis, such as wood, cotton and grain dusts, etc. Beyond ACHIH, there are few, if any, bioaerosol standards elsewhere. In turn, therefore, although there have been exposure assessments reported for certain environments and for certain types of bioaerosols for the purposes of research, there has been no compliance monitoring. As a result, there is a general paucity of data on human exposures to bioaerosols that would be needed upon which to base the development of scientifically based standards. In addition, the search for sampling methods consistent with the scientific rationale for exposure assessment has been fraught with many technical difficulties, some of which still remain unresolved. That said, however, the increase in interest in bioaerosols as an important class of environmental risk factor has stimulated the new generation of research that is being carried out to fill the remaining knowledge gaps that currently exist.

19.3   Technical issues for bioaerosol sampling

The nature and diversity of bioaerosols underscore the magnitude of the challenge that must be met in order to develop a scientific rationale for both standards and standards. *Viable* organisms are metabolically active ('living') and may be either *culturable* or *nonculturable*. Culturable organisms may reproduce, and hence grow, when subjected to appropriate environmental conditions (e.g. temperature, relative humidity) and in the presence of appropriate nutrients. Nonculturable organisms are less conducive to such reproduction or growth. By contrast, *nonviable* organisms are not metabolically active, and are so are not capable of reproduction. In turn, they are not culturable. For the purpose of sampling and measurement, a broad classification defines:

- *Culturable biological agents*, the viable bacteria and fungi in a sample that can be grown in a laboratory culture, and hence be amplified and made visible, so that their number – and ultimately their original airborne concentration – may be expressed in terms of the number of *colony-forming units* (CFU) (see the example in Figure 19.1).
- *Countable biological agents*, the pollens, grains, spores, bacteria, etc. – viable or nonviable – that can be identified and counted when viewed under the optical microscope (see the examples in Figure 19.2).

This simple breakdown provides the first guidance for the sampling and methodology for practical bioaerosol sampling. In the first, culturable organisms should be collected onto a substrate containing a medium that enables the organism to survive, and permits subsequent culture when the appropriate environmental conditions are applied (i.e. temperature and relative humidity during incubation). A range of collection media is available, depending on specific requirements. These include *general and enriched media* that allow the culture of a wide spectrum of biological organisms, and *differential media* that contain specific chemical components in order to permit the selective culture of specific organisms. Agar is widely used, supplemented as required by appropriate buffers and/or nutrients and other ingredients.

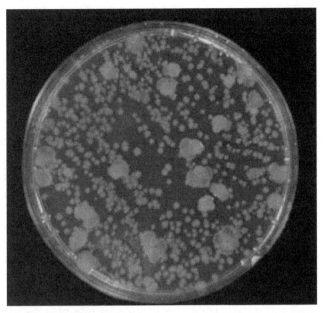

Figure 19.1 *Example of a cultured bioaerosol sample, showing colonies formed from individual viable bioaerosol particles deposited on an agar-coated Petri dish*

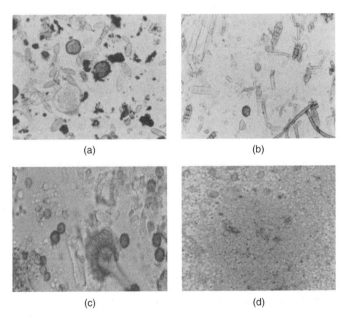

Figure 19.2 *Examples of noncultured bioaerosol sample as viewed under a microscope: (a) typical summer dry daytime air spores; (b) typical grain dust spores; (c) typical coarse moldy hay spores from top stage of a cascade impactor; (d) typical finer moldy hay spores from second stage of a cascade impactor. Photographs courtesy of the late John Lacey, Reprinted from Aerosol Science for Industrial Hygienists, Vincent. Copyright (1995), with permission from Elsevier*

It is a mixture of polysaccharides derived from red algae that forms a gel at temperatures below about 40°C. Alternatively, for some modes of sampling, the organisms may be collected in an appropriate liquid medium. These must be compatible with the organisms being sampled, with the requirement being that the bioaerosols collected in this way are unable to multiply. Typically they may include distilled water or buffered salt solutions, gelatin-phosphate solution, tryptose-saline solution, 5 % (by volume) skimmed milk in distilled water (Hensel and Petzoldt, 1995). By contrast, nonculturable organisms may be collected onto an inert surface, such as a glass slide, greased surface or a flat filter, which may subsequently be observed by light microscopy so that the collected organisms can be identified and counted. For bioaerosol sampling in general, Willeke and Macher (1999) identified three indices of performance:

- *Inlet sampling efficiency*, the effectiveness with which particles are aspirated from the atmospheric environment of interest without significant biases (of the types extensively discussed elsewhere in this book).
- *Particle removal efficiency*, the effectiveness with which particles, once aspirated, are captured at the filter or substrate of interest (also as discussed elsewhere).
- *Biological recovery efficiency*, the effectiveness by which the collected particles may be conveyed to the assay system for quantitative analysis without changing their viability, biological activity or physical integrity.

It is the latter which distinguishes the sampling of bioaerosols from that for most other aerosol types. Some organisms are less resilient than others, and may be more or less damaged or degraded during the process of sampling itself. The desiccation of biological material deposited on a conventional filter or substrate as air continues to pass during sampling is a primary cause of sample degradation. The degree of degradation varies greatly from one bioaerosol type to another. For example, pollen grains and microbial spores are more stable than bacteria in the presence of light, excess heat, cold, relative humidity, and certain toxic gases, as well as to the physical stresses that occur during sampling. From all the preceding, it becomes clear that bioaerosols may only be adequately assessed once the principles of not only aerosol sampling (including, where appropriate, particle size-selective classification), but also sample viability and culture preparation, are properly understood and applied accordingly.

19.4   Early bioaerosol sampling

An early description of bioaerosol sampling was given by Charles Darwin (1909–1914) in his account of the voyage of his ship The Beagle. He described collecting

'. . . a little packet of this brown-coloured fine dust, which appeared to have been filtered from the wind by the gauze of the vane at the masthead.'

The sample was subsequently examined by Professor C.G. Ehrenberg (about whom Darwin wrote 'I must take this opportunity of acknowledging the great kindness with which this illustrious naturalist has examined many of my specimens'), who made drawings on the basis on sample analysis by light microscope. These revealed organisms that are recognised today as common mold spores. Even earlier, Carnelly *et al.* (1887) had reported research into the levels of mold and bacteria in schools and public housing, even to the point of recommending a general level of 600 CFU m^{-3} as a limit value for human exposures. They employed a bioaerosol sampling device that had been developed in Germany a few years before by Hesse, which they described as comprising a cylindrical sampling tube coated on the inside with beef agar gel. Typical sampling flow rate, was of the order of about 20 to 30 Lpm. Another, even older, sampling method, dating as far back as the early 1800s, was the *open-plate deposition* (OPD) method. Interestingly, the mold and bacteria levels in residences that were reported from around that time based on this method, indicated that bioaerosol levels were not very different from what are often found today, even though the favored growth media have changed over the years. The simple OPD approach is still occasionally used today, although it is not recommended for any measurement where the desired end-point is a bioaerosol concentration (Willeke and Macher, 1999).

One the earliest impaction-based device used for bioaerosol sampling was the *aeroconiscope*. According to one report (Gregory, 1973), this was introduced by Maddox and Cunningham as early as the 1840s. But it was described definitively in 1870 by Maddox. This device depended on the external wind to provide the impaction of airborne particles onto a sticky slide that could subsequently be examined under a microscope. The sampler was attached to a movable vane that assured the plate would always face directly into the wind. The aeroconiscope was popular for a while and was used in a number of health-related studies reported during the 1800s. However, the results were of variable quality because of the wind speed dependence of the performance of the device, well-known now from studies of the impaction process. In the late 1870s, Miquel developed an improved impaction-based sampler in which air was directed onto a glycerine-coated slide at about 20 Lpm through a slit-shaped orifice (Miquel, 1879). Air movement was achieved by means of a water-driven pump. Not surprisingly, this device was found to be much more efficient and more consistent than the aeroconiscope. Indeed, it became the forerunner of many of the aerosol sampling devices – including bioaerosol samplers – that are widely used today.

Modern bioaerosol samplers have evolved as knowledge of aerosol mechanics and sampling science has advanced over the years, initially through a better understanding of particle collection mechanisms such as impaction and then through more recent recognition of the importance of aspiration efficiency and its relevance to health-related aerosol measurement.

19.5   Criteria for bioaerosol sampling

Previous chapters have presented a framework for health-related aerosol sampling that recognises the important of particle inhalability and penetration into, and deposition in, the various regions of the respiratory tract. Such physical particle size-selective criteria are relevant to bioaerosols every bit as much as for chemical aerosols. This is implicit in the performance indices for bioaerosol sampler performance identified by Willeke and Macher. In their 1994 review, Griffiths and DeCosemo identified samplers whose physical sampling performance characteristics – in terms of aspiration efficiency – had been determined, and those where such information had not yet been obtained. Around the same time, Grinshpun *et al.* (1994) reported experimentally obtained inlet characteristics for several commercially available bioaerosol samplers. That said, the explicit adoption of health-related particle size-selective criteria to the practical sampling of bioaerosols has so far been slow. It is important that, in due course, they should be incorporated into the culture of bioaerosol sampling in the same way that it has for other types of aerosol. Apart from strengthening the scientific basis of exposure assessment, this would also reduce intersampler biases. Meanwhile, however, a large number of bioaerosol samplers have been developed, many of which are available commercially and are widely used, with varying – and sometimes unknown – physical performance characteristics. A selection will be described below.

19.6   Inertial samplers

The aforementioned Maddox sampler from the 1800s was the first bioaerosol sampler – indeed, perhaps the first aerosol sampler for any purpose – that employed the inertial properties of particles moving in air relative to a stationary surface in order to cause them to be deposited onto on a surface. In that case, the air was moving relative to a stationary collecting surface. A later instrument operating along similar lines, and still in use today, is the *Rotorod* sampler (May *et al.*, 1976). In the operation of this instrument the collecting surface is moved rapidly in relation to the relatively stationary air. The particles are deposited onto sticky tape placed on the forwards-facing surfaces of the rotating arms that could be removed post-sampling and placed on glass slides for microscope analysis. It has been used mainly for the collection and estimation of the airborne concentration of spores. Like the much earlier Maddox sampler, the Rotorod is largely qualitative (Crook, 1995a) because the equivalent sampling flow rate can only be approximated. For more quantitative measurements, such samplers have been largely superceded by aspirating devices such as impactors with single or multiple jets.

19.6.1   Passive samplers

Earlier, a range of passive, nonaspirating samplers were described that had been developed for collecting coarse 'nuisance' particles in the ambient atmosphere (see Chapter 17). Passive samplers small enough to be used for personal aerosol assessment in occupational settings were also described (see Chapter 14). For most of these samplers, inertia – aided by the external wind – along with gravity provided the primary agencies of particle transport to collection surfaces. In 1946, Durham proposed a very simple passive sampling system for atmospheric allergens involving mainly the gravitational deposition of particles onto an oil-coated flat slide. Recognising the interferences associated with falling debris and rain, a cover was provided so that the aerosol to be collected was convected into the region of the sampling surface under the influence of the horizontal wind. Durham proposed this sampling system as a candidate standard allergen sampling system, although there is no evidence that it was widely adopted as such. Later, Tauber (1974) described a somewhat more sophisticated passive sampler designed along similar lines to some of the earlier ones, intended specifically for the collection of large pollen grains. As shown in Figure 19.3, particles entered through the circular orifice at the top under the influence of a combination of inertial and gravitational forces and entered the quiescent air in the cylindrical container, where most of them fell under the influence of gravity and were collected and retained on the glycerol-coated collection surface at the bottom of the container.

19.6.2   Single-stage impactors

In a given bioaerosol sampling situation, it is often the case that the particle sizes of the organisms of interest fall within defined ranges. Single-stage impactors may therefore be designed with defined

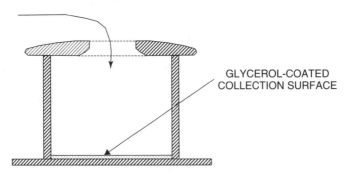

Figure 19.3   *Diagram to illustrate the principle of operation of the Tauber pollen collector (Tauber, 1974)*

Figure 19.4 *Example of a slit-type impactor: the 7-day Volumetric Spore Trap. Photograph courtesy of Burkard Manufacturing Co. Ltd, Rickmansworth, UK*

deposition efficiency that is sufficient to ensure efficient collection of the particles of interest. The Miquel sampler mentioned above was the first step along these lines as long ago as the 1870s. Since then, much more refined instruments have been built and used. In the *Hirst spore trap* (Hirst, 1952), the aerosol was aspirated horizontally at 10 Lpm through a single narrow slit that was arranged to always be facing into the wind by virtue of the wind vane that carried the main sampler assembly. The aspirated particles were collected by impaction onto a vertically mounted glass slide greased with petroleum jelly. The slide was mounted on a motorised stage and translated during sampling so that time resolution of the deposited particles was made possible. Later a hand-held slit-type impactor was described by Clarke and Madelin (1987) where, again, particles were impacted onto a coated slide, this time stationary. Several samplers based on the principle of the Hirst spore trap are commercially available today. One is the *7-day recording volumetric spore trap* (Burkard Air Sampling Co. Ltd, Rickmansworth, UK), where the particles are impacted onto an adhesive-coated transparent plastic tape placed on a clockwork-driven drum. This is a compact unit with a built-in pump and was designed to sample airborne particles such as fungus spores and pollens continuously for periods of up to 7 days without attention. This instrument is shown in Figure 19.4. Another sampler along similar lines is the *AeroTrap*™ (Brandt Instruments, Prairieville, LA, USA), otherwise known as the *Bioslide*™ (A.P. Buck, Inc., Orlando, FL, USA). This time the slit faces upwards and the aspirated particles are deposited downwards onto a moving horizontal slide. In other variants of the slit-sampler idea, particles are impacted onto surfaces – Petri dishes, glass slides, rotating drums, moving tape, etc. – which are coated with agar media, thus facilitating subsequent culture. One popular model is the *airborne bacteria sampler* (ABS) originally developed by the UK Medical Research Council in the 1930s and still commercially available from a number of sources (e.g.

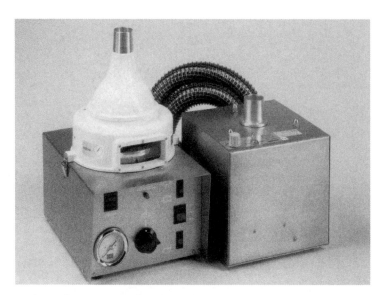

Figure 19.5   *Airborne bacteria sampler. Photograph courtesy of Casella London Ltd, Bedford, UK*

Casella CEL, Bedford, UK; BGI Inc., Waltham, MA, USA). It is shown in Figure 19.5. Aerosol is aspirated through up to four slits located at the top of the sampling head and particles are deposited by impaction onto a 14–16 mm diameter, agar-coated Petri dish placed on a slowly rotating turntable whose speed of rotation may be varied to suit the bioaerosol concentration in a given application. It can operate at a sampling flow rate up to as high as 700 Lpm, allowing a large sample to be recovered in a short sampling period. Griffiths *et al.* (1993) and Upton *et al.* (1993) evaluated the performance of this instrument in wind tunnel experiments and reported that aspiration efficiency dropped off very sharply with particle size, falling to about 0.1 for $d_{ae} = 10$ µm for sampling in air moving at 4 m s^{-1}.

Multiple-jet single-stage impactors have also appeared for the sampling of a specific single bioaerosol particle size range, in which the aspirated aerosol was drawn through a plate containing many, usually circular, impaction jets. These are often referred to as 'sieve samplers'. But as Willeke and Macher pointed out, such samplers operate strictly on the principles of impaction and that the term 'sieve' is therefore a misnomer. A significant advantage of these is that the particles are deposited much more uniformly on the collection substrate, which is especially convenient for culturing and subsequent evaluation. A leading example is the *Andersen N-6* (Thermo Electron Corporation, Waltham, MA, USA), featuring 400 impactor jets in a single flat circular plate and intended for operation at 28.3 Lpm (see Figure 19.6). Particles are deposited onto an agar-coated plate by impaction with $_{50}d_{ae} = 0.65$ µm. Smaller particles pass through and are collected onto a backing filter. For practical purposes, this sampler is a single-stage variant of the Andersen static cascade impactor mentioned in previous chapters. So, as far as inlet efficiency is concerned, the experimental wind tunnel data of Barrett *et al.* (1984) for the *nonviable* version are applicable, along with similar data for the *viable* version by Griffiths *et al.* and Upton *et al.* These data all agreed that sampling efficiency for this instrument was close to unity over quite a wide range of particle sizes up to $d_{ae} = 20$ µm for low wind speeds. Sampling efficiency fell markedly for higher wind speeds. A number of variations on the popular Andersen N-6 design are also available, including the *Aerotech 6* (Aerotech Laboratories, Phoenix, AZ, USA) and the *BioStage*® *single-stage bioaerosol impactor* (SKC Inc., Eight Four, PA, USA). The latter is essentially similar to

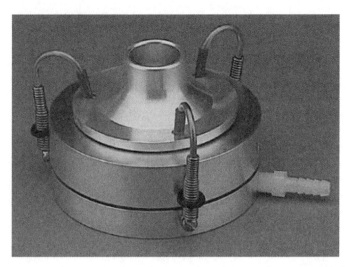

Figure 19.6   *Andersen N-6 single-stage, multi-orifice sampler. Photograph courtesy of Thermo Electron Corporation, Waltham, MA, USA*

the other two but also features a lower flow rate version (14.15 Lpm and 200 impactor jets). These, among others, are identified by the National Institute for Occupational Safety and Health (2004) in its recommended method (Method 0800) for bioaerosol sampling for culturable organisms, including bacteria, fungi and thermophilic actinomycetes.

19.6.3   Cascade impactors

In general, cascade impactors allow the classification of aerosols into specific ranges of particle aerodynamic diameter. These are useful in bioaerosol sampling – as they are for other aerosol types – in that they provide information about aerosol concentrations relevant to deposition in the various regions of the human respiratory tract. In addition, for bioaerosols, they enable discrimination between different types of organism on the basis of their aerodynamic sizes. The most widely used samplers are drawn again from the Andersen range of instruments. One is the 28.3 Lpm *Andersen two-stage viable cascade impactor* with $_{50}d_{ae}$-values at 0.83 and 6.28 μm, respectively, another the 28.3 Lpm *Andersen six-stage viable cascade impactor* with $_{50}d_{ae}$-values in the range from 0.58 to 6.24 μm (Thermo Electron Corporation, Waltham, MA, USA). These instruments are shown in Figure 19.7 and the quoted $_{50}d_{ae}$-values are the ones determined experimentally by Buttner *et al.* (1997) and summarised by Willeke and Macher.

The 2 Lpm *eight-stage Marple personal cascade impactor* first introduced by Rubow *et al.* (1987) is most commonly used for the sampling of inert aerosols, but has also been adapted for the sampling of bioaerosols (Crook, 1995a). In the conventional use of this sampler, particles – including those of biological origin – are collected onto the polycarbonate membrane impaction substrates and may be viewed directly by optical microscopy (or electron microscopy, if desired). Earlier, Crook *et al.* (1988) had described how viable spores collected in this way could be recovered by rinsing them into a suspension, injected onto agar plates and then cultured. Elsewhere, Macher and Hansson (1987) described how the collection substrates in the original Marple sampler could be replaced by small trays containing gelatin medium in order to improve the collection and sustenance of viable particles.

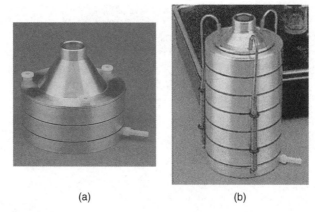

(a)   (b)

Figure 19.7   *Two versions of the Andersen viable cascade impactor: (a) the 28.3 Lpm Andersen two-stage viable cascade impactor with $_{50}d_{ae}$-values at 0.83 and 6.28 μm, respectively; (b) the 28.3 Lpm Andersen six-stage viable cascade impactor with $_{50}d_{ae}$-values in the range from 0.58 to 6.24 μm. Photographs courtesy of Thermo Electron Corporation, Waltham, MA, USA*

19.6.4   Impingers

In 1922, Greenberg and Smith introduced the *impinger*, in which aspirated air was impacted onto an impaction surface immersed in a liquid medium. A typical layout for this sampler was shown earlier in Figure 15.3. The recommended sampling flow rate was 28.3 Lpm, and so it required a large pump. Therefore it was – and still is – used as a static sampler. In the impinger, the mechanism of particle collection was inertial, similar to that in a conventional impactor except that the impacting particles were dispersed into the liquid medium. Most such samplers are constructed from glass, and have become known as 'all-glass impingers' (AGI). The original Greenberg–Smith concept was developed as sampler for inert dusts but was later adapted for use as a bioaerosol sampler (Henderson, 1952). In applications of the original impinger to bioaerosols, concerns soon emerged about the effect on particle viability and structural integrity of the impingement of the jet directly onto the surface. So modified designs were developed, one in which the jet was introduced at an angle, thus reducing the velocity at which particles arrived at the impingement surface (Shipe *et al.*, 1959), and another where the jet-to-surface distance was increased from 4 to 30 mm (May and Harper, 1957), similarly reducing the impingement velocity. The latter has since been widely used and is available as the 12.5 Lpm *AGI-30* (Ace Glass Inc., Vineland, NJ, USA). This sampler usually contains about 125 ml of an appropriate aqueous liquid such as deionised water or phosphate buffer. Ehrlich *et al.* (1966) showed that the overall collection efficiency of the AGI-30 was significantly greater than a typical single slit-type impactor sampler. More recently, however, Willeke *et al.* (1998) described a new version which, although it was externally very similar to the AGI-30, featured a redesigned jet nozzle configuration where collection was now by a combination of both inertial and centrifugal forces. The latter was achieved by virtue of an asymmetrical jet design that introduced coherent swirling motions into the fluid. In addition to the additional particle collection mechanism, it was expected that the loss by re-entrainment of already collected particles would be reduced since there were fewer bubbles generated. The new sampler was initially referred to as the 'swirling aerosol collector', but was later available commercially as the 12.5 Lpm *BioSampler®*, for which the recommended sampling flow rate was 12.5 Lpm (SKC Inc., Eighty Four, PA, USA). It is shown in Figure 19.8. Willeke *et al.* conducted experiments to test the relative performances of the new sampler and the old AGI-30, using 20 ml of deionized water as the collecting medium along with inert

ASPIRATED AEROSOL ⟶

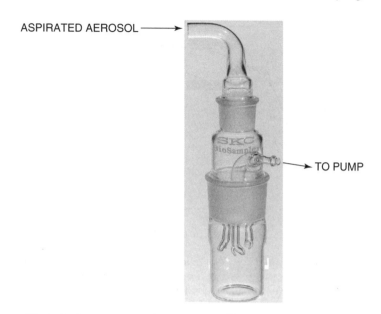

TO PUMP ⟶

Figure 19.8   *The modified all-glass impinger, the BioSampler®. Photograph courtesy of SKC Inc., Eighty Four, PA, USA*

monodisperse polystyrene latex test particles. The results are shown in Figure 19.9, and show that that the collection efficiency of the BioSampler was much greater across the whole range of particle sizes tested. Later, the BioSampler and AGI-30 were tested against *B. subtilis* and *P. fluorescens* biological test organisms as a function of sampling flow rate, in terms of the efficiencies of both collection and recovery of the collected organisms (Lin *et al.*, 2000). These results are summarised in Figure 19.10. Again the greater collection efficiency of the BioSampler is evident. Furthermore, it was shown that the efficiency of recovery of the organisms was greater for the BioSampler, most clearly for the *B. subtilis*. In addition (not shown), by the use of a more viscous, less volatile liquid in the form of white mineral oil, it was shown that the BioSampler could be used without bias in a sampling run for as long as 8 h (compared with only about 30 min for a standard evaporating liquid such as deionised water).

In the 1930s, the much smaller *midget impinger* appeared and, once miniaturised sampling pumps became readily available from the 1970s onwards, this became a good candidate for use as a personal sampler. Typically, this version was intended for sampling at flow rates from 1 to 3 Lpm and with liquid capacity in the range from 10 to 25 ml. It is currently available from a number of sources (e.g. SKC Inc., Eighty Four, PA, USA; Zefon International Inc., Ocala, FL, USA). An even smaller *micro impinger* has also appeared, intended for sampling at 0.1 Lpm and with about 1 ml of collecting fluid (Crook, 1995a).

Following the rationale of the cascade impactor described above for bioaerosols, May (1966) also developed an all-glass *multi-stage liquid impinger* (MSLI) for which a five-stage version is still available today (Copley Scientific Ltd, Nottingham, UK) that may be operated at flow rates from 30 to 100 Lpm (where of course the $_{50}d_{ae}$-values for the five stages shift accordingly). As pointed out by Crook, one problem with this instrument is that, since it is made from glass, manufacturing tolerances are large such that the $_{50}d_{ae}$-values for the stages cannot be specified accurately.

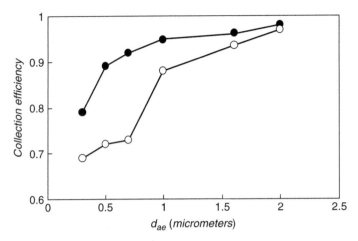

Figure 19.9   *Collection efficiency as a function of particle aerodynamic diameter (d_{ae}) for the BioSampler® and AGI-30 impingers, respectively, as measured in experiments using inert particles, with both samplers operated at a 12.5 Lpm: ○ AGI-30; ● BioSampler (Willeke et al., 1998). Adapted from Aerosol Science and Technology, 28, 439–456. Copyright (1998). Mount Laurel, NJ. Reprinted with permission*

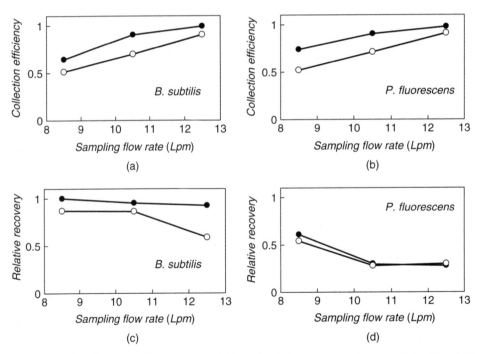

Figure 19.10   *Overall collection efficiency (a and b) and relative recovery (c and d) for B. subtilis and P. fluorescens test bioaerosols respectively, as functions of sampling flow rate for the BioSampler® and AGI-30 impingers (Lin et al., 2000). Adapted from Aerosol Science and Technology, 32, 184–196. Copyright (2000). Mount Laurel, NJ. Reprinted with permission*

19.7 Centrifugal samplers

19.7.1 Cyclones/wetted cyclones

As discussed in earlier chapters, cyclones are widely used for aerosol sampling in general. However, versions have been developed specifically for the sampling of bioaerosols. One example is the *Aerojet-General glass cyclone sampler* (Decker *et al.*, 1969; Buchanan *et al.*, 1972), capable of sampling flow rates as high as 1000 Lpm. The principle is essentially the same as for the cyclones described earlier, but here the internal walls of the cyclone are wetted with an appropriate liquid delivered into the inlet during sampling in order to irrigate the walls and wash particles deposited there down to the collecting vessel at the bottom. White *et al.* (1975) showed that the collection efficiency for this sampler was of the same order as for the AGI-30.

19.7.2 Centrifuges

The Canadian federal body, Public Works and Government Services Canada (PWGSC), has developed guidelines for fungal contamination that specifies the use of the *Reuter centrifugal sampler* (RCS) (GAP EnviroMicrobial Services, London, Ontario, Canada) (see Nathanson, 2005). The concept of this instrument is shown in Figure 19.11. Air is aspirated through the inlet at the top and enters the impeller drum axially. The aspirated air is thus set bodily in rotary motion and the airborne particles are driven by the resultant centrifugal force towards an agar surface on a plastic strip mounted at the perimeter

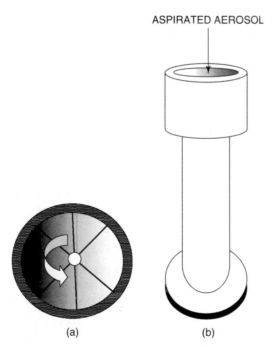

ASPIRATED AEROSOL

(a)   (b)

Figure 19.11 *The Reuter centrifugal sampler (GAP EnviroMicrobial Services, London, Ontario) (see Nathanson, 2005): (a) schematic plan view of the inlet, showing the impeller; (b) general layout of the device. In use, the agar-covered plastic strip is mounted on the inner perimeter of the cylindrical impeller housing*

of the drum. The deposit can then be analyzed in the usual way. The sampler is battery operated and capable of being hand held. So, it is convenient and portable enough to be carried around by the operator to different locations. Sampling flow rate is about 100 Lpm, but some questions have been raised about its low sampling efficiency (Macher and First, 1983).

19.8   'Total' and inhalable bioaerosol

Samplers for 'total' aerosol and inhalable aerosol of the type described earlier are also useful starting points for bioaerosol sampling if the goal is to perform particle size-selective health-related sampling. They have special potential for the sampling of large particles, as is the case for many pollens and other airborne allergens. Here, personal inhalable aerosol samplers are preferred in the light of the CEN/ISO/ACGIH criteria that have been extensively reviewed elsewhere in this book. The 2 Lpm *IOM personal inhalable aerosol sampler* and the 4 Lpm *Button personal inhalable aerosol sampler* have been suggested as promising candidates, especially when used with gelatin and sterilised polycarbonate filters from which micro-organisms may readily be removed into suspension, along with internal wall deposits, for subsequent analysis. Such filters are said to provide better micro-organism survivability when compared with other filter types. However, as cautioned by Griffiths and DeCosemo, such sampling systems afford little protection to already-collected organisms against the desiccation that occurs due to extended periods of air passing through the filter.

19.9   Other samplers

Crook (1995b) reviewed a number of other types of sampler most of which have not been widely used but nonetheless are of some historic interest. Electrostatic precipitation was employed in one instrument, the *Litton large volume electrostatic air sampler* (*LVEAS*). In this sampler, the collection surface was a rotating metal disk into the center of which a liquid medium was delivered by means of a peristaltic pump. As the disk was rapidly rotated, the liquid spread out as a film that covered the whole circular surface of the plate, thus irrigating the collecting surface. The liquid, along with the collected particles, flowed into a circular groove at the edge of the disk which was then pumped – by means of a second peristaltic pump – to a reservoir. Sampling flow rates with this system could be as high as 1000 Lpm. Samples collected in this way were analyzed in the same way as samples collected by other similarly irrigated samplers.

Thermal precipitation has also been considered for bioaerosol sampling. But the low sampling flow rates associated with this type of sampler mean that sampling for bioaerosols at low concentration was not feasible. Consequently, it has not been widely used.

References

Barrett, C.F., Ralph, M.O. and Upton, S.L. (1984) Wind tunnel measurements of the inlet efficiency of four samples of suspended particulate matter, Final Report on CEC Contract No. 6612/10/2, Report EUR 9378 EN, Commission of European Communities, Luxembourg.

Buchanan, L.M., Harstad, J.B., Phillips, J.C., Lafferty, E., Dahlgren, C.M. and Decker, H.M. (1972) Simple liquid scrubber for large-volume air sampling, *Applied Microbiology*, 23, 1140–1144.

Buttner, M.P., Willeke, K. and Grinshpun, S.A. (1997) Sampling and analysis of airborne microorganisms. In: *Manual of Environmental Biology* (Eds C.J. Hurst, G.R. Knudsen, M.J. McInerey *et al.*), American Society for Microbiology, Washington, DC, pp. 629–640.

Carnelly, T., Haldane, J.S. and Anderson, A.M. (1887) The carbonic acid, organic matter and micro-organisms in air, more especially of dwellings and schools, *Philosophical Transactions of the Royal Society of London B*, 178, 61–111.

Clarke, A.F. and Madelin, T. (1987) Technique for assessing respiratory health hazards from Hay and other source materials, *Equine Veterinary Journal*, 19, 442–447.

Cox, C.S. and Wathes, C.M. (Eds) (1995) *Bioaerosols Handbook*, CRC Press, Boca Raton, FL.

Crook, B. (1995a) Inertial samplers: biological perspectives. In: *Bioaerosols Handbook* (Eds C.S. Cox and C.M. Wathes), CRC Press, Boca Raton, FL, pp. 247–267.

Crook, B. (1995b) Non-inertial samplers: biological perspectives. In: *Bioaerosols Handbook* (Eds C.S. Cox and C.M. Wathes), CRC Press, Boca Raton, FL, pp. 269–283.

Crook, B., Griffin, P., Topping, M.D. and Lacey, J. (1988) An appraisal of methods for sampling aerosols implicated as causes of work-related respiratory symptoms. In: *Aerosols – their Generation, Behaviour and Implications*, The Aerosol Society, Bristol, pp. 327–333.

Darwin, C.R. (1909–1914) *The Voyage of The Beagle*, Collier and Son, New York.

Decker, H.M., Buchanan, L.M., Frisque, D.E., Filler, M.E. and Dahlgren, C.M. (1969) Advances in large volume air sampling, *Contamination Control*, 8, 13–17.

Douwes, J., Thorne, P., Pearce, N. and Heederik, D. (2003) Bioaerosol health effects and exposure assessment: progress and prospects, *Annals of Occupational Hygiene*, 47, 187–200.

Durham, O.C. (1946) The volumetric incidence of atmospheric allergens, IV: a proposed standard method of gravity sampling, counting and volumetric interpolation of results, *Journal of Allergy*, 17, 79–86.

Erhlich. R., Mettler, S. and Idoine, L.S. (1966) Evaluation of slit sampler in quantitative studies of bacterial aerosols, *Applied Microbiology*, 14, 328–330.

Greenberg, L. and Smith, G.W. (1922) A new instrument for sampling aerial dust, US Bureau of Mines Report of Investigations 2392.

Gregory, P.H. (1973) *Microbiology of the Atmosphere*, 2nd Edn, John Wiley & Sons, Inc., New York, NY.

Griffiths, W.D. and DeCosemo, G.A.L. (1994) Inlet characteristics of bioaerosol samplers, *Journal of Aerosol Science*, 25, 1425–1458.

Griffiths, W.D., Upton, S.L. and Mark, D. (1993) An investigation into the collection efficiency and bioefficiency of a number of aerosol samplers, *Journal of Aerosol Science*, 24, S541–S542.

Grinshpun, S.A., Change, S.W., Nevalainen, A. and Willeke, K. (1994) Inlet characteristics of bioaerosol samplers, *Journal of Aerosol Science*, 25, 1503–1522.

Henderson, D.W. (1952) An apparatus for the study of airborne infection, *Journal of Hygiene (Cambridge)*, 50, 53–68.

Hensel, A. and Petzoldt, K. (1995) Biological and biochemical analysis of bacteria and viruses. In: *Bioaerosols Handbook* (Eds C.S. Cox and C.M. Wathes), CRC Press, Boca Raton, FL, pp. 335–360.

Hirst, J.M. (1952) An automatic volumetric spore trap, *Annals of Applied Biology*, 39, 257–265.

Larsson, K.A. (1993) Prediction of the pollen season with a cumulated activity method, *Grana*, 32, 111–114.

Lin, X., Reponen, T., Willeke, K., Wang, Z., Grinshpun, S.A. and Trunov, M. (2000) Survival of airborne microorganisms during swirling aerosol collection, *Aerosol Science and Technology*, 32, 184–196.

Macher, J.M. (Ed.) (1999) *Bioaerosols: Assessment and Control*, American Conference of Government Industrial Hygienists, Cincinnati, OH.

Macher, J.M. and First, M.W. (1983) Reuter centrifugal air sampler: measurement of effective airflow rate and collection efficiency, *Applied Environmental Microbiology*, 45, 1960–1962.

Macher, J.M. and Hansson, H.C. (1987) Personal size separating impactor for sampling microbiological aerosols, *American Industrial Hygiene Association Journal*, 48, 652–655.

Maddox, R.L. (1870) On the apparatus for collecting atmospheric particles, *Monthly Microscopical Journal*, 1, 286–290.

May, K.R. (1966) Multistage liquid impinger, *Bacteriological Review*, 30, 559–570.

May, K.R. and Harper, G.J. (1957) The efficiency of various liquid impinger samplers in bacterial aerosols, *British Journal of Industrial Medicine*, 14, 287–297.

May, K.R., Pomeroy, N.P. and Hibbs, S. (1976) Sampling techniques for large windborne particles, *Journal of Aerosol Science*, 7, 53–62.

Miquel, M.P. (1879) Etude sur les poussierres organisees de l'atmosphere, *Annales d'Hygiene Publique et de Medecine Legale*, January 226.

Nathanson, T. (2005) Fungal contamination guidelines: interpreting the analysis. Paper on Indoor Air Quality, Office of Greening Government Operations, Ottawa, Ontario.

National Institute for Occupational Safety and Health (2004) Bioaerosol sampling (indoor air), Method 0800. In: *Manual of Analytical Methods*, 4th Edn, NIOSH, Cincinnati, OH.

Rubow, K.L., Marple, V.A., Olin, J. and McCawley, M.A. (1987) A personal cascade impactor – design, evaluation and calibration, *American Industrial Hygiene Association Journal*, 48, 532–538.

Shipe, E.L., Tyler, M.E. and Chapman, D.N. (1959) Bacterial aerosol samplers. II, Development and evaluation of the Shipe sampler, *Applied Microbiology*, 7, 349–354.

Spieksma, F.T. (1991) Aerobiology in the nineties: aerobiology and pollinosis, *International Aerobiology Newsletter*, 34, 1–5.

Tauber, H. (1974) A static non-overload pollen collector, *New Phytologist*, 73, 359–369.

Upton, S.L., Mark, D., Douglas, E.J. and Griffiths, W.D. (1993) A wind tunnel evaluation of the sampling efficiencies of some bioaerosol samplers. In: *Aerosols – Their Generation, Behaviour and Applications*, The Aerosol Society, Bristol, p. 156.

White, L.A., Hadley, D.J., Davids, D.E. and Naylor, R. (1975) Improved large volume sampler for the collection of bacterial cells from aerosol, *Applied Microbiology*, 229, 335–339.

Willeke, K. and Macher, J.M. (1999) Air sampling. In: *Bioaerosols: Assessment and Control* (Ed. J.M. Macher), American Conference of Government Industrial Hygienists, Cincinnati, OH, pp 11.1–11.25.

Willeke, K., Lin, X. and Grinshpun, S.A. (1998) Improved aerosol collection by combined impaction and centrifugal motion, *Aerosol Science and Technology*, 28, 439–456.

20

Direct-reading Aerosol Sampling Instruments

20.1 Introduction

For most of the sampling instruments described in this book, the sampled aerosol is collected in a form, for example on a filter or some other substrate, such that it may subsequently be assessed gravimetrically by weighing it on an analytical balance or processed for chemical analysis. Such instrumentation is suitable when time-averaged measurement can be justified. However, there are occasions where short-term – approaching 'real-time' – measurement is required: for example, where (a) the aerosol in question is thought to be particularly hazardous and where an immediate alert to high concentrations is required or (b) monitoring is desired in order to examine the effects of adjustments in process or dust control. So whereas time-weighted averaged aerosol concentrations, like those obtained using the samplers described in the preceding chapters, provide information like that shown in Figure 20.l(a), there is sometimes a need for information like that shown in Figure 20.1(b).

Many practical direct-reading instruments are aspirating samplers along the lines of many of those described already, but where the sampled aerosol is detected in real time. There is a range of physical options for direct-reading aerosol detection and measurement, all based on sensing the particles – either individually or in ensembles – using some sort of transducer which can provide a representative electric signal that can be read out or recorded and referred to some appropriate calibration. The majority of such instruments fall into one of five categories: optical, electrical, molecular, mechanical and nuclear. As before, the emphasis is placed on instrumentation of the type that finds application in occupational and ambient environmental settings, some developed for use as routine instruments for practical aerosol research and other, more specialised ones for use in aerosol research.

One of the earliest practical direct-reading instruments for aerosol measurement appears to be the '*Hazard*' device described by Drinker and Hatch as long ago as 1934. In this instrument, an aspirated sample was collected by impaction onto a moving continuous ribbon, which was subsequently passed through an optical detection system that operated on the principle of light attenuation. Since then, many instruments have appeared that provide aerosol information in real time.

Aerosol Sampling: Science, Standards, Instrumentation and Applications   James Vincent
© 2007 John Wiley & Sons, Ltd

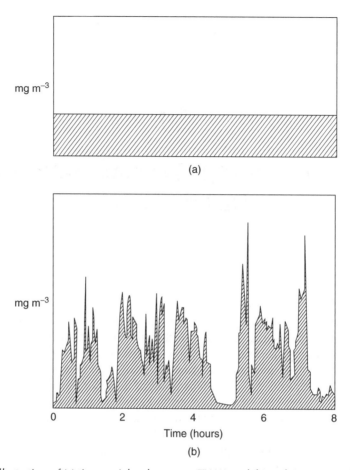

Figure 20.1   *Illustration of (a) time-weighted average (TWA) and (b) real time aerosol concentrations*

20.2   Optical aerosol-measuring instruments

Optical techniques provide an important means by which an aerosol may be assessed in real time. They have the great advantage over many other types of instrumentation that the measurement can be made remotely and without disturbing the aerosol. Available optical techniques fall into two basic categories, one based on measurement of the interaction between a light beam and the aerosol as a whole, and the other based on interactions with single, individual particles.

20.2.1   Physical background

The optical properties of aerosols are concerned with the interaction of electromagnetic radiation with individual suspended particles and with ensembles of particles. If a particle has different dielectric properties to those of the surrounding medium, as reflected in their relative refractive indices, then it represents a dielectric inhomogeneity. Physically, the problem may be treated in terms of a plane electromagnetic wave incident on a particle whose geometric surface defines the boundary enclosing a medium having different refractive index. The mathematical theory of such interactions was first described in the 1800s by James Clerk Maxwell, the solutions of which include all the well-known phenomena such as reflection, diffraction and refraction which, lumped together, constitute *light scattering* and *absorption*. The first theory of light scattering by small particles in air was published by Lord Rayleigh in the late 1800s, applying to very small particles of size much less than the wavelength of the radiation (λ). A significant advance, in terms of its relevance to aerosols, came later when Mie (1908) extended Maxwell's theory to larger particles. The overall picture of the physics of light scattering by aerosols has since been described comprehensively by Van de Hulst (1957) and Kerker (1963). All this provides the foundations of the two approaches to aerosol measurements that have emerged. The first is based on the attenuation of light as energy is removed from an incident beam, expressed in terms of either *transmission* or, conversely, *extinction*. The second is based on consideration of the scattered light itself. The two ideas are embodied in Figure 20.2. Both were reviewed by Hodkinson (1966).

For the first approach it is easily shown that:

$$\frac{I}{I_0} = \exp(-c_P Q t) \tag{20.1}$$

where I and I_0 are the intensities of the transmitted and incident light, respectively, t is the path length of the beam through the aerosol, and c_P is the projected area concentration of the aerosol. In this, Q is the *particle extinction coefficient*, containing the actual physics of the interaction between the light and the particle. In order to use Equation (20.1) as the basis of aerosol concentration measurement by photometry, it is necessary to know Q as well as to be able to estimate the relationship between c_P and the metric of aerosol concentration which is of true primary interest, usually mass concentration. For the former, Q is a complicated function of particle geometric diameter d and light wavelength λ, as well as particle shape and composition. It is not readily accessible for most particles of interest in the real world. If the extinction approach is to be used in a monitoring instrument, it is therefore usually necessary to calibrate the instrument for each application.

One important factor that comes from the practical application of the theory of light extinction to an aerosol concentration measuring instrument is to ensure that the light that is transmitted to the detector contains only the light that has not interacted with particles. However, since light is scattered from the incident beam in all directions, including the forwards direction, it is possible for some of the

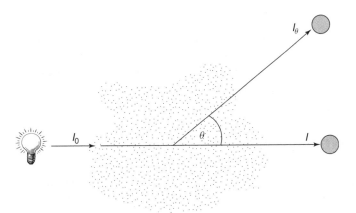

Figure 20.2 *Diagram to illustrate the concepts of light extinction and light scattering: I is the transmitted light intensity, and I_θ is the light intensity scattered at angle θ*

scattered light to reach the detector, in which case Equation (20.1) no longer strictly holds. In designing a practical instrument based on the principle of light extinction, the geometry of the light beam and the detecting optics should therefore be designed to eliminate – or at least minimise – such unwanted contributions to the detected light. Alternatively, it may be possible to correct for such effects (Wind and Szymanski, 2000).

The light scattering approach to aerosol measurement requires consideration of the angular distribution of the light that is scattered from a collimated light beam. For this the relevant physical property is the *particle scattering coefficient*, corresponding to the extinction coefficient (Q) in the first case considered. This is given by $S(\theta)$ the ratio of the light scattered by a single particle per solid angle at an angle (θ) to the forwards direction to the light geometrically incident on the particle. The intensity of the scattered light received at the angle θ is therefore given by:

$$I_\theta = k \left[\frac{c_P S(\theta)}{\sin \theta} \right] I_0 \tag{20.2}$$

where I_0 is again the intensity of the incident beam and c_P the projected area concentration. Here, k is a constant that contains the details of the geometry of the detecting optics, including the volume within the aerosol onto which the light is incident and the solid angle within which it is received by the detector. Like the extinction coefficient Q, $S(\theta)$ is very complicated and not universally accessible for practical situations.

Optical instruments operating on the basis of the detection of the scattered – rather than transmitted – light are more sensitive at lower concentrations. This is because it is easier to detect a change in a small light intensity against a dark background than to detect a small change in an intensity which is already bright. For this reason, light scattering photometry has been a more popular option for workplace aerosol monitoring. Furthermore, because so much information about the aerosol is contained in the angular distribution of the scattered light, there are many more options on which to base an instrument. Hodkinson classified light scattering instrumentation according to the angle at which the scattered light was detected, and identified a range of possibilities for practical optical aerosol measurement. Gebhart (1993) later identified the useful particle volume scattering coefficient, S_V, where, for a given angle and particle refractive index:

$$S_V = \frac{scattered\ light\ flux}{(\pi d^3/6)} \tag{20.3}$$

in which the denominator represents the individual particle volume. S_V is plotted in Figure 20.3 as a function of the dimensionless particle diameter, $\pi d/\lambda$, and it is seen that there is a range of particle size – for d from about 0.3 to 3 μm for visible light – where S_V is approximately constant. In this region, therefore, the scattered light intensity from an ensemble of particles is in turn approximately proportional to overall particle volume, and hence to particle mass.

20.2.2 Transmission/extinction monitoring

Inspection of Equation (20.1) reveals that the change in light intensity, and hence the ability to distinguish between I and I_0, increases with the product $c_P t$. So, for low concentrations, a long path length is desired. In modern workplaces, aerosol concentrations are generally not high enough to provide sensitive extinction measurement for a feasible optical path length. So it is not surprising to find that instruments based on the principles of light extinction have not been widely used in the monitoring of general workplace aerosols. In one exception, in the 1950s, one such instrument was used in coal mines

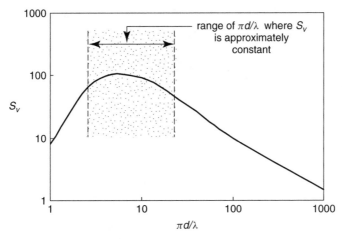

Figure 20.3 *Relationship between the volume scattering coefficient (S_V) and particle size parameter ($\pi d/\lambda$) (Gebhart, 1993). Adapted with permission from Gebhart, Aerosol Measurement (Eds. Willeke and Baron). Copyright (1993) Van Nostrand Reinhold*

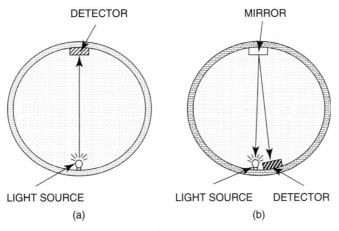

Figure 20.4 *Schematic to illustrate applications of light attenuation-based monitors in stack sampling: (a) single-pass; (b) two-pass*

in the Soviet Union for the monitoring of dust concentrations thought to be high enough to constitute an explosion risk (see Hodkinson, 1966). There appears to have been little else.

The few industrial applications that have been reported have been for situations where aerosol concentrations were very high, for example in the monitoring of aerosol emissions from industrial processes flowing through exhaust chimneys and ducts. Over the years, a number of instruments have been commercially available and applied for this purpose. Two typical layouts are shown schematically in Figure 20.4. Figure 20.4(a) indicates the layout for a simple, single-pass arrangement where the light source and detector are placed on opposite sides of the duct. Figure 20.4(b) shows an alternative arrangement where the light source and detector are mounted adjacent to one another on one side of the stack

(a)

(b)

Figure 20.5 *Typical packaged opacity meters for stack aerosol monitoring: (a) single-pass; (b) two-pass. Photographs courtesy of Environmental Monitor Service Inc., Yalesville, CT, USA*

and the light is reflected back across the stack by means of a retro-reflector on the opposite side. The latter is advantageous because the path length – t in Equation (20.1) – is doubled. One drawback with this arrangement is that the windows of the optical system are exposed to the aerosol and so are likely to be contaminated by particle deposition. In a practical system, it is necessary to design the optics either to minimise this effect or to allow easy cleaning of the surfaces. Versions of such *transmissometers* or *opacity monitors* are available commercially from several sources. Two are shown in Figure 20.5, from a range of packaged instruments where the light beam is collimated and electronically modulated in order to make the measurements insensitive to ambient light, one a single-pass and the other a double-pass instrument (e.g., Environmental Monitor Service Inc., Yalesville, CT, USA). Such instruments need to be calibrated for each application on a regular basis by comparison with gravimetric measurements of in-stack aerosol concentration obtained from samples collected using isokinetic probes.

The largest area of practical application of the transmission/extinction principle has been in the monitoring of atmospheric visibility. Here the path lengths of interest are much longer. Again practical instruments are available from several sources. Long range transmissometers consist of a light source and a detector which are usually placed from 0.5 to 10 km apart. Again, for this instrument, the light from the source is modulated in order to increase the sensitivity of detection in the presence of ambient background light. The aerosol concentration in such cases is traditionally obtained in terms of the *visibility*, expressed either as the *extinction* (in km^{-1}) or the *range* (in km).

20.2.3 Light scattering photometry

A much wider range of light scattering instrumentation has appeared on the market for aerosol concentration measurement. Unlike the transmission/extinction instruments described above, most instruments in this category are sampling devices where the light scattering is used as the means to assess the concentration of what has been aspirated. A major area of application has been in the monitoring of the fine respirable aerosol fraction in workplaces.

In Europe, two contrasting approaches were reported. In Britain, the *Safety in Mines Light Scattering Instrument (SIMSLIN)* that emerged during the 1970s was based on the horizontal elutriator system of the MRE Type 113A gravimetric sampler described in Chapters 14 and 15. Like the MRE, the SIMSLIN was intended for use in coal mines. The aerosol was aspirated and passed between the plates of a horizontal elutriator, with dimensions and sampling flow rate designed to produce a penetration characteristic matching the original British Medical Research Council (BMRC) respirable aerosol curve. In the MRE itself, the penetrating aerosol was deposited onto a filter so that, at the end of a sampling shift, a time-weighted average of the respirable aerosol mass concentration could be obtained by direct weighing of the filter. In the corresponding SIMSLIN, the aerodynamically selected aerosol entered an optical sensing zone in which scattered infrared light from a diode laser ($\lambda = 0.85 \ \mu m$) was collected in the angle 12 to $20°$ from the forwards direction and focused onto a photodiode detector. For these conditions, experiments in the laboratory with test aerosols generated from coal dusts showed that the scattered light flux, given by the output from the photodiode, was reasonably proportional to particle mass concentration over the particle size range of the respirable fraction. It was also found that the signal from the photodiode was only weakly dependent on particle refractive index for the range of mineral dust aerosols expected to be encountered in coal mines. So it looked promising as the basis of a direct-reading sampling instrument that would provide accurate measurements of the desired health-related aerosol fraction. Extension of the SIMSLIN concept led to the development of the *optical scattering instantaneous respirable dust indication system (OSIRIS)* (Leck, 1983), intended to provide the respirable dust concentration data in a form that could be telemetered to a central monitoring station. Although substantial research and development resources were devoted to both the SIMLSIN and OSIRIS instruments, they were never widely deployed in the field. Part of the reason was that the dust control regulations for British coal mines have long continued to be written in terms of time-weighted average mass concentrations requiring full-shift sampling and gravimetric assessment of collected dust.

In the German coal mining industry, a different approach was adopted in the *TM-Digital* (Armbruster and Breuer, 1983). This instrument is illustrated schematically in Figure 20.6(a), where it is shown how

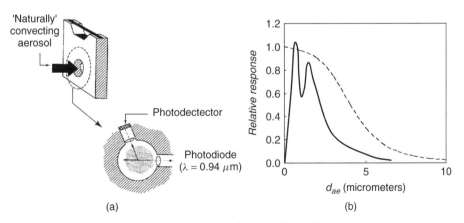

Figure 20.6   *TM-Digital respirable dust photometer: (a) diagram of basic lay-out; (b) optical response (continuous curve) shown in comparison with the latest respirable aerosol curve (dashed curve). Reprinted from Aerosol Science for Industrial Hygienists, Vincent. Copyright (1995), with permission from Elsevier*

the aerosol entered the sensing region as a result of the natural convection in the workplace atmosphere, and so required no pump and flow control system. The TM-Digital was therefore, in effect, a 'passive', nonaspirating dust sampler. It was assumed that the particles entering the selection zone would undergo no significant aerodynamic particle size selection. In the sensing region, scattered light from a parallel beam of monochromatic infrared light ($\lambda = 0.94$ µm) was detected at an angle of 70° to the forward-facing direction. Based on experiments in a laboratory chamber with coal dust test aerosols, the optical response of the TM-Digital was determined as a function of particle aerodynamic diameter (d_{ae}) for an aerosol typical of that found in mining applications. As shown in Figure 20.6(b), it is seen to be broadly similar in shape to the respirable aerosol curves shown in Chapter 11, albeit shifted somewhat towards smaller particle sizes. One drawback with the practical use of the TM-Digital arose from the fact that – unlike in the SIMSLIN – the sensing region was fully exposed to the workplace total aerosol. This meant that the optical surfaces tended to become contaminated by the deposition of particles, with the result that the performance calibration tended to 'drift' when the instrument was used unattended over long periods. The TM-Digital remains commercially available today (Helmut Hund GmbH, Wetzlar, Germany).

In the USA, a related instrument was the *respirable aerosol monitor* (*RAM*) developed originally by MIE Inc. (Bedford, MA, USA). In its original version, aerosol was aspirated with the aid of a pump and then passed through a cyclone which allowed the respirable fraction to penetrate to the optical sensing zone. Pulsed infrared light ($\lambda = 0.88$ µm) scattered in the angular range 45 to 90° from the forwards direction was detected by a photodiode. It was predicted from light scattering theory that the photodiode output should be approximately proportional to the respirable aerosol mass concentration in the sensing zone of the instrument. Typical response data for an earlier version, the RAM-1, are shown in Figure 20.7, based on the results of experiments by Rubow and Marple (1983) in a laboratory

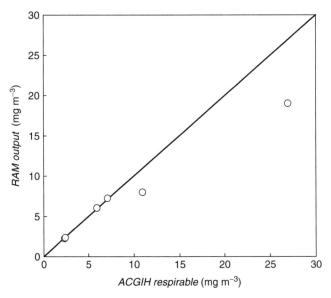

Figure 20.7 *Performance of the original RAM-1 respirable dust photometer shown in relation to the corresponding respirable fraction as defined by the 1968 ACGIH criterion, from experiments in a laboratory test chamber (data reported by Rubow and Marple, 1983). Adapted with permission from Rubow and Marple, Aerosols in the Mining and Industrial Work Environments, pp. 777–795. Copyright (1983) Ann Arbor Science Publishers*

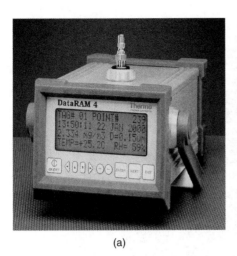

(a)   (b)

Figure 20.8   *Two current commercially available optical photometers for respirable aerosol: (a) DataRAM^{TM}; (b) Personal DataRAM^{TM}, shown in customary mode of use. Photographs courtesy of Thermo Electron Corporation, Waltham, MA, USA*

chamber using Arizona road dust of the type that is widely used for the testing of air cleaners. The output of the instrument was compared with actual mass concentrations for respirable aerosol based on the 1968 ACGIH definition, from measurements of particle size distributions obtained using a cascade impactor. The data shown were found to be in reasonable agreement for lower aerosol concentrations, but the RAM-1 values appeared to fall short for higher concentrations, suggesting some nonlinearity in the instrument's response. The RAM has appeared in a number of forms over the years and has been extensively characterised and upgraded. The instrument shown in Figure 20.8(a) is a recent fully automated digital version with built-in data-logging capability (the *DataRAM 4*^{TM}) that can provide both the concentration and median particle size, along with other environmental information such as temperature and humidity. When used with an appropriate pre-selector (not shown), the instrument can provide direct measurement of respirable, thoracic, PM_{10} or $PM_{2.5}$ fractions. An important extension was the *Mini-RAM*^{TM} which, like the TM-Digital, was an essentially 'passive' device, having no pump. Also like the TM-Digital, and unlike the RAM itself, it contained no aerodynamic pre-selector, but responded directly to the respirable aerosol fraction. It was developed in a miniaturised version that could be used as a personal direct-reading aerosol sampler, and is available today as the *Personal DataRAM* [see Figure 20.8(b)]. These later versions of the successors of the original RAM are now available commercially from Thermo Electron Corporation (Waltham, MA, USA).

Several related instruments are also currently available. The *DUSTTRAK*^{TM} is a versatile light scattering photometer in which scattered light is detected at 90° from the incident beam (TSI Inc., St Paul, MN, USA). It is a battery-operated, hand-held device that can be carried around by the operator and provide both real-time mass concentration read-out and data logging capability. By the use of appropriate cyclone or impactor-based pre-selectors, it can provide aerosol concentrations for the respirable, PM_1, $PM_{2.5}$ and PM_{10} particle size fractions. The *HAZ-DUST 1* also detects light scattered at 90° (SKC Inc., Eighty Four, PA, USA). This too is a small, hand-held, battery-operated instrument – indeed, it is so compact it may be used as a personal sampler. It is shown in this mode of use in Figure 20.9. Here, an appropriate sampling head for the aerosol fraction of interest is mounted in the breathing zone, and the pump and electronic processing unit is worn on the belt. The sensor is located in the tube close to

Figure 20.9 *A commercially-available portable optical photometer for general aerosol measurement, the HAZ-DUST® (SKC Inc., Eighty Four, PA, USA). It is shown here as part of a personal sampling system, with a sampling head for selecting a specific aerosol fraction, in this case a cyclone for respirable aerosol, and a miniature optical sensor, both located close together in the breathing zone. A 37 mm plastic cassette is located just behind the sensor in order to collect the sampled aerosol for subsequent analysis (if desired). The pump and electronic processing unit is worn on the belt. Photograph courtesy of SKC Inc., Eighty Four, PA, USA*

the sampling head. A 37 mm plastic cassette may be placed just behind the sensor in order to provide the opportunity for subsequent gravimetric and chemical assessment of the sampled aerosol.

Black smoke was long used as a metric for ambient air pollution, especially in the UK and elsewhere in Europe and the British Standards Institution method (British Standards Institution, 1969) was recommended and widely used as a standard method for analyzing filters post-sampling. Direct-reading versions, operating on essentially similar principles, appeared during the 1980s (Hansen *et al.*, 1984) and have been commercially available ever since (e.g. GIV GmbH, Breuberg, Germany).

The individual optically based direct-reading sampling instruments described in the preceding were introduced by way of illustration of how the principles of light scattering can provide useful real-time information about the mass concentrations of workplace and ambient atmospheric aerosols. The list is far from exhaustive.

20.2.4   Optical particle counters

The light scattering photometers described in the previous section were all based on the concept of angular scattering of light from an ensemble of particles, the signal from which can be related – directly

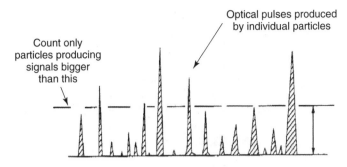

Figure 20.10 *Illustration of the principle of optical particle sizing and counting*

or indirectly – to a relevant index of concentration. By careful design of the sensing zone and choice of sensitive detection instrumentation, the same principles have been extended to enable the detection and assessment of individual particles. If the light scattered from an individual particle can be detected and registered electronically, it may not only be counted but also sized – that is, placed into a given size band (or electronic bin) based on the magnitude of the impulsive electrical signal arising from the detection of the scattered light by the particles. This general principle is illustrated in Figure 20.10. By such means, instruments can be designed that are capable of providing an overall particle size distribution. Many such devices have been described in the literature and found applications in occupational and environmental hygiene.

The optical configuration for a typical forward-scattering instrument is shown schematically in Figure 20.11. This formed the basis of the series of particle counting instruments commonly referred to for many years as the '*Royco*', For this family of instruments, particles were aspirated and enter an illuminated sensing zone from which scattered light was collected in the angular range 10–30°. A much wider effective scattering angle may be achieved by the use of an elliptical collecting mirror in the manner shown schematically in Figure 20.12, and this formed the basis of another, related, popular

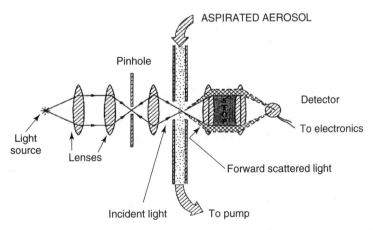

Figure 20.11 *Diagram to show the optical configuration and principle of operation of a typical forwards-scattering optical particle counter (e.g. Royco). Reprinted from Aerosol Science for Industrial Hygienists, Vincent. Copyright (1995), with permission from Elsevier*

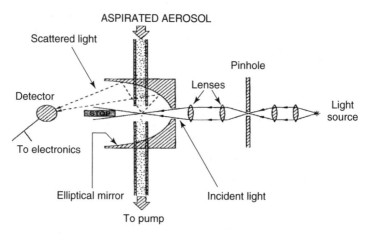

Figure 20.12 *Diagram to show the optical configuration and principle of operation of a typical wide-angle optical particle counter (e.g. Climet). Reprinted from Aerosol Science for Industrial Hygienists, Vincent. Copyright (1995), with permission from Elsevier*

family of particle counting instruments known as the '*Climet*' in which scattered light was collected in the angular range 15–105°. Two independent experimental studies were reported in the early 1980s to characterise the responses of popular versions of these two instruments, for particles in the size range from about 0.2 μm to about 10 μm (Mäkynen *et al.,* 1982; Chen *et al.,* 1984). The results were in broad agreement, indicating for both instruments a monotonic – almost linear – rise in response for increasing particle size in this range. Chen *et al.* noted that the smoothness of these responses was '... probably the result of the wide angle of integration with respect to scattering angle and wavelength.' Versions of both the Royco and Climet instruments are still commercially available (Hach Ultra Analytics, Grants Pass, OR, USA and Climet Instruments Company, Redlands, CA, USA, respectively). Typically they provide particle counts in the ranges of nominal particle size <0.3, 0.3–0.5, 0.5–0.7, 0.7–1, 1–2 and 2–5 μm. Related particle counting instruments based on the principles of light scattering are available from other sources. In general, for such instruments, since such responses are dependent on the type of particle, it is common practice to calibrate instruments of this type using polystyrene latex (PSL) test spheres of well-defined diameter and density. In that event, any instrument calibrated in this way provides particle size information that is 'PSL-equivalent'. Instruments like those described above have been widely used by occupational hygienists, mainly to provide quick, semi-quantitative information about changing aerosol conditions. One especially interesting application of one, the Royco, was as a counter for respirable airborne fibers in the workplaces of the asbestos textile industries. Here it was shown that, by setting the instrument to count particles with size > 5 μm, a reasonable correlation could be obtained with 'respirable' fiber counts as determined using the membrane filter and microscopy method described earlier (Addingly, 1966).

Fibrous aerosols are of considerable importance because of their severe implications to human health. For these, the criteria and methods for sampling and assessment are distinctly different to those for nonfibrous aerosols. It has long been recognised as a significant challenge to develop a direct-reading instrument which can distinguish and count fibres in the appropriate 'respirable' size range (National Institute for Occupational Safety and Health, 1979) and in the presence of other, nonfibrous particles. The *fibrous aerosol monitor* (*FAM*) proposed by Lilienfeld *et al.* (1979) was the first such instrument. In this ingenious device, the basic principle of operation was that airborne fibrous particles were subjected to a combined unidirectional and oscillating electric field. The larger unidirectional field aligned each fiber and the smaller oscillating component introduced a 'rocking' motion about its axes of alignment. The alignment and the rocking motion were dependent on the applied fields but independent of fiber dimensions (Lilienfeld, 1985). In the physical arrangement of the instrument, the electric fields were arranged so that the fiber rocking motion took place in the plane at right angles to an incident light beam, and light scattering associated with the rocking motion of the fiber was detected in this plane (Lilienfeld, 1987). It was found to be greater for long fibers and virtually nonexistent for spherical particles, so that particles could be electronically discriminated on the basis of their length. In practice, it was necessary to calibrate the FAM side-by-side with one of the more conventional membrane filter-based reference methods. This instrument was available commercially for a few years but has since been discontinued.

Apart from optical particle counters like those described where particle size is determined directly from the intensity of the pulse of scattered light, there are also optically based particle counting instruments which work on different principles. These include, for example, instruments where operation depends on the dynamic properties of particles as detected using phase-Doppler, time-of-flight and imaging techniques (as reviewed by Rader and O'Hern, 1993). One instrument that has become prominent is the *aerodynamic particle sizer (APS™)* (TSI Inc., St Paul, MN, USA), first mentioned earlier in Chapter 3. It was first proposed by Wilson and Liu (1980) and has since appeared in a succession of improved versions which have been widely used for both laboratory and field aerosol characterisation. The principle of operation of the APS is shown schematically in Figure 20.13(a) which shows how the aspirated particles are focused into a narrow beam by means of an annular separate sheath of clean air, and the beam is introduced into the sensing zone through an acceleration nozzle. Each particle is detected optically as it passes successively through a pair of narrowly focused divided beams from the split beam of a single HeNe laser. The resultant two scattered light pulses allows for the determination of particle velocity on the basis of its 'time-of-flight' between the two beams. The difference in velocity between the particle and the air is a direct measure of the particle's ability to respond to changes in the motion of the surrounding air, and so is related directly to particle aerodynamic diameter. A photograph of a current packaged instrument is shown in Figure 20.13(b). The APS is an important instrument in the context of much of what has been discussed elsewhere in this book. Other related optical particle counters to appear have included the *aerosizer* (Amherst Process Instruments, Amherst, MA, USA) (e.g. Dahneke, 1973) and the *electric single particle aerodynamic relaxation time analyzer (E-SPART)* (e.g. Renninger *et al.*, 1981).

Although the optical response, reflecting the light-particle interaction, is a primary performance feature of optical particle counting instruments like those described above, there are other important practical factors that need to be taken into account. In particular there is the question of defining the sensing zone into which the particles of interest may enter and from which the scattered light is received by the detector. This is determined by the design of the collector/detector optics. Here the technical challenge lies in making the sensing volume large enough such that there is always a good chance of finding a particle there, but not so large that there will be more than one particle at any given time. This may be achieved firstly by constraining the aerosol flow in the form of a thin particle beam in a sheath of clean air and then ensuring that the light-collecting optics receives light from a region which is small enough to contain only one particle. If more than one particle is present in the sensing zone at the same time, they will be counted and sized together and recorded as if they were a single, larger particle. Thus, in some practical situations, aerosol concentrations may be great enough that, even at low sampling flow rates, such 'coincidences' may preclude the use of a particular instrument. In some instruments (e.g. the APS), dilution of the sampled aerosol has been employed in order to alleviate this problem.

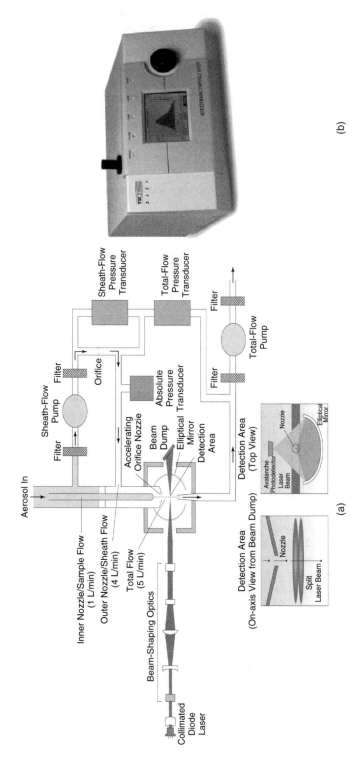

Figure 20.13 *The aerodynamic particle sizer (APS): (a) schematic to describe the principle of operation; (b) photograph of a current packaged instrument, TSI Model 3321. Pictures courtesy of Gilmore Sem, TSI Inc., St Paul, MN, USA*

20.2.5   Particle size and shape

The light scattered from an aerosol contains a great deal of information about the concentration, size, shape and composition of the particles. The challenge has always been to find ways to extract such information. Reid *et al.*, (1992) proposed a system that derived both particle size and shape information from the pattern of light scattered by individual aerosol particles as they passed through a laser beam. In the original version of this system, the shape parameter was determined from measurement of the degree of cylindrical symmetry using three photodetectors placed around the optical axis. The measured intensities at the three detectors were representative of the three-dimensional light scattering pattern, and were combined to provide a measure of particle symmetry and the degree of particle alignment with the flow through the sensing zone. The arrangement was successful in differentiating between particles of simple shape in the range of geometric particle size from about 1 to 10 μm. Further work refined the basic idea (e.g. Kaye *et al.*, 2000), leading up to the *aerosol shape analysis system* (ASAS). One recent commercial outcome from this effort is the four-detector *Aspect* (Biral, Bristol, UK), an instrument capable of classifying individual particles by geometric diameter in the range from 0.5 to 20 μm and *asymmetry factor* from 0 to 100. Here the asymmetry factor is a dimensionless quantity that is zero for perfectly spherical particles and is 100 for very long fibers. This instrument is potentially very useful for building a 'fingerprint' of the measured aerosol that goes beyond mere particle size distribution.

20.3   Electrical particle measurement

The diverse physical properties of aerosols enable a range of further options for particle detection and characterisation. The electrostatic charge carried by aerosol particles, and which can be placed on particles in a controlled way through corona discharge techniques, provides a powerful alternative to the optical approaches discussed above. There are two primary objectives of electrical measurement, the electric charge distribution in an aerosol and the particle size distribution.

Johnston (1983) described a *split-flow electrostatic elutriator* in which particles were sampled into a rectangular duct between a pair of conducting plates, and were then transported towards the upper or lower plate depending on the magnitude and polarity of the charge carried and on the applied voltage between the plates. Particles were classified according to their size, using a Royco optical particle counter set to the appropriate particle size channel, at the exits of the upper and lower halves of the duct, respectively, and for a range of applied voltages. By analysis of the two penetration curves using the *method of tangents* described earlier by Hurd and Mullins (1962) for a simple electrostatic elutriator, the distribution and magnitude of the charge on the sampled aerosol were determined. However, the process was laborious, prompting Wake *et al.* (1991) to develop an automated version. The Johnston split-flow electrostatic elutriator was used in research to characterise the state of charge of aerosols in a range of workplaces (Johnston *et al.*, 1985), driven by concerns that such properties may have a significant bearing on the lung deposition of particles inhaled by people and on the performances of sampling and control devices. However, no commercial versions of this instrument were ever developed.

Hochrainer (1985) suggested that it would be better to use a device where the aerosol was introduced into the inter-electrode space through an entrance slit whose location was well defined, thus rendering the method of tangents redundant and so providing a direct measure of the electrical mobility of the particles (and hence their charge). Such instruments are now referred to as '*electrical mobility spectrometers*'. Two versions have been developed, both based on cylindrical geometry and both remain popular to this day. The first is the *electrical aerosol analyzer* (EAA) shown schematically in Figure 20.14(a). Here the particles entering through the narrow annular slit at the top of the inter-electrode space, and those which are not deposited by electrical forces on either electrode, pass through to a particle counter that

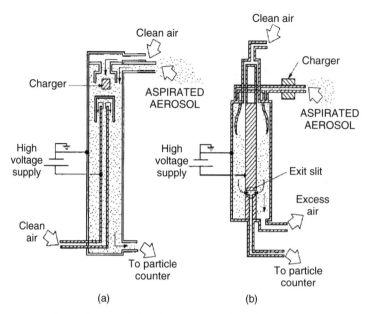

Figure 20.14 *Diagrams to show the principle of operation of (a) the electrical aerosol analyzer (EAA) and (b) the differential mobility analyzer (DMA). Reprinted from Aerosol Science for Industrial Hygienists, Vincent. Copyright (1995), with permission from Elsevier*

is capable of detecting particles in the size range of interest. Alternatively the EAA may be used as a source of classified fine particles for the purpose of aerosol experimentation. The second such instrument is the *differential mobility analyzer* (*DMA*), shown schematically in Figure 20.14(b), in which all those particles are detected which arrive at the central electrode exactly at the location of the exit slit shown in the diagram. For both the EAA and the DMA, the electrical mobility distribution of the aerosol entering both instruments is provided directly from the curve of particle counts versus applied voltage. In the *scanning DMA* the applied voltage is varied automatically. If the distribution of electric charge is known as a function of particle size, then the results may be used to determine the particle size distribution. Such instruments are best suited to fine particles, typically less than 1 μm in diameter, where the effects of gravity and inertia during aspiration and classification are negligible. In 1998, Chen *et al.* described a version of the DMA designed specifically for the characterisation of particles in the range of diameter from 3 to 50 nm, and this has important potential for practical applications at the present time when there is a high level of interest in ultrafine aerosols. Different versions of the DMA have been developed commercially for different ranges of particle sizes – the *Long DMA* for particles with diameter ranging from 10 nm to 1 μm and the *Nano-DMA* for particles with diameter ranging from 2 to 150 nm (TSI Inc., St Paul, MN, USA).

20.4   Condensation nuclei/particle counters

Instruments like those described above require the detection of very small particles. This is typically achieved using another class of particle counters whose operation depends on the condensation of liquid vapor onto the very small particles acting as nuclei, following the theory of heterogeneous

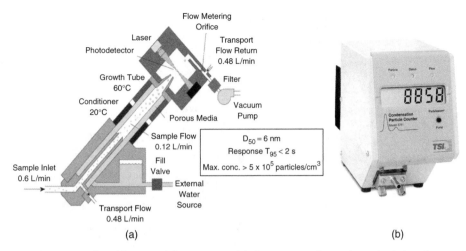

Figure 20.15 *Water condensation particle counter: (a) diagram to show the principle of operation of the condensation nuclei/particle counter; (b) photograph of a current packaged instrument, TSI Model 3781. Pictures courtesy of Gilmore Sem, TSI Inc., St Paul, MN, USA*

nucleation and particle growth (e.g. as reviewed by Hinds, 1999 and many others). The principle of operation of the *condensation particle counter* (*CPC*) – sometimes called the *condensation nuclei counter* (*CNC*) – involves firstly the sampling of aerosols and their introduction into a region which is saturated with water or some other appropriate substance (e.g. alcohol). The atmosphere is then made to supersaturate, preferably by convective cooling since supersaturation by expansion will require a nonuniform flow, usually considered undesirable from the sampling point of view. This causes molecules from the vapor phase to condense onto the very small particles acting as nuclei. The particles then grow to become large enough for detection by conventional optical particle counting techniques. The instrument was first described by Agarwal and Sem (1980). Versions have been on the market ever since, and several models are available today, including ones for the sampling and detection of ultrafine particles (TSI Inc. St Paul, MN, USA). By way of illustration, a typical current CPC is shown in Figure 20.15, where the model in question uses water as the growth liquid, wetting the porous media that surrounds the sampled aerosol stream. The CPC is widely used in fine particles aerosol research. In particular, it has become an important auxiliary instrument for use in counting ultrafine particles selected according to their size by DMA-type instruments. One interesting version of this device, the *Portacount*® (TSI Inc., St Paul, MN, USA) is frequently employed by occupational hygienists in the fit testing of personal respiratory protection equipment. As shown in Figure 20.16, this instrument incorporates a small, portable CPC that, in use, measures fine ambient atmospheric particles both inside and outside of the respirator when it is being worn by the subject, and provides a measure of the level of protection that is being afforded, in particular the quality of the fit between the respirator and the face of the wearer.

20.5 Mechanical aerosol mass measurement

For most of the direct-reading instruments described so far, the physics of the particle detection process is such that the mass concentration of the aerosol of interest is usually not provided directly. This means that they are not ideally matched to the needs of many aerosol sampling applications. There are,

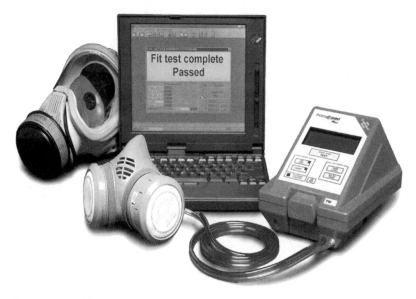

Figure 20.16   *The Portacount®, a portable version of the condensation nuclei counter. Photograph courtesy of Gilmore Sem, TSI Inc., St Paul, MN, USA*

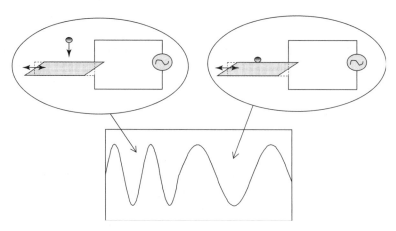

Figure 20.17   *Schematic to illustrate the principle of the piezoelectric mass balance concept. The AT-cut piezoelectric crystal oscillates in the transverse mode. The frequency of oscillation decreases when the effective mass of the crystal increases upon arrival of the particle*

however, some further physical options which can be applied to what are generally referred to as '*mass balances*'. These share the common feature that the aerosol of interest must first be sampled and collected efficiently onto an appropriate surface. One such device is the *piezoelectric mass balance*, the principle of which is shown in Figure 20.17. The main sensor is the piezoelectric crystal (e.g. 'AT-cut' quartz). When an oscillating potential difference is applied between the coated conducting surfaces placed as electrodes on each face of the crystal, the crystal vibrates mechanically in its transverse mode. The resonant frequency for these mechanical oscillations is a strong function of the mass of the crystal – by analogy with a simple physical spring–mass system. So any change in mass of the crystal produces a change in resonant frequency of the form:

$$\Delta f = k f^2 \Delta m \qquad (20.4)$$

where Δf is the change (decrease) in frequency from its value at f corresponding to a change (increase) in mass Δm, and where k is a coefficient which describes the sensitivity of mass detection, dependent on the type of crystal, its geometry and size, etc. The arrival of a particle at the vibrating surface will result in such a change in mass. A summary of the piezoelectric mass balance concept and its applications was given by Ward and Buttry (1990). From Equation (20.4), it can be seen that if Δf and f can be detected and measured in the external driving circuitry, then Δm may be obtained, provided that k can be determined by means of appropriate calibration. Therefore it follows that monitoring Δf over a short time interval will provide information about corresponding changes in the mass present on the surface, and hence on the sampled aerosol concentration. So an instrument based on this idea can in principle provide real time information about aerosol concentration. Electrostatic deposition has been shown to be an effective deposition mechanism, as featured in instruments described by Olin *et al.* (1971) (see Figure 20.18). Extensive development of the piezoelectric mass balance idea, leading up to commercial application, was reported by Sem and Tsurubayashi (1975) and Sem *et al.* (1977). More recently, the same idea was applied to the development of a personal aerosol sampler (Wilson *et al.*, 1997). Although piezoelectric mass balances operating on this principle were commercially obtainable for several years (e.g. from TSI Inc., St Paul, MN, USA), they appear to be no longer available. In some other applications of the piezoelectric mass balance idea, impaction has also been employed for particle

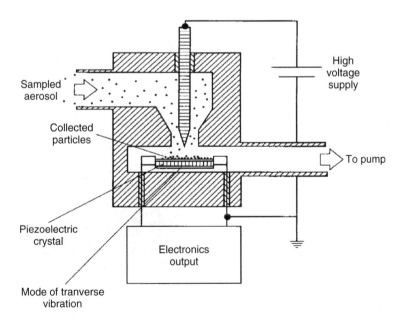

Figure 20.18 *Diagram to show a typical configuration for a piezoelectric mass balance aerosol sampling system in which the particles are deposited by electrostatic precipitation (after Olin et al., 1971). Reprinted from Aerosol Science for Industrial Hygienists, Vincent. Copyright (1995), with permission from Elsevier*

collection. In one example, piezoelectric mass balances were built into the stages of a cascade impactor (Carpenter and Brenchley, 1972; Fairchild and Wheat, 1984: O'Brien *et al.*, 1986;), and versions of these are available commercially (California Measurements Inc., Sierra Madre, CA, USA) and have been used quite extensively in studies of indoor and outdoor air quality, pharmaceutical research and occupational hygiene investigations.

There are some basic practical limitations associated with the piezoelectric mass balance approach. Firstly, the accurate detection of mass requires that the deposited particles are rigidly attached to the active crystal surface and are able to remain so during the rapid accelerations experienced during the mechanical oscillation of the crystal. Although this is the case for very fine particles, for which the short-range adhesion forces are large, it is less so for larger particles. Particles which adhere less well to the oscillating surface are less well detected, leading to significant undersampling of mass. The second limitation is that, if the crystal becomes heavily loaded, the response of the crystal as given by Equation (20.4) may become nonlinear and unpredictable. These and other factors such as temperature and humidity have been widely reported (see Daley and Lundgren, 1975; Lundgren *et al.*, 1976).

A somewhat related sampling instrument that has become prominent in recent years is the *tapered element oscillating microbalance* (*TEOM®*) first reported by Patashnick and Hemenway (1969). The principle of operation is shown in Figure 20.19(a). The device features a hollow glass tapered tube with a conductive coating. One end of the tube is anchored, while the other end supports a filter through which the sampled air is drawn and on which the particulate material is deposited. The tube oscillates about its anchor point, driven electromechanically by the external circuitry. As the mass on the filter – and hence on the end of the tapered element – increases, the frequency of the oscillation decreases. Again, the rate of change may be related to the mass rate at which the aerosol is being sampled, and hence to aerosol concentration. It is available commercially in several versions (Thermo Electron Corporation, Waltham, MA, USA), and these have been widely applied in recent years for the sampling of aerosols

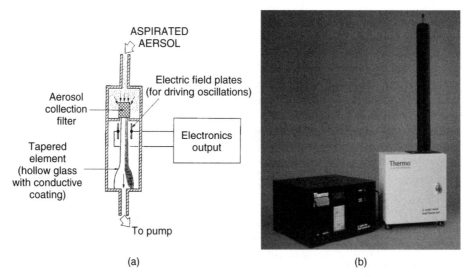

(a)   (b)

Figure 20.19   *The tapered element oscillating microbalance (TEOM): (a) schematic to show the principle of operation; (b) the packaged instrument. (a) Reprinted from Aerosol Science for Industrial Hygienists, Vincent. Copyright (1995), with permission from Elsevier. (b) Photograph courtesy of Thermo Electron Corporation, Waltham, MA, USA*

in both occupational and ambient environmental settings. One such packaged instrument is shown in Figure 20.19(b). In another recent interesting version, the TEOM concept has been incorporated into a package that is compact yet robust enough that, when used with a cyclone-based respirable aerosol pre-selector, it can be deployed as a personal respirable dust monitor, specifically with a view to use in underground mining situations (Volkwein *et al.*, 2004a, b).

20.6 Nuclear mass detectors

The attenuation of nuclear radiation as it passes through matter provides yet another alternative for the determination of aerosol concentration. As for the mass balances described above, this approach has been applied for particulate matter that has been deposited onto a substrate. Negatively charged beta-particles are a preferred form of radiation because their absorption by matter lies in the range where sensitive measurement can be made for deposits typical of what are found during practical aerosol sampling. The principle is illustrated in Figure 20.20. Beta-particles are emitted during the radioactive decay of isotopes such as ^{14}C or ^{147}Pm. The physical nature of their interaction with matter is such that the efficiency (the *cross-section*) of the absorption is proportional to the ratio between the atomic number (the number of protons in the nucleus) and the atomic weight for the substance in question. In principle, since this ratio does not vary much between the elements it follows that, to a good approximation, the interaction relates uniquely to mass.

For a beam of beta particles passing through an attenuating medium, the intensity falls according to the familiar exponential law, this time in the form:

$$I = I_0 \exp(-\mu X) \tag{20.5}$$

where μ here is the mass absorption coefficient (e.g. in $cm^2\ g^{-1}$) and X is the mass thickness ($g\ cm^{-2}$). In practice, the particles need to be deposited onto a collecting surface such as a filter or an impaction surface. The earliest report of a sampling instrument operating on this principle seems to be that by Nader and Allen (1960). But subsequent progress towards a practical device for use in the working environment owes much to the contributions of Macias and Husar (1976) and Lilienfeld (1970, 1975).

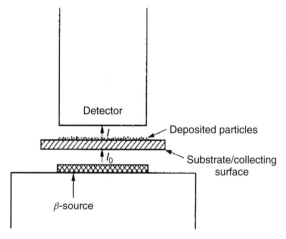

Figure 20.20 *Diagram to show the principle of operation of the beta-attenuation mass balance. Reprinted from Aerosol Science for Industrial Hygienists, Vincent. Copyright (1995), with permission from Elsevier*

In the beta-attenuation approach, one complication is that it is necessary to take into account the attenuation of the radiation not just by the collected particulate material itself but also by the collection substrate. This means that, in its application in a direct-reading instrument, it is necessary to correct for the contribution of the latter, for example, by alternately scanning areas of clean and mass-deposited substrate. Other complications arise from the effects associated with the geometry of the beta source and detection system. Careful calibration against known gravimetrically assessed reference samples is therefore required. Commercial instruments based on this principle have appeared over the years. In one, the *continuous ambient particulate monitor* (Thermo Electron Corporation, Waltham, MA, USA), the aspirated particles are deposited onto a moving filter tape which – by the use of two detectors – allows for the measurement of beta attenuation through both the clean tape and the part where particles have deposited and then provides particle mass concentration as a function of time.

20.7   Surface area monitoring

Knowledge gained from recent inhalation toxicology and epidemiology suggests that an appropriate metric for exposures to ultrafine particles may be one based on particulate surface area (as opposed to mass, as is the case for most aerosol exposures). With this in mind, there has been a search for direct-reading instrumentation that will provide aerosol measurements based on this index. In one instrument, the *AEROTRAK™ 9000 Nanoparticle Aerosol Monitor* (*NAM*) indicates the surface area of nanoparticle aerosols that deposit in the lung in accordance with the curves published by the International Commission for Radiological Protection (International Commission for Radiological Protection 1994) for the tracheobronchial and alveolar regions of the human respiratory tract (TSI Inc., St Paul, MN, USA). In this instrument (see Figure 20.21), fine particles in the size range of interest are charged by diffusion and are then detected by means of an electrometer. Importantly, according to the manufacturer, the

Figure 20.21   *Photograph of the AEROTRAK™ 9000 Nanoparticle Aerosol Monitor (NAM) for the measurement of the surface area concentrations of nano-sized aerosols corresponding to tracheobronchial and alveolar deposition in the human lung. Photograph courtesy of Gilmore Sem, TSI Inc., St Paul, MN, USA*

instrument does not measure just the total surface of the sampled aerosol but can be set to provide the surface area of fractions relating to regional lung deposition (e.g. tracheobronchial or alveolar). Since the charge imparted to particles in the manner indicated is directly related to surface area, the current imparted to the electrometer is also proportional to the surface area of the particles of interest.

20.8 Analytical chemical methods

As mentioned elsewhere in this book, there are situations where it is desired to measure the carbonaceous component of aerosols, for example in relation to black smoke and diesel particulate matter. Although interest in the former index of atmospheric aerosol has waned in recent years, the latter remains important. One direct-reading instrument relevant to such applications involves application of the stepwise, combustion-based (know as the '*thermal-CO$_2$*') approach by which to oxidise collected carbon-containing particulate matter to quantitate – and differentiate between – organic carbon and 'soot' components. This instrument is commercially available (Thermo Electron Corporation, Waltham, MA, USA).

20.9 Bioaerosol monitoring

There is great current interest in the real time detection and measurement of bioaerosols, driven in large measure by concerns about possible intentional releases of highly infectious agents into indoor and outdoor spaces. Conventional bioaerosol sampling methods for such agents require extended periods of sampling followed by even more extended assays. In certain instances, like those alluded to, such delays may not be acceptable. The technical problem is very difficult, not least because of the diversity of the biological agents that need to be assessed. More specifically, identification of biological organisms within a few minutes and at extremely low levels has been identified as a very difficult task, and urgent efforts have taken place during the past decade and are still going on (e.g. Ghosh and Prelas, 2002). Two different optical approaches are considered promising, the first involving fluorescence technology and the second involving an extension of the concept of simultaneously measuring particle size and shape.

20.9.1 Fluorescence technology

Many molecules of biological material, when exposed to ultraviolet light, exhibit specific excitation and emission spectra that represent their *intrinsic fluorescence*. Hairston *et al.* (1997) noted that nicotinamide adenine dinucleotide phosphate [NAD(P)H] is an indicator of bacterial activity, making it a good candidate for discriminating particles of biological origin from inert material. The fact that NAD(P)H is excited at a light wavelength of 340 nm and fluoresces in the emission band from 420 to 580 nm with a peak at 525 nm provides specific guidance for the detection of bioaerosols. Based on this, Hairston *et al.* described a new instrument for the real time detection of bioaerosols by means of the simultaneous measurement of particle size and the aforementioned intrinsic fluorescence. The first part involved direct application of the technology already applied in the development of the aerodynamic particle sizer (APS) and described earlier in this chapter. The second part involved adaptation of techniques applied elsewhere in liquid flow cytometry of the type that is widely used in experimental biology for the study of micro-organisms (e.g. Ho and Fisher, 1993). The result was a prototype instrument that became known as the '*fluorescent aerodynamic particle sizer*' (*FLAPS*). It contained two separate two laser systems, the first a HeNe laser that provided the pair of beams to provide the particle size-sensitive functions of the original APS and the second a HeCd ultraviolet laser emitting at a wavelength of 325 nm to stimulate the intrinsic fluorescence for particles of biological origin. For the latter the collecting optics were set up with a system of filters that served to transmit the desired wavelength band to a photomultiplier yet block the undesired light scattered from the system at other wavelengths. A prototype instrument was shown to perform well in the laboratory for the detection of *B. subtilis* and a commercial version of the FLAPS is now available (TSI Inc., St Paul, MN, USA).

20.9.2   Particle size and shape for bioaerosols

The earlier description of the *aerosol shape analysis system* (*ASAS*) provides options for the detection of bioaerosols. Although it is not claimed that the ASAS technology can identify specific agents, the articulation of a 'particle size/shape fingerprint' is a significant step towards the identification of particles having specific size and shape properties – as is the case for many micro-organisms. In the practical use of an instrument such as the *Aspect* mentioned earlier, the results obtained may be examined by sophisticated analytical algorithms in order to test for systematic changes in the aerosol which is being observed, facilitating a first indication of whether any such change correlates with the presence of an unwanted biological organism. Some 'false positives', which may be associated with specific aerosol types whose fingerprints are distinctive (e.g. soots and smokes, certain bioaerosols such as pollens, etc.), may be eliminated by correlation procedures (Biral, 2002).

20.9.3   Hybrid systems

Both the approaches described above provide results that are ambiguous. The 'fluorescent aerosol particle sizer' approach allows identification of the presence of biological material but not specific biological species. The 'particle size and shape' approach provides a great deal of information that might be useful towards implicating specific species, but does not actually contain any biological information at all. In addition, both approaches may be prone to producing 'false positives'. The problem is summarised in Figure 20.22 where it is shown how a combination of measurement of particle size, particle shape and particle nature (biological versus nonbiological) can provide options for rapid detection and identification

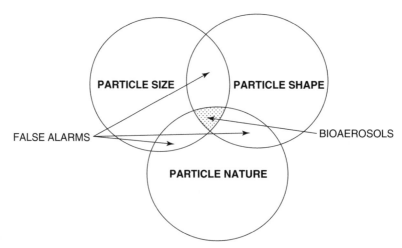

Figure 20.22   *Diagram to indicate how a combination of measurement of particle size, particle shape and particle nature (biological versus nonbiological) can provide options for rapid detection and identification of bioaerosols*

of bioaerosols and narrow down the discrimination process (Foot *et al.*, 2004). The new *VeroTect* instrument based on these ideas, with software containing a sophisticated discrimination algorithm, has been chosen for incorporation into the *Integrated Sensor Management System* (*ISMS*) for the UK Ministry of Defence (Biral, Bristol, UK).

References

Addingly, C.G. (1966) Asbestos dust and its measurement. *Annals of Occupational Hygiene*, 9, 73–82.

Agarwal, J.K. and Sem, G.J. (1980) Continuous flow, single particle-counting condensation nucleus counter, *Journal of Aerosol Science*, 11, 343–357.

Armbruster, L. and Breuer, H. (1983) Dust monitoring and the principle of on-line dust control. In: *Aerosols in the Mining and Industrial Work Environments* (Eds V.A. Marple and B.Y.H. Liu), Ann Arbor Science Publishers, Ann Arbor, MI, pp. 689–699.

Biral (2002) Shedding light on real-time biodetection. In: *Chemical Biological Warfare Review*, Biral, Portishead.

British Standards Institution (1969) Methods for the measurement of air pollution, 2. Determination of concentration of suspended particulate matter, British Standard 1747: Part 2.

Carpenter, T.E. and Brenchley, D.L. (1972) A piezoelectric cascade impactor for aerosol monitoring, *American Industrial Hygiene Association Journal*, 33, 503–510.

Chen, B.T., Cheng, Y.S. and Yeh, H.C. (1984) Experimental responses of two optical particle counters, *Journal of Aerosol Science*, 15, 457–464.

Chen, D.-R., Pui, D.Y.H., Hummes, D., Fissan, H., Quant, F.R. and Sem, G.J. (1998) Design and evaluation of a nanometer aerosol differential mobility analyzer (Nano-DMA), *Journal of Aerosol Science*, 29, 497–509.

Dahneke, B. (1973) Aerosol beam spectrometry, *Nature Physical Science*, 244, 54–55.

Daley, P.S. and Lundgren, D.A. (1975) The performance of piezoelectric crystal sensors used to determine aerosol mass concentrations, *American Industrial Hygiene Association Journal*, 36, 518–532.

Drinker, P. and Hatch, T. (1934) *Industrial Dust*, McGraw-Hill, New York.

Evans, B.M.T., Yee, E., Roy, G. and Ho, J. (1994) Remote detection and mapping of bioaerosols, *Journal of Aerosol Science*, 25, 1549–1566.

Fairchild, C.I. and Wheat, L.D. (1984) Calibration and evaluation of a real-time cascade impactor, *American Industrial Hygiene Association Journal*, 45, 205–211.

Foot, V.E., Clark, J.M., Baxter, K.L. and Close, N. (2004) Characterising single airborne particles by fluorescence emission and spatial analysis of elastic scattered light. In: *Optically Based Biological and Chemical Sensing for Defence* (Eds J.C. Carrano and A. Zukauskas), Proceedings of the SPIE 5617, The International Society for Optical Engineering, Bellingham, WA, pp. 292–299.

Gebhart, J. (1993) Optical direct-reading techniques: light intensity systems. In: *Aerosol Measurement* (Eds K. Willeke and P.A. Baron), Van Nostrand Reinhold, New York, Chapter 15.

Ghosh, T.K. and Prelas, M.A. (2002) Sensors and detection systems for biological agents. In: *Science and Technology of Terrorism and Counterterrorism* (Eds T.K. Ghosh, M.A. Prelas, D.S. Viswanath and S.K. Loyalka), Marcel Dekker, New York.

Hairston, P.P., Ho, J. and Quant, F.R. (1997) Design of an instrument for real-time detection of bioaerosols using simultaneous measurement of particle aerodynamic size and intrinsic fluorescence, *Journal of Aerosol Science*, 28, 471–482.

Hansen, A.D.A., Rosen, H. and Novakov, T. (1984) The aethalometer – an instrument for the real-time measurement of optical absorption by aerosol particles, *Science of the Total Environment*, 36, 191–196.

Hinds, W.C. (1999) *Aerosol Technology*, 2nd Edn, John Wiley & Sons, Inc., New York.

Ho, J. and Fisher, G. (1993) Detection of BW agents: flow cytometry measurements of *Bacillus subtilis* (BG) spore fluorescence, Defense Research Establishment Suffield Memorandum 1421.

Hochrainer, D. (1985) Measurement methods for electric charges on aerosols, *Annals of Occupational Hygiene*, 29, 241–249.

Hodkinson, J.R. (1966) The optical measurement of aerosols. In: *Aerosol Science* (Ed. C.N. Davies) Academic Press, London, pp. 287–357.

Hurd, F.K. and Mullins, J.C. (1962) Aerosol size distributions from ion mobility, *Journal of Colloid Science*, 17, 91–100.

International Commission for Radiological Protection 1994 *Human Respiratory Tract Models for Radiological Protection*, ICRP Publication 66, Pergamon Press, Elmsford, NY.

Johnston, A.M. (1983) A semi-automatic method for the assessment of electric charge carried by airborne dust, *Journal of Aerosol Science*, 14, 643–655.

Johnston, A.M., Vincent, J.H. and Jones, A.D. (1985) Measurements of electric charge for workplace aerosols, *Annals of Occupational Hygiene*, 29, 271–284.

Kaye, P.H., Barton, J.E., Hirst, E. and Clark, J.M. (2000) Simultaneous light scattering and intrinsic fluorescence measurement for the classification of airborne particles, *Applied Optics*, 39, 3738–3745.

Kerker, M. (1963) *Electromagnetic Scattering*, Pergamon Press, Oxford.

Leck, M.J. (1983) Optical scattering instantaneous respirable dust indication system, In: *Aerosols in the Mining and Industrial Work Environments* (Eds V.A. Marple and B.Y.H. Liu), Ann Arbor Science Publishers, Ann Arbor, MI, pp. 701–717.

Lilienfeld, P. (1970) Beta absorption impactor aerosol mass monitor. *American Industrial Hygiene Association Journal*, 31, 722–729.

Lilienfeld, P. (1975) A new ambient particulate mass monitor using beta attenuation. Paper presented at the 68th Annual Meeting of the Air Pollution Control Association, Boston, MA, Paper 75–65–2.

Lilienfeld, P. (1985) Rotational electrodynamics of airborne fibers, *Journal of Aerosol Science*, 16, 315–322.

Lilienfeld, P. (1987) Light scattering from oscillating fibres at normal incidence, *Journal of Aerosol Science*, 18, 389–400.

Lilienfeld, P., Elterman, P. and Baron, P. (1979) The development of a prototype fibrous aerosol monitor, *American Industrial Hygiene Association Journal*, 40, 270–282.

Lundgren, D.A., Carter, L.D. and Daley, P.S. (1976) Aerosol mass measurement using piezoelectric crystal sensors, In: *Fine Particles* (Ed. B.Y.H. Liu), Academic Press, New York, pp. 485–510.

Macias, E.S. and Husar, R.B. (1976) Atmospheric particulate mass measurement with beta attenuation mass monitor, *Environmental Science and Technology*, 10, 904–907.

Mäkynen, J., Hakulinen, J., Kivisto, T. and Isehtimaki, M. (1982) Optical particle counters: resolution and counting efficiency, *Journal of Aerosol Science*, 13, 529–535.

Mie, G. (1908) A contribution to the optics of turbid media, especially colloidal metallic suspensions, *Animals of Physics*, 25, 377–445.

Nader, J.S. and Allen, D.R. (1960) A mass loading and radioactivity analyzer for atmospheric particulates. *American Industrial Hygiene Association Journal*, 21, 300–307.

National Institute for Occupational Safety and Health (1979) USPHS/NIOSH membrane filter method for evaluating airborne asbestos fibers, Criteria for a recommended standard – occupational exposure to cotton dust, NIOSH Technical Report.

O'Brien, D.P., Baron, P. and Willeke, K. (1986) Size and concentration measurement of industrial aerosol, *American Industrial Hygiene Association Journal*, 47, 386–392.

Olin, J.G., Sem, G.J. and Christenson, D.L. (1971) Piezoelectric aerosol mass concentration monitor, *American Industrial Hygiene Association Journal*, 32, 791–800.

Patashnick, H. and Hemenway, C.L. (1969) Oscillating fibre microbalance, *Review of Scientific Instruments*, 400, 1008–1011.

Rader, D.J. and O'Hern, T.J. (1993) Optical direct-reading techniques: *in situ* sensing. In: *Aerosol Measurement* (Eds K. Willeke and P.A. Baron), Van Nostrand Reinhold, New York, pp. 345–380.

Reid, K., Martin, A.T. and Clark, J.M. (1992) Particle shape discrimination using the aerosol shape analysis system (ASAS), *Journal of Aerosol Science*, 23, S325–S328.

Renninger, R.G., Mazumder, M.K. and Testerman, M.K. (1981) Particle sizing by electrical single particle aerodynamic relaxation time analyser, *Review of Scientific Instruments*, 52, 242.

Rubow, K.L. and Marple, V.A. (1983) Instrument evaluation chamber: calibration of commercial photometers. In: *Aerosols in the Mining and Industrial Work Environments* (Eds V.A,. Marple and B.Y.H. Liu), Ann Arbor Science Publishers, Ann Arbor, MI, pp. 777–795.

Sem, G.J. and Tsurubayashi, K. (1975) A new mass sensor for respirable dust measurements, *American Industrial Hygiene Association Journal*, 36, 791–799.

Sem, G.J., Tsurubayashi, K. and Homma, K. (1977) Performance of the piezoelectric microbalance respirable aerosol sensor, *American Industrial Hygiene Association Journal*, 38, 580–588.

Van de Hulst, H.C. (1957) *Light Scattering by Small Particles*, John Wiley & Sons, Inc., New York.

Volkwein, J.C., Thimons, E., Dunham, D., Patashnik, H. and Rupprecht, E. (2004a) Development and evaluation of a new personal dust monitor for underground mining applications. In: *Proceedings of the 29th International Technical Conference on Coal Utilization and Fuel Systems*, Clearwater, FL, Volume II, pp. 1355–1375.

Volkwein, J.C., Vinson, R.P., McWilliams, L.J., Tuchman, D.P. and Mischler, S.E. (2004b) Performance of a new personal respirable dust monitor for mine use, National Institute for Occupational Safety and Health Publication No. 2004151, United States Department of Health and Human Services Centers for Disease Control, Pittsburgh, PA.

Wake, D., Thorpe, A., Bostock, G.J., Davies, J.K.W. and Brown, R.C. (1991) Apparatus for measurement of the electrical mobility of aerosol particles: computer control and data analysis, *Journal of Aerosol Science*, 22, 901–916.

Ward, M.D. and Buttry, D.A. (1990) *In situ* interfacial mass detection with piezoelectric transducers, *Science*, 249, 1000–1007.

Wilson, J.C. and Liu, B.Y.H. (1980) Aerodynamic particle size measurement by laser-Doppler velocimetry, *Journal of Aerosol Science*, 11, 139–150.

Wilson, L.W., Hepher, M.J., Reilly, D. and Jones, J.D.C. (1997) Development of a personal dust monitor with a piezoelectric quartz crystal sensor, *Measurement Science and Technology*, 8, 128–137.

Wind, L. and Szymanski, W.W. (2000) Forward scattering corrections for optical extinction measurements in aerosol media, *Journal of Aerosol Science*, 31, S400–S401.

Part D

Aerosol Sampler Applications and Field Studies

21

Pumps and Paraphernalia

21.1 Introduction

The major portion of this book has dealt with the parts of aerosol sampling systems that select and collect particles in defined particle size fractions – that is the sampling head, including the inlet and inner cavity or conduit leading down to the filter or sensing zone. These embody all the considerations of the air flow and particle transport in the sampling system and govern the basic performance of each and every sampler, defining the extent to which it meets the practical needs of occupational or environmental hygiene. That is the extent of aerosol sampling science. However, the realisation of practical workable sampling systems that can produce the desired results in terms of aerosol characterisation requires additional technical equipment and methodology, including air moving apparatus, flow measurement, filter and collection substrates, and analytical apparatus and procedures – pumps and paraphernalia. Aerosol samplers cannot be taken into the field without these. Here a short review is provided of the additional considerations that need to be taken into account in the design and practical application of aerosol sampling systems. Further details are given in the air sampling instruments handbook published by the American Conference of Governmental Industrial Hygienists (ACGIH), currently available in its 9th edition (American Conference of Governmental Industrial Hygienists, 2001).

21.2 Air moving systems

Active sampling for aspirating samplers requires an air mover that can deliver the desired flow rate against the pressure drop not only across the sampling head itself but also – in most samplers – across the filter. Since, as was seen in the first half of this book, the performances of aerosol samplers are strongly dependent on external and internal flow conditions, it is important that the aspirating flow closely matches the design flow rate, and that the flow rate is accurately maintained throughout the period during which sampling takes place. This therefore also requires an accurate means by which the flow rate can be controlled and monitored. Air moving systems vary a great deal in capacity and size, depending on the application. They may be broadly categorised in terms of size, portability and flow rate, including: mains-powered, nonportable, high-volume (up to the order of $m^3\ min^{-1}$) for area

Aerosol Sampling: Science, Standards, Instrumentation and Applications   James Vincent
© 2007 John Wiley & Sons, Ltd

sampling, mains-powered, portable, high/medium-volume (up to 100 Lpm) for area sampling, battery-powered, portable, low-volume (up to 10 Lpm) for area sampling, and battery-powered low-volume (up to 5 Lpm) for personal sampling.

21.2.1 Pumps

In their chapter in the ACGIH air sampling instruments handbook, Monteith and Rubow (2001) identified five types of air moving apparatus that provide options for sampling: diaphragm pumps, piston pumps, rotary vane pumps, blowers and air ejectors. Of these, blowers come into the category more commonly known as 'fans'. Although they can deliver very large volumetric flow rates, they can operate only against relatively small pressure drops. They have been used in large sampling devices with very high flow rates (e.g. of the order of $m^3 min^{-1}$), including for example the wide ranging aerosol classifier (WRAC) described in Chapter 18. Air ejectors have had only limited use in aerosol sampling – for example in dilution systems that were developed for use in stack sampling (see Chapter 16). The other three types of air mover have been more popular for general air sampling applications in occupational and environmental hygiene.

In the diaphragm pump, a flexible metal or elastomer diaphragm, driven by a reciprocating arm on an eccentric, electrically driven drive, successively draws air in through one port and exhausts it through the other port in the second half of the cycle of motion (see Figure 21.1). The inlet and outlet flows are facilitated by nonreturn valves located at each port. Diaphragm pumps operating on this principle are available commercially for flow rates ranging from very small up to as high as 200 Lpm. Because there is no need for sealing or lubrication, they are oil free and so produce very little contamination. The piston pump concept shown in Figure 21.2 is similar in that it too depends on a reciprocating action and a system of inlet and outlet nonreturn valves. But now the piston operates within a cylinder. It may be lubricated and sealed by oil, or may be oil-free (e.g. by the use of Teflon piston seals). One disadvantage of oil lubrication is that oil mist may be generated, passing through the outlet and so entering the area from which the air is being sampled. The advantage of this device over the diaphragm pump is that a greater volume of air is moved in each stroke, and a higher differential pressure is generated. Commercial versions are available with flow rates from about 1 Lpm up to several 100 Lpm. The rotary vane pump shown in Figure 21.3 operates under a different principle. Here, a cylinder rotates eccentrically within another cylinder, and contains blades that 'catch' the air in the cavity between the two cylinders and transport it towards the outlet. Because of the eccentricity, the blades need to be capable of radial motion in order to continuously adjust to the changing width of the air space enclosed. In an approach used in some rotary vane pumps, including the one shown, the blades take the form of flat plates that slide within slots in the inner cylinder. Contact between the freely sliding blades and the outer cylinder is maintained by the outwards-directed centrifugal force. Rotary vane pumps are available commercially in both oil-lubricated and oil-free versions. Again the generation of oil mist for lubricated versions may be problematical. A related problem may arise for dry rotary vanes, where solid particles may enter the exhausted airflow arising from the friction associated with the contact between the blades and the inner wall of the outer cylinder. For such pumps the seal is made by the use of graphite vanes, so that the aerosol that is released contains carbon dust. Again, a wide range of flow rates can be achieved, up to several 100 Lpm.

Pumps like those described are available from a wide range of sources, for which Monteith and Rubow provided a very comprehensive listing, current as of 2001.

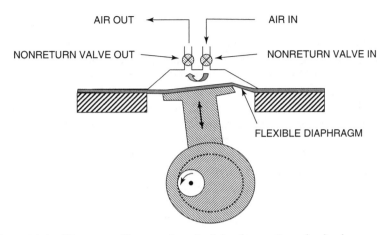

Figure 21.1 *Diagram to illustrate the principle of operation of a diaphragm pump*

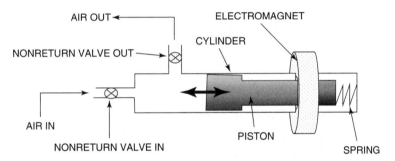

Figure 21.2 *Diagram to illustrate the principle of operation of a piston pump*

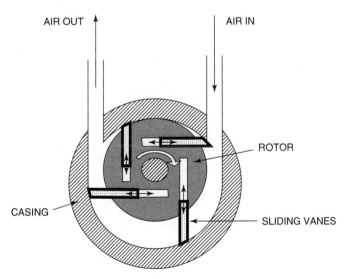

Figure 21.3 *Diagram to illustrate the principle of operation of a rotary vane pump*

21.2.2 Personal sampling pumps

The emergence of pumps sufficiently miniaturised that they could be used for personal sampling was an important milestone in the history of aerosol sampling, in the first instance for occupational hygiene applications – starting in the 1970s – but nowadays, increasingly, for assessing the exposures of people to aerosols in ambient and living environments. The first personal sampling pump was described by Sherwood and Greenhalgh (1960), in which a small DC motor, powered by a miniature mercury cell, was used to drive a PTFE diaphragm. The first prototype (Mark 1) ran continuously for 8 h at just 0.5 Lpm, and its successor (Mark 2) was able to run for as long as 30 h. The latter weighed about 0.5 kg. Both prototypes are shown in the photograph in Figure 21.4. As Sherwood and Greenhalgh described it, the pump was fitted into a (then) typical bicycle lamp casing small enough that it could be carried in the pocket of a traditional lab coat or worn on a belt as shown. A succession of further improvements, including a rechargeable NiCd battery pack, led eventually to the first commercial personal sampling pump, produced in the UK by the Casella company (now Casella CEL Ltd, Bedford, UK). Over the years, improvements continued, and features were added – for example timers, automatic start–stop sequences, digital read-out of flow rate, etc. – and personal sampling pumps of a wide range of types are now available from a wide range of sources. The vast majority are of the diaphragm type. Again, a comprehensive list of sources and available models has been given by Monteith and Rubow.

More recently there has also emerged a new generation of pumps, generally within the categories mentioned, but even lighter and more compact that previously, and with many additional features ('bells and whistles') to make them more convenient in use to the hygienist. Many can provide higher flow

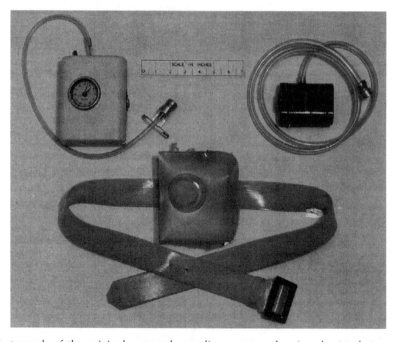

Figure 21.4 *Photograph of the original personal sampling system, showing the Mark 1 and Mark 2 proto-types, and carrying belt (Sherwood and Greenhalgh, 1960). Reproduced with permission from Sherwood and Greenhalgh, Annals of Occupational Hygiene, 2, 127–132. Copyright (1960) British Occupational Hygiene Society*

rates for longer periods, and are able to cope with higher pressure differentials. On many the sampling flow rate is controlled electronically. Many have programmable capability in order to provide pre-set sampling start and stop times, or allow for more complicated sampling sequences. This is especially useful for personal sampling pumps where the hygienist might want to conduct sampling for specific worker operational patterns. Other pumps operating at very low flow rate are even more compact. Although most of these were developed originally for the sampling of gases and vapors, they also have the potential for application with low flow rate personal aerosol samplers, including for example the 'Baby-IOM' described in Chapter 14.

21.2.3 Pulsation damping

One significant problem with pumps of the reciprocating type – diaphragm and piston – lies in the fact that flow pulsations are an inevitable consequence of the principle of operation. This has to be considered in the light of the strong dependence of the various indices of sampler performance – aspiration efficiency, penetration, sampling efficiency – on instantaneous sampling flow rate, along with a number of other important variables. So it is to be expected that temporal variations in sampling flow rate would have a bearing on sampler performance. As early as 1972, LaViolette and Reist described the phenomenon for a range of personal sampling pumps available at the time, driven by special interest in the performance of the 10 mm nylon (Dorr–Oliver) cyclone sampler of the type widely used as a personal sampler for collecting respirable aerosol. Using a hot-wire anemometer to measure the air velocity at the sampler entry, they obtained oscillograms that showed large fluctuations about the desired mean flow, also indicating considerable differences from one pump type to another. Such pulsations were considered to be great enough to warrant technical intervention to reduce or eliminate the flow variations. Even prior to the 1972 work of LaViollette and Reist, pulsation dampers had been proposed and used for this purpose. Each typically comprised a cavity that would be placed in the sampling line between the sampling head and the pump, providing the capacity to absorb the pressure variations associated with the flow fluctuations, containing elements with energy-absorbing elastic walls, for example in the form similar to rubber finger stalls. But, as LaViollette and Reist noted, the early pulsation dampers for pumps were bulky and so were especially problematical for personal sampling situations.

In a later study, also directed at the 10 mm nylon cyclone, Bartley *et al.* (1984) carried out more measurements of sampling flow fluctuations for a number of pumps popular at the time. The pumps in question were equipped with pulsation dampers of the type that were available at the time. Flow pulsations, although less than those noted by LaViolette and Reist, were still evident. Bartley *et al.* noted especially that the magnitude and harmonic content of the fluctuations at the pump itself were quite different from those at the cyclone inlet, associated with damping and resonance effects associated with the filter, filter cassette and air volumes contained within the overall sampling system. In terms of the effect of the flow pulsations on sampler performance, it is the flow properties at the sampler that would be the ones of primary interest. With this in mind, Bartley and his colleagues investigated the bias in aerosol sampling efficiency for a number of widely used personal sampling pumps. Some typical results from experiments to measure cyclone penetration as a function of particle aerodynamic diameter (d_{ae}), for both pulsating and smoothed flows, are shown in Figure 21.5. They show clearly the effect of the pulsating flow. In particular, penetration was significantly reduced across the whole range of particle size studied. Bartley *et al.* used such curves to estimate mass biases for such pulsation-related differences for the 10 mm nylon cyclone used with the pumps studied and for realistic ranges of typical particle size distributions, revealing negative mass biases in the range from 1 to 15 %, tending to be greater for coarser particle size distributions. The effects of the pulsating flow on cyclone performance were subsequently also studied by Berry (1991), confirming the order of magnitude of such biases.

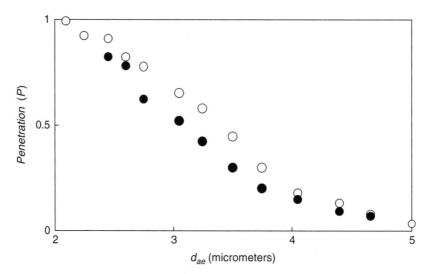

Figure 21.5   *Typical set of experimental results for the penetration (P) of a 10 mm nylon cyclone as a function of particle aerodynamic diameter (d_{ae}) for a typical personal sampling pump, operated at 2 Lpm and showing data for smooth flow (●) and 34 Hz pulsating flow (O), respectively (Bartley et al., 1984). Adapted with permission from Bartley et al., American Industrial Hygiene Association Journal, 45, 10–18. Copyright (1984) American Industrial Hygiene Association*

Over the years pump manufacturers have made great strides in the development of effective pulsation damping devices, and flow pulsation is no longer regarded as a problem in modern commercially available pumps of all the types used for aerosol sampling. Indeed, pulsation-free flow is now an expectation of any commercially available sampling pump, and is often a requirement in standardised methods. For example, the UK Health and Safety Executive, in its most recent document MDHS14/3 on aerosol sampling states that '... for cyclone samplers, pulsation damped flow is particularly important and an external pulsation damper must be used if the pump does not contain an integral damper.' (Health and Safety Executive, 2000). In the USA, the Occupational Safety and Health Administration (OSHA) has published similar guidelines.

21.3   Flow rate

The delivery of a well-defined sampling flow rate that is constant over the full duration of a sampling run is an essential requirement of all pumps that are used in aerosol sampling. The setting of the correct flow rate for the sampler in question, and in turn its accurate measurement, are critical to the establishment of optimal performance of the sampler. In addition, the sampling flow rate determines the air volume that has been aspirated during the whole sampling period, and so enables the mass of particulate matter to be converted into a mass concentration. From such considerations it is clear that the critical volumetric flow rate is that which passes through the sampling head. This is where the particle size-based selection that governs performance takes place. Elsewhere in the sampling line the actual volumetric flow rate may change due to changes in static pressure, as occurs for example during the passage of air through a filter.

21.3.1 Flow control

A simple – and popular – flow control device is the *critical orifice*. Here the principle of operation is based on the fact that the air flowing through an orifice in a duct or tube will reach the speed of sound if the orifice downstream pressure (nearest the pump) is less than half the upstream static pressure (nearest the sampler inlet). When that situation is reached, the application of a still higher negative pressure at the pump will not change the flow rate through the sampler. So what is needed is to ensure that the pump provides a sufficiently high negative pressure downstream of the orifice. Usually in practice the critical orifice is placed in the sampling line downstream of the collecting filter and ahead of the pump inlet. During sampling it is acknowledged that the build up of particulate matter on the filter will lead to increased pressure drop across the filter. However, it is a general rule that, provided the pressure drop across the filter remains small compared with the pressure drop across the critical orifice, the volumetric air flow upstream of the filter does not change markedly. The critical orifice is therefore seen as a simple yet effective means of controlling the air flow rate to the desired value. In this way different orifices may be chosen to deliver different aspiration flow rates.

Modern technology has provided the opportunity for flow control by means of electronic feedback control methods, some where the pump speed itself is held constant regardless of resistance and others where the mass flow rate is sensed and controlled. The latter is especially useful because it takes account of changes in static pressure – and in turn volume, via the gas laws – as the air moves through the sampling line, and ensures a constant flow rate at the inlet of the sampling head (which, of course, remains at atmospheric pressure).

21.3.2 Flow measurement

The simplest means of flow rate monitoring during sampling is by the *rotameter* that is placed in the sampling line downstream of the filter and (usually) ahead of the pump. A rotameter consists of a float inside a tapered vertical tube whose cross-sectional area increases from the bottom upwards. The float moves under the influence of the upwards fluid mechanical drag forces due to the air flow around it in the tube and the gravitational force that acts on it in the downwards direction. The float comes to rest where these forces balance out, and the position in the tube where this is achieved may be calibrated in terms of the air flow entering the tube. For many commercially available pumps, the rotameter is integral to the pump which is purchased. In truth, however, rotameters are not regarded as primary flow measuring devices, and are not particularly accurate. At best, perhaps, they should be regarded only as flow 'indicators'. In this role they can certainly inform the hygienist that the flow has changed significantly while sampling is taking place and that flow rate is in the 'right ball-park'. In preparing a sampling system for practical use, it is common practice to make direct measurement of the flow rate actually entering the sampler – this, after all, is the place where the airflow needs to be known.

Hygienists are taught from the beginning how to apply the *primary standard procedure* for flow rate measurement based on the principle of positive displacement. In its simplest form it is referred to as the 'bubble flowmeter'. It employs a vertical glass column, the top of which is connected directly to the sampling device so that the aspirated air is drawn upwards through the column. The bottom of the column is dipped into a soap solution such that a soap film (or bubble) is drawn into the column and attaches itself to the inner wall. The soap film seals the column of air which is being aspirated, and so rises through the column as the air is displaced. By tracking the soap film as it rises, and timing it as it passes between graduated markings on the glass wall of the column, indicating the actual volume displaced, the volumetric flow rate may be determined directly. Since the soap film has virtually no mass and there is virtually no friction, this method provides an unambiguous, unbiased measure of the aspirated flow rate. For samplers with complex inlets such that a connection between the glass tube and the inlet is not possible, then the geometry of this system needs to be modified so that the whole sampler is placed within a sealed vessel with two tubes, one connected to the outlet side of the sampler and then externally to the pump and the other tube connecting the air volume inside the vessel to the bubble column. The net result in terms of the air flow path is the same as for the simple version. The vertical tube, volumetric measurement and timing may be incorporated into a single compact, automated instrument, and a number of such instruments are commercially available, most of them operating on the basis of optical detection of the position of the soap film and electronic timing of its time of travel between markers. The bubble flowmeter is very good for low flow rate samplers and so the systems described are very good for personal samplers. They are less good for high flow rate samplers where the displacement is so great that the bubble cannot easily be tracked. Alternative air volume displacement measurement may be achieved in such cases using an accurate gas meter.

In practical aerosol sampling, the operator should employ 'best practice'. Here, in principle, the correct sampling flow rate is established and accurately measured before sampling commences. Then it is measured again at the end of the sampling period before the filter or particle collecting substrate collecting is removed for analysis. Samples taken where the flow rate has changed significantly (e.g. has fallen by more than, say, 10 %) during the sampling period should be discarded since neither the sampler's particle size-selective performance nor the sampled air volume would be known.

21.4   Collection media

In many direct-reading aerosol sampling instruments, particles are detected while they remain airborne (e.g. by light scattering) and so at no stage need to be separated from the air. In other direct-reading instruments they are collected onto a filter or substrate and are continuously analyzed, for example by beta-attenuation, or by a mechanical mass balance such as a piezoelectric or tapered element. Beyond direct-reading instruments, however, most aerosol samplers – which represent the great majority of *all* samplers – require the collection of particles onto a filter or substrate over a defined sampling period so that the collected material can be analyzed subsequently and separately.

21.4.1   Filters

The purpose of a filter in a sampling system is to efficiently collect all the particles that reach it. It should be placed as close as possible to the sampling head itself in order to minimise losses, so what is collected on the filter is as close as possible to the aerosol that is of interest. For most samplers the filter is disk-shaped, but rectangular filters are available for use in some devices (e.g. the high-volume sampler shown in Figure 17.10). The filter should be mounted within a filter holder arrangement that makes a good seal with the body of the sampling system such that *all* the air that has been aspirated does in fact pass through the filter. The filter itself consists of a porous medium that is capable of efficiently removing particles from the air (e.g. by impaction, interception, etc.) and retaining them in a form that can subsequently be analyzed, either visually (under the microscope), gravimetrically (by weighing on a balance), or analytically (by wet chemistry or by analytical instrumentation of the types that are now widely available). These differing requirements provide criteria by which a particular type of filter may be chosen for a particular application.

Lippmann (2001) described the character of filtration media of the type that are used in aerosol sampling, identifying them as '... porous structures with definable external dimensions such as thickness and cross-section normal to the flow.' Such filters differ considerably in terms of the flow pathways through the bulk of the media. *Fibrous filters* illustrated schematically in Figure 21.6 represent an

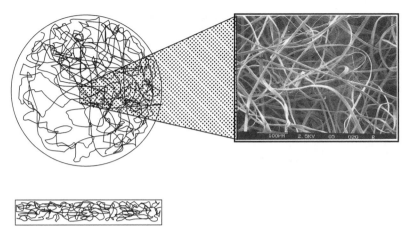

Figure 21.6 *Diagram (not to scale) to illustrate the nature of a fibrous filter. Inset: Electron micrograph of a typical fibrous filter mat, scale as shown*

important family of media, comprising fibers of, for example, glass or quartz, woven together to form a dense mat of elements which become the surfaces onto which particles may be deposited by combinations of impaction, gravitational settling, interception, diffusion and – in some cases – electrostatic forces. The air passes between these fibrous elements, flowing *externally* around each one. *Membrane filters* take the form of thin sheets – usually circular – of gel material characterised by interconnecting pores through which the air may flow. Such filter media may be characterised usefully in terms of 'pore size'. Now the air passes *internally* through the tortuous path between the successive pore-like elements. The gel material may take the form of cellulose ester, nylon, polyvinyl chloride (PVC) or Teflon. One special form of membrane filter, the Nuclepore® filter, has a distinctively different structure in which the air path is through straight cylindrical tubes that have been 'drilled' in the originally continuous polycarbonate sheet by a process of neutron bombardment and etching. A third category, less widely used, includes *granular bed filters* which, for air sampling applications comprises granular material packed into a sheet and sintered in order to form a permanent structure. Thin disks of silver granules sintered in this way are available in the form known as the '*silver membrane filter*'.

21.4.2 Filtration efficiency

As noted in earlier chapters, the various collection mechanisms of aerosols onto surfaces are strongly dependent on particle size. The physics of filtration has been discussed in full by Brown (1993) and others, and a useful concise review has been given by Hinds (1999). By way of illustration, Figure 21.7 indicates the collection efficiency of a filter as a function of particle diameter (d) for a situation typical of filtration in the context of aerosol sampling. Efficiency is shown in terms of the specific contributions of the primary individual mechanisms, and it is seen in general that the efficiency is dominated at larger particle sizes by gravitational settling, impaction and inertial interception, and at smaller sizes by diffusion. Also shown is the contribution from the interception of diffusing particles, less strongly dependent on particle size. Not shown is the contribution from electrostatic forces that will be present if the particles are electrically charged, where the specific efficiency of collection will tend to increase with particle size. In terms of filtration efficiency for the types of filters used to collect particles in aerosol sampling, the overall collection efficiency is the most important. Here, as the result of the

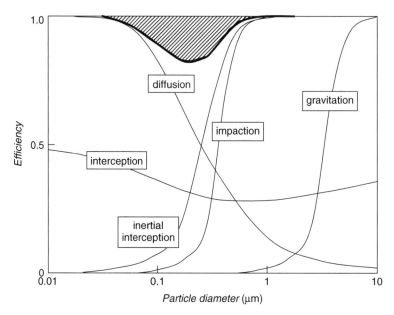

Figure 21.7 *Graph to show the relative roles of the various particle collection mechanisms indicated as functions of particle diameter (d) for a filter typical of the ones used for particle collection during aerosol sampling and for a single face velocity. Adapted with permission from Hinds, Aerosol Technology: Properties, Behavior and Measurement of Airborne Particles, 2nd Edn. Copyright (1999) John Wiley & Sons, Ltd*

competition between the various mechanisms, efficiency is high for large and small particles. But it is least in the transition region around $d = 0.2$–0.3 µm. The shaded area in Figure 21.7 represents the particles that are *not* collected. As shown in Figure 21.8, the depth and position of this minimum penetration efficiency shift with the air velocity at the face of the filter (the *face velocity*), becoming shallower for smaller face velocity and shifting to larger particle size. Relations like these have studied experimentally – and confirmed – for actual filters of the type used in aerosol sampling applications in occupational and environmental hygiene.

The preceding is useful, especially for fibrous filters but it describes a quite idealised situation. In reality, filter performance is complicated by the actual microstructure of the filter media. Whereas for fibrous filters the particles are deposited with equal efficiency throughout the body of the media, for membrane and Nuclepore® filters particles are deposited proportionately much closer to the front face. As noted by Lippmann, this tendency for membrane and Nuclepore filters provides the advantage that particles are more readily available for observation by optical and electron microscopy. Also, on-filter analytical methods can be used without interferences associated with the filter media itself. This brings with it the disadvantage that the 'carrying' capacity of the filter media is significantly reduced such that, as particles accumulate, the pressure drop rises more sharply than for fibrous filters. In addition, there is a greater tendency for particles to become removed from the filter by blow-off or other mechanical means.

21.4.3 Mass stability

For a high proportion of aerosol sampling, a goal is to measure the total mass that has been collected onto a filter. This is routinely achieved by weighing the collecting filter before and after sampling,

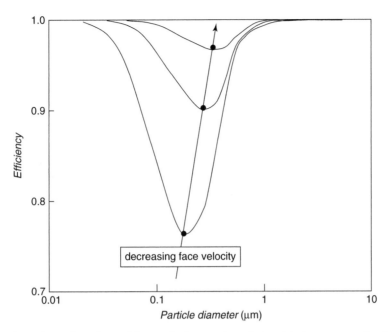

Figure 21.8 *Graph to show the effect of face velocity on the relationship between collection efficiency and particle diameter (d) for a filter typical of the ones used for particle collection during aerosol sampling. Adapted with permission from Hinds, Aerosol Technology: Properties, Behavior and Measurement of Airborne Particles, 2nd Edn. Copyright (1999) John Wiley & Sons, Ltd*

with the expectation – ideally – that the mass of aerosol collected is simply the difference between the two masses. However this apparently simple process is fraught with difficulties, firstly associated with the intrinsic accuracy with which the pre- and post-sampling masses can be differentiated even under ideal conditions, especially in the light of the increasingly small aerosol concentrations that are being assessed, and secondly associated with the external influences that cause mass instability of the filter and its contents. The latter are especially problematical, leading to variability that is difficult to control. A major external influence is the adsorption of water molecules onto the filter. Others include handling and electrostatic charge effects.

In order to accurately measure a small collected mass, it is important to know the *limit of detection (LOD)* and *limit of quantification (LOQ)* of the measurement method used. These define the envelope for the practical implementation of gravimetric sample analysis. In most industrial hygiene situations, the LOD is defined as '... a stated limiting value designating the lowest concentration that can be detected and that is specific to the analytical procedure used', and the LOQ is '... a stated limiting value designating the lowest concentration that can be quantified with confidence and that is specific to the analytical procedure used' (DiNardi, 1995). The American Society of Testing and Materials (ASTM, 2000) is more specific, defining the LOD as three times the standard deviation as estimated from repeated weighings of blank filters, and the LOQ as 10 times the same standard deviation. The balance itself may also contribute to weighing imprecision. Some balances are designed from the outside to be capable of measuring smaller masses than others. Major departures from original calibrations can be avoided by proper maintenance and regular calibration to accepted standards, for example as recommended in the USA by the National Institute of Standards and Technology (NIST).

Several laboratory studies have been carried out to investigate the variability of filter masses. In 1974, Mark reported a study of the moisture uptake by a number of filters, including a glass fiber filter and several membrane filters, tracking the filter masses over a period of time during which the relative humidity was also monitored. Changes in the masses of most of the membrane filters were seen to follow quite closely the fluctuations in relative humidity. The mass of the glass fiber filter was much less sensitive, appearing to be relatively steady. Later, Vaughan *et al.* (1989) made a closer examination of filter variability, performing repeated measurements of the masses of two contrasting filters of the type widely used by occupational hygienists, a 25 mm glass fiber filter (Whatman GF/A) and a 25 mm membrane filter made from PVC-acrylonitrile (Gelman DM800). The experiments were performed in an air-conditioned laboratory using two electronic balances, one with readability down to 0.001 mg and the other with readability down to 0.01 mg. The results from multiple weighings are summarised in Table 21.1. They show that the LOD-values were always significantly greater than the readability of the balance used. Interestingly, they show that the LOD-values were not very strongly dependent on the readability of the balance used, indicating that the variability does indeed come mainly for external factors such as moisture uptake. In the Vaughan *et al.* study, the variability of the glass fiber filter did not appear to be significantly different from that of the membrane filter for the two filters tested.

Recent interest in sampling for the inhalable aerosol fraction, in particular the widespread use of the IOM personal inhalable aerosol sampler (see Chapter 14), has extended the discussion about filter stability. The IOM sampler is notable for the fact that the filter is contained within a cassette such that it is the whole filter-cassette assembly that is weighed before and after sampling. So the entity whose mass stability is at issue is the combined filter-cassette system. This has been studied by a number of workers. Smith *et al.* (1998) cited field studies carried out the US National Institute for Occupational Safety and Health (NIOSH) that suggested considerable mass variability deriving predominantly from the cassettes, and not so much from the filters used. For example, for blanks of the plastic cassettes which had been supplied with the original versions of the IOM sampler, mass variability was seen to be considerably greater than Vaughan *et al.* had observed for glass fiber filters in the laboratory using a balance with the same readability. In response, therefore Smith *et al.* set out to investigate the mass stability of combinations of plastic and stainless steel IOM cassettes, containing glass fiber or PVC filters. They found that when plastic cassettes were exposed to dry conditions inside a desiccator, their masses were reduced by 1.5 mg in 4–5 days, and when they were subsequently left out in the weighing environment, their masses showed a similar rate of increase. Stainless steel cassettes, however, were relatively stable under local conditions of both desiccation and room humidity. In later studies, Li and Lundgren (1999) and Lidén and Bergman (2001) reached similar conclusions about the amount

Table 21.1 *Summary of limits of detection (LOD) for two filter types and for two electronic balances with different readabilities, from repeat measurements in an air-conditioned laboratory (from Vaughan et al., 1989). The LOD-values are also converted into equivalent limiting values for the measurement aerosol concentration for 4-h time-weighted average (TWA) samples*

Filter type	Designation	Balance readability	LOD (mg)	Equivalent aerosol concentration limit for 4-h TWA sampling (mg m^3)
Glass fiber	Whatman GF/A	0.001	0.018	0.038
		0.01	0.07	0.143
Membrane	Gelman DM800	0.001	0.016	0.033
		0.01	0.07	0.141

Table 21.2 *Summary of limits of detection (LOD) and limits of quantitation (LOQ) for different combinations of filter and cassette as used with the IOM personal inhalable aerosol sampler and using an electronic balance with readability 0.01 mg, obtained from repeated weighings of the combinations indicated (from Paik and Vincent, 2002)*

Filter type	Designation	Cassette	LOD (mg)	LOQ (mg)
Glass fiber	Whatman GF/A	Plastic	0.16	0.54
		Ni-coated plastic	0.16	0.53
		Stainless steel	0.19	0.65
Membrane	PVC	Plastic	0.16	0.53
		Stainless steel	0.12	0.42
	Teflon	Plastic	0.13	0.43
		Stainless steel	0.12	0.39

of moisture lost or absorbed by plastic IOM cassettes. Paik and Vincent (2002) performed a new experimental study in which three types of IOM sampler cassette (plastic, nickel-coated plastic and stainless steel) in various combinations with three filter types (glass fiber and PVC and Teflon membrane filters). All the weighings were carried out using an electronic balance with readability 0.01 mg. In the experiments, the filters were initially exposed to a ^{210}Po source (to eliminate static electricity) before placing them into their respective cassettes. Then they were desiccated overnight in a chamber containing silica gel at approximately 20 % relative humidity. After initial desiccation, the filter-cassette assemblies were conditioned for 8–12 h in the balance room environment, weighed for the first time, then transported and exposed for 1 h in another laboratory, after which they were brought back and reconditioned to the balance room environment and reweighed. This procedure was adopted in an effort to – partially at least – simulate what happens during sample handling during field investigations. The results for the LOD and LOQ, respectively, are summarised in Table 21.2. They show that, in general, the differences in mass stability between the various combinations were quite small. But they confirm that the LOD-values were much greater than Vaughan *et al.* had noted for glass fiber and membrane filters alone, and hence confirmed the important contribution of the cassette to the overall variability. Lidén and Bergman came to a similar conclusion. Importantly, such results indicate a seemingly intrinsic limitation in making gravimetric measurements at low aerosol concentrations at levels sometimes experienced in modern workplace environments. For example, an 8 h personal sampler for an aerosol whose average concentration was, say, 0.1 mg m^{-3}, taken at a flow rate of 2 Lpm, would yield a collected mass of about only 0.1 mg. This is well below the LOQ – even the LOD – for filter-cassette combinations like those described.

21.4.4 Choices and applications

The choice of filter type and material are strongly dependent on the application and on the intended method of analysis. The basic collection efficiency of the filter media is only a serious concern when the aerosol to be collected has a significant proportion of the mass in particle sizes below 1 μm. So for most gravimetric sampling applications, where the collected particulate matter is to be weighed on a balance, it is usually neglected. However, more care should be taken when making a choice for sampling fine aerosols (e.g. welding fume). These considerations apart, where samplers are to be analyzed gravimetrically, the two primary concerns of interest are the mass stability of the filter and the ability of the filter to collect and retain (possibly) large amounts of particulate material. Traditionally, therefore, the glass fiber filter has consistently been the primary choice for such applications.

On the other hand, where it is required to analyze the sample microscopically, the particles need to be collected onto a single plane in a relatively flat surface. This calls for the membrane filter. For the sampling of airborne asbestos fibers, standardised methods call for specific membrane filter types. For example, the method recommended by NIOSH for the optical detection and counting of respirable asbestos fibers by light microscopy calls for a mixed cellulose ester (MCE) membrane filter with a pore size from 0.45 to 1.2 μm. Such a filter has the ability to be cleared – by exposure to an atmosphere of acetone vapor – so that the particles can be observed without interference from the structural background of the filter media itself. Membrane filter samples can also be used for scanning or transmission electron microscopy, and this is especially useful if it is desired to use a single filter for both optical and electron microscopic analysis (e.g. Kohyama and Kurimori, 1996). However, for even greater visibility of particles at this level of resolution, the Nuclepore® filter is widely used.

Some quantitative analytical methods can be performed on samples while they remain *in situ* on the original collecting filter. One, for example, is the analysis of samples by quantitative X-ray diffraction, an important tool for the determination of many crystalline substances, including – importantly – crystalline silica (quartz). Davis and Johnson (1982) examined a number of filter types, and found that the quality of the results from X-ray diffraction were dependent on background scatter and on particle loading. Teflon filters were found to be superior for filter loadings less than about 0.2 mg m^{-3} and quartz and glass fiber filters were superior for loadings greater than about 0.3 mg m^{-3}. Infrared spectrophotometry is also used for direct on-filter analysis of samples, and this approach has been used for the quantitative analysis of cristobalite (Shinohara, 1996), a polymorph of quartz, having the same chemistry but with different crystalline structure and similarly important in relation to human exposure. Shinohara successfully employed a polypropylene membrane filter, noting its high transparency and lack of interferences in the infrared spectral range of interest. Direct on-filter analysis of bulk asbestos has also been carried out by infrared spectrophotometry.

Many analytical procedures require the recovery of the particulate material from the filter and its transfer onto or into some other media. In some instances, the particulate material may be removed from the filter directly. However, it is more common that the filter is destroyed during the sample recovery process. Here, the choice of type of filter media is important in optimising the recovery process, and so is dependent on the chemical species that is to be recovered. So, for example, cellulose paper filter may be chosen if the intention is to recover inorganic materials by low temperature plasma ashing, digestion in a strong acid, or incineration in a muffle furnace. Membrane filters may be chosen if the aim is to recover the collected material by dissolving the filter in a solvent. Glass fiber filters may be chosen if the sample is to be recovered by leaching or dissolving the sampler from the filter. The choice goes beyond just the recovery process. It also must take into account possible background interferences that might subsequently influence with the desired analysis, in particular if the filter were to contain trace amounts of the substance of interest (e.g. Mark, 1974: Gelman *et al.*, 1979).

Thimble-type filters of the sintered granular bed type are used in some stack sampling applications. Disk-shaped silver membrane filters are also used for some specialised applications.

21.4.5 Substrates

Continuous, nonporous, solid surfaces are employed as collecting surfaces in some aerosol sampling systems. The simplest are the glass plates that have been widely used in many of the devices mentioned in earlier chapters, including thermal precipitators and konimeters. For these, the glass is perfectly flat and there are no chemical artifacts. Here the main issue of concern is the quality of the adhesion of particles to the surface after collection. The problem can be particularly acute for larger particles. Here, therefore, it has been customary to provide a retentive collective surface by applying an appropriate grease in order to increase adhesion and so reduce particle losses by blow-off or re-entrainment.

Cascade impactors are quite widely used in occupational and environmental hygiene. For these, in the light of the high local air velocities that are usually encountered during particle collection, the properties of collection substrates are particularly important. In some cascade impactors, especially the many prototypes that have appeared over the years, the substrates of interest were in fact the actual impaction plates themselves. But in some commercially available cascade impactors, separate substrates are supplied by the manufacturer. For example, the popular Marple-type personal cascade impactor (see Chapter 18) is supplied with thin substrates made from Mylar, stainless steel or glass fiber. When used dry, such substrates are prone to particle loss by blow-off or re-entrainment. Turner and Hering (1987) explored a range of greasing options for impactor substrates, and showed that the effectiveness of the coatings were strongly dependent both on the type of grease and on the substrate itself. Cascade impactor manufacturers usually recommend a number of greases that have been shown by experience to perform effectively. One is the very pure Apiezon-L grease that is widely used elsewhere as a vacuum grease. Usually it is mixed with an appropriate solvent and the solution is then sprayed onto the substrate and conditioned in order to stabilise it before use. Silicone grease has also been widely used, sprayed directly from a canister.

Other types of substrate for impactors have included oil-soaked glass frits, which Turner and Hering found to be very effective in retaining collected particles. Reischl and John (1978) proposed an oil-soaked sintered metal disk, and this too was found to be very effective. Where space permits, polyurethane foam substrates have been used and shown to have very good retentive properties, even when used dry.

Cassettes have already been mentioned as component substrates, including where the cassette is integral with the filter itself, and the whole catch of particulate material is weighed, both that on the filter and that retained elsewhere within the cassette. In this way, the mass stabilities of both the cassette and the filter are important, as noted above. The retention properties of the cassette are less important because any re-entrained particulate matter will end up on the filter anyway, and so will still be assessed.

21.5   Analysis of collected samples

21.5.1   Handling and transport of samples

The practice of aerosol sampling requires manual handling of the various relevant items, including the sampler itself and associated pumping apparatus, and especially the sample filter (or substrate), first as it is loaded into the sampler and then, later, as it is removed and brought back to the laboratory for analysis. All the various steps provide opportunities for mishaps, or sampler damage or modifications. These problems may all be minimised by the adoption of careful and consistent practices. They are compounded when the samples need to be transported to another location for analysis. Today, this is an increasing issue as more and more analyses of air samples are carried out by specialised central laboratories, usually ones accredited for performing the analyses in question. For these, the analysis of a sample, by whatever means, is backed up by appropriate quality assurances. In addition such laboratories have developed very good procedures by which samples that have been collected in the field can be conveyed with the minimum of risk of damage. In the USA, OSHA provides strict guidelines for such shipping and handling, specifically for the shipment of samplers to its Salt Lake Technical Center, but also to other accredited laboratories.[1] Apart from requesting certain information about the

---

[1] See OSHA website at http://www.osha.gov/dts/osta/otm/otm_ii/otm_ii_4.html.

contents of the sample and factors which might represent interferences in the subsequence analysis, OSHA also provides appropriate containers designed to ensure continued integrity of the sample during transportation. In addition, importantly, the guidelines provide for a shipping procedure that maintains the chain of legal accountability or custody.

21.5.2   Gravimetric methods

As already mentioned, gravimetric methods entail determination of the whole mass of the sample that has been collected, and that involves weighing the filter (or other substrate) before and after sampling has taken place. The weighings take place by the use of an appropriate analytical balance, most of which nowadays are of the electronic variety. Such balances are rated first in terms of their readability, and modern instruments have readabilities as low as 0.001 mg. However, as already noted, the variability in the masses of filters and other substrates will always be such that the readability can never be regarded as the LOD. Balances are rated secondly in terms of their tare weight capability – that is, the maximum mass they can carry. Many modern balances can have low readability and yet still carry large tare weights. This is especially useful if the balance is to be used to measure a small mass change in a relatively large filter or cassette associated with aerosol collection. Mark (1990) demonstrated the practical feasibility that, using a modern electronic balance, sampling cassettes weighing even as much as tens of grams could be placed on the scales yet measurement of changes of as little as a fraction of a milligram could be carried out without loss of sensitivity.

Appropriate balance, balance room and conditioning facilities are necessary for reliable gravimetric measurements. Several sets of guidelines have been proposed. Those of the Environmental Protection Agency (EPA) are somewhat typical, providing guidelines for filter conditioning, weighing facilities and procedures for accurate and reproducible measurements of particulate mass.[2] These specify that the balance to be used should be placed on a vibration-free surface and should be electrically grounded. It should be located in a controlled environment where the mean temperature is from 20 to 23°C and should not vary by more than ±2 °C over a 24 h period. Mean relative humidity should be maintained at 30–40 %, and should not vary by more than ±5 % over 24 h. EPA specifies that samples should be conditioned in the same controlled environment as the balance for a minimum of 24 h prior to weighing, and that samples should be electrically neutralised prior to weighing, for example by exposure to ions from a small radioactive beta-source such as ^{210}Po. In addition to the samples to be weighed, field blanks should also be placed in the weighing sequence and carried through the process of pre- and post-sampling weighings, including loading into – and unloading from – a sampling head. The results of these may subsequently be used to assess the LOD and LOQ specific to the sampling exercise in question, and hence play a role as a check on quality control. Other approaches have been used and often found to be successful, for example, where sampling and gravimetric analyses need to be carried out at remote (e.g. industrial) sites where idealised conditions are not available. For these, it is often satisfactory to condition the samples overnight in the room – not necessarily conditioned – where the balance is kept. Other approaches have involved placing samples overnight in a desiccator and weighing them *immediately* after being removed the following morning.

21.5.3   Chemical analysis

Many samples need to be processed in order to extract information about chemicals of specific interest, and here there is a large battery of options for methods and instrumentation. In the first instance

---

[2] See EPA website at http://www.epa.gov/ttn/amtic/files/ambient/pm25/qa/balance.pdf.

there are 'wet' chemical methods by which collected samples may be recovered and – using appropriate reagents – analyzed for specific constituents. Then there is a wide range of quantitative physical instrumentation by which to directly analyze recovered material. These may include, for example, atomic absorption (AS) spectrophotometry for metals at high to moderate concentrations, inductively coupled plasma atomic emission spectrometry (ICP-AES) or mass spectrometry (ICP-MS) for metals at low and ultra-low concentrations, X-ray diffractometry (XRD) for crystalline silica and other minerals, etc. Most such instruments are available commercially. They can be quite expensive, and require considerable expertise for sample preparation and operation. So, as already mentioned, many researchers and practitioners make the decision to send samples for analysis to specialised accredited laboratories that can maintain the facilities and staff needed to operate such equipment consistent with quality assurance and control requirements.

Each sampled substance of interest requires its own combination of recovery and analytical procedures. There are far too many to describe in a book of this type. Aerosol scientists and all those engaged one way or another with aerosol sampling need to be aware of the options that are available. The most comprehensive account is given in the NIOSH *Manual of Analytical Methods,* 4th edition, a regularly updated resource that lists all substances of interest and describes a complete measurement procedure for each, including details of sampling filters and flow rates and measurement instrumentation (National Institute for Occupational Safety and Health, 2004).

References

American Conference of Governmental Industrial Hygienists (2001) *Air Sampling Instruments for Evaluation of Atmospheric Contaminants*, 9th Edn, (Eds B.S. Cohen and C.S. McCammon), ACGIH, Cincinnati, OH.

American Society of Testing and Materials (2000) *ASTM Standard practice for controlling and characterizing errors in weighing collected aerosols*, ASTM D 6552, ASTM International, West Conshohocken, PA.

Bartley, D.L., Breuer, G.M., Baron, P.A. and Bowman, J.D. (1984) Pump fluctuations and their effect on cyclone performance, *American Industrial Hygiene Association Journal*, 45, 10–18.

Berry, R.D., (1991) The effect of flow pulsations on the performance of cyclone personal respirable dust samplers, *Journal of Aerosol Science*, 22, 887–899.

Brown, R.C. (1993) *Air Filtration: An Integrated Approach to the Theory and Applications of Fibrous Filters*, Pergamon Press, Oxford.

Davis, B.L. and Johnson, L.R. (1982) On the use of various filter substrates for quantitative particulate analysis by X-ray diffraction, *Atmospheric Environment*, 16, 273–282.

DiNardi, S.R. (1995) *Calculation Methods for Industrial Hygiene*, Van Nostrand Reinhold, New York.

Gelman, C., Mehta, G.V. and Meltzer, T.H. (1979) New filter compositions for the analysis of airborne particulate and trace metals, *American Industrial Hygiene Association Journal*, 40, 926–932.

Health and Safety Executive (2000) General methods for sampling and gravimetric analysis of respirable and inhalable dust, MDHS 14/3, Health and Safety Executive, London.

Hinds, W.C. (1999) *Aerosol Technology: Properties, Behavior and Measurement of Airborne Particles*, 2nd Edn, John Wiley & Sons, Ltd, New York.

Kohyama, N. and Kurimori, S. (1996) A total sample preparation method for the measurement of airborne asbestos and other fibers by optical and electron microscopy, *Industrial Health*, 34, 185–203.

LaViolette, P.A. and Reist, P. (1972) An improved pulsation dampener for use with mass respirable sampling devices, *American Industrial Hygiene Association Journal*, 33, 279–282.

Li, S.N. and Lundgren, D.A. (1999) Weighing accuracy of samples collected by IOM and CIS inhalable samplers, *American Industrial Hygiene Journal*, 60, 235–236.

Lidén, G. and Bergman, G. (2001) Weighing imprecision and handleability of the sampling cassettes of the IOM sampler for inhalable dust, *Annals of Occupational Hygiene*, 45, 241–252.

Lippmann, M. (2001) Filters and filter holders. In: *Air Sampling Instruments for Evaluation of Atmospheric Contaminants*, 9th Edn (Eds B.S. Cohen and C.S. McCammon), American Conference of Governmental Industrial Hygienists, Cincinnati, OH.

Mark, D. (1974) Problems associated with the use of membrane filters for dust sampling when compositional analysis is required, *Annals of Occupational Hygiene*, 17, 35–40.

Mark, D. (1990) The use of dust collecting cassettes in dust samplers, *Annals of Occupational Hygiene*, 34, 281–291.

Monteith, L.E. and Rubow, K.L. (2001) Air movers and samplers. In: *Air Sampling Instruments for Evaluation of Atmospheric Contaminants*, 9th Edn, (Eds B.S. Cohen and C.S. McCammon), ACGIH, Cincinnati, OH.

National Institute for Occupational Safety and Health (2004), *Manual of Analytical Methods*, 4th Edn, NIOSH, Cincinnati, OH.

Paik, S. and Vincent, J.H. (2002) Filter and cassette mass instability in ascertaining the limit of detection of inhalable airborne particulate, *American Industrial Hygiene Association Journal*, 63, 698–702.

Reischl, G.P. and John, W. (1978) The collection efficiency of impaction surface, *Staub Reinhaltung der Luft*, 38, 55–58.

Sherwood, R.J. and Greenhalgh, D.M.S. (1960) A personal air sampler, *Annals of Occupational Hygiene*, 2, 127–132.

Shinohara, Y. (1996) Direct quantitative analysis of respirable cristobalite on filter by infrared spectrophotometry, *Industrial Health*, 34, 25–34.

Smith, J.P., Bartley, D.L. and Kennedy, E.R. (1998) Laboratory investigation of the mass stability of sampling cassettes from inhalable aerosol samplers. *American Industrial Hygiene Association Journal*, 59, 582–585.

Turner, J.R. and Hering, S.V. (1987) Greased and oiled substrates as bounce-free impaction surfaces, *Journal of Aerosol Science*, 18, 215–224.

Vaughan, N.P., Milligan, B.D. and Ogden, T.L. (1989) Filter weighing reproducibility and the gravimetric detection limit, *Annals of Occupational Hygiene*, 33, 331–337.

22

Field Experience with Aerosol Samplers in Workplaces

22.1 Introduction

The history of health-related aerosol standards for workplace exposures owes much to the growth of knowledge and development of sampling instruments, going back to the early 1900s and continuing to the present day. Some, but not all, has run parallel to the history of aerosol sampling for ambient atmospheric aerosol. So two 'schools of thought' have evolved. Many of the players – the aerosol scientists – have been the same, but others have specialised in one area or the other. Over the years, there have been many shifts in sampling philosophy and criteria, accompanied by advances in sampling devices and strategies, leading in turn to adjustments to standards and exposure limits. Occupational epidemiology has been a major stimulant. Many of the diseases associated with aerosol exposures at work have long latency periods, where clinically diagnosable outcomes may appear only long after the relevant exposure history began – for example, pneumoconiosis associated with respirable dust inhaled during coal mining, lung cancer associated with inhaled dust during nickel primary production, and many others. The changes in exposure assessment methodology over time have complicated the historical occupational exposure databases required to gain a full understanding of the epidemiology of such diseases. Field experiences with aerosol sampling instruments like those described in preceding chapters have provided important steps along the road to implementation of steadily improving methodologies and equipment. Of particular interest are the field studies aimed at providing exposure histories described in terms of single, most-relevant indices. Important ingredients towards achieving this goal have been the side-by-side studies that have been carried out in the field in order to make direct comparisons between instruments and sampling approaches under real-world conditions.

The sampler testing protocol EN 13205 developed by the Comité Européen de Normalisation (CEN) entitled 'Workplace atmospheres – assessment of performance of instruments for measurement of airborne particle concentrations' (Comité Européen de Normalisation, 2002), includes guidance not only on the laboratory testing of aerosol samplers but also field studies (see Chapter 3). In such field studies for a given aerosol fraction, the protocol requires the identification of an existing aerosol sampler that has been shown to accurately and consistently collect the fraction of interest, and then its operation in the field alongside candidate other samplers for the same fraction. Here, therefore, the chosen sampler is a

Aerosol Sampling: Science, Standards, Instrumentation and Applications   James Vincent
© 2007 John Wiley & Sons, Ltd

reference instrument, against which all the others are to be compared. EN13205 specifies that such comparisons should be carried out for as wide a range of conditions as possible pertaining to the field site(s) in question. This relatively recent protocol firmly underlines the importance – long recognised – that field studies of aerosol samplers considered to be candidate samplers for specific aerosol fractions are an extremely important part of the process of their validation.

Many of the aerosol samplers described in the preceding chapters have been applied by occupational hygienists in aerosol exposure studies in many countries, as well as in routine surveys to assess compliance with exposure standards. Some of the samplers listed have been used quite routinely in the field, others on a more restricted basis. Some have been commercially available for many years, some only recently, and others only in the form of research prototypes. Overall, a considerable amount of experience has been gained about not only their performance characteristics but also their convenience of use. Much of this has appeared in published reports. Here, examples are given of some of the key field studies that were subsequently influential in the longer run of occupational exposure assessment, occupational epidemiology and standards setting, marking important milestones in the various transitions that have taken place as perspectives and priorities have shifted.

22.2   Personal and static (or area) sampling

In earlier chapters, both static and personal aerosol sampling have been mentioned. But it has long been suggested by occupational hygienists that the two approaches might not produce the same results for aerosol concentration, even if the same sampling instrument were to be used. In recent decades, however, personal sampling has become the preferred occupational hygiene culture in most countries, acknowledging that this is the most effective – perhaps *only* – way to enable the assessment of the actual exposures of people to both aerosol and gaseous contaminants. This became practical only after personal sampling pumps became widely available. Sherwood and Greenhalgh (1960) were the first to describe a prototype of a small sampling pump that was miniaturised to the extent that it – and its associated battery power-pack – could be carried on the body of a worker for as long as a full working shift of up to 8 h. In addition to producing the first technical personal sampling system, Sherwood and Greenhalgh also produced the first evidence from field studies of the systematic differences between the results for static and personal sampling respectively. They showed that personal sampling *always* provided higher concentrations than corresponding static measurements. More recently, Cherrie (1999) reviewed the available data for such field comparisons, including the early Sherwood and Greenhalgh results. They are summarised in Table 22.1 where the comparisons included results for asbestos and man-made

Table 22.1   *Results of field studies to compare aerosol concentrations as measured by personal and static (or area) sampling, respectively, for various samplers and indices of concentration (as summarised by Cherrie, 1999)*

Aerosol type	Industrial setting	Ratio of personal to static sampling
Asbestos fibers	Store simulation	8.5
Para-aramid fibers	Spinning, carding, etc.	1.5
Man-made mineral fibers	Workroom simulation	1.2
'Total' aerosol	Uranium dust processing	5.0
'Total' aerosol	Soft paper production	1.8
Inhalable aerosol	Flax spinning	2.6
Inhalable aerosol	Cotton primary production	4.5

mineral fiber producing and using operations, uranium dust processing, soft paper production, cotton primary production and flax spinning. In some of these studies, the same sampler type was used as both a personal and a static sampler, and in others the samplers were different. For the latter, for example, in cotton production, the IOM inhalable aerosol sampler was chosen as the personal sampler, and its results were compared with those using a 50 Lpm static sampler that was recommended for many years for cotton dust sampling by the UK Health and Safety Executive HSE (1980). Regardless of sampler type, however, personal or static, the results in Table 22.1 show that the personal samplers consistently provided greater aerosol concentrations than the chosen static sampler. Cherrie noted that the largest ratios were found in large, well-ventilated work areas, and the smallest ratios in small, poorly ventilated areas. Similar results have been found for gaseous contaminants. They all confirm what occupational hygienists had long previously suspected.

The reasons for the differences between personal and static sampling remain unclear. For aerosol sampling, it was argued for many years that the presence of the human body when a personal sampler is worn on the lapel changes the aerodynamic sampling properties of the sampler. This would appear to be plausible from blunt sampler theory like that described in Chapter 5. However, as noted quite recently by Paik and Vincent (2002), the role of the bluff body appears to be weaker than was originally thought. Several other reasons for the differences have been considered, mostly speculative. One is that workers tend, on average, to be located closer to sources than is possible for fixed-point static samplers. Another is that individual people generate 'personal dust clouds' by virtue of their own specific activities, which they carry with them as they go about their activities. However, this latter – rather vague – concept is not very helpful in elucidating the underlying reasons for the difference.

22.3 Relationship between 'total' and inhalable aerosol

The evolution of standards for a new coarse aerosol particle size-selected fraction based on what can be inhaled through the nose and/or mouth of humans was discussed extensively in Chapter 10. The discussion drew attention to the fact not only that particles of different aerodynamic diameters are inhaled with different aspiration efficiency, but also that the corresponding aspiration efficiencies of aerosol samplers intended for sampling the coarse fraction would depend on sampler design and operating parameters, and would therefore differ from one sampler to another.

Historically, it has been widely assumed that a sampler without a pre-selector would aspirate particles with efficiency close to unity. Samplers for 'total' aerosol were therefore designed with a wide range of shapes, dimensions, and operated at a wide range of sampling flow rates without particular regard to their performance. It became a tacit assumption that the establishment of – and compliance with – aerosol standards for this coarse fraction was determined by the sampler chosen to do the job. However, the introduction of inhalability criteria from the 1980s onwards provided the opportunity to rationalise the definition and measurement of the coarse aerosol fraction in terms of a single, truly scientific, health-based criterion. This spurred the emergence of a new family of samplers designed specifically to collect the inhalable fraction, and – as described in Chapter 14 – a small number of such instruments have since become commercially available. Most such samplers were intended for use as personal samplers.

With the emergence of the new scientific framework for occupational aerosol exposure assessment, along with appropriate instrumentation, serious discussions began on how to implement the new sampling criterion for the inhalable fraction. The UK HSE made its move as early as 1989 when, under the (then) new regulations for the Control of Substances Hazardous to Health (COSHH), occupational exposure limits (OELs) for aerosol that had previously been thought of in terms of 'total' aerosol were redefined en masse in terms of 'total inhalable particulate' – what we now know as the inhalable fraction. It remains that way today. The change was facilitated by the fact that the seven-hole personal sampler was already available commercially in Britain and had been identified at the time by HSE scientists as satisfactorily collecting the inhalable fraction. Indeed, this sampler is still listed as an option in HSE's more recent *Methods for the Determination of Hazardous Substances* (MDHS 14/3) (Health and Safety Executive, 2000).

Elsewhere, standards setting organisations moved more slowly. The American Conference of Governmental Industrial Hygienists (ACGIH) noted that changing the criterion for sampling, and hence recommending adoption of new instrumentation not yet familiar to the professional occupational hygiene community, could lead to unforeseen consequences in the application of its TLVs. Three potential areas of difficulty were cited. The first was a quasi-political one – that suitable instruments should be available in a variety of models and designs from more than one manufacturer, thus providing occupational hygienists with a fair choice of instrument and supplier. The second concerned the potential impact of the change in sampling criterion – and in turn range of sampling instruments – on the compliance (or otherwise) by industry with existing standards. At the time it was not known how the proposed changes would influence measured exposures, except to broadly predict that, based on all the laboratory test data like those described in Chapter 14, measured aerosol concentration levels would inevitably rise. Thirdly, in the light of the expected changes in measured exposure concentrations for any given workplace situation, concern was expressed about how the new values might relate to the large banks of historical exposure data that had been accumulated over many years using earlier generations of sampler, and where it might be desired to use such data alongside new data for future epidemiological and standards setting purposes. In the light of such concerns, during the early 1990s ACGIH encouraged industry participation in research aimed at assessing the impact of the change in sampling criterion, specifically with respect to replacing the measurement of workers' exposures to 'total' aerosol by ones made using a sampler recognised as being capable of collecting the inhalable fraction. More specifically still, it was intended that this would involve the conduct and publication of results from side-by-side sampling studies in the field, for exposure assessment for 'total' aerosol using the *37 mm closed-face plastic cassette* – the sampler most commonly used for many years by occupational hygienists in the USA – and the *IOM personal inhalable aerosol* sampler, respectively. The latter was considered at the time as being an appropriate 'reference' sampler for the inhalable fraction (Bartley, 1998). The result during the years that followed was a large number of reports of field studies for various combinations of samplers.

22.3.1 Side-by-side comparative studies

The earliest studies to compare personal samplers for 'total' and inhalable aerosol samplers for actual workplace exposures were reported at the 1991 Inhaled Particles Symposium in Edinburgh by Lillienberg and Brisman, and were subsequently published in 1994. They carried out an exposure assessment survey in Swedish bakeries in which flour dust was collected by both 37 mm open-face and IOM samplers, where each participating worker wore the two samplers side-by-side in his/her breathing zone for full-shift sampling. It is not insignificant to note that this required each worker also to wear two sampling pumps which meant that, although personal sampling pumps had been increasingly miniaturised over the years, there was a considerable burden placed on the participating workers. The Lillienberg and Brisman study yielded 29 separate data records, each one representing a pair of samples providing full-shift time-weighted average concentrations for 'total' and inhalable aerosol exposures, respectively. It was shown that the IOM sampler consistently collected significantly more particulate matter than the 37 mm open-face sampler, ordinary least squares regression yielding a bias of close to 2. Soon afterwards, in a paper presented at the 1994 American Industrial Hygiene Conference and Exposition in

New Orleans, Shen *et al.* reported similar results for workers' exposures in an American borate/boric acid plant. This time, however, the sampler chosen for 'total' aerosol in the side-by-side comparison was the 37 mm closed-face sampler. From 58 data records, the IOM sampler was found to collect significantly more than the 37 mm closed-face sampler, with ordinary least squares regression again yielding a bias of greater than 2.

Such studies proliferated in the years that followed, accompanied by refinements in methodology. Tsai *et al.* (1996a) described a study of workers in the nickel primary production industry, confirming the trends that had started to become apparent from the reports of Lillienberg and Brisman and of Shen *et al.* They also described more refined statistical regression procedures for determining the biases between samplers in given exposure scenarios. The aims of the regressions were to estimate the ratio $E_{inhalable}/E_{total}$ for given similarly exposed groups (SEGs) of workers, where $E_{inhalable}$ was the exposure as measured for inhalable aerosol and E_{total} was the exposure as measured for 'total' aerosol. $E_{inhalable}$ was obtained using the IOM sampler (so that $E_{inhalable} \equiv E_{IOM}$) and E_{total} using the 37 mm closed-face sampler (so that $E_{total} \equiv E_{37}$). It was assumed that: (a) the goal was to find the relation between the two measures of exposure so that E_{37}-values could be converted to E_{IOM}-values; and (b) the relationship should be linear and pass through the origin of coordinates such that:

$$E_{IOM} = S \cdot E_{37} \tag{22.1}$$

where S was the slope of the relationship representing the bias between the instruments. In addition, it was assumed that the errors of individual exposure measurements were random, meaning that replicate samples of the same exposure should be normally distributed.[1] Equation (22.1) assumes trivially that, when no aerosol was present, both samplers should have collected zero particulate mass. Tsai *et al.* considered several regression methods, but opted for weighted least squares (WLS) regression since this best preserved the assumption of the normality of errors and also reduced the effect of excessive 'leverage' associated with small numbers of large exposure values for either sampler. By inspection of residuals when the WLS method was applied, it was found that the optimum weighting was achieved when both E_{IOM} and E_{37} were multiplied by the ratio $(1/E_{37})^2$. In this way, the results were considered to be equivalent to the univariate weighted analysis of the E_{IOM}/E_{37} ratios for the individual sampler pairs. Finally, with regard to outliers, it was acknowledged that, in real-world aerosol sampling, there were many possible events or factors that might cause outrageous results that were not consistent with the actual sampling process itself. So it was decided arbitrarily that sample pairs should also be removed from consideration if the ratio of E_{IOM}/E_{37} for each sample pair lay outside the range from 2/3 to 7. In any event, in the application of this criterion, the impact of the removal of any such data pairs was inspected and found to be small. Other workers adopted alternative statistical approaches. For example, Vinzents *et al.* (1996) obtained S in Equation (22.1) from the ratio of the medians estimated from the cumulative distributions of exposure for the two samplers.

As already stated, much initial interest was devoted to the comparison between the 37 mm closed-face sampler and the IOM sampler (see Figures 14.6 and 14.20), driven by the fact that the 37 mm sampler had been very widely used by occupational hygienists in the USA. A large number of such inter-sampler comparisons have been reported since the early 1990s and three typical yet contrasting sampler sets of data are shown by way of illustration in Figures 22.1–22.3. They are taken from studies at workplaces in nickel mining (Tsai *et al.*, 1995), lead smelting (Spear *et al.*, 1997) – both summarised by Werner *et al.* (1996) – and carbon black manufacturing (Kerr *et al.*, 2002). The data are shown as scatter plots representing the relationships between E_{37} and E_{IOM}. Also shown on each graph is the 1:1

---

[1] Not to be confused with the expected log-normal distribution of exposures.

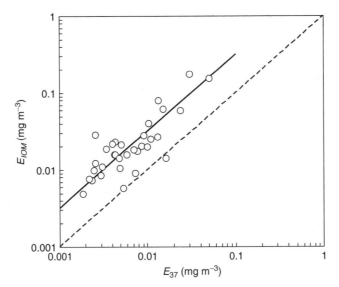

Figure 22.1 *Comparison between workers' exposures to overall airborne dust as measured using the IOM personal inhalable aerosol sampler (E_{IOM}) and the 37 mm closed-face cassette sampler (E_{37}) in a nickel mine (Tsai et al., 1995, summarised by Werner et al., 1996). The solid line represents the best fit from linear regression and the dashed line represents 1:1. Reproduced with permission from Werner et al., The Analyst, 121, 1207–1214. Copyright (1996) Royal Society of Chemistry*

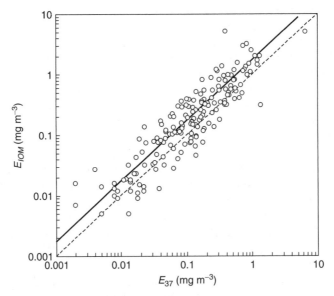

Figure 22.2 *Comparison between workers' exposures to airborne lead as measured using the IOM personal inhalable aerosol sampler (E_{IOM}) and the 37 mm closed-face cassette sampler (E_{37}) in a lead smelter (Spear et al., 1997, summarised by Werner et al., 1996). The solid line represents the best fit from linear regression and the dashed line represents 1:1. Reproduced with permission from Werner et al., The Analyst, 121, 1207–1214. Copyright (1996) Royal Society of Chemistry*

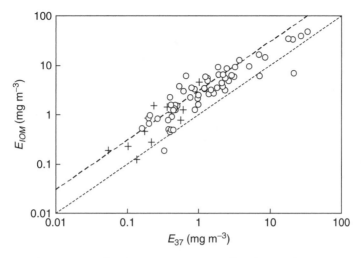

Figure 22.3 *Comparison between workers' exposures to overall airborne dust as measured using the IOM personal inhalable aerosol sampler (E_{IOM}) and the 37 mm closed-face cassette sampler (E_{37}) in a carbon black production facility (from Kerr et al., 2002). The solid line represents the best fit from linear regression and the dashed line represents 1:1. Adapted from Applied Occupational and Environmental Hygiene, Personal Sampling for Inhalable Aerosol Exposures of Carbon Black Manufacturing Industry Workers, 17, (10), pages 681–692. Copyright 2002. ACGIH® Cincinnati, OH. Reprinted with permission*

line representing the situation where both samplers are in perfect agreement. The dashed line represents the regression line representing the estimated bias between the two samplers. It is seen that the inhalable aerosol concentration results were consistently greater than for 'total' aerosol. This tendency was entirely consistent with what had been learned about the performances of these samplers in the laboratory studies reported earlier in Chapter 14. Data from these studies like these (from Tsai *et al.*, 1995, 1996a and b for aerosol exposures at primary nickel production, alloy production and electroplating facilities; Wilsey *et al.*, 1996 for machining fluids aerosols at an armaments plant; Vinzents *et al.*, 1996 for aerosol exposures at woodworking, welding lead battery and aluminum foundry facilities; Demange *et al.*, 2002 for iron production; and Calzavara *et al.*, 2003 for asphalt and roofing production) are summarised in Table 22.2. This table embodies a number of important features, in particular:

- Depending on the aerosol fraction of health-related interest, the collected samples were analyzed for total collected particulate mass (i.e. from gravimetric weighing of the collecting substrates – the entire cassette for the IOM sampler and the filter for the 37 mm closed-face sampler – before and after sampling) or for specific collected species (i.e. as measured by analytical instrumentation – for example, atomic absorption for lead or inductively coupled plasma atomic emission spectrophotometry for nickel). In most cases where results for both were given, the data were quite close. The exception, where there was a marked difference, was for asphalt and roofing (Calzavara *et al.*, 2003), where the S-value for the benzene-extracted fraction was close to unity, much smaller than for overall aerosol, suggesting that this fraction was contained within very small particles.
- The IOM samplers consistently collected more particulate mass than the 37 mm closed-face cassette, where the bias was generally greatest for situations where the aerosol was coarsest (e.g. mining) and least where the aerosol was finest (e.g. welding), with S ranging from close to unity to as high as greater than 3.

Table 22.2 *Summary of results of comparative studies for 37 mm closed-face plastic cassette sampler for 'total' aerosol versus the IOM personal inhalable aerosol sampler, both operated at a sampling flow rate of 2 Lpm and used side-by-side as personal samplers on actual workers in real-world workplace situations, where S is as given by Equation (22.1). For the regression analyses: OLS, ordinary least squares; WLS, weighted least squares; A, average of the individual sample pair ratios; RM, ratio of medians of the cumulative exposure distributions for the IOM and 37 mm sampler, respectively. Note: for woodworking shop B, the diameter of the sampler entry was 5.6 mm, as opposed to 4 mm for all the other studies cited*

Industry	Component analyzed	n	Regression	S	Reference
Borates/boric acid	Total collected particulate	58	OLS	2.20	Shen *et al.*, 1994
Nickel mine		32		3.20	
Nickel mill		21		2.72	
Nickel smelter A		35		1.65	Tsai *et al.*, 1995
Nickel smelter B	Ni	23	WLS	2.84	
Nickel refinery		36		2.12	
Nickel alloy plant		46		2.29	Tsai *et al.*, 1996a
Nickel electroplating A		21		2.02	Tsai *et al.*, 1996b
Nickel electroplating B		21		3.01	
Nickel mine		30		3.64	
Nickel mill		20		2.61	
Nickel smelter A		39		1.97	Tsai *et al.*, 1995
Nickel smelter B	Total collected particulate	23	WLS	2.43	
Nickel refinery		37		2.50	
Nickel alloy plant		45		1.94	Tsai *et al.*, 1996a
Nickel electroplating A		25		2.77	Tsai *et al.*, 1996b
Nickel electroplating B		25		3.29	
Lead smelter	Pb	151	WLS	1.77	Spear *et al.*, 1997
Armaments machine shop	Total collected machining fluids particulate	23	WLS	2.96	Wilsey *et al.*, 1996
Woodworking shop A	Total collected particulate	10		1.79	
Woodworking shop B		40		1.79	
Welding shop	Total collected particulate			0.95	
	Al	15	RM	1.36	Vinzents *et al.*, 1996
Lead battery	Total collected particulate			2.36	
	Pb	11		1.29	
Aluminum foundry	Total collected particulate				
Carbon black production	Total collected particulate	15		2.90	
Carbon black packing and shipping		60	WLS	3.27	Kerr *et al.*, 2002
Iron	Total collected particulate			1.62	Demange *et al.*, 2002
	Fe	54	A	1.39	
Asphalt production/roofing manufacturing	Total collected particulate			1.93	Calzavara *et al.*, 2003
	Benzene-extracted	58	OLS	0.92	

Table 22.3 *Suggested factors that may be used in practical situations for converting personal exposures to 'total' aerosol obtained using the 37 mm closed-face plastic cassette sampler to the equivalent inhalable aerosol exposure as obtained using the IOM sampler (Werner et al., 1996)*

Aerosol classification	Examples of industry sector or operation	Suggested conversion factor
Dust	Mining Ore and rock handling Handling and transport of bulk aggregates Textiles Flour and grain handling, etc.	2.5
Mist	Machining Paint spray Electroplating	2.0
Hot processes	Metal smelting and refining Alloy production Metals recovery, etc.	1.5
Fumes and smokes	Welding Garages	1.0

In the context of changing aerosol standards for coarse aerosols in workplaces, in particular the trend towards switching from exposure assessment based on the measurement of 'total' aerosol to one based on the inhalable fraction, Werner *et al.* (1996) reviewed the observed biases between the 37 mm closed-face and IOM samplers. Recognising the importance of such biases to the occupational hygiene community in North America in particular, they proposed some correction factors that might be used to convert exposures to 'total' aerosol to the corresponding exposure to the inhalable fraction. These are listed in Table 22.3, and were intended to provide occupational hygienists with 'ball-park' estimates of 37 mm closed-face versus IOM sampler biases for particular categories of field conditions. In addition to allowing for the correction of 'total' aerosol exposures, for example as might be required in some future standard, they also allowed assessment of the practical impact of switching to a new standard based on the inhalability criterion.

Such biases are also of interest where other samplers have been in use for sampling 'total' aerosol, driving similar inter-sampler comparisons of a number of personal samplers under field conditions, in nearly all which the IOM sampler was the chosen reference sampler for the inhalable fraction. Some of these from these additional field studies are summarised in Table 22.4. Here, following Equation (22.1), we have:

$$E_{IOM} = S \cdot E_{sampler} \qquad (22.2)$$

One comparison of particular interest is that between the IOM sampler and the CIS sampler (see Figure 14.8) that also performed well in the Bartley analysis. Some results for this comparison, from the Kerr *et al.* (2002) carbon black study, are shown in the table and graphically in Figure 22.4. They support the Bartley conclusion that – at least for the particular industry studied – the CIS sampler is a fair alternative to the IOM sampler for collecting the inhalable fraction. From this and other knowledge about the performance characteristics of this sampler, it is likely that a similar comparison might be found for most other industrial situations. Another device of related interest is the seven-hole sampler

Table 22.4   *Summary of results of comparative studies for various samplers versus the IOM personal inhalable aerosol sampler, where all samplers except for the Respicon® were operated at a sampling flow rate of 2 Lpm, where S is as given by Equation (22.2). The Respicon® was operated at 3.5 Lpm. All samples were used side-by-side as personal samplers on actual workers in real-world workplace situations. For the regression analyses: OLS, ordinary least squares; WLS, weighted least squares; LT, ordinary least squares on log-transformed results*

Industry	Sampler tested	Component analyzed	n	Regression	S	Reference
Bakeries	37 mm open-face sampler	Total collected particulate	29	OLS	1.82	Lillienberg and Brisman, 1994
Carbon black production	CIS sampler	Total collected particulate	7	WLS	1.02	Kerr *et al.*, 2002
Carbon black packing/shipping			17		0.92	
Various metal mines	Seven-hole sampler	Total collected particulate	27	OLS	2.13	Terry and Hewson, 1996
	Single-hole sampler		22	OLS	3.22	
Nickel refinery	Respicon®	Total collected particulate	107	LT	1.79	Koch *et al.*, 2002

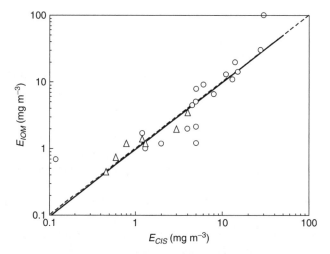

Figure 22.4   *Comparison between workers' exposures to overall airborne dust as measured using the IOM personal inhalable aerosol sampler (E_{IOM}) and the CIS personal sampler (E_{CIS}) in a carbon black production facility (from Kerr et al., 2002). The solid line represents the best fit from linear regression and the dashed line represents 1:1. Adapted from Applied Occupational and Environmental Hygiene, Personal sampling for inhalable aerosol exposures of carbon black manufacturing industry workers, 17(10), pages 681–692. Copyright 2002. ACGIH®, Cincinnati, OH. Reprinted with permission*

[see Figure 14.7(b)]. Although this did not fare as well in the Bartley analysis, it has – as already noted – continued to be regarded in some quarters as an acceptable sampler for the inhalable fraction. However, as shown in Figure 22.5, the results of Terry and Hewson (1996) for worksites in an Australian

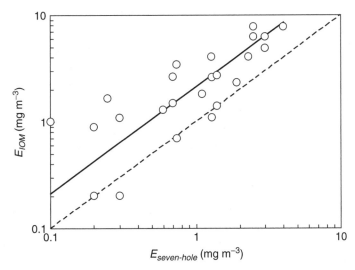

Figure 22.5 *Comparison between workers' dust exposures as measured using the IOM personal inhalable aerosol sampler (E_{IOM}) and the seven-hole personal sampler ($E_{seven-hole}$) at worksites in an Australian mining industry facility (from the data of Terry and Hewson, 1996). The solid line represents the best fit from linear regression and the dashed line represents 1:1*

mining industry facility showed that this sampler did not perform well in relation to the inhalable fraction as measured using the IOM sampler, undersampling significantly. Table 22.4 also summarises results reported by Lillienberg and Brisman for the 37 mm open-faced sampler in bakeries, and by Koch *et al.* (2002) for the Respicon sampler (see Figure 15.36) in a nickel refinery. For both of these samplers, significant biases in relation to the IOM sampler were evident.

The Button sampler (see Figure 14.24) arrived later as a candidate for collecting the inhalable fraction, and so did not feature in the field studies reported above. However a small amount of workplace experience with this instrument has been obtained. Hauck *et al.* (1997) described a short field study in which it was used as a personal sampler alongside the 37 mm closed-face plastic cassette. Here, however, the button sampler was operated at a sampling flow rate of 2 Lpm, as opposed to the 4 Lpm recommended for this instrument. At the lower flow rate, the results for the Button sampler were found to agree quite well with those for the 37 mm sampler. These data, however, did not go so far as to confirm the Button as a practical sampler for the inhalable fraction and further field studies with this instrument are clearly needed at the design flow rate.

Brown *et al.* (1995) conducted field trials at a hard metals tool-fabrication plant, two platinum plants and two ferrous metals, in which they deployed their novel electret-based personal sampler (see Figure 14.30) in side-by-side experiments with the IOM and seven-hole sampler, respectively. Although the masses collected by the electret sampler were always much less than those collected by the aspirated IOM and seven-hole samplers, for each plant there was fair linearity between the results for the electret sampler and those for each of the aspirated samplers. However, the quantitative relationship between the two types of sampler varied greatly from one plant to the other, reflecting differences in particle size distribution and other factors that might have influenced the particle electrical mobility that governed the performance of the electret sampler. Although these results were somewhat encouraging, the electret sampler does not appear to have been developed further or used subsequently in occupational aerosol exposure assessments.

22.3.2   The practical impact of changes from 'total' to inhalable aerosol measurement

An important objective of carrying out the simultaneous side-by-side measurements of personal exposures to 'total' and inhalable aerosol was to enable assessment of the practical impact of changing the aerosol sampling criterion. This was assessed by inspection not only of the direct relationship between the respective concentrations, as discussed above, but also the exposure statistical exposure distributions obtained in those studies. Figure 22.6 shows a typical example, taken from the data first shown in Figure 22.1. The bias already noted between the two samplers is clearly evident. The overall trend is that, when plotted on log-probability axes, each data set falls quite close to a straight line, reflecting the expected log-normal distribution. Discussion of the practical implications is carried out easiest by reference to the hypothetical example shown in Figure 22.7 (Werner *et al.*, 1996). It shows a case where the bias between the samplers in question, expressed as S in Equation (22.1), is 2. This is consistent with much of what was observed in practice. The figure shows that, for an OEL of $1 \, \mathrm{mg \, m}^{-3}$, the proportion of samples exceeding the OEL – the *exceedance* – is about 5 % for the original 'total' aerosol measurement and shifts to about 10 % when the metric of exposure is changed to the inhalable fraction. So, as expected, the exceedance does increase as a result of the change. But in practical terms, most occupational hygienists would regard the impact on compliance as relatively small. More generally, therefore, it may be concluded that, although the inter-sampler bias is indeed significant, the impact on the extent to which there is compliance with the new standards is only marginally affected in workplace situations where aerosol levels are already well controlled. However, the situation becomes more complicated when not only the basis of sampling changes but so too does the OEL itself. There have been several such cases in recent years. For example, for nickel-containing aerosols, in the late 1990s, ACGIH changed the basis of the TLV from 'total' to inhalable aerosol and at the same time

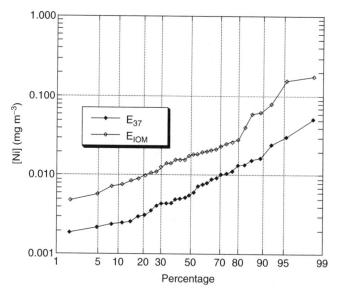

Figure 22.6 *Comparison between workers' exposures to overall airborne dust as measured using the IOM personal inhalable aerosol sampler (E_{IOM}) and the 37 mm closed-face cassette sampler (E_{37}) in a nickel mine (from the same data shown in Figure 22.1), shown here in terms of the respective cumulative distributions of the two measures of exposure. Reproduced with permission from Werner et al., The Analyst, 121, 1207–1214. Copyright (1996) Royal Society of Chemistry*

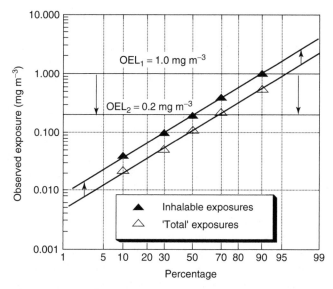

Figure 22.7 *Hypothetical example of the comparison between workers' exposures to inhalable and 'total' aerosol, respectively, shown in terms of the distributions of the two measures of exposure, used as a basis for discussing the impact of changing sampling criteria – and in turn sampling instruments – on the level of compliance with occupational exposure limits (OELs). Reproduced with permission from Werner et al., The Analyst, 121, 1207–1214. Copyright (1996) Royal Society of Chemistry*

lowered the exposure limits for the species that it listed. This represents a new situation – what has been referred to by some industry occupational hygienists as a 'double-whammy'. If, for example, the OEL in Figure 22.7 is reduced to 0.2 mg m^{-3}, then the original exceedance would now be as great as 30 %, which most occupational hygienists would *not* regard as satisfactory. But, worse, the new exceedance would be significantly greater still, now about 50 %.

22.4 Converting particle counts to particle mass

In many industries, aerosol sampling went through a long phase in the 1990s where particle size selective sampling involved the collection of samples where particles were collected onto glass substrates and were subsequently analyzed – usually by optical microscopy – in terms of the number concentrations of particles conforming to some particle size criterion (e.g. particle physical size less than 5 or 10 μm). For many years and for many industries, standards and limit values reflected this approach.

22.4.1 Respirable aerosol exposures in the coal industry

The history of respirable aerosol sampling and standards had its roots in the mining industries. The British coal mining industry was prominent in many important developments in the period following the Second World War, including the evolution of sampling concepts and instruments and their application in the field. This effort was part of a much larger epidemiological program known as the '*Pneumoniosis Field Research*' (*PFR*) that incorporated both dust exposure assessment and medical and physiological surveys. The latter included X-radiography, anthropometric measurements, lung function testing and respiratory symptoms questionnaires (Rogan *et al.*, 1967). The PFR was conducted at 25 underground coal mines in 14 regions across the UK, from Scotland in the north to North and South Wales in the west to Kent in the south, chosen to reflect the ranges of coal types and environmental conditions found at large in the industry. In the coal industry, coal type is identified in terms of *coal rank*, where 'high rank' includes high carbon content coals such as anthracite and bituminous and 'low rank' includes lower carbon content coals such as lignite and sub-bituminous. The PFR was a study of the exposure and health histories of 30 000 coalminers, carried out in the form of 5-yearly environmental and medical surveys, the first ending in 1958, the second in 1963, and so on.

Bedford and Warner had noted as early as 1943 that standards for coal mine dust exposures would best be represented by the mass concentration of aerosol for particles of physical size less than 5 μm. However, they proposed that, for convenience, and in the absence at the time of a sampler that could directly provide the desired particle size classification, OELs should be expressed in terms of the number concentration of particles in the size range from 1 to 5 μm. This was adopted. It was only later that the concept of the aerodynamic selection of respirable aerosol, in the form of the horizontal elutriator, and subsequent gravimetric analysis, would emerge. In the meantime, the thermal precipitator, which had first arrived in the 1930s (Watson, 1936–1937), became the mainstay of respirable dust sampling in British coal mines, where collected samples were size-selected and counted by optical microscopy. Aerosol concentrations were expressed in terms of the number of particles in the size range from 1 to 5 μm per cubic centimeter of air (p cm^3). The instrument in question, the *standard thermal precipitator (STP)*, was shown earlier in Figure 15.4. Data were also obtained using the *long-running thermal precipitator (LRTP)* in which particles had been aerodynamically selected prior to deposition. From the use of such instruments, aerosol concentration data expressed in terms of particle count were prevalent during the first and second PFR surveys. By the late 1950s, it was becoming apparent that the mass concentration of respirable dust exposures was a more appropriate index of exposure (Rivers *et al.*, 1960), confirming what had earlier been noted by Bedford and Warner. Meanwhile the practical limitations of the thermal precipitator under arduous field conditions were also by now well apparent. So the search for a particle size-selective gravimetric sampler was intensified, leading to the MRE respirable dust sampler (see Figure 15.6) which became used increasingly in the PFR during the mid-1960s, eventually taking over completely by the late 1960s. By the third PFR survey period, starting in 1964, it was clear that it would be necessary to develop correction factors to convert the earlier particle number concentrations to equivalent respirable mass concentrations. This was first approached by Rogan *et al.* on the basis of the particle size distributions obtained from the microscopic analysis of STP samples for particles in the range 1–5 μm. They estimated particle size distributions in terms of the simple form:

$$n(d) = \beta \exp[\beta(1-d)] \tag{22.3}$$

in which β was the *exponential size distribution parameter*. In this expression, smaller values of β were indicative of coarser aerosols. This particle size distribution information was then used to make estimates of the relative mass-to-number ratio for each colliery, leading ultimately to corresponding estimates of the mean particle mass concentration. The results for the 25 PFR collieries are shown in Table 22.5. Here it is noted that the estimated mean mass concentrations were expressed in arbitrary units, such that they were useful only for comparative purposes. However, they are interesting in showing the clear differences in particle size distribution from one colliery to the next and, in turn, the differences in conversion factor that would be required to convert STP measurements to appropriate respirable mass concentrations. One interesting feature of the results was that the β-values were found to be lowest, and hence the estimated mass-to-number ratios the highest, for the high-rank collieries in the South Wales region. Collieries 19 and 20 in particular were most notable for the fact that the coal type was anthracite, the highest rank of all.

Table 22.5 *Summary of respirable dust measurements made in British coal mines during the late 1950s and early 1960s, carried out during the first two surveys of the Pneumoconiosis Field Research (PFR): conversion of respirable particle count concentrations (particles with diameter between 1μm and 5μm) from standard thermal precipitator samples to equivalent mass concentrations (from Rogan et al., 1967)*

Region	Colliery number	Mean measured concentration (respirable particles cm^{-3})	Mean β-value	Estimated mean concentration (respirable mass m^{-3}) (arbitrary units)	Estimated mass -to-number ratio (arbitrary units)
Scotland	1	160	0.78	128	0.80
	2	170	0.72	146	0.86
	3	120	0.71	104	0.87
	4	110	0.63	106	0.96
	5	190	1.03	114	0.60
Northumberland	6	180	0.71	157	0.87
	7	270	0.65	256	0.95
Cumberland	8	230	0.54	248	1.08
Durham	9	300	0.65	282	0.94
	10	350	0.57	364	1.04
Yorkshire	11	210	0.60	210	1.00
	12	220	0.58	227	1.03
	13	300	0.60	300	1.00
Lancashire	14	490	0.53	534	1.09
North Wales	15	240	0.48	278	1.16
Nottinghamshire	16	540	0.54	583	1.08
Staffordshire	17	390	0.60	390	1.00
Warwickshire	18	150	0.50	170	1.13
South Wales	19	110	0.41	139	1.26
	20	130	0.44	159	1.22
	21	210	0.47	248	1.18
	22	140	0.42	175	1.25
	23	380	0.45	456	1.20
	24	130	0.50	147	1.13
Kent	25	150	0.54	162	1.08

The early studies reported by Rogan *et al.* were followed up by more sampling campaigns in the field in which both STPs – and, later, LRTPs – and MRE samplers were placed side-by-side at underground work sites in the 25 PFR mines. The first results were reported by Hamilton *et al.* (1967) and more by Dodgson *et al.* (1971), providing a comprehensive set of experimentally derived conversion factors – known as '*mass number indices*' (*MNIs*) – that could be used for conversion of all the earlier thermal precipitator data to equivalent MRE respirable aerosol mass concentrations. The MNIs were expressed in terms of the ratio of the gravimetric respirable aerosol concentration (in mg m^{-3}) to particle counts obtained from thermal precipitator samples (in 1000 p cm^{-3}). Dodgson *et al.* reported a number of important correlations. Figure 22.8(a) shows the observed relationship between MNI and the particle size distribution parameter (β) first introduced by Rogan *et al.* and Figure 22.8(b) the one between MNI and coal rank, expressed here as the percentage carbon content of the coal. MNI-values ranged widely from as low as about 7 to as high as 35 mg m^{-3}/1000 p cm^{-3}, increasing with decreasing β and being greatest for the highest rank coal mines. These trends are consistent with those suggested by Rogan

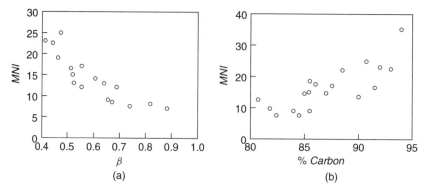

Figure 22.8 *Observed relationships between mass number index (MNI) and other aspects of airborne dust in coal mines: (a) particle size distribution parameter (β) first introduced by Rogan et al. (1967); (b) coal rank, expressed in terms of carbon content (data taken from Dodgson et al., 1971). Adapted with permission from Dodgson et al., Inhaled Particles Volume III, pp. 757–781. Copyright (1971) British Occupational Hygiene Society*

et al. Dodgson and his colleagues also reported differences in MNI within collieries between different work areas.

The most important single outcome of this large body of work was the enabling of the creation of long histories of dust exposures for the mineworkers expressed in terms of definitive respirable mass concentration. The cited works therefore played a pivotal role in the epidemiological studies of coalworkers' pneumoconiosis, providing data that were crucial in the development of new standards for coalminers' dust exposures which, in the years following implementation, led directly to new environmental control strategies that were demonstrably very effective in reducing the prevalence of lung disease. We can now look back and appreciate that the PFR was one of the finest examples of occupational epidemiology for aerosol-related disease, incorporating aerosol science into the comprehensive science-based exposure assessment that enabled unprecedented direct associations between dust exposure and lung disease that not only underpinned reliable new standards for the coal industry but also provided a valuable model for occupational epidemiology elsewhere.

22.4.2 Respirable aerosol exposures in hard rock mining

Dust exposures during hard rock mining (e.g. metal ores) have long been of concern due to the potential for workers' exposures to crystalline silica – or quartz – leading to the very serious lung disease of silicosis. It is in this industry sector where the *konimeter* sampler (see Figure 15.1) was widely used, especially in Canada and South Africa. As described in Chapter 15, this instrument aspirated a small 'snap' sample and particles were deposited by impaction onto a glass substrate, and the sampled particles were subsequently size-selected and counted by optical microscopy. The konimeter was an important basis of exposure assessment and regulation in Canadian hard rock mining during the long period from 1940 to 1982. Throughout that period, aerosol exposures were expressed in terms of the number concentration of particles, typically in the size range below about 10 μm, per cubic centimeter of air (p cm^{-3}), and this was the basis of the standards. In the early 1980s, however, attention was switched to more explicit measurement of the mass concentration of crystalline silica as contained within the respirable fraction that would be collected by a personal cyclone of the type also described in Chapter 15. Similar to the experience in the British coal industry, it became an important priority to rationalise the long-term exposure histories of workers in terms of the new – more relevant – index of exposure. Verma *et al.* (1987) carried out side-by-side field comparisons of three konimeter models, all of which had been used at various times in the Canadian and South African hard rock mining industries. The models in question were known as the '*Devers*', the '*Witwatersrand*' and '*Gathercole*' konimeter, respectively. In addition, they examined a version of the Witwatersrand konimeter with a pre-impactor (the 'Witwatersrand-I'). The four konimeter types were tested at a uranium ore mine and a gold mine, and the results are shown in Figure 22.9. The approach taken by Verma *et al.* was to compare the Gathercole directly with the three other types, with the intention that all konimeter results would be adjusted to be equivalent to those for the Gathercole. The results of the linear individual regressions are shown also in Figure 22.9. The results show that there were clear biases between the different konimeter models and that those biases were quite dependent on the type of mine.

Soon afterwards, Verma *et al.* (1989) reported a second important study in which more side-by-side measurements were made in workplaces, this time to determine directly the relationship between the old and new indices of exposure. The purpose was to assemble a complete set of exposure data from 1940 onwards that could be expressed in terms of the mass concentration of respirable crystalline silica, leading to estimates of the lifetime cumulative mass exposures of workers. That involved establishing a basis by

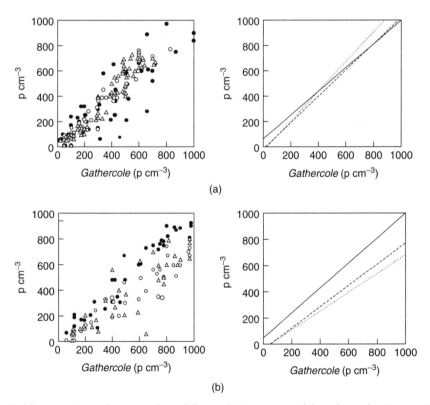

Figure 22.9 *Field comparisons between four different konimeter models, where the Devers (● and ___), Witwatersrand (O and – – –) and Witwatersrand-I (△ and ......) models were compared with the Gathercole; raw data to the left, results of regression analysis to the right: (a) for a gold mine; (b) for a uranium mine (Verma et al., 1987). Adapted with permission from Verma et al., Annals of Occupational Hygiene, 31, 451–461. Copyright (1987) British Occupational Hygiene Society*

which all the old konimeter data could be converted into equivalent respirable quartz mass concentrations. In Canada, for the latter, the gravimetric sampler of interest was the 10 mm nylon personal cyclone (see Figure 15.8). As Verma and his colleagues noted, the relationship was complicated by the fact that the two types of instrument measured entirely different quantities. The konimeter provided only particle counts without regard to composition, and each sample was taken over a very short time interval. On the other hand, the cyclone provided a sample taken over several hours that was analyzed (e.g. by infrared spectrophotometric methods) in terms of just the silica content of the aerodynamically selected respirable aerosol. Also as noted in their earlier 1987 study, four different types of konimeter were used, and these too needed to be rationalised. Additionally, they also examined the effects associated with 'between-microscopist' subjective particle counting differences for konimeter samples and how these related to results obtained using an automated scanning device. Finally, after all such factors had been standardised, all the konimeter results were converted to 'scanner-equivalent' values, and these were the ones used in the direct comparisons with the cyclone data. As in the 1987 study, results were obtained for both gold and uranium mines, respectively. The comparisons are summarised in Figure 22.10, where the shaded areas are drawn to represent the scope of nearly all the original data. When the raw data had been plotted on linear axes by Verma *et al.* in the original paper, the results were seen to be sharply nonlinear for low respirable silica concentrations. Verma *et al.* explored options for mathematical functions to describe the relationship in question, and proposed:

$$\text{konimeter counts} = \alpha_0[1 - \exp(\alpha_1 \cdot \text{cyclone mass})] + \alpha_2 \cdot \text{cyclone mass} \tag{22.4}$$

where the α-values here were coefficients that needed to be obtained by inspection of the original data. Table 22.6 summarises the values that were estimated from regression analysis, taking into account the influence of ore type. The resultant curves for the gold and uranium mines, respectively, are shown alongside the data summaries in Figure 22.10. Elsewhere, Hewson (1996) carried out a similar exercise at worksites in the Western Australian metals extraction industries. He too used field studies to provide data for converting konimeter particle count data to equivalent respirable mass concentration data.

22.4.3   Inhalable aerosol exposures in the nickel industry

Ontario, Canada, is the home of some of the largest nickel mining and primary production operations in the world. For many years, occupational aerosol exposures there were measured in terms of particle counts from samples obtained using the konimeter, not only in deep underground mining but also in the milling, smelting and refining primary production processes at the surface. Samples were usually taken within workers' breathing zones. In workplaces in these industrial sectors, workers have historically been exposed to nickel-containing aerosols. Meanwhile, a record has been established over the years of serious aerosol-related ill-health, including nasal and lung cancers. This has driven the long history of aerosol measurement in these workplaces. In 1990 this body of work was reviewed comprehensively by the International Committee on Nickel Carcinogenesis in Man under the chairmanship of Sir Richard Doll.

Like the other industries mentioned above, the nickel industry too has seen shifts in exposure assessment methods and strategies, and in occupational exposure standards. For many years, as in other mining-related industries in Canada, it was customary to measure workplace aerosols in terms of particle counts for samples obtained using the konimeter. The version of choice was usually the Gathercole model featured in the Verma *et al.* (1987) study. Most of the measurements between the 1950s and mid 1970s were made using this sampling instrument and the data were therefore recorded in terms of p cm^{-3}. In the 1970s, some gravimetric measurements of ambient 'total' aerosol were made using a high-volume sampler known as the '*Hurricane*' (Gelman Instrument Company, Ann Arbor, MI, USA),

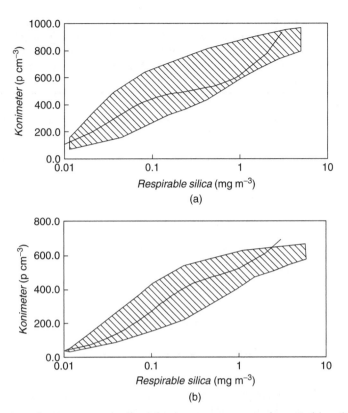

Figure 22.10 *Comparison between standardised konimeter counts and respirable silica concentration as obtained from 10-mm nylon personal cyclone samples, summary of the data reported by Verma et al. (1989) for: (a) Canadian gold mines; (b) Canadian uranium mines. The hatched areas encompass nearly all the data points reported in the original paper. The solid line describes the nonlinear regression Equation (22.4) of Verma et al.*

Table 22.6 *Summary of the regression analysis describing the relationship between standardized konimeter data and the corresponding gravimetric data for respirable silica as sampled using the 10 mm nylon cyclone (Verma et al., 1989)*

Coefficient	Gold mining	Uranium mining
α_0	459	437
α_1	-27.41	-9.31
α_2	159	86.3

having a very high sampling flow rate of over 3500 Lpm. During the late 1970s, personal sampling of workers' exposures to 'total' aerosol was carried out using the 37 mm closed-face cassette. Then during the 1990s, as described earlier in this chapter, sampling campaigns were carried out to make personal measurements of inhalable aerosol, using the IOM sampler, and comparing them with simultaneous

Table 22.7   *Summary of historical data for airborne dust, as measured by various methods, in a nickel smelter. The earlier data are summarised from Ramachandran and Vincent (1999) and the 1994 data from Tsai et al. (1995). For the konimeter, Hurricane and 37 mm cassette, the data are expressed as arithmetic means and the IOM data as medians. Figures in parentheses indicate the number of data records and those in italics represent the estimated range of data falling within the 95 percentile*

Period	Konimeter (breathing zone, p cm^{-3})	Hurricane (workplace air, 'total' aerosol, mg m^{-3})	Hurricane (workplace air, mg Ni m^{-3})	Personal ('total' aerosol, mg m^3)	Personal ('total' Ni, mg m^{-3})	Personal (inhalable aerosol, mg m^{-3})	Personal (inhalable Ni, mg m^{-3})
				37 mm cassette		IOM sampler	
1956–1963	959 (24) *765–1153*	–	–	–	–	–	–
1964–1966	561 (15) *458–664*	17 (1)	10 (1)	–	–	–	–
1967–1971	623 (27) *530–716*	16.5 (2) *0–98*	7 (2) *0–56*	–	–	–	–
1972–1975	529 (18) *456–602*	53.1 (27) *7–99*	37 (11) *12–62*	–	–	–	–
1976–1979	–	1.31 (17) *0.8–1.8*	–	1 (11) *0–28*	0.4 (11) *0–10*	–	–
1994	–			1 (33) *0–7*	0.2 (33) *0–2*	2 (33) *0.5–10*	0.4 (33) *0–1.5*

measurements using the 37 mm sampler. The overall challenge was to link all these measurements in order to create a long and consistent record based on a true health-related metric of exposure. Since, by the late 1990s OELs for nickel-containing aerosols were being expressed in terms of the inhalable fraction, the ultimate goal was to create long-term exposure histories for workers according to that metric. The problem is illustrated in Table 22.7 for a set of data that was available for nickel-containing aerosols in a nickel smelter, summarised from Tsai *et al.* (1995) and Ramachandran and Vincent (1999). The table shows how, as discussed above, the metric of the measured aerosol concentration has shifted over the years. Importantly, it also shows that the number of data records in most of the various categories, especially when the size of time interval is taken into account, were quite sparse such that the possibilities for statistical analyses of the types that might be used for exposure assessment were limited. Although rough estimates of the appropriate conversation factors are possible from Table 22.7 simply by comparing data between columns for the periods where more than one data set were available, this is only satisfactory in the absence of anything better. Later, Ramachandran (2001) conducted a more sophisticated analysis. In the first part, he located some comparative data from side-by-side experiments in the smelter using the konimeter and a respirable aerosol sampler, and found an average value for the ratio of respirable overall aerosol mass concentration to konimeter counts of 0.0021 mg m^{-3}/p cm^{-3}, consistent with the range of values (0.0002–0.003 mg m^{-3}/p cm^{-3}) published by Verma *et al.* in 1989 for the hard rock miners. To relate the respirable dust measurements to the corresponding inhalable values, he used particle size distribution information that had been reported by Tsai (1995) for the same workplaces and made a fair estimate of the desired conversion factor, reporting a value of 0.1. By combining these two conversions, a coefficient was obtained by which to convert konimeter data to corresponding personal inhalable aerosol exposures. Meanwhile, in order to incorporate the Hurricane-based measurements of 'total' aerosol, Ramachandran used a set of side-by-side data for the Hurricane and 37 mm closed-face cassette samplers, respectively, that had been obtained – again – by occupational hygienists in the smelter during the 1970s. These were converted to inhalable aerosol concentrations using the factors obtained later from the side-by-side comparisons of the 37 mm cassette and the IOM sampler. As a result of all these manipulations, a preliminary matrix upon which to base the long-term histories of workers' exposures to overall inhalable aerosol in the smelter was created. The problem remained that, even after rationalisation into a single, most-appropriate metric, the raw data were sporadic and uncertain.

To improve matters further, Ramachandran invoked specialised statistical procedures, the purpose of which was to refine the original knowledge of the exposures by narrowing their probability distributions, and hence sharpen the level of confidence with which they could be viewed. More specifically, this involved an approach named after the British mathematician, Thomas Bayes, who lived during the 1700s and developed ideas that were not widely used until as recently as the 1950s. Details of modern applications of 'Bayesian' methods to the refinement of sparse data sets have been given by, for example, Little and Rubin (1987). In its application to the particular aerosol exposure assessment problem addressed by Ramachandran, important inputs to the Bayesian-based data refinement process were the professional and scientific judgments of carefully selected, appropriately qualified experts with respect to relevant aspects of the aerosol formation, release and dispersion – that is, the physical factors relating to human exposure. For each expert, quantitative opinions on those aspects, given based on specific historical plant information that had been provided, were communicated by questionnaire. The results that were returned were incorporated into a physical model of exposure in order to make 'prior' estimates of the probability distributions of exposures. For each case of interest, the prior was combined within the Bayesian framework, leading to a new probability distribution known as the 'posterior', again for each expert. This was the desired endpoint. By way of illustration, the resultant history of inhalable aerosol exposures (for overall airborne dust) in the nickel smelter is shown in Figure 22.11 for the period from 1960 until 1979, based on the input of one expert. The open circles show the rationalised inhalable aerosol results from the first part of the analysis. The closed triangles show the refined inhalable aerosol exposures based on the input from the single expert. These results show, not surprisingly,

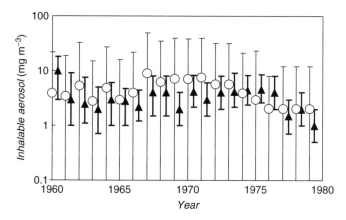

Figure 22.11 *Example of the history of inhalable aerosol exposure concentrations in a large nickel smelter over the period from 1960 to 1979. The open circles are measured data that have been taken over the years using different sampling instrumentation and rationalised in terms of the inhalable fraction. The closed triangles are the results of refinement to the original data set using the Bayesian framework as applied by Ramachandran (2001) for the input of a single expert. Adapted with permission from Ramachandran, Annals of Occupational Hygiene, 45, 651–667. Copyright (2001) British Occupational Hygiene Society*

that exposures declined somewhat over the more recent years of the period in question. Even more striking was the fact that the confidence intervals in the results after refinement, as reflected in the error bars, were sharply less. Such research suggests some powerful new tools that are becoming available for the optimisation of aerosol exposure data sets that, in their original form, are sparse, biased and very uncertain – a scenario that is all too familiar for historical aerosol exposures.

22.5 Field experience with samplers for respirable aerosol

22.5.1 Gravimetric mass sampling

Many respirable aerosol samplers have been described in this book for which the objective has been the collection of the respirable aerosol fraction onto a filter such that the collected mass can be determined by weighing or by some appropriate analytical method (e.g. for a particular chemical species sub-fraction). As has been discussed extensively, the definition of what is respirable has either defined the chosen sampling instrument or has been defined by it. Either way, the many such instruments that have found widespread application have exhibited a wide range of performance characteristics. In addition to the performance studies that have been carried out in the laboratory (again see Chapter 15), there has been a long history of comparative field studies of different sampling instruments in industrial workplaces. Some of them involved field comparisons between samplers that aimed to collect, nominally at least, the same specific respirable fraction. But, although these were useful in terms of helping occupational hygienists to make specific choices, most such studies were of limited interest in the wider historical context. However, some were of more general interest. One body of work in the USA was focused on the 10 mm Dorr–Oliver nylon cyclone that has been used extensively for many years in USA coal mines and other industrial workplaces. It was originally chosen and operated on the basis of the proximity of its performance to the earlier 1968 ACGIH criterion. But, in the light of the intense activity elsewhere in the development of respirable aerosol exposure histories for mineworkers for epidemiological purposes – most notably in the British PFR – it was of great interest to be able to link the exposures measured in the USA with those measured in Europe. More specifically, it was desirable to link personal exposures obtained using the 10 mm nylon cyclone with those obtained using the MRE sampler which was not only static (i.e. was an area or fixed point sampler) but also was designed to conform to a different definition of respirable aerosol, the BMRC curve. With this in mind, Tomb *et al.* (1973) carried out side-by-side field studies in underground coal mines and showed that respirable aerosol concentrations equivalent to the BMRC definition could be obtained by multiplying measured 10 mm nylon cyclone concentrations by 1.38 when the cyclone sampler was operated at 1.7 Lpm. Later, Bartley *et al.* (1983) carried out a numerical assessment of the performance of the same cyclone sampler and found that it could be adapted to collect mass corresponding to the BMRC definition of respirable aerosol by operating it at a flow rate of 1.2 Lpm and by multiplying the collected mass by 0.91. Treaftis *et al.* (1984) confirmed this adaptation in a series of tests carried out in a range of coal mine environments with the cyclone operated in this way and operated side-by-side with the MRE sampler.

Later still, Groves *et al.* (1994) carried out a field study at large-scale abrasive blasting operations in a chemical plant, making side-by-side comparisons of three personal respirable aerosol cyclones, one the 10 mm Dorr–Oliver designed to collect respirable aerosol according to the earlier 1968 ACGIH definition (operated at 1.7 Lpm), the second a version of the Higgins–Dewell sampler designed to collect respirable aerosol according to the 1952 BMRC definition (at 1.9 Lpm), and the third a modified Higgins–Dewell cyclone designed to match the later 1985 ACGIH definition (at 2.3 Lpm). The samplers were all tested together as area samplers, mounted together in close proximity on a tripod-mounted boom assembly. At the same time, particle size distributions were measured using a Marple-type miniature eight-stage cascade impactor (see Figure 18.12) placed on the same rig, for which the results were used to estimate the mass concentration for each sampling run corresponding to each of the three respirable aerosol fractions of interest. The results are summarised in Table 22.8. They show that the actual concentrations and those calculated from the particle size distributions were quite consistent. Importantly, they show that the rank order of the concentrations measured was consistent with what would have been expected on the basis of the $_{50}d_{ae}$-value for the three respirable aerosol curves respectively. The ratio of the measured 10 mm nylon cyclone and the corresponding BMRC results (from the Higgins–Dewell cyclone) was 1.83, higher than the 1.38 obtained by Tomb *et al.* from their coal mine studies, although for a very different field scenario. However, the general directions of the trends observed by Groves *et al.*, supported as they were by the cascade impactor results, were generally reassuring in terms of the broad agreement between measured and expected respirable aerosol concentrations. This work confirmed that changes, even apparently small ones, in sampling criteria can produce consistent and significant biases in measured aerosol concentrations.

22.5.2 Workplace comparisons between optical and gravimetric aerosol samplers

Occupational hygienists have long acknowledged that traditional aerosol sampling, where particles are collected onto filters over prolonged periods of duration up to a full working shift and are analyzed subsequently in the laboratory, was laborious and costly. However, most standards have long been – and still are – expressed in terms of such time-weighted average exposure measurement. So the search for direct reading instrumentation, of the type described in Chapter 20 that can yield data commensurate with those obtained using the traditional approach, has continued over many years. Devices operating on the principles of light scattering have featured strongly. Side-by-side field studies with these alongside more conventional samplers have provided the ultimate validation test for such instruments.

Table 22.8 *Summary of results of field study at blasting abrasive operations at a chemical plant, comparing three different types of cyclone side-by-side, and showing corresponding mass concentrations estimated from particle size distribution data obtained using an eight-stage Marple-type cascade impactor (Groves et al., 1994). The table also shows expected and actual rank order of the masses collected by the three cyclones*

	10 mm nylon cyclone	Calculated from particle size distribution	Higgins–Dewell cyclone	Calculated from particle size distribution	Modified Higgins–Dewell cyclone	Calculated from particle size distribution
Flow rate (Lpm)	1.7		1.9		2.3	
Applicable respirable aerosol curve		1968 ACGIH		1952 BMRC		1985 ACGIH
$_{50}d_{ae}$-value (μm)		3.5		5		4
Sampled concentration (mg m^{-3})	1.97	1.96	3.62	2.64	2.30	2.36
Expected rank order of mass collected		3		1		2
Actual rank order of mass collected		3		1		2

Coal mine dust exposures provided considerable stimulus for such activity in the early days. In their work already cited to characterise mine dusts, Dodgson *et al.* also described side-by-side comparisons in a variety of underground coal mines between gravimetric respirable aerosol concentrations obtained using the MRE sampler and an optical device referred to as a '*Tyndalloscope*'. The latter appears to have been an early version of the German TM-Digital device referred to in Chapter 20. The results shown in Figure 22.12 revealed that the measured scattered light intensity for the Tyndalloscope (I) increased quite linearly with respirable mass (M), but with a slope that was markedly dependent on the colliery, and hence on coal rank. This tendency reflected the different optical properties of the coal dust particles having difference size distributions, shape characteristics and chemical composition (including carbon content). Linear regression analysis of the results in Figure 22.12 yielded the useful relationship:

$$M = \frac{I}{(1.46 + 0.13 \cdot \%\text{noncoal})} \tag{22.5}$$

As Dodgson *et al.* noted, the scatter in the data was such that the possibility of other functional dependences could not be precluded. At the very least, however, it was shown that, if calibrated for each given workplace aerosol against an appropriate gravimetric reference instrument (e.g. the MRE), instruments like the Tyndallometer could provide useful data about full-shift, time-weighted average, respirable aerosol levels.

Such research clearly indicated the feasibility of applying direct reading, light scattering based respirable aerosol monitors. In time other such instruments started to emerge. In 1983, Ford *et al.* reported

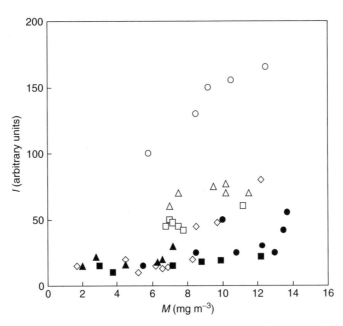

Figure 22.12 *Comparison between scattered light intensity (I) for a Tyndallometer and the gravimetric respirable mass concentration (M) obtained using the MRE sampler, both instruments placed side-by-side in a variety of British coal mines, each mine represented by a separate symbol (Dodgson et al., 1971). Adapted with permission from Dodgson et al., Inhaled Particles Volume III, pp. 757–781. Copyright (1971) British Occupational Hygiene Society*

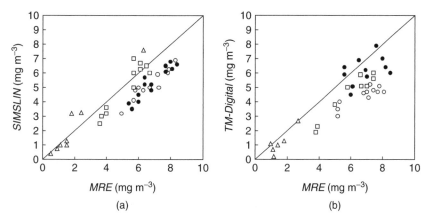

Figure 22.13 *Comparison between output from two direct-reading optical instruments and respirable aerosol measured gravimetrically using the MRE sampler: (a) for the British SIMSLIN; (b) for the German TM-Digital; summarising the results of field trials reported by Ford et al. (1983) for four different locations in coal mines. Adapted with permission from Ford et al., Aerosols in the Mining and Industrial Work Environments, pp. 759–775. Copyright (1983) Ann Arbor Science Publishers*

experiments in underground coal mines to examine the electrical outputs of the SIMSLIN and TM-Digital instruments in comparison with what was collected by MRE sampler. The latter was taken to be a reference sampler for the respirable fraction as defined by the BMRC convention. As noted earlier in Chapter 20, in the SIMSLIN the respirable aerosol was pre-selected according to the BMRC curve by the use of a horizontal elutriator similar to that of the MRE itself. By contrast, the TM-Digital was a pumpless, passive instrument without any mechanical pre-selection of particles. Both instruments provided continuous and cumulative average outputs, and had previously been calibrated against a 'standard' coal dust aerosol in a wind tunnel. In the field studies the samplers were all operated for up to 4 h. The results are shown in Figure 22.13. In general, agreement with the 1:1 calibration line was fair, although there were clearly significant differences between worksites. From the physics of light scattering by aerosols it is reasonable to expect dependences on particle composition and particle size distribution. However, Ford and his colleagues were not able to discern any clear such correlations.

The Mini-RAM light scattering based instrument was first mentioned in Chapter 20. In the version described there, it was an essentially 'passive' sampler, in which the aerosol of interest entered the sensing zone by virtue of external air motions. An 'active' version has also been used in which air entered the sensing zone by virtue of a pump, first passing through a 10 mm nylon cyclone that aerodynamically selected the respirable fraction. Both instruments provided read-out directly in terms of $mg\,m^{-3}$. In one field study reported by Lehocky and Williams (1996), the active version was used side-by-side with a set of 10 mm nylon cyclones used as gravimetric respirable samplers in fixed-point static mode. The experiments were carried out at worksites inside a number of coal-fired power stations, with the sampling instruments stationed close to coal transporting and handling operations. The results were supplemented by later work by the same group for the same arrangement and at the same locations (Middendorf *et al.*, 1999). The later study also included results for the Mini-RAM operated in the passive mode. The full sets of results are plotted together in Figure 22.14. The trends in the relationships for both Mini-RAM modes of operation were generally fairly linear within the expected variability for such data. Both versions consistently provided somewhat excessively high values for respirable aerosol concentration, as was borne out by subsequent statistical analysis.

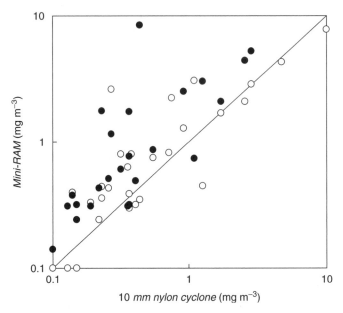

Figure 22.14 *Comparison between results obtained at workplaces in a coal-fired power station using two versions of the Mini-RAM sampler and the 10-mm nylon personal respirable aerosol sampler:* • *Passive Mini-RAM;* ○ *Active Mini-RAM (from the data of Lehocky and Williams, 1996 and Middenhorf et al., 1999)*

The studies cited above, and others not mentioned, certainly showed that light scattering based instruments can yield consistent data for time-weighted average, full-shift respirable aerosol concentrations. However the field experience indicated that attention needs to be given to the calibration of such instruments for each application, in full recognition of the influences of aerosol type and particle size distribution. Occupational hygienists should be encouraged but cautious.

22.6   Classification of workplace aerosols

22.6.1   Particle size distribution

Particle size distribution is a major factor in the inhaled dose that is received by humans exposed to aerosols, determining the way in which the deposition of inhaled particles is distributed throughout the respiratory tract. It also governs the amount of particulate matter collected by sampling instruments dedicated to specific aerosol fractions, and dictates their relative performances. In addition it is very influential in the performances of control systems for airborne particulate material. At an important practical level, the United States National Institute for Occupational Safety and Health (NIOSH, 1996), in its guidelines for respiratory protection, specifies that certain filters

'... should be used only when the mass median aerodynamic diameter (MMAD) is known to be greater than 2 μm. If this diameter is less than 2 μm or is unknown, a (high-efficiency) filter should be used.'

This places a specific responsibility on the occupational hygienist to acquire the knowledge needed to make that decision. For some or all of the above reasons, occupational hygienists have long sought to determine the particle size distributions of workplace aerosols.

In the earlier years, microscopic analyses of samples collected onto slides and other substrates were predominant. As described earlier in this chapter, Rogan *et al.* (1967) obtained particle size distributions from the counting and sizing of particles collected from coal mine atmospheres onto the slides of thermal precipitators and used the information obtained to link the thermal precipitator results with the gravimetric results obtained using later generations of respirable aerosol sampler. One sampling system described by Menichini (1986) involved a cascade impactor with a small number of stages (four) but where the collection substrates comprised glass slides for each of which collected particles could be counted and sized by optical microscopy. The particle size distribution was determined by combining the particle counts/sizes for each stage and the impaction properties of the four stages. Menichini applied this sampling system to the evaluation of oil mist droplets in industrial scenarios where various types of oil were being used for lubricating molds in concrete and glass casting operations. More generally, however, most of the workplace particle size distribution measurements reported in the literature have been carried out using cascade impactors usually with larger numbers of stages, where the amounts – usually masses – of particulate material collected on the various stages were used to generate the desired particle size distributions.

During the 1980s, a major multi-nation European project was carried out in Europe to evaluate a range of sampling instruments in the mining industries (Vincent, 1991). Six laboratories were involved, including the German Bergbau-Forschungsinstitut and Silikose-Forschungsinstitut (Essen and Bochum, respectively), the French Centre d'Etudes et Recherche des Charbonnages (Verneuil en Halatte), the Belgian Instituut voor Reddingswezen, Ergonomie en Arbeidshygiene (Hasselt), the Italian Istituto di Medicina del Lavoro (Milan), and the British Institute of Occupational Medicine (Edinburgh). Each laboratory evaluated one or more cascade impactors and their abilities – or otherwise – to provide aerosol fractions consistent with other samplers dedicated to specific fractions. The project involved both laboratory and field studies in underground mines; coal mines in Germany, France, Belgium and UK, and a pyrites mine in Italy. The cascade impactors tested were the 10 Lpm IOM static inhalable dust spectrometer (SIDS), the 40 Lpm precision cascade impactor (PCI) and the 2 Lpm personal inhalable dust spectrometer (PIDS), all three of which were described earlier in Chapter 18, and all three of which were capable of providing particle size distributions over the full inhalable range. For each workplace sample for each instrument, the particulate masses collected in the entries and on the various impactor stages, were applied to the estimation of best-fit double-mode log-normal functions, and the results were then used to calculate the particulate masses contained in the various fractions corresponding to other instruments place side-by-side. The latter included samplers for the inhalable, respirable and 'alveolar deposition' fractions (IOM static inhalable aerosol sampler, MRE and TBF50, respectively, see Chapters 14 and 15). It was reported that, in general, the SIDS, PCI and PIDS produced results that consistently correlated quite well with the results obtained using the dedicated samplers. Overall, this work provided encouragement to the view that cascade impactors could be used as versatile samplers, each one capable of providing information about more than one fraction.

When the culture of aerosol exposure assessment in workplace began to shift towards personal sampling from the 1970s onwards, interest also grew in cascade impactors that were sufficiently miniaturised to allow their use as personal samplers. The Marple personal cascade impactor (see Figure 18.12) was one important development in the late 1980s and since has been commercially available for several years. It has been widely used by occupational hygienists for characterising workplace aerosols in relation to all the rationales outlined above. In one example of its field application, Bullock and Laird (1994) reported results from a study where it was used to determine particle size distributions for aerosol exposures in the paper and wood products industry. Some of the results are shown in Figure 22.15. They are fairly typical of many of those reported for this instrument. For aerosols which would be expected to be

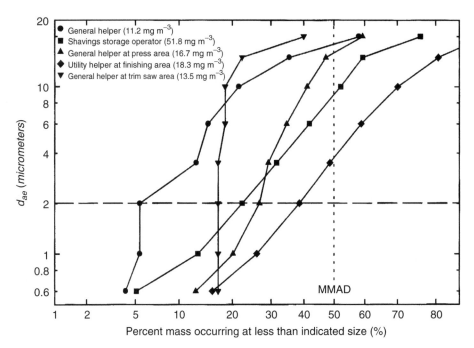

Figure 22.15 *Set of typical cumulative particle size distributions obtained using the Marple personal cascade impactor, for workers' exposures in a wood products facility (Bullock and Laird, 1994). Reproduced with permission from Bullock and Laird, American Industrial Hygiene Association Journal, 55, 836–840. Copyright (1994) American Industrial Hygiene Association*

quite coarse, they indicate the limitation of particle size distribution measurement at the high end of the distribution, where there is no information beyond the $_{50}d_{ae}$ for the top impactor stage. Li *et al.* (2001) also used this sampler in the wood products industry, and reported significant differences in particle size distribution when it was used side-by-side with other types of cascade impactor. They specifically noted the biases introduced by dependencies of aspiration efficiency on sampler orientation and external windspeed. The performance of the Marple sampler in other 'coarse aerosol' situations in the field was discussed by Nieuwenhuijsen *et al.* (1998) in their studies of aerosols in agricultural operations such as field crop farming (land planing, disking, ploughing, fertilising, planting, harvesting, baling, etc.), fruit and nut farming (hand and mechanical harvesting) and dairy farming (milking, feeding and manure removal). They observed that most of the particulate material was collected on the top stages, noting in particular that it was therefore difficult to estimate the MMAD. Application of the Marple personal cascade impactor has been more successful, however, in other industrial situations, for example in the ferrous industries where Perrault *et al.* (1992) reported MMAD-values from about 7 to 13 μm with geometric standard deviations less than 2.0, placing the ranges of particle size firmly within the range of capability of the instrument. More recently a modified version of the Marple cascade impactor was proposed, in which a porous foam-based top stage was added to significantly extend the range of particle size selectivity. This sampler was used successfully in a preliminary study to characterise aerosol particle size distributions in workplaces of a lead battery manufacturing facility and a foundry (Wu, 2005).

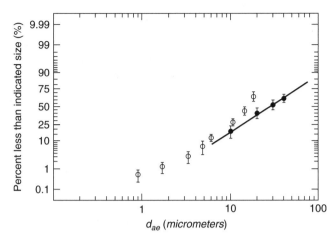

Figure 22.16 *Typical cumulative particle size distributions obtained using the IOM personal inhalable dust spectrometer (PIDS) obtained for a worker at a pig breeding farm (Vinzents, 1994). The open circles indicate raw cascade impactor data and the closed circles the data for the inhalable fraction obtained from the mass collected on the entry stage. Reproduced with permission from Vinzents, American Industrial Hygiene Association Journal, 55, 977–980. Copyright (1994) American Industrial Hygiene Association*

The earlier 2 Lpm PIDS sampler mentioned above (see also Figure 18.11) was similarly designed to overcome the limitation in particle size range evident in other cascade impactor-based aerosol spectrometer. As described in Chapter 18, the inlet of this instrument was designed to aspirate the inhalable fraction, by reference to the IOM personal inhalable aerosol sampler. The top (entry) stage also incorporated a cassette to enable recovery of everything that passed through the inlet and failed to reach the first impactor stage, the particle size-dependent efficiency of this inlet stage having previously been obtained experimentally (Gibson *et al.*, 1987). Although the PIDS was never available commercially, prototypes featured in a number of field studies. In one of them, Vinzents (1994) described its application in aerosol characterisation in pig breeding facilities. Figure 22.16 contains the cumulative particle size distribution for a typical sample, indicating first the distribution obtained from using just the raw data for the eight impactor stages and then the extended particle size distribution after the inlet deposit had been taken into account. For the latter, Vinzents assumed a log-normal distribution for the aspirated aerosol and estimated the mass of aerosol with d_{ae} less than 40, 30, 20 and 10 μm, respectively, by combining this distribution with the entry curve for the entry stage and integrating. The magnitude of the effect of the entry section for particles at the coarse end of the particle size distribution was clearly evident. In another application, Spear *et al.* (1998) used the PIDS in their study of the exposures to airborne lead of workers at a large lead smelter. Personal samplers were taken at worksites representative of ore storage and milling, the sinter plant, the blast furnace and the drossing area. The particle size distributions were recovered using an inversion algorithm described by Ramachandran *et al.* (1996a). The retrieved distributions were used to determine the concentrations in the inhalable, thoracic and respirable fractions as defined by the harmonised CEC/ISO/ACGIH criteria and the relationships between them. The results summarised in Table 22.9 reflect the degrees of coarseness (or fineness) of the aerosols in the four work areas indicated. In this way, the particle size distribution data were reduced to indices that could be easily understood in terms of the regional and overall respiratory dose to the exposed workers.

Table 22.9   *Summary of five indices of particle size distribution for workers' exposures to airborne lead at worksites in a large lead smelter, obtained using the PIDS sampler and expressed in terms of the CEN/ISO/ACGIH criteria for the inhalable, thoracic and respirable fractions (Spear et al., 1998)*

Plant area	n	% inhalable	% thoracic	% respirable	% thoracic/ respirable	% respirable/ inhalable
Ore storage/mill	10	59.8	15.7	5.1	25.6	8.5
Sinter plant	16	53.6	6.3	1.6	11.5	3.1
Blast furnace	12	57.9	19.5	12.4	33.6	20.8
Drossing area	8	53.8	7.7	2.7	14.3	5.1

22.6.2   Combined particle size measurement and chemical speciation

The PCI mentioned in the European five-nations research also featured in a field study by Verma *et al.* (1994) in Canadian surface and underground metal ore mining operations. The purpose was to provide information to facilitate the grouping of dust exposures associated with various tasks in the various workplaces in order to derive konimeter count versus respirable mass relationships to use in developing extended worker exposure histories suitable for use in epidemiology. The PCI was chosen because, with its $_{50}d_{ae}$-value for the top stage being as high as 32 μm, the useful particle size range of this instrument was considerably greater than for other cascade impactors readily available at that time. Another advantage of the PCI over many other such samplers was its high flow rate (40 Lpm) which, under most field conditions, was able to provide adequate amounts of particulate material at each stage to make it possible to carry out accurate analyses for chemical speciation. Verma and his colleagues were specifically interested in free crystalline silica. Particle size distributions were represented simply in terms of the cumulative mass plotted as a function of the $_{50}d_{ae}$-values for the various stages. The PCI was found to work consistently well for the aerosols found at the locations studied. Interestingly it was shown that the free crystalline silica content tended to increase as the particle size decreased, suggesting that the silica content was contained predominantly in the smaller particles.

The observation of Verma *et al.* about the limited range of particle size of most commercially available cascade impactors is an important one because workplace aerosols may range from very coarse to very fine – perhaps more so than those commonly found in the ambient atmosphere. This is implicit in the emergence of the inhalability criterion for the coarser fraction. With this in mind, Kerr *et al.* (2001) proposed a modified version of the static 28.3 Lpm Andersen Mk II cascade impactor in which a new top stage was added, incorporating a modified entry and a plug of porous plastic foam filtration media (see Figure 18.8). As for the modified Marple sampler mentioned above, the aim was to significantly extend the range of particle size selectivity. This sampler was used in a study to characterise aerosol exposures in workplaces of the primary nickel production industry (Vincent *et al.*, 2001). In particular, the large sampling flow rate provided amounts of collected material at each stage sufficient to permit chemical analysis in order to recover nickel containing species groups considered to be specifically relevant to health, namely water soluble, sulfidic, oxidic and metallic (Doll *et al.*, 1990). It should be noted that these species groups are not so much actual nickel compounds but, rather, are groups of compounds broadly as defined but are specific to the sequential leaching analytical procedure described by Zatka *et al.* (1992). They are generally consistent with the chemical fractions – namely, soluble inorganic compounds, nickel subsulfide, insoluble inorganic compounds and elemental nickel – identified by ACGIH in its current list of TLVs. The modified Andersen sampler was used to characterise nickel-containing aerosols at a range of workplaces in the mills and smelters of two companies and a refinery of one of them. Material was recovered from the entry and impactor stages of the sampler and analyzed for the nickel species indicated, from which – using the inversion procedure described earlier by Ramachandran *et al.* (1996a), modified only by insertion of the appropriate particle size dependent collection kernels for the various stages of the instrument – particle size distributions were obtained for each of the chemical species groups. In addition, the health-related inhalable, thoracic and respirable fractions were determined. A typical result, for an example sample taken at one of the smelters, is shown in Figure 22.17, showing a summary of the distributions of both particle size and chemical species fractions. It shows high proportions of sulfidic and oxidic species, and low soluble and metallic species, in all particle size fractions. This was not surprising for the smelter stage of the nickel primary production process. Interestingly, here and for all the other samples reported, there did not appear to be any significant differences in the distribution of chemical species from one particle size fraction to the next. Graphs like those in Figure 22.17 may be usefully regarded as 'fingerprints' of the exposures that could occur at the workplaces studied. They contain information simultaneously about the physical dose of particles to the different parts of the respiratory tract and about the chemical, and hence toxicological, effect that might be stimulated there. In short, this is a simple portrayal of most of what might be needed in order to assess the exposure, dose and – ultimately – the hazard. To illustrate this point, Vincent *et al.* modeled the potential exposures at each workplace studied and used the results to reduce each exposure scenario to the 'equivalent sulfidic fraction' (ESF). This was aided greatly by the uniformity of the chemical species across particle size fractions. Thus it was possible to write down the expression:

$$ESF = \frac{TLV_{sulfidic}}{TLV_{sulfidic}} \cdot (\% \text{ sulfidic}) + \frac{TLV_{sulfidic}}{TLV_{insoluble}} \cdot (\% \text{ metallic} + \% \text{ oxidic}) + \frac{TLV_{sulfidic}}{TLV_{soluble}} \cdot (\% \text{ soluble})$$

(22.6)

Table 22.10 *Ranking of plant areas/worksites according to the mean level of equivalent exposure to sulfidic nickel (from Vincent et al., 2001)*

Company and Process	Plant area	ESF (%)	Mean inhalable nickel exposure concentration ($mg\,m^{-3}$)	Mean equivalent exposure to inhalable sulfidic nickel ($mg\,m^{-3}$)
Company A				
Smelter	Matt processing	86	1.20	1.03
	Matt crushing	89	0.47	0.42
	Converter aisle	72	0.44	0.31
Refinery	Top blown rotary converter	58	0.49	0.28
Smelter	Flash furnace	87	0.31	0.27
	Air cleaner	65	0.21	0.14
Refinery	Packaging	53	0.14	0.07
Mill	All plant areas	82	0.03	0.03
Company B				
Smelter	Granulation	92	0.17	0.16
	Matt end	75	0.18	0.14
	Converter aisle	66	0.09	0.06

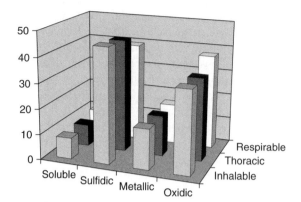

Figure 22.17   *Typical summary presentation of the distributions of particle size and chemical species fractions, this for a single sample taken at a nickel smelter (Vincent et al., 2001). Reproduced with permission from Vincent et al., Journal of Environmental Monitoring, 3, 565–574. Copyright (2001) Royal Society of Chemistry*

In this expression, each species is weighted according to the TLV assigned to it by ACGIH, under the assumption that the TLVs in question relate to similar health effects (e.g. lung cancer). Table 22.10 shows some results for ESF estimated from fingerprints like those shown in Figure 22.17. The companies are labeled A and B, and a number of the plant locations that were studied are identified. In this table, the calculated ESF-values were combined with the mean measured inhalable nickel exposure levels at the places indicated to produce equivalent inhalable sulfidic nickel exposures, leading in turn to rank ordering of possible level of hazard at the workplaces indicated.

The preceding discussion has been focused on the combined measurement of chemical speciation and particle size distribution measurement. The literature in this area remains relatively small. By contrast, however, there is a very large literature of reported field studies where aerosol samples for specific size fractions have been analyzed with respect to one or more chemical species. There is of course of great interest in relation to health effects, especially since many OELs are prescribed in terms of specific such fractions (e.g. lead, etc.). For the most part, such studies belong outside the scope of this book.

22.7   Diesel particulate matter

Ever since they became associated with lung cancer and other serious ill-health, diesel particulate matter and toxic substances contained within it have been of special interest in relation to sampling in both workplaces and the ambient atmosphere. Confined workplaces where there is substantial use of diesel-powered vehicles or machines, including underground mines, are of specific major concern. Sampling presents some significant challenges. As described in Chapter 15, a sampler was proposed – and is now commercially available – with an impactor-based pre-selector that could be used in conjunction with a personal respirable cyclone, providing a sharp 'cut' at $d_{ae} = 0.9$ μm, thus physically separating the coarser dust and finer diesel particulate components. Research carried out in North America under the auspices of the Diesel Emissions Evaluation Program (DEEP) has been reported where results from this particle size selective approach were compared with those from others based on chemical speciation methods (Watts, 2000). The field study was carried out in a number of Canadian underground mines, where the samplers actually had a 'cut' at $d_{ae} = 0.8$ μm. A second approach employed the *respirable combustible dust* (RCD) method in which respirable aerosol passing through a cyclone chosen

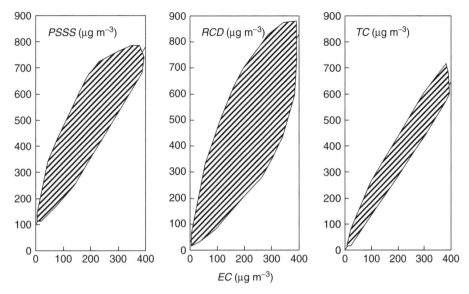

Figure 22.18 *Summaries of comparisons between measurement methods for diesel particulate (from the data of Watts, 2000). PSSS, particle size selective sampling (gravimetric); RCD, respirable combustible dust (gravimetric); TC, total carbon (analytical); EC, elemental carbon (analytical)*

to match the CEN/ISO/ACGIH convention was subjected to controlled combustion such that the mass removed – or burned off – was the RCD fraction of interest. Here, because the RCD comprised all the combustible material collected onto the filter, and included all carbonaceous material, it was apparent that the RCD approach is not generally appropriate to coal mining situations where the coarser dust also contains carbon. It was entirely satisfactory for the metal ore mines where the field study was conducted. The third approach studied involved the measurement of *elemental carbon* (EC) in collected samples, using a thermal-optical method recommended by the National Institute for Occupational Safety and Health (NIOSH, 1994). This method also allows the estimation of *total carbon* (TC). In the DEEP project, comparisons were made between measurements using the particle size selective approach along with measurements of RCD, EC and TC, in all of which EC was taken to be the reference by virtue of the fact that, in general, it exhibits the least interferences and biases. In fact, EC is the most sensitive and specific marker for diesel particulate in most environmental settings. The results of the method inter-comparisons are summarised in Figure 22.18 and provide means to relate the other methods to EC and to indicate the intrinsic variability in the correlations. The gravimetric particle size-selective sampling method was seen to perform quite well in the field trials.

22.8 The future of workplace aerosol measurement

Cherrie (2003) reviewed the status of the science underpinning occupational hygiene, noting the importance of the pioneering work of Sherwood and Greenhalgh in stimulating a new generation of workplace personal sampling, for both aerosols and gaseous contaminants. The introduction of the personal sampling pump was identified as a critical step. However, Cherrie also noted the apparent shift in modern thinking – even policy – about occupational exposure assessment in a direction away from actual measurements of the type made for aerosols using samplers like those described in this book. For example, in the UK, the Health and Safety Executive's National Database contains more than 80 000 individual measurements for about 400 different substances. Most of those measurements were made in the years between 1985 and 1990, and the rate of acquisition of data from new samplers has declined quite sharply in more recent years. Part of the reason may be the fact that fewer industrial companies are nowadays carrying out measurements of chemical exposures of their workers. Noting the same trend, Kromhout (2002) suggested that the emergence of predictive models may have contributed in some way to the decline. No information was given to support or deny whether such trends may be present in other countries, including the USA. Meanwhile, Cherrie noted the paradox that, as instruments for making accurate and valid exposure assessments have become more available and at lower cost, their use in the real world appears to be declining. His commentary ended with the apt observation that

'.. if we are to continue to efficiently manage the risks from chemicals then we must all make more use of the personal sampling techniques pioneered in the 1960s.'

Well said!

References

Bartley, D.L. (1998) Inhalable aerosol samplers, *Applied Occupational and Environmental Hygiene*, 13, 274–278.

Bartley, D.L., Bowman, J.D., Breuer, G. and Doemeny, L.J. (1983) Accuracy of the 10 mm cyclone for sampling respirable coal mine dust, In: *Aerosols in the Mining and Industrial Work Environments* (Eds V.A. Marple and B.Y.H. Liu), Ann Arbor Science Publishers, Ann Arbor, MI, pp. 897–906.

Bedford, T. and Warner, C.G. (1943) Chronic respiratory disease in South Wales coal miners II – environmental studies, UK Medical Research Council Special Report Series, No. 244, HMSO, London.

Brown, R.C., Hemingway, M.A., Wake, D. and Thompson, J. (1995) Field trials of an electret-based passive dust sampler in metal-processing industries, *Annals of Occupational Hygiene*, 39, 603–622.

Bullock, W.H. and Laird, L.T. (1994) A pilot study of the particle size distribution of dust in the paper and wood products industry, *American Industrial Hygiene Association Journal*, 55, 836–840.

Calzavara, T.S., Carter, C.M. and Axten, C. (2003) Air sampling methodology for asphalt fume in asphalt production and asphalt roofing manufacturing facilities: total particulate sampler versus inhalable particulate sampler, *Applied Occupational and Environmental Hygiene*, 18, 358–367.

Cherrie, J.W. (1999) The effect of room size and general ventilation on the relationship between near and far-field concentrations, *Applied Occupational and Environmental Hygiene*, 14, 539–546.

Cherrie, J.W. (2003) Commentary: the beginning of the science underpinning occupational hygiene, *Annals of Occupational Hygiene*, 179–185.

Comité Européen de Normalisation (2002) *Workplace atmospheres – assessment of performance instruments for measurement of airborne particle concentrations*, CEN Standard EN 13205.

Demange, M., Görner, P., Elcabache, J.M. and Wrobel, R. (2002) Field comparison of 37 mm closed-face cassettes and IOM samplers, *Applied Occupational and Environmental Hygiene*, 17, 200–208.

Dodgson, J., Hadden, G.G., Jones, C.O. and Walton, W.H. (1971) Characteristics of the airborne dust in British coal mines, In: *Inhaled Particles* (Ed. W.H. Walton), Unwin, Old Woking, Vol.III, pp. 757–781.

Doll, R.M., Andersen, A.A., Cooper, W.C., *et al.* (1990) Report of the International Committee on Nickel Carcinogenesis in Man, *Scandinavian Journal of Work Environment and Health*, 16 (Suppl. 1), 1–82.

Ford, V.F.W., Minton, R. and Mark, D. (1983) Comparative trials in coal mines of the TM-Digital and SIMSLIN dust monitors, In: *Aerosols in the Mining and Industrial Work Environments* (Eds V.A. Marple and B.Y.H. Liu), Ann Arbor Science Publishers, Ann Arbor, MI, pp. 759–775.

Gibson, H., Vincent, J.H. and Mark, D. (1987) A personal inspirable dust spectrometer for applications in occupational hygiene research, *Annals of Occupational Hygiene*, 31, 463–479.

Groves, W.A., Hahne, R.M.A., Levine, S.P. and Schork, M.A. (1994) A field comparison of respirable dust samplers, *American Industrial Hygiene Association Journal*, 55, 748–755.

Hamilton, R.J., Morgan, G.D. and Walton, W.H. (1967) The relationship between measurements of respirable dust by mass and number in British coal mines, In: *Inhaled Particles and Vapours II* (Ed. C.N. Davies), Pergamon Press, Oxford, pp. 533–546.

Hauck, B., Grinshpun, S., Reponen, A., Reponen, T., Willeke, K. and Bornschein, R. (1997), Field testing of new aerosol sampling method with a porous curved surface as inlet, *American Industrial Hygiene Association Journal*, 58, 713–719.

Health and Safety Executive (1980) *Cotton dust sampling*, Guidance Note EH 25, HSE, HMSO, London.

Health and Safety Executive (2000) General methods for sampling and gravimetric analysis of respirable and inhalable dust, In: *Methods for the Determination of Hazardous Substances*, MDHS 14/3, HSE, HMSO, London.

Hewson, G.S. (1996) Estimates of silica exposure among metalliferous miners in Western Australia, *Applied Occupational and Environmental Hygiene*, 11, 868–877.

Kerr, S.M., Muranko, H.J. and Vincent, J.H. (2002) Personal sampling for inhalable aerosol exposures of carbon black manufacturing industry workers, *Applied Occupational and Environmental Hygiene*, 17, 681–692.

Kerr, S.M., Vincent, J.H. and Ramachandran, G. (2001) A new approach to sampling for particle size distribution and chemical species 'fingerprinting' of workplace aerosols, *Annals of Occupational Hygiene*, 45, 555–568.

Koch, W., Thomassen, Y. and Vincent, J.H. (2002) Evaluation of the Respicon as a personal aerosol sampler for use in industrial environments, *Journal of Environmental Monitoring*, 4, 657–662.

Kromhout, H. (2002) Design of measurement strategies for workplace exposure – authors reply, *Occupational and Environmental Medicine*, 59, 788–789.

Lehocky, A.H. and Williams, P.L. (1996) Comparison of respirable samplers to direct-reading real-time aerosol monitors for measuring coal dust, *American Industrial Hygiene Association Journal*, 57, 1013–1018.

Li, S.N., Lundgren, D.A., Rovell-Rixx, D. and Ray, A.E. (2001) Effect of impactor inlet efficiency on the measurement of wood dust size distribution, *American Industrial Hygiene Association Journal*, 62, 19–27.

Lillienberg, L. and Brisman, J. (1994) Flour dust in bakeries – a comparison between methods, In: *Inhaled Particles VII* (Eds J. Dodgson and R.I. McCallum), Pergamon Press, Oxford, pp. 571–575.

Little, R.J.A. and Rubin, D.B. (1987), *Statistical Analysis with Missing Data*, John Wiley & Sons, Ltd, New York.

Menichini, E. (1986) Particle size distribution of oil mist in the workplace, *Annals of Occupational Hygiene*, 30, 349–363.

Middendorf, P.J., Lehocky, A.H. and Williams, P.L. (1999) Evaluation and field calibration of the Mini-RAM PDM3 aerosol monitor for measuring respirable and total coal dust, *American Industrial Hygiene Association Journal*, 60, 502–511.

National Institute for Occupational Safety and Health (1994) NIOSH Manual of Analytical Methods, Fourth Edition: Elemental Carbon (Diesel Particulate), NIOSH Method 5040, DHHS Publication No. 94–112.

National Institute for Occupational Safety and Health (1996) *NIOSH Guide to the Selection and Use of Particulate Respirators; II Detailed Guidelines for Use*, DHHS Publication No. 96–101.

Nieuwenhuijsen, M.J., Kruize, H. and Schenker, M.B. (1998) Exposure to dust and its particle size distribution in California agriculture, *American Industrial Hygiene Association Journal*, 59, 34–38.

Paik, S.Y. and Vincent, J.H. (2002) Aspiration efficiencies of disc-shaped blunt nozzles facing the wind, for coarse particles and high velocity ratios, *Journal of Aerosol Science*, 33, 1509–1523.

Perrault, G., Dion, C., Ostiguy, C., Michaud, D. and Baril, M. (1992) Selective sampling and chemical speciation of airborne dust in ferrous foundries, *American Industrial Hygiene Association Journal*, 53, 463–470.

Ramachandran, G. (2001) Retrospective exposure assessment using Bayesian methods, *Annals of Occupational Hygiene*, 45, 651–667.

Ramachandran, G. and Vincent, J.H. (1999) A Bayesian approach to retrospective exposure assessment, *Applied Occupational and Environmental Hygiene*, 14, 547–557.

Ramachandran, G., Johnson, E.A. and Vincent, J.H. (1996a) Inversion techniques for cascade impactor data, *Journal of Aerosol Science*, 27, 1083–1097.

Ramachandran, G., Werner, M.A., Spear, T.M. and Vincent, J.H. (1996b) On the assessment of particle size distributions in workers' aerosol exposures, *The Analyst*, 121, 1225–1232.

Rivers, D., Wise, M.E., King, E.J. and Nagelschmidt, G. (1960) Dust content, radiology and pathology in simple pneumoconiosis of coal workers, *British Journal of Industrial Medicine*, 20, 87–90.

Rogan, J.M., Rae, S. and Walton, W.H. (1967) The National Coal Board's Pneumoconiosis Field Research – an interim review, In: *Inhaled Particles and Vapours II* (Ed. C.N. Davies), Pergamon Press, Oxford, pp. 493–507.

Shen, P.T., Culver, B.D., Taylor, T.H., Granken, E.P. and Sattaur, J. (1994) Unpublished paper presented at the 1994 American Industrial Hygiene Conference and Exposition, New Orleans.

Sherwood, R.J. and Greenhalgh, D.M.S. (1960) A personal air sampler, *Annals of Occupational Hygiene*, 2, 127–132.

Spear, T.M., Werner, M.A., Bootland, J., Harbour, A., Murray, E.P., Rossi, R. and Vincent, J.H. (1997) Comparison of methods for personal sampling of inhalable and total lead and cadmium-containing aerosols in a primary lead smelter, *American Industrial Hygiene Association Journal*, 58, 893–899.

Spear, T.M., Werner, M.A., Bootland, J., Muray, E., Ramachandran, G. and Vincent, J.H. (1998) Assessment of particle size distributions of health-relevant exposures of primary lead smelter workers, *Annals of Occupational Hygiene*, 42, 73–80.

Terry, K.W. and Hewson, G.S. (1996) Comparative assessment of dust sampling heads used in western Australian mines - implications for dust sampling practice, In: *Occupational Hygiene Solutions* (Ed. G.S. Hewson), Proceedings of the 15th Annual Conference of the AIOH, Australian Institute of Occupational Hygienists, Tullamarine, Victoria, Australia.

Tomb, T.F., Treaftis, H.N., Mundell, R.L. and Parobeck, P.S. (1973) Comparison of respirable dust concentrations, measured with MRE and modified personal gravimetric sampling equipment, United States Bureau of Mines Report of Investigations No. 7772.

Treaftis, H.N., Gero, A.J., Kacsmar, P.M. and Tomb, T.F. (1984) Comparison of mass concentrations determined with personal respirable coal mine dust samplers operating at 1.2 liters per minute and the Casella 113A gravimetric sampler (MRE), *American Industrial Hygiene Association Journal*, 45, 826–832.

Tsai, P.J. (1995) Health-related exposures of the nickel industry workers, PhD Thesis, University of Minnesota, Minneapolis, MN.

Tsai, P.J., Vincent, J.H., Wahl, G. and Maldonado, G. (1995) Occupational exposure to inhalable aerosol in the primary nickel production industry, *Occupational and Environmental Medicine*, 52, 793–799.

Tsai, P.J., Vincent, J.H., Wahl, G.A. and Maldonado, G. (1996a) Worker exposure to inhalable and 'total' aerosol during nickel alloy production, *Annals of Occupational Hygiene*, 40, 651–669.

Tsai, P.J., Werner, M.A., Vincent, J.H. and Maldonado, G. (1996b) Exposure to nickel-containing aerosol in two electroplating shops: comparison between inhalable and 'total' aerosol, *Applied Occupational and Environmental Hygiene*, 11, 484–492.

Verma, D.K., Sebestyen, A. and Muir, D.C.F. (1987) An evaluation of konimeter performance: II. A field comparison of konimeters, *Annals of Occupational Hygiene*, 31, 451–461.

Verma, D.K., Sebestyen, A., Julian, J.A., Muir, D.C.F., Schmidt, H., Bernholz, C.D. and Shannon, H.S. (1989) Silica exposure and silicosis among Ontario hardrock miners: II. Exposure estimates, *American Journal of Industrial Medicine*, 16, 13–28.

Verma, D.K., Sebestyen, A., Julian, J.A., Muir, D.C.F., Shaw, D.S. and MacDougall, R. (1994) Particle size distribution of an aerosol and its sub-fractions, *Annals of Occupational Hygiene*, 38, 45–58.

Vincent, J.H. (1991) Joint investigations of new generations of sampler for airborne dust in mines, Industrial Health and Safety, Commission of the European Communities, Synthesis Report EUR 13414 EN, Luxembourg.

Vincent, J.H., Ramachandran, G. and Kerr, S.M. (2001) Particle size and chemical species 'fingerprinting' in primary nickel primary production industry workplaces, *Journal of Environmental Monitoring*, 3, 565–574.

Vinzents, P.S. (1994) Mass distribution of inhalable aerosols in swine buildings, *American Industrial Hygiene Association Journal*, 55, 977–980.

Vinzents, P.S., Thomassen, Y. and Hetland, S. (1996) A method for establishing tentative occupational exposure limits for inhalable dust, *Annals of Occupational Hygiene*, 39, 795–800.

Watson, H.H. (1936–1937) A system for obtaining, from mine air, dust samples for physical, chemical and petrological examination, *Transactions of the Institution of Mining and Metallurgy*, 46, 155–240.

Watts, W.F. (2000) Diesel particulate matter sampling methods: statistical comparison, Report of Diesel Emissions Evaluation Program (DEEP) Project, DEEP Secretariat, CAMIRO Mining Division, Sudbury, ON.

Werner, M.A., Spear, T.M. and Vincent, J.H. (1996) Investigation into the impact of introducing workplace aerosol standards based on the inhalable fraction, *The Analyst*, 121, 1207–1214.

Wilsey, P.W, Vincent, J.H., Bishop, M.J., Brosseau, L.M. and Greaves, I.A. (1996) Workers' exposures to inhalable and 'total' oil mist aerosol in a metal machining shop, *American Industrial Hygiene Association Journal*, 57, 1149–1153.

Wu, Y.-H. (2005) Application of particle size distribution measurement to the characterization of workplace aerosols, PhD Dissertation, University of Michigan, Ann Arbor, MI.

Zatka, V.J., Warner, J.S. and Maskery, D. (1992) Chemical speciation of nickel in airborne dusts: analytical method and results of am inter-laboratory test program, *Environmental Science and Technology*, 26, 138–144.

23

Field Experience with Aerosol Samplers in the Ambient Atmosphere

23.1 Introduction

The development of sampling criteria, standards and sampling methods for aerosols in the ambient atmosphere has followed a somewhat different path than for workplaces. Unlike the case of occupational aerosol measurement, there have not been the swings between the indices of particle number count and mass concentrations. With the exception of asbestos fibers, aerosol levels in the ambient atmosphere have nearly always been defined in terms of the mass concentration. Even where black smoke was the chosen index, it was expressed – by calibration of the optical method underpinning its determination – in terms of the mass concentration of soot. Furthermore, the question of personal versus static sampling did not surface until relatively recently. So most sampling for ambient aerosol has involved instruments placed at fixed points. As far as sampling criteria are concerned, whereas the international focus for occupational environments has converged for the most part on the Comité Européen de Normalisation/International Standards Organisation/American Conference of Governmental Industrial Hygienists (CEN/ISO/ACGIH) conventions, the emphasis for ambient atmospheric aerosol has moved increasingly towards the PM_{10} – and latterly $PM_{2.5}$ – philosophy first introduced by the US Environmental Protection Agency (EPA).

In the USA, the Environmental Protection Agency promulgated procedures – the Federal Reference Method (FRM) – for testing samplers, specifically most recently for samplers for the PM_{10} and $PM_{2.5}$ fractions (EPA, 1997). The formal documentation required not only laboratory tests but also the identification of one or more samplers as reference samplers, and their application in field studies alongside one or more of the candidate samplers for the same fraction. Europe followed along similar lines when the CEN published its protocol EN 12341 that described standardised procedures for testing samplers for the PM_{10} fraction that, too, placed considerable emphasis on field studies (CEN, 1998). So, on both sides of the Atlantic, field studies of prospective new samplers were therefore formally built into the standards infrastructure. Many other nations have since followed. That said, however, the need for field studies to validate emerging new aerosol samplers was acknowledged much earlier. A vast literature has grown over many years, describing many applications of samplers of diverse types in the ambient atmosphere, many of them of local interest only, for example providing data in relation to air pollution levels in specific cities or regions, without adding significantly to aerosol sampling science.

Aerosol Sampling: Science, Standards, Instrumentation and Applications   James Vincent
© 2007 John Wiley & Sons, Ltd

Here, therefore, as for samplers for the occupational setting in the preceding chapter, attention is focused on those field studies that have provided bridges between different approaches or otherwise elucidated the nature of aerosol sampling problems and solutions. A very wide range of samplers – wider even than samplers for occupational hygiene applications – has been used over the years, some of which have been described in earlier chapters. Of the others, most have physical and operational features in common with those that were described.

23.2 'Nuisance' dust

Samplers for large particles of dust or grit are important in relation to their nuisance as perceived by local populations, for which Vallack and Shillito (1998) described a model based on the rate of complaints received as a function of the particle deposition rate as measured using a deposition gauge. The dust gauges that were developed for collecting particles relevant to this problem took the form of nondirectional devices such as the British Standards Institution (BSI) deposition gauge (see Figure 17.2) and the 'inverted Frisbee' gauge (see Figure 17.4) and directional devices such as the CERL gauge (Figure 17.6). As already noted in Chapter 17, the problem is that collection efficiencies are very different from one sampler to the next. Vallack and Chadwick (1992) reported side-by-side comparisons between a variety of such samplers at a pair of field sites in the north of England, one at a village close to a large, coal-fired electricity generating station and the other at a village in a relatively 'clean' rural area. The study employed a pair of inverted Frisbees, with inner collection surfaces both dry and coated with oil, respectively. The aim of the latter was to increase the adhesion of deposited particles and so reduce re-entrainment, thus increasing collection efficiency. In addition, a BSI gauge and a related gauge described by the ISO were employed. Sampling took place for 1 month durations over an extended period of 2 years, thus incorporating two cycles of seasonal changes. Table 23.1 summarises the main results, expressed in terms of the rate of deposited mass per unit area (e.g. mg m^{-2} day^{-1}). In the first instance these confirmed that nuisance dust levels were indeed smaller at the pristine rural site than close to the power station. But, in addition, they confirmed trends that were expected from what was already known about the performances of such samplers from laboratory studies, as shown in the earlier Figures 17.3 and 17.5. That is, the inverted Frisbee consistently collected more particulate matter than the other two samplers, although the results for the dry frisbees were only marginally greater. Importantly, the oiled frisbees collected more than the dry ones. After compositional analysis of collected samples, Vallack and Chadwick explained some of the differences in terms of particle composition, noting that the inverted frisbees were more efficient for large particles in the soil/sand category.

Table 23.1 *Summary of results for a number of horizontal dust deposit gauges, presented as mean daily deposition rate, operated over a period of 2 years at two contrasting locations in the north of England (from Vallack and Chadwick, 1992)*

Sampler	Mean daily deposition rate (mg m^{-2} day^{-1}) over 2-year period	
	Location close to power station	'Clean' rural location
Dry inverted frisbee	69.0	25.0
Oiled inverted frisbee 1	76.2	32.8
Oiled inverted frisbee 2	71.4	36.9
British standard gauge	62.0	24.6
ISO gauge	61.4	25.4

Table 23.2 *Summary of results for a number of horizontal dust deposit gauges, presented as mean daily deposition rate, operated over a period of 2 years at two contrasting locations – same as for Table 23.1 – in the north of England (from Vallack, 1995)*

Sampler	Mean daily deposition rate (mg m^{-2} day^{-1}) over 2-year period	
	Location close to power station	'Clean' rural location
Dry inverted frisbee	34.2	23.0
Oiled inverted frisbee	89.1	42.7
Dry inverted frisbee with mesh insert	48.7	N/A
Dry inverted frisbee with foam insert	64.1	
British standard gauge	49.6	28.8

In a follow-up field study, Vallack (1995) explored both dry and oiled inverted frisbees, along with dry inverted frisbees with mesh and foam inserts, respectively. The latter contained further modifications of the original dry frisbee and were also aimed at improving collection efficiency by reducing re-entrainment losses. Table 23.2 summarises the results of this second field study. Here, the general trend was the same as for the earlier study (see Table 23.1) in that the nuisance dust levels were much greater in the vicinity of the power station. But, again, the results for the dry frisbee did not show the expected greater collection efficiency in relation to the BSI gauge. Indeed it was worse! Vallack noted the poor performance of the dry frisbee when used in the field, commenting pointedly that 'it is unfortunate that it has been adopted for routine monitoring by some organisations', and expressed the hope that the use of a dry frisbee without appropriate inserts or treatment would be discontinued.

The horizontal dust gauges featuring in the field studies of Vallack and his colleagues are all nondirectional. By contrast, the CERL vertical gauge (see earlier Figure 17.6) was directional, providing information that could assist in identification of sources of coarse nuisance dust or grit. Such samplers have all been used in the UK, along with samplers for finer aerosol fractions, in environmental assessments of large scale industrial operations. For example, they were quite recently employed together in the 2004 survey to characterise local background air quality in the vicinity of the very large 4000 MW Drax Power Station in the north of England, both prior to and during trials to introduce new coal/petroleum coke blends.[1] This has been a typical real-world application for samplers of this type.

Dust gauges like those described have been used quite widely in Europe and many other countries, including Australia and New Zealand. Soon after the CERL gauge appeared in the early 1960s, Macey (1967) reported experience with the instrument in Western Australia, and noted that the difference in climate between Western Australia and the UK, including the frequency of rainfall, produced significant differences in performance and operation of the sampler when used for sampling over extended periods. He therefore recommended a new approach to the calculation and interpretation of the raw data it provided to take such differences into account. It is interesting to note that the measurement of coarse nuisance dust and grit using such samplers has not been popular in the USA.

23.3 Total suspended particulate and black smoke

Apart from the criterion of the German Maximale Arbeitsplatzkonzentration (MAK) based on setting the inlet velocity of a sampler at 1.25 m s^{-1} (MAK, 1981), which has had only limited application,

---

[1] See http://www.draxpower.co.uk/files/test program 2004.pdf.

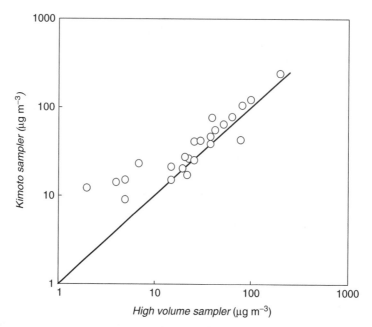

Figure 23.1 *Comparison between results for total suspended particulate (TSP) obtained using two different samplers, one a high volume sampler like that shown in Figure 17.10 and the other a Japanese 400 Lpm sampler (from the data of Reponen et al., 1996). The solid line represents perfect 1:1 agreement*

there has never been any internationally recognised, particle size-selective performance criterion for defining total suspended particulate (TSP). So standards for TSP in individual countries have been determined by chosen instruments, usually without regard to their sampling efficiency characteristics. This means that there has been no widely accepted basis for defining a reference sampler for TSP. This may account for the fact that, although such instruments have been extensively used in field studies to characterise air pollution in localities and regions throughout the world, and many reports of their applications have appeared in the peer-reviewed literature, there appears to be a paucity of reported field studies to compare the various TSP samplers that have been used. In one of those few studies, Reponen *et al.* (1996), compared results for TSP obtained at a single site near Kuopio, Finland, using two samplers, the first a high volume sampler similar to that in Figure 17.10 and another a Japanese high-volume, 400 Lpm, dichotomous sampler (Kimoto, Osaka, Japan, Model 130). The results are summarised in Figure 23.1, showing that these two instruments were in fair agreement with one another, especially for TSP concentrations above about 10 $\mu g\,m^{-3}$. However, such results are relevant only to the two samplers tested and for conditions pertaining to the Kuopio location. They are only of limited wider value.

Several methods have been proposed over the years for measuring 'black smoke' (BS). In the UK, BS was the primary index of ambient atmospheric aerosol for a long time from 1961 until the early 1990s, and this was the subject of the measurement method specified by the British Standards Institution (BSI, 1969). This produced a long history of records of local measurements. But it has been less relevant in more recent years, not least because carbonaceous aerosols that were once so predominant in the ambient air in Britain now represent a much smaller proportion of airborne particles. In turn, the original calibration of the optical 'smoke shade' method, carried out in the 1960s to provide the equivalent mass

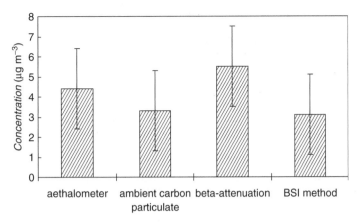

Figure 23.2 *Comparison between measurements (arithmetic means and standard deviations) of black smoke at a site beside a heavy-traffic urban road during October and November in the city of Antwerp, Belgium, using three direct-reading instruments together with the BSI method (Berghmans et al., 1996). Adapted from Journal of Aerosol Science, 27, Berghmans et al., S689–S690. Copyright (1996), with permission from Elsevier*

concentration, is no longer applicable. Other methods were also developed for the measurement of BS including, as mentioned in Chapter 17, one promulgated by the Organisation for Economic Cooperation and Development (OECD) which was similar to the BSI method. In addition, a number of direct-reading instruments have been used. Berghmans *et al.* (1996) conducted side-by-side studies of several measuring instruments for BS – including also the BSI method – at an urban roadside site in Antwerp, Belgium. The instruments tested included an aethalometer (GIV GmbH, Breuberg, Germany, Model AE-10 M), an ambient carbon particulate monitor (rp Air Quality Instrumentation, Thermo Electron Corporation, East Greenbush, NY, USA, Model 5400) and a beta-attenuation monitor (FAG Kugelfischer, Schweinfurt, Germany, Model FH 62 IN). The results are summarised in Figure 23.2. The ambient carbon particulate monitor showed the closest agreement with the BSI method. The beta-attenuation monitor provided the largest mean value, but this is not surprising since this instrument effectively actually measured *all* the collected particulate matter. It was seen that all the samplers – the direct-reading instruments and the BSI method – gave results that exhibited similar variability, and all four methods fell within the overall error bars.

The primary alternative to BS was TSP, and this was the approach adopted in many other countries, especially in the USA and some parts of Europe. TSP has traditionally been sampled using aspirating instruments like those described in Chapter 17. The available laboratory-based performance data for TSP samplers showed that their particle size selectivity varied greatly from one instrument to another (e.g. Barrett *et al.*, 1984). Since the BS and TSP approaches are so different, it is likely that correlations between BS and TSP would be of limited value only. That said, some field studies have been reported. One of the TSP samplers tested in the laboratory by Barrett *et al.* was the 45 Lpm *Kleinfiltergerat* ('small filter') GS050/3, recommended in Germany by the Verein Deutscher Ingenieure (VDI) since 1982 as a reference method against which to compare other samplers for TSP. In 1983, Laskus reported comparative measurements of TSP using this sampler and BS using the BSI approach over the period April 1980 to March 1981 at sites in the German city of Berlin, one in an industrial area and the other in the city suburbs. The results are summarised in Table 23.3. Here, the BS-values were obtained using the original calibration. It is seen that, for each location and each season, the TSP-values consistently exceeded the BS-values. This was not surprising since the smoke would have represented a finer

Table 23.3   *Summary of results of side-by-side comparisons of data collected at locations in the German city of Berlin for black smoke (BS) and total suspended particulate (TSP) as measured using the Kleinfiltergerat GS050/3 (from Laskus, 1983)*

Location	Period	Arithmetic mean (μg m^{-3})		Regression	TSP/BS (ratio of means)
		BS	TSP		
Suburbs	Summer	18	72	TSP $= 2.4 \cdot$ BS $+ 29.0$	4.00
	Winter	30	65	TSP $= 1.80 \cdot$ BS $+ 11.9$	2.17
Industrial	Summer	33	109	TSP $= 1.45 \cdot$ BS $+ 61.2$	3.30
	Winter	61	137	TSP $= 1.72 \cdot$ BS $+ 32.1$	2.25

sub-fraction of the coarser fraction collected by the Kleinfiltergerat to represent TSP, notwithstanding the low sampling efficiency of the latter for larger particles. Also not surprising, the measured levels at the industrial location were considerably higher than in the more pristine suburbs environment. Summer BS-levels were lower in relation to TSP than for the winter period at both locations, reflecting reduced usage in that period of carbonaceous fuel for domestic and industrial heating. Indeed, Laskus went further to suggest that the primary contributions to BS in the summer were of different origin, coming mainly from vehicle emissions. Laskus also reported gravimetric measurements of the masses collected by the BS sampling system, and noted that these produced concentration levels that were consistently greater than those provided by the BS optical measuring system. In his mind, this threw the general applicability of the calibration of the BS method into doubt, and he recommended TSP as a more consistent and reliable basis for standards. A more recent field study to compare BS and TSP was performed during the 1980s in Baghdad, Iraq, where this time a high volume ('Hi-Vol') sampler like that shown in Figure 17.10 was used to collect TSP (Kanbour *et al.*, 1990). No correlation was seen between BS and TSP and this was explained in terms of the fact that the TSP was dominated more by wind-blown dust and less by carbonaceous aerosols. Although this study was primarily of local interest, the results underlined the fact, by now becoming widely accepted, that both BS and TSP were inappropriate bases for international air quality guidelines. It was therefore timely, in the late 1980s, that new particle size-selective criteria emerged based on the nature of actual human aerosol exposure.

23.4   Black smoke and particle size fractions (PM$_{10}$ and PM$_{2.5}$)

As already mentioned, the BS index of atmospheric particulate persisted in Europe until quite recently. So, historic measurements of BS have remained of considerable interest there, especially in relation to the epidemiology needed for the development of improved standards. With this in mind, Heal *et al.* (2005) conducted a year-long comparative study of daily BS, PM$_{10}$ and PM$_{2.5}$-values at an urban rooftop site in the Scottish city of Edinburgh. For the purpose of comparison, BS-values were also obtained at a rural site to the south of the city. Sampling for the PM$_{10}$ and PM$_{2.5}$ was carried out using a pair of Partisol samplers (Thermo Electron Corporation, Waltham, MA, USA), the second with a sharp-cut cyclone below the PM$_{10}$ head to separate the finer fraction. BS was measured using the BSI method with a recent calibration (DETR, 1999). The results are summarised in Figure 23.3, represented in terms of the dimensionless ratios BS/PM$_{10}$, BS/PM$_{2.5}$ and PM$_{2.5}$/PM$_{10}$, and separated according to

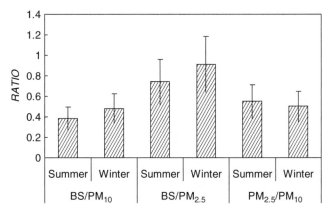

Figure 23.3 *Summary of comparisons between black smoke (BS), PM₁₀ and PM₂.₅ at an urban site in the city of Edinburgh, Scotland (Heal et al., 2005), shown in terms of the distributions of the daily measured values (medians and upper and lower quartiles) over a period of 1 year. The results are separated by season (summer and winter)*

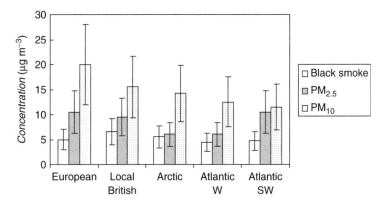

Figure 23.4 *Apportionment of the contributions to daily black smoke, PM₁₀ and PM₂.₅ concentrations for the city of Edinburgh from nearby and distant sources, based on a year long study of urban atmospheric aerosol carried out at an urban site (Heal et al., 2005). Adapted from Atmospheric Environment, 39, Heal et al., 3711–3718. Copyright (2005), with permission from Elsevier*

season (summer and winter). They show that the proportion of BS increased somewhat during the winter. Not surprisingly, BS as a proportion of PM_{10} was significantly less than as a proportion of $PM_{2.5}$. It was noted that, in the winter, BS and $PM_{2.5}$, became very close to one another. Based on daily measurements that had been made of wind speed and direction at the sampling site, the aerosol data were analyzed in terms of the contributions arriving from a variety of possible sources, including European, local British, Arctic, Atlantic W and Atlantic SW. Figure 23.4 summarises the results. Again it is seen that BS and $PM_{2.5}$ were close. The fact that BS was quite constant regardless of the source supported the view that BS – more than any of the other fractions – was dominated by local sources, and that long-range transport of BS was correspondingly lower than for the other fractions.

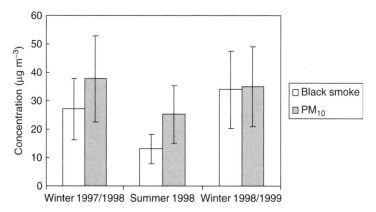

Figure 23.5 *Summary of comparisons between black smoke and PM₁₀ at a rural site in the small village of Žloukovice in the Czech Republic (from the data of Braniš and Domasová, 2003), shown in terms of the distributions of the daily measured values (medians and upper and lower quartiles) over a period of 18 months. The results are separated by season (summer and winter)*

Interest in BS has also persisted elsewhere in Europe. A recent field, study in the Czech Republic was reported by Braniš and Domasová (2003) where direct inter-sampler comparisons were made for BS and PM_{10}, respectively. A large data set was collected in the small village of Žloukovice during the winter 1997/1998, summer 1998 and winter 1998/1999. This site was chosen because there were no significant industries in the vicinity nor any significant road traffic, and the aerosol would therefore have been dominated by sources associated with domestic heating, either by coal or wood. Sampling for PM_{10} was carried out using a low flow rate, 10 Lpm instrument referred to as a 'Harvard impactor' that had been designed and calibrated to meet requirements of the US EPA for PM_{10} (Marple *et al.*, 1987; Turner *et al.*, 2000). The measurement of BS was carried out using the OECD method. The main results are summarised in Figure 23.5. The fact that BS and PM were so close during the winter months confirmed that most of the aerosol came from domestic heating. The weaker association during the summer suggested different sources. Among other studies of BS and PM_{10}, measurements of both indices were an important feature of a large epidemiologic project – Pollution Effects on Asthmatic Children in Europe (PEACE) – involving 14 laboratories in 10 European countries (Hoek *et al.*, 1997). But direct side-by-side comparisons between BS and PM_{10} were not reported.

Overall, although emphasis was shifting in later years towards standards and measurement based on the PM_{10} and $PM_{2.5}$ criteria, the results from studies like those described above suggest that BS and the health-related particle size-selective fractions are not uniformly correlated. There are still some experts who feel that there should be a continuing separate role for the measurement of BS in aerosol sampling in the ambient atmosphere.

23.5   Transition to particle size-selective sampling

As was the case for workplace aerosol exposure assessment, the transition of standards for ambient atmospheric aerosol from ones based on the old TSP concept to new ones based on new, finer fractions representative of particle penetration into the human respiratory tract pointed to the need to assess the impact of the change on compliance or exceedance. More specifically, field studies were needed in order to establish the relationship between the old TSP and the new PM_{10}. In the USA in the early 1980s, however, EPA was still considering the possibility of a PM_{15} criterion to represent the exposures of people to inhaled particles capable of penetrating below the larynx and into the lung (see Chapter 11). With this in mind, Watson *et al.* (1983) examined the masses collected by a number of sampling devices

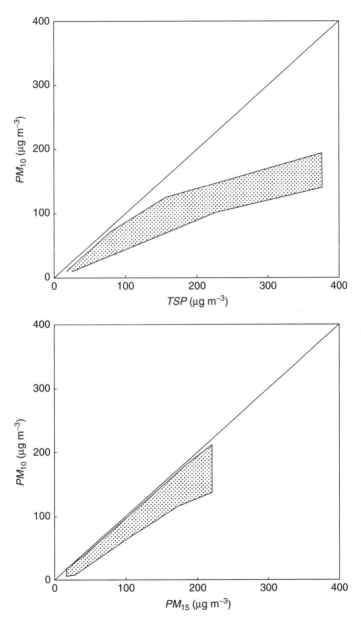

Figure 23.6 *Summary of results of side-by-side comparisons of PM_{10}, PM_{15} and TSP at six locations in the US (see Table 23.4), where the shaded areas contain most of the individual data points (from the data of Rodes and Evans, 1985)*

placed side-by-side with primary particle size-selectivity falling broadly within the performance envelope of the PM_{15} criterion that was being considered at the time. They compared the collected masses with the corresponding masses of TSP collected using high volume samplers. The reported data were taken from field measurements at various sites made under the auspices of the EPA Inhalable Particulate Matter Sampling Network and which first appeared in EPA reports. The average ratio between PM_{15} and TSP for many hundreds of samples was found to range from about 0.60 to 0.75 with standard deviation from about 0.07 to 0.19. The individual samplers tested were found on average to collect masses to within $\pm 5\%$ of one another. From their results and analysis, Watson *et al.* estimated that inlets matching the emerging new PM_{10} could be expected to collect between 80% and 90% of the mass collected by the PM_{15} samplers. In a somewhat later study for six contrasting locations at cities widely dispersed in the USA within the same EPA network, Rodes and Evans (1985) compared PM_{10}-values with those for TSP and PM_{15}, respectively, using samplers that had been considered by EPA to be acceptable for meeting the criteria that were relevant at the time. Again, a large amount of data was collected and these are summarised in Figure 23.6. Not surprisingly, they show that the PM_{10}-values were consistently smaller than those for TSP. This time, as shown in the results of regression analyses summarised in Table 23.4, the ratio of PM_{10} to TSP was about 0.5, and was reasonably consistent across the eight locations. The ratio of PM_{10} to PM_{15} was between 0.7 and 0.9, again generally quite consistent across locations, and in good agreement with what had been reported by Watson *et al.*

Although countries in the European Community considered for a while the possibility of adopting particle size-selective criteria based on the earlier recommendations of the International Standards Organisation (ISO, 1983), interest soon veered towards PM_{10}. Under the terms of a joint agreement with the EPA, the Dutch National Institute of Public Health and Environmental Hygiene, Bilthoven, The Netherlands, conducted a 1 year study to examine the relationship between PM_{10} and TSP concentrations at contrasting sampling sites. These included Vlagtwedde, a typical rural environmental in the northeastern part of the country, and Bilthoven, a suburban environment with meteorological features reflecting contributions from mesoscale pollution transport from industrial areas of neighboring countries (van der Meulen *et al.*, 1987). PM_{10} was collected using a 167 Lpm dichomotomous sampler (Sierra Andersen 241) and TSP using a 2800 Lpm sampler described as the 'Electronica'. It was found that the results for each location and each sampler were log-normally distributed, consistent with what has been seen elsewhere. Indeed, the geometric standard deviations for these data were of the order of from 2 to 3, roughly on a par with those found for most workplace aerosol situations. Overall, the respective PM_{10} and TSP-values were about 15% higher at the suburban Bilthoven site than at the more pristine rural Vlagtwedde site. From linear regression of all the data, the overall ratio of PM_{10} to TSP concentration

Table 23.4 *Summary of results of side-by-side comparisons of PM_{10}, PM_{15} and TSP at six locations in the USA (from Rodes and Evans, 1985)*

City	Mean PM_{10} ($\mu g\ m^{-3}$)	Mean PM_{15} ($\mu g\ m^{-3}$)	Mean TSP ($\mu g\ m^{-3}$)	Mean $\dfrac{PM_{10}}{PM_{15}}$	Mean $\dfrac{PM_{10}}{TSP}$
Birmingham, AL	50.9	59.7	102.0	0.834	0.506
Buffalo, NY	37.1	41.9	75.5	0.870	0.478
Philadelphia, PA	35.4	40.2	71.0	0.891	0.496
Phoenix, AZ	38.9	52.1	99.2	0.748	0.381
Pittsburgh, PA	–	41.4		0.961	0.577
Rubidoux, CA	71.7	92.0	130.4	0.805	0.485

was estimated to be about 0.7, compared with the lower value of 0.5 previously estimated for USA sites by Rodes and Evans. van der Meulen and his colleagues suggested that this difference may have been associated with the fact that some of the USA data had been taken in regions more arid than those generally encountered in Europe, leading to a larger contribution from the coarse mode of atmospheric aerosol.

23.6 PM$_{10}$

As mentioned above, the arrival of the PM$_{10}$ criterion in the USA in the 1980s, and later in Europe and elsewhere, carried with it the specific requirement to carry out comparative field studies in order to validate candidate samplers for that fraction. The usefulness of this requirement was the subject of an important comprehensive field study, carried out by EPA scientists in 1983 and 1984, in which comparisons were made at four separate USA locations for a range of PM$_{10}$ inlets mostly representative of what were commercially available at the time (Rodes *et al.*, 1985). The study also included one prototype inlet developed at the EPA and not commercially available. The samplers listed in Table 23.5 identify both the inlets and the base sampling systems, and these were generally representative of the types described in Chapter 17. They were also generally characteristic of most of the PM$_{10}$ samplers that are used today. Four of the instruments, designated by 'DCT', were dichotomous samplers and so, in addition to PM$_{10}$, also provided measures of PM$_{2.5}$. The locations listed in Table 23.6 were chosen – on the basis of information obtained from the EPA Inhalable Particulate Matter Sampling Network – to provide a range of particle size distributions that would be representative of most anticipated applications. In the field studies at the listed sites, particle size distributions were estimated from data obtained using the wide ranging aerosol classifier (WRAC, see earlier Figure 18.15) and were found to adequately meet the desired characteristics. A very large database was generated from this research. A summary of the main findings for the pooled data is given in Table 23.7. Most of the sampler pairs suggested quite good agreement. Rodes and his colleagues commented that,

'If the mass concentration measurements of all the *PM*$_{10}$ samplers agree with one another within 10 % under field conditions, it is reasonable to assume that the FRM approach adequately predicts sampler performances.'

A small number of the samplers tested fell outside such acceptability. That the Rodes *et al.* paper prompted a flurry of vigorous commentary in the pages of the journal, notably involving parties

Table 23.5 *Summary of inlets and sampling systems that were used in the 1983/1984 EPA intersampler comparisons (from Rodes et al., 1985). The designation DCT identifies dichotomous inlets that provided PM$_{2.5}$ in addition to PM$_{10}$*

Inlet	Source	Sampler	Flow rate (Lpm)
GMW-9000	General Metal Works	GMW Accu-Vol	1130
SA-321	Sierra Andersen	GMW Accu-Vol	1130
SA-321A	Sierra Andersen	GMW Accu-Vol	1130
SA-254	Sierra Andersen	SA Med Flo	113
GMW-9100	General Metal Works	GMW 554M	113
SA-246b-DCT	Sierra Andersen	Sierra 244E	16.7
GMW-9200-DCT	General Metal Works	Sierra 244E	16.7
FS-85-16-DCT	Flow Sensors	Sierra 244E	16.7
EPA prototype-DCT	–	Sierra 244E	16.7

Table 23.6   *Summary of the locations where intersampler comparisons were carried out, indicating the anticipated general level of concentration and degree of coarseness or fineness of the aerosol (from Rodes et al., 1985)*

Location	Concentration (μg m^{-3})	Particle size	Notes
Durham, NC	Low (<50)	Fine and Coarse	–
East St Louis, IL	Medium (50–80)	Fine and Coarse	–
Phoenix, AZ	High (>80)	Coarse	Site close to major road construction with significant fugitive dust
Rubidoux, CA	High (>80)	Fine and Coarse	–

Table 23.7   *Summary of regression analysis of pooled data (from all four sites) for PM$_{10}$ in relation to the various intersampler comparisons (from Rodes et al., 1985) (see Table 23.5)*

Sampler comparison y	Sampler comparison x	Number of samples (n)	$y = ax + b$ a	$y = ax + b$ b
FS-85-16-DCT	SA-246b-DCT	48	0.94	2.37
EPA-DCT	SA-246b-DCT	68	1.09	−2.95
GMW-9100	SA-246b-DCT	45	0.87	−2.59
FS-85-16-DCT	GMW-9200-DCT	47	1.08	−1.26
GMW-9100	GMW-9200-DCT	44	1.00	−5.89
GMW-9000	FS-85-16-DCT	46	0.90	−1.42
GMW-9100	FS-85-16-DCT	45	0.95	−6.03
SA-254	FS-85-16-DCT	46	1.09	−8.45
GMW-9100	EPA-DCT	45	0.80	−0.15
SA-254	EPA-DCT	55	0.93	−1.72
GMW-9100	SA-321	45	0.64	6.80
GMW-9100	GMW-9000	45	1.05	−4.04
SA-321A	SA-246b-DCT	18	0.96	1.36
SA-321A	EPA-DCT	18	0.89	1.69

concerned with the manufacturer of one or more of those samplers, was an indication of the importance of this work at a time when the new PM$_{10}$ standards were just becoming established. A follow-up field study was carried out to examine some of the questions that had been raised (Purdue *et al.*, 1986), and the results led to the suggestion that manufacturers of inlets, especially those inlets that had compared less favorably in the main inter-sampler evaluation, should develop appropriate inlet maintenance procedures, in particular to overcome problems of re-entrainment or particle bounce for larger, grittier particles. Of course, all of the samplers tested have since been superseded and improved new versions appeared in the years that followed. That said, this body of work was a milestone in the establishment of procedures by which samplers for ambient atmospheric aerosol could be transferred from the drawing board and laboratory to applications in the real world.

Other such studies followed. A number of samplers for PM$_{10}$ were compared side-by-side in the field during the PEACE study mentioned above (Hoek *et al.*, 1997). Five laboratories out of the 10-nations project participated in this sampler comparison exercise, each one using its own chosen, commercially available sampler alongside the 10 Lpm Harvard sampler agreed as an overall reference sampler for PM_{10} (Air Diagnostics and Engineering Inc., Naples, ME, USA). Sampling was carried out during the winter of 1993/1994 in the cities of Amsterdam (The Netherlands), Teplice (The Czech Republic), Pisa (Italy), Oslo (Norway) and Budapest (Hungary), and at sites chosen not to be immediately influenced by heavy traffic or industry. The 'local' samplers were listed by Hoek *et al.* as the Sierra Andersen 241 dichotomous sampler, the Strohlein Andersen PM_{10} sampler, the Andersen PM_{10} high volume sampler, the Sierra Andersen 245 dichotomous sampler and the MPSI 100 beta-gauge monitor with a PM_{10} inlet, for the five sites respectively. The results for the five comparisons are plotted together in Figure 23.7, where it is seen that all the 'local' samplers performed in generally very good agreement with the Harvard reference sampler. This was confirmed by the results of the linear regression that Hoek *et al.* carried out for each data set, summarised in Table 23.8. In an earlier field study, Chow (1995) had examined the performances of a number of PM_{10} samplers with different designs, and reported variability in the average collected masses as great as 50 %. Hoek *et al.* argued, however, that such differences were the result of inconsistent operating procedures, including the greasing of the inside walls of the sampler inlets so that, from the results shown in Figure 23.7, a more optimistic view was presented.

Some field studies focused on more limited ranges of samplers, a number of them aimed at the introduction of direct-reading instrumentation. The tapered element oscillating microbalance (TEOM, see Chapter 20) has been of considerable interest in recent years because it provides data that correlate directly with the mass of particulate matter collected. Price *et al.* (2003) conducted field studies in the city of Sunderland in the north of England to compare PM_{10} as measured using a conventional gravimetric sampler in the form of a 16.7 Lpm Partisol Plus 2025 and two versions of the TEOM (Thermo Electron Corporation, Waltham, MA, USA). One of the TEOMs was operated at 50 °C and the other with a

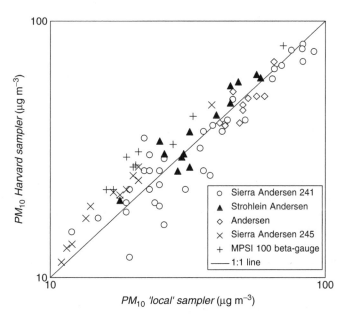

Figure 23.7 *Field comparisons for side-by-side reference sampler for PM_{10} (the Harvard impactor) and chosen 'local' samplers for PM_{10} at sites in Amsterdam (O), Teplice (▲), Pisa (◇), Oslo (×) and Budapest (+) (from the data of Hoek et al., 1997). 'Local' samplers are as shown, and are as listed by Hoek et al*

Table 23.8 *Summary of the main results of the regression analysis for the results plotted in Figure 23.7 for the field comparisons between the Harvard reference sampler for PM_{10} and the corresponding 'local' PM_{10} samplers used in the cities indicated (from Hoek et al., 1997)*

Location	Sampler	Regression
Amsterdam	Sierra Andersen 241 dichotomous sampler	$PM_{10Harvard} = 0.93 \cdot PM_{10local} - 1.4$
Teplice	Strohlein Andersen PM_{10} sampler	$PM_{10Harvard} = 1.19 \cdot PM_{10local} - 1.3$
Pisa	Andersen PM_{10} high volume sampler	$PM_{10Harvard} = 1.30 \cdot PM_{10local} - 19.3$
Oslo	Sierra Andersen 245 dichotomous sampler	$PM_{10Harvard} = 1.23 \cdot PM_{10local} - 0.9$
Budapest	MPSI 100 beta-gauge monitor	$PM_{10Harvard} = 1.06 \cdot PM_{10local} - 6.0$

Table 23.9 *Comparisons between samplers for PM_{10} at the kerbside of a busy road junction in the city of Sunderland, using the Partisol Plus 2025 gravimetric sampler and two versions of the direct reading tapered element oscillating microbalance (TEOM). the second one with the SES sampling conditioning system (Price et al., 2003). The results show monthly averages for the inter-sampler correlations*

Month	$\dfrac{\text{TEOM}}{\text{Partisol}}$	$\dfrac{\text{TEOM}}{\text{TEOM (SES)}}$	$\dfrac{\text{TEOM (SES)}}{\text{Partisol}}$
November	0.589	0.879	0.502
December	0.746	0.952	0.680
January	0.403	0.867	0.608
February	–	–	–
March	0.668	0.931	0.664
April	–	0.936	0.650
May	0.737	0.918	0.845
June	0.794	0.932	0.841
July	–	–	–
August	0.887	0.975	0.854

sample equilibrium system (SES) operating at the lower temperature of $30\,^{\circ}\mathrm{C}$. For the latter, the SES also incorporated a dryer that continuously conditioned both the sample and the bypass flows of the TEOM. The intention of this second version was to allow greater retention of semi-volatile components. The three instruments were deployed in a sampling campaign where they were placed side-by-side over a period of 9 months at the kerbside of a busy road junction. The results of the inter-sampler comparisons are summarised in Table 23.9. They show that the TEOM in both versions consistently undersampled with respect to PM_{10}, but less so for the version with the SES, the more so in the winter months. Such differences suggested that there was greater retention of moisture by the gravimetric sampler. In another field study conducted at four locations in Taiwan, a commercial direct-reading, beta-attenuation instrument was studied side-by-side with two high volume gravimetric samplers for PM_{10} (Chang *et al.*, 2001). The beta-gauge featured a cyclone inlet matching the PM_{10} criterion when operated at 18.9 Lpm. The average daily value of the ratio of beta-gauge to gravimetric PM_{10} was found to be about 1.08 for both gravimetric samplers, with standard deviation about ± 0.06, except for conditions where the deliquescence point was exceeded so that, due to increased moisture absorption by some particles, the ratio rose to greater than about 1.20. A similar field study was carried out at a rural site in Finland where a 38.3 Lpm gravimetric PM_{10} reference sampler was compared side-by-side with a different direct-reading, beta-attenuation monitor over the period from September to December (Salminen and Karlsson, 2003). The mean daily ratio of beta-attenuation to gravimetric sampler concentrations was found to be 1.06 with standard deviation ± 0.12. Although the beta-attenuation monitor slightly overestimated the PM_{10}, the results were close enough to be regarded as the same for all practical purposes. In addition, the monitor was found to be capable of measuring mass concentrations to as low as 2 $\mu g\,m^{-3}$. The studies cited, and others like them, demonstrated that direct-reading instruments like those described have considerable potential for practical use in the measurement of ambient PM_{10} concentrations.

23.7 $PM_{2.5}$

The EPA reference method developed over the years for PM_{10} provides excellent guidance for corresponding reference methods for the measurement of $PM_{2.5}$. In the Rodes *et al.* field study, the four dichotomous samplers also provided $PM_{2.5}$ in addition to PM_{10}, and excellent agreement was found between them. In the first instance it was not surprising that the $PM_{2.5}$ comparisons were more consistent than those for PM_{10} because the finer aerosol fraction would have been much less sensitive to effects on aspiration efficiency associated with external wind speed. Rodes *et al.* noted that, although the dichotomous samplers were more difficult to operate than the simpler ones dedicated just to PM_{10}, '... the experience gained... resulted in repeatable and predictable samples.' In Europe, however, some difficulties were noted. In preliminary inter-sampler studies for $PM_{2.5}$ carried out in the field in several European countries, it was reported that inter-sampler differences could be as high as 30 %. These were attributed to the fact that the chemical composition of $PM_{2.5}$ is markedly different than for PM_{10}, in particular with regard to the unstable, semi-volatile components, such as ammonium nitrate and organic compounds, present in $PM_{2.5}$ (Commission of European Communities, 2004). Losses of semi-volatile matter already noted for PM_{10} would be even more pronounced for $PM_{2.5}$, and be strongly influenced by seasonal and geographical factors. For example, while negligible losses were reported for one springtime study in Scandinavia, losses of up to 70 % were reported for the same aerosol sampling system in a study during the winter in Central Europe.

23.8 Personal exposures to PM_{10} and $PM_{2.5}$

Concentrations of atmospheric aerosol have traditionally been made using fixed point sampling devices placed at locations chosen to be relevant to populations. In this way, characterisation of atmospheric aerosol has been relatively simple and economic. However, the underlying rationale for sampling has usually been the estimation of human exposures. In occupational hygiene, personal sampling became the preferred – and widely used – approach back in the 1970s once sampling pump and sampling head technologies had delivered sufficiently miniaturised equipment. The difficulties associated with measuring personal aerosol exposures among the general population have already been noted. However, the 'disconnect' between measures like those described above and the actual exposures of people has persisted. Epidemiologists in particular have been anxious to obtain measurements of ambient aerosol levels that more closely reflect actual human exposure. In recent years, therefore, from the early 1990s onwards, efforts have been made to bridge the gaps between fixed point and personal aerosol measurement, and so account for the widely differing outdoor and indoor environments which most people inhabit during their daily lives. Colome *et al.* (1992) described one of the first such studies in which they compared outdoor and indoor residential PM_{10} levels in Orange County, CA, USA. They employed a 4 Lpm

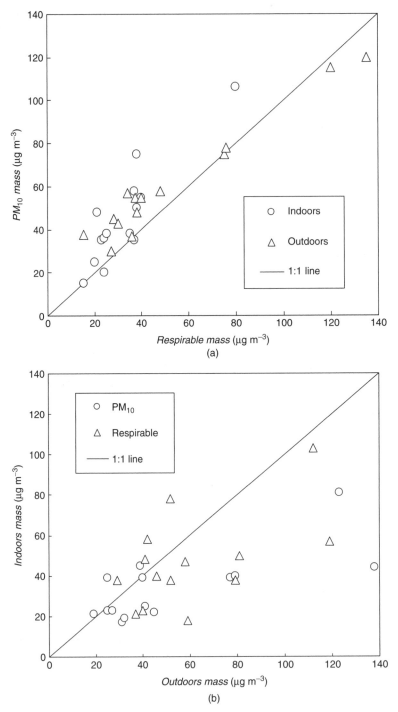

Figure 23.8 *Comparisons between paired samples taken outdoors and indoors in residences in Orange County, CA, USA and using two types of sampler: (a) PM₁₀ mass concentration as a function of respirable (cyclone) mass concentration for both indoors and outdoors; (b) indoors mass concentration as a function of outdoors mass concentration for both PM₁₀ and respirable (cyclone) samples (Colome et al., 1992). Adapted from Atmospheric Environment, 26A, Colome et al., 2173–2178. Copyright (1992), with permission from Elsevier*

PM_{10} sampler of the type described by Marple *et al.* and a 2 Lpm cyclone of the type widely used for occupational personal respirable aerosol sampling but used here in the first instance as a static sampler. The results in Figure 23.8(a) describe the direct comparison between daily 24-h, time-weighted average concentrations for the two samplers for both outdoor and indoor environments combined. As seen, the PM_{10} results were consistently somewhat greater than those for respirable. The differences were quite small, indicating the predominance of fine particles in all the environments studied. Figure 23.8(b) shows the results of the corresponding comparison between the outdoor and indoor aerosol levels, and much greater scatter in the data points is now seen. But, notwithstanding the differences in particle size-selection characteristics of the two samplers, the general tendency was for the outdoor concentrations to be greater than the indoor ones. However it was noted that, since the residences studied were occupied by nonsmokers and contained no pets, associated with the fact that one or more occupant was an asthma sufferer, it was likely that these indoor aerosol levels would have been lower than for most of the general population. Personal sampling was carried out for a small number of residents in the homes studied. Here, in the absence at the time of a commercially available personal sampler for PM_{10}, the participating individuals wore small, 2 Lpm cyclones of the type that had been used for collecting the respirable fraction in the fixed point studies, carrying them for the full 24 h on each sampling day. The results correlated much better with the in-residence concentration values. Colome *et al.* explained this trend in terms of the fact that the subjects spent a high proportion of their time indoors, as was – and is – the case for most people in modern western societies.

Later, Liu *et al.* (2004) reported a study in the city of Beijing, China, that described outdoor and indoor levels of a range of aerosol fractions, including not only PM_{10} and $PM_{2.5}$, but also PM_1. The measurements were carried out in as many as 49 public places, including computer rooms, offices, classrooms, restaurants, dormitories, supermarkets, libraries, etc., most of which were described as 'nonsmoking'. It was winter, so the windows of the locations studied were usually closed. The measurements themselves were made using a light scattering based direct-reading sampler ('Dustmate', Sibata Company, Japan), which provided simultaneous values not only for the PM_{10}, $PM_{2.5}$ and PM_1 but also TSP. From the large body of data, regression analysis was used to obtain the indoor-to-outdoor concentration ratios for these four fractions for the different types of environment studied. The main results are summarised in Table 23.10. In general it is seen that indoor concentrations were consistently lower than those outside, which was broadly consistent with the findings of Colome *et al.* In cases where the ratio was close to or greater than unity, there was usually a rational explanation – for example, heating and cooking in restaurants.

The studies cited are good examples to illustrate the importance of taking into account indoor, as well as outdoor, aerosol levels in assessing the overall aerosol exposures of the general population, also taking into account the types of indoor environment and types and intensities of potential aerosol sources in them. It is generally acknowledged that round-the-clock personal exposure assessment should be the

Table 23.10 *Summary of results of analysis for the ratio of indoor-to-outdoor aerosol concentration ratio for each of the aerosol fractions and indoor locations indicated, for the city of Beijing, China during the winter (Liu et al., 2004)*

	Computer rooms	Offices	Dormitories	Restaurants	Classrooms	Libraries	Supermarkets
PM_{10}	0.520	0.534	0.498	0.724	0.549	0.480	0.758
$PM_{2.5}$	0.770	0.705	0.460	1.080	0.839	0.358	1.065
PM_1	1.003	0.692	0.422	1.099	0.991	0.503	1.050
TSP	0.400	0.484	0.508	0.693	0.634	0.303	0.308

ultimate goal, and some research studies have recently begun to emerge which address this issue. One good example is the Williams *et al.* report in 2003 of a major 1-year investigation of airborne particulate matter in the Research Triangle area of North Carolina, USA. It was sponsored by EPA under the auspices of Research Triangle Park Particulate Matter Panel Study, and was driven by the needs of epidemiology into fine particle health effects in a population of people whose health was already compromised. The research involved not only indoor and outdoor measurements of PM_{10} and $PM_{2.5}$ but also a careful study of personal exposures to $PM_{2.5}$. Thirty-eight individuals participated in this part. An important ingredient was the introduction of a new personal sampling system that went a long way towards minimising the inconvenience to the participant. The central feature of this system was a light-weight nylon vest designed to compactly carry all the personal sampling equipment, retaining the sampling head firmly in the breathing zone. Each participant wore the vest at all times except during sleeping, bathing or dressing, at which times the vest – and the still-running sampling equipment – were placed as close as possible nearby. Personal $PM_{2.5}$ measurements were made using the 2 Lpm $PM_{2.5}$ version of the PEM sampler shown earlier in Figure 15.28. The same sampler was used to collect residential indoor, outdoor and ambient $PM_{2.5}$, while PM_{10} PEM samplers were used to collect corresponding PM_{10}. Dichotomous samplers were also employed to provide not only PM_{10} and $PM_{2.5}$ but also $PM_{10-2.5}$. The results of the overall data are summarised in Table 23.11, pooled across all the subjects, residences, seasons and cohorts studied. More detailed regression analysis of the data showed that the personal exposures to $PM_{2.5}$ were significantly higher than the fixed point measures. Of particular interest, personal $PM_{2.5}$ was greater than ambient $PM_{2.5}$ by a factor of close to 2. But the association was weak ($R^2 = 0.16$), suggesting the strong influence of indoor sources. Williams and his colleagues were careful to point out, however, that such a relationship would be highly dependent on the types of indoor spaces inhabited by the subjects at various times during the day, in particular the type of housing and ventilation. This general tendency between personal and static sampling for aerosol exposures in ambient and living environments has been confirmed in other studies (e.g. Soutar *et al.*, 1999; Gulliver and Briggs, 2004). It is comparable with what has been known for many years in occupational hygiene settings. In particular, it appears to be a general rule that aerosol concentrations obtained by personal aerosol sampling are consistently greater than those obtained by fixed point, static sampling. Again, as for occupational exposures, vague references to the 'personal aerosol cloud' (Williams *et al.*, 2003) are not particularly instructive.

Table 23.11 *Summary of aerosol mass concentration data (in $\mu g\ m^{-3}$) in the categories indicated, taken in the Research Triangle area of North Carolina, USA, pooled over all subjects, residences, seasons and cohorts (from Williams et al., 2003)*

	Number of samples	Mean	Minimum	Maximum
Personal $PM_{2.5}$	712	23.0	3.4	70.1
Indoor $PM_{2.5}$	761	19.1	2.3	80.1
Outdoor $PM_{2.5}$	761	19.3	5.0	43.7
Ambient $PM_{2.5}$	746	19.2	5.0	44.9
Indoor PM_{10}	761	27.7	4.4	70.6
Outdoor PM_{10}	761	30.4	7.9	46.4
Ambient PM_{10}	752	31.4	4.8	51.5
Indoor $PM_{10-2.5}$	761	8.6	0	111.8
Outdoor $PM_{10-2.5}$	761	11.1	0	86.9
Ambient $PM_{10-2.5}$	210	10.0	0	62.3

23.9 Classification of ambient atmospheric aerosols

23.9.1 Particle size distribution

The measurement of particle size distribution for atmospheric aerosols has been carried out for many years, driven mainly by the needs of research. Such information may be valuable in terms of diagnosing aerosol sources or origins. Although there are no standards that require specific knowledge of particle size distribution, such knowledge is useful prior to setting up field studies, at the very least to indicate the degree of coarseness or fineness of the aerosol to be collected, hence guiding the choice of specific sampling instrument. More particularly, the determination of a quantitative particle size distribution using a single measuring instrument, can enable the estimation – from just one sample – of any number of individual particle size fractions like those described above.

Chapter 18 described how a detailed, continuous particle size distribution may be obtained using instruments such as cascade impactors, especially when used with an appropriate inversion algorithm. There have been many accounts of such measurements made in the field for ambient atmospheric aerosols. The WRAC that featured in the 1985 field study of Rodes *et al.* was a large, trailer-mounted, four-stage, parallel-impactor sampler. It was used there specifically to help in the characterisation of sampling sites where comparisons would be made between samplers for PM_{10}. Rodes and his colleagues fitted the data for the masses collected on each of the stages of the WRAC to a bi-modal log-normal distribution for each of the four cities listed in Table 23.6, and these distributions were referred to those for the 'typical' urban particle size distribution defined earlier by Lundgren and Paulus (1975). Particle size distribution measurement also featured strongly in the more recent intensive field campaign INTERCOMP2000 (Hitzenberger *et al.*, 2004). This was a large project carried out under the auspices of sub-project AEROSOL within the program EUROTRAC-2 (Müller *et al.*, 2004). An important aspect of this work was the set of side-by-side comparisons that were made between a number of cascade impactors that had become widely used for atmospheric aerosol measurement. Of particular interest were their relative abilities to provide accurate measures of PM_{10} and $PM_{2.5}$. The field measurements were carried out during a short 2-week period in April 2000 at a single rural location 45 km northeast of the German city of Leipzig. Three types of cascade impactor were used: three versions of the Berner low-pressure impactor (the nine-stage version operating at 30 Lpm, the seven-stage at 70 Lpm, and the six-stage with heated inlet at 75 Lpm, respectively), the 11-stage 30 Lpm micro-orifice uniform deposit impactor (MOUDI) and the 11-stage 28 Lpm electrical low pressure impactor (ELPI). For each sampler, the fraction of interest – PM_{10} or $PM_{2.5}$ – was estimated on the basis of the particulate mass passing beyond the stage with $_{50}d_{ae}$ closest to 10 and 2.5 μm, respectively. The results are summarized in Table 23.12. They are encouraging in that they were generally quite consistent from one sampler to the next. A notable exception was the ELPI, which, as recalled from Chapter 18, was a direct-reading cascade impactor with real-time read-out of particle masses on the various stages. By contrast, the Berner and MOUDI instruments were all gravimetric. Hitzenberger *et al.* commented that the relatively high PM_{10} concentration recorded by the ELPI was probably the result of moisture uptake on the impactor stages (which would not have been observed for the gravimetric samplers after the pre-weighing sample conditioning).

Particle size distribution measurement has been an important component of the study of ultrafine aerosols in ambient air, and data are now appearing in the literature obtained from field studies with direct-reading instruments like the differential mobility analyzer (DMA), described in Chapter 20. Figure 23.9 shows some results reported by Harrison *et al.* (2000) based on data from a DMA located at an urban sampling site during the month of June in the city of Birmingham, UK. It illustrates how such sampling can be useful in identifying the presence of a significant ultrafine fraction. Here, for example, Harrison and his colleagues noted in particular a mid-morning burst in particle number concentration

Table 23.12 *Summary of averaged results for PM$_{10}$ and PM$_{2.5}$ ambient aerosol concentrations as estimated from the particle size distribution measurements obtained using the cascade impactors listed (Hitzenberger et al., 2004). The field study was carried out at a field site near Leipzig, Germany under the auspices of the European sampling campaign INTERCOMP2000*

Sampler	PM$_{10}$ (μg m^{-3})	PM$_{2.5}$ (μg m^{-3})
30 Lpm Berner low pressure impactor, nine-stage	16.74	13.22
70 Lpm Berner low pressure impactor, seven-stage	14.89	12.49
75 Lpm Berner low pressure impactor, six-stage, heated inlet	14.42	–
30 Lpm MOUDI, 11-stage	17.53	14.67
28 Lpm ELPI	–	26.51

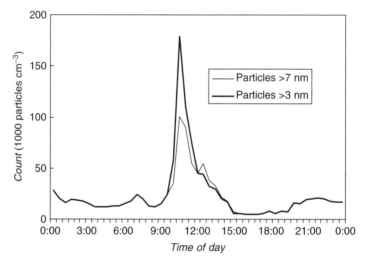

Figure 23.9 *Particle counts for ultrafine aerosol sampled by differential mobility analyzer (DMA) in the city of Birmingham, UK during the month of June, showing results for particles >7 nm and >3 nm, respectively (Harrison et al., 2000). Adapted with permission from Harrison et al., Philosophical Transactions of the Royal Society of London, 358, 2567–2580. Copyright (2000) The Royal Society of London*

in the range of particle diameter from 3 to 7 nm. Such data reflected formation and growth processes in the fine fraction of atmospheric aerosol, and were a sharp reminder of the transient nature of such aerosol, changing rapidly with time and strongly dependent on emissions of pollutants, especially from motor vehicles, and meteorological conditions.

23.9.2   Chemical composition

The composition of atmospheric aerosols varies greatly throughout the particle size range, from location to location, and from one set of meteorological conditions to another. Primary standards are expressed in terms of mass concentrations within specified particle size fractions so that measurement of airborne concentrations is usually based on gravimetric sampling. This is generally justifiable for PM_{10}. Field studies to investigate the composition within PM_{10} have generated data for a wide range of chemical species. Those of most widespread interest have been organic and elemental carbon, sulfate, nitrate and ammonium, but others have focused on mercury and other metals. The measurements of Maenhaut *et al.* (2004) for the distribution of chemical components in PM_{10} samples collected at a site in Hungary during Summer 2003 may be considered as somewhat typical of aerosol characteristic of continental background. They showed that the primary components of PM_{10} were, on average, organic matter (46 %), crustal matter (24 %) and sulfate (18 %). Other matter amounted to just 12 %.

Brook *et al.* (1997) reported measurements of TSP, PM_{10}, $PM_{2.5}$ and sulfate (SO_4^{2-}) that had been made during the period from 1984 to 1993 at a number of Canadian cities. The results are summarised in terms of the mean values in Figure 23.10. Brook and his colleagues noted in particular the lower sulfate concentrations for cities to the west of the Great Lakes, and the fact that they were highest in southern Ontario. This would appear to suggest industrial origins. In addition, the high values of TSP and (not shown here) its greater variability for the prairie cities of Winnipeg, Calgary and Edmonton suggested a higher level of coarse aerosol likely to be crustal in origin. Earlier, Thurston *et al.* (1994) had shown a significant relationship between SO_4^{2-} levels and the rate of respiratory-related hospital admissions in the city of Toronto, clearly identifying the SO_4^{2-} as a significant causal agent. In 1996, Lippmann and Thurston reviewed the data base for SO_4^{2-} levels and health effects in the USA. Earlier field data for SO_4^{2-} had been tainted by the 'sulfate artifact', associated with the collection of SO_2 vapor by the fiber glass filters that had been used at the time and its subsequent on-filter transformation to SO_4^{2-}. But later work provided the means to correct those earlier data, enabling them to be used in epidemiologic studies (Burnett *et al.*, 1994). Based on their review, Lippmann and Thurston suggested SO_4^{2-} as the best single index for routine, health-related aerosol monitoring in the ambient environment, correlating better with respiratory ill-health than the other metrics, and being stable and easily analyzable.

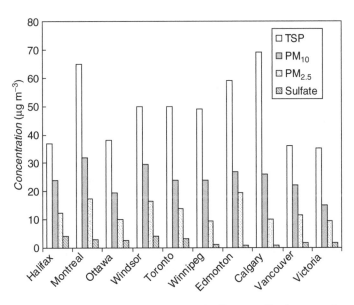

Figure 23.10 *Summaries of results for TSP, PM_{10}, $PM_{2.5}$ and sulfate (SO_4^{2-}) for a number of Canadian cities as indicated, taken from samples obtained during 1984–1993 (from the data reported by Brook et al., 1997)*

23.9.3   Bioaerosols

Sampling for bioaerosols has, perhaps, the longest history of all aerosol-related field studies, nearly all of it in the ambient outside air. Charles Darwin's description of his bioaerosol sampling exercise aboard *The Beagle* may have been one of the earliest such reports (see Chapter 19). For bioaerosols in the ambient atmosphere, the range of particles of interest is potentially even more diverse than for chemical species, compounding the degree of difficulty in trying to develop a coherent sampling rationale along the lines of that which has evolved for inert aerosols in workplaces and the ambient atmosphere. As noted by Lacey and Venette (1995) in their review of the sampling that has been carried out in cities, much of the attention has been given to sampling for pollens and fungal allergens, for which most sampling was carried out on the roofs of buildings in order to avoid the effects of very adjacent sources. Over the years, most such sampling has involved the use of sedimentation methods (in which particles were collected passively onto greased slides and plates), aspirating spore traps, Andersen samplers and rotorod-type, rotating-arm samplers. Here, however, even as early as the 1940s, Durham (1944) showed poor correlation between sedimentation and aspirating sampler methods. Because of results like these, sedimentation methods have fallen out of general use. A long sequence of spore measurements using aspirating spore traps has been made in cities in countries within the European Community (e.g. Bagni *et al.*, 1976; Spieksma *et al.*, 1980; and D'Amato *et al.*, 1988). Sampling for bacteria in outside air has also been of continuing interest, and again much of the measurement for the purpose of determining general background levels has been carried out on rooftops. However, in some locations considered to be especially relevant to potential human exposures to bacteria from specific potential sources (e.g. waste disposal facilities, sewage treatment plants, etc.), sampling has been carried out closer to ground level. For all bioaerosol types, a wide variety of sampling methods has been employed. However, there appears to have been little to link them together towards the development of a coherent single rationale.

References

Bagni, N., Charpin, H., Davies, R.R., Nolard, N. and Stix, E. (1976) City spore concentrations in European Economic Community, *Clinical Allergy*, 6, 61–68.

Barrett, C.F., Ralph, M.O. and Upton, S.L. (1984) Wind tunnel measurements of the inlet efficiency of four samplers of suspended particulate matter, Final Report on CEC Contract No. 6612/10/2, Report EUR 9378 EN, Commission of European Communities, Luxembourg.

Berghmans, P., Pauwels, J., Roekens, E. and Bogaert, R. (1996) Comparison of methods for the concentration measurement of black (carbonaceous) aerosol in ambient air, *Journal of Aerosol Science*, 27, S689–S690.

Braniš, M. and Domasová, M. (2003) PM_{10} and black smoke in a small settlement: case study from the Czech Republic, *Atmospheric Environment*, 37, 83–92.

British Standards Institution (1969) Methods for the measurement of air pollution, 2. Determination of concentration of suspended particulate matter, British Standard 1747: Part 2.

Brook, J.R., Dann, T.F. and Burnett, R.T. (1997) The relationship among TSP, PM_{10}, $PM_{2.5}$ and inorganic constituents of atmospheric particulate matter at multiple Canadian locations, *Journal of the Air and Waste Management Association*, 47, 2–19.

Burnett, R.T., Dales, R.E., Raizenne, M.E., Krewski, D., Vincent, R., Dann, T. and Brook, J. (1994) Effects of low ambient levels of ozone and sulfates on the frequency of respiratory admissions to Ontario hospitals, *Environmental Research*, 65, 172–194.

Chang, C.T., Tsai, C.J., Lee, C.T., Change, S.Y., Cheng, M.T. and Chein, H.M. (2001) Differences in PM_{10} concentrations measured by beta-gauge monitor and hi-vol sampler, *Atmospheric Environment*, 35, 5741–5748.

Chow, J. (1995) Measurement methods to determine compliance with ambient air quality standards for suspended particles, *Journal of the Air and Waste Management Association*, 45, 320–382.

Colome, S.D., Kado, N.Y., Jaques, P. and Kleinman, M. (1992) Indoor–outdoor air pollution relations: particulate matter less than 10 μm in aerodynamic diameter (PM_{10}) in homes of asthmatics, *Atmospheric Environment*, 26A, 2173–2178.

Comité Européen de Normalisation (1998) Air quality: determination of the PM_{10} fraction of suspended particulate matter. Reference method and field test procedure to demonstrate reference equivalence of measurement methods, CEN Standard EN 12341.

Commission of European Communities (2004) Commission decision of 29th April 2004 concerning guidance on a provisional reference method for the sampling and measurement of $PM_{2.5}$, *Official Journal of the European Union*, L 160/53, Annex: Guidance on $PM_{2.5}$ measurement under Directive 1999/30/EC

D'Amato, G., Mullins, J., Nolard, N., Spieksma, F.T.M. and Wachter, R. (1988) City spore concentrations in the European Economic Community (EEC). VII. Oleaceae (Fraxinus, Ligustrum, Olea), *Clinical Allergy*, 18, 541–547.

DETR (1999) Instruction manual: UK smoke and sulphur dioxide network, Report AEAT-1806, AEA Technology, Harwell.

Durham, O.C. (1944) The volumetric incidence of atmospheric allergen: II. Simultaneous measurements by volumetric and gravity slide methods, results with ragweed pollen and *Alternaria* spores, *Journal of Allergy*, 15, 226–235.

Environmental Protection Agency (EPA) (1997) Ambient air monitoring reference and equivalent methods, *United States Federal Register*, 40 CFR Parts 50, 53 and 58

Gulliver, J. and Briggs, D.J. (2004) Personal exposure to particulate air pollution in transport microenvironments, *Atmospheric Environment*, 38, 1–8.

Harrison, R.M., Shi, J.P., Xi, S., Khan, A., Mark, D., Kinnersley, R. and Yin, J. (2000) Measurement of number, mass and size distribution in the atmosphere, *Philosophical Transactions of the Royal Society of London*, 358, 2567–2580.

Heal, M.R., Hibbs, L.R., Agius, R.M. and Beverland, I.J. (2005) Interpretation of variations in fine, coarse and black smoke particulate matter concentrations in a northern European city, *Atmospheric Environment*, 39, 3711–3718.

Hitzenberger, R., Berner, A., Galambos, Z., Maenhaut, W., Cafmeyer, J., Schwartz, J., Müller, K., Spindler, G., Wieprecht, W., Acker, K., Hillamo, R. and Mäkelä, T. (2004) Intercomparison of methods to measure the mass concentration of the atmospheric aerosol during INTERCOMP2000 – influence of instrumentation and size cuts, *Atmospheric Environment*, 38, 6467–6476.

Hoek, G., Welinder, H., Vaskovi, E., Ciacchini, G., Manalis, N., Røyset, O., Reponen, A., Cyrys, J. and Brunekreef, B. (1997) Interlaboratory comparison of PM10 and black smoke measurements in the PEACE study, *Atmospheric Environment*, 31, 3341–3349.

International Standards Organisation (1983) Air quality – particle size fraction definitions for health-related sampling, Technical Report ISO/TR/7708–1983 (E) ISO, Geneva, revised 1992.

Kanbour, F.I., Altai, F.A., Yassin, S., Harrison, R.M. and Kitto, A.-M. N. (1990) A comparison of smoke shade and gravimetric determination of suspended particulate matter in a semi-arid climate (Baghdad, Iraq), *Atmospheric Environment*, 24, 1297–1301.

Lacey, J. and Venette, J. (1995) Outdoor air sampling techniques, In: *Bioaerosols Handbook* (Eds C.S. Cox and C.M. Wathes), CRC Press, Boca Raton, FL, pp. 407–471.

Laskus, L. (1983) Comparative measurements of suspended particulates using several gravimetric methods and the British smoke sampler, *Science of the Total Environment*, 31, 23–40

Lippmann, M. and Thurston, G.D. (1996) Sulfate concentrations as an indicator of ambient particulate matter air pollution for health risk evaluations, *Journal of Exposure and Analytical and Environmental Epidemiology*, 6, 123–146.

Liu, Y., Chen, R., Shen, X. and Mao, X. (2004) Wintertime indoor air levels of PM_{10}, $PM_{2.5}$ and PM_1 at public places and their contributions to TSP, *Environment International*, 30, 189–197.

Lundgren, D.A. and Paulus, H.J. (1975) The mass distribution of large particles, *Journal of the Air Pollution Control Association*, 25, 1227–1231.

Macey, H.H. (1967) Experience with CERL dust pollution gauge, *Atmospheric Environment*, 1, 637–643.

Maenhaut, W., Raes, N., Chi, X., Wang, W., Cafmeyer, J., Ocskay, R. and Salma, I. (2004) Chemical composition and mass closure of the atmospheric aerosol at Kpuszta, Hungary, in Summer 2003, *Journal of Aerosol Science*, 35, S799–S800.

Marple, V.A., Rubow, K.L., Turner, W. and Spengler, J.D. (1987) Low flow rate sharp cut impactors for indoor air sampling: design and calibration, *Journal of the Air Pollution Control Association*, 37, 1303–1307.

Maximale Arbeitsplatzkonzentration (MAK) (1981) *Mitteilungen der Senatskomission zur Prüfung gesundheitsschädlicher Arbeitsstoffe*, Maximale Arbeitsplatzkonzentration, Harald Boldt, Boppard.

Müller, K., Spindler, G., Maenhaut, W., Hitzenberger, R., Wieprecht, Baltensberger, U. and ten Brink, H.M. (2004) INTERCOMP2000, a campaign to assess the comparability of methods in use in Europe for measuring aerosol composition, *Atmospheric Environment*, 38, 6459–6466.

Price, M., Bulpitt, S. and Meyer, M.B. (2003) A comparison of PM_{10} monitors at a kerbside site in the northeast of England, *Atmospheric Environment*, 37, 4425–4434.

Purdue, L.J., Rodes, C.E., Rehme, K.A., Holland, D.M. and Bond, A.E. (1986) Intercomparison of high-volume PM_{10} samplers at a site with high particulate concentrations, *Journal of the Air Pollution Control Association*, 36, 917–920.

Reponen, A., Ruuskanen, J., Mirme, A., Pärjälä, E., Hoek, G., Roemer, W., Hosiokangas, J., Pekkanen, J. and Jantunen, M. (1996) Comparison of five methods for measuring particulate matter concentrations in cold winter climate, *Atmospheric Environment*, 30, 3873–3879.

Rodes, C.E. and Evans, E.G. (1985) Preliminary assessment of 10 µm particulate sampling at eight locations in the United States, *Atmospheric Environment*, 19, 293–303.

Rodes, C.E., Holland, D.M., Purdue, L.J. and Rehme, K.A. (1985) A field comparison of PM_{10} inlets at four locations, *Journal of the Air Pollution Control Association*, 35, 345–354.

Salminen, K. and Karlsson. V. (2003) Comparability of low-volume PM_{10} sampler with beta-attenuation monitor in background air, *Atmospheric Environment*, 37, 3707–3712.

Soutar, A., Watt, M., Cherrie, J.W. and Seaton, A. (1999) Comparison between a personal PM_{10} sampling head and the tapered element oscillating microbalance (TEOM) system *Atmospheric Environment*, 33, 4373–4377.

Spieksma, F.T.M., Charpin, H., Nolard, N. and Stix, E. (1980) City spore concentrations in the European Economic Community (EEC), *Clinical Allergy*, 10, 319–329.

Thurston, G.D., Ito, K., Hayes, C.G., Bates, D.V. and Lippmann, M. (1994) Respiratory hospital admissions and summertime haze air pollution in Toronto, Ontario: consideration of the role of acid aerosols, *Environmental Research*, 65, 271–290.

Turner, W.A., Olson, B.A. and Allen, G.A. (2000) Calibration of sharp-cut impactors for indoor and outdoor particle sampling, *Journal of the Air and Waste Management Association*, 50, 484–487.

Vallack, H.W. (1995) A field evaluation of frisbee-type deposit gauges, *Atmospheric Environment*, 29, 1465–1469.

Vallack, H.W. and Chadwick, M.J. (1992) A field comparison of dust deposit gauge performance at two sites in Yorkshire, *Atmospheric Environment*, 26A, 1445–1451.

Vallack, H.W. and Shillito, D.E. (1998) Suggested guidelines for deposited ambient dust, *Atmospheric Environment*, 32, 2737–2744.

van der Meulen, A., van Elzakker, B.G. and van den Hooff, G.N. (1987) PM_{10}: results of a one-year monitoring survey in The Netherlands, *Journal of the Air Pollution Control Association*, 37, 812–818.

Verein Deutscher Ingenieure (1982) Particulate matter measurement, Measurement of mass concentration in ambient air, Standard method for the comparison of non-fractionating methods, VDI 2463:8, Dusseldorf.

Watson, J.G., Chow, J.C. and Shah, J.J. (1983) The effect of sampling inlets on the PM_{10} and PM_{15} to TSP concentration ratios, *Journal of the Air Pollution Control Association*, 33, 114–119.

Williams, R., Suggs, J., Rea, A., Leovic, K., Vette, A., Croghan, C., Sheldon, L., Rodes, C., Thornburg, J., Ejire, A., Herbst, M. and Sanders, W. (2003) The Research Triangle Park Particulate Matter Panel Study: PM mass concentration relationships, *Atmospheric Environment*, 37, 5349–5363.

Index

Page references in **bold** indicate tables and references in *italics* indicate figures.

Aerosol Sampling: Science, Standards, Instrumentation and Applications James Vincent
© 2007 John Wiley & Sons, Ltd